DISTILLERS GRAINS
Production, Properties, and Utilization

DISTILLERS GRAINS

Production, Properties, and Utilization

Edited by

KeShun Liu and Kurt A. Rosentrater

CRC Press
Taylor & Francis Group
Boca Raton London New York

CRC Press is an imprint of the
Taylor & Francis Group, an **informa** business

AOCS
PRESS

CRC Press
Taylor & Francis Group
6000 Broken Sound Parkway NW, Suite 300
Boca Raton, FL 33487-2742

First issued in paperback 2019

© 2012 by Taylor & Francis Group, LLC
CRC Press is an imprint of Taylor & Francis Group, an Informa business

No claim to original U.S. Government works

ISBN-13: 978-1-4398-1725-4 (hbk)
ISBN-13: 978-0-367-38256-8 (pbk)

Visit the Taylor & Francis Web site at
http://www.taylorandfrancis.com

and the CRC Press Web site at
http://www.crcpress.com

Contents

PART I   Introduction, History, Raw Materials, and Production

PART II   Properties, Composition, and Analytics

PART III   Traditional Uses

PART IV   Further Uses

PART V   Emerging Uses

PART VI   Process Improvements

Figures

Preface

In the United States and elsewhere, the fuel ethanol industry has experienced a surge in growth during the last decade. This has been due, in large part, to increasing demand for motor fuels as well as government mandates for renewable fuels and fuel oxygenates. At this point in time, most of the fuel ethanol produced in the United States is made from corn. Elsewhere, however, other crops (such as sugarcane and various small grains) are commonly used. As demand for transportation fuels continues to increase, it is anticipated that the fuel ethanol industry in the United States and other countries will continue to grow, at least in the near term, and increasing amounts of corn will be used to achieve this.

In the United States, the majority of fuel ethanol manufacturers use the dry grind method, in which whole grain kernels are ground and the starch in grain flour is fermented into ethanol and carbon dioxide. This leaves the rest of the grain constituents (protein, lipids, fiber, minerals, and vitamins) relatively unchanged chemically, but concentrated. These residual components are sold as distillers grains, most commonly in the form of distillers dried grains with solubles (DDGS). Owing to the tremendous growth in fuel ethanol production in the United States, the supply of distillers grains has greatly increased in recent years. Like fuel ethanol, DDGS has quickly become a global commodity for trade.

Marketability and suitable uses of DDGS are key to the economic viability of fuel ethanol production. As the industry grows, the importance of distiller grains has also increased.

Thus, there has been a growing body of literature on distillers grains. There has also been a need for research into best practices for optimal coproduct utilization. However, a comprehensive compilation of and discussion about distillers grains for scientific communities, and the feed and ethanol industries has been lacking. In line with these developments, *Distillers Grains: Production, Properties, and Utilization* has been compiled. Considerable efforts have been made to cover this important fuel ethanol coproduct with a broad scope, and to provide readers with the most up-to-date technical information available. Thorough discussions on all aspects relating to DDGS have been provided, ranging from the historic perspective to the current status of the fuel ethanol industry, from structure and composition of feedstock grains to physical and chemical properties of DDGS, from dry grind processing to various types of end uses, and from analytical methodologies and mycotoxin occurrence to new methods for improving production efficiency and DDGS quality.

The book is divided into six major parts and consists of 26 chapters in total. Part I has six chapters and covers introduction, perspectives, history, structure and composition of raw grains, and manufacturing processes from corn as well as other grains. Part II deals with physical characteristics, chemical composition, and methodologies for analysis of DDGS; in addition, two chapters provide in-depth discussions on DDGS lipids and mycotoxin occurrence. Part III consists of four chapters, and covers traditional use of DDGS as livestock feed (beef cattle, dairy cattle, swine, and poultry). Part IV has three chapters and discusses further uses of DDGS, including feed for fish, feed for other animals, as well as food ingredients for human consumption. Some emerging opportunities are covered in Part V, including use of DDGS in bioplastics and as feedstocks for bioenergy (thermochemical conversion, anaerobic digestion, pretreatment for cellulosic ethanol production, and extraction and use of DDGS lipids for biodiesel production). The last part covers process improvements, including new and improved enzymes for fuel ethanol production, and fractionation of DDGS. It ends with a chapter on concluding thoughts.

This book brings together cutting-edge information on many aspects of DDGS. The chapters have been authored and coauthored by more than 35 scientists with specific expertise in their respective fields, and we appreciate their diverse approaches and viewpoints on DDGS. To ensure high quality, each chapter has been peer reviewed by at least two external professionals who are experts in their fields. We appreciate their valuable inputs and many constructive suggestions for improvement. Furthermore, extensive lists of references in each chapter should be helpful to readers who wish additional information for a specific topic.

Our special thanks are extended to Jodey Schonfeld, Publications Director, AOCS Press (Urbana, Illinois, USA) for getting our book project initiated, to AOCS Books and Special Publications Committee for their timely and constructive reviews of the book proposal; to John Lavender, Senior Vice-President of Publishing; Randy Brehm, Senior Editor; and Patricia Roberson; Jennifer Spoto, Rachael Panthier; Kevin Craig; and Jessica Vakili, all from CRC Press, Taylor & Francis Group LLC (Boca Raton, Florida, USA) for implementing all aspects of the book project (coordination, production, copyrights, cover design, and marketing), and to Soniya Ashok, Project Manager at Newgen Publishing and Data Services (Chennai, India) for managing the production phrase (editing, design, typesetting, proof trafficking and corrections) of the book. Their support and assistance, along with close cooperation from all the authors, are crucial elements toward successful execution of this project. Thanks are also expressed to our professional colleagues and spouses for their encouragement and support.

This book has been written to serve as a timely and comprehensive reference for animal and food scientists, feed and food technologists, ethanol plant managers and technicians, nutritionists, academic and governmental professionals, college students, and others who are interested in the technical aspects of fuel ethanol production from grains and the resulting coproducts. We will be pleased if this book contributes to a better understanding of distillers grains in terms of production, properties, and end uses, and helps in promoting and expanding utilization of this important biofuel coproduct.

KeShun Liu
Kurt A. Rosentrater

Editors

KeShun Liu, Ph.D. is a research chemist with the United States Department of Agriculture, Agricultural Research Service (USDA-ARS), National Small Grains and Potato Germplasm Research Unit, Aberdeen, Idaho, conducting research on developing plant-based ingredients for fish feed and other uses. His expertise is in chemistry, processing, and value-added utilization of grains, oilseeds, legumes and their by-products (including biofuel coproducts such as distillers grains). He received a bachelor degree in horticulture from Anhui Agricultural College (Hefei, China) and master and doctoral degrees in food science from Michigan State University, and did post-doctoral work at Coca-Cola Co. and the University of Georgia. Prior to joining USDA-ARS, he was an employee at Monsanto Co. and the University of Missouri-Columbia. Thus, he has over 26 years of research experience at academic institutions, private industries, and governmental agencies in the field of food science and technology. Over the years, he has authored or co-authored more than 90 publications, wrote, edited or co-edited three (not counting this one) reference books, including *Soybeans: Chemistry, Technology, and Utilization* (1997); *Asian Foods: Science and Technology* (1999); and *Soybeans as Functional Foods and Ingredients* (2004); organized or co-organized over 35 symposia for scientific meetings; and given more than 86 technical presentations to domestic and international audiences. He has also been an active member of American Oil Chemists Society (AOCS), Institute of Food Technologists (IFT), and American Association of Cereal Chemists, and has served in various leadership capacities for AOCS and IFT divisions, and received, the honors of AOCS Award of Merit and AOCS fellow.

Kurt A. Rosentrater, Ph.D., is a bioprocess engineer with the United States Department of Agriculture, Agricultural Research Service (USDA-ARS), at the North Central Agricultural Research Laboratory in Brookings, South Dakota. His research program focuses on increasing the value and utility of distillers grains, the coproducts from fuel ethanol manufacturing. He is actively developing value-added uses for these materials, including feed, food, and industrial applications. His expertise is in value-added product development, alternative recycling and reprocessing strategies for food and organic waste streams, modeling and simulation of processing systems, plant layout, and process design. Prior to his work at the NCARL, he was an assistant professor at Northern Illinois University in DeKalb, Illinois, in the Department of Engineering and Industrial Technology, where he taught in the areas of research methods, manufacturing systems, engineering mechanics, and design. Before this, he worked for a design-build firm and was responsible for process and equipment design, as well as plant and site layout for agri-industrial manufacturing facilities. He attended Iowa State University where he received his BS, MS, and PhD degrees in agricultural and biosystems engineering. He has authored or co-authored over 285 research publications, of which 90 have been peer-reviewed scientific journal articles. He is widely sought as a speaker, both domestically and internationally, and has given more than 70 invited presentations based on his research. He is an active member of the American Society of Agricultural and Biological Engineers, the American Association of Cereal Chemists, the Institute of Food Technologists, the American Society for Engineering Education, the American Physical Society, and Sigma Xi Honorary Research Society.

Contributors

Michael E. Barnes
McNenny State Fish Hatchery
South Dakota Department of Game
  Fish and Parks
Spearfish, South Dakota

Amy B. Batal
Department of Poultry Science
University of Georgia
Athens, Georgia

Kristjan Bregendahl
Sparboe Farms
Litchfield, Minnesota

Michael L. Brown
Department of Wildlife and Fisheries Sciences
South Dakota State University
Brookings, South Dakota

John Caupert
National Corn-to-Ethanol Research Center
Edwardsville, Illinois

Alfredo Dicostanzo
Department of Animal Science
University of Minnesota
St. Paul, Minnesota

Nicholas R. DiOrio
Department of Technology
Northern Illinois University
DeKalb, Illinois

Alvaro D. Garcia
Department of Dairy Science
South Dakota State University
Brookings, South Dakota

Michael J. Haas
Sustainable Biofuels and Co-Products
Eastern Regional Research Institute
United States Department of Agriculture,
  Agricultural Research Service
Wyndmoor, Pennsylvania

Conly L. Hansen
Department of Nutrition and Food Sciences
Utah State University
Logan, Utah

Kevin B. Hicks
Sustainable Biofuels and Co-Products
Eastern Regional Research Institute
United States Department of Agriculture,
  Agricultural Research Service
Wyndmoor, Pennsylvania

Arnold R. Hippen
Department of Dairy Science
South Dakota State University
Brookings, South Dakota

Milan Hruby
Danisco Animal Nutrition
Woodbury, Minnesota

Klein Ileleji
Department of Agricultural & Biological
  Engineering
Purdue University
West Lafayette, Indiana

Paula Imerman
College of Veterinary Medicine
Iowa State University
Ames, Iowa

David B. Johnston
Sustainable Biofuels and Co-Products
Eastern Regional Research Institute
United States Department of Agriculture,
  Agricultural Research Service
Wyndmoor, Pennsylvania

Kenneth F. Kalscheur
Department of Dairy Science
South Dakota State University
Brookings, South Dakota

KeShun Liu
National Small Grains and Potato Germplasm
    Research Unit
United States Department of Agriculture,
    Agricultural Research Service
Aberdeen, Idaho

Robert A. Moreau
Sustainable Biofuels and Co-Products
Eastern Regional Research Institute
United States Department of Agriculture,
    Agricultural Research Service
Wyndmoor, Pennsylvania

Nathan S. Mosier
Department of Agricultural & Biological
    Engineering
Purdue University
West Lafayette, Indiana

K. Muthukumarappan
Department of Agricultural and Biosystems
    Engineering
South Dakota State University
Brookings, South Dakota

Nhuan P. Nghiem
Sustainable Biofuels and Co-Products
Eastern Regional Research Institute
United States Department of Agriculture,
    Agricultural Research Service
Wyndmoor, Pennsylvania

Andrew W. Otieno
Department of Technology
Northern Illinois University
DeKalb, Illinois

John L. Richard
Consultant
Cave Creek, Arizona

Kurt A. Rosentrater
North Central Agricultural Research Laboratory
United States Department of Agriculture,
    Agricultural Research Service
Brookings, South Dakota

Gravis W. Schaeffer
Columbia Environmental Research Center
United States Geological Survey
Yankton, South Dakota

Gerald C. Shurson
Department of Animal Science
University of Minnesota
St. Paul, Minnesota

Radhakrishnan Srinivasan
Department of Agricultural and Biological
    Engineering
Mississippi State University, Mississippi

Charlie Staff
Distillers Grains Technology Council
University of Louisville
Louisville, Kentucky

Hans H. Stein
Department of Animal Sciences
University of Illinois
Urbana, Illinois

Robert A. Tatara
Department of Technology
Northern Illinois University
DeKalb, Illinois

Nancy Thiex
Oscar E. Olson Biochemistry Laboratories
South Dakota State University
Brookings, South Dakota

Jill K. Winkler-Moser
Food and Industrial Oil Research
United States Department of Agriculture,
    Agricultural Research Service
Peoria, Illinois

Cody L. Wright
Department of Animal and Range Sciences
South Dakota State University
Brookings, South Dakota

Yanhong Zhang
National Corn-to-Ethanol Research Center
Edwardsville, Illinois

Reviewers

Carl J. Bern
Department of Agricultural and Biosystems
    Engineering
Iowa State University
Ames, Iowa

Akwasi A. Boateng
Sustainable Biofuels and Co-Products
Eastern Regional Research Center
United States Department of Agriculture,
    Agricultural Research Service
Wyndmoor, Pennsylvania

Rodney Bothast
Southern Illinois University
Edwardsville, Illinois

David C. Bressler
Department of Agricultural
    Food and Nutritional Sciences
University of Alberta
Edmonton, Alberta, Canada

Hugh Chester-Jones
Southern Research and Outreach Center
University of Minnesota
Waseca, Minnesota

N. Andy Cole
Conservation and Production Research
    Laboratory
United States Department of Agriculture,
    Agricultural Research Service
Bushland, Texas

Michael A. Cotta
Bioenergy Research Unit
United States Department of Agriculture,
    Agricultural Research Service
National Center for Agricultural Utilization
    Research
Peoria, Illinois

Mohan Dasari
Feed Energy Company
Des Moines, Iowa

Bruce Dien
Bioenergy Research Unit
National Center for Agricultural Utilization
    Research
United States Department of Agriculture,
    Agricultural Research Service
Peoria, Illinois

Annie Donoghue
Poultry Production and Product Safety
    Research Unit
United States Department of Agriculture,
    Agricultural Research Service
Fayetteville, Arkansas

Oladiran Fasina
Department of Biosystems Engineering
Auburn University
Auburn, Alabama

B. Wade French
North Central Agricultural Research
    Laboratory
United States Department of Agriculture,
    Agricultural Research Service
Brookings, South Dakota

Alvaro Garcia
Department of Dairy Science
South Dakota State University
Brookings, South Dakota

Feng Hao
Department of Food Science and Human
    Nutrition
University of Illinois
Urbana, Illinois

Marvin Paulsen
Department of Agricultural and Biological
  Engineering
University of Illinois
Urbana, Illinois

Tomas Persson
Department of Biological and Agricultural
  Engineering
University of Georgia
Griffin, Georgia

Pratap C. Pullammanappallil
Department of Agricultural and Biological
  Engineering
University of Florida
Gainesville, Florida

Walter E. Riedell
North Central Agricultural Research
  Laboratory
United States Department of Agriculture,
  Agricultural Research Service
Brookings, South Dakota

Michael B. Rust
Northwest Fisheries Science Center
National Oceanic and Atmospheric Administration
Seattle, Washington

Wendy M. Sealey
Bozeman Fish Technology Center
United States Fish and Wildlife Service
Bozeman, Montana

Vijay Singh
Department of Agricultural and Biological
  Engineering
University of Illinois
Urbana, Illinois

Radhakrishnan Srinivasan
Department of Agricultural and Biological
  Engineering
Mississippi State University
Mississippi State, Mississippi

Robert Thaler
Department of Animal and Range Sciences
South Dakota State University
Brookings, South Dakota

Nancy Thiex
Oscar E. Olson Biochemistry Laboratories
South Dakota State University
Brookings, South Dakota

Kent E. Tjardes
Land O'Lakes Purina Feed
Omaha, Nebraska

Mehmet C. Tulbek
Northern Crops Institute
North Dakota State University
Fargo, North Dakota

Shon Van Hulzen
POET Plant Management
Sioux Falls, South Dakota

Park Waldroup
Poultry Science
Department of University of Arkansas
Fayetteville, Arkansas

James C. Wallwork
Center for Food Safety and Applied
  Nutrition
United States Food and Drug
  Administration
College Park, Maryland

Janitha Wanasundara
Agriculture and Agri-Food Canada
Saskatoon, Canada

Hui Wang
Center for Crops Utilization Research
Iowa State University
Ames, Iowa

Tong (Toni) Wang
Department of Food Science and Human
  Nutrition
Iowa State University
Ames, Iowa

Donald Ward
Grain Processing & Applications
Americas Genencor, a Danisco Division
Cedar Rapids, Iowa

Jill K. Winkler-Moser
Food and Industrial Oil Research
United States Department of Agriculture,
    Agricultural Research Service
Peoria, Illinois

Christine Wood
North Central Agricultural Research
    Laboratory
United States Department of Agriculture,
    Agricultural Research Service
Brookings, South Dakota

Cody Wright
Department of Animal and Range Sciences
South Dakota State University
Brookings, South Dakota

Temur Yunusov
NFI Iowa LLC
Osage, Iowa

Part I

Introduction, History, Raw Materials, and Production

1 Toward a Scientific Understanding of DDGS

Kurt A. Rosentrater and KeShun Liu

CONTENTS

1.1 INTRODUCTION

Recently, many people have asked what the fuel ethanol industry is going to do about the growing piles of nonfermented leftovers. Actually this question has been around for quite some time. As early as the 1940s, one report stated that "Grain distillers have developed equipment and an attractive market for their recovered grains" (Boruff, 1947), while another report described that "Distillers are recovering, drying, and marketing their destarched grain stillage as distillers dried grains and dried solubles" (Boruff, 1952). So it appears that a viable solution had already been developed as far back as the 1940s. And by the early 1950s, there was already a considerable body of published literature on both ethanol manufacturing as well as the use of distillers grains as animal feeds (see Chapter 3 for more information).

Over the course of the last century, some aspects of ethanol and distillers grain processing have changed, but others have not. For example, the production process that is currently used in modern fuel ethanol manufacturing plants remarkably resembles that of the 1940s (Figure 1.1); back then approximately 17 lb (7.7 kg) of distillers feed was produced for every 1 bu (56 lb; 25.4 kg) of grain that was processed into ethanol (which is very similar to today), but over 700 gal (2650 L) of water was required to produce this feed (Boruff et al., 1943; Boruff, 1947, 1952)—this was over two orders of magnitude higher than in modern plants!

1.2 GROWTH OF A MODERN INDUSTRY

Since Henry Ford's time, there had been interest in using grain-based alcohol as a transportation fuel. Prohibition, however, severely constrained development in the United States. After Prohibition ended, both the beverage and fuel ethanol industries grew. But it was not until the price of petroleum escalated during the Oil Crises of the 1970s that the fuel ethanol industry truly began to grow in the United States. And grow it has. At the end of 2010, over 13.1 billion gal/y (49.7 billion L/y) of fuel ethanol were produced, and 204 fuel ethanol manufacturing plants were operating in the United States. These are primarily located in midwestern states (coinciding with the U.S. Corn Belt), because this is where the raw materials (mainly corn) are mostly grown. Due to the growth in demand for ethanol, new plants are now being constructed outside the corn-producing regions of the United States as well. By the time this book is published, the statistics will undoubtedly have changed. Updated information can be found at RFA (2010) and its website (www.ethanolrfa.org).

3

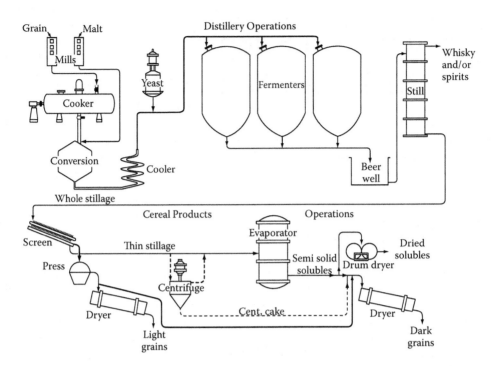

FIGURE 1.1 Process flow diagram for a grain alcohol distillery, ca. late 1940s. (Reprinted from Industrial and Engineering Chemistry, 1947, American Chemical Society. With permission.)

In recent years, as the price of gasoline has risen to near record levels again, there has been a contemporaneous surge in interest in renewable energy, particularly biofuels. This has been reflected in public discourse as well as scientific research on a variety of feedstocks (not just corn), in terms of production, logistics, processing, and conversion. The peer-reviewed literature is replete with new studies; books are also growing in number. Some of the most recent works include, but are not limited to, Cardona et al. (2009), Drapcho et al. (2008), Haas (2010), Ingledew et al. (2009), McNeil and Harvey (2008), Minteer (2006), Mousdale (2008), Mousdale (2010), and Olsson (2007).

While it can be debated whether ethanol production (and use) has a positive or negative energy balance, whether it leads to deforestation in the Amazon rain forest, and whether this approach to meeting fuel demand is sustainable in the long run, there is no question that millions of tons of nonfermented residues are currently available to the feed industry, primarily in the dry form of distillers dried grains with solubles (DDGS), and to a lesser extent in the wet form of distillers wet grains (DWG). During the past several years, ethanol coproducts have become major feed ingredients in North America, and DDGS has become a global agricultural commodity as well. Because the coproducts are now available in such great quantities, a tremendous amount of research has been conducted in recent years to determine their suitability in various livestock diets, as well as best practices for their use.

DDGS is a heterogeneous granular material, and it varies from plant to plant, as well as over time within any given plant. Thus DDGS can be challenging to use, both in terms of nutrient properties, but also physical characteristics. There is a growing body of literature (both scientific and anecdotal) attempting to determine optimal utilization for various animals. Most of this information has been in the form of conference papers, extension publications, magazine articles, as well as other forms of information available on the Internet. Many articles have been published for livestock producers—the end users of the coproducts. And each year, more peer-reviewed journal articles are published as well. To date, however, a comprehensive compilation and discussion of this information for the scientific and technical community has been lacking. This work aims to address that gap. The overall objective of this book is to provide a thorough summary of all aspects of DDGS, ranging from the corn kernel itself all the way through end use. The broad topic of distillers grains intersects a variety of disciplines, including physics, chemistry, microbiology, biology, animal science, as well as engineering and economics. This book will tie these diverse areas together.

1.3 CONCLUSIONS

As the industry continues to evolve, new manufacturing processes will undoubtedly change how ethanol is produced, and will also impact the resulting DDGS. Ethanol production cannot be successful without the sale of distillers grains. As the industry moves forward, it is important to have a solid reference base to serve as a strong technical guide on numerous aspects relating to distillers grains. And hopefully this book will serve useful in that regard.

REFERENCES

Boruff, C. S. 1947. Recovery of fermentation residues as feeds. *Ind. Eng. Chem.* 39(5): 602–607.

Boruff, C. S. 1952. Grain distilleries. *Ind. Eng. Chem.* 44(3): 491–493.

Boruff, C. S., B. Smith, and M. G. Walker. 1943. Water for grain alcohol distilleries. *Ind. Eng. Chem.* 35(11): 1211–1213.

Cardona, C. A., O. J. Sanchez, and L. F. Gutierrez. 2009. *Process Synthesis for Fuel Ethanol Production.* Boca Raton, FL: CRC Press.

Drapcho, C. M., N. P. Nhuan, and T. H. Walter. 2008. *Biofuels Engineering Process Technology.* New York: McGraw-Hill Companies, Inc.

Haas, B. P. 2010. *Ethanol Biofuel Production.* Hauppauge, NY: Nova Science Pub. Inc.

Ingledew, W. M., D. R. Kelsall, G. D. Austin, and C. Kluhspies. 2009. *The Alcohol Textbook,* 5th ed., eds. W. M. Ingledew, D. R. Kelsall, G. D. Austin, and C. Kluhspies. Nottingham, UK: Nottingham University Press.

McNeil, B., and L. Harvey. 2008. *Practical Fermentation Technology.* West Sussex, England: Wiley.

Minteer, S. 2006. *Alcoholic Fuels.* Boca Raton, FL: CRC Press.

Mousdale, D. M. 2008. *Biofuels: Biotechnology, Chemistry, and Sustainable Development.* Boca Raton, FL: CRC Press.

Mousdale, D. M. 2010. *Introduction to Biofuels.* Boca Raton, FL: CRC Press.

Olsson, L. 2007. Biofuels (Advances in Biochemical Engineering Biotechnology). New York: Springer.

RFA (Renewable Fuels Association). 2010. Biorefinery locations. Available online: www.ethanolrfa.org/bio-refinery-locations/. Accessed February 10, 2011.

2 Overview of Fuel Ethanol Production and Distillers Grains

Kurt A. Rosentrater

CONTENTS

2.1   INTRODUCTION

Modern societies face many challenges, including growing populations, increased demands for food, clothing, housing, consumer goods, and the concomitant raw materials required to produce all of these. Additionally, there is a growing need for energy, which is most easily met by use of fossil fuels (e.g., coal, natural gas, and petroleum). In 2008, the overall U.S. demand for energy was 99.3×10^{15} Btu (1.05×10^{14} MJ); 84% of this was supplied by fossil sources (U.S. EIA, 2009). Transportation fuels accounted for 28% of all energy consumed during this time, and nearly 97% of this came from fossil sources. Domestic production of crude oil was 4.96 million barrels per day, whereas imports were 9.76 million barrels per day (nearly two-thirds of the total U.S. demand) (U.S. EIA, 2009). Many argue that this scenario is not sustainable in the long term, for a variety of reasons (such as the need for energy independence and global warming), and other alternatives are needed.

Biofuels, which are renewable sources of energy, can help meet some of these increasing needs. They can be produced from a variety of materials, including cereal grains (such as corn, barley, and wheat), oilseeds (such as soybean, canola, and flax), sugary crops (such as sugar canes), legumes (such as alfalfa), perennial grasses (such as switchgrass, miscanthus, prairie cord grass, and others), agricultural residues (such as corn stover and wheat stems), algae, food processing wastes, and other biological materials. Indeed, the lignocellulosic ethanol industry is poised to consume large quantities of biomass. At this point in time, however, the most heavily used feedstock for biofuel production in the United States is corn grain, because industrial-scale alcohol production from corn starch is readily accomplished, and at a lower cost, compared to other available biomass substrates. The most commonly used process for the production of fuel ethanol from corn is the dry grind process (Figure 2.1), the primary coproduct of which is distillers dried grains with solubles (DDGS).

The goals of this chapter are twofold: (1) to briefly describe the current fuel ethanol industry, how the industry has evolved to its current state, and controversies surrounding this industry; and (2) to

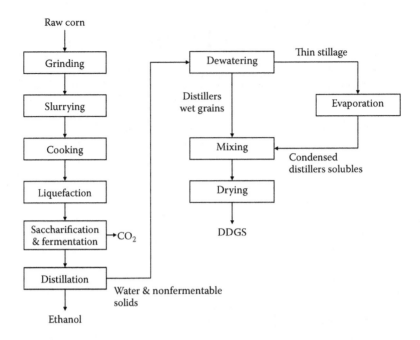

FIGURE 2.1 Simplified flow chart of dry grind fuel ethanol production.

briefly discuss ethanol and coproduct manufacturing practices, explain the importance of coproducts, and describe how coproduct quality and potential uses are expanding as the industry continues to evolve.

2.2   THE FUEL ETHANOL INDUSTRY—CURRENT AND PAST

Ethanol is a combustible material that can readily be used as a liquid fuel. It has, in fact, been used for this purpose for more than 150 years (see Chapter 3), although up until recent times the industry has been quite small. The modern corn-based fuel ethanol industry, however, has reached a scale that can augment the nation's supply of transportation fuels. In 2008, for example, ethanol displaced more than 321 million barrels of oil (Urbanchuk, 2009), which accounted for nearly 5% of all oil imports.

To help meet the increasing demand for transportation fuels, the number of ethanol plants has been rapidly increasing in recent years, as has the quantity of fuel ethanol produced (Figure 2.2). At the beginning of 2011, however, that number had risen to 204 plants with a production capacity of nearly 51.1 billion L/y (13.5 billion gal/y) (www.ethanolrfa.org). And, over the next several years, the Renewable Fuel Standard (RFS) mandates the use of 15 billion gal/y (56.8 billion L/y) of renewable biofuels (i.e., primarily corn-based ethanol). Because the industry is dynamic and still growing, the current production numbers will likely be outdated by the time this book is published. As production volume increases, the processing residues (known collectively as "distillers grains"—Figure 2.3) will increase in tandem (as shown in Figure 2.2). It is anticipated that over 40 million metric tonnes (t) of distillers grains (wet and dry) will eventually be produced by the U.S. fuel ethanol industry.

Over the last decade, while many new fuel ethanol plants have been built, considerable innovations have occurred in the industry, not only in production processes used and final products produced, but also in terms of optimizing raw materials, water, and energy consumed. Some of these innovations have arisen with the advent of dry grind processing. Due to many advantages, including

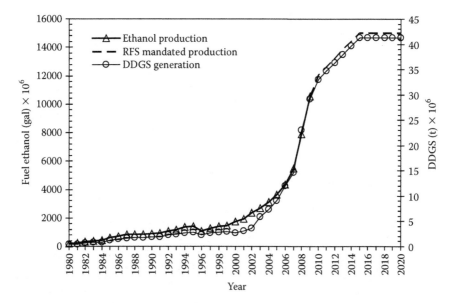

FIGURE 2.2 Trends in fuel ethanol production over time, and estimated DDGS generation; RFS denotes levels mandated by the RFS. (Adapted from RFA, 2009b. *Industry Resources: Co-products.* Renewable Fuels Association: Washington, DC. Available online: www.ethanolrfa.org/pages/industry-resources/coproducts/; RFA, 2009c. *Growing Innovation. 2009 Ethanol Industry Outlook.* Renewable Fuels Association. Washington, DC. Available at: www.ethanolrfa.org/pages/annual-industry-outlook/.)

FIGURE 2.3 An example of distillers dried grains with solubles (DDGS), as it is currently available from most fuel ethanol plants (author's photograph).

lower capital and operating costs (including energy inputs), most new ethanol plants are dry grind facilities (Figure 2.4) as opposed to the older style wet mills. In 2002, 50% of U.S. ethanol plants were dry grind; in 2004 that number had risen to 67%; in 2006 dry grind plants constituted 79% of all facilities; and in 2009 the fraction had grown to over 80% (RFA, 2009c).

Most ethanol plants are centered in the Midwest and north-central portions of the United States—which happens to coincide with the primary production region for corn (i.e., fermentation substrate)—the Corn Belt. But, in recent years, ethanol plants have been built in other states as well.

FIGURE 2.4 A typical dry grind corn-to-ethanol manufacturing plant. This plant has a nominal production capacity of approximately 120 million gal/y (450 million L/y) (author's photograph).

For example, in 2002, there were ethanol plants in 20 states; in 2009, however, there were plants in 26 states (Figure 2.5). Some of these plants are so-called destination plants, which are located near concentrated ethanol markets, and the raw corn is shipped to the plant. This scenario is the inverse of those plants that are located in the Corn Belt.

It is true that as the industry has grown, the concomitant consumption of corn has grown as well (Figure 2.6). In 2008, for example, over 30% of the U.S. corn crop was used to produce ethanol. When examining these numbers, however, it is important to be aware of several key points: exports have been relatively constant over time, there has been a slight decline in the corn used for animal feed, and the overall quantity of corn that is produced by U.S. farmers has been substantially increasing over time. Thus, it appears that the corn that is used to produce ethanol is actually arising mostly from the growing corn supply; it is also slightly displacing use for livestock feed. It is also important to note that the corn that is redirected away from animal feed is actually being replaced by DDGS and other ethanol coproducts in these animal feeds.

The fuel ethanol industry has gained considerable momentum over the last several years, although it is not really a new industry. Rather, the manufacture of ethanol from corn has been evolving for over 150 years. A brief summary of the history of the industry is warranted, and is provided in Table 2.1. Only recently has this industry become truly visible to the average citizen—at least to the current generation. This has been due, in part, to the growing demand for transportation fuels, escalating prices at the fuel pump, positive economic effects throughout rural America, as well as questions and controversies surrounding the production and use of corn ethanol.

2.3   THE FUEL ETHANOL INDUSTRY—QUESTIONS AND CONTROVERSIES

There have been many questions over the years regarding the sustainability of corn-based ethanol; these have been asked by the scientific community, policymakers, as well as the public. Many of these questions have focused on the production of the corn itself, resource and energy inputs versus outputs (i.e., the net energy balance and the life cycle of ethanol), economics, resulting impacts of manufacture, and the performance of ethanol in vehicles vis-à-vis gasoline. To address these, many studies have been conducted to examine the overall costs and benefits of this biofuel.

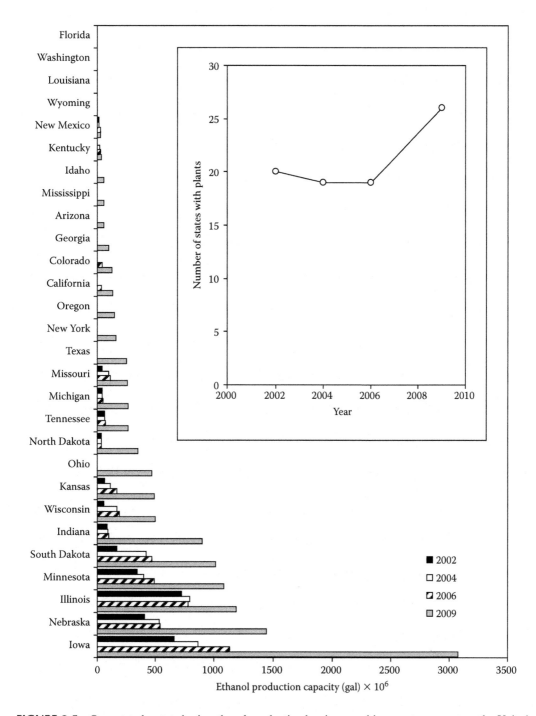

FIGURE 2.5 On a state-by-state basis, ethanol production has increased in recent years across the United States; Midwestern states (i.e., corn belt states) lead production. (Adapted from RFA, 2009a. *Biorefinery Locations.* Renewable Fuels Association: Washington, DC. Available online: www.ethanolrfa.org/bio-refinery-locations/; RFA, 2009b. *Industry Resources: Co-products.* Renewable Fuels Association: Washington, DC. Available online: www.ethanolrfa.org/pages/industry-resources/coproducts/.)

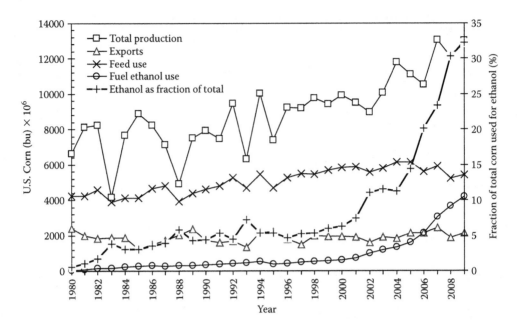

FIGURE 2.6 U.S. corn production (bu) and major categories of use since 1990. (Adapted from ERS, 2009. *Feed Grains Database: Yearbook Tables.* Economic Research Service, U.S. Department of Agriculture: Washington, DC. Available online: www.ers.usda.gov/data/feedgrains/.)

TABLE 2.1
Highlights in the History of Fuel Ethanol

Date	Event
1796	Johann Tobias Lowitz (b. 1757, d. 1804), a Russian chemist (born in Germany) obtained purified ethanol by filtering distilled ethanol through charcoal
1808	Nicolas-Theodore de Saussure (b. 1767, d. 1845), a Swiss chemist, determined the chemical formula of ethanol
1826	Samuel Morey (b. 1762, d. 1843), an American inventor, developed an engine that ran on ethanol or turpentine
1828	Michael Faraday (b. 1791, d. 1867), an English chemist, produced ethanol by acid-catalyzed hydration of ethylene
1840	Ethanol was widely used as lamp fuel in the United States; often it was blended with turpentine and called "camphene"
1858	Archibald Scott Couper (b. 1831, d. 1862), a Scottish chemist, was the first to publish the chemical structure of ethanol
1860	Nicholas Otto (b. 1832, d. 1891), a German inventor, used ethanol to fuel an internal combustion engine
1862	Congress enacted a $2.08/gal tax on both industrial and beverage alcohol to help fund the Civil War effort
1876	Nicholas Otto (b. 1832, d. 1891), a German inventor, developed the Otto Cycle, the first modern combustion engine; it combined four separate strokes with piston chambers for fuel combustion
1890–1900	France holds multiple car races to compare engine performance using ethanol, benzene, and gasoline fuels
1896	Henry Ford's (b. 1863, d. 1947) first automobile, the quadricycle, used corn-based ethanol as fuel
1899	Germany became a world leader in use of ethanol as a fuel; established an office to regulate alcohol sales, the Centrale fur Spiritus Verwerthung; it used taxes, tariffs, and subsidies to keep ethanol prices equivalent to those of petroleum

TABLE 2.1 (Continued)
Highlights in the History of Fuel Ethanol

Date	Event
1906	In Germany, approximately 72,000 distilleries produced 27 million gal of ethanol; between 10% and 30% of German engines ran on pure ethanol
	U.S. government ended the $2.08/gal tax on ethanol (funding for the Civil War was no longer necessary)
1908	Hart-Parr Company (Charles City, IA) manufactured tractors that could use ethanol as a fuel
	Henry Ford's (b. 1863, d. 1947) Model T used corn-based ethanol, gasoline, or a combination as fuel
1918	World War I caused increased need for fuel, including ethanol; demand for ethanol reached nearly 60 million gal/year
1919–1933	Prohibition in the United States
1919	In Brazil, the governor of Pernambuco state mandated that all governmental vehicles use ethanol as fuel
1920	As gasoline became more popular, Standard Oil added ethanol to gasoline to increase octane and reduce engine knocking
	Harold Hibbert (b. 1878, d. 1945) a chemist at Yale University, experimented with the conversion of cellulose into sugar, and garnered the interest of General Motors
1921	Tetraethyl lead developed as an octane enhancer and engine knock reducer
1921–1933	Several studies were conducted into the efficacy and safety of tetraethyl lead as a fuel additive, although many were not published, especially those that directly compared it to ethanol
1924–1925	Tetraethyl lead becomes suspected for the death of many workers in U.S. oil refineries (due to lead poisoning)
1930–1940	Corn-based "Gasahol" (generally 6% to 12% ethanol) becomes widely available in Midwest gasoline markets; see Figure 2.7
1930	To boost farm income during the Great Depression, the Nebraska legislature petitioned Congress to mandate blending 10% ethanol into gasoline, but it was defeated
1932	Iowa State University chemists tested efficacy of gasoline/ethanol blends, and convinced local retailers to offer blends of 10% ethanol with gasoline at service stations in Ames, IA
1933	A campaign to increase industrial uses for farm crops and commodities (including fuel ethanol) became popular across the Midwest, and was known as "Farm Chemurgy"
1936	Experimental ethanol manufacturing and blending with gasoline was initiated by the Chemical Foundation, which was backed by Ford; this program founded the Agrol Company in Atchison, KS
1937	As a result of this experiment, Agrol 5 (5% ethanol) and Agrol 10 (12.5%–17.0% ethanol) were widely sold at service stations throughout Midwestern states
1939	The Agrol plant was closed due to high costs, poor markets and economics
1940	The U.S. Army constructed and operated a fuel ethanol plant in Omaha, NE
1942	The Agrol plant reopened to manufacture materials for the war effort, especially aviation fuel and rubber; production was increased from 100 million gal/year to 600 million gal/year
1941–1945	World War II caused increased need for fuel, including ethanol; ethanol was also used to manufacture rubber; ethanol was produced at nearly 24 million gal/year in the United States
1945–1978	After World War II, a reduced need for fuel led to low fuel prices, and a decline in commercial ethanol in the United States; almost no ethanol was produced during this time period
1970s	Disruptions in the Middle East coupled with environmental concerns over leaded gasoline led to renewed interest in ethanol in the United States
1973	Arab Oil Embargo—U.S. dependence on imported oil becomes obvious; severe economic impacts; gas shortages and lines
1975	Brazil's government launched the National Alcohol Programme to utilize ethanol as an alternative to gasoline
	United States began to phase out use of lead in gasoline
1976	Brazil's government mandated that only ethanol could be used in Brazilian motorsports
1977	Brazil's government mandated that all fueling stations had to offer ethanol
1978	U.S. Congress passed the Energy Tax Act of 1978 (PL 95–618), which provided exemption to the 4 cent/gal federal fuel excise tax for gasoline blended with at least 10% ethanol

continued

TABLE 2.1 (Continued)
Highlights in the History of Fuel Ethanol

Date	Event
1978/1979	First distillation column for the production of fuel ethanol was invented by Dennis and Dave Vander Griend at South Dakota State University in 1978/1979; see Figure 2.8
1979	Brazil's government mandated all automakers to manufacture ethanol-powered vehicles
1980	Sugar prices escalated, oil prices plummeted, Brazil's conversion to ethanol collapsed
	U.S. Congress passed the Crude Oil Windfall Profit Tax Act of 1980 (PL 96–223) (which extended the blenders tax credit) and the Energy Security Act of 1980 (PL 96–294) (with loan guarantees), which encouraged energy conservation and domestic fuel production
	U.S. ethanol industry began to grow, but less than 10 commercial ethanol plants were operating and produced nearly 50 million gal/y
1983	Broin family built an ethanol plant on their farm in Wanamingo, MN; capacity of 125,000 gal/y
	U.S. Congress passed the Surface Transportation Assistance Act (PL 97–424), which raised the gasoline excise tax from 4 to 9 cent/gal and increased the exemption for 10% ethanol blends from 4 to 5 cent/gal
1984	U.S. Congress passed the Tax Reform Act (PL 98–369), which increased the tax exemption for 10% ethanol blends from 5 to 6 cent/gal
	U.S. ethanol industry continued to grow; 163 operational plants
1985	Due to poor ethanol prices, 89 of the 163 plants closed; this was due to very low oil prices; the remaining 74 plants produced 595 million gal/y
1986	The Broin ethanol plant in Wanamingo, MN passed the breakeven point
	Tetraethyl lead banned in U.S. gasoline
1987	Broin family purchased a bankrupt ethanol plant in Scotland, SD; capacity of 1 million gal/y
1988	U.S. Congress passed the Alternative Motor Fuels Act of 1988 (PL 100–494), which funded research and demonstration projects for fuels and vehicles
	Denver, CO mandated the use of oxygenates in fuels to reduce carbon monoxide emissions during the winter
1990	U.S. Congress passed the Omnibus Budget Reconciliation Act (PL 101–508), which extended the ethanol-blended fuel excise tax exemption from 1990 to 2000, but decreased the level from 6 to 5.4 cent/gal
	U.S. Congress passed the Clean Air Amendment of 1990, which created new gasoline standards to reduce fuel emissions, and required fuels to contain oxygenates
1991	Broin & Associates is formed to design and construct ethanol plants
1992	U.S. Congress passed the Energy Policy Act of 1992 (PL 102–486), which established 5.7%, 7.7%, and 85% ethanol blends, set a goal of 30% alternative fuel use in light-duty vehicles by 2010, increased the use of oxygenates in several U.S. cities, and mandated alternative fuel use in government vehicles
1993	Brazil's government passed a law that required all gasoline to be blended with 20%–25% ethanol
1995	Dakota Gold began marketing distillers grains for Broin ethanol plants
	ICM, Inc. founded by Dave Vander Griend to design and construct fuel ethanol plants
1998	U.S. Congress passed the Transportation Efficiency Act of the 21st Century (PL 105–178), which extended the excise tax exemption through 2007
1999	Bans on MTBE begin to appear in various states, because it was found to contaminate water sources
2000	EPA recommended that MTBE should be phased out of use
	Broin began to develop the Broin Project X (BPX), which is a process for producing ethanol without cooking the corn
2004	Broin's BPX process officially commercialized
	California, New York, and Connecticut ban the use of MTBE in motor fuels
	VeraSun Energy opened the largest dry grind ethanol plant in the United States; 120 million gal/y in Aurora, SD

TABLE 2.1 (Continued)
Highlights in the History of Fuel Ethanol

Date	Event
2005	U.S. Congress passed the Energy Policy Act of 2005 (PL 109–58), which required that biofuels must be increasingly mixed with gasoline sold in the United States; levels had to increase to 4 billion gal/y by 2006, 6.1 billion gal/y by 2009, and 7.5 billion gal/y by 2012; this established the first RFS
	Broin developed a prefermentation fractionation process (known as BFrac) that separates corn germ and hull from starch
2007	U.S. Congress passed the Energy Independence and Security Act of 2007 (PL 110–140), which aimed to improve vehicle fuel economy and increase use of alternative fuels by expanding the RFS to include 9 billion gal/y of biofuels in 2008, increasing up to 36 billion gal/y of biofuels by 2022; corn ethanol was slated to increase to a maximum level of 15 billion gal/y by 2015; the balance of the RFS must be met by cellulosic and advanced biofuels
	Broin changes company name to POET
	POET received $80 million grant from the U.S. Department of Energy to build the nation's first commercial cellulosic ethanol plant in Emmetsburg, IA (known as Project Liberty)
2008	VeraSun Energy declared bankruptcy
	Food versus fuel debate raged throughout the media; led to global discussion about biofuels

Note: The people, events, and developments presented in this table were based upon information found in several sources, including Buchheit (2002); CABER (2007); Kovarik (2003); Kovarik (2006); Looker (2006); NDDC (2006); U.S. EIA (2008); and Way (2008).

FIGURE 2.7   Ethanol was extensively used as a motor fuel additive prior to the end of World War II, photo ca. 1933. (From www.jgi.doe.gov/education/bioenergy/bioenergy_7.html.)

FIGURE 2.8 The first distillation column for the production of fuel ethanol was invented by Dennis and Dave Vander Griend at South Dakota State University in 1978/1979 (author's photograph).

As some questions have been answered (the preponderance of scientific evidence indicates that ethanol's positive attributes outweigh the negative), new criticisms have arisen in recent years. These include water consumption, land use change, greenhouse gas emissions, and the use of corn for ethanol instead of food (i.e., the food vs. fuel debate). These questions are leading to further studies of ethanol as a fuel and its impacts, not only nationally, but also globally. It is true that these types of analyses are complicated; but assessing the effectiveness and truth of each study is even more difficult. Results are completely dependent upon initial assumptions and system boundaries. Thus, answers to these new questions have not yet been completely definitive. It is clear, however, that demand for transportation fuels in the United States continues to increase. Ethanol is one means to provide additional fuel for the consumer, and it has had tremendous benefits to rural economies. For example, it has been estimated that during 2008 alone, the fuel ethanol industry contributed more than $65.6 billion to the U.S. GDP, and supported nearly 494,000 jobs (Urbanchuk, 2009).

Even though a thorough discussion of the findings, strengths, and shortcomings of each of these studies is beyond the scope of this book, it is important that the reader is at least aware of the work that has been published. Table 2.2 provides a listing of many of the published reports, and the reader is referred to them for more specific details.

TABLE 2.2
Some Published Research Discussing Shortcomings/ Benefits of Fuel Ethanol

Topic	Reference
General	Agrawal et al., 2007
	Cassman et al., 2006
	Goldemberg, 2007
	Pimentel, 2008
	Pimentel and Pimentel, 2008
	Robertson et al., 2008
	Wishart, 1978
Costs, efficiencies, economics	De La Torre Ugarte et al., 2000
	Gallagher et al., 2005
	Gallagher and Shapouri, 2009
	Laser et al., 2009
	Perrin et al., 2009
	Rendleman and Shapouri, 2007
	Rosenberger et al., 2002
	Shapouri and Gallagher, 2005
Environmental impacts	Cassman, 2007
	Dias De Oliveira et al., 2005
	Donner and Kucharik, 2008
	Farrell et al., 2006
	Hill et al., 2006
	Kim and Dale, 2005b
	Kim and Dale, 2009
	Naylor et al., 2007
	Pimentel, 1991
	Plevin and Mueller, 2008
	Tilman et al., 2009
Food versus fuel	Alexander and Hurt, 2007
	Cassman and Liska, 2007
	Harrison, 2009
	Pimentel et al., 2008
	Pimentel, 2009
	Pimentel et al., 2009
Invasive species	Raghu et al., 2006
Land use change	Fargione et al., 2008
	Foley et al., 2005

continued

TABLE 2.2 (Continued)
Some Published Research Discussing Shortcomings/
Benefits of Fuel Ethanol

Topic	Reference
	Kim et al., 2009
	Liska and Perrin, 2009
	Searchinger et al., 2008
Land/water usage, biodiversity	Chiu et al., 2009
	Green et al., 2005
	Groom et al., 2008
	Mubako and Lant, 2008
	Searchinger and Heimlich, 2009
	Service, 2009
Lifecycle analysis	Brehmer and Sanders, 2008
	Dewulf et al., 2005
	Hedegaard et al., 2008
	Kaltschmitt et al., 1997
	Kim and Dale, 2002
	Kim and Dale, 2004
	Kim, and Dale, 2005a
	Kim and Dale, 2006
	Kim and Dale, 2008
	Kim et al., 2009
	Liska and Cassman, 2008
	Liska and Cassman, 2009
	Liska et al., 2008
	Outlaw and Ernstes, 2008
	Plevin, 2009
Net energy balance	Chambers et al., 1979
	Da Silva et al., 1978
	Dale, 2007
	Hammerschlag, 2006
	Persson et al., 2009a
	Persson et al., 2009b
	Persson et al., 2010
	Pimentel, 2003
	Pimentel and Patzek, 2005
	Shapouri, 1998
	Shapouri et al., 2002
	Shapouri et al., 2003a, 2003b
	Weisz and Marshall, 1979

2.4 ETHANOL MANUFACTURING

Dry grind ethanol manufacturing typically results in three main products: ethanol, the primary end product; residual nonfermentable corn kernel components, which are sold as distillers grains; and carbon dioxide. A common rule of thumb is that for each 1 kg of corn processed, approximately 1/3 kg of each of the constituent streams will be produced. Another rule of thumb states that each bushel of corn (~56 lb; 25.4 kg) will yield up to 2.9 gal (11.0 L) of ethanol, approximately 18 lb (8.2 kg) of distillers grains, and nearly 18 lb (8.2 kg) of carbon dioxide. Of course, these will vary to some degree over time due to production practices, equipment settings, residence times, concentrations, maintenance schedules, equipment conditions, environmental conditions, the composition and quality of the raw corn itself, the location where the corn was grown, as well as the growing season that produced the corn.

The overall production process consists of several distinct unit operations. Grinding, cooking, and liquefying release and convert the starch so that it can be fermented into ethanol using yeast. After fermentation, the ethanol is separated from the water and nonfermentable residues by distillation. Downstream dewatering, separation, evaporation, mixing, and drying are then used to remove water from the solid residues and to produce a variety of coproduct streams (known collectively as distillers grains): wet or dry, with or without the addition of condensed soluble materials. More specific details will be provided in a subsequent chapter dedicated to ethanol and DDGS processing (Chapter 5).

Carbon dioxide, which arises from fermentation, is generated by the yeast, due to the metabolic conversion of sugars into ethanol. This byproduct stream can be captured and sold to compressed gas markets, such as beverage or dry ice manufacturers. Often, however, it is just released to the atmosphere because location and/or logistics make the sales and marketing of this gas economically unfeasible. Carbon dioxide release may eventually be affected by greenhouse gas emission constraints and regulations.

2.5 IMPORTANCE OF COPRODUCTS FROM ETHANOL PRODUCTION

Because the starch is converted into glucose, which is consumed during the fermentation process, the nonfermentable materials will consist of corn kernel proteins, fibers, oils, and minerals. These are used to produce a variety of feed materials, the most commonly produced is DDGS. DDG (i.e., without added solubles) can also be produced, but DDGS is often dried to approximately 10% moisture content (or even less at some plants), to ensure an extended shelf life, and then sold to local livestock producers or shipped by truck or rail to various destinations throughout the nation. And they are increasingly being exported to overseas markets as well. Distillers wet grains (DWG) has been gaining popularity with livestock producers near ethanol plants in recent years; in fact, it has been estimated that, nationwide, up to 25% of distillers grains sales are now DWG (RFA, 2009b). But, because the moisture contents are generally greater than 50% to 60%, their shelf life is very limited, and shipping large quantities of water is expensive. DDGS is still the most prevalent type of distillers grain in the marketplace.

Currently available DDGS typically contains about 30% protein, 10% fat, at least 40% neutral detergent fiber (NDF), and up to 12% starch. Composition, however, can vary between plants and even within a single plant over time, due to a number of factors, which have already been mentioned. For example, Table 2.3 summarizes composition of DDGS samples collected from five ethanol

TABLE 2.3
Average Composition (% db) of DDGS from Five Ethanol Plants in South Dakota (±1 Standard Deviation in Parentheses)

Plant	Protein	Lipid	NDF	ADF	Starch	Ash
1	28.33[b] (1.25)	10.76[a] (1.00)	31.84[b] (4.02)	15.56[a] (2.29)	11.82[a] (1.20)	13.27[a] (3.10)
2	30.65[a] (1.20)	9.75[a] (1.05)	39.90[a] (3.95)	15.21[a] (3.95)	9.81[a] (1.52)	12.84[a] (2.56)
3	28.70[a] (1.32)	10.98[a] (0.95)	38.46[a] (4.01)	17.89[a] (4.01)	11.59[a] (1.42)	11.52[a] (3.05)
4	30.65[a] (1.23)	9.40[b] (0.16)	36.73[a] (1.07)	15.28[a] (0.49)	9.05[b] (0.33)	4.13[b] (0.21)
5	31.78[a] (0.63)	9.50[b] (0.41)	38.88[a] (0.86)	17.24[a] (1.12)	10.05[a] (0.65)	4.48[b] (0.22)

Source: Adapted from Bhadra, R., K. A. Rosentrater, and K. Muthukumarappan. *Cereal Chemistry*, 86(4), 410–420, 2009.

Note: Statistically significant differences among plants for a given component are denoted by differing letters, $\alpha = 0.05$, LSD.

plants in South Dakota during 2008. On a dry basis, crude protein levels ranged from 28.3% to 31.8%; crude lipid varied between 9.4% and 11.0%; ash ranged from 4.1% to 13.3%. In terms of within-plant variability, the crude protein, crude lipid, and starch content all exhibited relatively low variation, whereas NDF, acid detergent fiber (ADF), and ash all had substantially higher variability DDGS composition is covered in detail in Chapter 8.

A question that should be asked is: does DDGS composition vary according to geographic region in the United States? To address this concern, samples of DDGS from 49 plants from 12 states were collected and analyzed for proximate composition and amino acid profile (UMN, 2009). These data are shown in Tables 2.4 and 2.5, respectively. Across all of the samples, dry matter content varied from 86.2% to 92.4%; protein varied from 27.3% to 33%. Crude fat content displayed even higher variability, and ranged from 3.5% to 13.5%; crude fiber ranged from 5.37% to 10.58%; and ash content varied from 2.97% to 9.84%. On average, there do not appear to be any geographic trends for any of the nutrient components (Table 2.4). In terms of amino acids, lysine ranged from 0.61% to 1.19%, but again, no geographic variability was apparent.

Currently, the ethanol industry's primary market for distillers grains, most often in the form of DDGS, and to a lesser degree in the form of DWG, has been used as livestock feed (the other coproducts are sold in much lower quantities than either DDGS or DWG, and are not available from all ethanol plants). Feeding ethanol coproducts to animals is a practical method of utilizing these materials because they contain high nutrient levels, and they are digestible, to varying degrees, by most livestock. Use of DDGS in animal feeds (instead of corn grain) helps to offset the corn that has been redirected to ethanol production, although this fact is not well publicized by the media. Over 80% of all distillers grains is used in beef and dairy diets, but swine and poultry diets are increasing their consumption as well (UMN, 2009). Over the years, numerous research studies have been conducted on coproduct use in livestock diets, for both ruminant and monogastric feeds. These will be topics of several subsequent chapters (Chapters 12 through 15).

Not only are coproducts important to the livestock industry as feed ingredients, but they are also essential to the sustainability of the fuel ethanol industry itself. The sale of distillers grains (all types—dry and wet) contributes substantially to the economic viability of each ethanol plant (sales can generally contribute between 10% and 20% of a plant's total revenue, but at times it can be as high as 40%), depending upon the market conditions for corn, ethanol, and distillers grains.

TABLE 2.4
Average Composition (% db) of DDGS Samples from 49 Ethanol Plants from 12 States

State	Plants Sampled	Dry Matter (%)	Crude Protein (%)	Crude Fat (%)	Crude Fiber (%)	Ash (%)
Minnesota	12	89.03	30.70	11.73	6.96	6.63
Illinois	6	89.72	29.98	11.48	7.26	5.60
Indiana	2	90.55	29.40	12.80	8.07	5.86
Iowa	7	88.92	31.23	10.27	7.57	5.76
Kentucky	3	90.57	29.43	9.77	9.28	4.47
Michigan	1	89.60	32.60	11.00	7.37	6.06
Missouri	2	87.90	30.45	10.25	7.17	5.39
Nebraska	4	89.02	30.40	11.35	8.13	4.23
New York	1	88.21	30.00	9.60	7.87	4.55
North Dakota	4	89.21	31.75	11.70	6.89	6.32
South Dakota	4	88.61	31.80	11.53	6.65	4.78
Wisconsin	3	89.68	31.70	11.63	7.59	5.77
Overall Average	49 (Total)	89.25	30.79	11.09	7.57	5.45

Source: Adapted from UMN. *The Value and Use of Distillers Grains By-products in Livestock and Poultry Feeds.* University of Minnesota: Minneapolis, MN, 2009.

TABLE 2.5
Average Essential Amino Acid Profiles (% db) of DDGS Samples from 49 Ethanol Plants from 12 States

State	Plants Sampled	Agrinine (%)	Histidine (%)	Isoleucine (%)	Leucine (%)	Lysine (%)	Methionine (%)
Minnesota	12	1.39	0.84	1.20	3.63	0.99	0.61
Illinois	6	1.37	0.82	1.15	3.45	0.94	0.63
Indiana	2	1.19	0.79	1.08	3.28	0.85	0.60
Iowa	7	1.34	0.86	1.20	3.63	0.95	0.61
Kentucky	3	1.35	0.79	1.09	3.33	0.89	0.66
Michigan	1	1.28	0.86	1.18	3.67	0.87	0.71
Missouri	2	1.35	0.83	1.18	3.68	0.89	0.73
Nebraska	4	1.46	0.88	1.18	3.61	1.05	0.65
New York	1	1.46	0.85	1.21	3.64	1.04	0.61
North Dakota	4	1.37	0.88	1.24	3.76	0.97	0.65
South Dakota	4	1.47	0.87	1.22	3.70	1.08	0.62
Wisconsin	3	1.45	0.86	1.24	3.75	1.07	0.59
Overall Average	49	1.37	0.84	1.18	3.59	0.96	0.64

State	Plants Sampled	Phenylalanine (%)	Threonine (%)	Tryptophan (%)	Valine (%)	Tyrosine (%)
Minnesota	12	1.59	1.17	0.24	1.62	1.20
Illinois	6	1.51	1.11	0.22	1.52	1.22
Indiana	2	1.45	1.04	0.21	1.44	—
Iowa	7	1.57	1.14	0.25	1.60	—
Kentucky	3	1.48	1.09	0.26	1.43	—
Michigan	1	1.52	1.15	0.25	1.57	—
Missouri	2	1.53	1.15	0.24	1.58	—
Nebraska	4	1.58	1.15	0.26	1.58	1.14
New York	1	1.63	1.11	0.20	1.59	1.19
North Dakota	4	1.62	1.19	0.25	1.67	—
South Dakota	4	1.67	1.19	0.23	1.63	1.35
Wisconsin	3	1.65	1.14	0.22	1.64	1.25
Overall Average	49	1.56	1.13	0.24	1.57	1.22

Source: Adapted from UMN. *The Value and Use of Distillers Grains By-products in Livestock and Poultry Feeds.* University of Minnesota: Minneapolis, MN, 2009.

This is the reason why these process residues are referred to as "coproducts," instead of "byproducts" or "waste products"; they truly are products in their own right along with ethanol.

The sales price of DDGS is important to ethanol manufacturers and livestock producers alike. Over the last decade, the price for DDGS has ranged from approximately $61.6/ton ($67.9/t) up to $165/ton ($181.8/t) (Figure 2.9). On an as-sold basis, both DDGS and corn ($1.7/bu to $6.55/bu) have exhibited great volatility over the last several years, due to a number of market-influencing factors. Drastic price increases in both corn and DDGS occurred during 2004, 2007, and 2008. Taking another look at this data, and placing each product on an equivalent basis ($/ton; Figure 2.10), DDGS and corn prices have historically paralleled each other very closely. This should not be surprising, as DDGS is most often used to replace corn in livestock diet formulations. How closely are the prices of DDGS and corn related? Figure 2.11 shows that the relationship has been quite strong over the last several years. Monthly data for each commodity shows a slightly stronger linear relationship, but even on a yearly basis, the relationship is still significant. Overall, this behavior is positive in nature; in other words, the higher the price of corn, the higher the price of DDGS. Moreover, it appears that

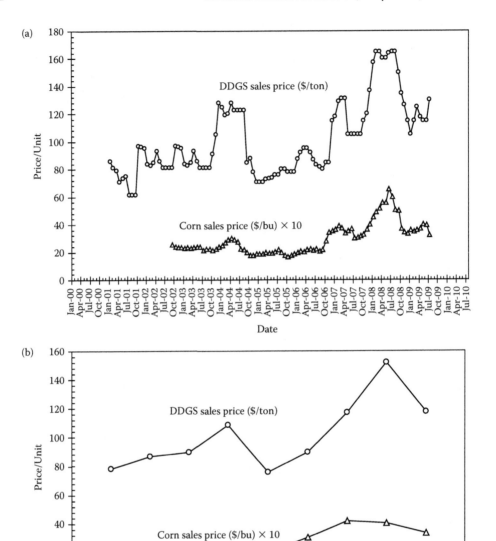

FIGURE 2.9 DDGS and corn sales prices over time on an as-sold basis: (a) monthly averages and (b) yearly averages. (Adapted from ERS, 2009. *Feed Grains Database: Yearbook Tables.* Economic Research Service, U.S. Department of Agriculture: Washington, DC. Available online: www.ers.usda.gov/data/feedgrains/.)

over time, DDGS has been declining in value vis-à-vis corn (Figure 2.12), at least slightly, although there is quite a bit of fluctuation in the markets. A ratio of less than 1.0 indicates that DDGS is sold at a discount compared to corn; this indicates that DDGS is considered less valuable than corn in the marketplace. Over time, a greater supply of DDGS has become available as more ethanol plants have come online. Usually as supply increases, price for a commodity decreases, due to traditional supply–demand behavior. However, by examining the historic data (Figure 2.13), it appears that this has not actually occurred for DDGS, at least not yet. This is probably a function of larger feed and grain market forces, which are reacting to demand for feed by the livestock industry, competing

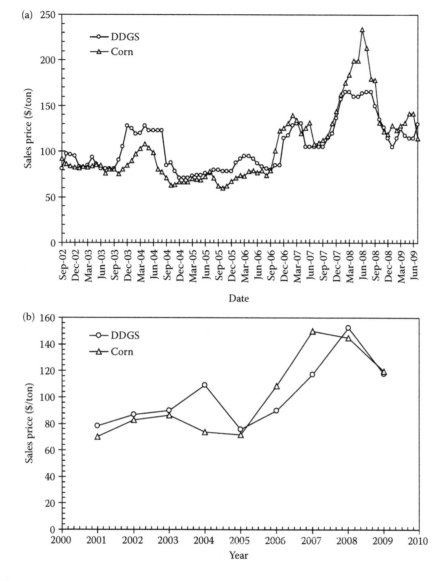

FIGURE 2.10 DDGS and corn sales prices over time on a similar quantity basis: (a) monthly averages and (b) yearly averages. (Adapted from ERS, 2009. *Feed Grains Database: Yearbook Tables.* Economic Research Service, U.S. Department of Agriculture: Washington, DC. Available online: www.ers.usda.gov/data/feedgrains/.)

demands for corn, as well as crude oil and transportation fuel prices. As the industry continues to expand, coproducts will continue to have a significant impact on the industry. The increased supply of distillers grains may eventually impact potential feed demand and sales prices, because ethanol coproducts must compete against other feed materials in the marketplace.

2.6 COPRODUCT CONSTRAINTS AND CHALLENGES

As the ethanol industry has grown, several constraints and challenges associated with the utility of distillers grains have arisen. Fortunately, several have already been addressed and the industry

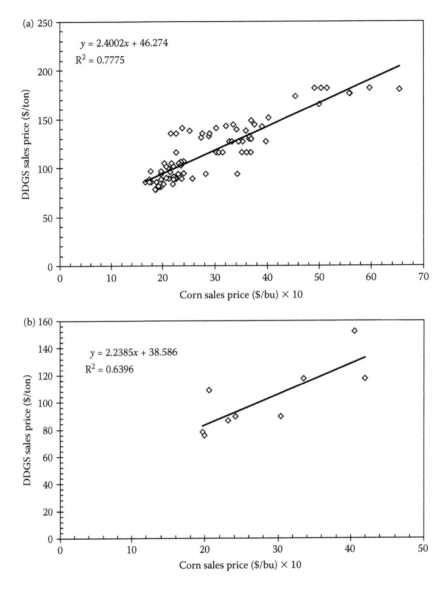

FIGURE 2.11 Relationship between DDGS sales price and corn sales price: (a) monthly averages and (b) yearly averages. (Adapted from ERS, 2009. *Feed Grains Database: Yearbook Tables*. Economic Research Service, U.S. Department of Agriculture: Washington, DC. Available online: www.ers.usda.gov/data/feedgrains/.)

has continued to move forward. First and foremost was how to use DDGS. The answer, of course, has been to utilize DDGS as a feed ingredient for livestock diets. Due to their ability to utilize high levels of fiber, ruminant animals have become the dominant consumers of DDGS. But, as producers and nutritionists increase their knowledge, through research and experience, the swine and poultry markets are also increasing their consumption as well.

Another impending challenge that arose a few years ago was the so-called 10 million tonne question. In other words, if the livestock industry would become saturated with "mountains of DDGS," then all of the unused material would probably have to be landfilled. A production level of 10 million metric tonnes of DDGS was perceived as unsustainable and more than the feed markets

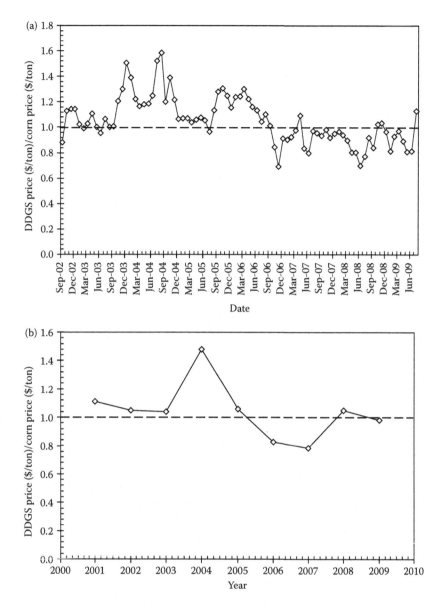

FIGURE 2.12 Ratio of DDGS sales price to corn sales price over time: (a) monthly averages and (b) yearly averages. (Adapted from ERS, 2009. *Feed Grains Database: Yearbook Tables.* Economic Research Service, U.S. Department of Agriculture: Washington, DC. Available online: www.ers.usda.gov/data/feedgrains/.)

could bear (Rosentrater and Giglio, 2005; Rosentrater, 2007). Fortunately, this situation has not occurred as the industry has grown. In fact, the opposite has happened—utilization in livestock diets has continued to increase. Predictions of the peak potential for DDGS use in domestic beef, dairy, swine, and poultry markets have estimated that between 40 and 60 million t could be used in the United States each year, depending upon inclusion rates for each species (Staff, 2005; Cooper, 2006; U.S. Grains Council, 2007). These are estimates of domestic use only, however; globally, the need for protein-based animal feeds continues to grow. Of the 23 million t of DDGS produced in 2008 (RFA, 2009b), 4.5 million t were exported to international markets (FAS, 2009); this

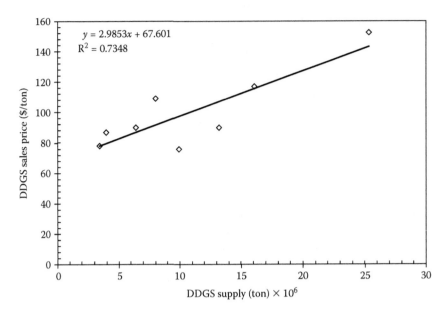

FIGURE 2.13 Relationship between DDGS supply and sales price. (Adapted from ERS, 2009. *Feed Grains Database: Yearbook Tables.* Economic Research Service, U.S. Department of Agriculture: Washington, DC. Available online: www.ers.usda.gov/data/feedgrains/.)

accounted for nearly 20% of the U.S. DDGS production that year. Furthermore, the potential for global exports is projected to increase for the foreseeable future (U.S. Grains Council, 2007). So, in fact, the opposite situation than that which was anticipated is occurring—that of being able to supply the growing demand.

A third challenge has been determining which analytical laboratory methods should be used to measure chemical constituents in DDGS (i.e., moisture, protein, lipid, fiber, and ash). This issue has become important for the trade and delivery of DDGS. The various methods that are available will produce differing results (some more so than others), and buyers and sellers need guaranteed analytical values. To address this, Thiex (2009) conducted an extensive multilaboratory examination of relevant methods, the results of which have led to recommendations endorsed as industry-preferred methods by the American Feed Industry Association, the Renewable Fuels Association, and the National Corn Growers Association. Information on analytical aspects is covered in Chapter 10.

There are still, however, several key issues that impact the value and utilization of distillers grains, both from the ethanol production standpoint, and from a livestock feeding perspective. Some of the most pressing topics include variability in nutrient content, quality, and associated quality management programs at each plant; lack of standardized metrics to quantify DDGS quality (beyond just the nutrient components and color); potential mycotoxin contamination; potential antibiotic and drug residues from processing; sulfur content; inconsistent product identity and nomenclature, especially as new coproducts result from the evolution of the ethanol production process; the large quantities of energy required to remove water during drying, coupled with the high cost of energy itself; costs associated with transporting DDGS to diverse and distant markets (especially fuel costs, rail leasing fees, etc.); international marketing and export challenges; and the need for more education and technical support for the industry. Poor flowability and material handling behavior are still challenges for many plants as well. All of these issues ultimately impact the end users—the livestock producers. The industry is currently working to address each of these, and in so doing will increase the utility and value of these coproducts.

2.7 IMPROVEMENT OF ETHANOL PRODUCTION, COPRODUCT QUALITY, AND END USES

The ethanol industry is very dynamic and has been continually evolving since its inception. A modern dry grind plant is vastly different from the inefficient and input-intensive gasohol plants of the 1970s. New developments and technology advancements, to name only a few, include new strains of more effective enzymes (see Chapter 24), higher starch conversions, cold cook technologies, improved drying systems, fractionation systems decreased energy consumption throughout the plant, increased water use efficiencies, and decreased pollutant emissions. Many of these improvements can be attributed to the design and operation of the equipment used in modern dry grind plants; but a large part is also due to computer-based instrumentation and control systems.

Much formal research as well as many informal trial and error studies have been devoted to adjusting existing processes in order to improve and optimize the quality of the coproducts that are produced. Ethanol companies have recognized the need to produce more consistent, higher quality DDGS that will better serve the needs of livestock producers.

The sale of DDGS and the other coproducts has been one key to the industry's success so far, and will continue to be important to the long-term sustainability of the industry. Although the majority of DDGS is currently consumed by beef and dairy cattle, use in monogastric diets, especially swine and poultry, continues to increase. Additionally, there has been considerable interest in developing improved mechanisms for feeding DDGS to livestock, especially in terms of pelleting/densification (Figure 2.14). This is a processing option that could result in significantly better storage and handling characteristics of the DDGS, and it would drastically lower the cost of rail transportation and logistics (due to increased bulk density and better flowability). Pelleting could also broaden the use of DDGS domestically (e.g., improved ability to use DDGS for rangeland beef cattle feeding) as well as globally (e.g., increased bulk density would result in considerable freight savings in bulk vessels and containers). Moreover, as feeding research continues, the use of DDGS for other species, such as fish (Figure 2.15) and companion animals, will begin to play larger roles as well.

There are also many exciting developments underway in terms of evolving coproducts (Figure 2.16), encompassing the development of more value streams from the corn kernel (i.e., upstream fractionation) as well as the resulting distillers grains (i.e., downstream fractionation). In fact, many plants have been adding capabilities to concentrate nutrient streams such as oil, protein,

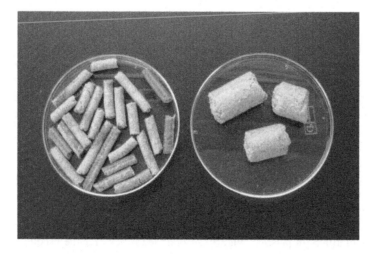

FIGURE 2.14 Pelleting is a unit operation that can improve the utility of DDGS, because it improves storage and handling characteristics, and allows more effective use in range land settings for beef cattle (author's photograph).

FIGURE 2.15 DDGS has been shown to be an effective alternative to fish meal as a protein source in Nile tilapia diets (and for other cultured fish species as well) (author's photograph).

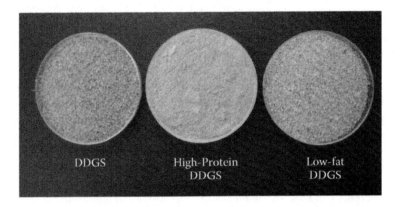

FIGURE 2.16 Examples of unmodified DDGS and some fractionated products that are becoming commercially available (author's photograph).

and fiber into specific fractions, which can then be used for targeted markets. For example, if the germ is removed from the corn kernel prior to ethanol processing (i.e., prefermentation), it can be used to produce food-grade corn oil. If the lipids are instead removed from the DDGS (i.e., postfermentation; Figure 2.17), they can readily be used to produce biodiesel, although they cannot be used for food-grade corn oil, because they are too degraded structurally. Another example is the removal of corn fiber, either from the kernel or from the DDGS; this type of fractionation leads to a DDGS with higher protein and fat levels. As these process modifications are developed, validated, and commercially implemented, improvements in the generated coproducts will be realized: higher quality DDGS and unique materials (such as high protein, low fat, and/or low fiber distillers grains) will be produced. Of course, these new products will require extensive investigation in order to determine how to optimally use them in livestock diets.

There are also new opportunities to increase the utility and value of DDGS. Some of these emerging possibilities include using DDGS (or components thereof) as ingredients in human foods (Figure 2.18), as biofillers in plastics (Figure 2.19), as feedstocks for the production of bioenergy (especially heat and electricity at the ethanol plant), and as substrates for the further production of ethanol or other biofuels (Figure 2.20). These topics will each be discussed in turn throughout the book.

FIGURE 2.17 Corn oil that has been extracted from DDGS can be used to manufacture biodiesel (author's photograph).

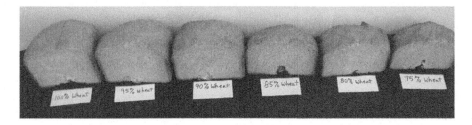

FIGURE 2.18 As a partial substitute for flour, DDGS can be used to improve the nutrition of various baked foods by increasing protein and fiber levels and decreasing carbohydrate (i.e., starch) content (author's photograph).

FIGURE 2.19 DDGS has been shown to be an effective filler in plastic products, replacing petroleum additives and increasing biodegradability (author's photograph).

FIGURE 2.20 DDGS can be thermochemically converted into biochar, which can subsequently be used to produce energy, fertilizer, or as a precursor to other biobased materials (author's photograph).

2.8 CONCLUSIONS

While it may be true that ethanol is not the entire solution to our transportation fuel needs, it is clearly a key component to the overall goal of energy independence. The industry has been rapidly expanding in recent years in response to government mandates, but also due to increased demand for alternative fuels. This has become especially true as the price of gasoline has escalated and fluctuated so drastically, and the consumer has begun to perceive fuel prices as problematic. As long as support for the expansion of the ethanol industry continues, there will be many opportunities for those involved in process and product development to help add value to various material streams in ethanol biorefineries—especially the coproducts known as distillers grains—as this industry continues to mature.

REFERENCES

Agrawal, R., N. R. Singh, F. H. Ribeiro, and W. N. Delgass. 2007. Sustainable fuel for the transportation sector. *Proceedings of the National Academy of Sciences* 104(12): 4828–4833.

Alexander, C., and C. Hurt. 2007. *Biofuels and Their Impact on Food Prices.* Bioenergy ID-346-W. West Lafayette, IN: Department of Agricultural Economics, Purdue University.

Andress, D. 2002. *Ethanol Energy Balances. Subcontract 4000006704 Report.* Washington, DC: Office of Biomass Programs, U.S. Department of Energy.

Bhadra, R., K. A. Rosentrater, and K. Muthukumarappan. 2009. Cross-sectional staining and the surface properties of DDGS and their influence on flowability. *Cereal Chemistry* 86(4): 410–420.

Brehmer, B., and J. Sanders. 2008. Implementing an energetic life cycle analysis to prove the benefits of lignocellulosic feedstocks with protein separation for the chemical industry from the existing bioethanol industry. *Biotechnology and Bioengineering* 102(3): 767–777.

Buchheit, J. K. 2002. *History of Ethanol Production. Rural Enterprise and Alternative Agricultural Development Initiative Report, No. 12.* Office of Economic and Regional Development, Southern Illinois University: Carbondale, IL.

CABER. 2007. *History of Ethanol.* Center for Advanced BioEnergy Research, University of Illinois at Urbana-Champaign. Available online: www.bioenergyuiuc.blogspot.com/2007/05/history-of-ethanol.html. Accessed February 10, 2011.

Cassman, K. G. 2007. Climate change, biofuels, and global food security. *Environmental Research Letters* 2(011002): DOI # 10.1088/1748–9326/2/1/011002.

Cassman, K. G., V. Eidman, and E. Simpson. 2006. *Convergence of Agriculture and Energy: Implications for Research Policy, QTA2006–3*. Ames, IA: Council for Agricultural Science and Technology.

Cassman, K. G., and A. J. Liska. 2007. Food and fuel for all: Realistic or foolish? *Biofuels, Bioproducts and Biorefining* 1(1): 18–23.

Chambers, R. S., R. A. Herendeen, J. J. Joyce, and P. S. Penner. 1979. Gasohol: Does it or doesn't it produce positive net energy? *Science* 206: 789–795.

Chiu, Y. -W., B. Walseth, and S. Suh. 2009. *Environmental Science and Technology* 43(8): 2688–2692.

Cooper, G. 2006. A brief, encouraging look at 'theoretical' distillers grains markets. *Distillers Grains Quarterly* 1(1): 14–17.

Da Silva, J. G., G. E. Serra, J. R. Moreira, J. C. Concalves, and J. Goldemberg. 1978. Energy balance for ethyl alcohol production from crops. *Science* 201(8): 903–906.

Dale, B. E. 2007. Thinking clearly about biofuels: Ending the irrelevant 'net energy' debate and developing better performance metrics for alternative fuels. *Biofuels, Bioproducts, and Biorefining* 1: 14–17.

De La Torre Ugarte, D. G., M. E. Walsh, H. Shapouri, and S. P. Slinsky. 2000. *The Economic Impacts of Bioenergy Crop Production on U.S. Agriculture, Agricultural Economic Report 816*. Washington, DC: USDA Office of the Chief Economist, U.S. Department of Agriculture.

Dewulf, J., H. Van Langenhove, and B. Van De Velde. 2005. Energy-based efficiency and renewability assessment of biofuel production. *Environmental Science and Technology* 39(10): 3878–3882.

Dias De Oliveira, M. E., B. E. Vaughan, and E. J. Rykiel, Jr. 2005. Ethanol as fuel: Energy, carbon dioxide balances, and ecological footprint. *BioScience* 55(7): 593–602.

Donner, S. D., and C. J. Kucharik. 2008. Corn-based ethanol production compromises goal of reducing nitrogen export by the Mississippi river. *Proceedings of the National Academy of Sciences* 105(11): 4513–4518.

ERS. 2009. *Feed Grains Database: Yearbook Tables*. Washington, DC: Economic Research Service, U.S. Department of Agriculture. Available online: www.ers.usda.gov/data/feedgrains/. Accessed February 10, 2011.

Fargione, J., J. Hill, D. Tilman, S. Polasky, and P. Hawthorne. 2008. Land clearing and the biofuel carbon debt. *Science* 319: 1235–1238.

Farrell, A. E., R. J. Plevin, B. T. Turner, A. D. Jones, M. O'Hare, and D. M. Kammen. 2006. Ethanol can contribute to energy and environmental goals. *Science* 311: 506–508.

FAS. 2009. Washington, DC: Foreign Agricultural Service, U.S. Department of Agriculture. Available online: www.fas.usda.gov/. Accessed February 10, 2011.

Foley, J. A., R. DeFries, G. P. Asner, C. Barford, G. Bonan, S. R. Carpenter, F. S. Chapin et al. 2005. Global consequences of land use. *Science* 309: 570–574.

Gallagher, P. W., H. Brubaker, and H. Shapouri. 2005. Plant size: Capital cost relationships in the dry mill ethanol industry. *Biomass and Bioenergy* 28: 565–571.

Gallagher, P. W., and H. Shapouri. 2009. Improving sustainability of the corn-ethanol industry. *Biofuels*, eds. W. Soetaert, and E. J. Vandamme, Ch. 12, 223–234. Chichester, West Sussex, UK: Wiley.

Goldemberg, J. 2007. Ethanol for a sustainable energy future. *Science* 315: 808–810.

Green, R. E., S. J. Cornell, J. P. W. Scharlemann, and A. Balmford. 2005. Farming and the fate of wild nature. *Science* 307: 550–555.

Groom, M. J., E. M. Gray, and P. A. Townsend. 2008. Biofuels and biodiversity: Principles for creating better policies for biofuel production. *Conservation Biology*: DOI # 10.1111/j.1523–1739.2007.00879.x.

Hammerschlag, R. 2006. Ethanol's energy return on investment: A survey of the literature 1990–present. *Environmental Science and Technology* 40: 1744–1750.

Harrison, W. 2009. The food versus fuel debate: Implications for consumers. *Journal of Agriculture and Applied Economics* 41(2): 493–500.

Hedegaard, K., K. A. Thyo, and H. Wenzel. 2008. Life cycle assessment of an advanced bioethanol technology in the perspective of constrained biomass availability. *Environmental Science and Technology* 42(21): 7992–7999.

Hill, J., E. Nelson, D. Tilman, S. Polasky, and D. Tiffany. 2006. Environmental, economic, and energetic costs and benefits of biodiesel and ethanol biofuels. *Proceedings of the National Academy of Sciences* 103(30): 11206–11210.

Kaltschmitt, M., G. A. Reinhardt, and T. Stelzer. 1997. Life cycle analysis of biofuels under different environmental aspects. *Biomass and Bioenergy* 12(2): 121–134.

Kim, S., and B. E. Dale. 2002. Allocation procedure in ethanol production system from corn grain. I. system expansion. *International Journal of Life Cycle Assessment* 7(4): 237–243.

Kim, S., and B. E. Dale. 2004. Cumulative energy and global warming impacts from the production of biomass for biobased products. *Journal of Industrial Ecology* 7(3–4): 147–162.

Kim, S., and B. E. Dale. 2005a. Life cycle assessment of various cropping systems utilized for producing bio-fuels: Bioethanol and biodiesel. *Biomass and Bioenergy* 29: 426–439.

Kim, S., and B. E. Dale. 2005b. Environmental aspects of ethanol derived from no-tilled corn grain: Nonrenewable energy consumption and greenhouse gas emissions. *Biomass and Bioenergy* 28: 475–489.

Kim, S., and B. E. Dale. 2006. Ethanol fuels: E10 or E85—Life cycle perspectives. *International Journal of Life Cycle Assessment* 11(2): 117–121.

Kim, S., and B. E. Dale. 2008. Life cycle assessment of fuel ethanol derived from corn grain dry milling. *Bioresource Technology* 99: 5250–5260.

Kim, S., and B. E. Dale. 2009. Regional variations in greenhouse gas emissions of biomass products in the United States—Corn-based ethanol and soybean oil. *International Journal of Life Cycle Assessment* 14: 540–546.

Kim, S., B. E. Dale, and R. Jenkins. 2009. Life cycle assessment of corn grain and corn stover in the United States. *International Journal of Life Cycle Assessment* 14: 160–174.

Kim, H., S. Kim, and B. E. Dale. 2009. Biofuels, land use change, and greenhouse gas emissions: Some unexplored variables. *Environmental Science and Technology*: DOI # 10.1021/es802681k.

Kovarik, W. 2003. Ethyl: The 1920's Environmental Conflict over Leaded Gasoline and Alternative Fuels. American Society for Environmental History Annual Conference, March 26–30. Available online: www.runet.edu/~wkovarik/papers/ethylconflict.html. Accessed February 10, 2011.

Kovarik, W. 2006. Ethanol's First Century. SVI International Symposium on Alcohol Fuels, Rio de Janeiro, Brasil, November 26–29. Available online: www.runet.edu/~wkovarik/papers/International.History.Ethanol.Fuel.html. Accessed February 10, 2011.

Laser, M., E. Larson, B. E. Dale, M. Wang, N. Greene, and L. R. Lynd. 2009. Comparative analysis of efficiency, environmental impact, and process economics for mature biomass refining scenarios. *Biofuels, Bioproducts, and Biorefining* 3: 247–270.

Liska, A., and K. G. Cassman. 2008. Towards standardization of life-cycle metrics for biofuels: Greenhouse gas emissions mitigation and net energy yield. *Journal of Biobased Materials and Bioenergy* 2(3): 187–203.

Liska, A., and K. G. Cassman. 2009. Response to Plevin: Implications for life cycle emissions regulations. *Journal of Industrial Ecology* 13(4): 508–513.

Liska, A. J., and R. K. Perrin. 2009. Indirect land use emissions in the life cycle of biofuels: Regulations vs science. *Biofuels, Bioproducts, and Biorefining*: DOI # 10.1002/bbb.153.

Liska, A. J., H. S. Yang, V. R. Bremer, T. J. Klopfenstein, D. T. Walters, G. E. Erickson, and K. G. Cassman. 2008. Improvements in life cycle energy efficiency and greenhouse gas emissions of corn-ethanol. *Journal of Industrial Ecology*: DOI # 10.1111/j.1530–9290.2008.105.x.

Looker, D. 2006. Ethanol: What might have been. *Agriculture Online* 2/14/2006. Available online: www.agriculture.com. Accessed February 10, 2011.

Lynd, L. R., and M. Q. Wang. 2004. A product-nonspecific framework for evaluating the potential of biomass-based products to displace fossil fuels. *Journal of Industrial Ecology* 7(3–4): 17–32.

Mubako, S., and C. Lant. 2008. Water resource requirements of corn-based ethanol. *Water Resources Research*: DOI: 10.1029/2007WR006683.

Naylor, R. L., A. J. Liska, M. B. Burke, P. Walter, S. D. Rozelle, and K. G. Cassman. 2007. The ripple effect: Biofuels, food security, and the environment. *Environment: Science and Policy for Sustainable Development* 49(9): 30–43.

NDDC. 2006. *Ethanol History—National Legislative History.* Bismarck, ND: North Dakota Department of Commerce. Available online: www.goefuel.com/ethanol/index.html. Accessed February 10, 2011.

Outlaw, J. L., and D. P. Ernstes. 2008. *The Lifecycle Carbon Footprint of Biofuels Conference Proceedings.* Miami Beach, FL, January 29.

Perrin, R. K., N. F. Fretes, and J. P. Sesmero. 2009. Efficiency in Midwest US corn ethanol plants: A plant survey. *Energy Policy* 37: 1309–1316.

Persson, T., A. Garcia y Garcia, J. O. Paz, J. W. Jones, G. Hoogenboom. 2009a. Maize ethanol feedstock production and net energy value as affected by climate variability and crop management practices. *Agricultural Systems* 100: 11–21.

Persson, T., A. Garcia y Garcia, J. O. Paz, J. W. Jones, G. Hoogenboom. 2009b. Net energy value of maize ethanol as a response to different climate and soil conditions in the southeastern USA. *Biomass and Bioenergy* 33: 1055–1064.

Persson, T., A. Garcia y Garcia, J. O. Paz, B. V. Ortiz, G. Hoogenboom. 2010. Simulating the production potential and net energy yield of maize-ethanol in the southeastern USA. *European Journal of Agronomy* 32: 272–279.

Pimentel, D. 1991. Ethanol fuels: Energy security, economics, and the environment. *Journal of Agricultural and Environmental Ethics* 4(1): 1–13.

Pimentel, D. 2003. Ethanol fuels: Energy balance, economics, and environmental impacts are negative. *Natural Resource Research* 12(2): 127–134.

Pimentel, D. 2008. Corn and other plants for transport biofuels. *Energy and Environment* 19(7): 1015.

Pimentel, D. 2009. Biofuel food disasters and cellulose ethanol problems. *Bulletin of Science, Technology, and Society* 29(3): 205–212.

Pimentel, D., A. Marklein, M. A. Toth, M. A. Karpoff, G. S. Paul, R. McCormack, J. Kyriazis, and T. Krueger. 2008. Biofuel impacts on world food supply: Use of fossil fuel, land and water resources. *Energies* 1: 41–47.

Pimentel, D., A. Marklein, M. A. Toth, M. N. Karpoff, G. S. Paul, R. McCormack, J. Kyriazis, and T. Krueger. 2009. Food versus biofuels: Environmental and economic costs. *Human Ecology* 37: 1–12.

Pimentel, D., and T. W. Patzek. 2005. Ethanol production using corn, switchgrass, and wood; Biodiesel production using soybean and sunflower. *Natural Resources Research* 14(1): 65–76.

Pimentel, D., and M. Pimentel. 2008. Corn and cellulosic ethanol cause major problems. *Energies* 1: 35–37.

Plevin, R. J. 2009. Modeling corn ethanol and climate: A critical comparison of the BESS and GREET models. *Journal of Industrial Ecology* 13(4): 495–507.

Plevin, R. J., and S. Mueller. 2008. The effect of CO_2 regulations on the cost of corn ethanol production. *Environmental Research Letters:* DOI: 10.1088/1748–9326/3/2/024003.

Raghu, S., R. C. Anderson, C. C. Daehler, A. S. Davis, R. N. Wiedenmann, D. Siberloff, and R. N. Mack. 2006. Adding Biofuels to the Invasive Species Fire? *Science* 313: 1742.

Rendleman, C. M., and H. Shapouri. 2007. *New Technologies in Ethanol Production, Agricultural Economic Report No. 842.* Washington, DC: USDA Office of Energy Policy and New Uses.

RFA. 2009a. *Biorefinery Locations.* Renewable Fuels Association: Washington, DC. Available online: www.ethanolrfa.org/bio-refinery-locations/

RFA. 2009b. *Industry Resources: Co-products.* Renewable Fuels Association: Washington, DC. Available online: www.ethanolrfa.org/pages/industry-resources/coproducts

RFA. 2009c. *Growing Innovation. 2009 Ethanol Industry Outlook.* Renewable Fuels Association. Washington, DC. Available at: www.ethanolrfa.org/pages/annual-industry-outlook

Robertson, G. P., V. H. Dale, O. C. Doering, S. P. Hamburg, J. M. Melillo, M. M. Wander, W. J. Parton, P. R. Adler, J. N. Barney, R. M. Cruse, C. S. Duke, P. M. Fearnside, R. F. Follett, H. K. Gibbs, J. Goldemberg, D. J. Mladenoff, D. Ojima, M. W. Palmer, A. Sharpley, L. Wallace, K. C. Weathers, J. A. Wiens, and W. W. Wilhelm. 2008. Sustainable biofuels redux. *Science* 322: 49–50.

Rosenberger, A., H. -P. Kaul, T. Senn, and W. Aufhammer. 2002. Costs of bioethanol production from winter cereals: The effect of growing conditions and crop production intensity levels. *Industrial Crops and Products* 15(2): 91–102.

Rosentrater, K. A. 2007. Ethanol processing coproducts—A review of some current constraints and potential directions. *International Sugar Journal* 109(1307): 1–12.

Rosentrater, K. A., and M. Giglio. 2005. What are the challenges and opportunities for utilizing distillers grains? *Distillers Grains Quarterly* 1(1): 15–17.

Searchinger, T., R. Heimlich, R. A. Houghton, F. Dong, A. Elobeid, J. Fabiosa, S. Tokgoz, D. Hayes, and T. -H. Yu. 2008. Use of U.S. croplands for biofuels increases greenhouse gasses through emissions from land-use change. *Science* 319: 1238–1240.

Searchinger, T., and R. Heimlich. 2009. Likely impacts of biofuel expansion on Midwest land and water resources. *International Journal of Biotechnology* 11(1–2): 127–149.

Service, R. F. 2009. Another biofuels drawback: The demand for irrigation. *Science* 326: 516–517.

Shapouri, H. 1998. Estimating the net energy value of corn-ethanol. *Fuel and Energy Abstracts* 39(3): 198.

Shapouri, H., and P. Gallagher. 2005. *USDA's 2002 Ethanol Cost-of-Production Survey, Agricultural Economic Report No. 841.* Washington, DC: USDA Office of Energy Policy and New Uses.

Shapouri, H., J. A. Duffield, and M. S. Graboski. 1995. *Estimating the Net Energy Balance of Corn Ethanol, Agricultural Economic Report NO. 721.* Washington, DC: U.S. Department of Agriculture.

Shapouri, H., J. A. Duffield, and M. Wang. 2002. *The Energy Balance of Corn Ethanol: An update, Agricultural Economic Report No. 813.* Washington, DC: U.S. Department of Agriculture.

Shapouri, H., J. A. Duffield, and M. Wang. 2003a. The energy balance of corn ethanol: An update. In *The Third International Starch Technology Conference—Coproducts Program Proceedings,* eds. M. Tumbleson, V. Singh, and K. Rausch, June 2–4, 2003, University of Illinois, Urbana, IL, p. 73.

Shapouri, H., J. A. Duffield, and M. Wang. 2003b. The energy balance of corn ethanol revisited. *Transactions of the ASAE* 46(4): 959–968.

Sheehan, J., A. Aden, K. Paustian, K. Killian, J. Brenner, M. Walsh, and R. Nelson. 2004. Energy and environmental aspects of using corn stover for fuel ethanol. *Journal of Industrial Ecology* 7(3–4): 117–146.

Sheehan, J., A. Aden, C. Riley, K. Paustian, K. Killian, J. Brenner, D. Lightle, R. Nelson, M. Walsh, and J. Cushman. 2002. *Is Ethanol from Corn Stover Sustainable?* Golden, CO: National Renewable Energy Laboratory, U.S. Department of Energy.

Staff, C. H. 2005. Question and answer. *Biofuels Journal* 3(4): 26–27.

Thiex, N. 2009. Evaluation of analytical methods for the determination of moisture, crude protein, crude fat, and crude fiber in distillers dried grains with solubles. *Journal of AOAC International* 92(1): 61–73.

Tilman, D., R. Socolow, J. A. Foley, J. Hill, E. Larson, L. Lynd, S. Pacala, J. Reilly, T. Searchinger, S. Somerville, and R. Williams. 2009. Beneficial biofuels—The food, energy, and environment trilemma. *Science* 325: 270–271.

UMN. 2009. *The Value and Use of Distillers Grains By-products in Livestock and Poultry Feeds.* University of Minnesota: Minneapolis, MN. Available online: www.ddgs.umn.edu/. Accessed February 10, 2011.

Urbanchuk, J. M. 2009. *Contribution of the Ethanol Industry to the Economy of the United States.* Wayne, PA: LECG.

U.S. EIA. 2009. *Annual Energy Review.* Washington, DC: Energy Information Administration, U.S. Department of Energy. Available online: www.eia.doe.gov/emeu/aer/. Accessed February 10, 2011.

U.S. EIA. 2008. *Ethanol Timeline.* Washington, DC: Energy Information Administration, U.S. Department of Energy. Available online: www.eia.doe.gov/kids/energy-cfm?page=H_ethanol

U.S. Grains Council. 2007. *An Independent Review of US Grains Council Efforts to Promote DDGS Exports.* Washington, DC: U.S. Grains Council. Available online: www.grains.org/ddgs-information. Accessed February 10, 2011.

Way, R. 2008. Wanamingo, MN: Birthplace of nation's first commercially viable ethanol plant. *Minnesota Post* 1/7/2008. Available online: www.minnpost.com/stories/2008/01/07/485/minnesotas_corn_ethanol_industry_blends_subsidies_politics_and_lobbying. Accessed February 10, 2011.

Weisz, P. B., and J. F. Marshall. 1979. High-grade fuels from biomass farming: potentials and constraints. *Science* 206:24–29.

Wishart, R. S. 1978. Industrial energy in transition: a petrochemical perspective. *Science* 199: 614–618.

3 Historical Perspective on Distillers Grains

Charlie Staff

CONTENTS

3.1 INTRODUCTION

The rapid growth of the fuel ethanol industry and recent amount of research on distillers grains and utilization have in most cases not indicated the interesting historical perspective of prior years. While the earlier years were of smaller beverage ethanol plants, the distillers grains products were very similar in many respects to those produced from giant fuel ethanol plants of today. Plus, the distillers grains coproducts are being utilized and fed to livestock in the same manner—to provide a valuable feed ingredient and an important plant revenue.

3.2 HISTORY OF DISTILLERS GRAINS

The reality is that distillers grains has been readily utilized in animal feeds for over 100 years in the United States. By common understanding and feed ingredient definition, distillers grains is the resulting product that remains after distillation of alcohol from grain that has been fermented by yeast. Therefore, grain, yeast, fermentation, and distillation are all required in the production of distillers grains. While true for distillers grains, it is not true for ethanol because it can be produced from many different sources (sugar cane, fruit, or mostly any carbohydrate source by fermentation or by chemical reaction from other organic compounds).

In early frontier days of the United States in the 1700s, clearing the frontier land, raising crops, and feeding animals were the major ways people survived. These frontiersmen brought with them from Europe most of the then-known agricultural practices and among these was the knowledge for producing distillers alcoholic beverages. Those knowledgeable found that this practice was quite helpful because being in a far rural area with few if any passable roads, transporting their harvested grain to a market was near to impossible. A much easier solution with higher financial rewards was fermenting the grain with yeast to produce alcohol and using simple pot stills distilling off the concentrated alcohol and placing it into wooden barrels (common packaging for commodities in this era). These new and more commonly used barrels were much easier to transport (usually using river or water systems) than loose grain. This had an even greater benefit in that after distillation there remained the liquid mixture (stillage), which was used to feed livestock and provided a second source of food and/or income. An example of this practice is President George Washington's restored distillery at his farm at Mount Vernon, Virginia. Details may be found at Distilled Spirits

Council of the United States website (www.discus.org/heritage/washington.asp). During the 1800s these on farm distilleries grew in number and size. The distillation equipment was expensive and knowledge about its use was limited so it became more convenient and useful for nearby farmers to bring their grain to those established distilleries and for a portion have their grain converted into distilled alcohol. This resulted in the farmers with distillation equipment to become larger; and therefore, larger numbers of livestock to consume the stillage, which at that time was called "slop." Very little is known about the composition or variability of "slop" during this time as there seems to be almost no written records. However, it was probably very similar to the author's experience when he visited the former Soviet Union in 1992 and toured two different "country distilleries." These were small with one continuous still, using grain (corn), freshly malted wheat (for enzymes), propagated yeast, and a feed lot with beef steers. The hot stillage, direct from the still, flowed in a pipe by gravity into a large tank that steers drank from. The condition of the steers looked remarkably good and I was told this and some grass or hay was all the feed that was required. Similar situations no doubt also occurred in the United States in the 1800s. The problem developed that the distilled product became such a profitable venture that distilleries became more numerous in number and some were even large enough to install more productive small continuous distillation stills. This was beginning to lead to a problem with handling of the increased amount of stillage. The population around the distilleries was increasing, larger space was needed for the livestock near the distillery, and managing livestock was less profitable, so in many cases what was not consumed by livestock went into the closest stream of water.

Apparently there was some drying of distillers grains in the late 1800s as it has been reported that distillers dried grains was widely used and was a popular feed ingredient in dairy rations prior to 1900 (Loosli and Turk, 1952). There seems to be few reported feeding studies to indicate the value of these distillery coproducts compared to other feeds available at the present time. Another reference indicates that "over twenty million bushels of grain, mostly maize, are used annually in the distilleries of the United States (Thomas, 1904). The annual output of distillers dried grains exceeds forty thousand tons and is largely exported to Germany for cattle feeding."

An early 1905 University of Pennsylvania feeding study indicates equal dairy cattle performance with cottonseed meal, tending to increase fat content of milk (Armsby and Risser, 1905). So it is quite evident that there was a substantial amount of distillers grains being dried and that the quality was good and at least equal to cottonseed meal available at that time. There seems to be very little record as to the process or equipment that was used during this time in the production and drying of distillers grains.

The period 1920–1933 was the United States' prohibition of all distilled alcoholic beverage and therefore the distilleries were closed except for a very few producing so-called medicinal products. In 1934, after prohibition, a complete change occurred in the distilling industry. A number of new distilleries were built, some very large for that period, such as the Hiram Walker Distillery in Peoria, Illinois and the Seagrams Distilleries in Lawrenceburg, Indiana and Louisville, Kentucky. These were both Canadian firms that had been producing distilled whiskies in Canada during the United States' prohibition period. Of course, these newer distilleries incorporated the latest technologies of that time period and all included drying equipment for producing dried distillers grains.

The first official definition for distillers dried grains was by the Association of Feed Control Officials (AAFCO) by 1913 and accepted in 1915 by the American Society of Animal Science Committee on terminology (Committee on Terminology Report, 1915). The terminology agreed to was: "The dried residue from cereals obtained in the manufacture of alcohol and distilled liquors. The product shall bear the designation indicating the cereal predominating." Today, the AAFCO definition used commercially to define the distillers grains products has been expanded to include sources from different grains, concentrated and dried solubles, wet and dried.

Boruff and Miller (1938) provide the earliest records as to the issues regarding the need to dry and recover a saleable feed product after distillation of alcohol (Boruff and Miller, 1938; Colley, 1938). By 1933 most states had sanitary laws governing stream pollution and most cities prohibited feeding of livestock in conjunction with the distillery. So, by necessity, these larger distilleries (20,000 bushels per day) installed the equipment to dry distillers grains even though at the time it was an expensive ($0.7–1.0 MM) and complicated process. Typically the process consisted of screening the hot stillage from the stills through metal or cloth screens and mechanically pressing the solids held on the screen (20% solids). The liquid portion that passed through the screens was sent to evaporators and concentrated to 15%–17% solids, which was then combined with the pressed solids and sent to rotary dryers. The authors state that the Hiram Walker Peoria, Illinois plant, was the first to incorporate basket centrifuges to further remove fine particulates from the screened effluent. This removal of a small percentage (approximately 10% solids) of gelled particles (sludge) by the centrifuges made it possible in the evaporators to increase the final evaporated solids from about 26% to 45%–50%, which resulted in a significant improvement in drying costs. The evaporated solubles (50% moisture) plus sludge from centrifuges (80% moisture) was added back and mixed with the wet pressed solids held on the screens (70% moisture). Included in the mixer were recycled dried grains from the rotary dryers. The product from the paddle mixer was typically uniformly fed to steam tube rotary dryers, which in the case of the Hiram Walker plant was five dryers, each 8 ½ ft in diameter and 26 2/3 ft long. Thus, while the scale is much smaller, it is interesting to note that some of the unit processes used were quite similar to those used in large fuel ethanol plants today. Of further interest are the commercial efforts during this period to solvent extract corn oil from dried distillers grains (Boruff and Miller, 1937; Walker, 1951). Also during this time there were considerable research and pilot testing of the anaerobic fermentation of distillery stillage to biogas with determination of gas yield and analysis of remaining fermentor sludge (Boswell and Hatfield, 1938).

During World War II in the early 1940s, there was a lack of protein feeds due to the sharp growth in animal production. The government realized that recovering and drying the stillage from the distilleries were ways of adding to availability of feed supply. Therefore, the government financed or helped finance approximately 80 distillers to establish facilities to recover and dry distillers grains. This additional feed supply was considered a major benefit to meeting the critical needs for expanded meat production. Development and commercial implementation of drying concentrated distillers solubles using double drum dryers were accomplished by Hiram Walker prior to World War II (Boruff and Weiner, 1944). Several of the larger distilleries installed the drum dryers because there were increasing demand and utilization by the poultry industry, and the dried solubles provided an attractive premium price over regular dried distillers grains. Also, they provided higher water-soluble vitamin and nutritional value equal to dried milk, which was especially important during World War II. It was thought that much of this nutritional value came from the very high content of inactive yeast cells, which was estimated by hemacytometer counts to be 3.5–4.5 billion cells/g. Comparing this to the same yeast culture dried would equate the dried solubles being 20% by weight dried yeast (Bauernfeind and Garey, 1944).

From initial sales in 1940, commercial sales of dried distillers solubles grew to approximately 20,000 tons in 1943 and were finding commercial usage in poultry, swine, calf, and dog rations. However, after 1947 the total sales of distillers grains declined due to less alcohol production because barreled whisky warehouses were finally being replenished after World War II. Plus the distillers dried solubles was more expensive to dry and it was becoming more difficult to obtain a significant premium price over distillers dried grains with solubles. Joseph E. Seagrams & Sons was reported to be the first to install a spray dryer in their Lawrenceburg, Indiana plant to commercially dry distillers solubles (Ridgway and Baldyge, 1947). Their use of boiler flue gas as the heat source reportedly led to issues of potential fires, excessive wear on atomizer nozzles, and excessive end product ash due to inefficient boiler fly ash separation. It is apparent that spray drying was not adopted by other members probably due to high capital costs and operational complexity, and the industry practice of drying solubles was soon abandoned altogether.

Typical composition of distillers light grains (without solubles), dark grains (with solubles), and dried solubles can be found in Tables 3.1 and 3.2. With the building of the larger beverage ethanol plants after prohibition and the following strong demand for ethanol during World War II,

TABLE 3.1
Chemical Composition of Distillers Feed from Corn, Rye, and a Typical Commercial Mash Bill

	Light Grains			Dark Grains (Grains with Solubles)			Solubles		
	Corn, 90% B. Malt 10%	Rye, 85% R. Malt 15%	Typical M. B.[a]	Corn, 90% B. Malt 10%	Rye, 85% R. Malt 15%	Typical M. B.[a]	Corn, 90% B. Malt 10%	Rye, 85% R. Malt 15%	Typical M. B.[a]
Moisture (%)	11	11	11	11	11	11	5	5	5
Protein (%)	30	18–26	28	28	29	28	25	36–38[c]	27
Fat (%)	11	6	10	9	3	8	6	1	5
Fiber (%)	11	15	12	7	8	7	1–3[b]	1–3[b]	1–3[b]
Ash (%)	2	2	2	6	6	6	8–9	8–9	8–9
N.F.E. (%)	35	44	37	39	43	40	52–55	44–49	51–54

Source: Boruff, C. S. *Our Products.* Second Conference of Feeds of the Beverage Industry, February 27, 1947, 12–13.

[a] Typical mash bill—70% corn, 20% rye, and 10% barley malt.
[b] Lower (1%) fiber content in solubles produced by use of centrifuges.
[c] Protein content of rye solubles decreases as mashing temperatures are increased.

TABLE 3.2
Vitamin Content (Micrograms per Gram) of Distillers Feed Products from Corn, Rye, and a Typical Plant Mash Bill

	Light Grains			Dark Grains (Grains with Solubles)			Solubles		
	Corn, 90% B. Malt 10%	Rye, 85% R. Malt 15%	Typical M. B.[a]	Corn, 90% B. Malt 10%	Rye, 85% R. Malt 15%	Typical M. B.[a]	Corn, 90% B. Malt 10%	Rye, 85% R. Malt 15%	Typical M. B.[a]
Thiamin	1	1	1	5	3	4	9	3	7
Riboflavin	3	3	3	9	8	9	18	14	17
Niacin	30	15	30	85	62	82	160	110	150
Pantothenic Acid	3	3	3	12	17	13	23	32	25
Biotin	—	—	—	—	—	1	—	—	2
Pyridoxin	—	—	—	—	—	—	—	—	9
Folic Acid	—	—	—	—	—	—	—	—	1
Choline	—	—	2000	—	—	4500	—	—	6500
P-aminobenzoic Acid	—	—	—	—	—	—	—	—	—
Unidentified essential factors	—	—	—	—	—	—	++++	—	++++

Source: Boruff, C. S. *Our Products.* Second Conference of Feeds of the Beverage Industry, February 27, 1947, 12–13.
[a] Mash bill of 70% corn, 20% rye, and 10% barley malt.

there was a significant increase in the production of both distillers dried grains and distillers dried solubles. The dried distillers grains increased in 1938 from approximately 150,000 tons to approximately 700,000 tons in 1945, while during the same years dried solubles became commercially available in 1939 and by 1945 approximately over 85,000 tons was being produced (Boruff, 1947). This increased availability of lower cost feed ingredients produced a number of animal feeding studies at universities and also larger beverage distilleries. Hiram Walker hired Dr. R. A. Rasmussen, an animal nutritionist, and Joseph E. Seagram (who had their own research farm), were both industry leaders and had active ongoing animal feeding studies. The animal research in this area had grown so extensively that in 1945, Seagrams Distillers Corporation hosted a meeting of industry, university, and government attendees to discuss the research being done and how to better coordinate and communicate the results in hopes that less duplication would occur. A list of presentations at this meeting is found in Figure 3.1. From this meeting came an industry consensus to establish a nonprofit organization "Distillers Feed Research Council (DFRC)" with Dr. Frank Shipman as its chairman. Dr. Philip Schaible was hired as Director of Research and William Stice as Educational Director. These men became very effective in establishing a

---

FIRST DISTILLERS GRAINS CONFERENCE AGENDA

Call to Order 7

Opening Address, Gen. Frank R. Schwengel 7

Address of Welcome, Mr. Owsley Brown 7

Harnessing Industry to War, Gen. Donald Armstrong 8

Permanent Chairman, Mr. H.F. Willkie takes Chair 11

Protein Feed Recovery Program of the United States Department

    of Agriculture, Mr. Walter Berger 12

Wet Feeding of Cattle, Dr. Frank M. Shipman 20

Dry Feeding of Cattle, Mr. R.S. Mather 23

Swine & Poultry Feeding, Mr. R.A. Rasmussen 26

Report of Industry on Dry Feed Recovery and Statistical Data on Number

    of Cattle Fed Wet Stillage, Mr. Don A. Fisher 33

Production and Distribution of Branded Feed Products, Mr. Charles P. Bur 39

The Feed Manufacture and Distillers By-Products, Mr. Syl Fisher 41

Luncheon Menu 43

The Grain Outlook, Mr. Wm. McArthur 44

The Farmer Looks to the Future, Mr. James Stone 53

Gen. Frank R. Schwengel Resumes the Chair 56

Resolution, Mr. Jos. A. Engelhard 57

Adjournment 58

Appointments to Permanent Standing Committee 59

Roster of those Registered at Conference 60

FIGURE 3.1 1945 Proceedings of the Conference on Feed and Other By-Product Recovery in Beverage Distilleries.

distillers grains research program that identified and funded research projects at major universities utilizing member dues, and an active promotional program to encourage feeding of distillers grains. Since its earliest inception, DFRC hosted annual distillers grains conferences, with the early meetings in Cincinnati, Ohio. Figures 3.2 and 3.3 are two of the many pictures we have in our files showing the active distillers grains promotional efforts at the 1947 conference and at subsequent meetings. The goal was always to bring together presentations of all the latest research on distillers grains utilization. Until 1996 these presentations were published by DFRC as written and bound conference proceedings. From 1945 to 1960 funds for the administration

FIGURE 3.2 Distillers Feed Research Council Meeting, January 24, 1950, Cincinnati, OH.

FIGURE 3.3 Distillers Feed Research Council Meeting, March 15, 1951, Cincinnati, OH.

of DFRC and its research programs originated from DFRC member companies, most of which were beverage alcohol producers. In 1960, DFRC became affiliated with and funded by Distilled Spirits Council of the U.S. (DISCUS) headquartered in Washington, DC. This affiliation ended in 1985. The loss of funding occurred due to a steadily declining whisky market (since 1968) and it became an economic necessity for DISCUS to separate DFRC as a totally independent organization and begin to again charge member dues. While this was a benefit in that DFRC could seek members outside of DISCUS members, it was a major problem for many small whisky producers to pay dues and remain members. By retaining large and some smaller whisky producers, with some early fuel ethanol producers, DFRC continued with reduced research funds exhibiting at animal shows and their prominent annual distillers grains conference. By 1996, with continuing loss of membership and resignation of the Director, the DFRC Board wisely sought out a new direction with a cooperative agreement with the University of Louisville, Louisville, Kentucky in which DFRC was to fund respective office costs and organizational management salaries. Also a different organizational name, Distillers Grains Technology Council (DGTC), was approved by the Board of Directors to indicate rapid industry technological and utilization changes occurring at that time. Clearly rapidly increasing petroleum price, lower corn prices, and encouragement of federal and states with tax subsidies were going to bring about the "second ethanol revolution" (gasohol being the first). Thus, profitability was back to producing fuel ethanol, which resulted in a greatly increased number of large dry mill ethanol facilities in the Midwest Corn Belt and increasing quantities of the coproduct, distillers grains. What to do with this growing amount of distillers grains was a deep concern in the industry. Research on feeding distillers grains to animals was well underway at universities and was being funded by other associations and government, so limited DGTC funds could be diverted from this area. What was lacking was the great need to communicate past and current research results on benefits of feeding distillers grains to animals. Therefore the efforts of DGTC since 1996 have been to increase presentations at animal meetings, exhibiting at dairy shows and animal science meetings and continuation of the annual symposium on distillers grains.

The energy crises in the United States during the late 1970s and subsequent ethanol subsidies brought on the first wave of building very large fuel ethanol plants (50+ MM gal/y) (Food Engineering, 1985). This was commonly referred to as the "gasohol era" (Reilly, 1979). Many of the large ethanol plants built during this period (1978–1988) were of questionable design, especially in regards to production of quality dried distillers grains. Problems with starch conversion, fermentation contamination, and inadequate drying capacity resulted in higher sugar, under-dried (high moisture), and over-dried (burnt-dark color-lower nutrition) distillers grains. This led to considerable market place complaints from nutritionists and feed formulators as to the variability in quality, and consequently resulted in a lost confidence of all distillers grains suppliers. This impacted the beverage alcohol distilleries as well that had developed well-established markets with known and trusted quality distillers grains, particularly in central and northeastern markets. It also created many upsets in the price structure of the distillers grains marketplace due to continued offerings of lower priced distressed products. This became such a problem to the members of DFRC that they proceeded to seek from AAFCO new ingredient definitions for distillers grains produced at fuel ethanol facilities. These had been approved to the level of tentative status when by 1998, most of the old gasohol plants had either shut down or made production changes to correct quality issues, at which time the renamed DGTC requested to AAFCO that the tentative definition for fuel ethanol distillers grains be eliminated from consideration for implementation. AAFCO Ingredient Definition Committee voted to approve this request and there have been no new distillers grains definitions added since 1982, all of which are found in the AAFCO Operations Manual, which can be purchased at (www.aafco.org). The distillers grains quality issues of the gasohol plants were used very effectively by the newly designed built (1993–2000) and equipped fuel ethanol plants in Minnesota and South Dakota. Utilizing lower drying temperatures and times, these plants produced dried distillers grains with a yellow to light brown color. These were effectively marketed as the "new generation" plants that offered "golden yellow" distillers grains as a means of differentiating them from the well-known quality problems of the earlier gasohol plant distillers grains.

In the early 1980s began the first commercial operation of a fuel ethanol plant located adjacent to a cattle feed lot using wet (30% solids) distillers grains as a source of feed. The Reeve Cattle Co. (Garden City, Kansas) continues to operate this ethanol distillery feed lot today and has expanded the production of the distillery and feed lot several times (Reeve, 1995). It seems ironic that this model is being successfully repeated on a much larger scale from the early 1800s where distilleries and feeding animals were closely located together.

3.3 CONCLUSIONS

Distillers grains has had a remarkable history of growth from the early, very small beverage ethanol plants to the mega gallon fuel ethanol plants of today. Environmental issues of the beverage plants in the 1940s forced the concentration and drying of the stillage so it could be stored, transported, and marketed. This however meant that animal research had to be collectively funded by these early beverage distillers to substantiate and convince the animal industry that distillers grains was a valuable, nutritious feed ingredient. This established a base of both processing and utilization knowledge for distillers grains, which in many ways benefitted the rapid growth and availability issues of the fuel ethanol era. Ideas that were adopted and/or abandoned in the past have been or may be tried in the future by the fuel ethanol plants as they seek out higher valued end products and markets.

Distillers grains composition and handling characteristics are highly dependent on type of grain used as a feedstock. The regional grain availability and transportation costs are a significant contributing factor in distillers grains variability for different plants, and this has sometimes been forgotten.

REFERENCES

Armsby, H. P., and A. K. Risser. 1905. Distillers dried grains vs. cottonseed meal as a source of protein. *Pennsylvania Agricultural Experiment Station Bulletin* 73.

Bauernfeind, J. C., and J. C. Garey. 1944. Nutrient content of alcohol fermentation by-products from corn. *Industrial and Engineering Chemistry* 36: 1 76–78.

Boruff, C. S. 1947. *Our Products—Animal Feed Products of the Distilling Industry.* Second Conference of Feeds of the Beverage Industry, February 27, 1947, 12–13.

Boruff, C. S., and D. Miller. 1937. Solvent extraction of corn oil from distillers grains. *Oil and Soap* 14: 312–313.

Boruff, C. S., and D. Miller. 1938. Complete recovery of distillery wastes. *Municipal Sanitation* 9: 259–261.

Boruff, C. S., and L. P. Weiner. 1944. *Feed By-products from Grain Alcohol and Whisky Stillage.* Proceedings of First Industrial Waste Utilization Conference, Purdue University, November 29–30, 1944.

Boswell, A. M., and W. D. Hatfield. 1938. Anaerobic fermentations. *State of Illinois Water Survey Bulletin No. 32:* 1–193.

Committee on Terminology Report. 1915. *Journal of Animal Science* 117.

Colley, L. C. 1938. Distillery by-products. *Industrial & Engineering Chemistry* June: 615–621.

Food Engineering. 1985. *Food Engineering* March: 156.

Loosli, J. K., and K. L. Turk. 1952. The value of distillers feeds for milk production. *Journal of Dairy Science* 35: 868.

Reeve, L. 1995. Lessons learned in feeding of high levels of distillers co-products over the last decade. *Distillers Feed Research Council Proceedings,* March 20–31, 1995, 71–72.

Reilly, J. P. 1979. Gasohol—future prospects. *Distillers Feed Research Council Proceedings,* March 29, 1979, 1–4.

Ridgway, J. W., and W. V. Baldyge. 1947. Progress on Spray Drying of Distillers Solubles. *Proceedings of the Third Industrial Waste Conference,* Purdue University, May 21–22, 1947, 128–137.

Thomas, F. H. 1904. *The Cereals in America.* New York: Orange Judd Co., 266.

Walker, D. D. 1951. Extraction of distillers dried grains in a soybean solvent extraction plant. *Journal of the American Oil Chemists' Society* 195–197.

4 Grain Structure and Composition

KeShun Liu

CONTENTS

4.1 INTRODUCTION

Cereals grains are the fruits of cultivated grasses. As members of the monocot family of *Gramineae*, cereal crops are mostly grown in the temperate and tropical regions of the world, and provide more food energy worldwide than any other type of crop. They are therefore staple crops. The principal cereal crops are corn (also known as maize), rice, wheat, barley, oats, rye, sorghum, triti-cale, and millets.

The word *cereal* is derived from *Ceres*, the name of the Roman goddess of harvest and agriculture. For thousands of years, cereals have been important food crops. Their successful production, storage, processing, and end utilization have undoubtedly contributed to development of modern civilization. In recent years, there emerges a new opportunity for production growth of cereals. As the world's petroleum supplies continue to diminish, there is a growing need for alternative fuels. One such substitute is fuel ethanol that is derived from renewal biomass, including sugary crops, starch-containing grains, and cellulosic materials. Although the current practice of producing fuel ethanol from grains and its phenomenal growth are debatable with respect to sustainability and a perceived link with increasing grain price, and thus are politically charged, there is no question that millions of tons of grains are now converted to fuel ethanol, with distillers dried grains with solubles (DDGS) (which contain nonfermented residues) as the main coproduct (Sanchez and Cardona, 2008; Balat and Balat, 2009; RFA 2010; Chapter 2 of this book). For ethanol production, corn is by far the most common cereal grain used in the United States (RFA, 2010). However, in other parts of the world, other grains such as sorghum, wheat, barley, and millets are also used (see Chapter 6).

Regardless of whatever the end uses are for cereal grains, as food, feed, fuel ethanol, or others, a knowledge of the structure and composition of cereal grains is necessary not only for understanding and optimizing plant growth and seed development to achieve the highest level of grain production, but also for developing improved methods of storage and handling for maximum preservation of grain quality, and methods of suitable processing for most efficient end utilization. This chapter is aimed at providing general information about structure and composition of cereal grains as well as the unique features of each cereal grain. These pieces of information are important for understanding the principles for bioethanol production from grains. They are also important for maximizing efficiency of ethanol production and improving quality of the DDGS coproduct. Additional information can be found in Kent and Evers (1994), Lasztity (1999), Evers and Millar (2002), Abdel-Aal and Wood (2005), and Delcour and Hoseney (2010). Information about individual grains can also be found in White and Johnson (2003) for corn, Khan and Shewry (2009) for wheat, Marshall and Sorrels (1992) for oat, Newman and Newman (2008) for barley, Champagne (2004) for rice, and Dendy (1994) for sorghum and millets.

4.2   GRAIN STRUCTURE

4.2.1   General Structure

The fruits of most plants contain one or more seeds, which can be easily separated from the rest of the fruit tissues when the fruits ripen. However, for *Gramineae* family, this is different: a fertilized egg cell in the ovary develops into a single seeded fruit. The fruit wall (pericarp) and seed coat are united and there is no simple method to separate the two. This type of fruit is characteristic for all grasses (including cereals). It is botanically known as *caryopsis*. Caryopses of cereals may unambiguously be called grains while the terms berry and kernel are generally used to describe other types of fruits. However, the word kernel or seed is frequently used to describe caryopses of cereals also, although strictly speaking, they do not have the same definition (Kent and Evers, 1994).

Although the structural parts of cereal grains are adapted differently among cereal species, a generalized structure can be described for all cereals (Figure 4.1). It mainly comprises embryo, endosperm, and tissues surrounding the embryo and endosperm. Some grain species also contain hulls (husks). Each of these structural parts also consists of different structural tissues.

4.2.1.1   Embryo

The embryo results from the fusion of male and female gametes. It is the most important grain component for the survival of the species as it is capable of developing into a plant of the filial generation Evers and Millar (2002). A mature embryo consists of an embryonic axis and scutellum. The

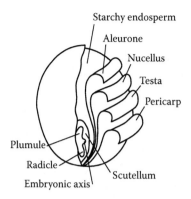

Starchy endosperm
Aleurone
Nucellus
Testa
Pericarp
Plumule
Radicle
Scutellum
Embryonic axis

FIGURE 4.1 Diagram of general structure of cereal grain, showing the relationships among the tissues. (Reprinted from Kent, N. L., and A. D. Evers, *Kent's Technology of Cereals,* 4th ed. Pergamon Press Ltd, Oxford, 1994.)

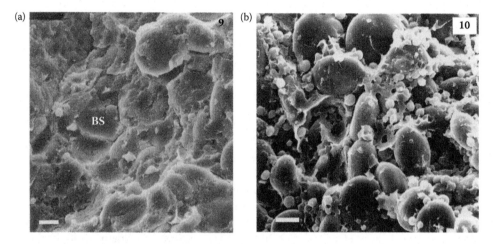

FIGURE 4.2 Scanning electron micrograph of hard (a) soft (b) wheat, showing the contents of endosperm cells. The white bar length is 10 μm. (Adapted from Hoseney, R. C. and P. A. Seib, Structural differences in hard and soft wheat. *The Baker's Digest,* 47(6): 26–28, 56, 1973.)

embryonic axis is the plant of the next generation. It further consists of primordial roots (known as radicle) and shoot with leaf initials (known as plumule). Scutellum, the name deriving from its shield-like shape, lies between the embryonic axis and endosperm. It consists mainly of parenchymatous cells, each containing nucleus, dense cytoplasm, and oil bodies, and functions as a secretory and absorptive organ to nourish the embryonic axis during germination. The word "germ" is frequently used by cereal chemists to describe the embryo. But strictly speaking, it refers to the embryo-rich fraction produced during milling.

4.2.1.2 Endosperm

The endosperm is the largest structural part of the grain. It is also the component with the greatest economic value in all grains. Nutrients stored in it are mostly insoluble. The major one is starch, followed by nonstarch polysaccharides (NSP), protein, lipid, minerals, and vitamins. The endosperm has two distinct tissues: starchy endosperm and aleurone. Starchy endosperm is a solid mass, consisting of cells that are packed with starch granules embedded in a protein matrix (Figure 4.2). These nutrients can be mobilized to support growth of the embryonic axis at the onset of germination. Starchy endosperm is commonly referred to as "endosperm" although strictly speaking it is only a major part of endosperm. Based on visual appearance, starchy endosperm of some cereals can be divided into two types, opaque (floury or mealy) endosperm and translucent (horny or vitreous) endosperm. In general, the opaque endosperm contains many air spaces. Starch granules in it are spherical in shape. Translucent endosperm is tightly compact, with few or no air spaces and its starch granules are polygonal. In wheat, some kernels have opaque endosperm while others have translucent endosperm. In corn, sorghum, and millets, both translucent and opaque areas are found in the starchy endosperm of a single kernel (Delcour and Hoseney, 2010).

Aleurone is the tissue surrounding the starchy endosperm, consisting of 1–3 layers of thick-walled block-like cells with dense contents and prominent nuclei. The number of layers present in aleurone is characteristic of the cereal species. Wheat, rye, oats, corn, and sorghum have one, while barley and rice have three. The size of aleurone cells is not a function of grain size; in wheat, for example, aleurone cells are approximately 50 μm cuboids, but in the much larger grain of corn, they are smaller. Pigmentation in the aleurone layer can give grains of some cereals a blue, red, or almost black appearance.

Unlike the tissue they surround, aleurone cells do not contain starch but have higher concentrations of protein, lipid, vitamins, and minerals than the starchy endosperm. They are extremely important in both grain development and germination. During grain development, aleurone cells divide to produce starchy endosperm cells, while during germination they are the site for synthesis of hydrolytic enzymes responsible for solubilizing the reserves. Furthermore, during roller milling, most of the aleurone layer is removed as part of the bran, leading to reduction of nutritional values of resulting flour. However, the removal of the aleurone layer can be reduced when decortication rather than roller milling is employed, as in production of pearled barley, milled rice, degermed corn, and wheat products that are milled by a process involving abrasion in the early stages (Dexter and Wood, 1996).

4.2.1.3 Tissues Surrounding the Endosperm and Embryo

The maternal tissue immediately surrounding the endosperm and embryo is known as nucellus or nucellar epidermis Evers and Millar (2002). It is the mass of tissues in which the endosperm and embryo have developed. Epidermal cells in many higher plants secrete a cuticle. For many cereals, a cuticle is present on the outer surface of the nucellar epidermis. Following fertilization, the embryo and endosperm expand at the expense of the nucellus, which is broken down except for a few remnants of the tissue and single layer of squashed empty cells. The compressed epidermis with its thin outer cuticle is all that remains of the nucellus. Nucellus is particularly prominent in sorghum but is apparently absent from mature oat grains. In corn only the cuticle persists. The nucellus is also called the hyaline layer or perisperm.

The outermost tissue of the seed is seed coat or testa (Figure 4.1). The nucellar epidermis is also regarded as a seed coat. The inner face of the testa lies adjacent to the cuticle of the nucellar epidermis. The testa consists of one layer in barley, oats, and rice, and two layers in wheat, rye, and triticale. In mature sorghum, no testa is present. In millet, if present, it is inconspicuous. The cuticle of the testa is thicker than that of the nucellar epidermis, and is responsible for the relative impermeability of the grain to water over most of its surface. It also plays a role in regulating gaseous exchange. Frequently, the testa accumulates pigmented corky substances in its cells during grain ripening and this may confer color on the grain and further contribute to impermeability Evers and Millar (2002).

In transverse section of caryopses featuring a crease, such as wheat grain, a discontinuity in the testa can be seen in that region. During grain development this facilitates transport of nutrient solution from the vascular strand to the nucellar projection and then to the endosperm. As the grain matures, impermeable material accumulates to form a pigment strand between the borders of the integuments, completing the waterproof layer around the grain. In red wheat, the pigment strand is dark brown but in white wheat it is less obvious and much paler. In corn, which does not have a crease, the tip cap area is the point of entry for nutrient transport from the plant to the developing grain. This path is sealed during maturation with a dark layer of dense cells called the hilar layer, which may be a possible source of black specks in cornmeal. This tip cap is the weakest point of the mature corn grain and is the point of water ingress when grains are steeped or tempered Evers and Millar (2002).

Pericarp refers to the caryopsis tissues that lie outside the seed coat and originate as components of the carpel wall. They are parts of the fruits but not of the seed. Therefore, pericarp is sometimes known as fruit coat. It is a multilayered structure consisting of several complete and incomplete layers. In all cereal grains, pericarp becomes dry at maturity and consists of largely empty cells. The innermost layer of the pericarp is the inner epidermis. In many cereals, this is an incomplete layer. Its cells are elongate, thus termed as "tube cells." The long axes of tube cells lie parallel to the long axis of the embryo. Outside the tube cell layer is the layer of cross cells. This layer takes its name from the fact that the long axes of its elongate cells lie at right angles to the grain's long axis (also the axis of the embryo). Cross cell shape varies among species and in different areas of the same grain. Over most surfaces of wheat, barley, triticale, and rye, they form a complete layer of cells about six times as long as their width, arranged in rows, but at the grain tips, they become squarer. Because of the fragmentary nature of the tube cell layer, it is the cross cells that adhere to the underlying cuticle of the testa. In corn, rice, sorghum, and pearl millets, there are large spaces among the distorted, more elongated cells. In oats, tube cells and cross cells are indistinguishable from other inner pericarp cells, which are all elongated and distorted with no common orientation. In rice all pericarp cells follow a cross cell orientation Evers and Millar (2002).

The mesocarp, the outside of cross cell layer, is a true layer in mature grains except in sorghum. The outermost layer of the pericarp and indeed of the caryopsis is the outer epidermis. It is one cell thick and adherent to the "hypodermis," which may be virtually absent, as in oats or some millets, one or two layers thick, as in wheat, rye, sorghum, and pearl millets, or several layers thick, as in corn. The outer epidermis has a cuticle, which probably helps control water loss in growing grains but generally becomes leaky on drying.

During seed development, pericarp serves to protect and support the growing endosperm and embryo. At this stage chloroplasts are present in the innermost layers (tube cells and cross cells) and starch accumulates as small granules in the central layers (mesocarp). By the time the grain matures, starch in the pericarp disappears and the cells in which the starch used to be are largely squashed or broken down. An exception to this feature can be found in some types of sorghum in which at least some of the cells and some of the starch granules remain.

The empty cells of the pericarp hold enough water to increase grain weight by 4%–5% after only a few minutes of immersion. They act as a reservoir for water that may ultimately enter the grain slowly through the micropyle. The absorbent nature of the pericarp is of great benefit in the milling of wheat, which is conditioned or tempered by water addition prior to milling. The grain is allowed to absorb water over a period of 12–24 h, during which time the bran layers are softened and become easily removed Evers and Millar (2002).

4.2.1.4 Lemma and Pales
All cereal caryopses are subtended on their parent plants by a pair of glumous bracts known as the lemma and pale, which in most types, surround and protect the immature fruit. In some species, nature caryopses are shed from the plant free from the structures, but in other species the lemma and pales remain as a hull, and are in close contact with the caryopsis even after it separates from the plant. Although the hull of all hulled species is formed from the lemma and pale, the mechanism of their connection with the fruit varies. The hulls of oats and rice are held in place by structural devices. The lemma and pale in rice have a rib and groove arrangement at their meeting margins. It is possible to free the fruits from them by using appropriate machines. By contrast, the lemma and pale of barley adhere to the pericarp and cannot be removed cleanly from the caryopsis by mechanical means Evers and Millar (2002). In hulless oats, hulls are less tightly adhering to the grain and thus thresh free during harvesting (Marshall and Sorrells, 1992). Hulless barleys are also cultivated (Bhatty 1999), but for malting purposes hulled varieties are preferred since hulls do not detach during mashing (Bamforth and Barclay, 1993; Newman and Newman, 2008).

It is usual for hulled grains to be traded with hulls in place and for analytical and compositional data to be presented for the grains as traded Evers and Millar (2002). Grain weights and the contributions of main morphological components are presented in Table 4.1. For hulled types (barley, oats, and rice), proportions are given for both hulled grains and caryopses. It is worth noting that the size of embryo relative to the whole grain varies with grain species. Wheat, barley, oats, and rice all have smaller embryos relative to the total grain mass, while sorghum, corn, and millets all have larger embryos.

4.2.2 Characteristics of Individual Cereal Grains

In spite of structural similarities, there are wide variations among cereal grains in size and shape. Comparison in size and shape among 12 grains are shown diagrammatically in Figure 4.3, and various weight and proportions of grain parts for some grains are given in Table 4.2. Shape, size, and mass are the most readily identifiable characteristics of the grains of individual cereal species, but within species, there is also considerable variation. Furthermore, morphology can be associated with quality parameters. For example, in rice, the ratio of length to width is a useful index of the nature of the endosperm, and more importantly it indicates the type of starch present; short grain rice, when cooked, tends to have a sticky texture. Corn grains vary from

TABLE 4.1
Typical Proportions (%) of Grain Parts in Some Cereals

Cereal	Hull	Pericarp and Testa	Aleurone	Starchy Endosperm	Embryo — Embryonic Axis	Scutellum
Wheat						
Thatcher	—	8.2	6.7	81.5	1.6	2.0
Vilmorin 27	—	8.0	7.0	82.5	1.0	1.5
Argentinian	—	9.5	6.4	81.4	1.3	1.4
Egyptian		7.4	6.7	84.1	1.3	1.5
Barley						
Whole grain	13	2.9	4.8	76.2	1.7	1.3
Caryopsis	—	3.3	5.5	87.6	1.9	1.5
Oats						
Whole grain	25		9.0	63.0	1.2	1.6
Caryopsis (groat)	—		12.0	84.0	1.6	2.1
Rye	—		10.0	86.5	1.8	1.7
Rice						
Whole grain	20		4.8	73.0	2.2	
Caryopsis						
Indian	—		7.0	90.7	0.9	1.4
Egyptian	—		5.0	91.7	3.3	
Sorghum	—		7.9	82.3	9.8	
Maize						
Flint	—	6.5	2.2	79.6	1.1	10.6
Sweet	—	5.1	3.3	76.4	2.0	13.2
Dent	—	6		82	12	
Proso millet	16	3	6.0	70	5	

Source: Data from Kent, N. L., and A. D. Evers, *Kent's Technology of Cereals*, 4th ed. Pergamon Press Ltd, Oxford, 1994.

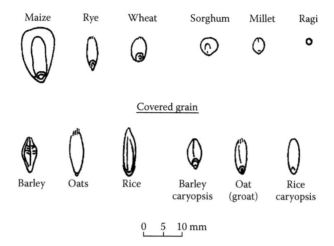

FIGURE 4.3   Grains of cereals showing comparative sizes and shapes. The caryopses of the three hulled grains, (barley, oates, and rice) are shown with and without their sorrounding pales. (Reprinted from Kent, N. L., and A. D. Evers, *Kent's Technology of Cereals,* 4th ed. Pergamon Press Ltd, Oxford, 1994.)

TABLE 4.2
Dimensions and Weight per 1000 Grains of Cereals

Cereal Grain	Dimensions		Weight per 1000 Grains	
	Length (mm)	Width (mm)	Range (g)	Average (g)
Barley	8–14	1.0–4.5	32–36	35
Corn	8–17	5–15	150–600	324
Millets (Pearl)	2	1.0–2.5	5–10	7
Oats	6–13	1.0–4.5		32
Rice	5–10	1.5–5.0		27
Rye	4–10	1.5–3.5	15–40	21
Sorghum	3–5	2–5	8–50	28
Triticale	10–12	2.5–3.0	28–48	36
Wheat	5–8	2.5–4.5	27–48	37

Source: Adapted from Kent, N. L., and A. D. Evers, *Kent's Technology of Cereals,*
4th ed. Pergamon Press Ltd, Oxford, 1994.

the near-spherical popcorn to flattened and angular flint maize. A depression on the distal face arising through contraction of endosperm is characteristic of dent corn and reflects the presence of a region of soft endosperm within a harder textured cup. Caryopses of members of the tribes *Triticease* (wheat, rye, barley, and triticale) and *Aveneae* (oats) can be distinguished by the presence of a crease, a reentrant region on the ventral side, extending along the grain's entire length, and deepest in the middle. It is most marked in wheat. In milling wheat into white flour, the crease presents the greatest difficulty for separation of starchy endosperm from other tissues.

Within a cultivar, sources of variation in grain size include growing conditions of the crop and the grain position within the inflorescence. In general the largest grains occur in the center of the inflorescence (Bremner and Rawson, 1978). Regardless of the size or shape of individual grains, the uniformity of their morphology is also important for processing. For example, more uniform wheat grains give more predictable milling performance, while malting performance of barley relies heavily on homogeneous samples. The uniformity can be readily achieved through sorting of grains by a physical parameter, such as the degree of shriveling.

4.2.2.1 Corn

Compared to other grains, corn caryopsis has a unique flat shape and low specific gravity. It has a blunt crown and pointed conical tip cap. Corn grain is also the largest among cereals, weighing

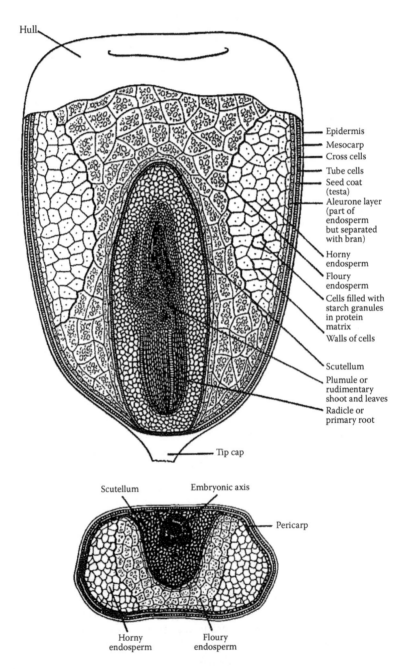

FIGURE 4.4 Longitudinal and cross sections of corn caryopsis (kernel). (Courtesy of the Corn Refineries Association, Washington, DC.)

250–300 mg each (yellow dent type). The caryopsis is attached to the cob by the pedicle, and contains a complete embryo and all other structural parts (Figure 4.4) for nutritional and enzymatic functions required for growth and development into a plant. The pericarp is thicker and more robust than that of the smaller grains. It is also known as hull or bran. The part of the hull overlying the embryo is known as the tip cap. The tip cap is the attachment point to the cob (Kent and Evers, 1994; White and Johnson, 2003).

Similar to other cereal grains, the largest fraction of the corn grain is the endosperm. Endosperm cells are packed with starch granules embedded in a combination matrix of amorphous protein Also embedded in this matrix are protein bodies composed almost entirely of the storage protein zein. In spite of the large size of the endosperm in corn, individual aleurone cells are small, comparable to those of oats and rice. Only one layer of them is present. In blue varieties, it is the aleurone cells that provide the coloration. By appearance, the starchy endosperm of a single corn kernel consists of two types of endosperm, floury and horny. The floury endosperm surrounds the central tissue and is opaque to transmitted light. Horny (also known as corneous) endosperm occurs as a deep cap surrounding a central core of floury endosperm (Figure 4.4). In the floury endosperm, many small starch granules (average 10 μm) occur. Protein (zein) also occurs in tiny granular form. In horny endosperm, the protein matrix is thicker and starch granules are compressed into polyhedral shapes. In dry milling of corn, the primary product is the isolated piece of endosperm, which is recovered by progressive grinding, sieving, and aspiration. The dry process causes floury endosperm breakage across the cell contents, releasing some free starch granules and producing a rough surface with many exposed starch granules and very little starch granules damage. In contrast, horny endosperms break more along cell wall lines but also across cells, with little release of starch granules but much granule damage (Kent and Evers, 1994; White and Johnson, 2003).

Based on shape and endosperm character, corn can generally be classified into five classes: flint corn, popcorn, flour corn, dent corn, and sweet corn. Flint corn has a rounded crown and the hardest kernels due to the presence of a large and continuous volume of horny endosperm. Popcorn is a small flint corn type. Flour corn generally also has a rounded or flat crown, but contains virtually all floury endosperm. Dent corn has a depressed crown that forms as the maturing kernel dehydrates. Among these corns, dent corn is the most abundantly grown, particularly in North and South America (White and Johnson, 2003).

Corn kernels also differ significantly in color from white to yellow, orange, purple, and brown. Color differences may be due to genetic differences in pericarp, aleurone, germ, and endosperm (Neuffer et al., 1997). However, only yellow or white dent corns are grown commercially. Some hybrids may have light tan or light orange pericarp.

In dry grind processing of grains for fuel ethanol production, yellow dent corn is mostly used as the feedstock. This is particularly true in the United States. Corn undergoes several basic steps, including dry grinding, cooking, liquefaction, saccharification, fermentation, distillation, and coproduct recovery and processing. The resulting coproduct is DDGS. DDGS consists of particles varying in size and composition. Close examination of DDGS has shown that the particles can be grouped into three classes: flakes, granules, and aggregated granules, each is determined by the composition of the structural part of corn that remains. The flakes come mostly from tip cap and broken seed coat of corn kernels. The granules are mostly nonfermentable materials that were left from ground endosperm and germ. The aggregated granules are glued together, apparently by condensed distillers solubles added during the final drying stage of the process (Liu, 2008). For more information on processing of corn into fuel ethanol, refer to Chapter 5. For more on DDGS chemical composition refer to Chapter 8; for more on DDGS physical properties refer Chapter 7.

4.2.2.2 Wheat

Wheat grain has a somewhat vaulted shape with embryo at one end, and a bundle of hairs, which is referred to as the beard or brush, at the other end. The most striking morphological characteristic of wheat is a deep crease or elongated reentrant region, which is parallel to its long axis but

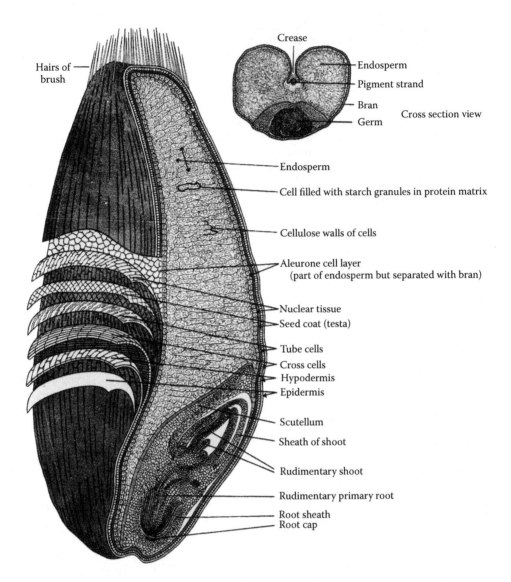

FIGURE 4.5 Longitudinal and cross sections of wheat caryopsis (kernel). (Courtesy of the North American Miller's Association, Washington, DC.)

opposite the embryo (Figure 4.5). At the inner margin of the crease there lies a layer of nucellus tissue between the testa and the endosperm. The aleurone layer consists of a single layer of cells of cubic shape. The starchy endosperm has round cells filled with starch granules embedded in a protein matrix (Figure 4.2). Two distinct populations of starch granules are present, large lenticular granules between 8 and 30 μm and near-spherical granules of less than 8 μm diameter. The former occupies about two-thirds of the starch mass. The endosperm is surrounded by the fused pericarp and seed coat (Hoseney, 1994; Khan and Shewry, 2009).

The texture of the endosperm has been a major criterion for classification of wheat, because this characteristic of the grain is related to the way the grain breaks down in milling. In hard wheat, fragmentation of the endosperm tends to occur along the lines of the cell boundaries, whereas the endosperm of soft wheat fractures in a random way. This phenomenon suggests a pattern of areas of mechanical strength and weakness in hard wheat, but fairly uniform mechanical weakness in soft wheat. Wheat hardness is related to the degree of adhesion between starch granules and the surrounding protein. Thus differences in endosperm texture must be related to differences in the nature of the interface between starch granules and the protein matrix in which they are embedded (Figure 4.2) (Simmonds et al., 1973). This is the biochemical basis for grain hardness. For the past three decades, the molecular genetic explanation of wheat grain hardness has become known, which is based on puroindoline proteins "a" and "b." When both proteins are in the functional wild state, grain texture is soft. When either one of the puroindolines is absent or altered by mutation, the result is hard texture. Durum wheat lacks puroindolines, so the texture is very hard (Morris, 2002).

The color of wheat grain can be another criterion for classification. Wheat grains have either a dark, orange–brown appearance or a light, yellowish color. The former is generally known as red wheat, while the latter known as white wheat. There was a time when many countries preferred white wheat. This preference has gradually disappeared. Red wheat varieties are popular nowadays. The main reason for this change is that white wheat is more susceptible to preharvest sprouting than the majority of the red wheat (Khan and Shewry, 2009).

During the first stage of milling, the outer layers of the wheat grain, that is the bran, are separated from the starchy endosperm. The fracture is located right under the aleurone layer. This means that bran is made up of the pericarp, the seed coat, and the aleurone layer. Wheat grain has been used for fuel ethanol production. Ojowi et al. (1997) showed that soft wheat, either soft white or soft red class, was preferred to hard wheat because soft wheat generally contains higher starch content.

4.2.2.3 Rice

The hull of rice is removed from the grain only by mechanical means, as it is locked by a "rib and groove" mechanism. The proportion of hulls in the rice grain averages about 20%. Once the hulls are removed, the outer epidermis of the pericarp is revealed as the outer layers of the caryopsis. Unlike the other small grain cereals, rice does not have a crease. It is laterally compressed, and this surface is longitudinally indented where broader ribbed regions of the pales restricted expansion during development. Distinctively, in all except one of the tissue layers (the tube cells) surrounding the endosperm, the cells are elongated transversely (in other grains only the cross cells are elongated in this direction). Aleurone cells are similar to those of oats, but the number of layers varies around the grain from one to three. Starch granules in the starchy endosperm cells are also similar to those of oats. Unusually, the embryo of rice is not firmly attached to the endosperm. Varieties of rice are classed according to grain width, length, and shape, described as short, medium, or long (Champagne, 2004). Unlike other grains, rice is almost exclusively used as human food throughout the world.

4.2.2.4 Barley

Most barley grains have adherent pales (hulls), which are removable only with difficulty. It amounts to about 13% of the grain mass on average. Some barleys are hulless. Regardless, barley grains are generally larger and more pointed than wheat. They have a ventral crease that is shallower than those of wheat and rye, and its presence is obscured by the adherent pales. Two to four (mostly three) aleurone layers are present, cells being smaller than those in wheat, about 30 µm in each direction. Blue color may be present due to anthcyanidin pigmentation (Newman and Newman, 2008).

4.2.2.5 Oats

The oat grain is featured by pales that are not removed during threshing. As they do not adhere to the groat (the term describes the actual caryopsis of the oat), they can be removed mechanically. Oats are traded with the hull in place. In this condition, they have an extremely elongated appearance and even with the hull removed groats are long and narrow.

Hulless oats is a type of oat that readily loses its husk during threshing, thus obviating the need for a special dehulling stage in milling. Comparatively, naked oats have higher protein and oil contents, and higher energy value. For hulled oats, groat contribution to the entire grain mass varies from 65% to 80%.

Oat pericarp has two layers while the testa comprises only a single layer with cuticle. In the endosperm there is a single aleurone layer. The wall of aleurone cells is not as thick as in wheat and rye. Conversely, the starchy endosperm cells have thicker walls than in wheat. Endosperm cells of oats have relatively higher protein and oil contents compared to those of other cereals (Marshall and Sorrels, 1992).

4.2.2.6 Rye

Rye grains are more slender and pointed than wheat grains, but they also have a crease and indeed share many of the features described for wheat. Rye grains may exhibit a blue–green cast due to pigment presence in the aleurone cells. Two populations of starch granules are present as in wheat, the large granules, seen under the microscope, often display an internal crack.

4.2.2.7 Sorghum

The grain of sorghum is near spherical with a relatively large embryo. It has no crease. There are two unusual features of the sorghum grain. First, its mesocarp consists of parenchymatous cells that still contain starch at maturity; the cells are not crushed as in other cereals. The starch granules in the mesocarp are up to 6 µm diameter, smaller than those in the endosperm. Second, the testa is absent. Instead, the nucellar layer is well developed, being up to 50 µm thick. Therefore, in some literature, the nucellar epidermis is referred to as the seed coat layers. In addition, cross cells do not form a complete layer; they are elongated vermiform cells aligned in parallel but separated by large spaces. Tube cells on the other hand are numerous and closer packed.

Pigmentation occurs in some sorghum. All tissues may be colored, but not all together. The starchy endosperm may be colorless or yellow and the aleurone may or may not contain pigments. The pericarp and nucellar layers may be clear, colored, or incomplete. The color of pericarp layers is mostly due to the presence of tannins, which are polyphenolic compounds. Because of unpalatability of the polyphenols, colored sorghums are certainly attacked less by birds than the white type (Dendy, 1994).

4.2.2.8 Pearl millet

Of the different millet species grown in the world, pearl millet is the most popular. Pearl millet has a small tear-shaped caryopsis that is threshed free of hulls. The kernels vary in color, including yellow, white, brown, and slate gray, but slate gray is the most common. The structure of pearl millet caryopsis is similar to that of other cereals. The germ in pearl millet is large (about 17%) in proportion to the rest of kernel. Its endosperm has both translucent and opaque endosperm, as do those of sorghum and corn. The translucent endosperm is void of air spaces and contains polygonal starch granules embedded in a protein matrix having well-defined protein bodies. The opaque endosperm contains many air spaces and spherical starch granules, also embedded in a protein matrix (Dendy, 1994).

4.2.2.9 Triticale

Triticale is a new cereal produced by crossing wheat (*Triticum*) and rye (*Secale*). Thus the morphology of its grain closely resembles its parent species. The caryopsis is generally larger than wheat, but also has a crease that extends its full length. It consists of a germ, an endosperm with aleurone as the outer layer, a seed coat, a pericarp, and the remains of the nucellar epidermis. The yellowish brown grain is featured by folds of the outer pericarp, apparently caused by shriveling of the grain, which is a major problem with triticale. Shriveling of the grain leads to low test weight, poor appearance, and unsatisfactory milling performance (Kent and Evers, 1994).

4.3 GRAIN COMPOSITION

4.3.1 GROSS COMPOSITION

The gross composition of all cereals is similar, being characterized by a small amount of water, abundant starch, sufficient protein and fiber, and relatively low amount of lipids. Minerals and vitamins are also present. Yet each species has its own distinguishing features. The differences between different cereal grains are considerable (Table 4.3).

Within a cereal species, variety and growing conditions play an important role in grain composition. Scott et al. (2006) characterized protein content and amino acid profile for a set of corn cultivars that were widely grown in different areas from the 1920s through 2001, and found that grain composition exhibited clear trends with time, with protein decreasing and starch increasing and that the grain protein content of modern hybrids responds to plant density and environment differently than the protein content of older varieties. These differences are consistent with a model in which protein content is modulated by different growth conditions. They also found that on a per tissue mass basis, the levels of the nutritionally limiting essential amino acids lysine, methionine and tryptophan dropped with time, while on a per protein basis, their levels were not significantly changed. The conclusion is that the development of modern hybrids has resulted in corn with reduced protein content, but the nutritional quality of this protein has not changed.

When grains are used for ethanol production by dry grind processing, variation in feedstock includes grain species, varieties, and blends. Even with the same species and varieties, there is variation in field condition and production year, which can lead to compositional differences of feedstock. With regard to species, corn is by far the most common cereal used for ethanol production in the U.S. However, in other parts of the world, other grains such as sorghum, wheat, pearl millets, and barley are also used. Due to differences in composition among grains the resulting DDGS are expected to differ in composition and feeding value. For example, Ortin and Yu (2009) compared wheat DDGS, corn DDGS, and blended DDGS from bioethanol plants and found great variation in chemical composition and nutritional values among them. Similarly, Kindred et al. (2008) showed the effect of variety and fertilizer nitrogen on alcohol yield (through biofermentation), grain yield, starch and protein content, and protein composition for winter wheat. Details about grain feedstock and their effect on DDGS quality is covered in Chapters 5 through 8.

TABLE 4.3
**Average Proximate Composition of Cereal Grains
(% Dry Matter)**

Cereal Grain	Protein	Oil	Starch	Ash	Total CHO[a]
Barley	10.9	2.3	53.4	2.4	84.4
Corn	10.2	4.6	69.5	1.3	83.9
Millets (Pearl)	10.3	4.5	58.9	4.7	80.5
Oats	11.3	5.8	55.5	3.2	79.7
Rice	8.1	1.2	75.8	1.4	89.3
Rye	11.6	1.7	71.9	2.0	84.7
Sorghum	11.0	3.5	65.0	2.6	82.9
Triticale	11.9	1.8	71.9	1.8	84.5
Wheat	12.2	1.9	68.5	1.7	84.2

Source: Adapted from Lasztity, R. *Cereal Chemistry.* Akademiai
     Kiado, Budapest, 1999.

[a] CHO = carbohydrate, total CHO is based on calculation ((100 −
     (% protein + % oil + % ash)).

TABLE 4.4
Chemical Composition of Embryo and Endosperm of Cereal Grains (% Dry Matter)

Cereal Grain	Embryo				Endosperm			
	Protein	Lipid	Ash	Total CHO[a]	Protein	Lipid	Ash	Total CHO[a]
Barley	21.9	16.2	6.7	55.2	9.0	1.0	0.6–0.8	89.1
Corn	17.3–19.0	31.1–35.1	9.9–11.3	34.6–41.7	6.9–10.4	0.7–1.0	0.2–0.5	88.1–92.2
Oats	36.2	—	—	—	9.6	5.9	1.8	82.7
Rice	17.7–23.9	19.3–23.8	6.8–10.1	42.2–56.2	7.3	0.4–0.6	0.4–0.9	91.5
Rye	37.2	13.4	5.8	43.6	8.8	1.0	0.8	89.4
Sorghum	—	—	—	—	8.3	0.9	0.4	90.4
Wheat	24.3–26.8	12.8	4.37–9.5	51.0–54.0	8.2–13.6	1.2	0.3	84.9–90.3

Source: Data in "corn" row adapted from Earle, F. R., J. J. Curtis, and J. E. Hubbard. *Cereal Chemistry,* 23, 504–511, 1946. The rest adapted from Lasztity, R. *Cereal Chemistry.* Akademiai Kiado, Budapest, 1999.

[a] CHO = carbohydrate, total CHO is based on calculation.

The chemical composition of cereal grains also varies greatly with structural parts (Table 4.4). The germ has the highest concentration of lipid. In fact, corn oil, an economically important commodity, comes mainly from its germ. The germ is also rich in protein, minerals, and lipid-soluble vitamins. However, its carbohydrate content is lower than in other parts of the kernel, particularly that of the endosperm. In contrast, starch is the main component of the endosperm. Protein content in endosperm is lower than that of germ. The endosperm is also lower in lipid, vitamin, and mineral contents. However, nutritional and functional significance of the endosperm is evident because of its largest proportion in a cereal grain by mass.

Besides germ and endosperm, two other structural parts that have distinct chemical composition are hulls and aleurone. The main components of hulls are NSP (nonstarchpolysaccharide), such as cellulose, hemicellulose, and lignin. No starch is found in hulls. Protein and lipid content are very low. Potassium and phosphorus are the main mineral components, although the major mineral in rice hulls is silica. Because of its nutritional insignificance, the hull is a poorly studied structural part of a grain. The aleurone layer has a quite different composition from the starchy endosperm, due to its physiological function during seed germination. It has higher protein content and generally does not contain starch, although considerable quantity of NSP is present. The aleurone layer is the main reservoir of many enzymes (such as amylases, proteases), which are responsible for degradation of starch and storage proteins during seed germination. The aleurone is rich in vitamins of B-group (thiamine, niacin, pyridoxine, riboflavin, and panthothenic acid) and minerals. It is also rich in phytate (myoinositol hexaphosphate), which is a major form (about 70%) of phosphorus in cereal grains (Schlemmer et al., 2009).

With regard to nutrient distribution within a kernel, all grains have similar patterns. Taking barley as an example, Liu et al. (2009) showed that protein was most concentrated in the outer areas, and decreased all the way toward the center core area. Starch showed an opposite trend, concentrated in the center and becomes less toward the outside of the grain. As for minerals, they are found chiefly in the outer layers of cereal grains (Liu et al., 2007). Lipid is found in small amounts scattered through the entire grain, but most of it is found in the embryo.

For dry grind processing of grains into fuel ethanol, the whole grain is used. Only starch is hydrolyzed to sugars that are then fermented into ethanol and carbon dioxide; all other chemical components are considered nonfermented residues and end up in a coproduct known as DDGS. Even for starch, a complete hydrolysis cannot be reached and thus DDGS has residual starch at about 5% (dry matter) (Liu, 2008). Since approximately two-thirds of the mass of the starting material (typically corn) is converted into carbon dioxide and ethanol it is normally expected that the concentrations of all unfermented nutrients, including protein, lipids, NSP, minerals, and vitamins, will be concentrated about threefold (Han and Liu, 2010). However, for some nutrients, such as Na, S, Ca, and inorganic P, the fold of increase is much higher than three, presumably due to exogenous addition of compounds containing the minerals or action of yeast phyase on grain phytate (Liu and Han, 2011). For a detailed discussion on changes in chemical composition during dry grind processing, from grain to DDGS, refer to Chapter 8.

4.3.2 STARCH

Starch is the most abundant carbohydrate component in cereal grains, and it is the main form of energy stored in grains. The amount of starch contained in a grain varies but is generally between 60% and 75% of the grain weight. In addition to its nutritive value, starch is important because of its effect on physical properties of many foods, such as the texture of cooked rice, the gelling of puddings, the thickening of gravies, and the setting of cakes. Starch is also an important industrial commodity, particularly in paper making. For bioethanol production from grains, starch is a key component since the whole production system (either wet milling or dry grind) centers around converting starch into glucose and fermenting glucose into ethanol (Bothast and Schlicher, 2005). Details on grain-based ethanol processing are covered in Chapter 5.

4.3.2.1 Chemistry

Starch is a polymer of α-D-glucose, and can have two distinguishable types, amylose and amylopectin. Amylose is a linear polymer of α-D-glucose linked through α-1, 4-glycosidic linkage. Its glucose unit ranges from 1000 to 4000, giving a molecular weight between 1.6×10^5 to 7.1×10^5 Daltons. Amylopectin is a branched polymer of glucose. In addition to α-1, 4-glycosidic linkage, it also has α-1,6-glycosidic linkage at branching points. Approximately every 20–25th glucose residue has a branch point. Amylopectin is a much larger molecule than amylose with a molecular weight of 1×10^7 to 1×10^9 Daltons. Its structure is built from about 95% α-1, 4- and 5% α-1, 6-linkages. The degree of polymerization is typically within the range 9,600 to 15,900 glucose units (Tester et al., 2004).

The ratio of the two polysaccharides varies according to the botanical origin of the starch. For normal starches, amylose typically makes up 20%–35% of the total starch. Over the years, mutants of many cereal species with amylose content significantly lower or higher than this normal range have been developed. High amylopectin cereal grains (amylose content <15% of total starch) are generally described as waxy since the endosperms of the first mutants discovered have a waxy appearance. For example, waxy corn is a starch variant of normal corn. It was first found in China in 1908. The trait is controlled by a single recessive gene. High amylose corn is a mutant that has an amylose content higher than 50% (Neuffer et al., 1997).

When suspended and cooked in water, amylose and amylopectin exhibit different properties due to physical structure and chemical bonding. When in solution, the linear structure of amylose assists in rapidly aligning with itself, resulting in more extensive hydrogen bonding and high gel strength. Thus, a high amount of energy is required to break these bonds and gelatinize starches with higher amylose content. The highly branched structure of amylopectin yields a large molecule, resulting in high viscosity. However, the branched amylopectin molecules cannot align as easily, thus have weaker hydrogen bonding and lower gel strength (Ellis et al., 1998). Such a difference in physicochemical properties between the two types of starch has something to do with grain-based ethanol production. For example, Sharma et al. (2007) evaluated the effect of two enzyme systems (granular starch hydrolyzing vs. conventional enzymes) and different ratios of amylose to amylopectin of corn starch on ethanol production. They found differences in ethanol yields between the two enzyme systems and among different amylose/amylopectin ratios. For pure starch and maize samples, the high amylopectin type resulted in higher ethanol concentrations than the high amylose type, by either enzyme system.

The waxy and normal types of cereal grains are readily distinguished by an iodine staining technique (Neuffer et al. 1997). Upon contact with iodine/potassium iodide solution, the normal grain typically stains deep blue because of the presence of amylose. However, for waxy types, which contain little or no amylose, they stain yellow to brownish red. One exception is sorghum, for which iodine is less indicative since its binding is reduced in lines with corneous endosperm. However, according to Pedersen et al. (2004), the iodine stain technique can be made applicable to sorghum grains, by heating samples at 95°C for 1 h to gelatinize starch first.

4.3.2.2 Starch Granules

Starch occurs as discrete granules in a broad array of plant tissues. In cereal grains, the granules are synthesized in plastids during seed development. The plastids that form starch are known as amyloplasts. In the cereals (corn, wheat, barley, rye, sorghum, and millet) with simple starch granules, each plastid contains one granule. In rice and oats, which contain compound starch granules, many granules are found in each amyloplast.

Starch granules are solid, optically clear bodies that appear white when seen as a bulk powder because of light scattering at the starch–air interface. Starch granules contain about 98%–99% amylose and amylopectin. The rest are some minor constituents, including lipids (such as lysophospholipids and free fatty acids), protein, and minerals (Tester et al., 2004).

TABLE 4.5
Properties of Starch Granules of Cereals

Cereal Grain	Range of Gelatinization Temperature (°C)	Shape	Granule Size (μm)	Special Features
Barley	51–60	Small, spherical	2–10	As wheat
		Large, lenticular	10–30	
Corn	62–72	Spherical	2–30	In floury endosperm
		Polyhedral	2–30, average 10	In flinty endosperm
Millets (Pearl)		Spherical	4–12	As corn
		Polyhedral	4–12	
Oats[a]	53–59	Compound, spherical	up to 60	Comprising up to 150 granuli
		Simple, polyhedral	2–5	
Rice[a]	68–78	Compound, polygonal	up to 50	Comprising up to 80 granuli
		Simple, polyhedral	2–5	
Rye	51–60	Large lenticular	10–40	As wheat, often displaying radical cracks, visible hilum
Sorghum	68–78	Spherical	16–20, average 15	As corn
Triticale	55–62	Large, lenticular	12–30	As wheat
		Small, spherical	1–10	
Wheat	51–60	Large, lenticular	15–35	Characteristic equatorial groove
		Small, spherical	1–10	Polyhedral where closely packed

Source: Data on "Range of Gelatinization Temperature (°C)" column adapted from Delcour and Hoseney (2010). Adapted from Kent, N. L., and A. D. Evers, *Kent's Technology of Cereals*, 4th ed. Pergamon Press Ltd, Oxford, 1994.

[a] Data are for isolated individual granules and not for compound granules.

(a)   (b)

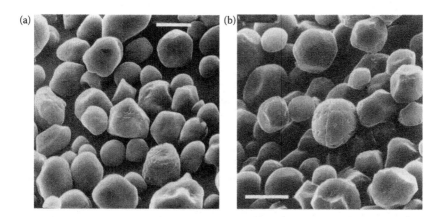

FIGURE 4.6 Scanning electron micrographs of isolated starch granules of corn (a) and pearl millet (b). The white bar on the left micrograph is 10 μm. (Reprinted from Robutti, J. L., et al. Modified opaque-2 corn endosperms. II. Structure viewed with a scanning electron microscope. *Cereal Chemistry* 51, 173–180, 1974.) The white bar on the right micrograph is 5 μm. (Reprinted from Delcour, J. A., and R. C. Hoseney, eds. *Principles of Cereal Science and Technology,* 3rd ed. AACC International, Inc., St. Paul, MN, 2010.)

There are large variations in granule size (1–100 μm in diameter), shape (round, lenticular, polygonal), size distribution (uni- or bi-modal), association as individual (simple) or granule clusters (compound), and composition (ratios of amylose:amylopectin, lipid, moisture, protein, and mineral content). These variations reflect the botanical origin. In the case of cereal grains, each species has characteristic features of starch granules, although some are similar (Table 4.5). For example, the *Triticeae* tribe (wheat, barley, rye, and triticale) all have two types and sizes of starch granules: the large (25–40 μm) lenticular (biconvex) and small (5–10 μm) spherical granules. Corn and sorghum are very similar to each other in shape and size. They average about 20 μm in diameter and their shape varies from polygonal to almost spherical (Figure 4.6). Pearl millet starch is also similar to that of corn and sorghum, except that its granules are smaller. Starch granules of rice and oats are similar in that they are small (2–5 μm) in size and polygonal in shape, and have some compound granules in which many granules fit together to produce large void holes. However, the two differ in size and shape of compound granules. Those of oats are large and spherical, and those of rice are smaller and polygonal.

Within the endosperm of a species, small differences in granule shape may arise as a result of granule packing conditions. These can be seen in grains as mealy (or opaque) and vitreous (or horny) regions. In mealy regions, packing is loose and granules adapt what appears to be their natural form. In horny regions, close packing causes granules to become multifaceted as a result of mutual pressure. Small indentations can also arise from other endosperm constituents such as protein bodies. For example, in the grain of corn and sorghum, starch granules in cells near the outside of the kernel (i.e., in the vitreous endosperm) tend to be polygonal, whereas those in cells from the center of the kernel (in the opaque endosperm) tend to be spherical.

In wet milling of grains, starch granules must be released from the matrix in order to recover starch. This can be accomplished by treating grains (such as corn) with an alkali (such as SO_2, which is also a reducing agent) at 45°C–50°C. This step is commonly known as steeping, which causes the matrix structure to weaken. When the treated endosperm is pulverized in water, the starch granules are released free of adhering matrix and can be separated by a gravitational procedure, such as centrifugation. The matrix fragments are recovered as a high protein fraction called gluten, which also contains endosperm cell walls and small starch granules (Kent and Evers, 1994). In dry grind processing for ethanol production, the whole grains (mostly corn) are dry milled, and then mixed and/or cooked with water. The starch in the flour matrix is converted into sugars by enzymes and fermented into ethanol and carbon dioxide by yeast (Bothast and Schlicher, 2005; Chapter 5 of this book).

4.3.2.3 Gelatinization

When starch is heated in the presence of water, such as in aqueous solution, it undergoes a series of changes known as gelatinization. This phenomenon is manifested as several changes in properties, including granule swelling and progressive loss of organized structure (detected as loss of birefringence and crystallinity), increased permeability to water and dissolved substances (including dyes), increased leaching of starch components, increased viscosity of the aqueous suspension, and increased susceptibility to enzymatic digestion. These changes are required and responsible for achieving certain unique characters of many foods containing starch, and in many cases for maximum conversion of starch to glucose in ethanol production based on starch or starch-containing feedstock.

When starch is placed in water at a room temperature, the granule is freely penetrated by water. With hydration, the starch can hold about 30% increase of its dry weight as moisture. The granule swells slightly, and this volume increase is generally considered to be about 5%. The volume change and water absorption are reversible, and heating the system to just below 60°C (the exact temperature depends on the variety of starch) will not bring about any other changes.

However, when the temperature is raised to above 60°C or higher, irreversible changes occur. At this high temperature, a few of the granules, usually the largest ones, swell very rapidly. As the temperature continues to rise, other granules undergo the same rapid swelling until, within a temperature range of 10°C–15°C, almost all have swollen. The process is known as starch gelatinization, which is defined as the loss of starch birefringence, as examined under polarized light. The temperature range over which the swelling of all the granules in the microscope field occurs is known as the gelatinization range. It includes initiation, midpoint, and termination in terms of temperature, is characteristic of the particular variety of starch being examined, and serves as a means of identification of starch (Table 4.5). The irreversible swelling is endothermic and can be quantified by thermal analysis techniques, such as differential scanning colorimetry.

Certain other changes occur simultaneously or shortly after the initial swelling of the starch granules. One of them is an increase in the clarity of the starch suspension, which may be measured quantitatively by a sudden change in light transmission. Another change is an increase in viscosity of the starch suspension. The viscosity increase results from substantial hydration and swelling of starch granules. With continued heating, the starch granules become distorted, and soluble starch is released into the aqueous system. The soluble starch and the continued uptake of water by the remnants of the starch granules are responsible for the viscosity increase. The changes, which occur after starch gelatinization, are termed pasting, during which solubilization of the starch continues. In excess water, the starch granule is not completely solubilized until the system reaches a temperature of 120°C or higher.

Then, there is a change that is of great significance in the baking of yeast breads and fuel ethanol production from starch or starch-containing feedstocks. It is the increase in susceptibility of the starch to enzyme attack. Although amylase can attack raw starch to a limited extent, dependent on both the type of starch and the source of the enzymes, its rapid action on starch follows gelatinization (Gallant et al., 1992; Oates, 1997; Tester et al., 2006). The extent of amylase action on starch depends on several factors, including medium temperature, pH, the source of particular amylases, etc.

In a typical dry grind process for fuel ethanol production, which also leads to production of DDGS, grains, mostly corn kernels, are ground in a hammer mill without removing other components, then slurried with water and α-amylase. The slurry is heated to high temperatures to gelatinize starch so that the amylase can randomly cleave the large α-1, 4-linked glucose polymers into shorter oligosaccharides known as dextrin. This phase is called liquefaction. After liquefaction, the mash is cooled, and glucoamylase is added to convert the liquefied starch into free glucose (fermentable sugars). This second phase is known as saccharification. Then, yeast is added to the mash to ferment the sugars into ethanol and carbon dioxide, and the remaining unfermented constituents are recovered and become DDGS (Power, 2003; Bothast and Schlicher, 2005). Refer to Chapter 5 for more in depth information on processing of corn into fuel ethanol.

Using the traditional amylase enzymes, without heating the slurry to gelatinize starch, the reaction would be very slow and result in slow and only partial macromolecule hydrolysis and very poor product gain (Oates, 1997). However, by heating the slurry, the process requires an intense amount of energy and special equipment, such as heat exchangers, steam jet cookers, and holding tanks. Although the traditional process is highly advanced, the process inevitably limits a plant's production capacity. To overcome this limitation, for the past decade, an effort has been to develop new types of amylolytic enzymes capable of hydrolyzing starch granules that have not been gelatinized (Kumar, 2008; Szymanowska and Grajek, 2009). The progress is marked by introduction of the commercial enzyme preparation called Stargen developed by Genecor in 2005. The preparation contains two amylolytic enzymes: α-amylase from *Aspergillus kawachi* expressed in *Trichoderma reesei* and glucoamylase from *Aspergillus niger*. These enzymes work synergistically below the gelatinization temperature. The omission of the energy-consuming process of thermal liquefaction allows reduction in operational costs. The new process is known as the raw starch hydrolysis method, also referred to as cold cooking or cold hydrolysis.

4.3.2.4 Gel Formation and Retrogradation

After gelatinization and pasting of starch, the amylose and amylopectin molecules may be considered "dissolved." When the system is allowed to cool without stirring, it will form a gel. A gel is a liquid system that has the properties of a solid. Some common examples are gelatins, pie fillings, and puddings. In gels, a small amount of solid material holds a large amount of water. Gelling of pasted starch is mainly due to the formation of intermolecular bonds, mostly hydrogen bonds. Amylose plays a more important role than amylopectin. For amylose, concentrations as low as 1.5% can form firm gels, while for amylopectin, the intermolecular association is too slight to be of any permanence unless the paste contains 30% or higher concentrations of starch.

The gel formation becomes progressively stronger during the first few hours after it is prepared. However, as a gel ages, the starch chains have a tendency to interact strongly with each other and force water out of the system. The gel shrinks. The squeezing of water out of a gel is called "syneresis." Longer storage gives rise to more interactions between the starch chains and eventually to formation of crystals. The increased association of the starch, with the accompanying changes in its properties, is termed retrogradation. When applied to starch, the term does not imply a backward movement of the starch molecules to their original position within the starch granule. This would be impossible because of the enormous disruption of the original orientation of the starch molecules during the swelling process. Instead, it means a return to a more orderly, partially crystalline state, but one quite different from its original.

4.3.2.5 Starch-Degrading Enzymes

Three enzymes are important for hydrolyzing starch to smaller molecules and eventually glucose (Kent and Evers, 1994; Delcour and Hoseney, 2010). They are α-amylase, β-amylase, and glucoamylase. Some of these enzymes (α- and β-amylases) are naturally present in cereal grains and become active during germination. All three enzymes can also be produced and isolated from microbial or other biological sources, and then added to starch or cereal flour as exogenous sources.

α-Amylase (EC 3.2.1.1) displays an endoaction and can hydrolyze the α-1,4 linkage of starch internally and randomly, yielding low molecular weight dextrins. It is optimally active at about pH 5.3, and contains calcium iron in its structure, which is necessary for its catalytic activity. When the starch has been gelatinized and pasted, α-amylase decreases the size of large starch molecules rapidly and thus reduces viscosity of a starch solution. However, given sufficient time, they also degrade granular starch (native, unheated starch) but slowly. Sound, intact cereal grains have low levels of the enzyme, but upon germination, the level of α-amylase increases many times.

β-Amylase (EC 3.2.1.2) is an enzyme having an exoaction. It can hydrolyze starch from the nonreducing chain end. The product removed through β-amylase action is maltose due to the hydrolysis of alternate α-1,4 linkages. β-amylase alone is basically inactive on granular starch but is capable of rapid action when the substrate is solubilized. β-amylase is found in sound, intact cereal grains and the level does not increase much as a result of germination. Its optimal pH is 5.5 and the enzyme is more susceptible to heat inactivation than α-amylase.

Even the combined action of α- and β-amylases cannot completely digest solubilized starch. Neither of them can catalyze hydrolysis of α-1,6 linkages and hence branch points of starch remain intact. Also those α-1,4 bonds close to branch points resist hydrolysis by the enzymes. Therefore, only about 85% of starch is converted to simple sugars. In order to increase yield of sugars in commercial processes, a debranching enzyme may be used. Glucoamylase (also known as amyloglucosidase, EC 3.2.1.3) from fungal sources is a popular debranching enzyme used commercially. It can catalyze hydrolysis of α-1,4 linkages at the nonreducing ends of the starch molecules and thus release glucose as an end product. Although slower, the enzyme can also hydrolyze α-1,6 linkages. Therefore, this enzyme can completely convert starch to glucose.

During fuel ethanol production by wet milling or dry grind processing, both α-amylase and glucoamylase are used to convert starch into glucose (Kelsall and Lyons, 2003). After starch is separated and heated (as in wet milling) or after grains are dry milled and flour is slurried and heated to gelatinize starch (as in dry grind), α-amylase is added. The enzyme acts randomly on the α-1,4 glucosidic linkages in amylose and amylopective but does not break the α-1,6 linkages of amylopectin. The resulting shorter straight chains (oligosaccharides) are called dextrins, while the shorter branched chains are called α-limit dextrins. The mixture of dextrins is much less viscous than pasted starch. The step is commonly known as liquefaction. Then a second enzyme, glucoamylase, is added. The enzyme converts both types of dextrins into individual glucose molecules. The step is known as saccharification.

4.3.3 PROTEINS

Protein is the second most abundant constituent of cereal grains, following starch. Depending on cereal species, variety, and agronomic conditions, the protein content can range from 5% to 20%. The type and amount of protein in cereal grains are important in terms of nutritional values as well as impacts on functional properties of food or feed containing the protein. For example, bread quality is mostly influenced by the type (proportion of gluten relative to the total protein) and content of wheat proteins in dough (Shewry, 2007).

During ethanol production from grains, proteins are either isolated before fermentation of starch into ethanol (such as in wet milling) or go through fermentation but remain in the coproduct, relatively unchanged (such as in dry grind processing). The relationship between the amount and the type of protein in a feedstock and the efficiency of bioethanol production has not clearly been demonstrated, but they have impact on nutritional quality of coproducts, such as distillers grains.

In term of classification, traditionally, cereal proteins have been classified into four types, according to their solubility, based on the classical work of T. B. Osborne in the early twentieth century on wheat. *Albumins* are proteins soluble in water. Their solubility is not affected by salt concentrations within a low range. However, these proteins can be coagulated upon heating. The classical example of this type of protein is ovalbumin (egg white protein). *Globulins* are proteins insoluble in pure water but soluble in dilute salt solutions, but insoluble at high salt concentrations. This class of proteins shows the classic salting in and salting out property. *Glutelins* are proteins insoluble in neutral aqueous solution, saline solutions, or alcohol but soluble in dilute acids or bases. *Prolamins* are proteins soluble in aqueous (70%–90%) ethanol. To add complexity to terminology, the glutelin of wheat is named glutenin, that of rice oryzenin, and that of barley hordenin. The prolamin of wheat is named gliandin, those of corn, sorghum, oats, and barley are zein, kafirin, avenin, and hordein, respectively.

This traditional classification is still used today, as it has stood the test of time. The method gives reproducible results and provides useful information about cereal proteins. However, it has some limitations. The fractions obtained show much complexity and are mutually contaminated. In addition, some proteins do not appear to fall into any of the four solubility groups. For example, wheat, barley, and rye contain glycoproteins, which are soluble in water but not coagulated by heat. Delcour and Hoseney (2010).

Most albumin and globulin proteins are physiologically active (enzymes). In cereals, they are concentrated in cells of aleurone and germ, but are lower in the endosperm. Nutritionally, these two types of proteins have a very good amino acid composition, relatively high in lysine, tryptophan, and methionine. The prolamins and glutelins are usually storage proteins. The cereal plant stores them in protein bodies. These proteins are limited to the endosperm in cereals, and in contrast to albumins and globulins, they are low in lysine, tryptophan, and methionine. Since albumin and globulin proteins are present in much lower quantity than the two storage proteins, the amino acid composition of whole cereal grain is thus influenced mostly by that of prolamins and glutelins.

Among the cereal flours, only wheat flour has the ability to form strong and cohesive dough that retains gas and produces a light, aerated baked product. Wheat proteins, and more specifically, the gluten proteins (a combination of glutenin and gliadin), are believed to be primarily responsible for that uniqueness of wheat flour. Thus, the storage proteins of wheat are unique because they are also functional proteins.

The proteins of cereal grains other than wheat do not have dough-forming properties to any extent. Rye and triticale probably come closer than the others, but their doughs are still weak. In many parts of the world, the so-called coarse grains, such as corn, sorghum, and pearl millet, are used to make dough-type products, such as chapathi of India, and tortilla of Central and South America. The dough produced is quite different from a wheat flour dough. The major cohesive force appears to be that created by the surface tension of water rather than by the cereal proteins.

In general, the protein content of cereal grains is estimated as the nitrogen content times 6.25. This factor is used for all cereals except for rice and wheat. The protein content of wheat is calculated as nitrogen content times 5.75, while the protein content in rice is calculated by multiplying the nitrogen content with 5.95 (Hoseney, 1994).

4.3.4 LIPIDS

Lipids present in cereals are complex as they consist of a large number of chemical classes and individual compounds. The distribution of the classes and compounds varies with not only cereal species, but also structural parts. Also, lipids can be bound to various other constituents in cereal and thus the same chemical entity can exhibit differences in solubility and other properties (Kitahara et al., 1997).

In general, the lipid content in cereals, based on the whole grain, is relatively low (about 3%, dry basis). Some grains, such as oats, contain higher content of lipids. Most lipids are concentrated in germ and aleurone. Corn is the major cereal used commercially for the production of oil. Rice bran oil and wheat germ oil are also in minor commercial production. With regard to lipid classes, the main components are triglycerides (about 70% of total lipid content). Phospho- and glycolipids are also significantly present. Minor components include free fatty acids, tocopherols, tocotrienols, and phytosterols (Moreau et al., 2001). Similar to lipids of oilseeds, the five main fatty acids of cereal lipids are palmitic, stearic, oleic, linoleic, and linolenic acids. From the outer surface to the inner core of seeds, the oil content decreased, while palmitic and stearic acids (in relative %) increased, oleic and linolenic decreased, and linoleic changed slightly (Liu 2011). Like protein, during dry grind processing, lipids in cereal grains undergo little change, and end up mostly in the coproduct (DDGS). Methods for removing lipids at either the frontend or backend during dry grind processing for ethanol production have been developed (see Chapters 5 and 8). The subject of lipids in DDGS is sufficiently covered in Chapter 9.

4.3.5 Nonstarch Polysaccharides

Nonstarch polysaccharides (NSP) are the constituents of grain cell wall materials. They are the main components of cereal grains, covering a large number of chemical compounds. These compounds consist of cellulose and noncellulosic carbohydrates. The latter includes a range of heteropolysaccharides without α-glucosidic linkages.

Similar to amylose, cellulose is a linear polymer of D-glucose units with 1,4 linkages. However, the steric configuration of the linkage on carbon 1 is β in cellulose, instead of α as in amylose. Because it is unbranched and has essentially a linear configuration, cellulose modules associate strongly with each other by forming hydrogen bonds between chains, and is quite insoluble. As partially crystalline microfibrils, it forms the basic structural materials of the cell walls of plant tissues. Hulls are the rich source of cellulose in cereal grains. Thus those cereals that are harvested with their hull intact (rice, barley, and oats) contain more cellulose than others. The pericarp of cereals is also quite rich in cellulose (up to 30%). The endosperm contains much less cellulose (only about 0.3%).

The noncellulosic carbohydrates can be further classified into neutral (containing mainly neutral sugar residues), acidic (containing mainly uranic acid residues, also referred to as pectic substances), and hemicellulose (Asp, 1996). Hemicellulose, sometimes, known as pentosans, are widely distributed in the plant kingdom and thought to make up the cell walls, along with cellulose and other substances, and serve as a cementing material that holds cells together. Chemically, they are quite diverse, varying in composition from a simple sugar such as is found in β-glucans, to polymers that may contain pentoses, hexoses, proteins, and phenolics. Sugars that are often reported to be constituents of cereal hemicellulose include D-xylose, L-arabinoes, D-galactose, D-glucose, D-glucuronic acid, and 4-O-methyl-D-glucuronic acid. Hemicelluloses are sometimes further divided into water soluble and water insoluble types. Due to the lack of uniform extraction and purification procedures, as well as lack of a definite test to show if one has a pure entity, the establishment of the true chemical nature of these fractions has been difficult.

Cell walls of different cereals have some common components. In general, cellulose is one component present in all cell walls. It is the material of the simplest and the youngest structure. In most cases, additional carbohydrates of varying complexity are deposited as a matrix, and some proteins are also included. Lignin is a common component of secondary thickening in the pericarp of all cereal grains. It is found in the pales but this is relevant to processing only in those grains of which they remain a part after threshing (i.e. oats, barley, and rice). The walls of nucellus and seed coat are generally unlignified. Walls of cereal endosperm (aleurone and starchy endosperm) consist predominantly of arabinoxylans and (1-3, 1-4) β-glucans, with smaller amounts of cellulose, heteromannans, protein, and estified phenolic acids. They are unlignified and contain little, if any, pectin and xyloglucan, or hydroxyproline-rich glycoprotein, all of which are common components of other primary cell walls.

However, cell wall composition is not consistent among species. This is particularly true with the composition of endosperm cell walls. For example, wheat has arabinoxylans as major components while barley and oats have cell walls rich in mixed linkage β-glucan. Recently, a great attention has been paid to β-glucan, due to their cholesterol lowering action (Chen and Huang, 2009). β-Glucans are also linear glucose polymers as is cellulose, but consist of about 30% of β-1,3 linkages and 70% of β-1,4 linkages, whereas cellulose has only β-1,4 linkages. It is considered as a soluble dietary fiber and can be used as functional food ingredients (Brennan and Cleary, 2005).

Beside the nutritional significance, information also exists as to the possible significance of differences in cereal cell wall composition as related to chemical and physical properties of cereal grains. Hicks et al. (2005) examined production of ethanol from barley instead of corn, but found that the presence of mixed linkage β-glucans caused high viscosity during processing. To solve this technical problem, they screened several commercial enzymes that can hydrolyze β-glucans and at the same time increase ethanol yield due to the fermentable glucose formed. Further information can be found in Chapter 6. A report also indicated that high β-glucan content in malt can cause difficulty in the filtering processes during brewing (Bamforth and Barclay, 1993).

Extraction of individual cell wall components is complex and unsuitable for routine analysis. Nevertheless an estimate of cell wall content is often required, particularly in relation to nutritional attributes of a product. In particular, NSP is a major component of dietary fiber. Dietary fiber does not constitute a defined chemical group but a combination of chemically heterogeneous substances, which cannot be digested by the endogenous secretions of human digestive tract. In recent years, the nutritional significance of dietary fiber has gained much attention (Anon., 2008; Cho and Samuel, 2009). The definition of dietary fiber is still evolving. It generally refers to NSP and lignin, but there is scientific evidence that this definition can be extended to include other indigestible food constituents such as resistant starch, oligosaccharides, resistant protein, polyphenols, and others (Goni et al., 2009).

Several analytical procedures have been developed to determine undigestible material as fiber, and each has its own terminology for fiber. These include crude fiber, acid detergent fiber (ADF), and neutral detergent fiber (NDF) (Bach Knudsen, 2001). Crude fiber is the residue left after boiling the defatted sample in dilute alkaline and then in dilute acid. The method recovers 50%–80% cellulose, 10%–50% lignin, and about 20% hemicellulose. ADF refers to the measured residue after samples are digested with a hot dilute sulfuric acid, containing a detergent cetyl trimethylammonium bromide. NDF is the residue left after extraction with a hot neutral solution of sodium dodecyl sulfate, also known as sodium lauryl sulfate. The goal is to separate the dry matter of feeds into those that are nutritionally available by the normal digestive process and those that depend on microbial fermentation for their availability. Therefore, the definition of dietary fiber differs from the above three in that it is not based on the method by which it is determined but on the value of the component.

During dry grind processing of grains into ethanol, NSP remains relatively unchanged chemically and all end up in DDGS (Han and Liu, 2010). Unlike other nonfermented components, such as proteins, lipids, minerals, and vitamins, which are nutritious as animal feed, high levels of NSP

TABLE 4.6
Mineral and Vitamin Composition (mg/100 g Dry Matter) for Several Cereal Grains

	Barley	Corn	Oats	Rice	Rye	Sorghum	Wheat
Minerals							
Phosphorus (P)	470	310	340	285	380	405	410
Potassium (K)	630	330	460	340	520	400	580
Calcium (Ca)	90	30	95	68	70	20	60
Magnesium (Mg)	140	140	140	90	130	150	180
Iron (Fe)	6	2	7	—	9	6	6
Copper (Cu)	0.9	0.2	4	0.3	0.9	0.5	0.8
Manganese (Mn)	1.8	0.6	5	6	7.5	1.5	5.5
Vitamins							
Thiamine (B1)	0.57	0.44	0.70	0.33	0.44	0.58	0.55
Riboflavin (B2)	0.22	0.13	0.18	0.09	0.18	0.17	0.13
Niacin (B3)	6.40	2.60	1.80	4.90	1.50	4.80	6.40
Pantothenic acid (B5)	0.73	0.70	1.40	1.20	0.77	1.00	1.36
Pyridoxine (B6)	0.33	0.57	0.13	0.79	0.33	0.60	0.53

Source: Adapted from Delcour, J. A., and R. C. Hoseney, eds. *Principles of Cereal Science and Technology*, 3rd ed. AACC International, Inc., St. Paul, MN, 2010.

in DDGS are undesirable for feeds of nonruminant animals. However, the carbohydrate portion of this coproduct may have higher value as a feedstock for additional fuel production. Efforts to convert NSP into simple sugars through a cellulosic approach (pretreatment and action of cellulosic enzymes) have been made (Kim et al., 2008). Detailed discussion of this is covered in Chapter 22.

4.3.6 MINERALS AND VITAMINS

About 95% of the minerals in caryopses of cereals consist of phosphates and sulfates of potassium, magnesium, and calcium (Table 4.6). Some of the phosphorus is present as phytate (Schlemmer et al., 2009). Important minor elements are iron, manganese and zinc, and copper. In addition, a large number of other elements are also present in trace quantities. Liu et al. (2007) reported that in hulled barley grain, major minerals in a decreasing order were P, K, Mg, S, and Ca, which had a level of 0.5 to 5.8 mg/g, while minor minerals, Ba, Cu, Mn, Fe, and Zn, also in a decreasing order, were in a range of 1 to 30 μg/g.

The content of minerals in hulls of barley, oats, and rice is higher than that in caryopses, and the hull is particularly rich in silica. Furthermore, within caryopses, minerals are concentrated in the aleurone layer (Liu et al., 2007). During dry grind processing of grains into fuel ethanol, minerals from the original feedstock are concentrated about threefold in DDGS over the original grain. However, Na, S, and Ca were found to have a much larger (more than three) increase over corn in some processing streams or DDGS, presumably due to exogenous addition of compounds containing these minerals during processing (Liu and Han, 2011).

Vitamins comprise a diverse group of organic compounds. They are necessary for growth and metabolism in the human and animal bodies, which are incapable of making them in sufficient quality to meet their needs. Most vitamins are known today by either their chemical name or identification as vitamin A, B, C, etc. Vitamins are sometimes classified according to solubility. Vitamins A, D, E, and K are fat soluble, but B and C vitamins are water soluble. Fat soluble vitamins are more stable to cooking and processing. Cereal grains contain significant quantities (in relation to daily requirements) of B vitamins (thiamin, riboflavin, pyridoxine nicotin acid, and pantothenic acid) and vitamin E (Table 4.6). Variation in content from one cereal to another is remarkably small except for niacin, the concentration of which is higher in barley, wheat, sorghum, and rice than in oats, rye, corn, and the millets. Vitamins are concentrated in the aleurone or the scutellum or both (Delcour and Hoseney, 2010).

4.4 CONCLUSION

This chapter covers general information about the structure and composition of cereal grains as well as unique features of each. Cereal grains are the fruits of cultivated grasses and members of *Gramineae* family. The fruit of a cereal is botanically known as caryopsis, featured by fusion of fruit wall and seed coat. A cereal grain generally consists of embryo, endosperm, and tissues surrounding the embryo and endosperm. Some species also have hulls. Each of these structural parts also consists of different structural tissues and has different chemical compositions. However, these structural parts are adapted differently among cereal species and there are wide variations in size and shape. In terms of composition, cereal grains are characterized by a small amount of water, abundant starch, sufficient protein and fiber, and relatively lower amounts of lipids. Minerals and vitamins are also present. However, there is considerable variation among species and among varieties within a species. During dry grind processing of grains into fuel ethanol, only starch is hydrolyzed to sugars, which are fermented into ethanol and carbon dioxide. All other chemical components remained relatively unchanged chemically and end up in a coproduct known as distillers dried grain with solubles (DDGS). Therefore, differences in grain species and feedstock composition will have impact on not only the yield and production efficiency of ethanol but also the composition and nutritional values of DDGS.

REFERENCES

Abdel-Aal E., and P. Wood, eds. 2005. *Specialty Grains for Food and Feed.* St. Paul, MN: AACC International.

Anon. 2008. Position of the American Dietetic Association: health implications of dietary fiber. *Journal of the American Dietetic Association* 108: 1716–1731.

Asp, N. -G. 1996. Dietary carbohydrates: classification by chemistry and physiology. *Food Chemistry* 57: 1–14.

Bach Knudsen, K. E. 2001. The nutritional significance of "dietary fiber" analysis. *Animal Feed Science and Technology* 90: 3–20.

Balat, M., and H. Balat. 2009. Recent trends in global production and utilization of bio-ethanol fuel. *Applied Energy* 86: 2273–2282.

Bamforth, C. W., and A. H. P. Barclay. 1993. Malting technology and the uses of malt. In *Barley: Chemistry and Technology*, eds. A. W. MacGregor and R. S. Bhatty, pp. 297–354. St. Paul, MN: American Association of Cereal Chemists.

Bhatty, R. S. 1999. The potential of hull-less barley. *Cereal Chemistry* 76: 589–599.

Bothast, R. J., and M. A. Schlicher. 2005. Biotechnological processes for conversion of corn into ethanol. *Applied Microbiology and Biotechnology* 67: 19–25.

Brennan, C. S., and L. J. Cleary. 2005. The potential use of cereal (1-3,1-4)-beta-D-glucans as functional food ingredients. *Journal of Cereal Science* 42: 1–13.

Bremner, P. M., and H. M. Rawson. 1978. The weights of individual grains of the wheat ear in relation to their growth potential, the supply of assimilate and interaction between grains. *Australian Journal of Plant Physiology* 5: 61–72.

Champagne, E. T., ed. 2004. *Rice Chemistry and Technology*, 3rd ed. St. Paul, MN: AACC International.

Chen, J. Z., and X. F. Huang. 2009. The effects of diets enriched in beta-glucans on blood lipoprotein concentrations. *Journal of Clinical Lipidology* 3: 154–158.

Cho, S. S., and P. Samuel, eds. 2009. *Fiber Ingredients: Food Applications and Health Benefits.* St. Paul, MN: AACC International.

Delcour, J. A., and R. C. Hoseney, eds. 2010. *Principles of Cereal Science and Technology,* 3rd ed. St. Paul, MN: AACC International.

Dendy, D. A. V., ed. 1994. *Sorghum and Millets Chemistry and Technology.* St. Paul, MN: AACC International.

Dexter, J. E., and P. J. Wood. 1996. Recent applications of debranning of wheat before milling. *Trends in Food Science and Technology.* 7: 35–41.

Earle, F. R., J. J. Curtis, and J. E. Hubbard. 1946. Composition of the component parts of the corn kernel. *Cereal Chemistry.* 23: 504–511.

Ellis, R. P., M. P. Cochrane, M. F. B. Dale, C. M. Duffus, A. Lynn, I. M. Morrison, R. D. M. Prentice, J. S. Swanston, and S. A. Tiller. 1998. Starch production and industrial use. *Journal of the Science of Food and Agriculture* 77: 289–311.

Evers, T., and S. Millar. 2002. Cereal grain structure and development: Some implications for quality. *Journal of Cereal Science* 36: 261–284.

Gallant, D. J., B. Bouchet, A. Bul´eon, and S. P´erez. 1992. Physical characteristics of starch granules and susceptibility to enzymatic degradation. *European Journal of Clinical Nutrition* 46: S3–S16.

Goñi, I., M. E. Díaz-Rubio, J. Pérez-Jiménez, and F. Saura-Calixto. 2009. Towards an updated methodology for measurement of dietary fiber, including associated polyphenols, in food and beverages. *Food Research International* 42: 840–846.

Han, J. C., and K. S. Liu. 2010. Changes in proximate composition and amino acid profile during dry grind ethanol processing from corn and estimation of yeast contribution toward DDGS proteins. *Journal of Agricultural and Food Chemistry* 58: 3430–3437.

Hicks, K. B., Senske, G. E., Hotchkiss, A. T., and O'Brien, D. 2005. *Screening of Commercial Beta-glucanases for Complete Hydrolysis of Barley Beta-glucan to Glucose.* Meeting abstract, Am. Chemical Soc. National Meeting, Washington DC, August 28–September 1.

Hoseney, R. C. 1994. *Principles of Cereal Science and Technology,* 2nd ed. St. Paul, MN: American Association of Cereal Chemists, Inc.

Hoseney, R.C., and P. A. Seib. 1973. Structural differences in hard and soft wheat. *The Baker's Digest.* 47(6): 26–28, 56.

Kelsall, D. R., and T. P. Lyons. 2003. Grain dry milling and cooking procedures: Extracting sugars in preparation for fermentation, Ch. 2. In *The Alcohol Textbook*, 4th ed. K. A. Jacques, T. P. Lyons, and D. R.
Kelsall. pp. 9–22. Nottingham, UK: Nottingham University Press.

Kent, N. L., and A. D. Evers. 1994. *Kent's Technology of Cereals*, 4th ed. Oxford: Pergamon Press Ltd.

Khan, K., and P. R. Shewry, eds. 2009. *Wheat: Chemistry and Technology*, 4th ed. St. Paul, MN: AACC
International.

Kim, Y., N. S. Mosier, and M. R. Ladisch. 2008. Process simulation of modified dry grind ethanol plant with
recycle of pretreated and enzymatically hydrolyzed distillers' grains. *Bioresource Technology* 99(12):
5177–5192.

Kindred, D. R., T. M. Verhoeven, R. M. Weightman, J. S. Swanston, R. C. Aguc, J. M. Brosnan, and R. Sylvester-
Bradley. 2008. Effects of variety and fertilizer nitrogen on alcohol yield, grain yield, starch and protein
content, and protein composition of winter wheat. *Journal of Cereal Science.* 48(1): 46–57.

Kitahara, K, T. Anaka, and T. Suganuma. 1997. Release of bound lipids in cereal starches upon hydrolysis by
glucoamylase. *Cereal Chemistry* 74: 1–6.

Kumar, M., 2008. *Native Grain Amylases in Enzyme Combinations for Granular Starch Hydrolysis.* Patent US
2007/017485 (WO208021050). Publication date: 21.02.2008.

Lasztity, R. 1999. *Cereal Chemistry.* Budapest: Akademiai Kiado.

Liu, K. S. 2008. Particle size distribution of distillers dried grains with solubles (DDGS) and relationships to
compositional and color properties. *Bioresource Technology* 99: 8421–8428.

Liu, K. S. 2011. Comparison of lipid content and fatty acid compositions and their distribution with seeds of 5
small grain species. *Journal of Food Science* 76: C334–C342.

Liu, K. S., F. T. Barrows, and D. Obert. 2009. Dry fractionation methods to produce barley meals varying in
protein, beta-glucan and starch contents. *Journal of Food Science* 74(6): C487–C499.

Liu, K. S., K. L., Peterson, and V. Raboy. 2007. A comparison of the phosphorus and mineral concentrations in
bran and abraded kernel fractions of a normal barley (*Hordeum vulgare*) cultivar versus four *Low Phytic
Acid (lpa)* isolines. *Journal of Agricultural and Food Chemistry* 55(11): 4453–4460.

Liu, K. S., and J. C. Han. 2011. Changes in mineral concentrations and phosphorus profile during dry-grind
process of corn into ethanol. *Bioresource Technology.* 102: 3110–3118.

Marshall, H. G., and M. E. Sorrells, eds. 1992. *Oat Science and Technology.* Madison, WI: American Society
of Agronomy, Inc. & Crop Sci. Soc. Am. Inc.

Moreau, R. A., V. Singh, and K. B. Hicks. 2001. Comparison of oil and phytosterol levels in germplasm acces-
sions of corn, teosinte, and Job's tears. *Journal of Agricultural and Food Chemistry* 49: 3793–3795.

Morris, C. F. 2002. Puroindolines: The molecular genetic basis of wheat grain hardness. *Plant Molecular
Biology* 48: 633–647.

Newman, R. K., and C. W. Newman. 2008. *Barley for Food and Health: Science, Technology, & Products.* St.
Paul, MN: AACC International.

Neuffer, M. G., E. H. Coe, and S. R. Wessler. 1997. *The Mutants of Maize.* Plainview, NY: Cold Spring Harbor
Press.

Oates, C. D. 1997. Towards an understanding of starch granule structure and hydrolysis. *Trends in Food Science
& Technology* 8: 375–382.

Ojowi, M. O., J. J. McKinnon, A. F. Mustafa, and D. A. Christensen. 1997. Evaluation of wheat-based wet
distillers grains for feedlot cattle. *Canadian Journal of Animal Science* 77: 447–454.

Ortin, W. G. N., and P. Q. Yu. 2009. Nutrient variation and availability of wheat DDGS, corn DDGS and blend
DDGS from bioethanol plants. *Journal of Agricultural and Food Chemistry* 89(10): 1754–1761.

Pedersen, J. F., S. R. Bean, D. L. Funneul, and R. A. Graybosch. 2004. Rapid iodine staining techniques for
identifying the waxy phenotype in sorghum grain and waxy genotype in sorghum pollen. *Crop Science*
44: 764–767.

Power, R. F. 2003. Enzymatic conversion of starch to fermentable sugars, Ch. 3. In *The Alcohol Textbook*,
4th ed., eds. Jacques, K. A., T. P. Lyons, and D. R. Kelsall, pp. 23–39. Nottingham, United Kingdom:
Nottingham University Press.

RFA (Renewable Fuels Association), 2010. *Climate of Opportunity: 2010 Ethanol Industry Outlook.* Washington
DC.

Robutti, J. L., R. C. Hoseney, C. E., and Wassom. 1974. Modified opaque-2 corn endosperms. II. Structure
viewed with a scanning electron microscope. *Cereal Chemistry* 51: 173–180.

Sanchez, O. J., and C. A. Cardona. 2008. Trends in biotechnological production of fuel ethanol from different
feedstocks. *Bioresource Technology* 99: 5270–5295.

Schlemmer, U., W. Frolich, R. M. Prieto, and F. Grases. 2009. Phytate in foods and significance for humans: Food sources, intake, processing, bioavailability, protective role and analysis. *Molecular Nutrition & Food Research* 53: S330–S375.

Scott, M. P., J. W. Edwards, C. P. Bell, J. R. Schussler, and J. S. Smith. 2006. Grain composition and amino acid content in maize cultivars representing 80 years of commercial maize varieties. *Maydica* 51(2): 417–423.

Sharma, V., K. D. Rausch, M. E. Tumbleson et al. 2007. Comparison between granular starch hydrolyzing enzyme and conventional enzymes for ethanol production from maize starch with different amylose:amylopectin ratios. *Starke* 59(11): 549–556.

Shewry, P. R. 2007. Improving the protein content and composition of cereal grain. *Journal of Cereal Science* 46: 239–250.

Simmonds, D. H., K. K. Barlow, and C. W. Wrigley. 1973. The biochemical basis of grain hardness in wheat. *Cereal Chemistry* 50: 553–562.

Szymanowska, D., and W. Grajek. 2009. Fed-batch simultaneous saccharification and ethanol fermentation of native corn starch. *ACTA Scientiarum Polonorum—Technologia Alimentaria* 8(4): 5–16.

Tester, R. F., X. Qi, and J. Karkalas. 2006. Hydrolysis of native starches with amylases. *Animal Feed Science and Technology* 130: 39–54.

Tester, R. F., J., Karkalas, and X. Qi. 2004. Starch—composition, fine structure and architecture. *Journal of Cereal Science* 39: 151–165.

White, P. J., and L. A. Johnson, eds. 2003. *Corn Chemistry and Technology,* 2nd ed. St. Paul, MN: AACC International.

5 Manufacturing of Fuel Ethanol and Distillers Grains—Current and Evolving Processes

Kurt A. Rosentrater, Klein Ileleji, and David B. Johnston

CONTENTS

Before examining all of the different possible uses of distillers dried grains with solubles (DDGS), a discussion regarding ethanol and DDGS production methods is warranted. This chapter will cover current production processes and some of the new practices that have been developed and are being adopted at plants, including front-end fractionation and back-end fractionation. Additionally, standard coproduct definitions will be discussed, as they govern the marketing and sales of these materials.

5.1 PRODUCTION PROCESSES AND PRODUCTS

While corn can be converted into ethanol by three commercial processes, namely wet milling, dry milling, and dry grind ethanol processing (Rausch and Belyea, 2006), only the last process, which constitutes the major growth in the U.S. fuel ethanol industry in recent years, will be thoroughly discussed in this chapter. As pointed out in a previous chapter, over 80% of U.S. ethanol plants currently use the dry grind process. Some plants are, however, beginning to implement dry milling and wet milling fractionation processes to produce more product streams at less cost than a wet milling facility (RFA, 2009).

5.1.1 Dry Grind Processing of Corn to Ethanol

The dry grind process has become the predominant method for production of fuel ethanol in recent years (in 2009 approximately 80% of all U.S. ethanol plants were dry grind) because of its lower investment and operational requirements, as well as advances in fermentation technology. The dry grind process (Figure 5.1) entails several key steps, including grain receiving, distribution, storage, cleaning, grinding, cooking, liquefaction, saccharification, fermentation, distillation, ethanol storage and loadout, centrifugation, coproduct drying, coproduct storage, and loadout. Additional systems that play key roles include energy/heat recovery, waste management, grain aeration, CO_2 scrubbing and extraction, dust control, facility sanitation, instrumentation and controls, and sampling and inspection. Figure 5.2 depicts an example of how all of these pieces may fit together in a commercial plant. During operation of a plant, there are many complex interactions between all of these components, and they all must work in concert for the plant to be efficient and effective.

A brief summary of dry grind ethanol production will be provided below. Additional detailed information on these processing steps can be found in Tibelius (1996), Weigel et al. (1997), Dien et al. (2003), Jaques et al. (2003), Bothast and Schlicher (2005), Rausch and Belyea (2006), and Ingledew et al. (2009). The reader is referred to these sources for more specific information. In addition, the effects of processing methods on physical properties are covered in Chapter 7, while the effects on DDGS composition are covered in Chapter 8.

While there are some "standard" configurations in terms of plant construction, each facility is actually unique—no two are identical. Design choices often depend upon individual client needs and requirements, operational flexibility, the ability to expand in the future, the creativity and imagination of engineers and designers, but most importantly, on cost.

5.1.1.1 Receiving, Cleaning, and Storage

As in traditional grain elevators, receiving operations introduce incoming grain into the storage facility. Corn is typically delivered to an ethanol plant in large wagons, hopper-bottom semitrucks, or even by rail cars at the large plants. The incoming grain is sampled at the receiving station (generally at or prior to the truck scale), and tested for moisture content, broken and foreign material, mold damage, and sometimes mycotoxin levels. Because of the extra expense associated with mycotoxin testing, plant managers typically monitor the harvest crop quality forecasts for mold and mycotoxin pressures, in order to determine how rigorous their mycotoxin testing should be. After testing, the inbound grain is then dumped into an underground receiving pit, and then transferred to mechanical distribution equipment (i.e., bucket elevators, spouts, and conveyors), which then transport the grain to storage. A key consideration with this operation is to maximize material throughput and minimize waiting time (especially at harvest). Receiving hopper (Figure 5.3) volumes can be up to 1200 bu (42.3 m³), and conveyor capacities are often greater than 20,000 bu/h (705 m³/h). Bucket elevators are used for vertical grain transfers (Figure 5.4), while drag or belt conveyors are used for horizontal transfers (although screw conveyors are often used for moving DDGS and distillers wet grains (DWG), but that will be discussed later).

In terms of grain storage, there are two options that are commonly used in the ethanol industry: concrete silos and steel bins (Figures 5.4 and 5.5). Concrete silos are most common for large-volume storage, and can be built with heights up to 150 ft (45.7 m), diameters up to 100 ft (30.5 m), and often have storage capacities of several hundred thousand bu (sometimes even greater than 1 million bu [35,239 m³; ~28,000 tons; ~25,400 t]) per silo. Steel bins, on the other hand, are more common for lower-volume storage (these are often found at small to midsize ethanol plants [~20 to 60 million gal/y]), and are often used for future facility expansions (even at the large ethanol plants [~80 to 150 million gal/y]. They are typically built with heights up to 115 ft (35 m), diameters up to 105 ft (32 m), and often have storage capacities more than 100,000 bu (3524 m³) (depending upon the diameter and height used). A typical rule of thumb is to have between 7 and 10 days of corn storage onsite at the ethanol plant.

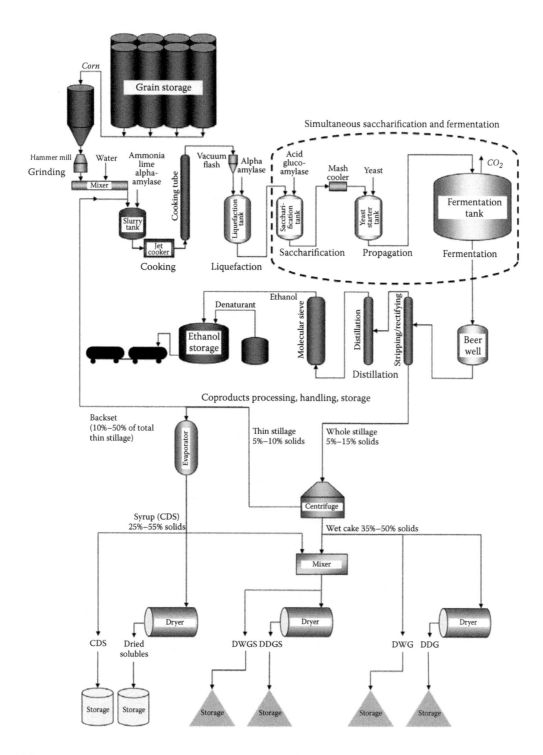

FIGURE 5.1 Flow chart of typical corn dry grind fuel ethanol and coproducts processing. (Adapted from Rosentrater, K. A. *International Sugar Journal* 109(1307): 1–12, 2007.)

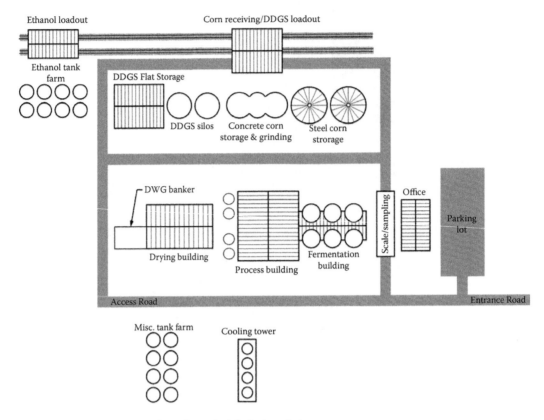

FIGURE 5.2   General plan view of a typical fuel ethanol plant.

FIGURE 5.3   A below-ground grain receiving hopper accepts incoming corn by truck or rail, and then the corn is transferred to a bucket elevator by a drag conveyor (author's photograph).

FIGURE 5.4 Corn is conveyed vertically by bucket elevators, and horizontally by drag conveyors. Note the drag scalper between the bucket elevator discharge and the distribution conveyor—it is used to clean the incoming grain. This is a 21-million gal/y plant (author's photograph).

FIGURE 5.5 A large-scale ethanol plant often uses both types of grain storage options. Large concrete silos are generally constructed with the original facility, and steel bins are then installed during future expansions. Note the small concrete silos used for DDGS storage. This is a 120-million gal/y (450 million L/y) plant (author's photograph).

After storage, the corn is then sent through at least one coarse cleaning operation, such as scalping and/or screening (Figure 5.6). This is done to remove broken corn kernels, fines, and chaff from the grain (i.e., screening), as well as foreign material (such as rocks, stems, cobs, leaves, and insects) (i.e., scalping). Also, large magnets (generally mounted on the grain spouts) are used to remove metal objects that may be in the grain stream. Removing these types of materials is very important for subsequent grinding and fermentation operations.

5.1.1.2 Grinding

In the conventional dry grind process, the entire corn kernel is ground into either a coarse meal or flour using either a hammer mill or a roller mill (Figure 5.7). Corn is ground before processing into ethanol in order to reduce particle size, which increases the surface area and thus exposes the starch in the endosperm so that it can more easily be accessed and transformed by the enzymes and yeast during subsequent processing steps. The resulting particle size can be affected by a number of factors, including the grinding equipment used (i.e., hammer mill or roller mill), the screen size (if a hammer mill is used), the roller or rotor speed, equipment wear, and the characteristics of the corn kernels themselves, such as hardness, shape, size, moisture content, existing stress cracks and fractures, and structural susceptibility to breakage. Typical particle size distributions of ground corn for ethanol production can range (lognormally) from 2.0 mm to less than 0.25 mm. Liu (2009a) found geometric mean diameter (d_{gw}) values for ground corn ranging from 0.430 to 0.516 mm from six ethanol plants; while Rausch et al. (2005) found a geometric mean value of 0.94 from nine ground

FIGURE 5.6 Two common types of grain cleaning equipment include drag scalpers (shown on the left) and rotary scalpers (shown on the right) (author's photograph).

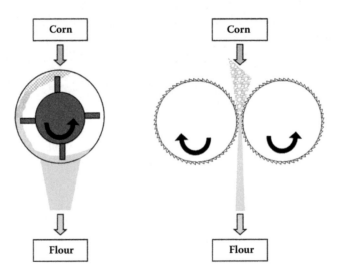

FIGURE 5.7 Schematic representation of size reduction (i.e., grinding). The image on the left illustrates the basic operation of a hammer mill, while the image on the right depicts a roller mill.

corn samples. Particle size has been shown to affect ethanol production. For example, Naidu et al. (2007) examined the effects of five ground corn particle size distributions (0.5, 2.0, 3.0, 4.0, and 5.0 mm). They found that particle size affected both ethanol yield and the concentration of soluble solids in the resulting thin stillage. The highest ethanol yield (12.6 mL/100 mL of beer) was attained at a particle size of 0.5 mm, which resulted in a soluble solids concentration of 25.1 g/L in the thin stillage. This was higher than the concentrations obtained for the other particle sizes, which had a mean concentration of 16.2 g/L.

Of the two main types of grinding systems, hammer mills are most common at ethanol plants because of high throughput capacities; but the trade off is high energy consumption (up to 600 hp [450 kW] can be required). Knives within the rotor assembly often travel at linear speeds between 9000 and 24,000 ft/min (2743 to 7315 m/min), depending on the diameter, and screen surface areas can be up to 6000 in^2 (3.9 m^2), especially for the larger hammer mills.

Roller mills are not often used in ethanol plants, but they are sometimes used in plants that use the dry milling process (which will be discussed later). These types of mills often have higher capital costs than hammer mills, but they will have greater throughput at a given horsepower. They generally produce coarser, but more uniform particle size distributions, and will not generate as many fines, nor will they heat the product as much. Rollers are generally between 9 and 12 in (23 and 30.5 cm) in diameter, and often travel at linear speeds between 1500 and 3000 ft/min (457 and 914 m/min). The grooved rollers rotate at slightly different speeds in order to grind the kernels. Additional details on corn milling can be found in Kelsall and Piggot (2009).

5.1.1.3 Cooking and Liquefaction

After grinding, the corn flour is mixed with water from the backset and process condensate (10% to 60% of the total liquid supplied, in fact, thus conserving water and energy) to form a slurry of approximately 30% solids. The starch, which has been exposed during milling, is then prepared for fermentation. The pH is adjusted to between 5.5 and 6.0 by the addition of ammonia (NH_3), lime ($CaCO_3$), or sulfuric acid (H_2SO_4), and the enzyme α-amylase is added at a rate between 0.04% and 0.08% (of the corn, on a dry basis, db). The slurry is then heated to between 80°C (176°F) and 95°C (203°F) for 15 to 20 min (although some plants may hold the slurry as long as 40 min) to begin starch gelatinization, which makes the starch more accessible and amenable to modification during subsequent processing steps, and allows the slurry to be pumped because the viscosity is reduced. The ground corn, backset, and cook water can also be added together simultaneously, which can result in an immediate temperature of ~85°C (185°F). The slurry is cooked using a jet cooker, which injects steam into the slurry (Figure 5.8). An actuator is used to adjust the mixing of the streams, and cooking temperatures between 120°C (248°F) and 140°C (284°F) are achieved.

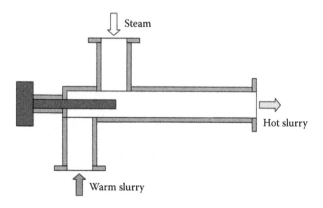

FIGURE 5.8 Schematic representation of a jet cooker.

During cooking, the slurry is held in a cooking column for 5 to 20 min, which fully gelatinizes the starch and breaks down the crystalline structure of the starch granules. After cooking, the slurry is then cooled to between 80°C (176°F) and 95°C (203°F) by flashing to atmospheric pressure in a vacuum condenser.

The mixture is then held in a liquefaction tank at 85°C (185°F) to 95°C (203°F) for 30 to 120 min, and more α-amylase enzyme is added at a rate between 0.05% and 0.08% (of the dry solids in the slurry). Caustic solution from the clean-in-place (CIP) system, or sulfuric acid, may be added to maintain the pH between 5.5 and 6.5. Lime and calcium may be added as nutrients for enzymes; backset may be added (generally 10% to 15%) for nutrients as well; and urea is often added as a subsequent nitrogen source (the yeast will require this nutrition during fermentation). The liquefaction step breaks the long starch polymers into short chains: the α-amylase enzyme hydrolyzes the starch polymers into short glucose chains called maltodextrins. A detailed discussion regarding enzymes, their importance to ethanol production, and their impact on DDGS is provided in Chapter 24.

5.1.1.4 Saccharification and Fermentation

The next step, saccharification, breaks down the short glucose chains into individual glucose molecules (known as dextrose). The enzyme α-amylase is used to cleave the α-1,4 glucosidic linkages, while the gluco-amylase is used to hydrolyze the α-1,6 glucosidic bonds (Ingledew et al., 2009). To accomplish this, mash from the liquefaction stage is cooled using plate heat exchangers to between 55°C (131°F) and 65°C (149°F). Another enzyme, gluco-amylase, is added either in its own tank or when the slurry is pumped to the fermentation vessels. Sulfuric acid is generally added to produce a pH between 4.0 and 4.5, which is optimal for this enzyme. The mash is kept at an elevated temperature for several minutes and then cooled to approximately 30°C (86°F) before being transferred into the fermentation tanks. In recent years it has become common for the saccharification step to occur in the fermentation tanks; this is known as simultaneous saccharification and fermentation (or SSF), and is indicated in Figure 5.1.

The mash is then transferred to fermentation tanks (Figure 5.9), which can have volumes greater than 2.0 million L (528,000 gal) each. Residence times in the fermentors may range from 40 to 72 h.

FIGURE 5.9   Fermentation tanks at a 120 million gal/y (450 million L/y) plant (author's photograph).

Generally speaking, pH levels are set at an initial pH of 4.0 to 5.0, and then they are allowed to run; sometimes they are maintained between 3.5 and 4.0. Temperatures are maintained at approximately 28°C (82°F) to 34°C (93°F) using external recirculating plate heat exchangers. In the fermentation vessels, gluco-amylase may be added; this enzyme converts the remaining dextrins into glucose. Yeast (*Saccharomyces cerevisiae*) is also added to the mash to consume simple sugars and convert glucose molecules into heat, ethanol, and carbon dioxide. Theoretically, 1 g of glucose will yield 0.51 g of ethanol and 0.49 g of carbon dioxide, following:

$$C_6H_{12}O_6 \Rightarrow 2C_2H_5OH + 2CO_2.$$

These maximums, however, will be reduced in reality (generally 90% to 95% of theoretical) due to a number of factors, including the growth of yeast biomass cells (between 5% and 10% of the glucose may actually be consumed for cell growth) during fermentation, as well as the possible production of other secondary products, including glycerol, acetic acid, lactic acid, propanol, butanol, furfural, or other unintended products. Furthermore, protease activities must be monitored during fermentation to ensure that the yeast are receiving adequate nutrition. When complete, the fermentation slurry will contain between 10% and 15% ethanol, as well as yeast cells, nonfermentable materials (i.e., proteins, fibers, oils, minerals) that are processed further and sold as distillers grains, secondary products, and water. A common rule of thumb states that for each 1 kg of corn processed, approximately 1/3 kg of each of ethanol, carbon dioxide, and distillers grains will be produced. Another rule of thumb states that each bushel of corn (~56 lb; 25.4 kg) will yield up to 2.9 gal (11.0 L) of ethanol, approximately 18 lb (8.2 kg) of distillers grains, and nearly 18 lb (8.2 kg) of carbon dioxide.

The carbon dioxide that is produced during fermentation is cleaned, and most often released to the atmosphere. Sometimes it is compressed and sold to beverage manufacturers, but plant location, logistics, and economics often make this option cost prohibitive for most plants. Generally, the carbon dioxide is subjected to a wet scrubbing system, in which the gas stream is thoroughly in contact with water in order to remove volatile organic compounds (VOCs). The water from the scrubber is then recycled through the ethanol plant as process water.

5.1.1.5 Distillation, Dehydration, and Ethanol Storage

The fermented liquid stream that leaves the fermentors is known as beer. The beer is transferred to a large holding tank called the beer well. The beer is then sent through a stripping/rectifier column to remove the ethanol. The overflow from the column is a mixture of ethanol and water, while the underflow is a liquid stream called whole stillage—it contains all the nonfermentable components of the corn as well as the yeast cells and water. The ethanol–water mixture is further processed through distillation columns and molecular sieves to remove the water. The mixture is continuously pumped through a heated multiple-column distillation system (often with 28 to 34 separate stages) that boils off the ethanol (ethanol has a boiling point of 78°C (172°F), whereas water has a boiling point of 100°C (212°F)). The superheated alcohol vapor, which is approximately 190 proof (95% pure ethanol and 5% water) then moves on to dehydration in a molecular sieve (with one or more columnar beds), which physically separates the remaining water from the ethanol vapor. Molecular sieves contain microporous beads (such as zeolyte), which adsorb the water vapor and result in 200 proof (100% pure) ethanol. The pure ethanol vapor is then condensed, cooled, blended with a denaturant (generally gasoline, so that it is not fit for human consumption, and thus not subject to beverage alcohol taxation) to produce the final liquid fuel ethanol product. This is then stored in tanks until it is loaded onto rail tanker cars for shipping to an ethanol/gasoline blending facility.

5.1.1.6 Coproduct Processing, Handling, and Storage

Conversion of nonfermentable residues into valuable animal feeds and other value-added coproducts is an important goal. The sale of all types of distillers products as livestock feed substantially contributes to the economic viability of ethanol manufacturing, and coproducts are thus vital components to each plant's operation (Rosentrater, 2006). Conversion of nonfermentable materials into coproducts begins with the whole stillage. The whole stillage (which contains the nonfermentable materials, in both dissolved and suspended form) generally has an overall solids content of approximately 5% to 15%, and can be processed into a range of distillers grains coproduct feed materials (Figure 5.1), depending upon the specific configuration of the ethanol plant. The whole stillage is centrifuged (most commonly with a decanter style or solid-bowl centrifuge) to remove water (the resulting process stream is known as thin stillage, and contains high concentrations of the water-soluble solids); the dewatered product, on the other hand, is known as wet cake, and it contains the suspended solids that were removed from the whole stillage. Wet cake is often sold directly as DWG to cattle feedlots that are within the locality of the ethanol plant. Wet cake, or DWG, can contain between 35% and 50% solids (although it often has 50% to 65%), and thus is still very high in moisture. Sometimes syrup or partially dried DDGS is mixed with the DWG to produce a modified product. More details about processing these coproduct streams can be found in Meredith (2003).

Often between 10% and 50% of the thin stillage (which generally contains between 5% and 10% solids content) is recycled through the process; this is known as "backset." The balance of the thin stillage is then processed through a multiple-effect evaporator system to produce condensed distillers solubles (CDS); this is referred to as "syrup" in the industry. During the evaporation process, the solids content of the thin stillage is increased to between 25% and 55%. CDS is a golden brown, thickened, highly viscous liquid. CDS is sold as a specific coproduct at only a few plants in the United States, and is usually fed to cattle as a mixed feed with silage. At most plants, it is recombined with the DWG, and this combination is then dried to about 10%–12% wet basis (wb), or over 90% solids content (to ensure a stable shelf life), to produce DDGS. This coproduct is a granular, bulk material, and is physically similar to other dry feed ingredients, such as soybean meal.

If the CDS is not added before the DWG is dried, then the resulting feed product is known as distillers dried grains (DDG). The majority of ethanol plants are set up to solely produce the dried coproducts (most currently produce DDGS, while only a few produce DDG). Dried products are stable, and can be either used locally or shipped via truck or rail for use by distant customers. Lately, however, there has been growing interest in local use of DWG, because it appears to offer some nutritional benefits to livestock, and the production of DWG substantially reduces the overall energy requirements for the ethanol plant vis-à-vis the production of DDG or DDGS.

Drying is one of the most energy-intensive operations of the ethanol plant, and can consume up to 1/3 of the plant's entire energy supply (Meredith, 2003a, 2003b). As a rule of thumb, approximately 0.06 kg of natural gas is required to evaporate 1 kg of water (or 1000 BTU per 1 lb of water) during drying. Two main types of drying systems are used in U.S. fuel ethanol plants: rotary drum dryers (Figure 5.10) and ring dryers (Figure 5.11). Each of these types of systems has several potential configurations. For example, both the rotary drum and the ring dryers can have partial or full combustion gas recycling systems, and both can be constructed with either a direct-fired heat source, or they can be indirectly fired (i.e., use heat exchangers to heat the drying air). Additionally, rotary dryers can use steam tube heating as a heat source instead of air. The equipment used and manner in which the DDGS is dried will greatly impact the resulting nutritional quality as well as physical properties (more about the influence of processing conditions will be discussed in Chapters 7 and 8).

At this point in time, rotary drying is the predominant type of drying system used in U.S. ethanol plants (Figure 5.10 shows a generic schematic for this process); for larger plants, usually more than one high-capacity dryer operates simultaneously in parallel. To begin the drying process, two product streams (DWG and CDS) are blended in a mixing chamber, screw conveyor, or paddle mixer, before being conveyed to the dryer. Additionally, a portion of freshly dried DDGS (at approximately 10% moisture content) from the rotary drum dryer is mixed in as well (this DDGS is actually recycled through the system again). The ratio of DWG, CDS, and freshly dried DDGS is such that the solids content of the blend is about 65% before entering the dryer. Air temperatures in the

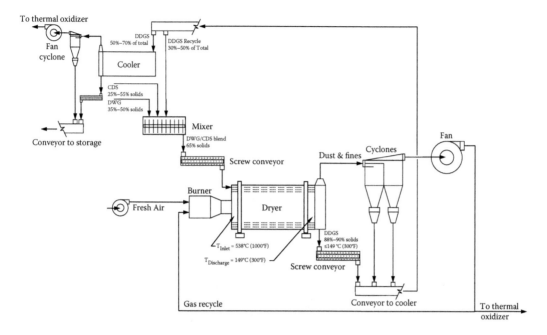

FIGURE 5.10 An example of a basic rotary drum drying process. (Adapted from Monceaux, D. A., and D. Kuehner. In *The Alcohol Textbook,* 5th ed., eds. W. M. Ingledew, D. R. Kelsall, G. D. Austin, and C. Kluhspies, pp. 303–322. Nottingham, UK: Nottingham University Press, 2009.)

dryer are very high: at the dryer inlet the temperature can be over 500°C (932°F), and at the dryer discharge over 100°C (212°F). Most rotary dryers operate in a cocurrent configuration, that is, the product and the airflow move in the same longitudinal direction inside the dryer. Upon exiting the dryer, the DDGS will generally be slightly lower (nearly 10°) in temperature than the air discharge temperature, and it will have approximately 10% to 12% moisture content (wb). Between 50% and 70% of the dried DDGS is then conveyed to storage. The other (i.e., recycled) portion is routed back to the mixer where it is blended with incoming DWG and CDS. Residence times in rotary dryers can range from approximately 10 to 20 min (although some particles can reside as long as 60 min, depending upon the moisture).

It is inside the rotary drum dryer that the DDGS particles are formed, as the drying air removes water vapor and the particles are mixed with other drying particles. The resulting DDGS characteristics will thus depend upon a number of drying factors: the ratio of DWG to CDS in the blend prior to entering the dryer, the process configuration used for blending (i.e. where in the process and how are the products blended), the ratio of recycled DDGS to stored DDGS, and drying parameters (such as drying temperatures, air flow rates, and drum rotation speed). Kingsly et al. (2010) examined the influence of some of these drying process variables on the resulting characteristics of DDGS. The drying process is also a granulation process, where wet solids are broken up (pulverized) and/or agglomerated into granulated, flowable materials (Probst and Ileleji, 2009). Having a liquid in the blend (i.e., CDS) during drying induces particle agglomeration, and results in spherical ball-shaped solids, which are often referred to as "syrup balls" in the industry. The size of agglomerates depends on the level of CDS in the blend, the physical process of introducing and blending the CDS with the DWG and recycled DDGS, dryer rotation speed, as well as the total number of dryer revolutions. Compared to newer plants, DDGS from older generation plants that use rotary dryers are typically darker in color, have a larger geometric mean particle size (d_{gw} of 3 mm or more), and have a larger particle size distribution. The large agglomerates occur in these older systems because of the way the CDS is introduced into the blend—often all at once, which precludes thorough mixing with

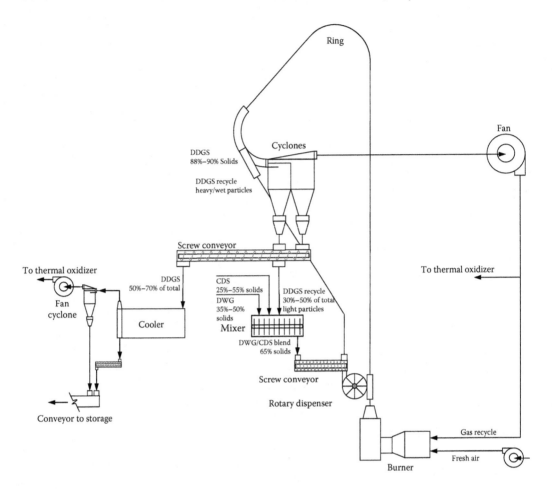

FIGURE 5.11 An example of a basic ring drying process (Adapted from Monceaux, D. A., and D. Kuehner. In *The Alcohol Textbook,* 5th ed., eds. W. M. Ingledew, D. R. Kelsall, G. D. Austin, and C. Kluhspies, pp. 303–322. Nottingham, UK: Nottingham University Press, 2009.)

the DWG and recycled DDGS. It should also be noted that the partial recycling of dried DDGS to reduce the energy requirements for drying, prevent plugging of the dryer, as well as speed up drying has the potential to cause product degradation due to denaturing of proteins (especially heat-sensitive amino acids) and other components that are repeatedly exposed to high temperatures.

Another type of rotary dryer configuration that is becoming commonly installed at newer dry grind ethanol plants consists of a pair of rotary drum dryers connected in series. In this configuration, the blend of DWG, CDS, and freshly recycled DDGS enters the first dryer, which dries the product to between 25% and 35% moisture content. The DDGS discharging from the second dryer, on the other hand, is between 10% and 12% moisture content. A portion of the DDGS from each dryer is recycled back through that specific dryer, along with the incoming material stream (DWG and CDS for the first dryer; partially dried DDGS and new CDS for the second dryer). Generally about four times more CDS is added to the second dryer versus the first. This newer configuration produces DDGS that is lighter (golden) in color, has a smaller particle size (generally less than 2 mm), and has a narrower particle size distribution. The decrease in particle size is correlated to a decrease in particle agglomeration during drying, because the addition of the syrup (which is the agglomeration agent) is accomplished in two stages instead of one, which reduces agglomeration formation and enhances bulk pulverization. Additionally, the second dryer

can often be operated at somewhat lower temperatures than the first; this also impacts the resulting DDGS properties (Kingsly et al., 2010).

In recent years, ring dryers have become used at many ethanol plants as well (Figure 5.11 shows
a generic schematic for this process). In this type of system, recycled DDGS is mixed with incoming CDS and DWG (as with rotary systems), but then the mixture travels through a disintegrator/
disperser (which provides some particle size reduction) where it is introduced into a moving hot air
stream. The wet material is then pneumatically conveyed through ductwork in the shape of a "ring,"
where the water is partially or totally evaporated from DDGS particles. After traveling around the
ring circuit, the air stream and entrained particles travel through a split manifold where the wetter
(and thus heavier) particles are separated from the lighter, drier particles. The wetter material falls
directly to the disperser for travel through the drying circuit again; the dry DDGS material, on the
other hand, is sent through cyclone separators to remove them from the air stream. The clean air is
then recycled through the system, while the DDGS is then either conveyed to storage, or recycled
through the drying system. Ring drying is a very fast process. Typical residence times in a ring
dryer are often only a few seconds in the air stream itself, while those particles that are recycled
can spend between 2 and 4 min in the system. Overall, a ring dryer will have higher electrical usage
compared to rotary dryers, most of which is used to power the primary fan. But these systems will
require lower heat (often between 5% and 10% less). Additionally, a ring dryer will have a smaller
"footprint," requiring less area for the system, but it will require a support tower because the system's components have a vertical orientation (Figure 5.12). Drying of DDGS has been discussed in
more depth in Meredith (2003a, 2003b) and Monceaux and Kuehner (2009).

FIGURE 5.12 Ring dryers require a support tower, but encompass a smaller footprint than rotary dryers
(author's photograph).

After drying, the DDGS (or DDG) is transferred to a storage structure (Figure 5.13)—most often a flat storage building (via drag conveyor), discharged into multiple piles, and allowed to cool to ambient temperature (this is known as "curing"). Flat storage buildings (Figure 5.14) are generally of sheet metal construction with steel frames and concrete floors; they are often at least 100 ft × 200 ft (30.5 m × 61 m) in size, and can store between 6000 and 8000 tons (5443 and 7257 t) of DDGS. At larger plants, however, concrete silos have become common (Figure 5.13). These can have diameters up to 55 ft (16.8 m), heights up to 110 ft (33.5 m), and often store up to 4000 tons (3629 t). Because DDGS does not generally flow well (as will be discussed in Chapter 7), screw-type reclaim unloaders must be used in these silos. If a large plant does have silos, the DDGS is still generally cured in a flat storage building first; otherwise flowability problems have been shown to become exacerbated. After storage, conveyors are used to transfer the DDGS to the loadout system, which includes a bulkweigher and rail loadout spouting (generally of flexible/retractable design). Capacities for loadout systems generally range from 200 to 400 tons/h (181 to 363 t/h). Sometimes loadout buildings are combined with grain receiving operations in a common structure (as in Figure 5.2).

FIGURE 5.13 Typical storage structures at a large-scale ethanol plant. The top image shows, from left, corn storage in concrete silos; steel corn receiving/DDGS loadout building; concrete DDGS silos; steel DDGS flat storage building; and on the far right, denatured ethanol rail loadout building. The image on the lower left illustrates a typical steel DDGS flat storage building. The image on the lower right illustrates a concrete DDGS storage silo (author's photograph (top)).

DWG, on the other hand, is placed in outdoor concrete bunkers (Figure 5.15). Depending upon the season and prevailing weather conditions, DWG is extremely susceptible to spoilage (due to the high moisture content and water activity), so ethanol plants try to sell the DWG before it degrades (often less than one week). If this occurs, then the product must be disposed of—generally by landfilling. When the product is sold, it must be shipped. Most often, local livestock producers

FIGURE 5.14 Distillers dried grains with solubles (DDGS) are most often piled into a flat storage building until cooled, at which time the DDGS will be transferred and loaded onto rail cars or semitrucks (photograph courtesy of USDA).

FIGURE 5.15 Distillers wet grains (DWG) being piled into a concrete bunker, awaiting loading onto livestock producers' trucks (author's photograph).

provide their own trucks (as opposed to the ethanol plants shipping the DWG), and they arrive at the plant when they need to procure the material for their animals. Due to its semisolid nature, and thus challenging material handling behavior, the DWG must be loaded onto semitrailers with a pay loader. Interestingly, the DWG and the CDS have completely different flow characteristics (semisolid vs. fluid), even though their solids contents can be very similar.

5.1.2 Effects of Processing Conditions on DDGS

Although other chapters in this book are dedicated to the chemical composition (Chapter 8) and physical properties (Chapter 7) of DDGS, it is appropriate to briefly discuss the effects that processing conditions may have on these attributes. Theoretically, all unit operations will ultimately have an influence on the resulting DDGS. Thus grinding, cooking, fermentation, distillation, centrifugation, evaporation, blending, and drying will all play a role, as will the blend ratio of CDS to DWG prior to drying. Moreover, the quality of the incoming corn will also influence the final coproducts as well. Practically speaking, however, some processing steps and parameters are more influential than others. Unfortunately, there have been very few studies to date that have tried to quantify these effects. Han and Liu (2010) collected samples of various process streams throughout the dry grind process (i.e., ground corn, raw slurry, cooked slurry, liquefied mash, fermented mash, whole stillage, DDG, DWGS, and DDGS), and determined the proximate compositions of each, in order to quantify chemical changes during processing. Additional discussions regarding the effects of processing on composition can be found in Chapter 8.

One of these studies, Liu (2009a), examined the particle size distributions (0.15 to 2.36 mm) of both raw ground corn and resulting DDGS samples from six commercial ethanol plants. This study found that the particle size of the raw corn and the DDGS were indeed highly correlated (r = 0.81), which logically makes sense. Overall, the geometric mean diameter (d_{gw}) of the ground corn ranged from 0.43 to 0.52 mm; that of the DDGS ranged from 0.48 to 0.70 mm. Thus, the grinding unit operation does impact the DDGS particles.

Another question that needs to be answered is: what effect does the addition level of CDS have upon the DDGS? To investigate the effects of CDS/DWG ratio on resulting chemical composition of DDGS, Cao et al. (2009) conducted a laboratory study. CDS was combined at five levels with DWG (0%, 23%, 27%, 40%, and 100% [i.e., no DWG]), and then the samples were oven dried to produce DDGS. As shown in Figure 5.16, the nutrient composition of the CDS played a progressively important role as the CDS increased in the blend. As the level of CDS grew, both the crude protein and the NDF values declined, specifically because the CDS was low in each of these components. Conversely, the crude fat and ash increased with increasing levels of CDS. Also of note, both P and the S (which are elements that are very important to livestock diet formulations) increased as the CDS increased. These results underscore the importance of blending a consistent level of CDS to the DWG at the ethanol plant in order to produce DDGS with consistent nutrient properties and low variation.

Along these lines, Kingsly et al. (2010) examined the effects of varying CDS addition levels on various physical and chemical properties of DDGS, but on a commercial scale. Three levels of CDS addition were used: 0 L/min (0 gal/min), 106 L/min (28 gal/min), and 212 L/min (56 gal/min). In this study, DDGS was manufactured using two rotary dryers in series; the first dryer used an inlet temperature of approximately 487°C (909°F), and an outlet temperature of 106°C (223°F); the second dryer had a varying inlet temperature (273°C (523°F), 377°C (711°F), and 499°C (930°F), in order to properly process the varying level of CDS), but the outlet temperature was nearly constant at 109°C (228°F); for all treatments, the speed of the recycling conveyor was held constant at 60% of the maximum. As shown in Figure 5.17, varying the level of CDS resulted in substantial changes in the DDGS properties—both the physical properties as well as the chemical composition. Moisture, geometric mean particle diameter (d_{gw}), bulk density, fat, and ash levels all increased linearly as the level of CDS increased. Protein and Hunter L scale (which is a color parameter that quantifies

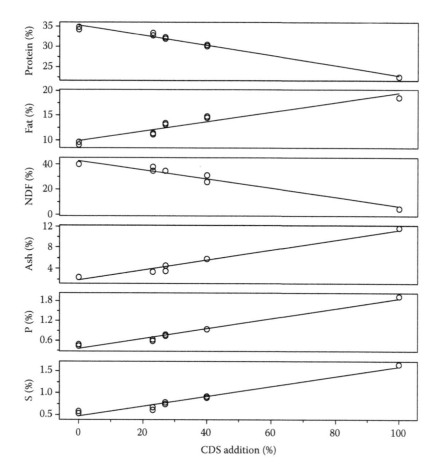

FIGURE 5.16 Effects of CDS addition rate on the resulting chemical composition of laboratory-produced DDGS. NDF is neutral detergent fiber. (Adapted from Cao, Z. L., J. L. Anderson, and K. F. Kalscheur. *Journal of Animal Science* 87: 3013–3019, 2009.)

lightness/darkness), not surprisingly, decreased as the CDS level increased; thus the DDGS darkened with increasing CDS levels.

Rosentrater and Wrenn (2009) conducted physical and flowability property analyses on DDGS samples that were produced under a series of varying conditions in order to investigate the effects of various manufacturing operations (specifically, ethanol processing [i.e., front-end processing conditions] and DDGS drying conditions [i.e., back-end processing conditions]) on the resulting properties of the DDGS. The experiments that focused on front-end variables were primarily concerned with how liquefaction, saccharification, and solids processing conditions (e.g., factors that influence fermentation performance) impacted the resulting DDGS. The trials that examined drying conditions, on the other hand, were primarily concerned with the flowrates of CDS and DWG, as well as dryer temperature. They measured DDGS angle of repose (AOR,°), aerated (i.e., loose) bulk density (ABD, kg/m³), Hausner Ratio (HR, –), and uniformity (–). Processing conditions that were varied included hammer mill screen size (0.06 to 0.11 in; 1.52 to 2.79 mm), jet cooker temperature (225°F to 235°F; 107°C to 113°C), slurry enzyme flowrate (0 to 103 gal/h; 0 to 390 L/h), mash tank enzyme flowrate (0 to 207 gal/h; 0 to 784 L/h), CDS flowrate (0 to 18 kg/h; 0 to 40 lb/h), DWG flowrate (0 to 57 kg/h; 0 to 126 lb/h), and dryer temperature (153°F to 243°F; 67°C to 117°C). During the experiments, several processing conditions were held constant, including a corn feed

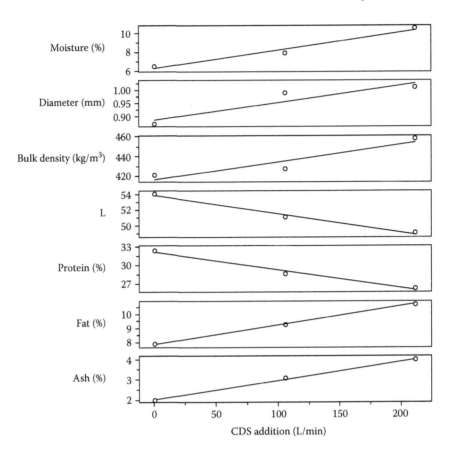

FIGURE 5.17 Effects of CDS addition rate on the resulting physical and chemical properties of commercially produced DDGS. L is the Hunter L (lightness–darkness) parameter. (Adapted from Kingsly A. R. P., K. E. Ileleji, C. L. Clementson, A. Garcia, D. E. Maier, R. L. Stroshine, and S. Radcliff. *Bioresource Technology* 101: 193–199, 2010.)

rate of 500 lb/h (227 kg/h), a process water temperature of 200°F (93.3°C), a process water density of 0.96 g/cm³ (60.5 lb/ft³), a jet cooker flow rate of 3 gal/min (11.4 L/min), a fermentor volume of 3600 gal (13,627 L), a fermentor temperature of 90°F (32.2°C), and a corn solids density of 1.47 g/cm³ (92.6 lb/ft³). In terms of front-end effects, they found that as hammer mill screen size increased (i.e., ground corn particle size increased), the bulk density decreased (Figure 5.18). Additionally, they found that as the enzyme flowrates increased, both the angle of repose and the bulk density of the DDGS increased. Regarding the back-end variables, they found that as the CDS level increased, the bulk density tended to increase as well, but the uniformity of the particles decreased (i.e., they became more nonuniform, and thus varied in shape).

5.1.3 Resource Consumption during Manufacture

After having discussed the production processes that are used to manufacture both ethanol and coproducts, as well as the influences that processing variables may have on the resulting coproduct properties, it is appropriate to briefly quantify how much energy and water are required for these processes. This topic has been one of interest, not only from the manufacturer's standpoint (because these requirements affect the bottom line for each plant), but also from a net energy

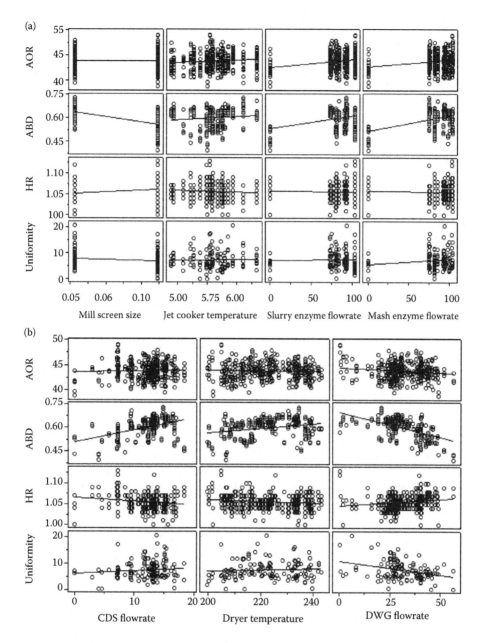

FIGURE 5.18 Effects of front-end (a) and back-end (b) processing conditions on the resulting physical properties of pilot scale-produced DDGS. AOR is angle of repose; ABD is aerated bulk density; HR is Hausner Ratio. (Adapted from Rosentrater, K. A., and B. A. Wrenn. 2009. *Examining the Effects of Ethanol Processing and Drying Conditions on the Physical Properties of Distillers Dried Grains with Solubles (DDGS)*. ASABE Paper No. 095563. St. Joseph, MI: ASABE.)

balance sustainability and lifecycle assessment standpoint (as mentioned in Chapter 2). Table 5.1 provides some estimates for the total energy and water that are required (on a per gallon of ethanol basis). The table illustrates that over time, energy and water use has been declining industry-wide. This trend is attributable to increases in efficiencies in the plants, which is partly due to new technologies being developed and deployed.

TABLE 5.1
Estimates of Resource Consumption during Ethanol Manufacturing

Year	Ethanol Yield (gal/bu)	Total Energy Use (BTU/gal)	Natural Gas Use (BTU/gal)	Electricity Use (kWh/gal)	Water Use (gal H_2O/gal)	DDGS Production (dry lb/gal)	Reference
1995	2.53	53,277	37,000	1.2			Shapouri et al. (1995)
1998					3.3–10.6 (5.8)		Keeney and Muller (2006)
2002	2.64	39,719			4.7		Shapouri et al. (2002); Wu (2008)
2003	2.75	45,900	35,000	1.09		6.55	Tiffany and Eidman (2003)
2004	2.63	48,539	39,031	1.07	6.80		Graboski (2002)
2005					3.5–6.0 (4.2)		Keeney and Muller (2006)
2006	2.80		34,000	0.75			Patzek (2007)
2008	2.62–2.96 (2.81)	17,706–44,034 (31,070)	16,000–36,883 (27,589)	0–1.57 (0.70)	2.65–4.90 (3.45)	4.7–8.1 (5.9)	Wu (2008)

Note: Values in parentheses indicate mean values.

5.2   FRONT-END FRACTIONATION

Front-end fractionation techniques offer opportunities to use various corn components in high-value food and industrial applications, as the separations occur before any of the ethanol unit operations. A few examples of experimental fractionated products are provided in Table 5.2, and some commercially developed products are listed in Table 5.3. Due to economic conditions in recent years, however, not all of these are actually available. These new processes have led to a burgeoning of coproducts of various compositions that are now available to livestock producers.

5.2.1   DRY FRACTIONATION OF CORN FOR ETHANOL PRODUCTION

Dry milling has historically been a process used for the production of food-grade corn flours, meals, and grits for various uses in the food and beverage industry. In dry milling of corn (Figure 5.19), the kernel is physically separated, using a dry process, into its primary components in which the germ (which contains the fat), tip cap, and pericarp (which contains fiber) are separated from the endosperm (which contains the starch) (Rausch and Belyea, 2006). This general process has also begun to be applied as a prefractionation method prior to fermentation in dry grind ethanol plants. In the ethanol industry this is known as dry fractionation, and a number of different dry milling technologies to fractionate corn are currently being retrofitted into existing dry grind fuel ethanol plants. The economics and value of dry fractionation lie in the fact that only the fermentable portion of the kernel, the endosperm, is ground and fermented into ethanol, while the other nonfermentable components (i.e., protein, fiber, oil, and tip cap) can be processed into other higher value products. Rausch and Belyea (2006) describe the dry milling process as follows: (1) corn kernel moisture content is first increased from approximately 15% to 22%, which causes differential swelling of the

TABLE 5.2
Some Experimentally Fractionated Ethanol Coproducts

Product	Type of Fractionation	Crude Protein	Crude Fat	Crude Fiber	ADF	NDF	Ash	Reference
Prefermentation								
DDGS	No fractionation	28.5	12.7		10.8		3.6	Singh et al. (2005)
DDGS	Quick germ process	35.9	4.8		8.2		4.0	Singh et al. (2005)
DDGS	Quick germ, quick fiber process	49.3	3.8		6.8		4.1	Singh et al. (2005)
DDGS	Enzymatic dry grind process	58.5	4.5		2.0		3.2	Singh et al. (2005)
DDGS	No fractionation	21.2	13.9					Martinez-Amezcua et al. (2007)
DDGS	Dry degerm defiber	23.8	8.7					Martinez-Amezcua et al. (2007)
DDGS	Quick germ, quick fiber process	28.0	5.4					Martinez-Amezcua et al. (2007)
Postfermentation								
DDGS	Elusieve	33.6	12.5			32.5		Srinivasan et al. (2005)
Pan DDGS	Elusieve	42.2	12.9			19.0		Srinivasan et al. (2005)
Enhanced DDGS	Elusieve	35.6	14.2			32.6		Srinivasan et al. (2005)
"Fiber"	Elusieve	19.3	7.05			53.3		Srinivasan et al. (2005)
DDGS	Elusieve	33.0	8.0		12.4	37.8	4.7	Rosentrater and Srinivasan (2009)
Pan DDGS	Elusieve	37.3	7.0		11.5	29.2	5.0	Rosentrater and Srinivasan (2009)
Enhanced DDGS	Elusieve	31.9	8.7		11.6	34.6	4.7	Rosentrater and Srinivasan (2009)
"Fiber"	Elusieve	25.2	6.6			42.3	4.9	Rosentrater and Srinivasan (2009)
Deoiled DDGS	Solvent extraction	34.0	2.7	8.4			4.8	Saunders and Rosentrater (2009)
Deoiled DDGS	Solvent extraction	34.5	3.5		12.9	45.0	5.2	Kalscheur et al. (2008)
Deoiled DDGS	Solvent extraction	33.3	2.1	9.7			4.8	Ganesan et al. (2009)

Note: All Nutrients Reported as % Dry Basis. ADF is acid detergent fiber; NDF is neutral detergent fiber.

germ relative to the other kernel components, and increases the resiliency of the germ; (2) the corn is then sent through a degerminator to separate the germ; this is an abrasion step that breaks the kernel into pericarp (i.e., bran), germ, and endosperm fragments; (3) additional steps remove pericarp and germ from the endosperm; and (4) an aspirator and a gravity table are used to further classify and purify the broken pieces of kernel into their respective components.

In comparison to the wet milling process, fractionation of the kernel components is not as complete in the dry milling process. The resulting distillers grains from plants that utilize dry milling fractionation generally has a higher concentration of protein than distillers grains produced using traditional dry grinding, but a lower fiber and oil content, because these components are separated out prior to fermentation. Additionally, dry milling fractionation processes save energy and increase fermentation

TABLE 5.3
Some Commercially Developed Fractionated Ethanol Coproducts

Manufacturer	Product	Type of Fractionation	Crude Protein	Crude Fat	Crude Fiber	ADF	NDF	Ash	Reference
FWS Technologies	Enhanced DDGS	Prefermentation	35.0–37.0	6.5			21.0	3.8	FWS (2009)
Poet	Dakota Gold BPX DDGS	No fractionation	28.2	10.8	7.1	10.0	26.1	4.8	Dakota Gold (2009a)
	Dakota Bran	Prefermentation	14.0	8.9	7.0	8.0	38.1	5.5	Dakota Gold (2009b)
	Dakota Germ—Corn Germ Dehydrated	Prefermentation	15.8	17.1	6.2	8.2	23.4	5.9	Dakota Gold (2009c)
	Dakota Gold HP DDG	Prefermentation	41.0	4.0	8.1	13.0	24.0	2.1	Dakota Gold (2009d)
Renessen	Enhanced DDG(S)	Pre/postfermentation	35.0–50.0	2.5–4.0		7.0–11.0	15.0–25.0		Stern (2007)
Solaris	Energia	Pre/postfermentation	30.0	2.5	8.2			2.5	Lohrmann (2006)
	Glutenol	Postfermentation	45.0	3.3	3.8	9.23	15.3	4.0	Lohrmann (2006)
	Neutra-Fiber	Prefermentation	6.8	1.5	17.1			0.6	Lohrmann (2006)
	NeutraGerm	Prefermentation	17.5	45.0	6.0			1.9	Lohrmann (2006)
	ProBran	Pre/postfermentation	9.5	2.0	16.6			1.0	Lohrmann (2006)

Note: All Nutrients Reported as % Dry Basis. ADF is acid detergent fiber; NDF is neutral detergent fiber.

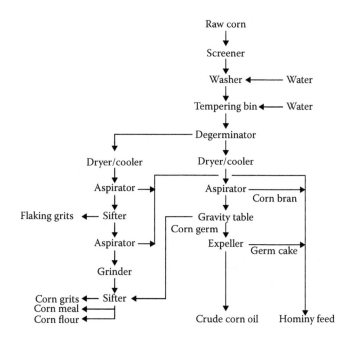

FIGURE 5.19 Flow chart of typical corn dry milling processing. (Adapted from Alexander, R. J. *Corn Chemistry and Technology,* 1st ed., eds. S. A. Watson and P. E. Ramstad, pp. 351–376. St. Paul, MN: American Association of Cereal Chemists, 1987, and Rausch, K. D., and R. L. Belyea. *Applied Biochemistry and Biotechnology* 128: 47–86, 2006.)

capacity since a highly concentrated stream of starch (which is converted into ethanol) is processed through the fermentors. The economics and value of germ and fiber separation prior to fermentation have been researched and are described in Singh and Eckhoff (1997) and Singh et al. (2002).

The dry fractionation process has a number of benefits, but it also has several disadvantages that need to be considered. Dry separation equipment does not cleanly separate the components without losing some starch. If starch is lost with these coproducts, it cannot be converted into ethanol, thus yields will be reduced. Additionally, the soluble components in the germ contain important nutrients for yeast metabolism. When the germ is removed using dry separations, these nutrients are removed as well. Without these being available to the yeast, metabolism can be significantly reduced and fermentation rates and yields decreased. Lastly, while the overall coproduct drying costs are reduced relative to the conventional process, electricity use is significantly increased for the various fractionation equipment.

Several variations of dry fractionation have been developed and commercially implemented (Table 5.3), including BFrac (www.poet.com), FWS (www.fwstl.com), Cereal Process Technologies (www.cerealprocess.com), and dry degerm defiber (Murthy et al., 2006) technologies. Singh and Johnston (2009) have provided an extensive discussion regarding dry fractionation techniques for dry grind plants, and the reader is referred to this source for more information.

5.2.2 WET FRACTIONATION OF CORN FOR ETHANOL PRODUCTION

In the corn wet milling process (Figure 5.20), the corn kernel is fractionated into individual components of starch, protein, fiber, germ, and soluble solids using an aqueous medium (Blanchard, 1992; Johnson and May, 2003). The process involves multiple stages, including chemical pretreatment using sulfites, followed by size and density separations to produce the isolated components. Wet milling is currently used to produce a significant portion (over 1 billion gallons [~ 3.8 billion L]

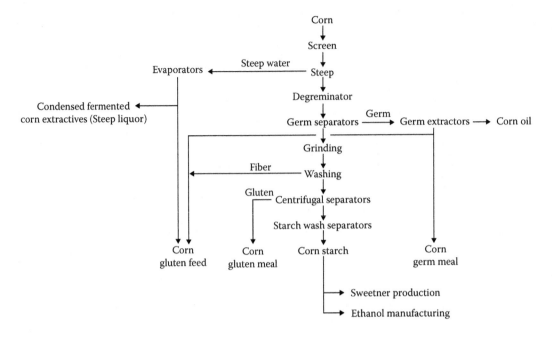

FIGURE 5.20 Flow chart of typical corn wet milling processing. (Adapted from Loy, D. D., and K. N. Wright, Nutritional properties and feeding value of corn and its by-products. In *Corn: Chemistry and Technology*, 2nd ed., eds. P. J. White, and L. A. Johnson, pp. 571–604. St. Paul, MN: American Association of Cereal Chemists, and Rausch, K. D., and R. L. Belyea. *Applied Biochemistry and Biotechnology* 128: 47–86, 2006.)

per year) of fuel ethanol; however, as previously mentioned, capital and operational costs are high relative to the dry grind process (Ramirez et al., 2009), so the majority of new facilities have been dry grind in recent years.

A number of prefractionation processes using wet milling unit operations have been developed for use in dry grind plants. As with the dry fractionation techniques, these processes separate out the higher value germ and can remove fiber as well. One process developed also incorporates an enzyme treatment that aids in the subsequent separation and fermentation processes (Ramirez et al., 2009). Unlike dry fractionation, however, wet fractionation also can remove the endosperm fiber and extract the soluble nutrients from the germ. Wet fractionation can also have higher starch recoveries relative to dry fractionation; however, all components must be hydrated, so the overall coproduct drying costs will be higher than traditional dry grind plants.

Several variations of wet fractionation have been developed, and include the quick germ process (Singh and Eckhoff, 1997), the quick germ quick fiber process (Wahjudi et al., 2000), and the enzymatic milling process (Wang et al., 2005). Singh and Johnston (2009) have provided an extensive discussion regarding wet fractionation techniques for dry grind plants, and for the sake of brevity, the reader is referred to this source for more information.

5.3 BACK-END FRACTIONATION

In addition to fractionating nutrients from the raw corn kernels prior to fermentation, another possibility that has been explored in recent years is the fractionation of nutrients from the coproducts themselves (i.e., postfermentation). To date, most of the interest in this type of fractionation has primarily revolved around the DDGS, the thin stillage, or the CDS. A few examples of experimental fractionated products are provided in Table 5.2, and some commercially developed products are listed in Table 5.3. As these tables show, the various fractionation processes can lead to coproducts with substantially unique nutrient compositions. A brief discussion about various back-end fractionation options follows; but more extensive discussions about each will be provided in subsequent chapters.

5.3.1 FRACTIONATION OF COMPONENTS FROM DDGS

The idea of removing chemical components from ethanol coproducts has actually been around for quite some time. An early example of concentrating nutrient streams was described in Wu and Stringfellow (1986), who found that a simple screening method (using sieves) could result in finer particle fractions with substantially higher protein content, as well as lower fiber, compared to the initial materials, for both corn DDG and DDGS. Composition of the screened DDG fractions depended on the mesh size (ranging from 20 through 80 mesh; 0.84 to 0.18 mm), and resulted in 14.5% to 49.3% protein, 8.4% to 9.3% fat, and 1.3% to 2.4% ash content (on a dry basis); screened DDGS, on the other hand, yielded protein contents ranging from 16.9% to 36.8%, fat contents of 7.4% to 12.0%, and ash contents of 2.0% to 6.5%.

In another early study, Wu and Stringfellow (1982) fractionated corn DDG (using moisture contents between 6% and 31%) and corn DDGS (using moisture contents between 5% and 31%). The materials were first milled and then screened on various sieves (ranging from 20 through 80 mesh; 0.84 to 0.18 mm). The original DDGS had 30% protein content, while the resulting fractions ranged in protein from 11% to 46%. Optimal DDGS processing conditions were found to be 21% moisture with two passes through the mill at 14,000 rpm. The original DDG, on the other hand, had a protein content of 25%, while the resultant fractions ranged in protein from 13% to 50%; optimal conditions were determined to be similar to those for the DDGS—21% initial moisture content but only one pass through the mill at 14,000 rpm. They also found that, for both the DDGS and the DDG, the fractions that had the highest protein also had the highest fat levels. And fiber appeared to be resistant to grinding, as the fractions with the largest particle size tended to have the highest fiber content.

Air aspiration can also be used to separate fiber from DDGS. Singh et al. (2002) examined samples from both corn-based dry grind ethanol plants in the Midwest, and beverage alcohol plants that used combined feedstocks consisting of corn, rye, and malted barley. An aspiration technique, which entailed placing DDGS on a 20 mesh (840 μm) screen and then aspirating with an air jet at a pressure of 2.8 atm, (284 kPa) was able to separate the pericarp from the germ fraction, though not extremely effectively. Overall, the aspirated DDGS samples contained (on a dry basis) 25.7% to 28.9% protein, 7.7% to 12.3% fat, 39.3% to 51.8% NDF, and 13.1% to 17.9% ADF.

Srinivasan et al. (2005) further expanded these concepts and developed a process (termed the "Elusieve" process) using a combination of elutriation (air classification/aspiration) and sieving to separate fiber particles from DDGS. First, DDGS was sieved into four size fractions, and then the three larger size fractions were subjected to air classification. The smallest size fraction (which was not subjected to air classification), the "pan" DDGS, represented nearly 40% of the original DDGS and had reduced fiber and increased protein content (by approximately 6% to 14%, and 7% to 8% points, respectively) compared to the original DDGS. The three larger size fractions, on the other hand, which were subjected to air classification, yielded heavier and lighter fractions. Air velocity played a key role in this classification. Overall, the resulting lighter fractions had higher fiber and lower protein and fat contents than the heavier fractions, depending on air velocity. The heavier fractions from air classification of the three larger size fractions were then combined together to produce an "enhanced" DDGS product that had nearly same protein content, but lower NDF content, than the original DDGS. The lighter fractions from air classification of the three larger size fractions were combined together to produce a "Fiber" product, which had higher NDF and lower protein and fat contents than the original DDGS. Liu (2009b) also examined the effects of combining sieving and air separation. This study found that, depending on the particle (i.e., sieve) size used, a maximum protein reduction of 56.4% and a maximum protein increase of 60.2% could be achieved. Additionally, oil could be reduced by as much as 81.4%, or increased by as much as 262.7%. The topic of mechanical fractionation will be covered in more depth in Chapter 25.

Another possibility for fractionation includes the removal of oil from the DDGS, which can readily be accomplished by solvent extraction. Singh and Cheryan (1998) were the first to report successful solvent extraction of corn oil from DDGS. Their optimal process entailed using a ratio of ethanol-to-DDGS of 6:1, and this resulted in extraction of nearly 50% of the oil. In general, if the oil is removed after ethanol processing, its quality is not high enough to be appropriate for food use; but it has been shown to be acceptable for conversion into biodiesel (more about this will be discussed in Chapter 23). The process of removing the oil will not only affect the DDGS composition, but it can also impact the physical nature of the particles. For example, Saunders and Rosentrater (2009) examined chemical and physical properties of commercially processed solvent extracted DDGS. Crude fat levels (2.7% db) were much lower, while crude protein (34.0% db) and crude fiber (8.4% db) contents were somewhat higher than traditional DDGS. The resulting DDGS exhibited water activity (0.24), thermal conductivity (0.07 W/m/°C), thermal diffusivity (0.17 mm^2/s), bulk density (482 kg/m^3; 30.1 lb/ft^3), and angle of repose (21.7°) values similar to unmodified DDGS. Color values were substantially lighter (Hunter L = 54.1 compared to ~40 to 50 for unmodified DDGS), however, because of fat-soluble pigment losses during processing. Similarly, Ganesan et al. (2009) examined physical and flowability properties of commercially produced solvent-extracted (2.1% db fat) and unmodified (9.3% db fat) DDGS to determine if fat level affected flowability behavior. The extracted DDGS had an angle of repose 4.3% lower and Carr compressibility 70% lower than unmodified DDGS. Jenike shear testing indicated that reduced fat DDGS had unconfined yield strength and Jenike compressibility values that were 15.7% and 40.0% lower, respectively, than unmodified DDGS, but had major consolidating stress and flowability index values that were 6.7% and 13.2% higher. Overall, a reduction in fat content slightly improved some flow properties, but both types of DDGS were ultimately classified as cohesive in nature.

5.3.2 FRACTIONATION OF COMPONENTS FROM LIQUID STREAMS

In recent years interest has grown in removing corn oil from liquid coproduct streams prior to the drying operation. This is not a new concept, however. Wu et al. (1981) examined fractionation of whole stillage using screening (20 mesh; 0.84 mm) and centrifugation (10,400 ×g for 10 min). Although the focus of their work was to concentrate protein, they were also able to produce streams that had fat contents concentrated up to 36% more than the level found in DDGS.

As the fuel ethanol industry has grown, commercial processes have been developed to remove corn oil from thin stillage, semiconcentrated thin stillage, and even CDS, in order to increase revenue for ethanol plants. This type of fractionation has been pursued because, although the resulting corn oil fractions cannot be used as food-grade oil, they can readily be converted into biodiesel. More on this will be covered in a Chapter 23.

Some of the processes that are currently being deployed include those of Mean Green BioFuels (www.meangreenbiofuels.com; Winsness, 2006), GS CleanTech (www.greenshift.com; Kirkbride McElroy, 2007; Cantrell and Winsness, 2009; Winsness and Cantrell, 2009), and ICM (www.icminc.com; ICM, 2009). In general, all of the processes that are currently available are based upon physical separation techniques, using various configurations of decanters and centrifuges, and have been reported to remove between 30% and 70% of the oil in the coproduct stream.

5.4 COPRODUCT DEFINITIONS

As discussed, there are a variety of fuel ethanol coproducts than can potentially be produced at a given plant. Each plant is unique, and employs various permutations of processing technologies. In terms of end use (i.e., feeding to livestock), it is important for livestock producers to understand specifically what they are purchasing. To facilitate the sale and fair trade of coproducts in the

TABLE 5.4

Common and Official (as Delineated by AAFCO, 2006) Coproduct Names and Definitions

Common Acronym	Official Name	Official Definition for Trade
DDGS	Corn Distillers Dried Grains with Solubles	"Is the product obtained after the removal of ethyl alcohol by distillation from the yeast fermentation of a grain or a grain mixture by condensing and drying at least ¾ of the solids of the resultant whole stillage by methods employed in the grain distilling industry. The predominating grain shall be declared as the first word in the name."
DDG	Corn Distillers Dried Grains	"Is obtained after the removal of ethyl alcohol by distillation from the yeast fermentation of a grain or a grain mixture by separating the resulting coarse grain fraction of the whole stillage and drying it by methods employed in the grain distilling industry. The predominating grain shall be declared as the first word in the name."
DWG (WDG)	Distillers Wet Grains	"Is the product obtained after the removal of ethyl alcohol by distillation from the yeast fermentation of a grain mixture. The guaranteed analysis shall include the maximum moisture."
CDS (syrup)	Corn Condensed Distillers Solubles	"Is obtained after the removal of ethyl alcohol by distillation from the yeast fermentation of a grain or a grain mixture by condensing the thin stillage fraction to a semi-solid. The predominating grain must be declared as the first word in the name."
DDS	Corn Distillers Dried Solubles	"Is obtained after the removal of ethyl alcohol by distillation from the yeast fermentation of a grain mixture by condensing the thin stillage fraction and drying it by methods employed in the grain distilling industry. The predominating grain must be declared as the first word in the name."

marketplace, the American Association of Feed Control Officials (AAFCO) has established official definitions (Table 5.4) for feed materials. These definitions have worked well for the last several decades, but this is changing as modern processing operations have evolved. Note that coproducts (especially DDGS) that are produced from new fractionation technologies may not necessarily be compatible with the definitions listed in Table 5.4. New definitions may need to be added as these newer processing technologies that modify DDGS from a typical dry grind process are deployed. This will be true for both the front-end as well as the back-end fractionation processes.

5.5 CONCLUSIONS

This chapter was intended to provide a broad overview of the production methods that are commonly used to manufacture both fuel ethanol and the concomitant coproduct streams from corn. Other grains can also be used; these will be discussed in the following chapter. The industry is dynamically evolving, so the fractionation techniques that have been discussed will most probably grow in use during the coming years. And newer technologies that will improve energy efficiency and decrease water use throughout the plant, as well as improve the ethanol yield, will undoubtedly continue to be refined. The ultimate impact of these modifications will be upon coproduct properties, and thus, utilization opportunities.

REFERENCES

AAFCO. 2006. *2006 Official Publication*. Oxford, IN: American Feed Control Officials Inc.

Alexander, R. J. 1987. Corn dry milling: Processes, products, and applications. In *Corn Chemistry and Technology,* 1st ed., eds. S. A. Watson and P. E. Ramstad, pp. 351–376. St. Paul, MN: American Association of Cereal Chemists.

Blanchard, P. 1992. Wet milling. *Technology of Corn Wet Milling and Associated Processes,* pp. 92–93. Amsterdam, the Netherlands: Elsevier Science Publishers.

Bothast, R., and M. Schlicher. 2005. Biotechnological processes for conversion of corn into ethanol. *Applied Microbiology and Biotechnology* 67(1): 19–25.

Cao, Z. L., J. L. Anderson, and K. F. Kalscheur. 2009. Ruminal degradation and intestinal digestibility of dried or wet distillers grains with increasing concentrations of condensed distillers solubles. *Journal of Animal Science* 87: 3013–3019.

Cantrell, D. F., and D. J. Winsness. 2009. *Method of Processing Ethanol Byproducts and Related Subsystems.* U.S. Patent No. 7,601,858. October 13, 2009.

Dakota Gold. 2009a. Dakota Gold BPX: Distillers Dried Grains with Solubles. Poet Nutrition: Sioux Falls, SD. Available online: www.dakotagold.com. Accessed November 1, 2010.

Dakota Gold. 2009b. Dakota Bran. Poet Nutrition: Sioux Falls, SD. Available online: www.dakotagold.com. Accessed November 1, 2010.

Dakota Gold. 2009c. Dakota Germ: Corn Germ Dehydrated. Poet Nutrition: Sioux Falls, SD. Available online: www.dakotagold.com. Accessed November 1, 2010.

Dakota Gold. 2009d. Dakota Gold HP: Distillers Dried Grains. Poet Nutrition: Sioux Falls, SD. Available online: www.dakotagold.com. Accessed November 1, 2010.

Dien, B. S., R. J. Bothast, N. N. Nichols, and M. A. Cotta. 2003. The U.S. corn ethanol industry: An overview of current technology and future prospects. In *The Third International Starch Technology Conference— Coproducts Program Proceedings,* eds. M. Tumbleson, V. Singh, and K. Rausch, pp. 10–21. June 2–4, 2003, Urbana, IL: University of Illinois.

FWS. 2009. The FWS corn fractionation system: Are you getting only a fraction of your profit potential? Available online: www.fwstl.com. Accessed November 1, 2010.

Ganesan, V., K. A. Rosentrater, and K. Muthukumarappan. 2009. Physical and flow properties of regular and reduced fat distillers dried grains with solubles (DDGS). *Food and Bioprocess Technology* 2: 156–166.

Graboski, M. S. 2002. *Fossil Energy Use in the Manufacture of Corn Ethanol.* St. Louis, MO: National Corn Growers Association.

Han, J. C., and K. S. Liu. 2010. Changes in proximate composition and amino acid profile during dry grind ethanol processing from corn and estimation of yeast contribution toward DDGS proteins. *Journal of Agricultural and Food Chemistry* 58: 3430–3437.

ICM. 2009. Oil recovery. ICM, Inc. Available online: www.icminc.com. Accessed November 1, 2010.

Ingledew, W. M., D. R. Kelsall, G. D. Austin, and C. Kluhspies. 2009. *The Alcohol Textbook,* 5th ed., eds. W. M. Ingledew, D. R. Kelsall, G. D. Austin, and C. Kluhspies. Nottingham, UK: Nottingham University Press, pp. 1–541.

Jaques, K. A., T. P. Lyons, and D. R. Kelsall. 2003. *The Alcohol Textbook,* 3rd ed. Nottingham, UK: Nottingham University Press, pp. 1–446.

Johnson, L. A., and J. B. May, 2003. Wet milling: The basis for corn biorefineries. In *Corn: Chemistry and Technology,* 2nd ed., eds. P. J. White, and L. A. Johnson. St. Paul, MN: American Association of Cereal Chemists, pp. 449–490.

Kalscheur, K., A. Garcia, K. A. Rosentrater, and C. Wright. 2008. *Ethanol Coproducts for Ruminant Livestock Diets.* Fact Sheet 947. Brookings, SD: South Dakota Cooperative Extension Service.

Keeney, D., and M. Muller. 2006. *Water Use by Ethanol Plants—Potential Challenges.* Minneapolis, MN: Institute for Agriculture and Trade Policy.

Kelsall, D. R., and R. Piggot. 2009. Grain milling and cooking for alcohol production: Designing for the options in dry milling. In *The Alcohol Textbook,* 4th ed., eds. K. A. Jacques, T. P. Lyons, and D. R. Kelsall, pp. 161–175. Nottingham, UK: Nottingham University Press.

Kingsly A. R. P., K. E. Ileleji, C. L. Clementson, A. Garcia, D. E. Maier, R. L. Stroshine, and S. Radcliff. 2010. The effect of process variables during drying on the physical and chemical characteristics of corn dried distillers grains with solubles (DDGS)—Plant scale experiments. *Bioresource Technology* 101: 193–199.

Kirkbride McElroy, A. 2007. Corn oil extraction opens new markets. *Distillers Grains Quarterly* 2(1): 7–8.

Liu, K. 2009a. Effects of particle size distribution, compositional and color properties of ground corn on quality of distillers dried grains with solubles (DDGS). *Bioresource Technology* 100: 4433–4440.

Liu, K. 2009b. Fractionation of distillers dried grains with solubles (DDGS) by sieving and winnowing. *Bioresource Technology* 100(24): 6559–6569.

Lohrmann, T. 2006. Rethinking ethanol coproducts. *Distillers Grains Quarterly* 1(4): 23–24.

Loy, D. D., and K. N. Wright. 2003. Nutritional properties and feeding value of corn and its by-products. In *Corn: Chemistry and Technology*, 2nd ed., eds. P. J. White, and L. A. Johnson, pp. 571–604. St. Paul, MN: American Association of Cereal Chemists.

Martinez-Amerizcua, C., C. H. Parsons, V. Singh, G. S. Murthy, and R. Srinivasan. 2007. Nutritional characteristics of corn distillers dried grains with solubles as affected by amounts of grains versus solubles and different processing techniques. *Poultry Science* 86: 2624–2630.

Meredith, J. 2003a. Dryhouse design: Focusing on reliability and return on investment. In *The Alcohol Textbook*, 4th ed., eds. K. A. Jacques, T. P. Lyons, and D. R. Kelsall, pp. 363–376. Nottingham, UK: Nottingham University Press.

Meredith, J. 2003b. Understanding energy use and energy users in contemporary ethanol plants. In *The Alcohol Textbook*, 4th ed., eds. K. A. Jacques, T. P. Lyons, and D. R. Kelsall, pp. 355–361. Nottingham, UK: Nottingham University Press.

Monceaux, D. A., and D. Kuehner. 2009. Dryhouse technologies and DDGS production. In *The Alcohol Textbook*, 5th ed., eds. W. M. Ingledew, D. R. Kelsall, G. D. Austin, and C. Kluhspies, pp. 303–322. Nottingham, UK: Nottingham University Press.

Murthy, G. S., V. Singh, D. B. Johnston, K. D. Rausch, and M. E. Tumbleson. 2006. Evaluation and strategies to improve fermentation characteristics of modified dry-grind corn processes. *Cereal Chemistry* 83(5): 455–459.

Patzek, T. W. 2007. A first-law thermodynamic analysis of the corn-ethanol cycle. *Natural Resources Research* 15(4): 255–270.

Ramirez, E., D. Johnston, A. J. McAloon, W. C. Yee, and V. Singh. 2008. Engineering process and cost model for a conventional corn wet milling facility. *Industrial Crops and Products* 27: 91–97.

Ramirez, E., D. Johnston, A. J. McAloon, and V. Singh. 2009. Enzymatic corn wet milling: Engineering process and cost model. *Biotechnology for Biofuels* 2: 2.

Rausch, K. D., and R. L. Belyea. 2006. The future of coproducts from corn processing. *Applied Biochemistry and Biotechnology* 128: 47–86.

Rausch, K. D., R. L. Belyea, M. Ellersieck, V. Singh, D. B. Johnston, and M. E. Tumbleson. 2005. Particle size distribution of ground corn and DDGS from dry grind processing. *Transactions of the ASAE* 48(1): 273–277.

RFA. 2009. *Growing Innovation. 2009 Ethanol Industry Outlook.* Renewable Fuels Association. Washington, DC. Available at: www.ethanolrfa.org/industry/outlook/. Accessed November 1, 2010.

Rosentrater, K. A. 2006. Some physical properties of distillers dried grains with solubles (DDGS). *Applied Engineering in Agriculture* 22(4): 589–595.

Rosentrater, K. A. 2007. Corn ethanol coproducts: Some current constraints and potential opportunities. *International Sugar Journal* 109(1307): 1–12.

Rosentrater, K. A., and R. Srinivasan. 2009. *Effect of Elusieve Fractionation on Physical Properties of Distillers Dried Grains with Solubles (DDGS).* ASABE Paper No. 095567. St. Joseph, MI: ASABE.

Rosentrater, K. A., and B. A. Wrenn. 2009. *Examining the Effects of Ethanol Processing and Drying Conditions on the Physical Properties of Distillers Dried Grains with Solubles (DDGS).* ASABE Paper No. 095563. St. Joseph, MI: ASABE.

Saunders, J., and K. A. Rosentrater. 2009. Properties of solvent extracted low-oil corn distillers dried grains with solubles. *Biomass and Bioenergy* 33: 1486–1490.

Shapouri, H., J. A. Duffield, and M. S. Graboski. 1995. *Estimating the Net Energy Balance of Corn Ethanol*, Agricultural Economic Report No. 721. Washington, DC: United States Department of Agriculture, Office of the Chief Economist, Office of Energy Policy and New Uses.

Shapouri, H., J. A. Duffield, and M. Wang. 2002. *The Energy Balance of Corn Ethanol: An Update*, Agricultural Economic Report No. 814. Washington, DC: United States Department of Agriculture, Office of the Chief Economist, Office of Energy Policy and New Uses.

Singh, V., and S. R. Eckhoff. 1997. Economics of germ preseparation for dry-grind ethanol facilities. *Cereal Chemistry* 74(4): 462–466.

Singh, V. and D. B. Johnston. 2009. Fractionation technologies for dry-grind corn processing. In *The Alcohol Textbook*, 5th ed., eds. M. W. Ingledew, D. R. Kelsall, G. D. Austin, and C. Kluhspies, pp. 193–207. Nottingham, UK: Nottingham University Press.

Singh, V., D. B. Johnston, K. Naidu, K. D. Rausch, R. L. Belyea, and M. E. Tumbleson. 2005. Comparison of modified dry grind corn processes for fermentation characteristics and DDGS composition. *Cereal Chemistry* 82: 187–190.

Singh, V., R. A. Moreau, K. B. Hicks, R. L. Belyea, and C. H. Staff. 2002. Removal of fiber from distillers dried grains with soluble (DDGS) to increase value. *Transactions of the ASABE* 45(2): 389–392.

Srinivasan, R., R. A. Moreau, K. D. Rausch, R. L. Belyea, M. D. Tumbleson, and V. Singh. 2005. Separation of fiber from distillers dried grains with solubles (DDGS) using sieving and elutriation. *Cereal Chemistry* 82: 528–533.

Stern, M. 2007. *The Renessen Corn Processing System: Rebalancing the Bioenergy/Feed Equation.* Innovation to Commercialization, IndEx 2007, Chicago, IL, February 22, 2007. Chicago, IL: Illinois Biotechnology Industry Organization.

Tibelius, C. 1996. Coproducts and near coproducts of fuel ethanol fermentation from grain. Agriculture and Agri-Food Canada—Canadian Green Plan Ethanol Program: Starchy Waste Streams Evaluation Project. Available online: res2.agr.ca/publications/cfar/index_e.htm. Accessed November 1, 2010.

Tiffany, D. G., and V. R. Eidman. 2003. *Factors Associated with Success of Fuel Ethanol Producers*, Staff Paper P03-7. Minneapolis, MN: Department of Applied Economics, University of Minnesota.

Wahjudi, J., L. Xu, P. Wang, V. Singh, P. Buriak, K. D. Rausch, A. J. McAloon, M. E. Tumbleson, and S. R. Eckhoff. 2000. Quick fiber process: Effect of mash temperature, dry solids and residual germ on fiber yield and purity. *Cereal Chemistry* 77: 640–644.

Wang, P., V. Singh, L. Xu, D. B. Johnston, K. D. Rausch, and M. E. Tumbleson. 2005. Comparison of enzymatic (E-mill) and conventional dry grind corn processes using a granular starch hydrolyzing enzyme. *Cereal Chemistry* 82: 734–738.

Weigel, J. C., D. Loy, and L. Kilmer. 1997. Feed co-products of the dry corn milling process. Iowa State University, Iowa Corn Promotion Board, Iowa Department of Agriculture, Renewable Fuels Association, National Corn Growers Association. Available online: www.iowacorn.org/ethanol/ethanol_17.html. Accessed November 1, 2010.

Winsness, D. J. 2006. Increase ethanol industry profits with corn oil extraction technology. *Ethanol Producer Magazine* February: 108–110.

Winsness, D. J., and D. F. Cantrell. 2009. *Method of Freeing the Bound Oil Present in Whole Stillage and Thin Stillage*. U.S. Patent No. 7,608,729.

Wu, M. 2007. *Analysis of the Efficiency of the U.S. Ethanol Industry 2007*. Washington, DC: Renewable Fuels Association.

Wu, Y. V., K. R. Sexson, and J. S. Wall. 1981. Protein-rich residue from corn alcohol distillation: Fractionation and characterization. *Cereal Chemistry* 58(4): 343–347.

Wu, Y. V., and A. C. Stringfellow. 1982. Corn distillers dried grains with solubles and corn distillers dried grains: Dry fractionation and composition. *Cereal Chemistry* 47: 1155–1180.

Wu, Y. V., and A. C. Stringfellow. 1986. Simple dry fractionation of corn distillers dried grains and corn distillers dried grains with solubles. *Cereal Chemistry* 63(1): 60–61.

6 Ethanol Production from Starch-Rich Crops Other than Corn and the Composition and Value of the Resulting DDGS

Robert A. Moreau, Nhuan P. Nghiem, Kurt A. Rosentrater, David B. Johnston, and Kevin B. Hicks

CONTENTS

6.1 INTRODUCTION AND COMPARISON OF CORN AND OTHER STARCH-RICH FEEDSTOCKS

Because corn (maize, *Zea mays*) is the predominant feedstock for fuel ethanol in the United States and in many other countries, most of the chapters in this book focus on ethanol production and DDGS composition from corn. However, corn is not the only starch-rich crop that has been used as a feedstock for fuel ethanol. This chapter will describe eleven other starch-rich crops that have been used or have potentials to be used as feedstock for fuel ethanol. A useful website that includes information and references comparing various types of DDGS is hosted by the University of Minnesota (University of Minnesota, 2009). Another recent chapter described the commercial development of fuel ethanol plants using starch-rich and cellulosic feedstocks (Nghiem, 2008).

Mention of trade names or commercial products in this publication is solely for the purpose of providing specific information and does not imply recommendation or endorsement by the U.S. Department of Agriculture.

This chapter will focus mainly on the chemical composition of starch-rich crop feedstocks (Table 6.1) and the composition of their stillage and DDGS (Table 6.2). We also compared the international production values for all of the crops and noted the major countries where each is produced (Table 6.3).

It should be noted that there has been considerable controversy in recent years about using corn and other edible grains as feedstocks to make fuel ethanol (Congressional Budget Office, 2009). This food versus fuel debate was exacerbated by recent upturns in the price of petroleum products. Most experts agree that the long-term solution to the food versus fuel problem is to develop improved feedstocks that will not compete with food production. Such feedstocks as sugar cane bagasse, switchgrass, miscanthus, and hybrid poplar have been suggested, but it is important to note that the growth of these feedstocks could compete with food production and the use of these feedstocks to make ethanol does not result in the production of a high-protein animal feeds such as DDGS, which could result in a "fuel versus feed" issue. At the moment the cost of making fuel ethanol from corn and other starch-rich grains is considerably cheaper than making fuel ethanol from lignocellulosic feedstocks. The authors hope that during the next decade processes will be developed to make economically and environmentally sustainable biofuels from a mixture of acceptable feedstocks in a way that optimizes the use of the world's land mass for the sustainable and affordable production of food, feed, fuel, and fiber. Until cellulosic ethanol become economically competitive with ethanol from starch, ethanol from grains will continue to be a bridge to more advanced biofuels.

Even though corn is the focus of most of the chapters in this book, we feel that it is important to briefly discuss corn in this chapter and use it as a benchmark to compare the composition of feedstocks and DDGS of other starch-rich crops. Modern corn hybrids have been bred to give high yields and to contain consistently high levels of starch (~70%), as well as possessing many other desirable agronomic and compositional traits (Table 6.1). Corn is the #1 crop in the world in terms of grain production (Table 6.3), followed by rice and then wheat (also far ahead of soybeans with a world production of 219 MMT, million metric tons, in 2007).

The primary components in corn and other grains are carbohydrates, proteins, and lipids (Tables 6.1 and 6.2). Among the carbohydrates, starch and crude fiber are reported in Tables 6.1 and 6.2. Proteins are polymers of various combinations of 20 amino acids (12 amino acids are considered essential amino acids, since they are not synthesized by men and other mammals and must be included in the diet). The amino acid composition of corn, sorghum, wheat, and barley is shown in Table 6.4. It is interesting to note that corn and sorghum grains are known to be low in lysine, but the DDGS produced from these grains contains significantly higher amounts of lysine due to the contribution of yeast proteins (Table 6.4). Also, the same is true with wheat and barley, whose grains contain higher amounts of lysine, than do corn and sorghum—the DDGS from wheat and barley shows a dramatic additional increase in lysine over that in the whole grains, due again to the presence of yeast proteins (Table 6.4). Further discussion about the yeast effect on DDGS proteins can be found in Chapter 8.

The main lipids in starch-rich crops and all plants are triacylglycerols. Each triacylglycerol is hydrolyzed during digestion to yield three fatty acids and glycerol, and the nutritional value of these lipids is mainly due to the fatty acid composition (Table 6.5). It is interesting to note that the most abundant fatty acid in corn and the other four crops is linoleic acid, which is an 18 carbon fatty acid with two double bonds (Table 6.5).

It is important to note that considerable research and development has been invested over many decades by industry and other researchers to improve the agronomic and compositional traits of corn (maize) (Farnham et al., 2003). For this reason, #2 yellow dent corn obtained from different parts of the country or from different production years can have very similar compositions of starch, protein, and other components. Small grains, on the other hand, and particularly grains like barley, rye, and triticale, have received very little attention in this regard, so that one might expect much greater variation in yields and composition of these grains harvested in different years or locations. For this reason, values given in Tables 6.1 through 6.5 reflect a work in progress, rather than data from improved and optimized commercial lines. As research on them progresses, more consistent

TABLE 6.1

Composition of Starch-Rich Feedstocks for Fuel Ethanol, % Dry Basis

Crop	Starch	β-Glucan	Protein	Crude Fat	Fiber	References
Corn	73.4 (67.8–74.0)	0	9.1 (9.1–11.5)	4.4 (3.9–5.8)	nr	Nghiem (2008)
Sorghum (milo)	73.9	0	10.9	3.2	2.2 (CF)	Wu et al. (1984)
Wheat, hard	64.8	0	15.7	1.3	3.4 (CF)	Wu et al. (1984)
Wheat, hard	61.1	nr	13.83	nr	nr	Zhao et al. (2009)
Wheat, soft	65.2	nr	11.51	nr	nr	
Wheat, waxy	59.1	nr	15.05	nr	nr	
Barley, Hulled[a]	56.38	4.26	7.92	nr	nr	Griffey et al. (2010)
Hull-less[b]	63.48	4.42	8.41	nr	nr	
Barley, various	50–65	1.9–10.7	8.1–21.2	0.9–3.3	nr	Bhatty (1993)
Barley						
Hulled	64.7	6	17.7	2.5	5.7 (ADF)	Ingledew et al. (1995)
Hull-less	68.7	5.5	18.5	2.6	2.3 (ADF)	
Hull-less + high β-glucan	70.0	6.8	15.5	2.3	2.0 (ADF)	
Hull-less + waxy starch + high β-glucan	62.7	8.8	18.7	2.9	2.4 (ADF)	
Oats, whole	48.2	nr	11.3	4.4	13.2 (CF)	Wu (1990)
Groats	61.8	nr	16.0	6.3	2.0 (CF)	
Groats	57.1–60.4	4.3–5.8	14.6–19.6	4.6–7.8	nr	Doehlert et al. (2001)
Oats, hulled	50.8	3.0	10.6	5.0	nr	Thomas and Ingledew (1995)
Oats, hull-less	59.8	5.7	16.3	6.4	nr	
Rye	57.0–62.5	nr	8.0–10.4	nr	nr	Hansen et al. (2004)
	nr	2.26	nr	nr	nr	Henry (1987)
Triticale	67.87	nr	10.33	nr	nr	Kučerova (2007)
Rice, brown long grain	77.2	nr	7.94	2.92	3.5 (CF)	USDA (2010)
Short, white	79.15		6.50	0.52	2.8 (CF)	
Field Peas, dehulled	77.8	nr	nr	nr	nr	Nichols et al. (2005)
Pearl Millet	65.3–70.39	nr	9.73–13.68	6.8	nr	Wu et al. (2006)
Cassava	63.3	nr	2.4	0.3	4.5 (CF)	Stevenson and Graham (1983)
Cassava	75–85	nr	1.5–3.0	0.2	3–4 (CF)	Shetty et al. (2007)
Sweet Potato	55.7	nr	6.9	0.2	13.2 (TDF)	USDA (2010)

[a] Mean of 6 hulled barley lines.

[b] Mean of 18 hull-less barley lines.

TABLE 6.2
Composition of DDGS and Stillage after Fermentation of Starch-Rich Feedstocks, % Dry Basis

Feedstock	Starch	β-Glucan	Protein	Crude Fat	Fiber	References
Corn	5.1	nr	31.3	11.9	10.2 (CF)	Belyea et al. (2004)[a]
	5.48	nr	30.8	11.4	7.41 (CF)	UMN (2009)
Sorghum (milo)	5.2–5.2	nr	39.9–45.1	10.8–12.0	7.5–9.2 (CF)	Corredor et al. (2006)
Wheat hard	5.7	nr	45.3	12.3	11.6 (CF)	Wu et al. (1984)
Wheat hard	nr	nr	35.59	7.66	5.56 (CF)	UMN (2003)
Wheat hard	6.3	nr	39.3	4.98	48.1 (NDF)	Nuez-Ortin and Yu (2009)
Wheat, waxy	0.43	nr	37.4	nr	nr	Zhao et al. (2009)
Wheat, soft	1.01	nr	32.3	nr	nr	
Wheat, hard	0.85	nr	38.8	nr	nr	
Barley						
Hulled	nr	nr	35.3	7.2	8.2 (CF)	Gibreel et al. (2009)
Hull removed	nr	nr	40.4	8.1	2.7 (CF)	
Hull-less	nr	nr	39.9	8.6	1.6 (CF)	
Hull-less	nr	1.7	24.2	nr	33.9 (ADF)	Ingledew et al. (1995)
Hull-less + high β-glucan	nr	1.9	34.5	nr	22.6 (ADF)	
		2.6	36.4	nr	22.7 (ADF)	
Hull-less + high β-glucan + waxy starch	nr	3.2	34.8	nr	21.4 (ADF)	
Hulled with EDGE process	1.64	0.2	21.75	4.43	39.36 NDF	Nghiem et al. (2010)
Oats	1.0	nr	18.8	9.7	25.4 (CF)	Wu (1990)
Rye, stillage	1.52 ± 0.08	nr	27.72 ± 0.20	3.32 ± 0.01	28.24 ± 0.72 (TDF)	Wang et al. (1998)
Rye, stillage	1.08	nr	25.00	2.64	32.84 (TDF)	Wang et al. (1999)
Pearled Rye, stillage	2.18	nr	29.72	3.22	18.66 (TDF)	
Triticale, stillage	1.67 ± 0.52	nr	34.54 ± 0.71	3.93 ± 0.06	32.61 ± 1.03 (NDF)	Wang et al. (1998)
Rice, stillage	32.6[b]	nr	55.8	8.2	2.1 (CF)	Minowa et al. (1994)
Field peas	0.79	nr	41.4	0.4	nr	Nichols et al. (2005)
Pearl Millet	3.45	nr	30.7	19.2	4.28 (CF)	Wu et al. (2006)
Sweet potato	55.3[b]	nr	27.2	1.7	11.4 (CF)	Minowa et al. (1994)

Abbreviations: ADF, acid detergent fiber; CF, crude fiber; NDF, neutral detergent fiber; nr, value not reported; TDF, total dietary fiber.

[a] Mean of samples from 49 dry grind ethanol plants.

[b] Nonfibrous carbohydrate.

TABLE 6.3
World Production and Major Producing Countries of Starch-Rich Crops

Crop	MMT	Major Countries
Grain Crops		
Corn[a]	784	United States (332), China (152), Brazil (52)
Rice[a]	650	China (185), India (141), Indonesia (57)
Wheat[a]	607	China (110), India (54), United States (54)
Barley[a]	136	European Union (58), Russia (16), Canada (12)
Sorghum (milo)[b]	59	United States (10), India (8), Nigeria (8)
Pearl Millet[a]	32[c]	India (11), Nigeria (8), Niger (3)
Oats[b]	25	Russia (5), Canada (3), United States (2)
Triticale[b]	14	Poland (4), Germany (3), France (2)
Rye[b]	13	Russia (4), Poland (3), Germany (3)
Field Peas[b]	12	Canada, France, Russia
Root Crops		
Cassava[b]	185	Thailand (77), Vietnam (14), Indonesia (6)
Sweet Potato[d]	125	China (100), Nigeria (3), Uganda (3)

[a] FAO 2007.
[b] FAO 2005.
[c] All millets.
[d] usda.mannlib.cornell.edu/MannUsda.

levels of starch, protein, and other components will likely result. These data in the tables therefore represent current status and a starting point for future improvements.

6.2 SORGHUM (MILO)

Sorghum (*Sorghum bicolor*) has a world production of about 59 MMT, with highest productions in the United States, India, and Nigeria (Table 6.3). Sorghum has been used as a fuel ethanol feedstock in the United States at many ethanol plants. Because the composition of the sorghum grain is similar to that of corn (Table 6.1), it can be blended with corn and used in ethanol plants designed for corn. The chemical composition of sorghum DDGS is also very similar to that of corn DDGS (Table 6.2) and the feed value of corn and sorghum DDGS was reported to be similar in feeding studies with beef and lambs (Lodge et al., 1997; Harborth et al., 2006). Decortication is an abrasive dehulling process that has been used to remove nonfermentable fiber from sorghum grain and the resulting DDGS is more valuable, especially for nonruminant feeds because they are higher in protein and lower in fiber (Corredor et al., 2006). The amino acid composition of sorghum grain and DDGS is similar to that of corn grain and DDGS (Table 6.4). Although there are few published reports of sorghum DDGS composition, it is possible that as new and improved regional cultivars of sorghum are developed (Buffo et al., 1998), when fermented they may produce DDGS with enhanced compositions of protein and other nutrients. Also, as with any grain, the proportion of protein can be increased by conventional breeding, but this is usually achieved at the expense of starch yields, and starch yields are usually considered to be the primary trait for ethanol production.

6.3 BARLEY

Barley (*Hordeum vulgare*) is the #4 cereal, behind corn, rice, and wheat with a worldwide production of 136 MMT in 2007 (Table 6.3). The countries that produce the highest amounts of barley are Russia, Canada, and Spain.

TABLE 6.4
Crude Protein and Amino Acid (AA) Composition of Selected Grains and their DDG or DDGS (% Dry Sample Weight)

	Corn Grain[a]	Corn DDGS[a]	Sorghum Grain[a]	Sorghum DDGS[a]	Wheat Grain[b]	Wheat DDG[b]	Barley Grain[c]	Barley DDG[c]
Crude protein	8.0	27.5	9.8	31.0	15.7	29.2	13.4	32.6
Amino Acid								
Arginine	0.39	1.16	0.32	1.10	0.87	2.11	0.77	1.71
Histidine	0.23	0.72	0.23	0.71	0.37	0.75	0.32	0.71
Isoleucine	0.28	1.01	0.37	1.36	0.52	1.05	0.46	1.24
Leucine	0.95	3.17	1.25	4.17	1.09	2.21	0.89	2.39
Lysine	0.24	0.78	0.20	0.68	0.43	1.08	0.54	1.21
Methionine	0.21	0.55	0.18	0.53	0.18	0.42	0.23	0.56
Phenylalanine	0.38	1.34	0.47	1.68	0.72	1.41	0.69	1.90
Threonine	0.26	1.06	0.29	1.07	0.49	1.49	0.45	1.09
Tryptophan	0.09	0.21	0.07	0.35	nr	nr	nr	nr
Valine	0.38	1.35	0.48	1.65	0.74	1.19	0.68	1.65
Alanine	0.58	1.94	0.86	2.90	0.58	1.44	0.54	1.34
Aspartic acid	0.55	1.83	0.60	2.17	0.88	1.97	0.76	1.86
Cysteine	0.16	0.53	0.18	0.49	nr	nr	0.31	0.40
Glutamic acid	1.48	4.37	1.92	6.31	5.11	7.26	3.93	9.54
Glycine	0.31	1.02	0.29	1.03	0.68	1.50	0.51	1.15
Proline	0.70	2.09	0.77	1.40	1.72	2.71	1.30	3.45
Serine	0.38	1.18	0.37	2.50	0.77	1.58	0.57	1.31
Tyrosine	0.27	1.01	0.25	nr	0.54	1.03	0.45	1.09

Note: Abbreviations, nr, value not reported.

[a] Stein, H. H. 2008. Use of distillers co-product in diets fed to swine, in Using Distillers Grains in the US and international livestock and poultry industries, Midwest Agribusiness and Trade Research Information Center, Iowa State U, Ames, Iowa. Available online: www.matric.iastate.edu/DGbook/chapters/chapter4.pdf. Accessed November 2, 2010.

[b] Wu, Y. V., K. R. Sexson, and A. A. Lagoda. 1984. Protein-rich residue from wheat alcohol distillation: Fractionation and characterization. *Cereal Chemistry* 61: 423–427.

[c] Wu, Y. V. 1986. Fractionation and characterization of protein rich material from barley after alcohol distillation. *Cereal Chemistry* 63: 142–145.

TABLE 6.5
Concentration of Fatty Acids in DDGS (Relative %)

	Linoleic	Oleic	Palmitic	Others
Corn[a]	49.7	25.3	16.1	8.9
Wheat[a]	56.1	11.6	21.4	11.1
Hulled barley[a]	50.9	11.6	23.3	14.0
Hulless barley[a]	51.4	12.3	24.0	12.5
Sorghum[b]	44.1	28.2	21.8	6.0

[a] Gibreel, A., J. R. Sandercock, L. Lan, L. A. Goonewardene, R. T. Zijlstra, J. M. Curtis, and D. C. Bressler. 2009. Fermentation of barley by using *Saccharomyces cerevisiae*: Examination of barley as a feedstock for bioethanol production and value-added products. *Applied and Environmental Microbiology* 75: 1363–1372. All fermentations were jet cooked before adding amylase.

[b] Wang, L., C. L. Weller, V. L. Schlegel, T. P. Carr, and S. L. Cuppett. 2007. Comparison of supercritical CO_2 and hexane extraction of lipids from sorghum distillers grains. *European Journal of Lipid Science and Technology* 109: 567–574.

Until recently, barley had been used only sparingly for fuel ethanol production because several of its attributes made it more costly to produce fuel ethanol from this feedstock compared to corn or most other starch-rich crops. The first negative factor is the presence of silica in the hulls (2%–6% of the weight of the hulls), which is abrasive and causes damage to grain handling and processing equipment. For this reason thicker and harder processing equipment needs to be used for barley than for corn. Two other strategies have been used to try to overcome the problems caused by the abrasiveness of barley grain. The first is to remove the hull by abrasive techniques such as pearling or scarification (Wang et al., 1997a; Flores et al., 2007). This process works well but it can also remove some of the starch, thus reducing ethanol yields per bushel of grain. Another strategy has been to use hull-less (sometimes spelled "hulless") or "naked" cultivars of barley, which lose their hulls in the field and during harvesting, so little or no abrasive hull is attached when the grain enters the ethanol plant.

The second negative attribute for barley is the fact that most barley previously used for fuel ethanol production had low levels of starch and gave low amounts of ethanol per bushel of grain. Most barley used in the past had been either "feed" grade or off-grade malting barley, both of which may have low levels of starch and very high levels of neutral detergent fiber (NDF). To solve the low-starch problem, breeding programs have been initiated recently to produce higher-starch and lower NDF "energy" varieties of both hulled and hull-less barley varieties (Griffey et al., 2010). Many of these newer barley varieties, especially the hull-less ones, contain lower levels of fiber and higher levels of starch that are more desirable for fuel ethanol production (Table 6.1). As a general rule, most hull-less varieties contain higher levels of starch, protein, and β-glucan and lower levels of fiber than similar hulled varieties (Table 6.1) so that hull-less varieties are receiving more attention as ethanol feedstock. Still another innovative approach to solve this problem has been to use roller milling and fractionation to separate whole barley kernels into high-starch fractions for ethanol production and low starch fractions for various food and feed applications (Flores et al., 2005). Fractions containing well over 70% starch and low levels of NDF that produce high yields of ethanol can be produced in this manner.

The third attribute of barley that makes fuel ethanol production difficult is the presence of β-glucans, which are present throughout the barley kernel and especially in the endosperm. β-Glucans are water-soluble mixed linkage β-1,3- and β-1,4-linked glucans that create very high viscosity when dissolved in aqueous solutions. β-Glucans derived from barley or oats can be valuable components in human diets because they are a healthy soluble fiber that can reduce

levels of serum LDL-cholesterol by 10%–15%. The presence of β-glucans in ethanol production, however, is not desirable. Early attempts to produce fuel ethanol from barley resulted in
extremely high mash viscosity, which created difficulties in mixing, pumping, saccharification,
and fermentation. The resulting DDGS from such processes contains high levels of β-glucans,
which are not well tolerated in swine and poultry diets and therefore decrease the value of barley
DDGS as a feed component for these species. The use of β-glucanase enzymes during mashing
and saccharification of barley (Ingledew et al., 1995) was shown to lower viscosity to acceptable
levels and is now standard practice in the industry. While these enzymes decreased viscosity,
they did not increase ethanol yields (Ingledew et al., 1995). This is because β-glucanase only
partially hydrolyzes the β-glucan polymers to produce short oligomers, which are not fermentable by *Saccharomyces cerevisiae*. These β-glucan oligomers therefore end up as a component
in the resulting DDGS. Researchers have recently developed a barley "EDGE" (Enhanced Dry
Grind Enzymatic) ethanol process that results in several process improvements and gives higher
ethanol yields. The EDGE process employs a second enzyme, a β-glucosidase, in combination
with β-glucanases, which hydrolyzes the glucan oligomers to glucose, and this glucose, along
with the glucose from barley starch can be fermented to ethanol (Nghiem et al., 2010). Using
the EDGE process, a barley variety that contains 65% starch and 5% β-glucan should provide
equivalent ethanol yield to a grain containing 70% starch. Recently, barley has been converted
to fuel ethanol using advanced GSHE "granular starch hydrolyzing enzyme" technology (Li et
al., 2007; Gibreel et al., 2009). This technology allows starch to be efficiently converted without
first gelatinizing the starch by exposure to high (jet cooking) temperatures. This simple process
is less complicated and requires less energy and processing equipment than conventional starch
conversion technologies. The GSHE process gives high ethanol yields and high quality DDGS.
The technology and similar ones are beginning to be used in the United States on corn grain and
in Europe for small grains.

In the past, barley DDGS has been criticized for having too much β-glucan and too much fiber
for monogastric animal diets, but the levels of β-glucan, NDF, and other components in DDGS will
depend upon the composition of the feedstock as well as the process (and enzymes) used to convert
it to ethanol. Comparison of the barley grain and DDGS compositions in Tables 6.1 and 6.2 reveals
the following trends:

- DDGS from hulled barley has high fiber levels and lower protein and fat levels. Its composition is more suitable for ruminant animals than monogastric types.
- DDGS from hull-less or dehulled varieties has lower fiber content than that from hulled
  varieties and also has higher protein and fat levels. All other factors being equal, this
  DDGS would be more valuable and suitable for monogastric animals than the DDGS from
  hulled varieties. It is of course also quite suitable for ruminant animals.
- DDGS made using the EDGE process contains low levels of β-glucans that are approximately one order of magnitude lower than that found in DDGS produced from barley ethanol production using β-glucanases without β-glucosidase.
- Based upon high levels of protein and low β-glucans, use of the EDGE process with hull-
  less or dehulled barley results in DDGS that should be more valuable than traditional
  hulled barley DDGS and well tolerated by monogastric animals.

The amino acid composition of barley DDGS is also quite good (Table 6.4) and is richer than corn
DDGS in lysine and some other essential amino acid levels. Because of the advances in barley variety improvement, improved ethanol conversion processes, and the positive attributes of hull-less
and dehulled barley DDGS, the first barley ethanol plant in the United States is scheduled to begin
production in Hopewell Virginia in 2010 (Robertson, 2010) beginning a new era in small grain
biorefining in the United States.

6.4 WHEAT

In Canada, Europe, and Australia, wheat (several *Triticum* species) is an important feedstock for fuel ethanol production. Zhao et al. (2009) recently reviewed the technology on wheat to ethanol conversion. In the United States, two ethanol plants ferment wheat starch (produced by wet milling), so they do not produce DDGS. In Europe and Australia, wheat is the primary feedstock being considered for expansion of the fuel ethanol industry. Recently Zhao et al. (2009) compared the composition and conversion efficiency of waxy, soft, and hard wheat into fuel ethanol. Ethanol yield was highly related to both total starch and protein content, but total starch was a better predictor of ethanol yield. Wang et al. (1998) showed that wheat can be sequentially abraded to yield pearled grains with a significant (12%) increase in starch content over the whole kernels.

Most wheat cultivars contain slightly more starch and slightly more protein than corn (Table 6.1). Wheat is the #3 crop in the world (following corn and rice) in terms of grain yield (607 MTT in 2007) (Table 6.3).

Compared to corn, wheat DDGS is potentially more valuable because it contains more protein and less fiber (Table 6.3). The nutrient variation and availability of wheat DDGS from bioethanol plants in Canada were recently surveyed (Nuez Ortin and Yu, 2009). The amino acid composition of wheat kernels and wheat DDGS is also better than for corn and sorghum (Table 6.4). Wheat is an important crop in western Canada, where virtually all new and existing grain ethanol plants use wheat as the main feedstock. A useful website that includes information and references about wheat DDG and DDGS is hosted by the University of Saskatchewan's Department of Animal and Poultry Science (University of Saskatchewan, 2010).

6.5 TRITICALE

Triticale (× *Triticosecale*) is a hybrid of wheat (*Triticum*) and rye (*Secale*) first bred in laboratories during the late 19th century. Triticale is a cereal grain with potential for use as an important feedstock for fuel ethanol production. The world annual production of triticale (14 MMT in 2007) is relatively small compared to that of the major crops such as corn and barley. The major producing countries are Poland, Germany, and France (Table 6.3). During the last 20 years, triticale production has increased.

Traditionally triticale has been used mostly for food and feed products. Triticale also has been used for fuel ethanol production. However, in terms of total production, its use for fuel ethanol production has been modest compared to other feedstocks such as corn and cane sugar. In Sweden, triticale has been used for fuel ethanol production together with wheat (Fields of Energy, 2006). Only recently, triticale has attracted considerable attention as a feedstock for fuel ethanol production as the direct result of the surge in demand for this renewable liquid fuel. Rye and triticale were studied as substitutes for wheat as fuel ethanol feedstocks due to their similar starch contents but much lower costs (Sosulski et al., 1997; Wang et al., 1997a, 1997b, 1998, 1999). Interest in using triticale for fuel ethanol production has been growing strong particularly in Canada. The Canadian government recently invested $15.5 million from the Agriculture and Agri-Food Canada's Agricultural Bioproducts Innovation Program in the Canadian Triticale Biorefinery Initiative (CTBI) (Christiansen, 2009; Agriculture and Agri-Food Canada, 2010).

Recent data on starch contents and other important characteristics related to ethanol production of triticale grown in the Czech Republic are summarized in Table 6.1 (Kučerova, 2007). According to actual data on ethanol production from triticale in Sweden, 2.65 kg triticale grain produces 1 L ethanol, 0.8 kg carbon dioxide, and 0.8 kg DDGS (Fields of Energy, 2006).

One advantage of triticale as a feedstock for ethanol production is its high content of endogenous α-amylase, which helps to reduce enzyme requirements and costs (Kučerova, 2007; Davis-Knight and Weightman, 2008). This was also studied by Pejin et al. (2009), who stated that in laboratory studies "in the case of triticale, technical enzymes were not needed for starch degradation." This would indicate that in addition to endogenous α-amylase, triticale must also contain endogenous glucoamylase or related enzymes. Since triticale is a cross between wheat and rye, it inherits some of the traits from rye and can grow under soil and environmental conditions that are not favorable to wheat. Unlike barley and rye, triticale mash does not have high viscosity and hence does not require application of viscosity-reducing enzymes like rye mash (see further discussion below in the section on rye). Addition of urea resulted in lower ethanol yields in some cases, for example, in a laboratory-scale fermentation using a mash containing 28.5 g of dissolved solids/100 mL of mash liquid, 89.4 ± 3.1% theoretical yield was obtained with 16 mM urea addition compared to 93.0 ± 0.1% theoretical yield without urea addition. However, the addition of 16 mM urea shortened the time required to complete the fermentation from 120 to 70 h. The final ethanol concentrations obtained were 14.7%– 15.1% (v/v). The chemical compositions of the fermentation stillages obtained in these and related experiments are summarized in Table 6.2 (Wang et al., 1998, 1999). These data show that triticale stillage is richer in protein and fat than rye stillage and when pearled prior to ethanol production, the resulting DDGS is even richer in protein and fat and lower in fiber. While the pearling process results in the loss of some fermentable starch, the resulting protein levels of over 40% may in some cases justify the loss of ethanol yields.

6.6   RYE

Rye (*Secale cereale*) has a world production of 13 MMT and the major producing countries are Russia, Poland, and Germany (Table 6.3). Like triticale, rye is also a cereal grain with potential for use as important feedstocks for fuel ethanol production. The total world production in 2008 was only 15 MMT for rye (FAO, 2005, 2007). During the last 20 years, rye production has declined.

Traditionally, rye has been used mostly for food, brewing, and feed products. Rye has also been used for fuel ethanol production. However, in terms of total production, use of rye for fuel ethanol production has been modest compared to other feedstocks such as corn and cane sugar. In Canada, rye has been used only occasionally as an adjunct in fuel ethanol manufacture (Ingledew et al., 1999). Poland is the only country where rye is an important feedstock for fuel ethanol production, where it accounts for more than 80% of the total fuel ethanol production (Czarnecki and Nowak, 2001). Only recently has rye attracted considerable attention as a feedstock for fuel ethanol production as the direct result of the surge in demand to meet Federal renewable fuel guidelines. Rye and triticale were considered as substitutes for wheat as fuel ethanol feedstocks due to their similar starch contents but much lower costs (Ingledew et al., 1999; Wang et al., 1999).

Some grain characteristics of rye grown in Denmark are summarized in Table 6.1 (Hansen et al., 2004). One of the problems of using rye as a fuel ethanol feedstock is its high pentosan and β-glucans contents, which result in very high viscosity of the mash. The β-glucan content of whole grain rye is about 2%, which is less than in barley and oats (Table 6.1; Henry, 1987).

Because of the high pentosan and β-glucans contents viscosity-reducing enzymes are needed, either in an additional preincubation step or during the simultaneous saccharification and fermenta- tion (SSF) (Ingledew et al., 1999; Czarnecki and Nowak, 2001; Balcerek and Pielech-Przybylska, 2009). The enzymes include pullulanase, cellulase, xylanase, β-glucosidase, pectinase, and pro- teinase. Addition of enzymes such as xylanase and pullulanase is found to also increase release of fermentable sugars and improve fermentation efficiency and ethanol yield (Balcerek and Pielech- Przybylska, 2009). Fermentation of the ground pearled grains reduces energy requirements for mash heating, mash cooling, and ethanol distillation (Sosulski et al., 1997; Wang et al., 1999). A mass balance for an ethanol fermentation using Canadian ground pearled rye was reported (Sosulski et al., 1997).

The composition of the whole rye stillage from two separate studies is shown in Table 6.2 (Wang et al., 1998, 1999). Wang et al. (1999) also reported that pearling rye, prior to ethanol production resulted in feedstock with higher starch content and results in DDGS with higher protein and fat and lower levels of fibrous polysaccharides (Table 6.2).

Whereas much has been studied and published on the nutritive value of corn DDGS, relatively little has been published on those from rye. One *in vitro* study (Mustafa et al., 2000) indicated that rye thin stillage and distillers grains from rye had some superior nutritional qualities in ruminant diets.

6.7 OATS

Oats (*Avena sativa*) have a world production of about 25 MMT, with Russia, Canada, and the United States being the major producing countries (Table 6.4). Like barley, a major attribute of oats is the presence of β-glucans (4%–6%) throughout oat kernels (Table 6.1). Another attribute of oats, which distinguishes it from other cereals, is its higher values of crude fat (4%–9%). The higher fat levels have been utilized to give baked goods a softer and moister texture than can be achieved with most other cereals, which all have lower levels of fat (Doehlert, 2001). Thomas and Ingledew (1995) showed that hull-less oats contained higher levels of starch than hulled varieties and produced more ethanol per unit of grain. They also showed that the use of cellulases or endo-glucanases was necessary to reduce mash viscosities, especially for very high gravity fermentations. Wu (1990) demonstrated the feasibility of fermenting ground oats and reported the composition of starch (1.0%), protein (18.8%), crude fat (9.7%), and crude fiber (25.4%) in the stillage (Table 6.2).

6.8 RICE

Rice (*Oryza sativa*) is the #2 cereal crop behind corn, with world production of 650 MMT in 2007 (Table 6.3). Rice has high levels of starch (~75%), no β-glucans, and low levels of fiber (Table 6.1). Although Minowa et al. (1994) demonstrated the feasibility of fermenting starch to ethanol, the high value of rice as a food has precluded it from serious consideration as a feedstock for fuel ethanol. Minowa et al. (1994) reported that stillage from rice fermentation contained starch (32.6%), protein (55.8%), fat (8.2%), and crude fiber (2.1%) (Table 6.2).

6.9 PEARL MILLET

Laboratory scale ethanol production from Pearl Millet (*Pennisetum glaucum*) was investigated by Wu et al. (2006). Because Pearl Millet (not to be confused with "pearled millet") can grow in semi-arid conditions where sorghum and corn cannot, it is being considered by some as a potential biofuel crop for those regions. In this study they examined four Pearl Millet cultivars with starch contents varying from 65.3% to 70.39% on a dry weight basis. The starting protein contents of the cultivars were significantly higher than corn and ranged from 9.72% to 13.68% on a dry weight basis. Crude fat contents were also high relative to corn and were as high as 6.80%. Mash concentrations from 20% to 35% dry mass were investigated by examining ethanol yields and fermentation efficiencies. Fermentation rates and efficiencies were found to be similar to corn. The resulting DDGS from Pearl Millet has a protein content of 30.74% and a fat content of 19.22%. For other components, see Table 6.2.

6.10 FIELD PEAS

Field peas (*Pisum sativum*) and other legumes have been considered as feedstocks for fuel ethanol. Laboratory ethanol production from field peas was investigated by Nichols et al. (2005). They used whole peas as well as air-classified fractionation to enrich starch content. The fractionation process increased the starch content from 46.2% (starting peas) to as high as 77.8% for the dehulled, 9 × 14,000 rpm pin-milled and air-classified fraction. The resulting DDGS from the field peas contained significantly higher levels of protein (42.8%) compared to corn DDGS, but has a much lower oil content (0.4%) (Table 6.2).

6.11 CASSAVA

Cassava (*Manihot esculenta*, also called manioc, Brazilian arrowroot, sagu, yucca, or tapioca) is an important tropical root crop, which had a world production of 185 MMT in 2005 and the major producing countries were Thailand, Vietnam, and Indonesia (Table 6.3). It is our understanding that cassava is being considered as a feedstock for fuel ethanol in China. Cassava root contains about 63% starch, 1.6% protein, and 4.5% fiber, when expressed on a dry weight basis (Table 6.1, Stevenson and Graham, 1983). Shetty et al. (2007) found higher starch content (Table 6.1) and gave complete details for the use of cassava as an alternative feedstock in the production of fuel ethanol. We are not aware of any published reports about the composition of stillage or DDGS from cassava but since the levels of protein are so low in the tuber, it is presumed that the protein in the DDGS would be low as well. Shetty et al. (2007) actually referred to this material as "DDS," dry distiller's solids.

6.12 SWEET POTATO

Sweet potatoes (*Ipomoea batatas*) are a second starch-rich crop that is being evaluated as a feedstock for fuel ethanol. Sweet potatoes have a world production of 125 MMT and the major producing countries are China, Nigeria, and Uganda (Table 6.3). Sweet potatoes contain about 55.7% starch, 6.9% protein, 0.2% fat, and 13.2% fiber, when expressed on a dry weight basis (Table 6.1). Minowa et al. (1994) reported that the stillage from sweet potatoes was comprised of 27.2% protein and 2% fat. Fermentation and ethanol yields of three sweet potato cultivars were also studied by Wu and Bagby (1987), who reported three stillage fractions (filter cake, centrifuged solids, and stillage solubles) for each cultivar, with protein levels ranging from 9.8% to 54.4%. There have been many press reports of sweet potato and "energy potato" biorefineries that are being planned or considered in various parts of the United States, especially the southern United States. As far as we are aware, there is little data published on the composition of the "energy" sweet potato varieties or the dried distiller's solids that would result from drying the stillage.

6.13 FINAL THOUGHTS: ADVANCED BIOFUELS AND
COPRODUCTS FROM GRAIN BIOREFINERIES

Corn continues to be "king" in the United States as the primary ethanol feedstock in 2010 and only time will tell about the future of other small grains and starchy feedstocks for these purposes. Many resources are currently being applied to develop economically and environmentally sustainable processes to convert nonfood lignocellulosic feedstocks into advanced biofuels by biochemical and thermochemical pathways. These technologies seem to be about 5 years away from economic feasibility and this has been so for more than 20 years. The use of winter barley as cover crops and then using the grain as ethanol feedstock on the East Coast is creating a new paradigm: the ability to produce a low-greenhouse gas "advanced" biofuel from a feedstock that we know how to efficiently convert to ethanol (Nghiem et al., 2010). Since the grain is grown on winter fallow land and doesn't interfere with summer crop rotations, it avoids the food versus fuel debate. Growing the winter barley also prevents the leaching of nutrients and sediment in winter soils into watersheds. The Chesapeake Bay Commission (2007) fully supports growing winter barley as a cover crop to clean up the Chesapeake Bay. This model of growing an additional winter crop can be repeated in many different parts of the United States and abroad. Each location may have a different crop but it can have the same effect. In Illinois, for instance, maybe that crop is Pennycress (Moser et al., 2009). Imagine the potential impact of winter energy cover crops on cleaning up the Mississippi River watershed and the dead zone in the Gulf of Mexico. Many of the starchy crops reported in this chapter plus others not yet imagined may play a large role in future production of advanced biofuels in the United States. If this is so, the DDGS or high-protein meals from those grains will be positioned to fill animal feeding needs in that region.

Two other important topics that were not addressed in the chapter were the topics of sustainability and the need for the development of biorefineries for the conversion of starch-rich grain to fuel ethanol, animal feeds, and to multiple higher value coproducts. There has been much discussion about converting corn dry grind ethanol plants into more efficient biorefineries and there is a need to continue this discussion as other starch-rich feedstocks are further developed. For instance, it is known that grains such as corn and barley contain valuable nutraceuticals and functional proteins and hydrocolloids that could be isolated in biorefineries as valuable coproducts. Whereas laboratory processes for such coproducts as phytosterols (Moreau et al., 1996; Lampi et al., 2004), tocopherols and tocotrienols (Moreau et al., 2007a, 2007b), lutein and zeaxanthin (Moreau et al., 2007c), functional polysaccharides (Doner et al., 1998; Yadav et al., 2007), and zein (Shukla and Cheryan, 2001) have been developed to isolate these potential coproducts from grains and by-products, none of the processes have been economically scaled-up to commercial levels in existing ethanol plants. With continued research, this will likely become a reality, enabling a more economically sustainable biofuels industry.

REFERENCES

Agriculture and Agri-Food Canada. 2010. Government of Canada makes significant investment in Alberta-Based Research Network. Available online: www.agr.gc.ca/cb/index_e.php?s1=n&s2=2009&page=n90306. Accessed February 10, 2011.

Balcerek, M., and K. Pielech-Przybylska. 2009. Effect of supportive enzymes on chemical composition and viscosity of rye mashes obtained by the pressureless liberation of starch method and efficiency of their fermentation. *European Food Research and Technology* 229: 141–151.

Bhatty, R. S. 1993. Nonmalting uses of barley. In *Barley Chemistry and Technology*, eds. A. W. MacGregor, and R. S. Bhatty, pp. 355–418. St. Paul: American Association of Cereal Chemists.

Belyea, R. L, K. D. Rausch, and M. E. Tumbleson. 2004. Composition of corn and distillers dried grains and solubles from dry grind ethanol processing. *Bioresource Technology* 94: 293–298.

Buffo, R. A., C. L. Weller, and A. M. Parkhurst. 1998. Relationships among grain sorghum quality factors. *Cereal Chemistry* 75: 100–104.

Chesapeake Bay Commission. 2007. Biofuels and the Bay—Getting It Right to Benefit Farms, Forests and the Chesapeake. A report of the Chesapeake Bay Commission. Available online: www.chesbay.state.va.us/Publications/BiofuelsAndTheBay1.pdf. Accessed November 3, 2010.

Christiansen, R. C. 2009. Canada invests $15.5 million in *triticale* research. Ethanol Producer Magazine, April 2009. Available online: www.ethanolproducer.com/article.jsp?article_id=5498&q=triticale&category_id=36. Accessed November 4, 2010.

Congressional Budget Office. 2009. The impact of ethanol use on food prices and greenhouse-gas emissions. Available online: www.cbo.gov/ftpdocs/100xx/doc10057/04-08-Ethanol.pdf. Accessed November 2, 2010.

Corredor, D. Y., S. R. Bean, T. Schober, and D. Wang. 2006. Effect of decorticating sorghum on ethanol production and composition of DDGS. *Cereal Chemistry* 83: 17–21.

Czarnecki, Z., and J. Nowak. 2001. Effects of rye pretreatment and enrichment with hemicellulolytic enzymes on ethanol fermentation efficiency. *Electronic Journal of Polish Agricultural Universities* 4(2): 12. Available online: www.ejpau.media.pl/volume4/issue2/food/art-12.html. Accessed November 2, 2010.

Davis-Knight, H. R., and R. M. Weightman. 2008. The potential of triticale as a low input cereal for bioethanol production. ADAS UK Ltd, Centre for Sustainable Crop Management, Project Report No. 434. Available online: www.hgca.com/document.aspx?fn=load&media_id=4586&publicationId=4631. Accessed November 2, 2010.

Doehlert, D. C., M. S. McMullen, and J. J. Hammond. 2001. Genotypic and environmental effects on grain yield and quality of oat grown in North Dakota. *Crop Science* 41: 1066–1072.

Doner, L. W, H. K. Chau, M. L. Fishman, and K. B. Hicks. 1998. An improved process for isolation of corn fiber gum. *Cereal Chemistry* 75: 408–411.

FAO. 2005. Food and Agriculture Organization of the United Nations, http://www.fao.org/es/ess/top/commodity.html?lang=en&item=71&year=2005. Accessed March 15, 2011.

Farnham, D. E., G. O. Benson, and R. B. Pearce. 2003. Corn perspective and culture. In *Corn Chemistry and Technology*, 2nd ed., eds. P. J. White, and L. A. Johnson, pp. 1–33. St. Paul, MN: American Association of Cereal Chemists, Inc.

Fields of Energy. 2006. SW Seed Limited. Available online: www.swseedco.com. Accessed November 2, 2010.

Flores, R. A., K. B. Hicks, D. W. Eustace, and J. G. Phillips. 2005. High-starch and high-ß-glucan barley fractions milled with experimental mills. *Cereal Chemistry* 82: 727–733.

Flores, R. A., K. B. Hicks, and J. Wilson. 2007. Surface abrasion of hulled and hulless barley: Physical characteristics of the milled fractions. *Cereal Chemistry* 84: 485–491.

Gibreel, A., J. R. Sandercock, L. Lan, L. A. Goonewardene, R. T. Zijlstra, J. M. Curtis, and D. C. Bressler. 2009. Fermentation of barley by using *Saccharomyces cerevisiae*: Examination of barley as a feedstock for bioethanol production and value-added products. *Applied and Environmental Microbiology* 75: 1363–1372.

Griffey, C., W. Brooks, M. Kurantz, W. Thomason, F. Taylor, D. Obert, R. Moreau, R. Flores, M. Sohn, and K. Hicks. 2010. Grain composition of Virginia winter barley and implications for use in feed, food and biofuels production. *Journal of Cereal Science* 51: 41–49.

Hansen, H. B., B. Moller, S. B Andersen, J. R. Jorgensen, and A. Hansen. 2004. Grain characteristics, chemical composition, and functional properties of *rye* (*Secale cereale* L.) as influenced by genotype and harvest year. *Journal of Agricultural and Food Chemistry* 52: 2282–2291.

Harborth, K. W., T. T. Marston, and D. A., Llewellyn. 2006. Comparison of corn and grain sorghum dried distillers grains and protein supplements for growing beef heifers, Kansas State University Beef Cattle Research 2006, Report of Progress 959. Available online: www.ddgs.umn.edu/articles-beef/2006-Harborth-%20Comparison%20of%20corn%20and--.pdf. Accessed November 2, 2010.

Henry, R. J. 1987. Pentosan and $(1{\rightarrow}3),(1{\rightarrow}4)$-β-glucan concentrations in endosperm and wholegrain of wheat, barley, oats and rye. *Journal of Cereal Science* 6: 253–258.

Ingledew, W. M., A. M. Jones, R. S. Bhatty, and B. G. Rosnagel. 1995. Fuel ethanol production from hull-less barley. *Cereal Chemistry* 72: 147–150.

Ingledew, W. M., K. C. Thomas, S. H. Hynes, and J. G. McLeod. 1999. Viscosity concerns with *rye* mashes used for ethanol production. *Cereal Chemistry* 76: 459–464.

Kučerova, J. 2007. The effect of year, site and variety on the quality characteristics and bioethanol yield of winter *triticale*. *Journal of the Institute of Brewing and Distilling* 113: 142–146.

Lampi, A.-M., R. A. Moreau, V. Piironen, and K. B. Hicks. 2004. Pearling barley and rye to produce phytosterol-rich fractions. *Lipids* 39: 783–787.

Li, M., P. Teunissen, G. Konieczny-Janda, K. B. Hicks, D. B. Johnston, J. Nghiem, and J. Shetty. 2007. *Viscosity Reduction and Efficient Conversion of Barley Starch into Bioethanol Using the Low Energy Intensive STARGEN™ Technology.* Proceedings for the 3rd European Bioethanol Technology Meeting. Detmold, Germany.

Lodge, S. L., T. J. Klopfenstein, D. H. Shain, and D. W. Herold. 1997. Evaluation of corn and distillers byproducts. *Journal of Animal Sciences* 75: 37–43.

Moreau, R. A., M. J. Powell, and K. B. Hicks. 1996. Extraction and quantitative analysis of oil from commercial corn fiber. *Journal of Agricultural and Food Chemistry* 44: 2149–2154.

Moreau, R. A., R. Flores, and K. B. Hicks. 2007a. The composition of functional lipids in hulled and hulless barley, in fractions obtained by scarification, and in barley oil. *Cereal Chemistry* 84: 1–5.

Moreau, R. A., K. E. Wayns, R. A. Flores, and K. B. Hicks. 2007b. Tocopherols and tocotrienols in barley oil prepared from germ and other fractions from scarification and sieving of hulless barley. *Cereal Chemistry* 84: 587–592.

Moreau, R. A., D. B. Johnson, and K. B. Hicks. 2007c. A comparison of the levels of lutein and zeaxanthin in corn germ oil, corn fiber oil, and corn kernel oil. *Journal of the American Oil Chemists Society* 84: 1039–1104.

Moser, B. R., G. Knothe, S. F. Vaughn, and T. A. Isbell. 2009. Production and evaluation of biodiesel from field pennycress (*Thlaspi arvense* L.). *Oil. Energy & Fuels* 23: 4149–4155.

Minowa, T., M. Murakami, Y. Dote, T. Ogi, S.-Y. Yokoyama. 1994. Effect of operating conditions on thermochemical liquefaction of ethanol fermentation stillage, *Fuel* 73: 579–582.

Mustafa, A. F., J. J. McKinnon, M. W. Ingledew, and D. A. Christensen. 2000. The nutritive value for ruminants of thin stillage and distillers' grains derived from wheat, rye, triticale and barley. *Journal of the Science of Food and Agriculture* 80: 607–613.

Nghiem, N. P. 2008. Biofuel feedstocks, Ch. 4. In *Biofuels Engineering Process Technology,* eds. C. M. Drapcho, N. P. Nghiem, and T. H. Walker, pp. 69–103. New York: McGraw-Hill, Inc.

Nghiem, N. P., K. B. Hicks, D. B. Johnston, G. Senske, M. Kurantz, M. Li, J. Shetty, and G. Konieczny-Janda. 2010. Production of ethanol from winter barley by the EDGE (enhanced dry grind enzymatic) process. *Biotechnology for Biofuels* 3: 8, doi:10.1186/1754-6834-3-8.

Nichols, N. N., B. S. Dien, Y. V. Wu, and M. A. Cotta. 2005. Ethanol fermentation of starch from field peas. *Cereal Chemistry* 2005 82: 554–558.

Nuez-Ortin, W. G. N., and P. Yu. 2009. Nutrient variation and availability of wheat DDGS, corn DDGS, and blend DDGS from bioethanol plants. *Journal of the Science of Food and Agriculture* 89: 1754–1761.

Pejin, D., L. J. Mojovic, V. Vucurovic, J. Pejin, S. Dencic, and M. Rakin. 2009. Fermentation of wheat and triticale hydrolysates: A comparative study. *Fuel* 88: 1625–1628.

Robertson, R. 2010. Virginia ethanol plant to open this summer. Southeast Farm Press, April 9. Available online: www.southeastfarmpress.com/grains/barley-ethanol-0409/. Accessed November 2, 2010.

Shetty, J. K., G. Chotani, D. Gang, and D. Bates. 2007. Cassava as an alternative feedstock in the production of renewable transportation fuel. *International Sugar Journal* 109: 3–11.

Shukla, R., and M. Cheryan. 2001. Zein: The industrial protein from corn. *Industrial Crops and Products* 13:171–192.

Sosulski, K., S. Wang, W. M. Ingledew, F. W. Sosulski, and J. Tang. 1997. Preprocessed barley, rye, and triticale as a feedstock for an integrated fuel ethanol-feedlot plant. *Applied Biochemistry and Biotechnology* 63–65: 59–70.

Stein, H. H. 2008. Use of distillers co-product in diets fed to swine, in Using Distillers Grains in the US and international livestock and poultry industries, Midwest Agribusiness and Trade Research Information Center, Iowa State U, Ames, Iowa. Available online: www.matric.iastate.edu/DGbook/chapters/chapter4. pdf. Accessed November 2, 2010.

Stevenson, M. H., and W. D. Graham. 1983. The chemical composition and true metabolisable energy content of cassava root meal imported into Northern Ireland. *Journal of the Science of Food and Agriculture* 34: 1105–1106.

Thomas, K. C., and W. M. Ingledew. 1995. Production of fuel alcohol from oats by fermentation. *Journal of Industrial Microbiology* 15: 125–130.

University of Minnesota. 2003. Nutrient composition and value of wheat DDGS. Available online: www.ddgs. umn.edu/othertypes/wheat/2003-wheat_ddgs.pdf. Accessed November 2, 2010.

University of Minnesota. 2009. Distillers grains by-products in livestock and poultry feeds, US profile comparison. Available online: www.ddgs.umn.edu/profiles.htm. Accessed November 2, 2010.

University of Saskatchewan. 2010. Department of Animal and Poultry Science, Wheat DDGS. Available online: www.ddgs.usask.ca. Accessed November 2, 2010.

USDA. 2010. National Nutrient Database for Standard Reference, Release 23. Available online: www.ars.usda. gov/Services/docs.htm?docid=8964. Accessed November 2, 2010.

Wang, L., C. L. Weller, V. L. Schlegel, T. P. Carr, and S. L. Cuppett. 2007. Comparison of supercritical CO_2 and hexane extraction of lipids from sorghum distillers grains. *European Journal of Lipid Science and Technology* 109: 567–574.

Wang, S., K. Sosulski, F. Sosulski, and M. Ingledew. 1997a. Effect of sequential abrasion on starch composition of five cereals for ethanol fermentation. *Food Research International* 30: 603–609.

Wang, S., K. C. Thomas, W. M. Ingledew, K. Sosulski, and F. W. Sosulski. 1997b. Rye and triticale as feedstock for fuel ethanol production. *Cereal Chemistry* 74: 621–625.

Wang, S., K. C. Thomas, W. M. Ingledew, K. Sosulski, and F. W. Sosulski. 1998. Production of fuel ethanol from *rye and triticale* by very-high-gravity (VHG) fermentation. *Applied Biochemistry and Biotechnology* 69: 157–175.

Wang, S., K. C. Thomas, K. Sosulski, W. M. Ingledew, and F. W. Sosulski. 1999. Grain pearling and very high gravity (VHG) fermentation technologies for fuel alcohol production from rye and triticale. *Process Biochemistry* 34: 421–428.

Wang, Y., M. Tilley, S. Bean, X. S. Sun, D. Wang. 2009. Comparison of method for extracting kafirin proteins from sorghum distillers dried grains with solubles. *Journal of Agricultural and Food Chemistry* 57: 8366–8372.

Wu, W. V. 1990. Recovery of protein-rich by products from oat stillage after alcohol distillation. *Journal of Agricultural and Food Chemistry* 38: 588–592.

Wu, X., D. Wang, S. R. Bean, and J. P. Wilson. 2006. Ethanol production from pearl millet using *Saccharomyces cerevisiae*. *Cereal Chemistry* 83: 127–131.

Wu, Y. V. 1986. Fractionation and characterization of protein rich material from barley after alcohol distillation. *Cereal Chemistry* 63: 142–145.

Wu, Y.V., and M.O. Bagby. 1987. Recovery of protein-rich byproducts from sweet potato stillage following alcohol distillation. *Journal of Agricultural and Food Chemistry* 35: 321–325.

Wu, Y. V., K. R. Sexson, and A. A. Lagoda. 1984. Protein-rich residue from wheat alcohol distillation: Fractionation and characterization. *Cereal Chemistry* 61: 423–427.

Yadav, M. P., D. B. Johnston, and K. B. Hicks. 2007. Structural characterization of corn fiber gums from coarse and fine fiber and a study of their emulsifying properties. *Journal of Agricultural and Food Chemistry* 55: 6366–6371.

Zhao, R., X. Wu, B. W. Seabourn, S. R. Bean, L. Guan, Y.-C. Shi, J. D. Wilson, R. Madl, and D. Wang. 2009. Comparison of waxy vs. nonwaxy wheats in fuel ethanol fermentation. *Cereal Chemistry* 86: 145–156.

Part II

Properties, Composition, and Analytics

7 Physical Properties of DDGS

Kurt A. Rosentrater

CONTENTS

7.1 INTRODUCTION

Many research studies have examined chemical properties and nutritional characteristics of distillers dried grains with solubles (DDGS), especially in terms of utilization as livestock feed ingredients, their digestibilities, and resulting animal performance. Much of this is actually discussed in other chapters of this book (see, Chapter 8 for a review of DDGS chemical composition, and Chapters 12 through 17 for discussions on feeding DDGS to various animals). Until just a few years ago, however, no information was available regarding the physical properties of DDGS. But that is beginning to change, though, as more research is being conducted and published. The scientific understanding of DDGS is increasing, and this chapter aims to encapsulate this knowledge.

DDGS is a heterogeneous granular material (also known as a bulk solid). It consists of a range of particle types, sizes, and shapes (Figure 7.1). These particles include corn kernel fragments (i.e., tip cap and pericarp tissues), nonuniformly crystallized soluble protein and lipid coatings on the surfaces of these fragments, and agglomerates, which form during drying (often referred to as "syrup balls").

The macroscale behavior of DDGS is ultimately a function of microscale particle–particle interactions. These microscopic relationships have a large influence on the utility of DDGS, both at the ethanol plant and at livestock production facilities, as they influence material handling, flowability, and storage behavior.

The objective of this chapter is to summarize what is currently known about the physical attributes of DDGS. Specifically, this chapter will provide typical ranges for DDGS properties (including various physical and flow properties) and definitions of these properties. It will also review relationships among soluble solids, moisture, and lipids, how these influence the nature of particle surfaces and resulting frictional properties, and ultimately how all of these affect flowability.

(a)   (b)

(c)

FIGURE 7.1 DDGS is a heterogeneous mixture of various particle sizes and shapes; (a) 10× magnification (scale gradations are in cm); (b) 60× magnification; (c) 200× magnification. Note the variation in color amongst particles. Also note the crystalline formations on the DDGS particles in (b) and (c) (author's photograph).

7.2 PHYSICAL PROPERTIES

As discussed in other chapters, DDGS often has a reputation for variability in chemical (i.e., nutritional) and physical properties, both among ethanol manufacturing plants, as well as at a given plant over time. Although all causative variables have not yet been fully quantified or even identified, variations in DDGS properties are generally attributed (often qualitatively) to a number of factors, including the characteristics of the raw corn itself, settings, conditions, additives, and chemicals used during processing, the proportion of condensed distillers solubles (CDS) added to the distillers wet grain (DWG) prior to drying, the type of dryer used, drying conditions (e.g., times and temperatures), cooling of the DDGS after drying (i.e., flat storage vs. vertical silo, final moisture content of the DDGS, time allowed for the product to cool prior to shipping, or even loading of the DDGS into railcars when it is still hot), and environmental conditions (especially ambient temperatures and relative humidities).

Even though there can be considerable nonuniformities in the properties of various DDGS samples, it is important to have baseline of information—providing this is one of the main goals of this chapter. Chemical data are essential for livestock diet formulations; they are also important for pursuing other potential end uses, such as human foods, bioenergy, and biofillers, to name a few. These topics are covered in depth in other chapters throughout the book. Physical properties, on the other hand, are critical for the design and operation of processing equipment (e.g., dryers, conveyors, mixers, pellet mills, extruders, etc.), processing facilities, and storage structures (e.g., flat storage buildings and vertical silos). As with other biomaterials, the physical characteristics of DDGS are intimately related to the chemical constituents, and the two cannot be completely separated from each other. Key physical properties include particle size, bulk density, angle of repose, moisture properties, thermal properties, and color. Table 7.1 summarizes much of the currently published data on DDGS physical properties. Details about these properties are discussed below.

TABLE 7.1
Typical Physical Properties of DDGS Samples from Commercial Ethanol Plants

Property	Range	Reference
Geometric mean diameter	0.21–1.38	Bhadra et al. (2009)
(d_{gw}, mm)	0.680–1.862	Clementson et al. (2009)
	0.434–0.949	Liu (2008)
	0.073–1.217	U.S. Grains (2008)
	0.256–1.087	U.S. Grains (2008)
	0.61–2.13	U.S. Grains (2008)
Geometric standard deviation	0.20–0.55	Bhadra et al. (2009)
(S_{gw}, mm)	0.418–1.494	Clementson et al. (2009)
	0.313–0.556	Liu (2008)
	0.26	U.S. Grains (2008)
	0.28	U.S. Grains (2008)
	1.56–2.75	U.S. Grains (2008)
Bulk density	490–600	Bhadra et al. (2009)
(BDA, kg/m³)	414.37–577.78	Clementson et al. (2009)
	389.3–501.5	Rosentrater (2006)
	365.22–504.58	U.S. Grains (2008)
	398.86–560.65	U.S. Grains (2008)
	493.37–629.53	U.S. Grains (2008)
Angle of repose	35.94–41.60	Bhadra et al. (2009)
(°)	26.5–34.2	Rosentrater (2006)
Moisture content	4.32–8.89	Bhadra et al. (2009)
(%, db)	13.2–21.2	Rosentrater (2006)
Water activity	0.53–0.63	Rosentrater (2006)
(–)		
Thermal conductivity	0.06–0.08	Rosentrater (2006)
(W/(m °C))		
Thermal diffusivity	0.13–0.15	Rosentrater (2006)
(mm²/s)		
Color—L	36.56–50.17	Bhadra et al. (2007)
(–)	40.0–49.8	Rosentrater (2006)
Color—a	5.20–10.79	Bhadra et al. (2007)
(–)	8.0–9.8	Rosentrater (2006)
Color—b	12.53–23.36	Bhadra et al. (2007)
(–)	18.2–23.5	Rosentrater (2006)

7.2.1 PARTICLE SIZE AND SHAPE

As with other granular materials, DDGS is composed of a distribution of particle sizes and shapes. Visual examples of this are illustrated in Figure 7.1. DDGS particles are not uniform, but rather tend to exhibit lognormal size distributions. These profiles are mathematically summarized using the geometric mean diameter (d_{gw}) and geometric standard deviation (S_{gw}) (American Society of Agricultura Engineers, ASAE, 2004). As shown in Table 7.1, the size of DDGS particles can often range from less than 0.1 mm to greater than 2 mm in diameter. Particle size and shape are functions of grinding (i.e., size reduction) prior to fermentation (Liu, 2009), due to type of grinder used, screen sizes used, knife sharpness/wear, etc., but they are also dependent upon the drying and cooling conditions after fermentation (Kingsly et al., 2010), and the agglomeration of fat, protein, and residual sugar molecules on the surface of the nonfermented solids, especially near glass transition temperatures (Bhadra et al., 2009b).

Particle size and shape are important because they affect other physical properties, such as bulk density, angle of repose, compressibility, heat transfer characteristics, and flowability properties (Ganesan et al., 2008b). Generally, the finer the particle size, the greater the surface area and number of contact points between particles, and the smaller (but more numerous) the interstitial air spaces between particles. This, consequently, can lead to greater compressibility, higher resulting cohesive bulk strength, and thus lower flowability.

7.2.2 BULK DENSITY

Bulk density (denoted as BD or P) is used to determine effective capacities for storage bins and containers. It is also fundamental to the structural design of these vessels. This parameter is defined as the mass of a granular material that will occupy a specific volume, such as a bin or rail car (Figure 7.2). Bulk density includes not only particle mass, but also the air entrained in the void spaces between the particles. As shown in Table 7.1, bulk density of DDGS from commercial fuel ethanol plants has been found to range from ~365 to 630 kg/m³ (~22.8 to 39.3 lb/ft³). Anecdotally, a bulk density of 30 lb/ft³ is common for DDGS throughout the industry.

Actually, there are two unique types of bulk density that are important: packed bulk density (BD$_P$) and aerated bulk density (BD$_A$). Aerated bulk density, also known as loose bulk density, is the easiest to measure and is determined by pouring a quantity of granular material into a container of known volume (BDA is provided in Table 7.1.). This parameter is representative of the bulk solid that has not been subjected to compression or packing. Packed bulk density, on the other hand, is the bulk density of the material after it has been compressed, and thus some of the entrained air has been displaced. This is representative of the material's actual bulk density in storage and transport, and is a more realistic quantity to use.

Bulk density is a function of particle size and shape, the mass of each individual particle, and how the particles are placed into the storage vessel. Lock and key interactions as the particles come into contact with each other determine the internal structure of the bulk solid, and influence how much packing occurs.

7.2.3 ANGLE OF REPOSE

Angle of repose is defined as the angle that forms between a horizontal plane and the slope of a pile (at rest) that has been formed by dropping the bulk material from some elevation (Figure 7.3). There are, in fact, two separate types of angle of repose: filling (Figure 7.4) and emptying (Figure 7.5); these occur when a material is poured into bins, trucks, railcars, or other storage structures, and when it is removed, or drained, from them. Filling and emptying angle of repose are not always

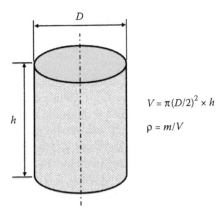

$$V = \pi (D/2)^2 \times h$$

$$\rho = m/V$$

FIGURE 7.2 Bulk density (ρ) is defined as the mass (*m*) that a given volume (*V*) will contain. In this example, the storage container is a cylindrical vessel, similar to a vertical storage silo.

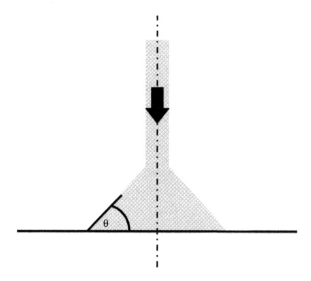

FIGURE 7.3   Angle of repose (θ) for a pile of granular material, such as DDGS, is the angle that forms between the slope of the pile and a horizontal plane.

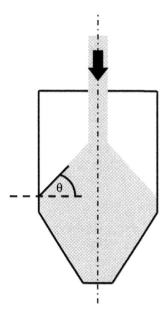

FIGURE 7.4   Illustration of angle of repose (θ) resulting from filling a storage vessel, such as a bin or rail car. Because boundary conditions are not the same as a flat surface, this value may be different from that of a pile on a flat floor.

equal due to the particle dynamics vis-à-vis potential energy versus kinetic energy changes when the particles flow from a given height compared to starting flow at rest.

In general, angle of repose is related to many of the flow properties of the material, and thus is an indirect indication of general flowability behavior. Angle of repose is a function of physical properties of the particles, such as size, shape, and porosity. It is also affected by the drying and cooling conditions used at the ethanol plant, especially when the sugar and fat molecules on the particle surfaces reach glass transition temperature, which influences the surface frictional properties, and is reflected in the stickiness, or cohesion, between particles. Bulk solids with an angle of repose

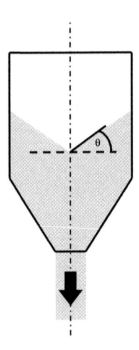

FIGURE 7.5 Illustration of angle of repose (θ) resulting from emptying a storage vessel, such as a bin or rail car. Because of dynamic flow conditions, and a lower potential energy compared to that during filling of the vessel, this value is often different from the filling angle of repose.

between approximately 25° and 35° are generally considered free flowing. Higher values, however, indicate poor flowability. As shown in Table 7.1, DDGS samples from commercial ethanol plants have exhibited angle of repose values ranging from ~26° to nearly 42°. Not surprisingly, DDGS often has flowability problems (this topic will be discussed in depth later in the chapter).

7.2.4 MOISTURE

In order to effectively utilize DDGS, drying is necessary to prevent microbial spoilage, ensure a long shelf life, and reduce shipping costs. Most ethanol plants currently dry DDGS to a moisture level of approximately 10% to 12%, or even less. This moisture content is typically recommended for feed products because it substantially reduces transportation costs and is microbiologically safe (Beauchat, 1981; Wang et al., 1997). As shown in Table 7.1, moisture content of DDGS samples from commercial fuel ethanol plants has been found to range from ~4.3% to 8.9% (dry basis—db), although some samples have been found up to nearly 21% (db).

Contemporaneous to the total quantity of water that is contained in the DDGS, the availability of that water to interact with microorganisms and participate in biological and biochemical reactions is also important. Water activity (a_w) quantifies the amount of "free" water (i.e., unbound water) available in materials for use by microorganisms and chemical agents, and hence is a measure of susceptibility to spoilage and deterioration. Products with no free water ($a_w = 0.0$) are not at risk for spoilage, whereas materials with 100% free water ($a_w = 1.0$) are at risk for rapid spoilage. Materials become safe from yeast growth below water activities of approximately 0.9, safe from bacterial growth below approximately 0.8, and safe from mold growth below approximately 0.6 (Barbosa-Canovas and Vega-Mercado, 1996). Water activity of DDGS samples from commercial fuel ethanol plants has been found to range from 0.53 to 0.63, which indicates that DDGS is largely shelf stable (Table 7.1).

7.2.5 THERMAL PROPERTIES

The thermal properties of DDGS govern heat storage and heat transfer in the granular material. As such, these attributes dictate behavior during processing at the plant (i.e., drying and cooling operations), during storage (i.e., product temperature vis-à-vis environmental temperature fluctuations), and during further value-added processing of DDGS (e.g., pelleting, extrusion, fabrication of bioplastics, cooking properties, etc.). Thermal properties can be summarized using the thermal diffusivity:

$$\alpha = \frac{k}{\rho c_\mathrm{p}} \tag{7.1}$$

where α is the thermal diffusivity (mm²/s), k is the thermal conductivity (W/(m°C)), ρ is the mass density (kg/m³), and c_p is the specific heat capacity (J/(kg°C)). DDGS samples from commercial ethanol plants have exhibited thermal conductivity values ranging from 0.06 to 0.08 W/(m°C), and thermal diffusivity values ranging from 0.13 to 0.15 mm²/s (Table 7.1).

7.2.6 COLOR

It is true that DDGS can exhibit considerable variability in both nutrient components as well as physical properties. This is easily apparent by visually examining DDGS samples (Figure 7.6). Color variations are readily visible across DDGS samples, and often range from "golden yellow" to "dark brown." These variations often arise due to differences in processing, handling, and storage conditions. However, much of the variability can often be attributed to the CDS addition level and drying conditions at the ethanol plant (Chapter 5). Color has been shown to be an indicator of nutritional quality, and can be related to amino acid digestibility (Goihl, 1993; Ergul et al., 2003; Batal and Dale, 2006; Fastinger and Mahan, 2006). Because of this, quantitative color measurement using the Hunter L-a-b three-dimensional color space (Figure 7.7) has become a frequently used quality control parameter. As shown in Table 7.1, DDGS samples from commercial ethanol plants have exhibited Hunter L values ranging from 36 to 50, Hunter a values ranging from 5 to 10, and Hunter b values ranging from 12 to 24. Anecdotally, the more yellow the DDGS, the better the nutrient quality and physical properties.

FIGURE 7.6 DDGS samples showing variability in nutrient components and physical properties. These often arise due to differences in processing, handling, and storage conditions (author's photograph).

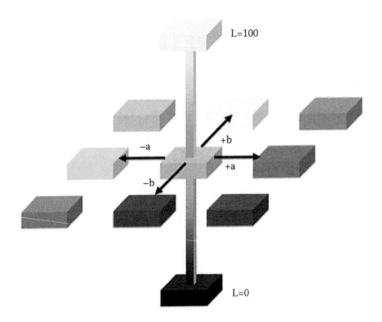

FIGURE 7.7   Three-dimensional color space used to define L-a-b color coordinates.

7.3   FLOWABILITY PROPERTIES

7.3.1   Principles

With the exponential growth of the U.S. fuel ethanol industry during the past several years (see Chapter 2), large quantities of distillers grains are now produced. To utilize these materials as livestock feeds, however, they must be transported greater distances via rail (often outside the corn belt), and then stored in bins and silos until final use. Unfortunately, discharge (i.e., unloading) of DDGS after reaching various destinations is often problematic (Figure 7.8), owing to agglomeration, caking, and bridging among particles that occurs during storage and transport. In fact, flowability has become one of the major issues that must be addressed for effective sales, marketing, distribution, and utilization of distillers grains (Janes, M. 2004. *VeraSun Energies*, personal communication; Cooper, T. 2005. *VeraSun Energies*, personal communication; Schlicher, 2005; Rosentrater, 2007). Because DDGS does not flow well from rail cars, workers often hammer the car sides and hopper bottoms to induce flow. This leads to severe damage to the rail cars themselves (Figure 7.9), repairs of which have become very expensive to ethanol manufacturing companies. Large rail carriers have even prohibited DDGS shipment on their own cars (UP, 2009). So ethanol companies either have to own or lease rail cars. Even though some anecdotal knowledge about flowability is present in the industry, it is often proprietary in nature, and is incomplete. Furthermore, only a few formal scientific studies have investigated handling and flow properties of distillers grains to date.

Flowability is defined as the ability of bulk granular solids and powders to flow. Basic flow of granular materials typically follows two main types of regimes: mass flow (Figure 7.10) and funnel flow (Figure 7.11). Mass flow, which is the type of flow that is desired, occurs when the entire bulk mass flows downward during bin discharge. Essentially, the first material placed in the bin (i.e., at the bottom) will be the first to leave. No arching or bridging forms between particles, and thus this type of flow is very predictable and controllable. Funnel flow, on the other hand, which is not desirable, often occurs when the bulk solid has flowability problems, such as with distillers grains. Essentially the granular mass exhibits bulk strength, with arching and bridging between particles, and thus mass flow is reduced or even prevented. A vertical flow channel (also known as a "rat hole") forms in the center of the granular mass and is surrounded by stagnant material (Figure 7.12).

FIGURE 7.8 Flow of distillers grains is often problematic, due to caking and bridging between particles, which frequently occurs during storage and transport (author's photograph).

FIGURE 7.9 Transporting distillers grains can lead to substantial rail car damage due to flowability problems (when trying to unload product), because sledgehammers are often used to induce flow (author's photograph).

Material flows down into this rat hole from the top of the bin, in successive layers, often unpredictably, uncontrollably, and sporadically. With this type of flow, the first material in the bin is often the last to discharge. Sometimes flow will even stop and not easily start again. Funnel flow can occur when the hopper bottom is not steep enough, when the discharge area is undersized, or when the bulk material has high cohesion between particles, which is often the case with distillers grains. So

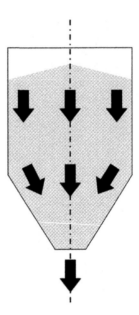

FIGURE 7.10 Illustration of mass flow bin discharge, which is desirable. This is indicative of free flowing materials.

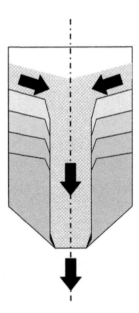

FIGURE 7.11 Illustration of funnel flow (rat-holing) bin discharge, which is not desirable. This is indicative of flowability problems.

the fundamental issue is determining what causes this cohesion, and thus flowability problems, in some situations but not others.

Flowability is, in fact, not an inherent natural material property in itself, but rather is the consequence of several interacting properties that simultaneously influence material flow. Environmental conditions during storage and the equipment used for handling, storing, and processing the material also play key roles. From studies of other granular materials, it has been determined that

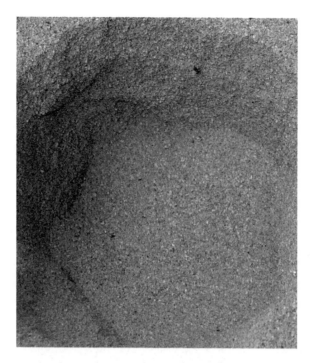

FIGURE 7.12 Funnel flow occurs when the central core discharges out of the storage vessel, but peripheral material does not. Note the vertical channel walls, which occur due to sticky particles and large frictional forces (author's photograph).

flowability problems may arise from a number of synergistically interacting factors, including product moisture, particle size distribution, storage temperature, relative humidity, storage time, compaction pressure distributions within the product mass, vibrations during transport (i.e., rail or truck shipping), and/or variations in the levels of these factors throughout storage and transport processes (Craik and Miller, 1958; Johanson, 1978; Moreyra and Peleg, 1981; Teunou et al., 1999; Fitzpatrick et al., 2004a, 2004b). In addition to the parameters listed above, other factors that may affect flowability include chemical constituents, (e.g., protein, fat, starch, and carbohydrate levels), as well as addition of flow conditioning agents (Peleg and Hollenbach, 1984; Prescott and Barnum, 2000).

Flow behavior is thus multidimensional in nature. Because of this, no single test can fully quantify a product's flowability; instead several tests are required. Toward this end, two main types of testing methodologies have been developed: Jenike (1964) and Carr (1965a, 1965b). Shear testers are the primary equipment used to measure the strength and flow properties of bulk materials (Figures 7.13 and 7.14). A shear test typically consists of two components: measurement of consolidation (i.e., compaction) and of bulk strength (Figure 7.15). The measured bulk strength depends on the degree of consolidation and how it was achieved (i.e., stress history). Stress history is strongly influenced by the strength of the bulk solid itself. Thus, each of these aspects is highly dependent upon the other. Reliable predictions of the strength of a bulk solid can be achievable only if the stress history and the directions of the major principal stresses during consolidation and failure are known (Figure 7.16).

In terms of shear testing, Jenike (1964) developed the fundamental methodologies for determining the strength and flow characteristics of powders and granular materials. To analyze flow in bins and hoppers, and to develop flow/no-flow criteria for various bulk solids, Jenike used the principles of plastic failure (Thomson, 1997). From a physical standpoint, the general principle

FIGURE 7.13   Typical Jenike shear tester setup (author's photograph).

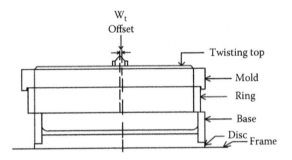

FIGURE 7.14   Schematic diagram of Jenike shear cell components.

is that granular flow is equivalent to failure of a solid material due to shear stress. In ideal, free flowing materials, resistance to flow is only the result of friction. In cohesive (sticky) materials, such as distillers grains, however, interparticle forces and binding are enhanced by compaction, which can produce mechanical strength in the bulk solid, and thus flowability problems (Peleg, 1983). Jenike's methodologies are benchmarks for determining design criteria for storage bins and silos, and have been used for characterizing many types of granular materials, including cement (Schrämli, 1967), lactose powder (York, 1975), sugar (Kamath et al., 1993), wheat flour (Kamath et al., 1994), confectionary sugar (Duffy and Puri, 1994), grains (Duffy and Puri, 1999), and milk

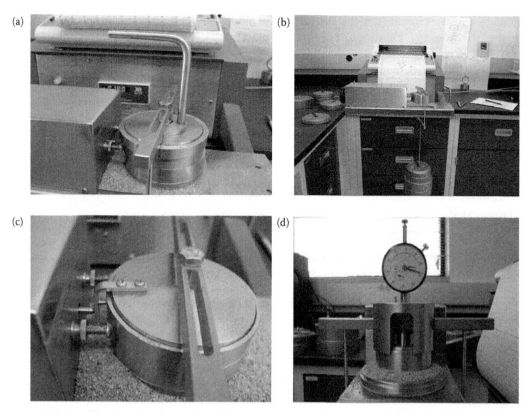

FIGURE 7.15 Experimental steps in a Jenike shear test: (a) preconsolidation (i.e., prior to adding weights); (b) consolidation (i.e., adding a specified load); (c) shearing of the material (i.e., determining frictional resistance); and (d) compressibility testing (author's photograph).

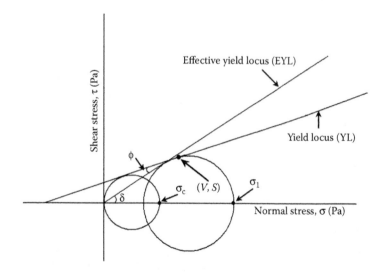

FIGURE 7.16 A Mohr circle failure plot is used to analyze Jenike shear cell experimental data. V denotes the normal load applied to the shear cell; S denotes the resulting shear stress in the sample; ϕ denotes the angle of internal friction; δ denotes the effective angle of internal friction, σ_1 is the major consolidation stress; σ_c is the unconfined yield strength.

powders (Fitzpatrick et al., 2004b). Further information about Jenike's methodology can be found in American Society for Testing and Materials (ASTM, 2006).

Carr (1965a, 1965b) also developed a series of laboratory procedures that permit the evaluation of flow properties of granular materials (Figures 7.17 and 7.18 and Table 7.2). This approach requires the determination of eight key properties: angle of repose, angle of fall, angle of difference, angle of spatula, bulk density, compressibility, uniformity, and dispersibility. These tests do not, however, account for consolidation or stress history (i.e., compaction over time), as the shear testing does. During storage of granular materials, pressure continuously acts throughout the duration of residence in the vessel, so consolidation can be an important factor to consider. Thus, it is important to consider Carr testing in conjunction with Jenike shear testing. Even so, the information determined by the Carr methodology is useful for designing bins and hoppers so that particle flow can be achieved, and provides information that is complimentary to that of the shear test data. Further information about Carr's methodology can be found in ASTM (1999).

Angle of Repose   Angle of repose is a bulk property of granular materials, as discussed in Section 7.2. But it is also used to assess flowability, according to the Carr methodology. Since it has already been thoroughly described, it will not be discussed further.

Angle of Fall   When a material lies at rest in a pile, it has a specific angle of repose. If the supporting surface experiences vibrations, impacts, or other movements, the material on the sloped sides of the pile will dislodge and flow down the slope. The new angle of repose that forms is referred to as the angle of fall. This parameter provides an indication of particle size, shape, uniformity, and cohesion, and thus the flowability of the material.

Angle of Difference   The angle of differences is calculated by subtracting the angle of fall from the angle of repose. This parameter is related to the internal cohesion of the granular particles. The larger the angle of difference, the more flowable the material.

Angle of Spatula   Angle of spatula provides an indication of the internal friction between particles. It is determined by inserting a flat blade into a pile of granular material and lifting vertically (Figure 7.19). The new angle of repose that the material forms relative to the horizontal blade

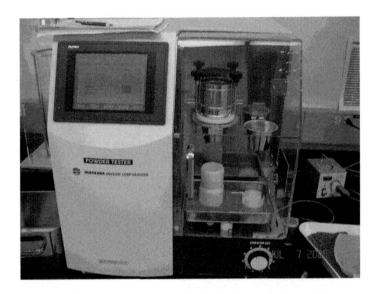

FIGURE 7.17   Typical semiautomated tester used to measure Carr indices (author's photograph).

FIGURE 7.18 Experimental steps in a Carr test: (a) angle of repose; (b) aerated bulk density; (c) packed bulk density; (d) angle of spatula; (e) angle of fall; and (f) dispersibility (author's photograph).

TABLE 7.2
Carr Classification of Granular Material Flowability by Carr Indices

Property	Index Value	Degree of Flowability/Floodability	Remediation Measures
Flow index	90–100	Very good	Bridge breaking measures not required.
	0–19	Very bad	Special apparatus and techniques are required to break the bridge/cake/agglomerations.
Flood index	80–100	Very high	Rotary seal must be used to prevent flushing.
	0–24	Won't flush	Rotary seal is not required to prevent flushing.

Source: Adapted from Carr, R. L. *Chemical Engineering* 72(3): 69–72, 1965.

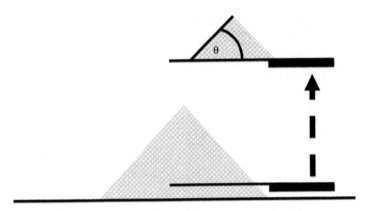

FIGURE 7.19 Illustration of angle of spatula (θ), formed when a flat blade is inserted into a pile of granular material and lifted vertically (this simulates movement of the supporting surface), is the new angle of repose resulting from this movement.

surface is known as the angle of spatula. This test simulates movement of the supporting surface and material handling behavior that will be encountered in real settings. Generally, bulk solids with an angle of spatula less than approximately 40° are considered free flowing.

Bulk Density   As with angle of repose, bulk density (aerated and packed) is an overall bulk property, but it is also used to assess flowability according to the Carr methodology. It has been thorough discussed in Section 7.2.

Compressibility   Some granular materials have a propensity to become tightly packed; others do not. After determining aerated and packed bulk densities, the compressibility of a material can be calculated as:

$$C = \left(\frac{\text{BD}_\text{P} - \text{BD}_\text{A}}{\text{BD}_\text{P}} \right) \times 100 \tag{7.2}$$

where C is compressibility (%), BD_P is the packed bulk density, and BD_A is the aerated bulk density. This parameter provides an indication of particle size, shape, uniformity, and cohesion, and thus the overall flowability of the material. Bulk solids with a compressibility value less than approximately 18% are considered free flowing. Hausner Ratio (HR) is related to compressibility, and is defined as:

$$\text{HR} = \frac{\text{BD}_\text{P}}{\text{BD}_\text{A}}. \tag{7.3}$$

Compressibility and Hausner Ratio are thus related by:

$$C = \left[1 - \frac{1}{\text{HR}} \right] \times 100. \tag{7.4}$$

Uniformity   The relative homogeneity of the size and shape of the particles within a bulk solid (i.e., uniformity) has a direct effect on a material's ability to flow. This can be quantified by determining the screen size that will allow 60% of a sample to pass through (which is generally a relatively large mesh size) and the screen size that will allow 10% to pass through (which is generally a relatively small mesh size). Uniformity is then calculated by dividing the 60% mesh size by the 10% mesh size. Thus, the smaller the uniformity value, the more homogeneous the particle sizes and shapes. A material that is more uniform will have a tendency to have better flowability than a material with a wide range of particle sizes.

Dispersibility   Dispersibility is a measure of the propensity for a granular material to form dust, and thus lose mass to the surrounding air. It is especially important during vertical flow (e.g., pile formation after spout or conveyor discharge). The more dispersible a material, the higher the potential for mass loss due to dust generation.

Although the Jenike and Carr procedures have been extensively used in many industries to understand and alleviate flow problems, to date few formal studies have used these to investigate the handling and flow characteristics of distillers grains. Determining the specific physical, chemical, and/or environmental factors, or interactions thereof, which cause flowability problems for these coproducts should be pursued aggressively, because solving this issue will have substantial economic implications for both the fuel ethanol and the livestock industries. Refer to Ganesan et al. (2008b) for a more in-depth discussion about flowability principles, especially as they relate to DDGS.

7.3.2   FLOW BEHAVIOR OF DDGS

In recent years, data on the flow properties of DDGS have become available, and various studies have examined multiple aspects of flowability both on a laboratory scale as well as a commercial scale. A summary of some of these data for DDGS samples collected from commercial ethanol plants is provided in Table 7.3.

For example, Ganesan et al. (2008a) produced laboratory-scale DDGS and examined the effect of five moisture content levels (10%, 15%, 20%, 25%, and 30% db) on the resulting physical and chemical properties of DDGS with four soluble solid levels (10%, 15%, 20%, and 25% db). To produce these materials, CDS was combined with DDG, and appropriate quantities of water were added to adjust moisture contents. Carr indices were used to quantify the flowability of the DDGS samples. The results showed that both soluble solid level and moisture content had negative effects on physical and flow properties (e.g., aerated bulk density, packed bulk density, and compressibility). According to dispersibility, flowability index, and floodability index, flowability declined significantly with an increase in moisture content for most of the soluble levels under consideration (as shown in Figure 7.20). The color values and protein content of the DDGS significantly decreased as soluble level increased as well.

In another study, Ganesan et al. (2008c) determined the Jenike flow properties of laboratory-produced DDGS using a Jenike shear tester. Measured properties included cohesion, effective angle of friction, angle of internal friction, yield locus, flow function, major consolidating stress, unconfined yield strength, and compressibility. This work investigated the effects of four levels of soluble solid content (10%, 15%, 20%, and 25% db) and five levels of moisture content (10%, 15%, 20%, 25%, and 30% db) on the resulting flow properties of DDGS. With an increase in soluble solid levels, the flow function curves shifted in a counterclockwise direction (i.e., toward the shear stress, or vertical) axis, which indicated worsening flowability. As soluble solid and moisture levels increased, the compressibility of the DDGS was found to increase, which indicated worsening flow. Overall, the DDGS was classified as a cohesive material, and was likely to produce cohesive arching and thus flowability problems.

Using data from these two studies, Ganesan et al. (2007) used exploratory data analysis to develop a mathematical model to predict the flowability of DDGS. A simple, yet robust model ($R^2 = 0.93$, standard error = 0.12) was developed by combining flow properties obtained from conventional Carr and Jenike tests via dimensional analysis and response surface modeling. In this model, compressibility, dispersibility, angle of internal friction, and effective angle of internal friction were

TABLE 7.3
Typical Flowability Properties of DDGS Samples from Commercial Ethanol Plants

Property		Consolidation	Range	Reference
Carr properties	Angle of repose (°)		35.94–41.60	Bhadra et al. (2009a)
			26.5–34.2	Rosentrater (2006)
	Aerated bulk density (g/cm³)		0.44–0.60	Bhadra et al. (2009a)
	Packed bulk density (g/cm³)		0.47–0.62	Bhadra et al. (2009a)
	Hausner ratio (–)		1.03–1.09	Bhadra et al. (2009a)
	Compressibility (%)		2.88–7.86	Bhadra et al. (2009a)
	Uniformity (–)		1.20–2.80	Bhadra et al. (2009a)
	Dispersibility (%)		36.40–55.53	Bhadra et al. (2009a)
Jenike properties	Effective angle of friction	Level 1	42.33–57.00	Bhadra et al. (2009a)
	(°)	Level 2	49.67–59.33	Bhadra et al. (2009a)
		Level 3	48.33–61.00	Bhadra et al. (2009a)
	Angle of internal friction	Level 1	32.33–47.00	Bhadra et al. (2009a)
	(°)	Level 2	32.67–46.00	Bhadra et al. (2009a)
		Level 3	25.00–54.33	Bhadra et al. (2009a)
	Unconfined yield strength	Level 1	3.23–18.00	Bhadra et al. (2009a)
	(kPa)	Level 2	1.52–6.56	Bhadra et al. (2009a)
		Level 3	0.44–1.24	Bhadra et al. (2009a)
	Major consolidation stress	Level 1	17.75–31.31	Bhadra et al. (2009a)
	(kPa)	Level 2	4.93–10.40	Bhadra et al. (2009a)
		Level 3	1.48–2.24	Bhadra et al. (2009a)
	Jenike flow function	Level 1	1.43–5.57	Bhadra et al. (2009a)
	(–)	Level 2	1.45–5.69	Bhadra et al. (2009a)
		Level 3	1.36–4.16	Bhadra et al. (2009a)
	Compressibility (1/cm)		2.94–22.30	Bhadra et al. (2009a)

nonlinearly related to Hausner Ratio (i.e., ratio of packed bulk density to loose bulk density), and regions of good flow, fair flow, and poor flow were identified. However, this model was developed exclusively based on coproduct materials from one ethanol plant and, as DDGS flow properties often differ among plants, it was suggested to use this methodology and approach to develop similar models to predict the flowability of DDGS from other plants.

Bhadra et al. (2009a) measured the flowability characteristics of DDGS samples from five commercial ethanol plants in the north central region of the United States. Carr and Jenike tests were performed and the resulting data were mathematically compared with Ganesan's previously developed empirical computer model. The largest DDGS particles had an average geometric mean diameter (d_{gw}) of 1.19 mm, while the lowest particle size had an average d_{gw} of 0.5 mm. Soluble solid levels ranged from 10.5% to 14.8% (db). The effective angle of friction was 42° to 57°. Additionally, some physical and flow properties exhibited fairly high linear correlations with each other, including aerated and with each packed bulk densities (r = 0.97), geometric standard deviation and Carr compressibility (r = 0.71), and geometric standard deviation and Hausner Ratio (r = −0.70). Overall flowability assessment validated the Ganesan et al.'s model and indicated that the commercial DDGS samples did have the potential for flow problems, although no samples exhibited complete bridging.

To examine the behavior of a new, nutrient-fractionated DDGS product, Ganesan et al. (2009) quantified physical and flow properties of commercially produced, unmodified (9.3% db fat) and fractionated, reduced fat (2.1% db) DDGS to determine if lipid level affected flowability. The compressive modulus of reduced fat DDGS was 28.2% higher than unmodified DDGS, but shear stress resistance was the same (0.03 kg/m²). Carr testing indicated that the reduced fat DDGS had an angle

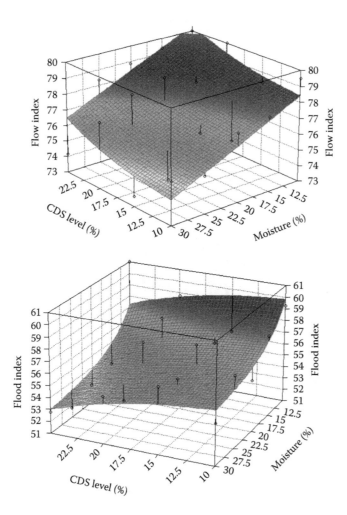

FIGURE 7.20 CDS level and moisture content in DDGS have nonlinear interactions. As the moisture content increases, flow index declines (i.e., flowability worsens), although floodability declines as well. As CDS level increases, flow index increases slightly (i.e., flowability improves), while floodability declines. (Adapted from Ganesan et al. *Cereal Chemistry,* 85(4), 464–470, 2008a.)

of repose 4.3% lower, and Carr compressibility 70% lower, than unmodified DDGS. However, in terms of uniformity and dispersibility, reduced fat DDGS was 100% and 41.5% greater than regular DDGS, respectively. Jenike shear testing revealed that reduced fat DDGS had unconfined yield strength and Jenike compressibility values that were 15.7% and 40.0% lower, respectively, than unmodified DDGS, and had major consolidating stress and flowability index values that were 6.7% and 13.2% higher, respectively. For regular DDGS, the flow function curve was located somewhat closer to the shear stress (vertical) axis, which indicated slightly worse flowability than the reduced fat DDGS. Overall, the reduction in the fat content showed slight improvement in some flow properties, especially compressibility. But both types of DDGS were ultimately classified as cohesive in nature. Additionally, exploration of the data using the previously developed computer assessment tool appeared to show that the reduced fat DDGS may have somewhat better flow characteristics.

Bhadra et al. (2009b) investigated the cross-sectional and surface natures of DDGS particles from five commercial ethanol plants, and how these characteristics interacted with overall physical and flow properties as well as composition. This study examined the distribution patterns of chemical components within cross sections, within section edges (i.e., surface layers), and on surfaces using standard staining techniques. Crude protein in the bulk DDGS samples ranged from 28.33% to 30.65% db, crude fat ranged from 9.40% to 10.98% db, and neutral detergent fiber (NDF) was 31.84% to 39.90% db. Moisture contents ranged from 4.61% to 8.08% db, and d_{gw} values were 0.37 to 0.52 mm. Cross-sectional staining of individual particles, however, indicated protein levels ranging from 19.57% to 40.39%, and carbohydrate levels from 22.17% to 43.06%, depending on the particle size examined and the production plant from which the DDGS was sampled. Staining of DDGS particles indicated a higher amount of surface layer protein (see Figure 7.21 for an example) compared to carbohydrate thickness on DDGS particles that had a lower flow function index (which indicated poor flowability). Additionally, surface fat staining (Figure 7.22) suggested that higher surface fat also occurred in samples with poor flow problems.

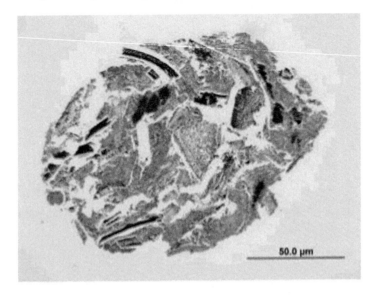

FIGURE 7.21 Cross-sectional image of a DDGS particle. Carbohydrate is denoted by dark stain, while protein by light stain. (Adapted from Bhadra et al. *Cereal Chemistry,* 86(4), 410–420, 2009b.)

FIGURE 7.22 Surface fat globules on DDGS particles (note the small spheres). (Adapted from Bhadra et al. *Cereal Chemistry,* 86(4), 410–420, 2009b.)

7.4 CONCLUSIONS

DDGS is similar to other dry, granular feed materials (such as soybean meal and corn gluten meal) in many respects. Currently, there are several challenges that affect the value and utility of DDGS, including variability and poor flowability. But more information is becoming available on the physical properties of DDGS, as well as the effects that processing parameters at the ethanol plants (especially CDS addition and drying operations) and the influence of storage conditions (including time, temperature, and relative humidity) have on these properties. Physical properties (especially moisture, particle size, shape, hardness, and stickiness) impact other physical attributes (such as thermal properties, bulk density, and angle of repose) and thus influence end use functionality. These properties affect material handling, storage behavior, as well as value-added processing operations. Physical properties are only one facet of DDGS, however; the others are chemical composition and nutritional quality (including digestibility). The following chapter discusses chemical composition of DDGS in depth, while nutritional quality is covered in the chapters on animal feeds (Chapters 12 through 17).

REFERENCES

ASAE (American Society of Agricultural Engineers). 2004. *S19.3, Method of Determining and Expressing Fineness of Feed Materials by Sieving.* ASAE Standards: Standards, Engineering Practices, Data, 51st ed. St Joseph, MI: ASAE.

ASTM (American Society for Testing and Materials). 1999. *ASTM D6393. Standard Test Method for Bulk Solids Characterization by Carr Indices.* ASTM Standards, W. Conshohocken, PA.

ASTM (American Society for Testing and Materials). 2006. *ASTM D6128. Standard Test Method for Shear Testing of Bulk Solids Using the Jenike Shear Cell.* ASTM Standards, W. Conshohocken, PA.

Barbosa-Canovas, G. V., and H. Vega-Mercado. 1996. *Dehydration of Foods.* New York: International Thomson Publishing.

Batal, A. B., and N. M. Dale. 2006. True metabolizable energy and amino acid digestibility of distillers dried grains with solubles. *Journal of Applied Poultry Research* 15: 89–93.

Beauchat, L. R. 1981. Microbial stability as affected by water activity. *Cereal Foods World* 26(7): 345–349.

Bhadra, R., K. Muthukumarappan, and K. A. Rosentrater. 2007. *Characterization of Chemical and Physical Properties of Distillers Dried Grain with Solubles (DDGS) for Value Added Uses.* Paper No. 077009. 2007 ASABE Annual International Meeting, Minneapolis, MN.

Bhadra, R., K. Muthukumarappan, and K. A. Rosentrater. 2009a. Flowability properties of commercial distillers dried grains with solubles (DDGS). *Cereal Chemistry* 86(2): 170–180.

Bhadra, R., K. A. Rosentrater, and K. Muthukumarappan. 2009b. Cross-sectional staining and surface properties of DDGS particles and their influence on flowability. *Cereal Chemistry* 86(4): 410–420.

Carr, R. L. 1965a. Evaluating flow properties of solids. *Chemical Engineering* 72(3): 163–168.

Carr, R. L. 1965b. Classifying flow properties of solids. *Chemical Engineering* 72(3): 69–72.

Clementson, C. L., K. A. Rosentrater, and K. E. Ileleji. 2009. *A Comparison of Measurement Procedures Used to Determine the Bulk Density of Distillers Dried Grains with Solubles (DDGS).* Paper No. 095761. 2009 ASABE Annual International Meeting, Reno, NV.

Craik, D. J., and B. F. Miller. 1958. The flow properties of powders under humid conditions. *Journal of Pharmacy and Pharmacology* 10: 136–144.

Duffy, S. P., and V. M. Puri. 1994. *Effect of Moisture Content on Flow Properties of Powders.* ASAE Paper No. 944033. St. Joseph, MI: ASAE.

Duffy, S. P., and V. M. Puri. 1999. Measurement and comparison of flowability parameters of coated cottonseeds, shelled corn and soybeans at three moisture contents. *Transactions of the ASAE* 42(5): 1423–1427.

Ergul, T., C. Martinez Amerzcua, C. M. Parsons, B. Walters, J. Brannon, and S. L. Noll. 2003. Amino acid digestibility in corn distillers dried grains with solubles. *Poultry Science* 82(Supplement 1): 70.

Fastinger, N. D., and D. C. Mahan. 2006. Determination of the ileal amino acid and energy digestibilities of corn distillers dried grains with solubles using grower-finisher pigs. *Journal of Animal Science* 84: 1722–1728.

Fitzpatrick, J. J., S. A. Barringer, and T. Iqbal. 2004a. Flow property measurement of food powders and sensitivity of Jenike's hopper design methodology to the measured values. *Journal of Food Engineering* 61(3): 399–405.

Fitzpatrick, J. J., T. Iqbal, C. Delaney, T. Twomey, and M. K. Keogh. 2004b. Effect of powder properties and storage conditions on the flowability of milk powders with different fat contents. *Journal of Food Engineering* 64(4): 435–444.

Ganesan, V., K. Muthukumarappan, and K. A. Rosentrater. 2008a. Effect of moisture content and soluble level on the physical, chemical, and flow properties of distillers dried grains with solubles (DDGS). *Cereal Chemistry* 85(4): 464–470.

Ganesan, V., K. A. Rosentrater, and K. Muthukumarappan. 2007. Modeling the flow properties of DDGS. *Cereal Chemistry* 84(6): 556–562.

Ganesan, V., K. A. Rosentrater, and K. Muthukumarappan. 2008b. Flowability and handling characteristics of bulk solids and powders—A review with implications for DDGS. *Biosystems Engineering* 101: 425–435.

Ganesan, V., K. A. Rosentrater, and K. Muthukumarappan. 2008c. Flow properties of DDGS with varying soluble and moisture contents using Jenike shear testing. *Journal of Powder Technology* 187: 130–137.

Ganesan, V., K. A. Rosentrater, and K. Muthukumarappan. 2009. Physical and flow properties of regular and reduced fat distillers dried grains with solubles (DDGS). *Food and Bioprocess Technology* 2: 156–166.

Goihl, J. 1993. Color, odor good indicators of DDGS nutritional value. *Feedstuffs* 65(21): 11.

Jenike, A. W. 1964. *Storage and Flow of Solids*. Bulletin No. 123. Utah Engineering Station. Logan, UT: University of Utah.

Johanson, J. R. 1978. Know your material—How to predict and use the properties of bulk solids. *Chemical Engineering*, Desk book issue: 9–17.

Kamath, S., V. M. Puri, H. B. Manbeck, and R. Hogg. 1993. Flow properties of powders using four testers—Measurement, comparison and assessment. *Powder Technology* 76: 277–289.

Kamath, S., V. M. Puri, and H. B. Manbeck. 1994. Flow property measurement using the Jenike cell for wheat flour at various moisture contents and consolidation times. *Powder Technology* 81(3): 293–297.

Kingsly, A. R. P., K. E. Ileleji, C. L. Clementson, A. Garcia, D. E. Maier, R. L. Stroshine, and S. Radcliff. 2010. The effect of process variables during drying on the physical and chemical characteristics of corn dried distillers grains with solubles (DDGS)—plant scale experiments. *Bioresource Technology* 101: 193–199.

Liu, K. 2008. Particle size distribution of distillers dried grains with solubles (DDGS) and relationships to compositional and color properties. *Bioresource Technology* 99: 8421–8428.

Liu, K. 2009. Effects of particle size distribution, compositional and color properties of ground corn on quality of distillers dried grains with solubles (DDGS). *Bioresource Technology* 100: 4433–4440.

Moreyra, R. and M. Peleg. 1981. Effect of equilibrium water activity on the bulk properties of selected food powders. *Journal of Food Science* 46:1918–1922.

Peleg, M. 1983. Physical characteristics of food powders. In *Physical Properties of Foods*, eds. M. Peleg, and E. Bagley, pp. 293–323. New York: AVI Publishing Company.

Peleg, M., and A. M. Hollenbach. 1984. Flow conditioners and anticaking agents. *Food Technology* 38(3): 93–102.

Prescott, J. K., and R. A. Barnum. 2000. On powder flowability. *Pharmaceutical Technology*. Oct: 60–84.

Rosentrater, K. A. 2006. Some physical properties of distillers dried grains with solubles (DDGS). *Applied Engineering in Agriculture* 22(4): 589–595.

Rosentrater, K. A. 2007. Ethanol processing coproducts—A review of some current constraints and potential directions. *International Sugar Journal* 109(1307): 1–12.

Schlicher, M. 2005. The flowability factor. *Ethanol Producer Magazine* 11(7): 90–93, 110–111.

Schrämli, W. 1967. On the measurement of the flow properties of cement. *Powder Technology* 1: 221–227.

Teunou, E., J. J. Fitzpatrick, and E. C. Synnott. 1999. Characterization of food powder flowability. *Journal of Food Engineering* 39(1): 31–37.

Thomson, F. M. 1997. Storage and flow of particulate solids. In *Handbook of Powder Science and Technology*, pp. 389–486. New York: Chapman and Hall.

UP. 2009. Biofuels & co-products. Union Pacific Railroad. Available online: www.uprr.com/customers/ag-prod/ethanol/overview.shtml. Accessed February 10, 2011.

U.S. Grains. 2008. Physical & chemical characteristics of DDGS. *DDGS User Handbook*. Available online: www.grains.org. Accessed February 10, 2011.

Wang, L., R. A. Flores, and L. A. Johnson. 1997. Processing feed ingredients from blends of soybean meal, whole blood, and red blood cells. *Transactions of the ASAE* 40(3): 691–697.

York, P. 1975. The use of glidants to improve the flowability of fine lactose powder. *Powder Technology* 11: 197–198.

8 Chemical Composition of DDGS

KeShun Liu

CONTENTS

8.1 INTRODUCTION

Distillers dried grains with solubles (DDGS) is a major type of distillers grains in the current market. By definition, distillers grains are cereal coproducts of the distillation process. There are two main sources of distillers grains. The traditional source is from brewers where beverage ethanol, such as beer or spirits, is produced, and a growing source is from fuel ethanol plants. Fuel ethanol can be produced from sugary, starchy, or cellulosic materials and has been highly concentrated to remove water and blended with other compounds to render the alcohol undrinkable. There are two major industrial methods for producing fuel ethanol from grains: wet milling and dry grind.

Wet milling involves soaking grains (mostly corn) in water (steeping) under carefully controlled conditions. The softened grain is then milled, and its structural or chemical components are separated by screening, centrifuging, and washing, to generate fractions of starch, germ, fiber, and gluten. The isolated starch is fermented by yeast into ethanol, processed into sweeteners (such as high fructose corn syrup), or sold as it is. The other fractions are further processed into different products, including oil (from germ), gluten meal (from gluten), and gluten feed (by mixing fiber and evaporated steepwater and drying the mixture). Both corn gluten meal and corn gluten feed are used for animal feed (Kent and Evers, 1994).

A detailed description of dry grind processing is covered in Chapter 5, as well as by Jacques et al. (2003), Bothast and Schlicher (2005), and Koster (2007). Briefly, as shown in Figure 8.1, grains (mostly corn) are dry milled, and then mixed and/or cooked with water. The starch in the flour is converted into sugars by enzymes and fermented into ethanol and carbon dioxide by yeast. After ethanol is distilled out, all residual components remain and end up in an intermediate stream known as whole stillage, which is further processed and recovered into several types of coproducts. Typically, whole stillage is centrifuged into thin stillage and distillers wet grains (DWG). Thin stillage is then evaporated to become condensed distillers solubles (CDS). For increasing shelf life and improving drying and handling, both DWG and CDS are subsequently mixed and then dried to produce DDGS. Thus, DDGS is the coproduct of dry grind processing of corn or other starchy grains into fuel ethanol. It is the major form of distillers grains available to the feed industry. For this reason, in this chapter, the terms "distillers grains" and "DDGS" are used interchangeably.

Dry grind ethanol production represents the majority of ethanol processing in the United States, and all newly constructed ethanol plants employ some variation on the basic dry grind process because such plants can be built at a smaller scale for a smaller investment (RFA, 2010). Although DDGS has been in the market for a century (see Chapter 3), its surge in global supply in recent years has stimulated many new investigations into this important coproduct of biofuel production. In particular, chemical composition of DDGS has been a great interest to researchers in animal science, ethanol producers, and traders in the feed industry. This chapter focuses on major and minor nutrients in DDGS in terms of concentrations (quantity), composition (quality), changes during the

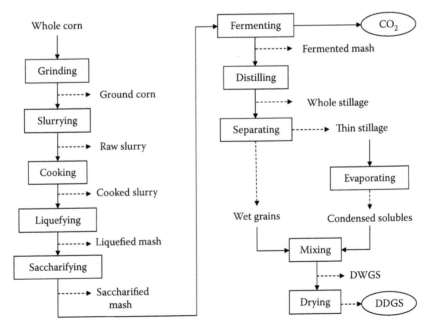

FIGURE 8.1 Schematic diagram of a conventional dry grind ethanol production from corn.

entire dry grind process, and the underlying causes for high compositional variation of DDGS. Additional topics, including particle size of DDGS, its relationship with chemical composition, and wet fractionation of DDGS, are also discussed. As for mycotoxins in DDGS, refer to Chapter 11.

8.2 CHEMICAL COMPOSITIONS

During dry grind processing, since approximately two-thirds of the mass of the starting material (typically corn) is converted into carbon dioxide and ethanol, it is normally expected that the concentrations of all unfermented nutrients, such as oil, protein, and minerals, will be concentrated about threefold. Thus DDGS can be a rich source of significant amounts of crude protein, amino acids, phosphorus, and other nutrients for animal feed. The main problem with use of DGGS as a feed ingredient is the high variability of nutrient concentrations among different DDGS sources.

There are many reports on the general composition of DDGS and their variability. Some are in published literature (Cromwell et al., 1993; Spiehs et al., 2002; Belyea et al., 2004, 2006; Skiatkiewicz and Koreleski, 2008; Kim et al., 2008; Liu, 2008, 2009a). Others are posted in various websites of state agricultural extension offices (USA), trade, or commodity organizations. Variation in the chemical composition of CDS is also reported (Belyea et al., 1998). As shown in Tables 8.1 through 8.3, nutrient contents in DDGS vary with reports; they differ not only among production plants but also among years of production from the same plant or even among batches. It is not surprising that nutrient composition often differs from standard reference values as reported in NRC (1994, 1998). Thus it has been recommended that a complete chemical analysis of each source of DDGS should be done on a regular basis.

8.2.1 GENERAL COMPOSITION

In an early study, Cromwell et al. (1993) evaluated the physical, chemical, and nutritional properties of DDGS from nine different sources (two from beverage and seven from fuel alcohol production systems). They found that considerable variability in nutrient content and physical properties existed among DDGS samples. When converted to a dry matter basis, crude protein ranged from 26.0% to 31.7%, fat from 9.1% to 14.1%, ash from 3.7% to 8.1%, acid detergent fiber (ADF) from 11.4% to 20.8%, and neutral detergent fiber (NDF) from 33.1% to 43.9%. The coefficient of variation (CV, also known as relative % of standard deviation) for these nutrients ranged from 5.3% to 27.7%. The average value for protein, oil, ash, total carbohydrate, ADF, and NDF were 29.7%, 10.7%, 5.3%, 54.3%, 15.9%, and 38.8%, respectively. The average dry matter content was 90.5% with a CV of 1.8% (Table 8.1).

A decade later, Spiehs et al. (2002) evaluated the nutrient content and variability of DDGS in a total of 118 samples from ten fuel ethanol plants during 1997, 1998, and 1999. They found that the average values (% dry matter) for protein, oil, ash, crude fiber, ADF, and NDF were 30.2, 10.9, 5.8, 8.8, 16.2, and 42.1, respectively. The CV ranged from 6.4% for protein to 28.4% for ADF (Table 8.1). The average dry matter content was 88.9%, with a CV of 1.7%. These values were not substantially different from those of Cromwell et al. (1993). Both showed higher variation in ash content and lower variation in dry matter content.

Later, Belyea et al. (2004) analyzed as many as 235 DDGS samples from a fuel ethanol plant in Minnesota, USA, and found that the average values (% dry matter) for protein, oil, ash, crude fiber, and ADF were 31.4, 12.0, 4.6, 10.2, and 16.8, respectively. They also reported the average content of residual starch as 5.3%. Thus, Belyea et al. (2004) gave higher average values of protein, oil, and crude fiber and lower values of ash than Spiehs et al. (2002). Liu (2008) showed that the average values of six DDGS samples from fuel ethanol plants for protein, oil, ash, and starch were 27.4, 11.7, 4.4, and 4.9, % dry matter, respectively. The lower estimate of the protein value as compared with the previous three studies might be due to use of 5.75 as a conversion factor from nitrogen instead of 6.25.

TABLE 8.1
General Composition of DDGS from Different Plants, Years, and Sources Reported in Different Sources

Source	Cromwell et al. (1993)			Spiehs et al. (2002)			Belyea et al. (2004)			Liu (2008)		
	Mean	Range	CV (%)[a]	Mean	Range[b]	CV (%)	Mean	Range[c]	CV (%)[c]	Mean	Range	CV (%)
No. of data points	9	9	9	118	10	118	235	5	5	6	6	6
Dry matter	90.5	87.1–92.7	1.8	88.9	87.2–90.2	1.7						
Protein	29.7	26.0–31.7	5.3	30.2	28.7–31.6	6.4	31.4	30.8–33.3	6.3	27.4	25.8–29.1	4.0
Oil	10.7	9.1–14.1	6.5	10.9	10.2–11.4	7.8	12.0	10.9–12.6	5.6	11.7	11.0–12.2	4.0
Ash	5.3	3.7–8.1	27.7	5.8	5.2–6.7	14.7	4.6	4.3–5.0	5.7	4.4	4.0–4.9	7.8
Starch							5.3	4.7–5.9	9.7	4.9	3.2–5.7	25.7
Total carbohydrate	54.3			53.1			52.1		5.2	56.5	55.7–57.9	1.6
Crude fiber				8.8	8.3–9.7	8.7	10.2	9.6–10.6	3.7			
Acid detergent fiber	15.9	11.4–20.8	21.1	16.2	13.8–18.5	28.4	16.8	15.4–19.3	9.3			
Neutral detergent fiber	38.8	33.1–43.9	10.0	42.1	36.7–49.1	14.3						

Note: Nutrient values are all expressed in or converted to % dry matter basis; bold indicates mean values.

[a] CV = coefficient of variation, also known as relative standard deviation.

[b] Range values for means of ten sample origins (locations).

[c] Range and CV (%) values for means of five sample groups (by year).

8.2.2 AMINO ACID COMPOSITION

In the early study, Cromwell et al. (1993) reported that among nine different sources from beverage or fuel alcohol production systems, when converted to dry matter basis, lysine varied from 0.48% to 0.97%, methionine from 0.49% to 0.61%, threonine from 0.99% to 1.28%, and tryptophan from 0.18% to 0.25%. Lys was the most variable among the 11 amino acids (AA) measured, with a CV = 18.71%. In the Spiehs et al. (2002) study, 119 DDGS samples were analyzed for ten essential amino acids. On a dry matter basis, the average lysine content was 0.85%, ranging from 0.72% to 1.02%. Again, lysine was the most variable among the ten amino acids, with an average CV = 17.3% (Table 8.2). Methionine values ranged from 0.49% to 0.69%, with an average value of 0.55%. Average tryptophan and threonine values were 0.25% and 1.13%, respectively. The mean values (% dry matter) for arginine, histidine, phenylalanine, isoleucine, leucine, and valine were 1.20, 0.76, 1.47, 1.12, 3.55, and 1.50, respectively. These values did not substantially differ from those reported by other researchers (Table 8.2). Cromwell et al. (1993) and Spiehs et al. (2002) only analyzed essential AA, but others (Batal and Dale, 2006; Kim et al., 2008; and Han and Liu, 2010) examined contents of both essential and nonessential AA. Like proximate composition, variation in individual amino acids exists among reports and among sample sources.

Although protein content in DDGS is increased about threefold over that in the grain feedstock, the protein quality (in terms of AA composition relative to total AA) is not substantially improved over the grain. Because corn is the primary grain used in fuel ethanol production, the resulting DDGS has a similar AA profile to corn even though fermentation yeast has some effect (to be discussed labor). In particular, the lysine level is low relative to crude protein in DDGS. Thus, the protein quality of DDGS is considered incomplete, relative to the amino acid requirements of animals.

In addition, due to its high susceptibility to heat damage, lysine content and digestibility are the main concerns in the use of DDGS as a feed component. Cromwell et al. (1993) reported that Lys concentration tended to be lowest in the darkest and highest in the lightest-colored DDGS, and correlation between the Hunter lab L score and Lys content was also significant. This was later confirmed by Fastinger et al. (2006) who reported that Lys content in five sources ranged from 0.48% to 0.76%, with the lowest Lys content in the darkest DDGS source. Furthermore, they also reported that apparent and true Lys digestibility evaluated on adult roosters was significantly lower in the dark colored than in other DDGS samples. Differences in other essential amino acid digestibility among sources were smaller but also significant. Concurrently, Batal and Dale (2006) observed considerable differences in the true amino acids digestibility among samples, and significantly lower total and digestible Lys content in the darker DDGS samples, and attributed it to the Maillard reaction, which occurs when a reducing carbohydrate such as glucose reacts with the epsilon amino group of Lys. The reaction may result in destruction of a significant amount of Lys during excessive heating. They further suggested that color analysis might be a quick and reliable method of estimating the amino acid, particularly Lys, digestibility of DDGS for poultry.

8.2.3 MINERALS

Many studies also documented mineral composition in DDGS (Spiehs et al., 2002; Batal and Dale, 2003; Belyea et al., 2006; Liu and Han, 2011). Like in other biological materials, major minerals in DDGS are Ca, P, K, Mg, S, and Na (Table 8.3). Mean concentrations ranged from 0.05% (dry matter) for Ca to 1.15% for K. The remaining four fell in between. Minor minerals in DDGS include Zn, Mn, Cu, Fe, Al, Se, etc. Their concentrations ranged between 6 ppm for Cu to 149 ppm for Fe, among reports.

High concentration and high variability of minerals are two issues that impact practical utilization and thus marketing of DDGS as animal feed (Spiehs et al., 2002; Batal and Dale, 2003; Belyea et al., 2006; Liu and Han, 2011). High concentration can lead to not only nutritional disorders but also excessive minerals in wastes. For example, excessive dietary S concentrations have been

TABLE 8.2
Amino Acid Composition of DDGS Reported in Different Sources

Source	Cromwell et al. (1993)			Spiehs et al. (2002)			Batal and Dale (2006)		Han and Liu (2010)			Kim et al. (2008)
	Mean	Range	CV (%)[a]	Mean	Range	CV (%)[a]	Mean	CV (%)[a]	Mean	Range	CV (%)[a]	
Number of data points	9	9	8	118	10	118	8	8	3	3	3	1
Essential												
Arg	**1.18**	0.95–1.33	9.70	**1.20**	1.11–2.17	9.1	**1.09**	14.68	**1.29**	1.16–1.40	9.45	**1.4**
His	**0.80**	0.65–0.93	12.65	**0.76**	0.72–0.82	7.8	**0.69**	8.70	**0.91**	0.82–1.01	10.41	**0.8**
Ile	**1.13**	1.06–1.26	5.89	**1.12**	1.05–1.17	8.7	**0.97**	6.19	**1.03**	0.91–1.25	18.85	**1.1**
Leu	**3.69**	3.05–4.40	12.56	**3.55**	3.51–3.81	6.4	**3.05**	4.59	**3.50**	3.18–3.91	10.62	**3.3**
Lys	**0.78**	0.48–0.97	18.71	**0.85**	0.72–1.02	17.3	**0.71**	22.54	**1.04**	0.88–1.15	13.63	**1.0**
Met	**0.57**	0.49–0.61	6.65	**0.55**	0.49–0.69	13.6	**0.54**	11.11	**0.72**	0.65–0.76	8.45	**0.6**
Phe	**1.61**	1.39–1.91	9.48	**1.47**	1.41–1.57	6.6	**1.31**	3.05	**1.50**	1.37–1.76	14.79	**1.4**
Thr	**1.13**	0.99–1.28	10.00	**1.13**	1.07–1.21	6.4	**0.96**	6.25	**1.17**	1.06–1.26	8.67	**1.1**
Trp	**0.22**	0.18–0.25	11.09	**0.25**	0.21–0.27	6.7	**0.20**	25.00				**0.2**
Val	**1.49**	1.30–1.64	7.18	**1.50**	1.43–1.56	7.2	**1.33**	5.26	**1.56**	1.40–1.80	13.72	**1.5**
Nonessential												
Ala							**1.78**	3.93	**2.07**	1.86–2.27	9.91	**1.9**
Asp							**1.75**	11.43	**1.97**	1.77–2.16	9.91	**1.7**
Cys	**0.59**	0.49–0.66	9.52				**0.56**	7.14	**0.57**	0.53–0.60	6.33	**0.5**
Glu							**3.49**	6.88	**5.48**	4.94–6.01	9.76	**3.3**
Gly									**1.19**	1.11–1.31	8.72	**1.1**
Pro							**1.99**	5.03	**2.19**	1.94–2.63	17.32	**2.0**
Ser							**1.09**	6.42	**1.45**	1.32–1.58	9.00	**1.2**
Tyr							**0.96**	9.38	**1.02**	0.87–1.29	22.65	**1.2**

Note: Nutrient values are expressed in or converted to % dry matter basis, except for Batal and Dale (2006) data, which were expressed as % wet (as is) basis; bold indicates mean values.

[a] CV = coefficient of variation.

[b] Range values for means of ten sample origins (locations).

TABLE 8.3
Concentrations of Minerals in DDGS Reported in Different Sources

Source	Parameters	K (mg/g)	P (mg/g)	Mg (mg/g)	S (mg/g)	Na (mg/g)	Ca (mg/g)	Fe (ug/g)	Zn (ug/g)	Mn (ug/g)	Cu (ug/g)
Spiehs et al. (2002)	Sample No.	118	118	118	118	118	118	118	118	118	118
	Minimum	6.9	7.0	2.5	3.3	1.2	0.3	75.3	44.7	10.7	4.7
	Maximum	10.6	9.9	3.7	7.4	5.1	1.3	156.4	312.0	21.3	7.6
	Mean	**9.4**	**8.9**	**3.3**	**4.7**	**2.4**	**0.6**	**119.8**	**97.5**	**15.8**	**5.9**
	CV (%)	14.00	11.70	12.10	37.10	70.50	57.20	41.10	80.40	32.70	20.40
Batal and Dale (2003)	Sample No.	12	12	12	12	12	12	12	12	12	12
	Minimum	6.7	5	2.1	5.8	0.9	0.1	67	44	9	3
	Maximum	9.9	7.7	3.3	11	4.4	7.1	325	88	48	18
	Mean	**9.1**	**6.8**	**2.8**	**8.4**	**2.5**	**2.9**	**149**	**61**	**22**	**10**
	CV (%)	12.08	12.29	14.28	25.00	60.00	93.00	57.70	21.30	50.00	43.00
Belyea et al. (2006)	Sample No.	9	9	9	9	9	9	9	9	9	9
	Minimum	9.31	7.10	2.99	3.44	0.60	0.25	90.0	75.0	15.6	4.9
	Maximum	12.40	9.43	3.79	8.27	2.30	0.34	109.0	170.0	19.3	6.8
	Mean	**11.22**	**8.52**	**3.48**	**5.76**	**1.30**	**0.28**	**98.7**	**113.7**	**17.0**	**5.6**
	CV (%)	9.60	8.71	7.70	25.06	40.99	11.14	5.87	36.52	7.04	10.92
Liu and Han (2011)	Sample No.	3	3	3	3	3	3	3	3	3	3
	Minimum	10.72	8.35	3.24	6.03	2.16	0.31	17.52	63.36	14.57	5.01
	Maximum	12.42	9.28	3.63	7.94	2.94	0.48	26.63	67.28	17.98	6.07
	Mean	**11.44**	**8.73**	**3.45**	**6.83**	**2.63**	**0.37**	**21.47**	**65.15**	**15.81**	**5.55**
	CV (%)	7.66	5.60	5.79	14.56	15.56	26.02	21.75	3.04	11.93	9.52

Note: Values are expressed in dry matter basis. CV = coefficient of variation, also known as relative standard deviation; bold indicates mean values.

associated with thiamine deficiency, which in turn causes polioencephalomalacia (PEM) in ruminants (Gould, 1998; Niles et al., 2002). It has also been linked, together with high nitrogen (N), to
increased odor production in manure (Spiehs and Varel, 2009). High phosphorus (P) concentration
in DDGS, which ranges from 0.5% to 1.0% (Spiehs et al., 2002), has been shown to cause increased
P excretion in livestock wastes (Koelsch and Lesoing, 1999; Spiehs and Varel, 2009), which in turn
increases the amount of land necessary to utilize manure P. Therefore, high DDGS inclusion in
rations for certain animals has been avoided because of potential problems with PEM and/or environmental concerns.

Variation in mineral contents (Table 8.3) is much larger than the composition of other nutrients
(Tables 8.1 and 8.2). For some minerals (such as S, Na, and Ca), the CV values (>25%) within
a single study were higher than others. Exogenous addition of some mineral compounds during processing may be an explanation. For example, ethanol plants may use sodium hydroxide to
sanitize equipment. They may also use it, along with sulfuric acid, to adjust the pH of mashes for
optimum enzyme activity during liquefaction and/or meeting yeast requirements during fermentation (Belyea et al., 2006). High variation in mineral contents makes accurate diet formulation
difficult because assumed concentrations could be different from actual concentrations. To prevent
the potential for underfeeding, producers often formulate diets on the assumption that mineral
concentrations are low. This practice results in overfeeding of nutrients, which can lead to not only
nutritional disorders but also in excess minerals in wastes. Because of excessive variation of elements in DDGS, they are frequently measured in order to develop a more complete nutrient profile
of DDGS.

8.2.4 Lipids

The lipids in DDGS originate from the feedstock for ethanol production (i.e., grain). In the United
States, the major feedstock is yellow dent corn, although sorghum and other grains are also used to
a limited extent. So, lipid profiles in distillers grains mostly resemble those in corn, except for the
about threefold increase in concentration. The major lipid is triglycerides while minor ones include
phytoesterols, tocopherols, tocotrienols, carotenoids, etc. (Winkler et al., 2007; Leguizamon et al.,
2009; Winkler-Moser and Vaughn, 2009; Majoni and Wang, 2010; Moreau et al., 2010a, 2010b).
Yet, unlike oil of the original feedstock, distiller grains was found to contain unusually higher
amounts of free fatty acids (6%–8% vs. 1%–2% in corn, based on extracted oil weight) (Winkler-
Moser and Vaughn, 2009; Majoni and Wang, 2010; Moreau et al., 2010b). Oil extracted from CDS
was also found to contain higher levels of free fatty acids (Moreau et al., 2010a).

Since the crude oil content in DDGS is around 10%, there is a renewed interest in removing
oil either before (frontend) (Singh et al., 2005) or after fermentation (backend) (Wang et al., 2008;
Cantrell and Winsness, 2009) (Chapter 5). Unlike oil removed at the frontend, oil removed at the
backend is no longer edible (mainly due to unusually high levels of free fatty acids) so it is mainly
for use as a feedstock of biodiesel production. Details on lipid composition in DDGS are covered in
Chapter 9 of this book.

8.2.5 Carbohydrates and Low Molecular Weight Organics

During dry grind processing, starch is converted to simple sugars, which are then fermented to
ethanol and carbon dioxide. However, other carbohydrates (CHO), such as cell wall CHO, remain
relatively unchanged chemically. DDGS also contains low molecular weight organic compounds
that are present in the original feedstock or produced during processing. Since starch conversion
cannot lead to completion under normal processing conditions, there are also some residual starch
and sugars in the coproduct (Liu, 2008).

Dowd et al. (1993) reported that the low molecular weight organics in the solubles of corn origin were lactic acid (10.40 g/L), glycerol (5.8 g/L), and alanine (free amino acid, 4.08 g/L), as well

as smaller amounts of ethanol, and various non-nitrogenous and nitrogenous acids, polyhydroxy alcohols, sugars, and glucosides. Wu (1994) measured various types of sugars in DDGS, distillers dried grains (DDG), and distillers dried solubles (DDS) by hydrolyzing samples with trifluoroacetic acid (TCA), followed by HPLC analysis. Analyses revealed that these ethanol coproducts contained many neutral sugars after TCA hydrolysis, including glycerol, arabinose, xylose, mannose, glucose, and galactose. DDS had the highest content of sugars (38.7%), followed by DDGS (38.0%) and DDG (35.8%). The sugar composition also differed among the three coproducts. For example, among the sugars in DDGS, the highest amount was glucose (11.9%), followed by xylose (8.5%), glycerol (7.8%), arabinose (6.4%), galactose (1.9%), and mannose (1.6%). Note that these sugars were not present in free form, but rather as a complex carbohydrate, commonly seen in cell walls. They became measurable after TCA hydrolysis.

Traditionally, DDGS is mainly used as animal feed. Compositional analysis is thus centered on key nutrients such as protein, oil, minerals, etc. Yet, with an increasing demand for fuel ethanol, DDGS is viewed as a potential feedstock for ethanol production by a cellulosic method (Chapter 22). Thus, Kim et al. (2008) developed a new analytical approach, which aimed at determining a more detailed chemical composition, especially for polymeric sugars, such as cellulose, starch, and xylan, which release fermentable sugars upon action by cellulosic enzymes. Not surprisingly, DDGS had higher water extractives than DDG (Table 8.4) since the solubles, as a part of DDGS, contained more simple sugars. Here, the ether extractives can be considered crude oil content, while water extractives can be considered soluble carbohydrates and related compounds such as glycerol and lactic acid. In this table, assuming that the complex carbohydrate consists of glucan (which includes residual starch and cellulose), xylan, and arabinan, by adding all four constituents plus water extractables, we can have a total carbohydrate (CHO) of 59.4% dry matter. This value was very close to the calculated amount of total CHO (59.0%) by subtracting the sum of protein, ether extractives, and ash from 100%. In either case (by measurement or by calculation), the total CHO content in DDGS was higher than those reported in Table 8.1.

TABLE 8.4
Cellulosic Biomass Compositional Analysis of Distillers Wet Grains (DWG) and DDGS

Constituents	DWG	DDGS
Dry matter (% wet basis)	35.3	88.8
Ether extractives	9.6	11.6
Crude protein	36.6	24.9
Ash	2.0	4.5
Total carbohydrate by calculation	51.8	59.0
Water extractives	8.8	24.7
Glucan	18.5	21.2
Cellulose	12.6	16.0
Starch	5.9	5.2
Xylan and arabinan	20.9	13.5
Xylan	14.9	8.2
Arabinan	5.5	5.3
Total carbohydrate measured	48.2	59.4

Source: Adapted from Kim, Y. et al. *Bioresource Technology* 99: 5165–5176, 2008.
Note: All values are expressed in dry matter basis except where otherwise noted.

8.3   PARTICLE SIZE AND ITS RELATIONSHIP WITH CHEMICAL COMPOSITION

DDGS is a heterogeneous particulate bulk consisting of all the nonfermentable components from corn kernels, after the majority of starch is hydrolyzed and depleted during ethanol production (Ileleji et al., 2007). Thus, the relative amounts of particles present, sorted according to size, would be a characteristic of a particular DDGS sample. Such a feature, commonly known as particle size distribution (PSD), has been widely used to describe many other powder materials, since it is an important quality parameter that helps in understanding the physical and chemical properties of a particular powder material (Barbosa-Canovas et al., 2005). See Chapter 7 for further details.

The way that PSD is expressed is usually defined by the method by which it is determined (Barbosa-Canovas et al., 2005). In general, a sieve analysis is the easiest method for particle size determination (ASAE Standards, 2003), where a powdery material is separated on sieves of different sizes, and PSD is defined in terms of mass frequency over discrete size ranges. It is based on an assumption that the particles are spheres that will just pass through a square hole in a sieve. In reality particles in most powder materials, such as DDGS, are irregular in shape, often extremely so. However, it does not diminish the value of particle size analysis. Based on sieve analysis, PSD is generally expressed in two ways: in the proportion of material retained on (or passed through) each sieve size (by a table or a graph) or as geometric mean diameter based on a statistical treatment (ASAE Standards, 2003). The former expression can be more easily understood by processors, but geometric mean diameter can be an effective way for comparing PSD of different samples on a statistical basis.

Particle size has been shown to affect the volume and acceptability of baked products incorporated with DDGS (Abbott et al., 1991). It could also affect digestibility by various species of animals (Wondra et al., 1995). Therefore, PSD data of DDGS are essential for many aspects, including formulation of animal feed, digestibility and nutrient availability, design of equipment and processing facilities, optimization of unit operations, storage, material handling systems, assessment of potential or flexibility for a particular nutrient enrichment by sizing, and of end product quality.

There is limited information in the literature on the PSD of DDGS and its relationship with chemical composition. Rausch et al. (2005a) compared the PSD of DDGS with those of ground corn measured in the same study and determine the relationship between the two, but no chemical properties were measured. In an attempt to provide information about DDGS PSD and its relationship with chemical composition, Liu (2008) studied 11 DDGS samples processed from yellow dent corn and collected from different ethanol processing plants in the U.S. Midwest area. PSD (by mass) of each sample was determined using a series of six selected U.S. standard sieves: No. 8, 12, 18, 35, 60, and 100, and a pan. The original sample and sieve sized fractions were measured for surface color and contents of moisture, protein, oil, ash, and starch. Total carbohydrate (CHO) and total nonstarch CHO were also calculated.

The study showed that the particle size of DDGS varied greatly within a sample, and PSD varied greatly among samples (Figure 8.2). The majority had a unimodal PSD, with a mode in the size class between 0.5 and 1.0 mm. The 11 samples had a mean value of 0.660 mm for the geometric mean diameter (d_{gw}) of particles and a mean value of 0.440 mm for the geometric standard deviation (S_{gw}) of particle diameters by mass. The study also showed that there was a great variation in chemical composition (Figure 8.3) and color attributes in whole and sieved fractions among DDGS from different plants. A few DDGS samples contained unusually high amounts of residual starch (11.1%–17.6%, dry matter basis, vs. about 5% for the rest), presumably resulting from modified processing methods. More importantly, although particle size and color parameters had little correlation with the composition of whole DDGS samples, distribution of nutrients as well as color attributes correlated well with PSD. In sieved fractions, protein content, color values of L and a were negatively correlated with particle size, while contents of oil and total CHO were positively correlated with particle size. This means that finer fractions were higher in protein concentration, but lower in oil and CHO, and lighter in color. Thus, there was a highly heterogeneous distribution of nutrients in sized fractions.

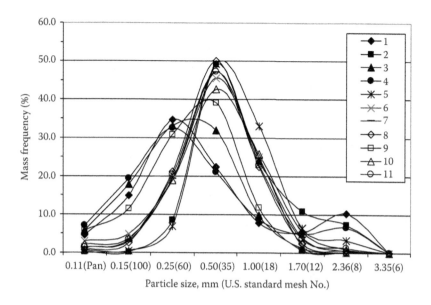

FIGURE 8.2 Particle size distribution of 11 DDGS samples collected from the U.S. Midwest region. Mass frequency was based on the proportion of materials retained on each sieve size, by weight. (Adapted from Liu, K. S. *Bioresource Technology* 99: 8421–8428, 2008. With permission.)

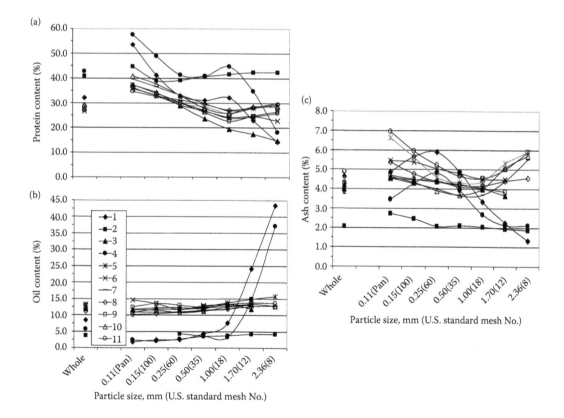

FIGURE 8.3 (a) Protein, (b) oil, and (c) ash contents (%, dry matter basis) in the original (whole) and sieve sized fractions of the 11 DDGS samples. (Adapted from Liu, K. S. *Bioresource Technology* 99: 8421–8428, 2008. With permission.)

FIGURE 8.4 Close-up photograph of a DDGS sample, showing different texture and shapes of particulate material. (Adapted from Liu, K. S. *Bioresource Technology* 99: 8421–8428, 2008. With permission.)

Liu (2008) further showed that the particles in DDGS can be grouped into three classes, flakes, granules, and aggregate granules (Figure 8.4). The flakes came mostly from tip cap and broken seed coats of corn kernels. The granules were mostly nonfermentable materials that were left from ground endosperm and germ. The aggregate granules are mostly granules glued together, apparently by solubles added during the final stage of the process. Because all three types of particulates varied in size and shape, sieving could cause changes of their proportions in sized fractions. Since the flakes were mostly fiber, while the granules and aggregates were mostly nonfiber components, shifts in their proportions led to change in composition in sieve sized fractions. Therefore, the study of Liu (2008) supports the idea of nutrient enrichment of DDGS through sieving and/or air clarification reported previously (Wu and Stringfellow, 1986; Srinivasan et al., 2005; Liu, 2009b) and in Chapter 25 of this book.

8.4 CHANGES IN CHEMICAL COMPOSITION DURING DRY GRIND PROCESSING

Some have reported compositional differences between the raw material (corn) and the end product (DDGS) of the dry grind process (Belyea et al., 2004). Others investigated compositional differences among different versions of the process—traditional versus modified methods (Wang et al., 2005; Singh et al., 2005). These studies provided some information about changes that occur during the process, but lacked details on changes that occur step by step during dry grind processing. Belyea et al. (2006) investigated mineral concentrations of primary process streams from the dry grind process, the first of its kind, although the documented changes were limited to minerals only. Recently, Liu and colleagues conducted a comprehensive study, with objectives to monitor changes in concentrations of various nutrients, composition of particular nutrients, and some physical properties during the entire dry grind process, from corn to DDGS. The study, documented in several reports (Han and Liu, 2010; Moreau et al., 2010; and Liu and Han, 2011), used three sets of samples that were provided from different commercial dry grind ethanol plants in Iowa. Each set consisted of ground corn, yeast, intermediate streams, and DDGS (Figure 8.1). Intermediate streams included raw slurry, cooked slurry, liquefied mass, saccharified mash, fermentation mash, whole stillage, thin stillage, CDS, DWG, and distillers wet grains with solubles (DWGS), although the total number of intermediate streams varied slightly among plants. Results of these reports are to be covered in detail in the following subsections.

8.4.1 Changes in General Composition

Protein, oil, and ash contents (on a dry basis) of the processing streams increased slightly at the beginning of the process, up to the saccharification step (Figure 8.5a). The increase of these components in cooked slurry as compared with ground corn was most likely due to using a portion of thin stillage as backset to slurry ground corn; the contents of protein, oil, and ash in thin stillage were much higher than ground corn. After fermentation, these nutrients were concentrated dramatically, about threefold over corn. The increase was mainly due to depletion of starch as it was fermented into ethanol and carbon dioxide. Distillation caused little changes in composition, but centrifugation did. Thin stillage was higher in oil and ash content but lower in protein content than DWG. This implies that in whole stillage, a larger portion of oil was emulsified in the liquid phrase, and the majority of ash was soluble, so that they went more in the liquid fraction than the solid fraction during centrifugation. Among all the stream samples, oil and ash were highest in thin stillage and its condensed form—CDS, while protein was highest in DWG. In addition, the ash content was so greatly reduced in DWG upon centrifugation that it was only slightly higher than that in ground corn. When the two were mixed together to become DWGS, the composition was averaged out and became similar to that of the whole stillage. There was a slight but significant (p < 0.05) difference in contents of protein, oil, and ash between DWGS and DDGS. This difference was most likely due

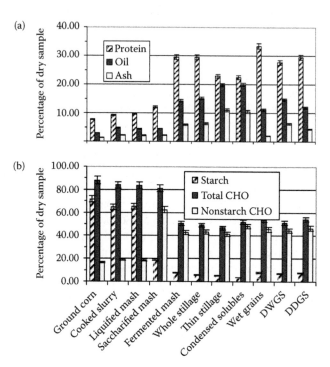

FIGURE 8.5 Changes in gross composition during dry grind ethanol processing from corn at a commercial plant in Iowa, USA. (a) contents of protein, oil and ash, and (b) contents of starch/dextrin, total CHO (carbohydrates), and total nonstarch CHO. (Adapted from Han, J. C., and K. S. Liu. *Journal of Agricultural and Food Chemistry* 58: 3430–3437, 2010. With permission.)

to the dynamics of drying, since part of the DDGS output was recycled, to be mixed with DWG and/ or CDS for improving operational performance (Kingsly et al., 2010).

Changes in starch/dextrin and total carbohydrate during dry grind processing of corn (Figure 8.5b) were opposite to those of protein, oil, and ash (Figure 8.5a). At the beginning of the process, starch and dextrin were relatively unchanged, although a decrease from corn to cooked slurry was noticeable. This decrease was apparently due to an increase in protein, oil, and ash contents discussed earlier. Upon saccharification, starch/dextrin decreased substantially and decreased further to about 6% after fermentation. It remained unchanged in the rest of processing streams. Enzymatic action and fermentation converted most of the starch to ethanol, but apparently could not achieve a complete conversion. Residual starch in coproducts was also reported elsewhere (Belvea et al., 2004; Pedersen et al., 2007; Liu, 2008). Concomitant with starch/dextrin change, the total CHO was relatively stable at about 83% until fermentation, where it decreased substantially to about 51%. This value fluctuated slightly in the rest of the processing streams (Figure 8.5b).

Total nonstarch carbohydrate refers to all carbohydrates excluding starch and dextrin. It includes cellulose, hemicellulose, and lignin, which are all cell wall components. It also includes soluble sugars and other low molecular organics (such as glycerol and lactic acid). In ground corn, starch was a major portion of total CHO and the total nonstarch CHO was around 17% of dry matter (Figure 8.5b). This value remained relatively unchanged until the step of saccharification, where it increased significantly due to conversion of starch and dextrose to simple sugars. Upon fermentation, depletion of soluble sugars caused some decrease in total nonstarch CHO, but the value was still about 43%, more than double the value in ground corn. This value fluctuated slightly in the remaining streams.

8.4.2 CHANGES IN AMINO ACID COMPOSITION

Amino acid composition is a major nutritional index of a protein ingredient. It is typically expressed as concentrations (% of sample weight, dry or fresh weight basis) or relative % (based on weight of total amino acids or protein in a given sample). DDGS proteins, like other proteins, contain essential and nonessential amino acids. In general, changes in AA concentrations, either essential or nonessential, followed the pattern of protein changes during the dry grind process (Table 8.5). Before fermentation, there was a slight change. Upon fermentation, concentrations of all AA increased 2.0–3.5-fold, resulting from starch depletion. When whole stillage was separated into thin stillage and distillers wet grains, AA concentrations, just like protein content, were higher in DWG than thin stillage. When the CDS and DWG were mixed to produce DWGS, the AA concentrations became close to that in whole stillage. There was some minor change upon drying into DDGS. The content of total amino acids was close to the protein content in each sample, but the difference between the two fluctuated between positive and negative values, depending on sample type and ethanol plant. This difference was presumably attributed to the difference in nonprotein nitrogen content among samples and the variation of two separate analytical methods. Note that Table 8.5 also includes yeast AA concentration.

Although the general trend in AA concentration followed that of protein content, the extent of change for each AA of a given downstream product as compared with that of ground corn varied with individual AA (Table 8.5). For example, upon fermentation, some amino acids increased in concentration significantly faster than others. Furthermore, when AA profile is expressed as relative % (based on total AA), it describes more protein quality than quantity. Unlike AA concentrations, the change in AA composition in terms of relative % (Table 8.6, converted from Table 8.5) did not follow the trend of protein change. Upon fermentation, some AA increased, others decreased, and still others remained unchanged.

8.4.3 CHANGES IN PROTEIN STRUCTURES

Reports on changes in protein structures are rather limited. Yu et al. (2010) investigated changes in protein molecular structures during bioethanol production, and found that proteins from original

TABLE 8.5

Amino Acid Concentration (% Dry Weight) during Dry Grind Ethanol Processing of Corn at Plant 1

Amino Acid (AA)	Milled Corn	Cooked Slurry	Liquefied Mash	Saccharified Mash	Fermented Mash	Whole Stillage	Thin Stillage	Condensed Solubles	Wet Grains	DWGS	DDGS	Yeast
Essential												
Arg	0.36 ± 0.03	0.37	0.34	0.39	1.28	1.22	1.03	0.98	1.34	1.24	1.32 ± 0.02	1.59 ± 0.06
His	0.32 ± 0.01	0.36	0.31	0.39	0.81	0.87	0.65	0.60	1.00	0.74	1.01 ± 0.00	0.87 ± 0.01
Ile	0.37 ± 0.03	0.29	0.35	0.30	1.05	0.77	0.48	0.69	0.99	0.96	0.91 ± 0.01	1.38 ± 0.01
Leu	1.24 ± 0.01	1.03	1.14	1.12	3.21	2.75	1.35	1.62	3.97	2.83	3.42 ± 0.04	2.37 ± 0.03
Lys	0.32 ± 0.00	0.32	0.32	0.31	1.13	1.01	0.81	0.97	1.17	1.05	1.09 ± 0.03	2.57 ± 0.01
Met	0.34 ± 0.08	0.45	0.33	0.37	0.67	0.69	0.60	0.56	0.68	0.48	0.76 ± 0.04	0.66 ± 0.03
Phe	0.66 ± 0.05	0.49	0.68	0.58	1.53	1.22	0.76	0.97	1.59	1.26	1.38 ± 0.07	1.44 ± 0.05
Thr	0.40 ± 0.01	0.40	0.38	0.42	1.15	1.05	0.77	0.83	1.29	1.05	1.19 ± 0.02	1.84 ± 0.02
Val	0.76 ± 0.01	0.69	0.75	0.70	1.57	1.31	1.03	1.23	1.55	1.40	1.47 ± 0.02	1.68 ± 0.00
Nonessential												
Ala	0.66 ± 0.01	0.66	0.66	0.66	2.05	1.81	1.29	1.45	2.20	1.87	2.09 ± 0.03	1.79 ± 0.02
Asp	0.60 ± 0.02	0.63	0.64	0.65	1.97	1.76	1.26	1.42	2.11	1.78	1.99 ± 0.03	3.35 ± 0.05
Cys	0.30 ± 0.05	0.34	0.30	0.35	0.59	0.55	0.49	0.46	0.64	0.48	0.58 ± 0.02	0.43 ± 0.03
Glu	1.80 ± 0.02	1.68	1.78	1.73	5.30	4.74	3.22	3.51	5.93	4.69	5.50 ± 0.06	6.33 ± 0.02
Gly	0.35 ± 0.02	0.34	0.36	0.37	1.20	1.08	0.93	1.03	1.14	1.13	1.16 ± 0.03	1.44 ± 0.01
Pro	0.68 ± 0.04	0.45	0.73	0.54	2.21	1.56	0.84	1.30	2.17	1.91	1.94 ± 0.03	0.68 ± 0.08
Ser	0.51 ± 0.01	0.49	0.48	0.53	1.41	1.26	0.87	0.92	1.60	1.27	1.44 ± 0.02	1.71 ± 0.03
Tyr	0.49 ± 0.17	0.38	0.50	0.35	1.16	0.78	0.56	0.76	1.00	0.74	0.91 ± 0.02	0.81 ± 0.15
Total AA	10.13 ± 0.28	9.35	10.04	9.73	28.28	24.46	16.93	19.32	30.38	24.88	28.15 ± 0.46	30.91 ± 0.31
Protein	7.70 ± 0.22	9.27 ± 0.13	9.77 ± 0.04	12.08 ± 0.02	29.43 ± 0.03	29.50 ± 0.48	22.90 ± 0.63	21.31 ± 0.24	33.40 ± 0.09	27.67 ± 0.37	29.47 ± 0.04	36.90 ± 0.28

Source: From Han, J. C., and K. S. Liu. *Journal of Agricultural and Food Chemistry* 58: 3430–3437, 2010.

Note: Means ± standard deviation. The rest are values of single measurement.

TABLE 8.6
Amino Acid Composition (Relative % of Individual Amino Acids in Each Sample) during Dry Grind Ethanol Processing of Corn at Plant 1

Amino Acid (AA)	Milled Corn	Cooked Slurry	Liquefied Mash	Saccharified Mash	Fermented Mash	Whole Stillage	Thin Stillage	Condensed Solubles	Wet Grains	DWGS	DDGS	Yeast
Essential												
Arg	3.55 ± 0.43	3.96	3.40	4.05	4.51	4.98	6.11	5.09	4.42	5.00	4.68 ± 0.13	5.15 ± 0.15
His	3.13 ± 0.17	3.85	3.09	4.05	2.86	3.57	3.84	3.13	3.28	2.98	3.59 ± 0.06	2.81 ± 0.07
Ile	3.61 ± 0.35	3.08	3.50	3.04	3.71	3.15	2.84	3.56	3.25	3.85	3.25 ± 0.02	4.46 ± 0.00
Leu	12.23 ± 0.25	11.00	11.32	11.49	11.36	11.23	7.95	8.40	13.07	11.38	12.16 ± 0.06	7.67 ± 0.18
Lys	3.19 ± 0.09	3.41	3.19	3.15	3.99	4.13	4.76	5.03	3.84	4.22	3.87 ± 0.05	8.31 ± 0.04
Met	3.30 ± 0.74	4.84	3.29	3.83	2.38	2.82	3.55	2.88	2.25	1.93	2.68 ± 0.10	2.14 ± 0.08
Phe	6.49 ± 0.32	5.28	6.79	5.97	5.40	4.98	4.47	5.03	5.22	5.05	4.90 ± 0.18	4.66 ± 0.20
Thr	3.96 ± 0.03	4.29	3.81	4.28	4.07	4.27	4.55	4.29	4.25	4.22	4.23 ± 0.01	5.95 ± 0.00
Val	7.50 ± 0.12	7.37	7.51	7.21	5.56	5.35	6.11	6.38	5.12	5.64	5.22 ± 0.00	5.43 ± 0.07
Nonessential												
Ala	6.56 ± 0.26	7.04	6.58	6.76	7.25	7.42	7.60	7.48	7.26	7.52	7.42 ± 0.01	5.79 ± 0.00
Asp	5.91 ± 0.33	6.71	6.38	6.64	6.97	7.19	7.46	7.36	6.95	7.16	7.06 ± 0.00	10.83 ± 0.06
Cys	2.95 ± 0.42	3.63	2.98	3.60	2.09	2.25	2.91	2.39	2.11	1.93	2.06 ± 0.02	1.38 ± 0.12
Glu	17.73 ± 0.66	17.93	17.70	17.79	18.73	19.40	19.03	18.15	19.53	18.85	19.52 ± 0.09	20.47 ±
Gly	3.43 ± 0.26	3.63	3.60	3.83	4.23	4.42	5.47	5.33	3.77	4.54	4.11 ± 0.05	4.66 ± 0.01
Pro	6.67 ± 0.23	4.84	7.30	5.52	7.81	6.39	4.97	6.74	7.16	7.66	6.90 ± 0.00	2.18 ± 0.25
Ser	5.02 ± 0.05	5.28	4.73	5.41	4.99	5.17	5.11	4.78	5.25	5.09	5.12 ± 0.03	5.52 ± 0.14
Tyr	4.82 ± 1.54	4.07	4.94	3.60	4.11	3.19	3.34	3.92	3.28	2.98	3.23 ± 0.00	2.60 ± 0.46

Source: From Han, J. C., and K. S. Liu. *Journal of Agricultural and Food Chemistry* 58: 3430–3437, 2010.

Note: Means ± standard deviation. The rest are values of single measurement.

grains had a significantly higher ratio of alpha helix to beta sheet than those of coproducts produced from bioethanol processing (1.38 vs. 1.03, $p < 0.05$). There were significant differences between wheat and corn (1.47 vs. 1.29, $p < 0.05$) but no difference between wheat DDGS and corn DDGS (1.04 vs. 1.03, $p > 0.05$). In terms of the ratio of protein amide 1 to 11 in the protein structure, the grains had a significantly higher value than their coproducts (4.58 vs. 2.84, $p < 0.05$). There was also a significant difference in the ratio between wheat DDGS and corn DDGS (3.08 vs. 2.21, $p < 0.05$), although no significant differences existed between wheat and corn (4.61 vs. 4.56, $p > 0.05$).

8.4.4 Changes in Fatty Acids and Functional Lipid Profiles

When the fatty acid composition of DDGS oil was expressed in relative %, linoleic acid was the major one (53.96%–56.53%), followed by oleic acid (25.25%–27.15%) and then palmitic acid (13.25%–16.41%), with low levels of stearic (1.80%–2.34%) and linolenic (1.15%–1.40%) acids (Moreau et al., 2010b). Although some minor yet significant difference existed in mean values of individual fatty acid among steps (fractions), all major fatty acids generally remained constant. This is also true among means of plants (Moreau et al., 2010b).

The major phytosterols in ground corn were sitosterol > campesterol > sitostanol > campestanol (Moreau et al., 2010b). Ten other minor phytosterols (stigmasterol, avenasterol, and others) and squalene were also detected but their total proportions ranged from 12% to 15% (based on total phytosterols mass). Since ergosterol, the major sterol in yeast (Redon et al., 2009), was not detected in any of the postfermentation samples, the contribution of yeast sterols to the total phytosterol pool was considered negligible. There were some differences in the sterol levels among samples collected at each step, but no obvious trends were observed. The proportions of the various phytosterols remained relatively constant among the nine fractions. Total phytosterol content in oil extracted from DDGS samples ranged from 1.5% to 2.5% and averaged around 2%. These data indicate that phytosterol content and composition remained relatively constant throughout dry grind processing.

Moreau et al. (2010b) also made quantitative analysis of the tocotrienols and tocopherols in the various fractions (Table 8.7) and confirmed previous reports that γ-tocopherol is the major tocopherol and γ-tocotrienol is the major tocotrienol in ground corn (Moreau et al., 2006) and in DDGS (Winkler et al., 2007), with small amounts of α- and δ-tocopherols and trace amounts of α- and δ-tocotrienols. Yet, in HPLC chronographs of all samples, there was an unknown peak that eluted between α-tocopherol and α-tocotrienol. In a previous paper, Moreau et al. (2010a) found the same unknown peak in ground corn and in "postfermentation" corn oil samples and suggested that it might be "α-tocomonoenol." This unknown tocol is labeled as α-T* in Table 8.7. Overall, levels of tocols and the proportions of the homologues remained relatively stable throughout the dry grind operation. However, there was an exception. Both δ-tocopherol and the unknown peak (α-T*) showed significant increase upon fermentation and remained relatively high thereafter. Since the total tocols included the values of these two compounds, they showed significant higher values for all fractions after the fermentation step. Thus, dry grind processing caused little changes to the majority of tocols and some increase of the remaining ones. The preservation of these important antioxidants may help maintain the oxidative stability of corn oil extracted from DDGS. Additional information on lipid changes can be found in Chapter 9.

8.4.5 Changes in Mineral Composition

Belyea et al. (2006) studied element concentrations in primary process streams from nine dry grind ethanol plants, after monitoring P concentrations and flows in corn wet milling streams (Rauch et al., 2005b). Samples included corn, ground corn, fermented mash (beer), DWG, CDS, and DDGS. They found that concentrations of most elements in corn were similar to published data and did not differ among processing plants. However, for the processing streams, there were differences in several element concentrations among processing plants. The concentrations of most elements in

TABLE 8.7
Tocopherol and Tocotrienol Composition of Nine Fractions from Three Dry Grind Ethanol Plants

Fraction no.	Plant	Alpha T	Alpha T*	Gamma T	Delta T	Alpha T3	Gamma T3	Delta T3	Total Tocols
1	8	12.74	46.22	109.93	16.19	12.08	24.90	1.35	223.42
1	9	19.80	32.13	84.92	9.96	11.09	19.48	0.98	178.37
1	10	23.16	31.35	85.80	9.65	13.04	25.94	1.48	190.41
	Mean[b]	18.57[b]	36.56[e]	93.55[b]	11.94[e]	12.07[b]	23.44[e]	1.27[cde]	197.4[f]
2	8	9.88	10.01	93.48	11.05	7.23	19.18	1.20	152.04
2	9	16.70	9.47	74.02	10.31	6.84	15.65	0.87	133.85
2	10	22.27	8.88	76.60	6.41	8.55	18.75	0.99	142.45
	Mean[b]	16.28[d]	9.45[g]	81.36[d]	9.25[f]	7.54[e]	17.86[f]	1.02[e]	142.78[h]
3	8	9.95	19.58	89.86	13.74	9.06	21.87	1.23	165.29
3	9	18.86	12.01	80.48	10.85	9.82	20.80	1.05	153.86
3	10	22.27	18.88	73.50	9.84	11.65	24.67	1.28	162.07
	Mean[b]	17.03[cd]	16.82[f]	81.28d	11.48[e]	10.17[cd]	22.45[e]	1.19[de]	160.41[g]
4	8	11.79	126.05	95.75	64.82	10.20	25.52	1.89	336.03
4	9	16.50	103.34	72.41	70.23	9.90	22.62	1.17	296.17
4	10	22.47	103.79	73.26	49.72	10.33	23.04	1.59	284.20
	Mean[b]	16.92[cd]	111.06[d]	80.47[d]	61.59[a]	10.14[cd]	23.73[de]	1.55[bcd]	305.47[c]
5	8	13.76	154.29	100.79	69.44	10.90	28.04	2.07	379.29
5	9	18.21	90.28	77.16	61.07	11.08	24.59	1.45	283.83
5	10	18.69	126.79	70.76	48.31	10.67	23.61	1.67	300.49
	Mean[b]	16.88[cd]	123.79[c]	82.9[cd]	59.60[b]	10.88[c]	25.42[c]	1.73[bc]	321.2[b]
6	8	9.33	188.08	91.09	70.66	12.74	33.29	2.48	407.68
6	9	15.98	113.45	77.01	50.41	15.37	32.63	2.64	307.49
6	10	17.80	201.42	67.74	54.01	13.30	30.70	3.16	388.14
	Mean[b]	14.37[e]	167.65[a]	78.61[d]	58.36[b]	13.80[a]	32.21[a]	2.76[a]	367.77[a]
7	8	9.62	93.63	88.05	12.55	12.09	30.72	2.15	248.81
7	9	15.95	123.11	68.84	34.86	13.40	28.28	2.76	287.19
7	10	17.25	155.53	59.94	23.70	13.61	29.12	2.41	301.55
	Mean[b]	14.27[e]	124.09[c]	72.28[e]	23.7[d]	13.03[a]	29.37[b]	2.44[a]	279.18[d]
8	8	14.06	40.29	112.53	27.93	6.71	18.25	1.06	220.82
8	9	21.52	30.79	89.95	28.51	7.31	17.25	0.88	196.21
8	10	28.82	50.84	92.90	27.55	8.01	19.20	0.96	228.29
	Mean[b]	21.47[a]	40.64[e]	98.46[a]	28.00[c]	7.34[e]	18.23[f]	0.97[e]	215.10[e]
9	8	13.21	112.21	104.00	22.14	9.62	25.29	1.91	288.37
9	9	16.18	87.01	73.87	28.32	7.52	20.30	1.41	234.61
9	10	24.05	197.89	84.52	24.39	12.79	29.33	2.35	375.32
	Mean[b]	17.81[bc]	132.37[b]	87.46[c]	24.95[d]	9.98[d]	24.97[cd]	1.89[b]	299.43[c]
Mean[c]	8	11.59[c]	87.82[b]	98.39[a]	34.28[a]	10.07[b]	25.23[a]	1.71[a]	269.08[a]
Mean[c]	9	17.74[b]	66.84[c]	77.63[b]	33.83[a]	10.26[b]	22.4[b]	1.47[b]	230.18[c]
Mean[c]	10	21.86[a]	99.49[a]	76.11[b]	28.17[b]	11.33[a]	24.93[a]	1.76[a]	263.66[b]

Source: Adapted from Moreau, R. A. et al. 2010b. *Journal of the American Oil Chemists Society.* DOI 10.1007/s11746-010-1674-y.

[a] Means of duplicate results, expressed as mg/100g extracted oil.
[b] Column means of three plants for each of nine fractions bearing different letters differ significantly at $p < 0.05$.
[c] Column means of nine fractions for each of three plants bearing different letters differ significantly at $p < 0.05$.

fermented mash were about three times those of corn, due to depletion of starch during fermentation. CDS had the highest element concentrations. However, the study did not include samples of cooked slurry, liquefied mash, or thin stillage. Liu and Han (2011) conducted a similar study but included all possible streams of dry grind processing, from corn to DDGS. They found that the changes in individual mineral content followed the changing pattern of protein, oil, and ash shown in Figure 8.5a. Fermentation caused the most dramatic increase in mineral content mainly due to depletion of starch. Upon centrifugation, more minerals went to the liquid fraction (thin stillage) than the solid fraction (DWG). They also showed that among processing streams, thin stillage (not CDS) had the highest levels of all minerals, while DWG had the lowest.

More importantly, several studies reported a larger fold (more than three) of increase over corn in Na, S, and Ca concentrations in some processing streams or DDGS, as compared to other minerals. Batal and Dale (2003) noticed that the content of most minerals in DDGS appeared generally consistent with a threefold concentration increase, but unusually larger range of values were noticed for Na and Ca. Belyea et al. (2006) also reported that in the stream of fermented mash, both Na and S had unusually higher increases in concentrations than other minerals. They attributed this to the addition of compounds which contained Na and S during the dry grind process. Liu and Han (2011) showed that Na, S, and Ca in fermented mash, whole stillage, and DDGS had much higher fold of increase over corn than other minerals (Table 8.8), presumably due to exogenous addition of compounds containing these minerals.

8.4.6 CHANGES IN VARIOUS FORMS OF PHOSPHORUS

The concentration and availability of phosphorous (P) in animal feed are the two most important factors that affect the retention of P in ingested feeds by animals and the amount of P excreted in wastes. P bioavailability is in turn determined by its chemical forms. Grains and their byproducts contain various forms of P, including inorganic P, phytate P, and the rest of P. Inorganic P (also known as phosphate P) has higher bioavailability than phytate P. The rest of P represents the sum of all P-containing compounds in a sample other than phytate P and inorganic P. It includes, for example, P found in DNA, RNA, proteins, lipids, and starch, and can be calculated by subtracting the sum of phytate P and inorganic P from total P.

Approximately two-thirds of total phosphorus in various grains is present as phytate or inositol hexaphosphate (Raboy, 1997), which is not well utilized by monogastric animals. Besides its low P availability, phytate has been shown to interact directly and indirectly with various dietary components to reduce their availability to animals (Morris, 1986). Theoretically, using low phytate ingredients in feeds should reduce P excretion, provided that available P levels in feeds containing these ingredients are appropriately adjusted downward.

Although reports on phosphorus content of both corn and DDGS are readily available (Spiehs et al., 2002; Batal and Dale, 2003), data on different forms of P as well as changes during dry grind processing of corn into DDGS are rather limited. Rausch et al. (2005b) monitored phosphorus concentrations and flow in corn wet milling streams but did not study streams of the dry grind process. Belyea et al. (2006) studied changes in element concentration in primary process streams from dry grind plants, but only total P content was measured (other forms of P were not).

Noureddini et al. (2009) analyzed total P in several streams of both dry grind and wet milling operations, and found that about 82% of total P in whole stillage went into the liquid fraction (thin stillage) and that CDS contained the highest phosphorus content (1.34%, dry matter) among selected streams of dry grind processing (corn, milled corn, whole stillage, CDS, DWG, DWGS, and DDGS). They further showed that about 59% of total P in whole stillage was phosphate P, and attributed the remaining P in whole stillage as phyate P. However, their HPLC analysis on this stream and its two centrifuged fractions did not reveal the presence of phyate. In contrast, Liu and Han (2011) showed that, in whole stillage, 48% of total P was phytate P, 25.19% was phosphate P, and the remaining 26.82% was contributed by the rest of P. When phosphate P and the rest of P are added together and

TABLE 8.8
Ratios of Streams versus Ground Corn (Fold of Increase) in Mineral Concentrations during Dry Grind Processing of Corn into Ethanol at Three Plants

Fraction Name	Plant No.	K	P	Mg	S	Na	Ca	Fe	Zn	Mn	Cu
Ground corn	1	1.00	1.00	1.00	1.00	1.00	1.00	1.00	1.00	1.00	1.00
	2	1.00	1.00	1.00	1.00	1.00	1.00	1.00	1.00	1.00	1.00
	3	1.00	1.00	1.00	1.00	1.00	1.00	1.00	1.00	1.00	1.00
	Mean	1.00[f]	1.00[f]	1.00[f]	1.00[f]	1.00[h]	1.00[e]	1.00[c]	1.00[e]	1.00[h]	1.00[d]
Cooked slurry	1	1.54	1.44	1.45	1.91	92.53	2.00	1.31	1.30	1.52	1.24
	2	1.54	1.45	1.47	1.52	87.76	2.10	1.09	1.23	1.41	1.20
	3	1.36	1.31	1.35	1.48	69.85	2.37	1.42	1.17	1.25	1.18
	Mean	1.48[e]	1.40[e]	1.43[e]	1.64[e]	83.38[g]	2.13[d]	1.27[c]	1.23[e]	1.39[g]	1.21[d]
Liquefied mash	1	1.50	1.26	1.38	1.67	84.74	2.09	0.97	1.08	1.30	1.08
	2	1.53	1.42	1.48	1.48	86.26	2.05	1.06	1.22	1.41	1.17
	3	1.29	1.28	1.26	2.27	69.82	2.24	0.95	1.15	1.25	1.17
	Mean	1.44[e]	1.32[e]	1.37[e]	1.80[e]	80.28[g]	2.11[d]	0.99[c]	1.15[e]	1.32[g]	1.14[d]
Fermented mash	1	4.03	3.70	3.70	7.40	250.45	5.23	2.00	3.09	3.61	3.27
	2	4.27	4.17	4.06	5.78	298.74	5.97	2.25	3.56	3.91	3.39
	3	4.22	3.95	3.94	7.30	394.92	8.77	3.26	3.51	3.90	3.63
	Mean	4.17[c]	3.94[c]	3.89[c]	6.82[bc]	314.70[d]	6.40[b]	2.45[b]	3.38[b]	3.81[d]	3.43[b]
Whole stillage	1	4.16	3.56	3.97	7.53	263.69	5.57	1.99	3.16	3.63	3.13
	2	4.43	4.01	4.17	6.17	520.33	6.05	2.58	3.57	4.58	3.31
	3	4.36	4.12	4.13	7.51	422.84	9.10	3.06	3.53	4.00	3.66
	Mean	4.31[c]	3.89[c]	4.09[c]	7.07[b]	402.29[c]	6.62[b]	2.50[b]	3.41[b]	4.07[c]	3.37[b]
Thin stillage	1	8.23	6.65	7.32	11.00	515.19	9.37	2.91	4.44	6.47	3.78
	2	8.08	6.85	7.33	7.24	859.08	9.03	3.33	4.32	5.97	3.87
	3	8.39	7.61	7.60	10.62	811.21	14.93	4.59	4.38	6.57	5.35
	Mean	8.23[a]	7.03[a]	7.41[a]	9.62[a]	728.49[a]	10.56[a]	3.55[a]	4.38[a]	6.34[a]	4.35[a]
Condensed solubles	1	7.12	5.63	6.33	11.30	483.41	8.72	3.74	4.01	5.62	3.31
	2	7.83	6.46	6.77	7.22	809.20	7.58	3.29	3.71	4.80	3.72
	3	8.41	7.20	7.14	10.70	805.83	12.44	3.61	3.08	4.77	5.24
	Mean	7.77[b]	6.41[b]	6.73[b]	9.74[a]	699.48[b]	9.13[a]	3.55[a]	3.60[b]	5.06[b]	4.11[a]
Wet grains	1	1.03	1.52	1.10	4.95	65.37	2.03	1.43	2.12	1.38	2.48
	2	1.18	1.67	1.38	4.46	141.97	3.30	1.98	2.63	2.23	2.82
	3	1.15	1.46	1.29	4.85	114.79	3.44	1.83	2.73	1.73	2.35
	Mean	1.12[f]	1.55[e]	1.25[e]	4.75[d]	107.38[f]	2.91[cd]	1.72[bc]	2.48[d]	1.77[f]	2.54[c]
DDGS	1	3.03	2.85	2.78	7.33	216.39	4.36	1.98	2.71	2.77	2.62
	2	2.80	2.99	2.95	5.62	278.32	5.20	3.03	3.17	3.47	3.05
	3	2.92	3.02	2.83	6.23	293.63	6.07	2.19	2.90	2.69	3.02
	Mean	2.92[d]	2.95[d]	2.85[d]	6.40[c]	262.78[e]	5.13[bc]	2.38[b]	2.92[c]	2.97[e]	2.90[c]

Source: From Liu, K. S. and J. C. Han. *Bioresource Technology* 102: 3110–3118, 2011.
Note: Column means of three plants for each of nine fractions bearing different letters differ significantly at p < 0.05.

considered as nonphytate P, the nonphytate P in DDGS was 56.30%. This value matches well with 54% reported by NRC (1994).

Liu and Han (2011) also determined the levels of different forms of P in all possible streams of dry grind processing, from corn to DDGS (Table 8.9). They found that, on average, milled corn contained 0.21% of phytate P, 0.03% of inorganic P, 0.05% of the rest of P, and 0.29% total P. There were

TABLE 8.9
Changes in Various Forms of Phosphorus (P) during Dry Grind Processing of Corn into Ethanol at Three Plants

Fraction Name	Plant No.	Phy P mg/g	Inorg P mg/g	Rest of P mg/g	Total P mg/g	Phy P/TP %	Inorg P/TP %	Rest P/TP %
Ground corn[a]	1	2.22[x]	0.26[x]	0.52[x]	3.01[x]	73.97[x]	8.76[x]	17.27[x]
	2	1.99[y]	0.23[y]	0.53[x]	2.75[x]	72.19[x]	8.47[x]	19.35[x]
	3	2.15[xy]	0.27[x]	0.49[x]	2.90[x]	74.11[x]	9.30[x]	16.59[x]
	Mean[b]	2.12[g]	0.26[f]	0.51[a]	2.89[g]	73	8.84[f]	17.74
	CV (%)	5.69	7.60	4.31	4.42	1.46	4.76	8.08
Cooked slurry	1	3.65	0.61	0.52	4.79	76.28	12.82	10.90
	2	3.19	0.81	0.54	4.54	70.30	17.85	11.85
	3	3.06	0.55	0.68	4.30	71.24	12.86	15.89
	Mean[b]	3.30[f]	0.66[e]	0.58[a]	4.54[ef]	72.61	14.51[e]	12.88[b]
	CV (%)	9.45	20.45	15.02	5.46	4.42	19.93	20.58
Liquefied mash	1	3.13	0.68	0.40	4.20	74.48	16.09	9.43
	2	3.25	0.71	0.54	4.50	72.28	15.79	11.93
	3	3.09	0.55	0.56	4.20	73.40	13.18	13.42
	Mean[b]	3.16[f]	0.65[e]	0.50[a]	4.30[f]	73.39	15.02[e]	11.59[b]
	CV (%)	2.69	12.72	18.08	4.00	1.50	10.67	17.38
Fermented mash	1	5.19	3.00	2.80	11.00	47.23	27.30	25.47
	2	5.42	2.73	3.85	12.00	45.13	22.78	32.09
	3	5.54	2.97	3.51	12.01	46.13	24.72	29.15
	Mean[b]	5.38[a]	2.90[b]	3.39	11.67[b]	46.16[b]	24.93[b]	28.9
	CV (%)	3.26	5.02	15.80	5.01	2.28	9.09	11.48
Whole stillage	1	5.24	2.95	2.92	11.11	47.18	26.54	26.28
	2	5.50	2.74	3.28	11.52	47.74	23.78	28.47
	3	6.37	3.28	3.34	12.99	49.05	25.24	25.71
	Mean[b]	5.71[b]	2.99[b]	3.18	11.88[b]	47.99[b]	25.19[b]	26.82
	CV (%)	10.40	9.11	7.18	8.35	2.00	5.47	5.43
Thin stillage	1	9.20	5.63	3.86	18.69	49.21	30.15	20.64
	2	9.40	6.62	2.99	19.01	49.47	34.81	15.72
	3	11.01	6.62	4.03	21.65	50.83	30.56	18.62
	Mean[b]	9.87	6.29	3.63	19.78	49.84	31.84	18.33
	CV (%)	10.03	9.05	15.43	8.22	1.74	8.11	13.51
Condensed solubles	1	7.67	6.08	3.25	17.01	45.11	35.76	19.13
	2	8.23	6.82	3.45	18.50	44.53	36.87	18.60
	3	9.96	7.42	3.38	20.76	48.00	35.74	16.26
	Mean[b]	8.62	6.77	3.36	18.75	45.88[ba]	36.12	18.00

continued

TABLE 8.9 (Continued)
Changes in Various Forms of Phosphorus (P) during Dry Grind Processing of Corn into Ethanol at Three Plants

	CV (%)	13.83	9.88	2.93	10.07	4.06	1.80	8.47
Wet grains	1	2.32	1.07	1.16	4.55	51.01	23.59	25.40
	2	2.59	0.91	1.72	5.23	49.63	17.48	32.89
	3	2.22	0.78	1.34	4.34	51.03	17.99	30.98
	Mean[b]	**2.38[g]**	**0.92[a]**	**1.41[b]**	**4.71[e]**	**50.56**	**19.69[a]**	**29.76**
	CV (%)	8.18	15.92	20.37	9.81	1.59	17.24	13.09
DDGS[a]	1	3.76[x]	3.04[x]	1.76[x]	8.56[y]	43.90[x]	35.55[x]	20.55[y]
	2	3.79[x]	2.56[y]	2.01[x]	8.35[y]	45.39[x]	30.72[y]	23.89[y]
	3	3.88[x]	2.64[y]	2.76[x]	9.28[x]	41.82[x]	28.45[y]	29.73[x]
	Mean[b]	**3.81[e]**	**2.75[b]**	**2.17**	**8.73[a]**	**43.70[a]**	**31.57**	**24.72**
	CV (%)	1.67	9.34	24.07	5.60	4.11	11.48	18.80
Mean of Fractions[a]	1	4.71[y]	2.59[y]	1.91[y]	9.21[z]	56.48[x]	24.06[x]	19.45[y]
	2	4.82[y]	2.68[xy]	2.1[xy]	9.60[y]	55.18[x]	23.17[y]	21.64[xy]
	3	5.25	2.79[x]	2.23[x]	10.27[x]	56.18[x]	22.00[z]	21.82[x]

Source: From Liu, K. S. and J. C. Han. *Bioresource Technology* 102: 3110–3118, 2011.

Note: Means of duplicate results. Concentration is expressed on dry weight basis. P = phosphorus, Phy = phytate, Inorg = inorganic, Rest P = the rest of P, TP = total P, CV = coefficient of variation.

[a] Column means for each of three plants bearing different xyz letters differ significantly at p < 0.05.

[b] Column means of three plants for each of nine fractions bearing different a–g letters differ significantly at p < 0.05.

little changes in P profile before saccharification. Upon fermentation, phytate P, inorganic P, and total P increased dramatically. Distillation caused little further change. Compared to thin stillage, DWG was much lower in all forms of P. Since CDS (resulting from concentrating thin stillage) was added back to DWG for drying, DDGS had similar composition and P profile as whole stillage.

Furthermore, Liu and Han (2011) found some interesting trends of P changes during dry grind processing. In terms of relative % of each form of P of the total P of, for the first few steps of the process, phytate P was about 73% of total P, while inorganic P and the rest of P collectively contributed the remaining 27%. Upon fermentation, phytate P decreased to about 46%, while inorganic P and the rest of P increased to about 27% each (doubled). In the remaining processing streams, contribution of phytate P toward total P fluctuated slightly. This observation indicates that upon fermentation phytate underwent some degradation, most likely due to the action of yeast phytase. In terms of fold of increase in concentration over corn, phytate P increased about 1.5-fold in cooked slurry and liquefied mash, similar to that of other minerals. However, upon fermentation, phytate P increased only 2.54-fold, as compared to three- to fourfold for most other minerals. Concomitantly, fold of increase for inorganic P concentration in these steps was much higher than that of most minerals: 2.58-fold in cooked slurry and 11.37-fold in fermented mash. In the final product (DDGS), the increase of phytate P was 1.80-fold that of corn, but inorganic P was 10.77-fold. The lower fold of increase in phytate P concentration over ground corn and much higher fold of increase in inorganic P, as compared to other minerals further indicated occurrence of phytate degradation during dry grind processing. Phytase is widely distributed in plants and microorganisms, including grains and fermentation yeast (Wodzinki and Ullah, 1996). Based on Shetty et al. (2003), it is possible that phytate is hydrolyzed during steps prior to fermentation by endogenous phytase of feedstock (corn) and/or during fermentation by yeast phytase. It is also possible that nonenzymatic hydrolysis of phytate occurs under harsh processing conditions (heat and/or pH changes) during the dry grind process. However, Liu and Han (2011) reasoned that activity of yeast phytase is most likely the major route for phytate degradation, since their study showed that only during fermentation were dramatic changes in phytate P and inorganic P observed in terms of both % relative to total P and fold of increase in concentrations over corn.

For years, the bioavailability of P in DDGS has been repeatedly shown to be significantly higher than that in corn (Amezcua et al., 2004; Pedersen et al., 2007). In one report for nutritional requirements of swine (NRC, 1998), the relative availability of P in corn was 14%. This value increased to 77% for DDGS produced from corn. Phytate degradation during the dry grind process, was supported by (1) observed decrease in % phytate P and concomitant increase in % inorganic P toward total P during fermentation; and (2) observed varying fold of increase over corn in concentrations of different forms of P, would account for improved P bioavailability in DDGS over corn. Most of the P in corn is bound in the phytate complex, so its bioavailability is very low. During dry grind processing, particularly the fermentation step, some of the bonds that bind P to the phytate complex have been hydrolyzed. This would make more P available for absorption and result in greater P bioavailability in DDGS compared to corn.

8.5 FACTORS AFFECTING CHEMICAL COMPOSITION OF DDGS

The average concentrations of various nutrients in DDGS available in the market can be quite variable with sources. This is evidenced by high CV values (Tables 8.1 through 8.3). For example, protein content in DDGS often has a CV in the range of 4.0%–6.4%. In comparison, protein in soybean meal generally has CV <2%. The nutrient composition of all feed ingredients varies, but using ingredients (such as DDGS) that are highly variable can reduce profitability of livestock operations because of increased feed costs and/or reduced production. Increased feed costs occur when diets are over supplemented to avoid the possibility of reduced production. Reduced production occurs when a diet does not contain adequate concentrations of a particular nutrient because a feed has less than anticipated concentrations of that nutrient. For example, Batal and Dale (2003) reported use of DDGS in laying hen formulas. In this incidence, use of the NRC (1998) value for sodium in the DDGS led to a severe deficiency with an almost total cessation of egg production. Subsequent analysis of the sodium content of the DDGS sample revealed 0.09% sodium in contrast to 0.48% as listed by the NRC (1998). Reformulation of the feed to provide intended level of sodium (0.18%) led to resumption of the level of egg production observed prior to initiation of the study. Therefore, in order to manage nutrient variation in DDGS, distilleries and ethanol plants need to provide average concentrations of major nutrients plus standard deviation by analyzing an adequate number of samples. Feed nutritionists need to properly characterize the composition of DDGS from respective suppliers the prior to incorporation into balanced rations.

The cause for varying DDGS composition can be multiple and has been the subject of many studies. They include, but are not limited to, differences in feedstock species, composition, processing methods (e.g., front-end or back-end fractionation), processing parameters (e.g., the amount of CDS added to DWG), effect of fermentation yeast, analytical methodology, etc. A better understanding of many causes for nutrient variations will help us develop strategies to control quality variation, and thus improve value-added utilization of DDGS.

8.5.1 GRAIN SPECIES, VARIETIES, AND BLENDS

For dry grind processing, variation in raw materials includes grain species, varieties, and blends. Even with the same species and varieties, there is variation in field conditions and production year, which can lead to compositional differences of feedstock. With regard to species, corn is by far the most common cereal grain used for ethanol production in the United States. However, in other parts of the world, other grains such as sorghum (Corredor et al., 2006), wheat (Ojowi et al., 1997; Nyachoti et al., 2005), pearl millet (Wu et al., 2006), and barley (Mustata et al., 2000), are also used.

Due to differences in composition among grains the resulting DDGS are expected to differ in composition and feeding value. Ortin and Yu (2009) compared wheat DDGS, corn DDGS, and blended DDGS from bioethanol plants and found great variation in chemical composition and nutritional values among them. Ethanol production from starch-rich crops other than corn and composition and values of the resulting DDGS are covered in Chapter 6 of this book.

Even with the same grain species, different varieties are sometimes used. For example, in Canada, many different classes and types of wheat can be used as feedstock for ethanol production. In general, soft wheat, either soft white or soft red class, is preferred to hard wheat because soft wheat generally contains higher starch content (Ojowi et al., 1997). Therefore, inconsistencies in the feedstock, ranging from variability in grain species and variety to variability in blends of different grains (corn, wheat, barley, etc.) are expected to have an effect on the nutritional characteristics of the DDGS produced. However, as discussed in the next section, some controversy exists in literature with regard to the effects of particle size and chemical composition of the same raw material (yellow dent corn) on those of DDGS.

8.5.2 Relationship between Ground Corn and DDGS in Particle Size and Composition

In the dry grind method for fuel ethanol production from corn, grinding is the first basic step as it reduces corn particle size by passing whole corn through a hammer mill containing screens with relatively small openings (3.2 to 4.8 mm diameter) (Bothast and Schlicher, 2005; Chapter 5 of this book). The resulting ground corn consists of a mixture of particles of different sizes. The particle size of ground corn is reported to affect fuel ethanol fermentation rate, ethanol yield, energy efficiency, and concentration of solids in thin stillage. Kelsall and Lyons (2003) reported that decreasing the particle size of ground corn from a relatively coarse grind (0.80 mm) to fine grind (0.48 mm) increased ethanol yield from 0.366 to 0.396 L/kg. However, they also suggested that decreasing the particle size excessively could have adverse affects on downstream processing steps, such as centrifugation. Guritno and Hague (1994) showed that reducing the particle size of different grains significantly increases energy consumption. Naidu et al. (2007) reported that the particle size of ground corn affected ethanol yield as well as concentration of solubles in thin stillage. In general, if particles are too large, starch granules are not gelatinized easily, enzyme access becomes limited, and all these lead to reduced production of fermentable sugars, resulting in lower ethanol yields. If particles are too small, the hammer mill needs more energy to grind corn, and finer particles also reduce centrifuge efficiency, and increase the amount of solids in thin stillage. Therefore, Kelsall and Lyons (2003) suggested that for efficient fermentation about 80% sample of ground corn should be 0.43 mm or larger, but particles larger than 0.84 mm or smaller than 0.25 mm should be no more than 10% each.

Although the effect of the particle size of raw material on ethanol fermentation has been generally established, its effect on the particle size of DDGS has not. Oftentimes, by intuition and common reasoning, corn processors believe that (1) PSD of ground corn affects that of DDGS; (2) the chemical and other physical properties of ground corn and DDGS are related to each other; and (3) variation in the composition of corn is a major cause of variation in the composition of DDGS. In many other processes (such as food processing and drug processing), raw material does show some effects on the final product quality. Yet, for fuel ethanol production by the dry grind process, information on the effect of raw material is not only limited but also contradictory.

Rausch et al. (2005a) compared PSD between ground corn and DDGS from dry grind processing. They found that mean geometric diameter (d_{gw}) values for nine ground corn samples and resulting nine DDGS samples were 0.94 and 0.92 mm, respectively, and thus concluded that the two types of samples were not significantly different from each other in particle size. They also found that the PSD of ground corn was not correlated (r = 0.35) to that of DDGS.

Belyea et al. (2004) compared the chemical composition of corn and DDGS produced in multiple years from a single plant and found that there were no significant correlations (r ranged of

−0.21–0.16) in nutrients (fat, protein, starch, crude fiber, and acid detergent fiber) between corn and DDGS. They concluded that variation in the chemical composition of DDGS was not related to the composition of corn used in fermentation but rather to variations in processing techniques. Later, the group (Belyea et al., 2006) studied element concentrations in primary process streams from nine dry grind plants. They found that the element content of corn was not different among plants, but the element content in several process streams varied with plants as well as streams. Since the nine plants used similar processing equipment to convert corn into ethanol and DDGS, they concluded that variations in element contents of DDGS and parent streams were due to processing conditions. Stein et al. (2009) found that energy and nutrient digestibility varied among sources of DDGS even when the DDGS was produced from ethanol plants that use corn grown within a narrow geographical region. Thus, factors other than corn growing region are responsible for the variability of these parameters in DDGS.

Yet, Liu (2009a) measured PSD, gross composition, and surface color of six ground corn samples and corresponding DDGS samples from different processing plants in the whole DDGS sample as well as in sieved fractions. In contrast to Rausch et al. (2005a), Liu (2009a) showed that the average geometric mean diameter (d_{gw}) of particles for the six ground corn and corresponding DDGS samples was 0.479 and 0.696 mm, respectively, and thus concluded that the d_{gw} of ground corn was significantly lower than that of corresponding DDGS. Change or improvement of processing methods over the span of several years might explain this discrepancy between the two studies. More importantly, the study of Liu (2009a) also showed that in terms of the geometric mean diameter (d_{gw}) of particles of the whole fraction and mass frequency of individual particle size classes, the relationship between ground corn and DDGS varied, but in terms of PSD, the two had a highly positive correlation (r = 0.807). Furthermore there were some positive correlations in contents of protein and nonstarch CHO and in L value (a color index) between corn and DDGS, while variations in nutrients and color attributes were larger in DDGS than in corn.

Results of Liu (2009a) disagreed with Belyea et al. (2004) and Rauch et al. (2005a) in certain aspects, but provided a scientific basis to partially support the common belief expressed by processors regarding relationships in quality parameters between corn and DDGS. Here, the support is partial since the study showed that raw material affected DDGS quality only to some extent, and that other factors, such as processing method and fermentation yeast, were also responsible for large variations in quality attributes of DDGS. The major reason causing the disagreement is that in both Belyea et al. (2004) and Rauch et al. (2005a), comparison and correlation were performed only in the whole sample between ground corn and DDGS, but in Liu (2009a), linear regression was also conducted for attributes measured in all sieved fractions between ground corn and DDGS.

8.5.3 Effects of Method Modifications

In a conventional dry grind process, corn is ground using a hammer mill; water is added to produce slurry that is to be fermented. Since northing is removed from the corn, the fermentable mash mainly consists of starch, protein, germ, and fiber fractions. Of these four fractions, only starch is fermentable, the other three fractions remain relatively unchanged and end up in the DDGS. Yet, over the years, several modified methods have been developed, with a main variation in fractionation either before (i.e., front end) or after (i.e., back end) the fermentation step to remove one or more nonfermentable fractions. As a result, not only ethanol production efficiency is improved, but also the chemical composition of DDGS is significantly altered (Singh et al., 2005; Wang et al., 2005; Khullar et al., 2009). The modified DDGS generally has higher protein, lower oil, and/ or lower fiber contents than conventional DDGS. For example, in one report (Singh et al., 2005), a modified process in a laboratory setting was found to reduce the fiber content of DDGS from 11% to 2% and increased the protein content of the DDGS from 28% to 58%. The following provides some discussion on the topic; additional information can be found in Chapter 5.

The front-end fractionation is further divided into wet and dry methods. The wet fractionation methods include the "quick germ" method, in which germ can be recovered from the mash by using a corn wet milling degermination process (Singh and Echhoff, 1996), and the "quick germ and quick fiber" method, in which corn was soaked in water, both germ and pericarp fiber were removed before fermentation (Singh et al., 1999; Ponnampalam et al., 2004). In addition, Wang et al. (2005) reported modifications based on front-end fractionation, which included treatment of corn slurry with enzymes and/or using a new granular starch hydrolyzing enzyme STARGEN 001. STARGEN 001 can convert starch into dextrins at low temperatures as well as hydrolyze dextrins into fermentable sugars. Robinson et al. (2008) evaluated nutritional composition of four types of distillers grains resulting from different processing methods, conventional DDGS, DDGS using BPX technology (raw starch hydrolysis), high protein distillers grains (with most fiber and germ removed prior to fermentation), and dehydrated dry milled corn germ. They found significant differences among the four samples for many of the nutrients measured.

For front-end dry fractionation, Murthy et al. (2005) reported a method for processing corn into ethanol as follows: The grain is tempered, degermed, and passed through a roller mill. Ground corn is sieved to separate germ and fiber fractions from the endosperm, and the remaining endosperm is processed by conventional dry grind ethanol methods to produce ethanol. Corredor et al. (2006) investigated the effect of decortication as a pretreatment method on ethanol production from sorghum, as well as its impact on the quality of the resulting DDGS. Eight sorghum hybrids with 0%, 10%, and 20% of their outer layers removed were used as raw materials for ethanol production. The decorticated samples were fermented to ethanol. Removal of germ and fiber before fermentation allowed for greater starch loading for ethanol fermentation and resulted in increased ethanol production. Ethanol yields increased as the percentage of decortication increased. They concluded that the decortication process resulted in DDGS with higher protein and lower fiber contents. Wang et al. (1999) carried out a similar work with rye and triticale as feedstocks for ethanol production and found that partial removal of outer grain solids by pearling in an alcohol plant would improve plant efficiency and decrease energy requirements for mash heating and cooling, and ethanol distillation. They did not look at the effect of pearling on DDGS quality.

Back-end fractionation refers to removal of oil, fiber, and other valuable components from ethanol coproducts at any stage after fermentation. It not only modifies DDGS composition, but also results in value-added products. The methods can also be divided into dry and wet. The back-end dry fractionation is limited to removing fiber from DDGS by sieving (Wu and Stringfellow, 1986), sieving and elutriation (Srinivasan et al., 2005), or sieving followed by winnowing (Liu, 2009b), while reports on wet methods of back-end fractionation focus mostly on removing oil from various types of coproducts. They include removing oil from DDGS through ethanol extraction (Singh and Cheryan, 1998), from thin stillage through centrifugation (Wang et al., 2008) and from CDS through heating to a high temperature and centrifugation with a disk stack centrifuge (Cantrell and Winsness, 2009). Conventional DDGS contains about 12% oil on dry matter basis. Although the presence of oil increases the energy density of DDGS as livestock feed, it may interfere with normal milk production by dairy cattle, and may impact fat deposits and meat quality in swine. Therefore, partial removal of oil from DDGS will improve its feed quality. More importantly, the oil recovered can be used as a feedstock for biodiesel production (Chpater 23). The back-end recovery process also requires much less capital to build and has lower operating costs than the front-end fractionation. The subject of DDGS wet fractionation is to be covered separately in Section 8.6 of this chapter, while the subject of DDGS dry fraction is to be covered in Chapter 25.

8.5.4 PROCESSING PARAMETERS

Even with the same dry grind method, changes in some processing parameters or practices can cause significant changes in chemical composition and nutritional properties of DDGS (Belyea et al., 2004, 2006; Kingsly et al., 2010). For example, during dry grind processing, some plants use sodium hydroxide for sanitation of processing equipment and sulfuric acid (sulfur) for pH adjustment. This not only causes larger variations for the contents of Na and/or S in DDGS, but also creates nutritional challenges when included in some animal diets (Belyea et al., 2006).

Much of the variation in nutrient contents is related to the drying step (Kingsly et al., 2010). Uneven mixing and variability in the quantity and quality of CDS added back to the DWG during drying will certainly affect the nutrient content of resulting DDGS. Fluctuation in the ratio of CDS to DWG entering the dryer occurs as the rates are often adjusted in order to improve the drying characteristics. The ratio may vary from batch to batch and from plant to plant (Belyea et al., 2004, 2006). Aggregation and lumping during the drying process often occur if the ratio is too high. Also, during drying, DDGS is subject to high temperature conditions, which may result in reduced protein quality despite the high overall crude protein content. More about drying is covered in Chapter 5.

8.5.5   EFFECTS OF FERMENTATION YEAST

Yeast is one of the least expensive raw materials for dry grind processing. Yet, it is an important ingredient for fuel ethanol production. A healthy and well-selected strain of yeast is needed for an efficient fermentation. It can also potentially affect the final product quality. As early as 1944, Bauerfeind et al. defined corn DDS as a grain-yeast concentrate comprising the water-soluble nutrients derived from the original grains and from the grain-yeast fermentation. Thus, there is no doubt that DDGS proteins come from corn and yeast. Yet, the effect of fermentation yeast on DDGS protein quantity and quality (AA profile) has not been well documented.

Based on literature search, there are at least four methods that are described to estimate yeast contribution toward distillers grains products. As discussed below, the estimate results vary greatly across methods. Since major factors affecting DDGS quality and market values are protein quantity (concentration) and quality (amino acid composition), and since yeast AA profile is better than that of corn, investigation into yeast effect and accurate estimation of yeast contribution will have a positive impact on the feed and ethanol industries, and at the same time increase our basic understanding of the processing system.

Bauerfeind et al. (1944) suggested that yeast cell (all dead) content can be estimated by hemacytometer counts of thin stillage, condensed solubles, or dried solubles. They reported that DDS contained about $4 \pm 0.5 \times 10^9$ cells per gram. When this figure was compared to that of dried yeast, the approximation was reached that 20% by weight of dried solubles is dried yeast. The method may not be applicable directly to DDGS unless a separation of solubles is carried out or a portion of solubles is first estimated or assumed.

Ingledew (1999) used an assumption and calculation approach to estimate the amount that yeast contributed toward DDGS by both mass and protein. He stated that in the late 1990s, the annual fuel ethanol production in North America was 7×10^9 L. Assuming that the fermentation mash contains 12% alcohol, the total mash would be 6×10^{10} L. Assuming that at the peak of fermentation, yeast count in a fermentor is 1.9×10^{11} cells per L mash, the total yeast cells in total annual mash production would be $6 \times 10^{10} \times 1.9 \times 10^{11} = 1.14 \times 10^{22}$. Assuming that 1 g dry yeast contains 4.87×10^{10} cells, the total annual mash would contain $(1.14 \times 10^{22})/(4.87 \times 10^{10}) = 2.34 \times 10^{11}$ g = 2.34×10^5 metric tons (MT) of yeast biomass. Assuming that average yeast contains 38% protein, the total annual mass would contain $2.34 \times 10^5 \times 38\% = 88,920$ MT yeast protein. Assume that 1000 L ethanol leads to 860 kg DDGS, the total annual DDGS would be $(7 \times 10^9$ L)/1000 L $\times 860$ kg = 6×10^6 MT. Assume that the average protein content in DDGS is 28%, the total annual DDGS protein was $6 \times 10^6 \times 28\% = 1.68 \times 10^6$ MT. Therefore, yeast contribution by mass $(2.34 \times 10^5)/(6 \times 10^6) = 3.9\%$; and yeast contribution by protein $88,920/(1.68 \times 10^6) = 5.3\%$. Since the approach was based on many assumptions, the accuracy of the final estimation is uncertain.

Belyea et al. (2004) calculated the average ratio of AA concentrations (based on dry sample weight) of DDGS versus yeast and suggested that yeast contribute up to 50% of DDGS protein. This approach apparently had some shortcomings: (1) it disregarded corn protein contribution; (2) it did not include nonessential AA; and (3) the average ratio in AA concentrations actually reflected the ratio of protein concentration of DDGS versus yeast.

Recently, Han and Liu (2010) proposed a multiple lineat regression model based on changes in amino acid profile in terms of relative % during the entire process of dry grind ethanol production. As discussed earlier, when amino acid composition was expressed on dry sample weight basis, the change in AA concentrations, either essential or nonessential, followed the pattern of protein changes during the dry grind process (Table 8.5). However, when expressed as relative % (based on total AA), the change of AA profile (Table 8.6), unlike AA concentration, did not follow the trend of protein change. Upon fermentation concentrations of some AA concentrations increased, others decreased, still others remained unchanged. This is because the expression of AA in terms of relative % to total AA focuses more on quality of protein rather than on quantity. More importantly, when amino acid composition is expressed as concentration in dry samples (Table 8.5), there was little information about influence of yeast AA composition on upper and downstream products (including DDGS). However, when AA is expressed as relative % (protein based) (Table 8.6), the influence of yeast AA on that of a downstream products becomes clear. For example, Arg in corn was 3.55%, in yeast 5.15%, so the trend was increasing. For example, Met in corn was 3.32%, in yeast, 2.14%, so the trend was decreasing. Although changing trends in AA composition depend mostly on the difference between yeast and corn AA compositions, there were some exceptions. For example, for Pro, there was no clear pattern of change during processing, but yeast had much lower value than corn (2.18% vs. 6.67%).

The above observations were based on data measured on stream products from Plant 1. The changing patterns of amino acid profile in stream products of dry grind process from two other Plants (2 and 3) were very similar to those found with samples of Plant 1 (data not shown). This confirms all the observations on change in AA composition during the dry grind process of corn with samples from Plant 1.

Thus, Han and Liu (2010) proposed that AA composition (% relative) of a stream (response variable) is a function of AA of corn (independent variable 1) and AA of yeast (independent variable 2), based on a multiple linear regression model:

$$Y = AX_1 + BX_2 + C$$

where Y = relative % of an amino acid in a stream; X_1 = relative % of the AA in ground corn; X_2 = relative % of the AA in yeast; A = a fixed value parameter indicating the extent of contribution by corn AA; B = a fixed value parameter showing the extent of influence by yeast AA; C = a fixed value parameter showing the intercept on the Y-axis.

According to the above proposed model, regression results show that parameters A, B, and C, varied greatly with the type of stream products, but only slightly with the sample source (Plant No.) (Table 8.10). When regression was conducted on the combined data set of three plants, results show that before fermentation, the value of parameter A was about 0.92, and B was around 0.05. After fermentation, value A was reduced to about 0.84 and value B increased to about 0.20. This implies that the average AA composition for fermented mash from all three plants would increase by a factor of 0.84 if AA composition of ground corn increased by 1% and AA composition of yeast remained fixed. Similarly, a 1% increase in yeast AA composition, with corn AA held fixed, would now increase mean AA of fermented mash by a factor of 0.20. Furthermore, upon centrifugation, the B value increased in thin stillage and CDS, but decreased in DWG. The A value changed accordingly, decreasing in thin stillage and CDS but increasing in DWG. The two parameters in both DWGS and DDGS became similar to those found in whole stillage and fermented mash.

Based on multiple linear regression of amino acid composition (% relative) of DDGS with those of ground corn and yeast as two independent variables, the yeast accounted for about 20% of effect on DDGS amino acid profile, while corn accounted for the remaining 80%. The method was also useful to estimate the effect of yeast on AA profile of intermediate streams. For example, before

TABLE 8.10

Multiple Linear Regression for Amino Acid Composition (Relative % of Individual Amino Acids in Each Sample) of Intermediate Streams and DDGS of Dry Grind Processing from Three Commercial Plants, with Milled Corn and Yeast AA Composition as Variants X_1 and X_2, Respectively

Sample Source	Parameter	Raw Slurry	Cooked Slurry	Liquefied Mash	Saccharified Mash	Fermented Mash	Whole Stillage	Thin Stillage	Condensed Solubles	Wet Grains	DWGS	DDGS	Average
Plant 1	A		0.783	0.967	0.840	0.840	0.745	0.428	0.535	0.924	0.783	0.843	0.769
	B		0.162	0.013	0.121	0.191	0.304	0.482	0.366	0.198	0.258	0.244	0.234
	C		0.332	0.116	0.240	−0.188	−0.295	0.531	0.580	−0.719	−0.244	−0.516	−0.016
	r^2		0.965	0.993	0.980	0.963	0.965	0.926	0.945	0.963	0.949	0.962	0.961
Plant 2	A	0.842	0.969	0.947		0.821	0.794	0.565	0.476	0.892		0.895	0.800
	B	0.151	0.000	0.048		0.177	0.224	0.330	0.433	0.199		0.156	0.191
	C	0.048	0.165	0.030		0.019	−0.109	0.617	0.545	−0.530		−0.320	0.052
	r^2	0.975	0.985	0.984		0.940	0.965	0.940	0.949	0.942		0.940	0.958
Plant 3	A		1.026	0.845		0.846	0.885	0.555	0.444	1.061	0.924	0.939	0.836
	B		−0.038	0.218		0.218	0.175	0.354	0.426	0.082	0.157	0.178	0.197
	C		0.062	0.136		−0.378	−0.361	0.535	0.767	−0.847	−0.484	−0.728	−0.144
	r^2		0.996	0.960		0.959	0.963	0.952	0.929	0.947	0.963	0.956	0.958
Combined	A		0.918	0.926		0.835	0.802	0.513	0.488	0.950	0.843	0.889	0.796
	B		0.046	0.060		0.196	0.238	0.390	0.407	0.165	0.214	0.195	0.212
	C		0.208	0.080		−0.180	0.240	0.571	0.622	−0.680	−0.343	−0.511	0.001
	r^2		0.977	0.978		0.954	0.963	0.936	0.940	0.949	0.955	0.951	0.956

Source: From Han, J. C., and K. S. Liu. *Journal of Agricultural and Food Chemistry* 58: 3430–3437, 2010.

Note: Regression was based on a multiple linear model $Y = AX_1 + BX_2 + C$, Where Y = relative % of an amino acid in an downstream product; X_1 = relative % of the AA in milled corn; X_2 = relative % of the AA in yeast; r^2 = the square of the correlation coefficient.

fermentation, yeast accounted for about 5% (due to recycle of thin stillage), but after fermentation its contribution increased to about 20%. This value was higher in thin stillage and CDS, but lower in DWG. Since yeast had a better AA profile than corn (particularly with regard to Lys), the higher the yeast effect, the better the DDGS AA profile as compared to that of corn.

Among the four methods described, the proposed model of multiple regression model by Han and Liu (2010) is believed to be the most accurate estimation for the effect of yeast on the AA profile of DDGS. First, it links DDGS AA profile as a function of both corn and yeast AA profiles. Second, it includes all amino acids. Third, it is based on relative %, rather than absolute concentrations of AA. The latter is affected by the protein content in the sample. Fourth, it can estimate the yeast effect on not only DDGS but also any intermediate streams. Finally, unlike the previous three methods, which focus on how much protein in DDGS is of yeast origin, the regression approach focuses on the impact of yeast and corn AA profiles on that of DDGS and intermediate streams.

8.5.6 Analytical Methodologies

The use of various analytical methods for DDGS has, in part, led to significant variation of reported compositional values among laboratories, and therefore created confusion for producers, marketers, nutritionists, regulatory bodies, and end users. For example, Ileleji et al. (2010) reported that the various methods that have been used for moisture determination of DDGS did not give identical results, and therefore caution should be exercised when selecting a method for determining moisture in DDGS. One key factor leading to the situation of using various methods is lack of standardized protocols for characterizing DDGS composition and quality. To address the issue, Thiex (2009) conducted a study to evaluate analytical methods for DDGS. The study was commissioned by the American Feed Industry Association (AFIA), Renewable Fuels Association (RFA), and National Corn Growers Association (NCGA), since with increasing production of DDGS, both fuel ethanol and animal feed industries are demanding standardized protocols for analytical methodology. Details of this work as well as recommended analytical methodologies for DDGS are covered in Chapter 10.

Another key factor leading to the current situation is lack of systematic studies that have investigated factors affecting oil analysis of DDGS. Although DDGS is derived from milled corn, and both milled corn and DDGS are dry particulate materials, the two have different chemical compositions and physical matrices. Therefore, their responses to factors affecting analytical methods are expected to be different. Considering oil analysis as an example, recently, Liu (2010) used AOCS Approved Procedure, Am 5–04 (AOCS 2005) for measuring crude oil content in both milled corn and resulting DDGS, and found that for the crude oil analysis by the AOCS method, particle size had no effect for milled corn but it had the most significant effect for DDGS among other factors. DDGS, with the larger particle size (compared to those of the original matrix), tends to have significantly lower measured values of crude oil content than samples with reduced particle sizes, when other analytical conditions are kept the same (Table 8.11). On average, the measured oil content in DDGS ranged from 11.11% (original matrix), to 12.12% (<0.71 mm) and to 12.55% (<0.50 mm). It is commonly believed that there is a strong relationship between surface area and solvent extraction efficiency. The smaller size of particles, the greater surface area, and thus the greater extraction efficiency would be. However, the effect of particle size on crude oil analysis cannot be fully explained by the increase in surface area of particles, since the same study showed that for milled corn, particle size had no significant effect (Table 8.11). It is presumably attributed to the differences in chemical composition and physical matrix between raw corn and DDGS.

8.5.7 Other Factors

Improper sampling can be another factor causing variation in DDGS composition, since only a very small portion of sample is typically used for chemical analysis. Aggregation and inconsistent physical

TABLE 8.11
Least Square Means of Crude Oil Content for Levels of Sample Type (milled corn or DDGS) × Plant No. (1, 2, or 3) × Particle size (original matrix, <0.71 mm, or <0.50 mm)

Level	Mean (%)	Level	Mean (%)
Milled corn, 1, Original matrix	2.892[h]	DDGS, 1, Original matrix	11.384[e]
Milled corn, 1, <0.71 mm	3.146[h]	DDGS, 1, <0.71 mm	12.819[b]
Milled corn, 1, <0.50 mm	3.114[h]	DDGS, 1, <0.50 mm	13.294[a]
Milled corn, 2, Original matrix	3.706[g]	DDGS, 2, Original matrix	11.409[e]
Milled corn, 2, <0.71 mm	3.693[g]	DDGS, 2, <0.71 mm	11.932[d]
Milled corn, 2, <0.50 mm	3.623[g]	DDGS, 2, <0.50 mm	12.428[c]
Milled corn, 3, Original matrix	2.116[i]	DDGS, 3, Original matrix	10.533[f]
Milled corn, 3, <0.71 mm	2.231[i]	DDGS, 3, <0.71 mm	11.624[e]
Milled corn, 3, <0.50 mm	2.220[i]	DDGS, 3, <0.50 mm	11.956[d]
Mean of 3 plant samples with a same particle size range			
Milled corn, original matrix	**2.90[d]**	DDGS, original matrix	**11.11[c]**
Milled corn, <0.71 mm	**3.02[d]**	DDGS, <0.71 mm	**12.12[b]**
Milled corn, <0.50 mm	**2.99[d]**	DDGS, <0.50 mm	**12.55[a]**

Source: Adapted from Liu, K. S. *Cereal Chemistry* 87(3): 243–249, 2010.
Note: Degrees of freedom = 4, standard error = 0.05373, % dry matter basis.
Column means bearing different letters differ significantly at $p < 0.05$.
Original matrix = control, unground samples; DDGS = distiller dried grains with solubles.

characteristics of the DDGS also make it challenging to obtain a truly representative sample from such a small quantity of material (Clementson et al., 2009). Thus, sampling of the material must include a large pooled sample, comprised of multiple samples per batch throughout the production process.

8.6 WET FRACTIONATION OF DDGS

The major components in DDGS are protein, oil, and carbohydrates. Fractionation of DDGS into value-added products may improve the economic viability of dry grind corn ethanol facilities in the wake of variable corn and ethanol prices. In Section 8.5.3, back-end fractionation was briefly covered, while dry fractionation of DDGS is covered separately in Chapter 25. This section covers only wet fractionation of DDGS, which is a part of an overall strategy of back-end fractionation. Additional information on both front-end and back-end fractionation can also be found in Chapter 5.

Brehmer et al. (2008) investigated protein separation techniques to produce higher value-added product options for distillers grains to improve the corn-ethanol industry. This would allow additional utilization of the cellulosic components and separation of the proteins from DDGS for use as chemical precursors. They treated several different corn feedstock layouts with second generation ammonia fiber expansion (AFEX) pretreatment technology and tested for protein separation options (protease solubilization). They concluded that the resulting system has the potential to greatly improve ethanol yields, lower bioprocessing energy costs, and can satisfy a significant portion of the organic chemical industry.

The same group also conducted research aimed at creating a high protein, high lysine product from the grain byproduct using alkaline protein extractions in conjunction with hydrolysis of the remaining fiber to sugars, which are then fermented to ethanol (Bals et al., 2009). They found that alkaline extractions improved the lysine content in protein products, although protein solubility did not exceed 45% of the total protein. However, oligomeric carbohydrates, starch, and other water solubles were also extracted, leading to a low purity protein product. They concluded that alkaline extraction is unlikely to be a useful tool for fractionation of distiller's grains. In a follow-up study, Datta et al. (2010) attempted to simultaneously extract proteins from and enzymatically saccharify cellulosic materials in DDGS, using food-grade biobased solvents (ethyl lactate, D-limonene, and methyl esters). This approach would produce a high-value animal feed while simultaneously producing additional sugars for ethanol production. Their preliminary experiments on protein extraction resulted in recovery of 15%–45% of the protein, with hydrophobic biobased solvents obtaining the best results. The integrated hydrolysis and extraction experiments showed that biobased solvent addition did not inhibit hydrolysis of the cellulose. However, only 25%–33% of the total protein was extracted from DDGS, and the extracted protein largely resided in the aqueous phase, not the organic solvent phase.

In a similar study, Cookman and Glatz (2009) reported protein extractions based on aqueous ethanol, alkaline-ethanol, and aqueous enzyme treatments, and found that all three methods extracted a significant amount of the protein from dried, defatted distillers grains (DDDG). Comparatively, the enzymatic extraction was effective for both milled and unmilled DDDG. The alkaline-ethanol extraction was similarly effective for milled but not unmilled distillers grains. Simple extraction with alcohol was not as effective. The AA profile of each protein extract was consistent and resembled that of zein. For the protease-assisted extractions, 95% the proteins were in the form of peptides smaller than 10 kDa.

The third research group (Wang et al., 2009) extracted kafirin proteins from sorghum DDGS with various extraction methods including use of acetic acid, HCl-ethanol, and NaOH-ethanol under reducing conditions, and achieved extraction yields of 44.2%, 24.2%, and 56.8%, respectively. They also found that extraction conditions affected purity and thermal properties of the extracted kafirin proteins. Acetic acid and NaOH-ethanol extracted protein with higher purity (98.94% and 94.88%, respectively) than kafirins extracted with the HCl-ethanol (42.32% protein), while the original material (sorghum DDGS) contained 35.47% protein. The acetic acid and HCl-ethanol based extraction methods tended to extract more high-molecular weight protein than the NaOH-ethanol based method, while the γg kafirins were found only in extracts from the NaOH-ethanol extraction system.

The use of solvent-based methods to recover DDGS proteins has been a relatively new attempt. Such an effort, although still very limited, may lead to efficient recovery and production of purified protein products that have high-value utilization. For example, the extracted proteins may be used for many bioindustrial applications such as adhesives and resins. The residue after protein extraction is a carbohydrate rich product, which is better suited for conversion to fermentable sugars. Although fractionation of DDGS into value-added products may serve to improve the economic viability of dry grind ethanol plants, its potential is yet to be seen. Since during the conventional dry grind process grain proteins are subjected to some harsh conditions, including heating, pH adjustment and fermentation, some of them are denatured or tightly bound to other molecules in the final DDGS products. Wet extractions, based on above reports, have so far had limited effect in obtaining protein products with high recovery yield. In addition, the use of harsh and heavy solvent systems, such as strong acid or alkaline solvents, makes the methods not only costly but also less environmentally friendly.

8.7 CONCLUSIONS

Marketability and suitable uses of DDGS are keys to the economic viability of fuel ethanol production. As the industry grows, the importance of distillers grains has also increased. In this chapter, several topics, which are crucial to the use of DDGS, have been discussed with updated information, ranging from nutrient levels in DDGS to their variations among reports and within the same report, and from compositional changes during the entire dry grind process to analysis of several key factors causing high compositional variation.

During dry grind processing, starch is converted to glucose and then to ethanol and carbon dioxide. All other components in the original grain feedstock remain relatively unchanged. As a result, all nutrients other than starch in DDGS are concentrated about threefold over the initial grains. Thus, DDGS is nutritious as a livestock feed ingredient, or even for aquafeed and pet food. Yet one key challenge for using DDGS as a feed ingredient is its large variation in nutrient levels compared to some other feed ingredients, such as soy meal. The main factor causing the large variation is the dry grind process itself since it is more complex than processes of other feed protein ingredients (such as oilseeds), and entails more steps and more variables in processing conditions by the same method. Using different grain species or their blends as feedstock adds another factor. For better utilization of DDGS, both suppliers and end users need to better understand factors that cause variation and develop strategies to decrease it.

REFERENCES

Abbott, J., J. Opalka, and C. F. McGuire. 1991. Dried distillers' grains with solubles: Particle size effects on volume and acceptability of baked products. *Journal of Food Science* 56(5): 1323–1326.

Amezcua, C. M., C. M. Parsons, and S. L. Noll. 2004. Content and relative bioavailability of phosphorus in distillers dried grains with solubles in chicks. *Poultry Science* 83: 971–976.

AOCS (American Oil Chemists Society) 2005. *Approved Procedure Am 5–04, Rapid Determination of Oil/Fat Utilizing High Temperature Solvent Extraction*. Urbana, IL: American Oil Chemists Society.

ASAE (American Society of Agricultural Engineers) Standards. 2003. *Methods for Determining and Expressing Fineness of Feed Materials by Sieving. S319.3*. St. Joseph, MI: American Society of Agricultural Engineers.

Bals, B., V. Balan, and B. Dale. 2009. Integrating alkaline extraction of proteins with enzymatic hydrolysis of cellulose from wet distiller's grains and solubles. *Bioresource Technology* 100: 5876–5883.

Barbosa-Canovas, G. V., E. Ortega-Rivas, P. Juliano, and H. Yan. 2005. *Food Powders, Physical Properties, Processing, and Functionality*. New York, NY: Kluwer Academic/Plenum Publishers.

Batal, A. B., and N. M. Dale. 2003. Mineral composition of distillers dried grains with solubles. *Journal of Applied Poultry Research* 12: 400–403.

Batal, A. B., and N. M. Dale. 2006. True metabolizable energy and amino acid digestibility of distillers dried grains with solubles. *Journal of Applied Poultry Research* 15: 89–93.

Bauernfeind, J. C., J. C. Garey, W. Baumgarten, L. Stone, and C. S. Boruff. 1944. Nutrient content of alcohol fermentation by-products from corn. *Industrial & Engineering Chemistry* 36: 76–78.

Belyea, R. L., K. D. Rausch, and M. E. Tumbleson. 2004. Composition of corn and distillers dried grains with solubles from dry grind ethanol processing. *Bioresource Technology* 94: 293–298.

Belyea, R. L., S. R. Eckhoff, M. A. Wallig, and M. E. Tumbleson. 1998. Variation in the composition of distillers solubles. *Bioresource Technology* 66: 207–212.

Belyea, R. L., T. E. Clevenger, V. Singh, M. Tumbleson, and K. D. Rausch. 2006. Element concentrations of dry-grind corn-processing streams. *Applied Biochemistry and Biotechnology* 134(2): 113–128.

Bothast, R. J., and M. A. Schlicher. 2005. Biotechnological processes for conversion of corn into ethanol. *Applied Microbiology and Biotechnology* 67: 19–25.

Brehmer, B., B. Bals, J. Sanders, and B. Dale. 2008. Improving the corn-ethanol industry: Studying protein separation techniques to obtain higher value-added product options for distillers grains. *Biotechnology and Bioengineering* 101: 49–61.

Cantrell, D. F., and D. J. Winsness. 2009. Method of processing ethanol byproducts and related subsystems. *U.S. Patent No. 7,601,858B2*. October 13.

Clementson, C., K. E. Ileleji, and R. L. Stroshine. 2010. Particle segregation within a pile of bulk distillers dried grains with solubles (DDGS) and variability of nutrient content. *Cereal Chemistry* 86(3): 267–273.

Cookman, D. J., and C. E. Glatz. 2009. Extraction of protein from distiller's grain. *Bioresource Technology* 100(6): 2012–2017.

Corredor, D. Y., S. R. Bean, T. Schober, and D. Wang. 2006. Effect of decorticating sorghum on ethanol production and composition of DDGS. *Cereal Chemistry* 83: 17–21.

Cromwell, G. L., K. L. Herkelman, and T. S. Stahly. 1993. Physical, chemical, and nutritional characteristics of distillers dried grains with solubles for chicks and pigs. *Journal of Animal Science* 71: 679–686.

Datta, S., B. D. Bals, Y. P. J. Lin, M. C. Negri, R. Datta, L. Pasieta, S. F. Ahmed, A. A, Moradia, B. E. Dale, and S. W. Snyder. 2010. An attempt towards simultaneous biobased solvent based extraction of proteins and enzymatic saccharification of cellulosic materials from distiller's grains and solubles. *Bioresource Technology* 101: 5444–5448.

Dowd, M. K. 1993. Low molecular weight organic composition of ethanol stillage from corn. *Cereal Chemistry* 70: 204–209.

Fastinger, N. D., J. D. Latshaw, and D. C. Mahan. 2006. Amino acid availability and true metabolizable energy content of corn distillers dried grains with solubles in adult cecectomized roosters. *Poultry Science* 85: 1212–1216.

Gould, D. H. 1998. Polioencephalomalacia. *Journal of Animal Science* 76: 309–314.

Guritno, P., and E. Haque. 1994. Relationship between energy and size reduction of grains using a three-roller mill. *Transactions of the ASAE* 37(4): 1243–1248.

Han, J. C., and K. S. Liu. 2010. Changes in proximate composition and amino acid profile during dry grind ethanol processing from corn and estimation of yeast contribution toward DDGS proteins. *Journal of Agricultural and Food Chemistry* 58: 3430–3437.

Ingledew, W. M. 1999. Yeast—Could you base a business on this bug? In *Biotechnology in the Feed Industry, Proceedings of Alltech's Fifteenth Annual Symposium*, eds. T. P. Lyons, and K. A. Jacques, pp. 27–47. Nottingham, UK: Nottingham University Press.

Ileleji, K. E., K. S. Prakash, R. L. Stroshine, and C. L. Clementson. 2007. An investigation of particle segregation in corn processed dried distillers grains with solubles (DDGS) induced by three handling scenarios. *Bulk Solids & Powder—Science & Technology* 2: 84–94.

Ileleji, K. E., A. A. Garcia, A. R. P. Kingsly, and C. L. Clementson. 2010. Comparison of standard moisture loss-on-drying methods for the determination of moisture content of corn distillers dried grains with solubles. *Journal of AOAC International* 93: 825–832.

Jacques, K. A., T. P. Lyons, and D. R. Kelsall. 2003. *The Alcohol Textbook,* 4th ed. Nottingham, UK: Nottingham University Press.

Kelsall, D. R., and T. P. Lyons. 2003. Grain dry milling and cooking procedures: Extracting sugars in preparation for fermentation, Ch. 2. In *The Alcohol Textbook*, 4th ed., eds. K. A. Jacques, T. P. Lyons, and D. R. Kelsall, pp. 9–22. Nottingham, UK: Nottingham University Press.

Kent, N. L., and A. D. Evers. 1994. *Kent's Technology of Cereals,* 4th ed. Oxford: Pergamon Press Ltd.

Khullar, E., E. D. Sall, K. D. Rausch, M. E. Tumbleson, and V. Singh. 2009. Ethanol production from modified and conventional dry-grind processes using different corn types. *Cereal Chemistry* 86(6): 616–622.

Kim, Y., N. S. Mosier, R. Hendrickson, T. Ezeji, H. Blaschek, B. Dien, M. Cotta, B. E. Dale, and M. Ladisch. 2008. Composition of corn dry-grind ethanol by-products: DDGS, wet cake, and thin stillage. *Bioresource Technology* 99: 5165–5176.

Kingsly, A. R. P., K. E. Ileleji, C. L. Clementson, A. Garcia, E. E. Maier, R. L. Stroshine, and S. Radcliff. 2010. The effect of process variables during drying on the physical and chemical characteristics of corn dried distillers grains with solubles (DDGS)—Plant scale experiments. *Bioresource Technology* 101: 193–199.

Koelsch, R., and G. Lesoing. 1999. Nutrient balance on Nebraska livestock confinement systems. *Journal of Animal Science* 77: 63–71.

Koster, H. 2007. Dry milling ethanol by-products for animal feed. *International Sugar Journal* 109: 201–206.

Leguizamon, C., C. L. Weller, V. L. Schlegel, and T. P. Carr. 2009. Plant sterol and policosanol characterization of hexane extracts from grain sorghum, corn and their DDGS. *Journal of the American Oil Chemists Society* 86(7): 707–716.

Liu, K. S. 2008. Particle size distribution of distillers dried grains with solubles (DDGS) and relationships to compositional and color properties. *Bioresource Technology* 99: 8421–8428.

Liu, K. S. 2009a. Effects of particle size distribution, compositional and color properties of ground corn on quality of distillers dried grains with solubles (DDGS). *Bioresource Technology* 100: 4433–4440.

Liu, K. S. 2009b. Fractionation of distillers dried grains with solubles (DDGS) by sieving and winnowing. *Bioresource Technology* 100: 6559–6569.

Liu, K. S. 2010. Selected factors affecting crude fat analysis of distiller dried grains with solubles (DDGS) as compared with milled corn. *Cereal Chemistry* 87(3): 243–249.

Liu, K. S., and J. C. Han. 2011. Changes in mineral concentrations and phosphorus profile during dry-grind process of corn into ethanol. *Bioresource Technology.* 102: 3110–3118.

Majoni, S., and T. Wang. 2010. Characterization of oil precipitate and oil extracted from condensed corn distillers solubles. *Journal of the American Oil Chemists Society* 87(2): 205–213.

Moreau, R. A., and K. B. Hicks. 2006. A reinvestigation of the effect of heat pretreatment of corn fiber on the levels of extractable tocopherols and tocotrienols. *Journal of Agricultural and Food Chemistry* 54: 8093–8102.

Moreau, R. A., K. B. Hicks, D. B. Johnston, and N. P. Laun. 2010a. The composition of crude corn oil recovered after fermentation via centrifugation from a commercial dry grind ethanol process. *Journal of the American Oil Chemists Society* 87: 895–902.

Moreau, R. A., K. S. Liu, J. K. Winkler-Moser, and V. Singh. 2010b. Changes in lipid composition during dry grind ethanol processing of corn. *Journal of the American Oil Chemists Society.* 88:435–442.

Morris, E. R. 1986. Phytate and dietary mineral bioavailability. In *Phytic Acid: Chemistry and Applications.* ed. Graf, E. pp. 57–76. Minneapolis, MN: Pilatus Press.

Murthy, G. S., V. Singh, D. B. Johnston, K. D. Rausch, and M. E. Tumbleson. 2006. Evaluation and strategies to improve fermentation characteristics of modified dry-grind corn processes. *Cereal Chemistry* 83: 455–459.

Mustafa, A. F., J. J. McKinnon, and D. A. Christensen. 2000. Chemical characterization and in situ nutrient degradability of wet distillers' grains derived from barley-based ethanol production. *Animal Feed Science and Technology* 83: 301–311.

Naidu, K., V. Singh, D. B. Johnston, K. D. Rausch, and M. E. Tumbleson. 2007. Effects of ground corn particle size on ethanol yields and thin stillage soluble solids. *Cereal Chemistry* 84(1): 6–9.

Niles, G. A., S. Morgan, W. C. Edwards, and D. Lalman. 2002. Effects of dietary sulfur concentrations on the incidence and pathology of polioencephalomalacia in weaned beef calves. *Veterinary and Human Toxicology* 44: 70–72.

Noureddini, H., M. Malik, J. Byun, and A. J. Ankeny. 2009. Distribution of phosphorus compounds in corn processing. *Bioresource Technology* 100: 731–736.

NRC (National Research Council). 1994. *Nutrient Requirements of Poultry,* 9th ed. Washington, DC: National Academy Press.

NRC (National Research Council). 1998. *Nutrient Requirements of Swine,* 10th ed. Washington, DC: National Academy Press.

Nyachoti, C. M., J. D. House, B. A. Slominski, and I. R. Seddon. 2005. Energy and nutrient digestibilities in wheat dried distillers' grains with solubles fed to growing pigs. *Journal of the Science of Food and Agriculture* 85: 2581–2586.

Ojowi, M. O., J. J. McKinnon, A. F. Mustafa, and D. A. Christensen. 1997. Evaluation of wheat-based wet distillers grains for feedlot cattle. *Canadian Journal of Animal Science* 77: 447–454.

Ortin, W. G. N., and P. Q. Yu. 2009. Nutrient variation and availability of wheat DDGS, corn DDGS and blend DDGS from bioethanol plants. *Journal of Agricultural and Food Chemistry* 89(10): 1754–1761.

Pedersen, C., M. G. Boersma, and H. H. Stein. 2007. Digestibility of energy and phosphorus in ten samples of distillers dried grains with solubles fed to growing pigs. *Journal of Animal Science* 85: 1168–1176.

Ponnampalam, E., D. B. Steele, D. Burgdorf, and D. McCalla. 2004. Effect of germ and fiber removal on production of ethanol from corn. *Applied Biochemistry and Biotechnology* 113: 837–842.

Raboy, V. 1997. Accumulation and storage of phosphate and minerals. In *Cellular and Molecular Biology of Plant Seed Development,* eds. B. A. Larkins, and I. K. Vasil, Vol. 4, pp. 441–447. Dordrecht: Kluwer Academic Pub.

Rausch, K. D., R. L. Belyea, M. R. Ellersieck, V. Singh, D. B. Johnston, and M. E. Tumbleson. 2005a. Particle size distribution of ground corn and DDGS from dry grind processing. *Transactions of the ASAE* 48(1): 273–277.

Rausch, K. D., L. M. Raskin, R. L. Belyea, R. M. Agbisit, B. J. Daugherty, T. E. Clevenger, and M. E. Tumbleson. 2005b. Phosphorus concentrations and flow in maize wet milling streams. *Cereal Chemistry* 82: 431–435.

Redon, M., J. M. Guillamon, A. Mas, and N. Rozes. 2009. Effect of lipid supplementation upon *Saccharomyces cerevisiae* lipid composition and fermentation performance at low temperature. *European Food Research and Technology* 228: 833–840.

RFA (Renewable Fuels Association). 2010. *Climate of Opportunity: 2010 Ethanol Industry Outlook.* Washington, DC: Renewable Fuels Association.

Robinson, P. H., K. Karges, and M. L. Gibson. 2008. Nutritional evaluation of four co-product feedstuffs from the motor fuel ethanol distillation industry in the Midwestern USA. *Animal Feed Science* and *Technology* 146: 345–352.

Shetty, J. K., B. Paulson, M. Pepsin, G. Chotani, B. Dean, and M. Hruby. 2008. Phytate in fuel ethanol production offers economical and environmental benefits. *International Sugar Journal* 110: 160–174.

Singh, N., and M. Cheryan. 1998. Extraction of oil from corn distillers dried grains with solubles. *Transactions of the ASAE* 41: 1775–1777.

Singh, V., and S. R. Eckhoff. 1996. Effect of soak time, soak temperature, and lactic acid on germ recovery parameters. *Cereal Chemistry* 73: 716–720.

Singh, V., D. B. Johnston, K. Naidu, K. D. Rausch, R. L. Belyea, and M. E. Tumbleson. 2005. Comparison of modified dry grind processes for fermentation characteristics and DDGS composition. *Cereal Chemistry* 82: 187–190.

Singh, V., R. A. Moreau, L. W. Doner, S. R. Eckhoff, and K. B. Hicks. 1999. Recovery of fiber in the corn dry-grind ethanol process: Feedstock for valuable coproducts. *Cereal Chemistry* 76: 868–872.

Skiatkiewicz, S., and J. Koreleski. 2008. The use of distillers dried grains with solubles (DDGS) in poultry nutrition. *World's Poultry Science Journal* 64: 257–265.

Spiehs, M. J., and V. H. Varel. 2009. Nutrient excretion and odorant production in manure from cattle fed corn wet distillers grains with solubles. *Journal of Animal Science* 87: 2977–2984.

Spiehs, M. J., M. H. Whitney, and G. C. Shurson. 2002. Nutrient database for distiller's dried grains with solubles produced from new ethanol plants in Minnesota and South Dakota. *Journal of Animal Science* 80: 2639–2645.

Srinivasan, R., R. A., Moreau, K. D. Rausch, and R. L. Belyea. 2005. Separation of fiber from distillers dried grains with solubles (DDGS) using sieving and elutriation. *Cereal Chemistry* 82: 528–533.

Stein, H. H., S. P. Connot, and C. Pedersen. 2009. Energy and nutrient digestibility in four sources of distillers dried grains with solubles produced from corn grown within a narrow geographical area and fed to growing pigs. *Asian-Australasian Journal of Animal Science* 22: 1016–1025.

Thiex, N. J. 2009. Evaluation of analytical methods for the determination of moisture, crude protein, crude fat, and crude fiber in distillers dried grains with solubles. *Journal of AOAC International* 92(1): 61–73.

Wang, H., T. Wang, L. A. Johnson, and A. L. Pometto. 2008. Effect of the corn breaking method on oil distribution between stillage phases of dry-grind corn ethanol production. *Journal of Agricultural and Food Chemistry* 56 (21): 9975–9980.

Wang, P., V. Singh, L. Xu, D. B. Johnston, K. D. Rausch, and M. E. Tumbleson. 2005. Comparison of enzymatic (E-Mill) and conventional dry-grind corn processes using a granular starch hydrolyzing enzymes. *Cereal Chemistry* 82(6): 734–738.

Wang, S., K. C. Thomas, K. Sosulski, W. M. Ingledew, and F. W. Sosulski. 1999. Grain pearling and very high gravity (VHG) fermentation technologies for fuel alcohol production from rye and triticale. *Process Biochemistry* 34: 421–428.

Wang, Y., M. Tilley, S. Bean, X. S. Sun, and D. H. Wang. 2009. Comparison of methods for extracting kafirin proteins from sorghum distillers dried grains with solubles. *Journal of Agricultural and Food Chemistry* 57(18): 8366–8372.

Winkler, J. K., K. A. Rennick, F. J. Eller, and S. F. Vaughn. 2007. Phytosterol and tocopherol components in extracts of corn distiller's dried grain. *Journal of Agricultural and Food Chemistry* 55: 6482–6486.

Winkler-Moser J. K., and S. F. Vaughn. 2009. Antioxidant activity of phytochemicals from distillers dried grain oil. *Journal of the American Oil Chemists Society* 86: 1073–1082.

Wodzinski, R. J., and A. H. Ullah. 1996. Phytase. *Advances in Applied Microbiology*. 42: 264–303.

Wondra, K. J., J. D. Hancock, K. C. Behnke, R. H. Hines, and C. R. Stark. 1995. Effects of particle size and pelleting on growth performance, nutrient digestibility, and stomach morphology in finishing pigs. *Journal of Animal Science* 73: 757–763.

Wu, X., D. Wang, S. R. Bean, and J. P. Wilson. 2006. Ethanol production from pearl millet using *Saccharomyces cerevisiae*. *Cereal Chemistry* 83: 127–131.

Wu, Y. V., and A. C. Stringfellow. 1986. Simple dry fractionation of corn distillers' dried grains and corn distillers dried grains with solubles. *Cereal Chemistry* 63(1): 60–61.

Wu, Y. V. 1994. Determination of neutral sugars in corn distillers dried grains, corn distillers dried solubles, and corn distillers dried grains with solubles. *Journal of Agricultural and Food Chemistry* 42: 723–726.

Yu, p. Q., Z. Y. Niu, and D. Damiran. 2010. Protein molecular structures and protein fraction profiles of new coproducts from bioethanol production: a novel approach. *Journal of Agricultural and Food Chemistry* 58: 3460–3464.

9 Lipids in DDGS

Jill K. Winkler-Moser

CONTENTS

9.1 INTRODUCTION

The word "lipids" generally describes a class of nonpolar organic molecules that are not soluble in water but are soluble in nonpolar organic solvents such as hexane, chloroform, and ether (Moreau, 2005). Lipids usually contain long chains of hydrocarbons, but may also contain oxygen, nitrogen, phosphorus, and sulfur. Plant seed oils, usually called vegetable oils, and animal fats, are composed of a mixture of many different lipids, but the main component is triacylglycerols (TAGs). TAGs (Figure 9.1a) make up anywhere from 85% to 98% of a crude (unrefined) plant seed oil or animal fat. The term "oil" is used for triacylglycerols that remain liquid at room temperature, while the term "fat" refers to triacylglycerols that are solid at room temperature. In corn, sorghum, and other seeds, TAGs are stored in the germ. They are composed of three fatty acids esterified to each hydroxyl position of a glycerol molecule. The five major fatty acids: palmitic, stearic, oleic, linoleic, and linolenic, which are 16- or 18-carbons in length, and have zero to three double bonds, make up the majority of plant TAGs. The percentage of each fatty acid in the TAGs varies depending on plant species and variety; these percentages are known as the "fatty acid composition." Fatty acids can be hydrolyzed from TAGs by heat and moisture, acid or alkali, or by enzymes called lipases, to form free fatty acids (FFA) and diacylglycerols, monoacylglycerols, or free glycerol, depending on how completely the fatty acids were hydrolyzed from the TAG molecule. Other lipids that are commonly found in vegetable oils include, but are not limited to: diacylglycerols, monoacylglycerols, FFA, phytosterols, tocopherols and tocotrienols (vitamin E), waxes, phospholipids, carotenoids, acyl alcohols, acyl aldehydes, and hydrocarbons (Moreau, 2005). Examples of the structures of phospholipids, carotenoids, tocopherols, and tocotrienols are shown in Figure 9.1b through e. The other lipids mentioned above are found in much lower quantities in crude vegetable oils. Whereas TAGs mainly function as carbon storage, lipids such as phospholipids, tocopherols, phytosterols, and carotenoids have other important functions within plant cells, and they are found in tissues other than just the seeds.

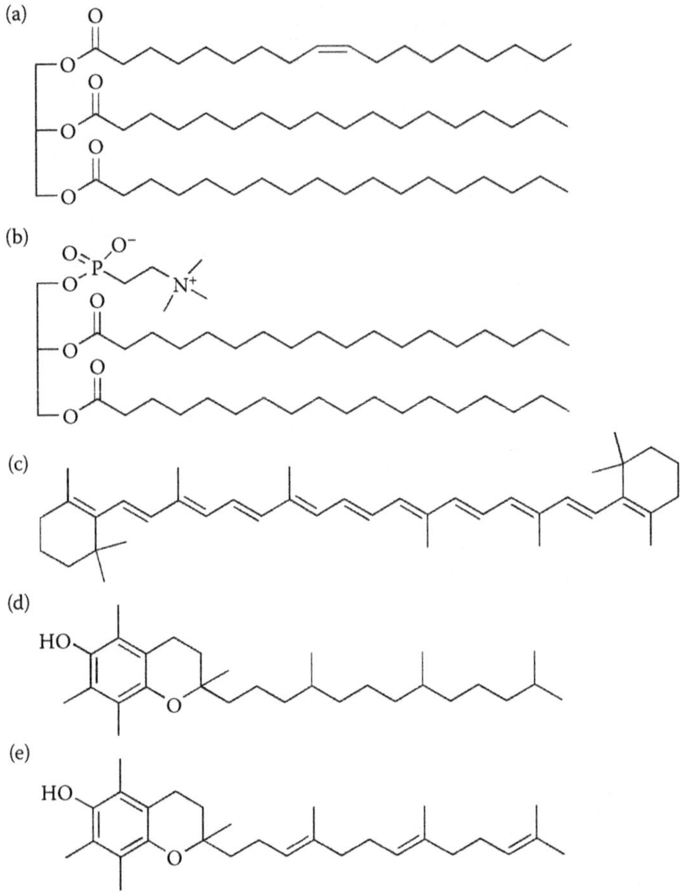

FIGURE 9.1 Structures of some common plant lipids: (a) a triacylglycerol; (b) phosphatidylcholine, a phospholipid; (c) β-carotene, a carotenoid; (d) α-tocopherol; and (e) α-tocotrienol.

9.2 LIPIDS IN DDGS

Over 13 billion gallons of ethanol were produced in the United States in 2010, generating over 30 million metric tons of distillers dried grains with solubles (DDGS) as a coproduct (Rosentrater, 2009; Renewable Fuels Association, 2011; U.S. Energy Information Administration, 2011). DDGS has, on average, 10% (by weight) oil (Belyea and Rausch, 2004; Rausch and Belyea, 2006), so over 3 million metric tons of oil are potentially available. Until recently, references to DDGS lipids in the literature were mostly in the context of reporting total fat content (total ether solubles) for the purpose of determining nutrient composition and energy (caloric) content. However, extraction of oil from DDGS to increase the number and value of coproducts from the dry grind ethanol process has been gaining interest (Watkins, 2007). In addition to the value of the oil itself, removing the oil increases the protein content of DDGS, making it a more valuable feed component (Rausch and Belyea, 2006). The high content and availability of oil in DDGS limits its use in the diets of some animals, such as swine (Shurson and Spiehs, 2002; Hutjens, 2008; Stein and Shurson, 2009), so a modified DDGS with lower fat and higher protein is desirable to increase the market for DDGS in nonruminant diets. Oil extracted from DDGS can be sold as a feedstock for biodiesel production, or if produced in a food-grade environment, as an edible oil. In addition, some of the individual lipids in DDGS oil are bioactive phytochemicals that have potential value as nutraceutical ingredients for the food and cosmetic industry. About eight decades ago, at least one distillery produced edible corn oil from DDGS: the Hiram Walker and Sons whiskey distillery in Peoria, Illinois built an oil extraction plant that processed 100 tons of distillers dried grains per day to produce approximately 14,000 pounds of crude oil per day (Boruff and Miller, 1937; Anonymous, 1977). The extracted oil was sold to vegetable oil refineries for refining and human consumption (Boruff and Miller, 1937). The Hiram Walker and Sons whiskey distillery was eventually sold to Archer Daniels Midland Company, and is now used to produce industrial ethanol, and the oil extraction has ceased. Today, oil from DDGS would not be considered suitable for human consumption unless the production plant, processes used, and final product were all certified food grade.

The lipids in DDGS originate from the feedstock for ethanol production, which in the United States is mainly corn, although sorghum and other grains are also used to some extent. About 95% of the commercial corn oil produced in the United States [2.5 billion pounds per year (USDA, 2009)], is extracted from corn germ isolated from corn wet milling plants (Anonymous, 2006). Approximately 85% of the total oil in a kernel of corn is found in the germ, the rest is found in the endosperm and hull fractions. In conventional dry grind fermentation, the entire corn kernel is ground, and as the starch and sugars are utilized for fermentation, the lipids remain un-utilized and partition between the thin stillage and the wet grains, and ultimately end up in the DDGS. Thus, the oil is concentrated from 3%–5% in whole corn to 8%–12% in corn DDGS (Belyea and Rausch, 2004; Rausch and Belyea, 2006). Since the lipids in DDGS are derived from the entire corn kernel, and have been subjected to the processing conditions from ethanol production, their composition and properties are different from commercial corn oil from corn germ. This chapter will summarize the available literature on the characteristics of the lipids in DDGS. The focus will be on lipids from either corn or sorghum DDGS, since these are the two feedstock sources on which literature is available. Additional information on lipids in various distillers grains functions, such as corn distillers dried grains with no added solubles (DDG) and oil that has been removed by centrifugation of whole or thin stillage is also included.

9.2.1 Glycerolipids and FFA

Standard methods for determination of total oil content (often called crude ether solubles, total fat, or total oil) in DDGS extract of ethyl ether, petroleum ether, or hexane soluble compounds using a Soxhlet or similar extractor, and then gravimetrically measure the extracted lipid weight (AOAC, 1984; AOCS, 1998). The major lipid component in the extracts is TAGs, which have a caloric value of 9 kcal/g. Total lipid content of corn DDGS ranges from 8% to 12% on a dry weight basis (Belyea and Rausch, 2004), and in sorghum DDGS is around 9% (Wang et al., 2005). Though the percentages of TAGs are not reported, Wang et al. (2007, 2008) showed by thin-layer chromatography (TLC) that TAGs are the main component in sorghum DDGS oil. The fatty acid compositions of corn germ oil and sorghum kernel oil TAGs are shown in Table 9.1. The fatty acid composition of TAGs from corn DDG and DDGS oil, as well as from oil extracted at every major step in the dry grind ethanol fermentation process, is similar to commercial corn oil or corn germ oil (Winkler et al., 2007; Majoni and Wang, 2010; Moreau et al., 2011), thus, processing conditions do not appear to significantly affect the fatty acid composition.

Crude vegetable oils typically have low content of FFA (0.5%–1.5%) because TAGs are stored and protected in protein-coated "oil bodies" in intact seeds and because there is very little lipase activity in seeds until germination (Quettier and Eastmond, 2009). However, FFA content is higher in sorghum DDGS and corn DDG (Wang, 2007; Winkler et al., 2007). The reason(s) for the high FFA content are not known, but potential causes are corn or yeast lipase activity, the high temperatures combined with high moisture content as well as acid and base used for pH adjustment, and high temperatures used during the evaporation of thin stillage and the drying of wet grains. As of yet there are no published reports on the FFA content of corn at each processing step of a dry grind operation; however, Noureddini et al. (2009) reported that hexane extracts of corn whole stillage, condensed distillers solubles, and DDG all had 7.5%–8% FFA. Winkler-Moser and Vaughn (2009) reported 6.8% FFA content in a Soxhlet extract (using hexane) of corn DDG. Wang et al. (2007)

TABLE 9.1
Fatty Acid Composition of Corn (*Zea mays*) Germ Oil and
Sorghum (*Sorghum bicolor*) Seed Oil Triacylglycerols

Fatty Acid (%)[a]	Corn	Grain Sorghum
16:0 (Palmitic)	9.2–16.5	15–25
18:0 (Stearic)	0–3.3	1.0–1.4
18:1 (Oleic)	20–42.2	30–42
18:2 (Linoleic)	39.4–65.6	36–51
18:3 (Linolenic)	0.5–1.5	1.6–2.3

Source: Firestone, D., ed. *Physical and Chemical Characteristics of Oils, Fats, and Waxes,* 2nd ed. Urbana, 2006.

[a] The number in front of the colon represents the number of carbons in the fatty acid, the number following the colon designates the number of double bonds.

reported FFA content in sorghum DDGS oil ranging from 6% to 32.9% of the oil (0.5%–4% of DDGS) depending on the extraction solvent and conditions. Much higher FFA content was observed when supercritical CO_2 was used for extractions rather than hexane in a recirculating solvent extractor. FFA content in supercritical CO_2 extracts increased with increasing temperature (40°C–70°C), but temperature had no impact on FFA extracted using recirculating hexane. The FFA in sorghum DDGS were composed of palmitic, oleic, and linoleic acids (Wang et al., 2007).

In terms of overall oil quality, high FFA content is usually considered detrimental for several reasons. High FFA content in crude oils leads to refining losses (Gupta, 2008), and contributes to low oil stability to oxidation (Frankel, 2005). Oils for human consumption are refined to remove FFA to a level of 0.5% or less (Gupta, 2008) because lipid oxidation results in the production of off-flavors and odors (Frankel, 2005). For biodiesel production, oils with FFA content greater than 1% require an acid pretreatment because the fatty acids form soaps during base-catalyzed esterification, which interfere with the separation of the glycerol from the fatty acid methyl esters (Freedman et al., 1984). On the other hand, given the amount of FFA that have been extracted from sorghum DDGS using supercritical CO_2 (Wang et al., 2007), it may be economically feasible to extract the FFA for use by chemical and cosmetic industries.

9.2.2 PHYTOSTEROLS

Phytosterols (plant sterols) are triterpenes, similar in structure to cholesterol, found ubiquitously in plants (Moreau et al., 2002). They exist as free sterols, acylated sterols, where the hydroxyl group is esterified to a fatty acid, as sterol glycosides, acylated sterol glycosides, or as esters to ferulic or *p*-coumaric acid (Figure 9.2). In vegetable oils, they are found mainly either in their free or acylated form (Piironen et al., 2000; Moreau et al., 2002). Like cholesterol in mammalian cells, phytosterols help to regulate the fluidity of plant cell membranes, though they also play a role in other cellular and developmental processes (Piironen et al., 2000). Phytosterols inhibit the intestinal absorption of dietary cholesterol and reabsorption of biliary cholesterol. The general consensus is that consumption of approximately 2 g/day of phytosterols as part of a low-fat diet can reduce total serum and low density lipoprotein (LDL) cholesterol by ~10% thereby reducing the risk of cardiovascular heart disease (Weststrate and Meijer, 1998; Gylling and Miettinen, 2005; Wu et al., 2009). Since the year 2000, the FDA has allowed a health claim for foods containing at least 0.65 g or 1.7 g/serving of phytosterol or phytostanol (fully saturated phytosterol) esters, respectively (FDA, 2009). Since then, the number of "functional food" products containing phytosterols has skyrocketed, in addition to numerous dietary supplements and pharmaceutical products (Ohr, 2006). There also have

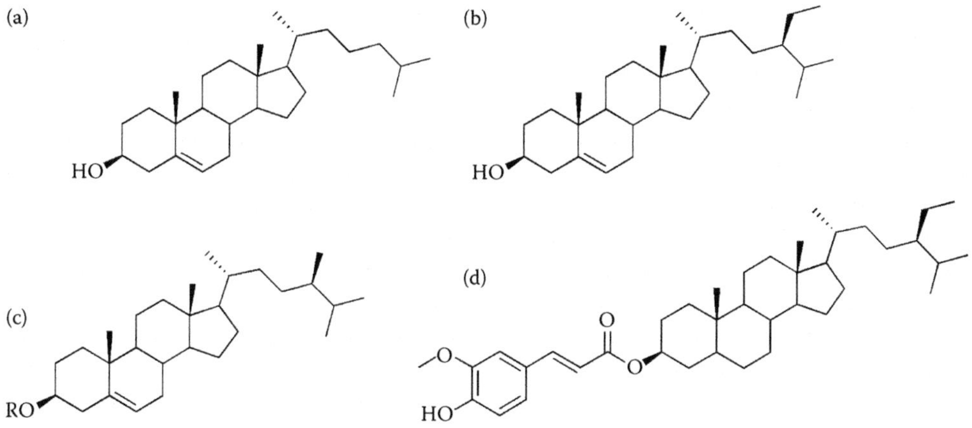

FIGURE 9.2 Structures of cholesterol and plant sterols: (a) cholesterol; (b) β-sitosterol; (c) steryl ester, R represents a long-chain fatty acid; and (d) sitostanyl ferulate, a stanyl ferulate (ferulate phytosterol ester or FPE).

been numerous reports that phytosterols may have anticancer properties; these studies have recently been reviewed (Woyengo et al., 2009). Currently, the two main sources for large-scale isolation of phytosterols are deodorizer distillates from the refining of vegetable oils, and tall oil, a byproduct of the wood pulp industry (Fernandez and Cabral, 2007).

Phytosterols (free and acylated) are present in most commercial vegetable oils, such as soybean, corn, sunflower, and canola, at concentrations ranging from 2 to 10 mg/g (Piironen et al., 2000; Phillips et al., 2002). In comparison to commercial corn oil extracted from corn germ, corn fiber oil and corn kernel oil have much higher phytosterol content (Moreau et al., 1996; Singh et al., 1999; Moreau and Hicks, 2005). Corn kernel and corn fiber oils also have a considerable amount of the ferulic acid esters of sterols (ferulate phytosterol esters, or FPE) that are virtually absent in commercial corn oil because FPE are derived from the aleurone layer of corn kernel pericarps (Moreau et al., 2000). Since corn DDG and DDGS are derived from whole corn kernels, the phytosterol content and composition of their oils are similar to corn kernel oil (Winkler et al., 2007, Moreau et al., 2011). Phytosterol content, including free sterols, acylated sterols, and FPE, of corn DDG and DDGS has been reported to range between 1.0 and 3.5 mg/g, depending on the DDGS source, extraction solvent, and technique; content in lipid extracts ranges from 9 to 22 mg/g (Singh et al., 2001; Winkler et al., 2007; Wong et al., 2008; Leguizamón et al., 2009; Moreau et al., 2011). Phytosterol content of sorghum DDGS ranges from 0.8 to 2.5 mg/g, again, depending on the extraction solvent and technique (Wang et al., 2007; Leguizamón et al., 2009). Leguizamón et al. (2009) also reported that total phytosterol content of grain sorghum DDGS is actually up to 50% higher when content of acylated steryl glycosides (ASG) and steryl glycosides (SG) is included in the analysis. However, only low levels of ASG and SG are extracted by conventional pressing and/or hexane extraction (Moreau et al., 2003). Moreau et al. (2011) analyzed phytosterol content and composition of lipid extracts from corn after each step in the dry grind process from three ethanol plants, and concluded that while the lipid concentration increased after fermentation, the content and composition of phytosterols in the lipid extracts were not significantly impacted by processing steps.

In most plants and in vegetable oils, the predominant phytosterols are sitosterol, campesterol, and stigmasterol. The quantity of fully saturated phytosterols (phytostanols) is usually less than 2% of the total sterol content (Phillips et al., 2002). However, corn DDG and DDGS lipid extracts have unusually high contents of sitostanol (12%–17% of total) and campestanol (7%–9% of total) compared to corn oil (Winkler et al., 2007; Moreau et al., 2011), and are similar in composition to corn fiber oil (Phillips et al., 2002). This was later determined to be due to FPE in the extracts, since corn FPE are mainly composed of phytostanols rather than phytosterols (Seitz, 1999). FPE are present in grains such as corn, rice, wheat, and rye (Seitz, 1989), but are not found in commercial vegetable oils other than rice bran oil. Due to the ferulic acid group, these compounds have antioxidant activity and have been shown to protect oils during frying (Xu and Godber, 2001; Nyström et al., 2005, 2007). The FPE from rice bran oil, which are collectively called gamma-oryzanol, also have cholesterol-lowering activity due to the sterol moiety (Seetharamaiah and Chandrasekhara, 1988; Rong et al., 1997). FPE content of corn DDG and DDGS is approximately 0.4–0.5 mg/g; content in lipid extracts ranges from between 2 and 4 mg/g depending on the solvent and technique used for extraction (Singh et al., 2001; Winkler et al., 2007). FPE have not been reported in sorghum DDGS, but presumably they are present at much lower quantities than in corn DDGS since only 0.01 mg/g were reported in sorghum kernels (Singh et al., 2003).

9.2.3 Tocopherols and Tocotrienols

Tocopherols and tocotrienols are structurally very similar; tocotrienols are the unsaturated analogues of tocopherols, containing three double bonds in the triterpenyl side chain (Figure 9.1d and e). Tocopherols and tocotrienols each have four isoforms (α-, β-, γ-, δ-), differing in the number and position of methyl groups on the chromanol ring (Figure 9.1d and e; Abidi, 2000). These eight vitamers comprise the family of compounds responsible for natural vitamin E activity (Sen et al., 2007). Tocopherols are important antioxidants in all plant tissues. They are found in vegetable oils at levels ranging from 300 to 1000 mg/kg and are the primary antioxidants protecting most oils (Kamal-Eldin, 2006). Tocopherols, especially α-tocopherol, are added to numerous food, cosmetic, and personal hygiene products. α-Tocopherol is often produced synthetically; the major source of natural tocopherols is deodorizer distillates from vegetable oil refining (Watanabe et al., 2004). Tocotrienols are not present in significant quantities in commodity vegetable oils except for palm oil (Elson, 1992); they are also found in rice bran oil, wheat germ oil, and barley oil (Ward et al., 2008). Tocotrienols have antioxidant activity similar to tocopherols in bulk oil systems (Schroeder et al., 2006). However, they are also gaining attention due to an increasing number of studies investigating the role that they may have in preventing cardiovascular heart disease and cancer, and in providing neuroprotection (Sen et al., 2007).

Winkler et al. (2007) analyzed tocopherol and tocotrienol content in DDG oil extracted using either hexane accelerated solvent extraction (ASE) or hexane Soxhlet, ethanol ASE or Soxhlet, and supercritical CO_2. Total tocopherol and tocotrienol content ranged from 730 to 1820 mg/kg oil (130–240 mg/kg DDG). The composition of tocopherols in the DDG oil was similar to corn germ oil, in that there was mostly γ-tocopherol (350–950 mg/kg) with lower amounts of α-tocopherol (90–260 mg/kg) and δ-tocopherol (21–58 mg/kg). The γ-tocopherol content was lower than what Moreau et al. (2006) reported in corn germ oil, however, this may have been due to loss from heat during processing and drying, since tocopherols are oxidized and degraded at high temperatures. The tocotrienol contents of corn DDG lipid extracts analyzed by Winkler et al. (2007) were similar to the reported content in corn kernel oil (Moreau and Hicks, 2005, 2006); levels of γ-tocotrienol, α-tocotrienol, and δ-tocotrienol in the DDG extracts were 210–460 mg/kg, 70–170 mg/kg, and 6–18 mg/kg, respectively. The tocotrienols are apparently preferentially located in the endosperm (Grams et al., 1970), which explains why high content is found in lipid extracts of DDGS but not in corn germ oil. Moreau et al. (2011) analyzed content and composition of tocopherols and tocotrienols in lipid extracts from corn after each step in the dry grind process from three ethanol plants. The tocopherol content increased significantly after saccharification and fermentation, indicating perhaps, that starch hydrolysis may have helped to release bound tocopherols. Total tocol (tocopherols plus tocotrienols) concentration decreased after evaporation of thin stillage, but was still higher in lipids extracted from DDGS than from the original ground corn. Wang et al. (2007) extracted lipids from sorghum DDGS using either recirculating hexane or supercritical CO_2. They reported total tocopherol content to range from 40 to 90 mg kg^{-1} DDGS. α-Tocopherol was the most prominent tocopherol, followed by γ-tocopherol; neither β- nor δ-tocopherol was detected.

9.2.4 WAXES AND POLICOSANOLS

Waxes are mixtures of compounds with a high (>40 °C) melting point including long-chain (>C20) alkanes, fatty acids, alcohols, wax esters, and other high-melting lipids; wax esters are composed of long chain alcohols esterified to long-chain fatty acids (Lee and Parish, 1998). Sorghum lipids contain considerable wax-like material composed mainly of saturated, long-chain (28- and 30-carbon) aldehydes, alcohols, and acids (Hwang et al., 2002). The long-chain alcohols, known as policosanols, are of commercial interest and are sold as nutritional supplements because they are believed to lower blood cholesterol (Gouni-Berthold and Berthold, 2002), reduce blood platelet aggregation (Arruzazabala et al., 1996), and improve muscle endurance (Kabir and Kimura, 1995). Recent studies indicate that there is still some controversy over whether policosanols really lower blood cholesterol or have cardioprotective effects (Jones et al., 2009; Kassis et al., 2009). Hwang et al. (2004) measured policosanol content of sorghum grains and DDGS. Policosanol content in sorghum DDGS ranged from 2.5 to 2.8 mg/g and was mainly composed of octacosanol (28:0) and triacontanol (30:0). In another study, policosanol content of sorghum DDGS ranged from 1.6 to 4.7 mg/g (Wang et al., 2007). Leguizamón et al. (2009) analyzed lipids extracted from whole corn and sorghum kernels and reported that they had similar content and composition of policosanols. However, policosanol content was much lower in lipids extracted from ground corn kernels or corn DDGS, while sorghum DDGS had approximately threefold higher policosanols compared to whole sorghum kernels. No explanation was given for the loss of policosanols in the ground corn and corn DDGS; there were differences in how the corn was ground (hammer mill) compared to the sorghum (coffee grinder) but this does not seem to adequately explain the difference in policosanol content. According to Bianchi and Avato (1984), corn kernel wax is mainly composed of wax esters (76 wt.% of wax components), with only about 2% long-chain alcohols. Wax content in crude corn oil has been reported to be around 0.5 mg/g (Majoni and Wang, 2010). A high-melting precipitate in oil extracted by centrifugation of corn condensed distillers soluble showed a wax content of 2.5 mg/g (Majoni and Wang, 2010). Further investigation is needed to determine the wax content, including policosanols, of corn DDGS.

9.2.5 CAROTENOIDS

Corn kernel and corn fiber oil are good sources of carotenoids, especially lutein and zeaxanthin (Moreau et al., 2007). Winkler-Moser et al. (2009) observed that a hexane extract of corn DDGS had 66 µg/g oil of carotenoids and xanthophylls, mainly composed of β-carotene, lutein, zeaxanthin, and β-cryptoxanthin. As with tocotrienols, carotenoids are concentrated in the endospum tissue of corn (Weber, 1987). Carotenoid content in the DDG oil was 33 times higher than reported in oil extracted from corn germ using hexane, 5.5-fold higher than in hexane extracted corn fiber oil, and 2.5-fold higher than found in hexane extracted ground corn (Moreau et al., 2007). Moreau et al. (2010) reported total carotenoid and xanthophylls content of DDGS lipid extracts from three ethanol plants ranging from 10 to 30 µg/g oil. They discovered that carotenoid and xanthophylls content was higher (25–50 µg/g) in whole stillage, wet distillers grains, thin stillage, and distillers soluble, but 25%–50% appeared to be destroyed by the drying process. β-Carotene and β-cryptoxanthin are precursors to Vitamin A (Bendich and Olson, 1989). Carotenoids have also been shown to have a number of beneficial physiological actions other than Vitamin A activity, including antioxidant activity, enhanced immune response, and chemoprotective activity against several types of cancer (Bendich and Olson, 1989). Lutein and zeaxanthin are both associated with reduced risk of cataracts and macular degeneration (Seddon et al., 1994). β-Carotene and carotenoids have both antioxidant and prooxidant activity *in vitro* (Frankel, 2005), and have also been shown to synergistically enhance the antioxidant activity of tocopherols and tocotrienols (Schroeder et al., 2006) in bulk oils and liposomes.

9.3   EXTRACTION OF LIPIDS FROM DDGS

Since several functional lipids described above, especially the phytosterols, are concentrated in the fiber portion of corn, considerable attention has been placed on removing the fiber from the rest of DDGS to create several different fractions and coproducts including, corn fiber oil (Singh et al., 1999, 2001, 2002, 2005; Srinivasan et al., 2005, 2006, 2007). Many of these methods are described in Chapters 5, 8, and 25. Therefore, these approaches will not be discussed here. The leading technologies used to extract oil for biodiesel production are briefly described in this chapter, but more detailed information about these technologies and the suitability of DDGS oil for biodiesel applications can be found in Chapter 23.

9.3.1   Oil Extraction for Biodiesel Production

TAGs from vegetable oils are the main feedstock for biodiesel production all over the world. Soybean oil is the principal triacylglycerol feedstock for biodiesel production in the United States, but animal fats, other commodity seed oils, and waste restaurant vegetable oils are also used as feedstocks to some extent. In 2008, U.S. biodiesel production hit a historic high level of over 670 million gallons per year, but dropped dramatically to around 300 million gallons per year in 2010, mainly because of the expiration of the biodiesel tax credit. (U.S. Energy Information Administration, 2011) TAGs are transesterified into fatty acid methyl esters using a base catalyst. Considering that over 30 million metric tons of DDGS are produced annually (Renewable Fuels Association, 2011; U.S. Energy Information Administration, 2011), the amount of oil in DDGS would be enough to feed current biodiesel production needs.

Technologies have been developed for the extraction of oil either pre- or postfermentation in a dry grind ethanol operation. Front-end extraction technology was developed through a joint venture between Cargill and Monsanto called Renessen, LLC. In the proprietary process, a dry grind plant is modified to include an initial fractionation where the high-oil corn fractions (i.e., the germ) are removed prior to fermentation (Watkins, 2007). According to their website, the high-oil fraction can then be used to make corn oil suitable either for human food or for biodiesel.

Another company, GreenShift (http://www.greenshift.com), has developed patented technology for the recovery of oil after fermentation in a dry grind ethanol plant. Oil is mechanically removed via centrifugation of thin or whole stillage. These technologies and the suitability of the oil obtained are described more fully in Chapter 23. According to the Greenshift website, their oil extraction technologies are currently being used in four dry grind ethanol plants across the United States. In the current industrial process, the total oil in corn feedstock is approximately equally distributed between the thin stillage and the distillers grains (Wang et al., 2008). Wang et al. (2008) found that flaking and extruding corn prior to fermentation increased the percentage of oil partitioned into the thin stillage to 90%. Oil extracted via centrifugation of thin stillage or condensed distillers solubles is often found to have a cloudy appearance due to lipids that have a melting point higher than 25°C and a "waxy" appearance. Majoni and Wang (2010) reported that the precipitate had a high concentration of FFA (37% of lipids). The FFA fraction of the precipitate had a high percentage of saturated palmitic acid (70%), which is the likely explanation for the high melting point. Other components in the precipitate included waxes and wax esters (2.5 mg/g), phytosterols (8.6 mg/g), and other neutral and polar lipids (triacylglycerols, diacylglycerols, and phospholipids).

9.3.2   Solvent Extraction of DDGS

Since oil is the third most abundant component in DDGS and is a concentrated source of energy, the animal feed industry needs to have accurate and universal procedures for oil analysis in order to ensure product quality and consistency. The most important feature of any oil analysis procedure is the ability to extract all of the oil. This ability is affected by factors unique to each food/feed and analytical procedure. In a study of the factors affecting crude oil analysis of DDGS, Liu (2010) identified sample origin (ethanol plant), extraction time, particle size, solvent type (hexane or petroleum ether), and drying time as having significant effects on crude oil content.

Petroleum ether and hexane are the two most common solvents used for extracting oil for determination of crude oil because they effectively remove the neutral lipids that contribute to caloric content (AOAC, 1984; AOCS, 1998). However, other solvents may be more favorable for large-scale oil extraction when the goal is either to obtain the oil cheaply, to use more environmentally friendly solvents, or to obtain higher quantities of lipid phytochemicals. Several studies have investigated the effects of different types of solvents as well as extraction conditions and techniques on the yield and composition of DDGS lipids. Singh and Cheryan (1998) published one of the first reports of oil extraction from corn DDGS. Oil was extracted using ethanol because it would be readily available in an ethanol plant, and thus could easily be recirculated after oil extraction. Using a stirred reactor at 50°C and ethanol-to-DDGS ratios (mL/g) ranging from 2 to 10, the impact on yield of oil, total solids, glycerol, and protein was determined. Oil extraction was optimized at an ethanol-to-DDGS ratio of 6 mL/g, but only about 50% of the available oil was extracted.

Winkler et al. (2007) compared the yield and composition of lipid phytochemicals in extracts from corn DDG using hexane ASE, hexane Soxhlet, ethanol ASE, ethanol Soxhlet, and supercritical CO_2. The highest amount of solids, phytosterols, FPE, tocopherols, and tocotrienols was extracted by ethanol Soxhlet extraction (32.7%), followed by ethanol ASE (17.6%). However, there was also considerable nonlipid material in the ethanol extracts. Hexane and supercritical CO_2 extraction resulted in oil yields of 11%–12.7%, and extracted similar levels of phytochemicals, although hexane resulted in slightly higher oil yields.

Wang et al. (2005) looked at the effects of hexane-to-solid ratio (ml/g), extraction temperature, and extraction time on the yield and composition of sorghum DDGS lipids using a bench-scale recirculating extraction system. They found that the higher the extraction temperature, from 45°C to 68°C, the higher the oil yields and the more quickly the extraction yields were maximized. Extraction yields also increase when hexane-to-solid ratios were increased from 2 ml/g to 5 mL/g, but they started to level off after a ratio of about 4 mL/g. Maximum extractability of 94.5% of total lipids was achieved with a solvent to solids ratio of 5 mL/g at a temperature of 68°C and an extraction time of 6 h. The effect of supercritical CO_2 extraction conditions on sorghum DDGS oil yield and composition was also examined by the same research group (Wang et al., 2008). Oil yields were maximized using an extraction pressure of 27.5 MPa, a mass ratio of CO_2 to DDGS of 45 kg/kg, an extraction time of 4 h and a temperature of 70°C. Sterol and policosanol content increased with increasing flow rate and extraction time, and FFA content was lowest with lower temperatures and higher extraction pressures. Supercritical CO_2 extraction had higher yield of total lipids as well as policosanols, phytosterols, FFA, and tocopherols compared to recirculated hexane extraction of sorghum DDGS (Wang et al., 2007).

9.4 CONCLUSIONS

The oil in DDGS has potential value as a third coproduct from the dry grind ethanol process. Triacylglycerols are valued in food and feed products, and as feedstock for biodiesel production. Lipid components such as triacylglycerols and several other phytochemicals, especially phytosterols, FPE, policosanols, tocopherols, and tocotrienols, are concentrated in DDGS because of the removal of starch from corn or sorghum kernels during fermentation and because of the fact that lipids from the entire grain, including the bran and fiber, are present as opposed to the germ components only. Therefore, DDGS is also a potential source for the extraction of these phytochemicals for use in functional foods, cosmetics, or dietary supplements. Regulatory issues regarding the safety of lipids extracted from DDGS for food use have to be considered, as well as the economics surrounding the cost of extraction and the value of the oil and its phytochemicals. However, because of the concentrations of bioactive phytochemicals in DDGS oils, they have potential marketability as functional foods. For example, DDG and DDGS oils have a similar phytosterol profile to corn fiber oil, which has been shown to reduce plasma cholesterol in Guinea pigs as well as hamsters (Ramjiganesh et al., 2011; Jain et al., 2008). Phytochemicals extracted from corn DDGS have also been shown to have antioxidant activity (Winkler-Moser and Vaughn, 2009). In addition, grain sorghum lipids have been shown to prevent proliferation of human colon cancer cells (Zbasnik et al., 2009). The properties and potential uses for the lipids in DDGS are only in the early stages of exploration.

REFERENCES

Anonymous. 1977. Continuous solvent extraction. *Journal of the American Oil Chemists Society* 54: 202A–207A.

Anonymous. 2006. *Corn Oil*, 5th ed., pp. 24. Corn Refiners Association, St. Louis. Available online: www.corn.org/CornOil.pdf. Accessed February 02, 2010.

Abidi, S. L. 2000. Chromatographic analysis of tocol-derived lipid antioxidants. *Journal of Chromatography* 881: 197–216.

AOAC. 1984. *Official Methods of Analysis of AOAC International,* 17th ed., ed. W. Horwitz. Gaithersburg, MD: AOAC International.

AOCS. 1998. *Official Methods and Recognized Practices of the American Oil Chemists' Society,* 5th ed., ed. D. Firestone. Champaign, IL: AOCS Press.

Arruzazabala, M. L., S. Valdes, R. Mas, L. Fernandez, and D. Carbajal. 1996. Effect of policosanol successive dose increases on platelet aggregation in healthy volunteers. *Physiological Reviews* 34: 181–185.

Belyea, R. L., K. D. Rausch, and M. E. Tumbleson. 2004. Composition of corn and distillers dried grains with solubles from dry grind ethanol processing. *Bioresource Technology* 94(3): 293–298.

Bendich, A., and J. A. Olson. 1989. Biological actions of carotenoids. *FASEB Journal* 3: 1927–1932.

Bianchi, G., and P. Avato. 1984. Surface waxes from grains, leaves, and husks from maize (*Zea mays* L.). *Cereal Chemistry* 61: 45–47.

Boruff, C. S., and J. Miller. 1937. Solvent extraction of corn oil from distillers grains. *Journal of the American Oil Chemists Society (Oil and Soap)* 14: 312–313.

Elson, C. E. 1992. Tropical oils: Nutritional and scientific issues. *Critical Reviews in Food Science and Nutrition* 31: 79–102.

FDA, Code of Federal Regulations, Title 21, 101.83, 2009

Fernandes, P., and J. M. S. Cabral. 2007. Phytosterols: Applications and recovery methods. *Bioprocess Technology* 98: 2335–2350.

Frankel, E. N. 2005. *Lipid Oxidation,* 2nd ed. Bridgwater: The Oily Press, PJ Barnes and Associates.

Freedman, B., E. H. Pryde, and T. L. Mounts. 1984. Variables affecting the yields of fatty esters from transesterified vegetable oils. *Journal of the American Oil Chemists Society* 61: 1638–1643.

Gouni-Berthold, I., and H. K. Berthold. 2002. Policosanol: Clinical pharmacology and therapeutic significance of a new lipid-lowering agent. *American Heart Journal* 143: 356–365.

Grams, G.W., Blessin, C.W., and Inglett, G.E. 1970. Distribution of tocopherols with the corn kernel. *Journal of the American Oil Chemists Society* 47: 337–339.

Gupta, M. K. 2008. *Practical Guide to Vegetable Oil Processing.* Urbana, IL: AOCS Press.

Gylling, H., and, T. A. Miettinen 2005. The effect of plant stanol- and sterol-enriched foods on lipid metabolism, serum lipids and coronary heart disease. *Annals of Clinical Biochemistry* 42: 254–263.

Hutjens, M. 2008. Distiller's grains opportunities. Illini Dairy Net 2002. Available online: www.livestocktrail.uiuc.edu/dairynet/paperDisplay.cfm?ContentID=632. Accessed July 7, 2009.

Hwang, K. T., S. L. Cuppett, C. L. Weller, M. A. Hanna, and R. K. Shoemaker. 2002. Aldehydes in grain sorghum wax. *Journal of the American Oil Chemists Society* 79: 529–533.

Hwang, K. T., C. L. Weller, S. L. Cuppett, and M. A. Hanna. 2004. Policosanol contents and composition of grains sorghum kernes and dried distillers grains. *Cereal Chemistry* 81: 345–349.

Jain, D., N. Ebine, X. Jia, A. Kassis, C. Marinangeli, M. Fortin, et al. 2008. Corn fiber oil and sitostanol decrease cholesterol absorption independently of intestinal sterol transporters in hamsters. *The Journal of Nutritional Biochemistry* 19: 229–236.

Jones, P. J. H., A. N. Kassis, and C. P. F. Marinangeli. 2009. Policosanols lose their lustre as cholesterol-lowering agents. *Journal of Functional Foods* 1: 236–239.

Kabir, Y., and S. Kimura. 1995. Tissue distribution of (8014C)-octacosanol in liver and muscle of rats after serial administration. *Annals of Nutrition & Metabolism* 39: 279–284.

Kamal-Eldin, A. 2006. Effect of fatty acids and tocopherols on the oxidative stability of vegetable oils. *European Journal of Lipid Science and Technology*. 58: 1051–1061.

Kassis, A. N., S. Kubow, P. J. H. Jones. 2009. Sugar cane policosanols do not reduce LDL oxidation in hypercholesterolemic individuals. *Lipids* 44: 391–396.

Liu, K. S. 2010. Selected factors affecting crude fat analysis of distiller dried grains with solubles (DDGS) as compared with milled corn. *Cereal Chemistry* 87(3): 243–249.

Leguizamón, C., C. L. Weller, V. L. Schlegel, and T. P. Carr. 2009. Plant sterol and policosanol characterization of hexane extracts from grain sorghum, corn, and their DDGS. *Journal of the American Oil Chemists Society* 86: 707–716.

Majoni, S., and T. Wang. 2010. Characterization of oil precipitate and oil extracted from condensed corn distillers soluble. *J. Am. Oil Chem. Soc.* 87: 205–213.

Moreau, R. A., M. J. Powell, and K. B. Hicks. 1996. Extraction and quantitative analysis of oil from commercial corn fiber. *Journal of the American Oil Chemists Society* 44: 2149–2154.

Moreau, R. A., V. Singh, A. Nuñez, and K. B. Hicks. 2000. Phytosterols in the aleurone layer of corn kernels. *Biochemical Society Transactions* 28: 803–806.

Moreau, R. A., B. D. Whitaker, and K. B. Hicks. 2002. Phytosterols, phytostanols, and their conjugates in foods: Structural diversity, quantitative analysis, and health-promoting uses. *Progress in Lipid Research* 41: 457–500.

Moreau, R. A., M. J. Powell, and V. Singh. 2003. Pressurized liquid extraction of polar and nonpolar lipids in corn and oats with hexane, methylene chloride, isopropanol, and ethanol. *Journal of the American Oil Chemists Society* 80: 1063–1067.

Moreau, R. A. 2005. Extraction and analysis of food lipids. In *Methods of Analysis of Food Components and Additives*, ed. S. Otles, pp. 97–110. Boca Raton, FL: CRC Press/Taylor and Francis.

Moreau, R. A., and K. B. Hicks. 2005. The composition of corn oil obtained by the alcohol extraction of ground corn. *Journal of the American Oil Chemists Society* 82: 809–815.

Moreau, R. A., and K. B. Hicks. 2006. Reinvestigation of the effect of heat pretreatment of corn fiber and corn germ on the levels of extractable tocopherols and tocotrienols. *Journal of Agricultural and Food Chemistry* 54: 8093–8102.

Moreau, R. A., D. B. Johnston, and K. B. Hicks. 2007. A comparison of the levels of lutein and zeaxanthin in corn germ oil, corn fiber oil, and corn kernel oil. *Journal of the American Oil Chemists Society* 84: 1039–1044.

Moreau, R.A., Hicks, K.B., Johnston, D.B., Laun, N.P. 2010. The composition of crude corn oil recovered after fermentation via centrifugation from a commercial dry grind ethanol process. *Journal of the American Oil Chemists Society* 87: 895–902.

Moreau, R. A., K. S. Liu, J. K. Winkler-Moser, and V. Singh. 2011. Changes in lipid composition during dry grind ethanol processing of corn. *Journal of the American Oil Chemists Society*. 88: 435–442.

Noureddini, H., S. R. P. Bandlamudi, and E. A. Guthrie. 2009. A novel method for the production of biodiesel from the whole stillage-extracted corn oil. *Journal of the American Oil Chemists Society*. 86: 83–91.

Nyström, L., M. Mäkinen, A. -M. Lampi, and V. Piironen. 2005. Antioxidant activity of steryl ferulate extracts from rye and wheat bran. *Journal of Agricultural and Food Chemistry* 53: 2503–2510.

Nyström, L., T. Achrenius, A. -M. Lampi, R. A. Moreau, and V. Piironen. 2007. A comparison of the antioxidant properties of steryl ferulates with tocopherol at high temperatures. *Food Chemistry* 101: 947–954.

Ohr, L. M. 2006. The (heart) beat goes on. *Food Technology* 60: 87–92.

Phillips, K. M., D. M. Ruggio, J. I. Toivo, M. A. Swank, and A. H. Simpkins. 2002. Free and esterified sterol composition of edible oils and fats. *Journal of Food Composition and Analysis* 15: 123–142.

Piironen, V., D. G. Lindsay, T. A. Miettinen, J. Toivo, and A. -M. Lampi. 2000. Plant sterols: Biosynthesis, biological function and their importance to human nutrition. *Journal of the Science of Food and Agriculture* 80: 939–966.

Quettier, A. -L., and P. J. Eastmond. 2009. Storage oil hydrolysis during early seedling growth. *Plant Physiology and Biochemistry* 47(6): 485–490.

Ramjiganesh, T., S. Roy, H. C. Freake, J. C. McIntyre, and M. L. Fernandez. 2002. Corn fiber oil lowers plasma cholesterol by altering hepatic cholesterol metabolism and up-regulating LDL receptors in Guinea pigs. *Journal of Nutrition* 132: 335–340.

Rausch, K. D., and R. L. Belyea. 2006. The future of coproducts from corn processing. *Applied Biochemistry and Biotechnology* 128: 47–86.

Renewable Fuels Association. 2010. Ethanol industry outlook 2011. Available online: http://www.ethanolrfa. org/. Accessed March 2011.

Rong, N., L. M. Ausman, and R. J. Nicolosi. 1997. Oryzanol decreases cholesterol absorption and aortic fatty streaks in hamsters. *Lipids* 32: 303–309.

Rosentrater, K. 2009. Distill dried grains with solubles (DDGS)—A key to the fuel ethanol industry. *Inform* 20: 789–800.

Schroeder, M. T., E. M. Becker, and L. H. Skibsted. 2006. Molecular mechanism of antioxidant synergism of tocotrienols and carotenoids in palm oil. *Journal of Agricultural and Food Chemistry* 54: 3445–3453.

Seddon, J. M., U. A. Ajani, R. D. Sperduto, R. Hiller, N. Blair, and T. C. Burton, et al. 1994. Dietary carotenoids, Vitamins A, C, and E, and advanced age-related macular degeneration. *Journal of the American Medical Association* 272: 1413–1420.

Seetharamaiah, G. S., and N. Chandrasekhara. 1988. Hypocholesterolemic activity of oryzanol in rats. *Nutrition Reports International* 38: 927–935.

Seitz, L. M. 1989. Stanol and sterol esters of ferulic and *p*-coumaric acids in wheat, corn, rye, and triticale. *Journal of Agricultural and Food Chemistry* 37: 662–667.

Sen, C. K., S. Khann, and S. Roy. 2007. Tocotrienols in health and disease: The other half of the natural vitamin E family. *Molecular Aspects of Medicine* 28: 692–798.

Shurson, J., and M. Spiehs. 2002. Feeding recommendations and example diets containing Minnesota-South Dakota produced DDGS for swine. Available online: www.ddgs.umn.edu/feeding-swine/exampleswinediets-revised.pdf. Accessed July 2009.

Singh, N., and M. Cheryan. 1998. Extraction of oil from corn distillers dried grains with solubles. *Transactions of the ASAE* 41: 1775–1777.

Singh, V., R. A. Moreau, L. W. Doner, S. R. Eckhoff, and K. B. Hicks. 1999. Recovery of fiber in the corn dry-grind ethanol process: a feedstock for valuable coproducts. *Cereal Chemistry* 76: 868–872.

Singh, V., R. A. Moreau, K. B. Hicks, R. L. Belyea, and C. H. Staff. 2001. Recovery of phytosterols from fiber removed from distiller dried grains with solubles (DDGS). Paper number 01–016012 Paper presented at the annual meeting of the American Society of Agricultural Engineers.

Singh, V., R. A. Moreau, K. B. Hicks, R. L. Belyea, and C. H. Staff. 2002. Removal of fiber from distillers dried grains with solubles (DDGS) to increase value. *Transactions of the ASME* 45: 389–392.

Singh, V., R. A. Moreau, K. B. and Hicks. 2003. Yield and phytosterol composition of oil extracted from grain sorghum and its wet-milled fractions. *Cereal Chemistry* 80: 126–129.

Singh, V., D. B. Johnston, K. Naidu, K. D. Rausch, R. L. Belyea, and M. E. Tumbleson. 2005. Comparison of modified dry-grind corn processes for fermentation characteristics and DDGS composition. *Cereal Chemistry* 82: 187–190.

Srinivasan, R., R. A. Moreau, K. D. Rausch, R. L. Belyea, and M. E. Tumbleson. 2005. Separation of fiber from distillers dried grains with solubles (DDGS) using sieving and elutriation. *Cereal Chemistry* 82: 528–533.

Srinivasan, R., V. Singh, R. L. Belyea, K. D. Rausch, R. A. Moreau, and M. E. Tumbleson. 2006. Economics of fiber separation from distillers dried grains with solubles (DDGS) using sieving and elutriation. *Cereal Chemistry* 83: 324–330.

Srinivasan, R., R. A. Moreau, K. D. Rausch, M. E. Tumbleson, and V. Singh. 2007. Phytosterol distribution in fractions obtained from processing of distillers dried grains with solubles using sieving and elutriation. *Cereal Chemistry* 84: 626–630.

Stein, H. H., and G. C. Shurson. 2009. Board invited review: The use and application of distillers dried grains with solubles in swine diets. *Journal of Animal Sciences* 87: 1292–1303.

U.S. Department of Agriculture, Economic Research Service: Oil Crops Yearbook. 2009. Available online: usda.mannlib.cornell.edu/MannUsda/viewDocumentInfo.do?documentID=1290. Accessed February 2010.

U.S. Energy Information Administration, Monthly Report, February 2011. Available online: http://www.eia. gov/mer/contents.html, accessed March, 2011.

Wang, L., C. L. Weller, and K. T. Hwang. 2005. Extraction of lipids from grain sorghum DDG. *Transactions of the ASAE* 48: 1883–1888.

Wang, L., C. L. Weller, V. L. Schlegel, T. P. Carr, and S. L. Cuppet. 2007. Comparison of supercritical CO_2 and hexane extraction of lipids from sorghum distillers grain. *European Journal of Lipid Science and Technology* 109: 567–574.

Wang, L., C. L. Weller, V. L. Schlegel, T. P. Carr, and S. L. Cuppett. 2008a. Supercritical CO_2 extraction of lipids from grain sorghum dried distillers grains with solubles. *Bioresource Technology* 99: 1373–1382.

Wang, H., T. Wang, L. A. Johnson, and A. L. Pometto III. 2008b. Effect of corn breaking method on oil distribution between stillage phases of dry-grind corn ethanol production. *Journal of Agricultural and Food Chemistry* 56: 9975–9980.

Ward, J. L., K. Poutanen, K. Gebruers, V. Piironen, A. -M. Lampi, L. Nyström, et al. 2008. The HEALTHGRAIN cereal diversity screen: Concept, results, and prospects. *Journal of Agricultural and Food Chemistry* 56: 9699–9709.

Watanabe, Y., T. Nagao, Y. Hirota, M. Kitano, and Y. Shimada. 2004. Purification of tocopherols and phytosterols by a two-step *in situ* enzymatic reaction. *Journal of the American Oil Chemists Society* 81: 339–345.

Watkins, C. 2007. Two fuels from one kernel. *Inform* 18: 714–717.

Weber, E. 1987. Carotenoids and tocols of corn grain determined by HPLC. *Journal of the American Oil Chemists Society* 64: 1129–1134.

Weststrate, J. A., and G. W. Meijer. 1998. Plant sterol-enriched margarines and reduction of plasma total- and LDL-cholesterol concentrations in normocholesterolaemic and mildly hypercholesterolaemic subjects. *European Journal of Clinical Nutrition* 52: 334–343.

Winkler, J. K., K. A. Rennick, F. J. Eller, and S. F. Vaughn. 2007. Phytosterol and tocopherol components in extracts of corn distiller's dried grain. *Journal of Agricultural and Food Chemistry* 55: 6482–6486.

Winkler-Moser, J. K., and S. F. Vaughn. 2009. Antioxidant activity of phytochemicals from distillers dried grain oil. *Journal of the American Oil Chemists Society* 86: 1073–1082.

Wong, A. 2008. Phytosterols in selected grain processing residues. *Electronic Journal of Environmental, Agricultural, and Food Chemistry* 7 6(June): 2948–2958. Available online: ejeafche.uvigo.es/index.php?option=com_docman&task=doc_view&gid=359&Itemid=33. Accessed February 14, 2011.

Wu, T., J. Fu, Y. Yang, L. Zhang, and J. Han. 2009. The effects of phytosterols/stanols on blood lipid profiles: a systematic review with meta-analysis. *Asia Pacific Journal of Clinical Nutrition* 18: 179–186.

Woyengo, T. A., V. R. Ramprasath, and P. J. H. Jones. 2009. Anticancer effects of phytosterols. *European Journal of Clinical Nutrition* 63: 813–820.

Xu, Z., and J. S. Godber. 2001. Antioxidant activities of major components of γ-oryzanol from rice bran using a linoleic acid model. *Journal of the American Oil Chemists Society* 78: 645–649.

Zbasnik, R., T. Carr, C. Weller, K. T. Hwang, L. Wang, S. L. Cuppett, and V. Schlegel. 2009. Antiproliferation properties of grain sorghum dry distiller's grain lipids in Caco-2 cells. *Journal of Agricultural and Food Chemistry* 57: 10435–10441.

10 Analytical Methodology for Quality Standards and Other Attributes of DDGS

Nancy Thiex

CONTENTS

10.1 INTRODUCTION

Distillers grains is the second largest processed feed ingredient in the United States, at an estimated 2010 production of 35 million tons (MT), with 9 MT going to exports (personal communication, Charles Staff, Distillers Grain Technology Council, Louisville, KY). As a feed ingredient, distillers grains are typically analyzed in laboratories for one of three purposes. This chapter is divided into three sections, accordingly: (1) for quality standards for trading purposes; (2) to determine nutrients for inclusion in livestock feeds; and (3) to determine contaminants that may limit inclusion in livestock feeds.

Analytical methods are categorized based on the level of validation to which the method has been subjected. Levels of method validation include single laboratory validation (SLV), multilaboratory validation, and those subjected to a full harmonized protocol collaborative study. The different levels of validation range in their degree of ruggedness; the most rigorous form of method validation is achieved through a full harmonized protocol collaborative study where statistical criteria for the data must be met in a minimum of eight laboratories.

SLV generally applies to a specific laboratory, technician, and equipment. Criteria for the elements to be validated by SLV are published by the International Union of Pure and Applied Chemistry (IUPAC) (Thompson et al., 2002), and other associations and federal agencies. These elements include applicability or scope of the method, selectivity, calibration, trueness or accuracy, precision, recovery, range, detection limit, limit of quantification, sensitivity, ruggedness, fitness for purpose, uncertainty, and others depending on the method. A method subjected to multilaboratory validation has been validated in two to seven labs and is intended to provide information on how a method is reproduced outside of the original lab. A method receives full collaborative study validation when eight or more labs provide acceptable data using the method. The criteria for this validation have been harmonized by the IUPAC, the International Standards Organization (ISO), and the AOAC International (AOAC) (Horwitz, 1992).

The production of fuel ethanol and distillers dried grains with solubles (DDGS) is covered in Chapter 5, while the chemical composition of feedstock (grains) and DDGS is treated in Chapters 4 and 8, respectively. In this chapter, the focus is on the specific methods used for the analysis of DDGS, including the type of method and the level of validation to which the method has been subjected. As much as possible, methods indicated for use in distillers grains in this chapter will be those which have been validated at the level of a full collaborative study and are internationally accepted methods for trade/commerce and regulation. Methods that have not been validated at some level appropriate for the intended use of the method for elements listed above (accuracy, precision, trueness, etc.) are unacceptable since the quality of the data generated has not been established. Such methods may include "proof of concept" assays that, although needed to advance analytical technology, do not yet meet the criteria necessary for widespread use.

10.2 ANALYTICS FOR TRADING PURPOSES (QUALITY STANDARDS)

For trading purposes, DDGS is tested for and marketed on quality standards for moisture, crude protein, crude fat, and crude fiber content. The methods most often used for measuring these analytes are Type 1 methods as defined by the Codex Alimentarius Commission, Committee of Methods of Analysis and Sampling Committee scheme (Codex Alimentarius Procedural Manual, 2010). They are "defining" or "empirical" methods, meaning the analytical results obtained are defined by the conditions specified for the method. Any change to the conditions of the method (extraction or drying times, extraction or drying temperatures, particle size, reagent type, reagent concentration, etc.) will bias results. When variations in test conditions for a specific analyte by an empirical analytical method are practiced in laboratories, the laboratories will report different concentrations for the analyte in the same laboratory sample. A lack of agreement in reported results leads to confusion and frustration for those who are dependent upon the results of analyses, whether they are buyers, sellers, producers, nutritionists, or regulators. However, standardizing laboratories is even more difficult than standardizing methods.

10.2.1   Moisture Analysis

Loss on drying methods are "defining" or "empirical" methods, estimating moisture based on evaporative loss on drying. When materials are heated to drive off water, other volatile substances present can be lost. If the temperature is excessive, additional weight loss can be assigned to the degradation of heat sensitive substances. Different combinations of temperature and length of heating are employed in various loss on drying methods. Loss on drying methods commonly used in feed laboratories include AOAC **934.01** (AOAC, 2005d), Loss on Drying at 95%–100°C (Vacuum Oven); AOAC **935.29** (AOAC, 2005e), Moisture in Malt (103–104°C/5 h); NFTA **2.2.2.5** (Undersander et al., 2002); Lab Dry Matter (105°C/3 h); and AOAC **930.15** (AOAC, 2005c), Loss on Drying (135°C/2 h). The Karl Fisher method, AOAC **2001.12** (AOAC, 2005z) is a specific titration for water (H_2O) and is a reference method that can be used to assess the loss on drying empirical methods.

10.2.2   Crude Protein Analysis

Crude protein is commonly determined in feed laboratories by either "combustion" or "Kjeldahl" methods. Both types of protein methods measure nitrogen content, but they are by very different techniques. With either procedure, crude protein is estimated as % N times 6.25. Since the methods do not measure true protein, rather nitrogen, they are termed "crude" protein methods. AOAC **990.03** (AOAC, 2005t) measures nitrogen by combustion, oxidizing nitrogenous compounds followed by conversion to nitrogen and thermal conductivity detection of the nitrogen. The Kjeldahl procedure can be broken into three main steps: digestion, distillation, and quantitation. Digestion of the test portion in sulfuric acid destroys organic matter and converts nitrogen sources to ammonium sulfate. Salt (potassium sulfate) is added to raise the boiling point during digestion and a catalyst (e.g., mercury, copper, selenium, or both copper and titanium) is added to speed digestion. The salt/acid ratio is critical and is dependent upon the choice of catalyst and digestion temperature (rack or block). The recommended ratio of salt/acid for digestions with Hg catalyst using AOAC **954.01** (AOAC, 2005i) is 0.6; for digestions with $CuSO_4/TiO_2$ catalyst using AOAC **988.05** (AOAC, 2005r) is 0.835; for digestions using selenium catalyst is 0.47; and for digestions with copper catalyst on a traditional rack using AOAC **984.13** (AOAC, 2005q) is 0.75 or 0.6 if performed on a block digestor using either AOAC **976.06** (AOAC, 2005o) or AOAC **2001.11** (AOAC, 2005y) (Thiex, 1994). Due to waste disposal issues, the use of catalyst (especially Hg catalyst) and the traditional 800 mL Kjeldahl flasks has lost popularity. AOAC **2001.11** measures nitrogen by the traditional Kjeldahl acid block digestion utilizing a copper catalyst to convert nitrogenous compounds to ammonia, which is distilled into boric acid and titrated.

Methods that measure true protein instead of crude protein are of renewed interest since the inclusion of melamine and other economic adulterants into protein supplements was discovered in recent years to falsely bias the crude protein results toward higher readings.

10.2.3   Crude Fat (Oil) Analysis

The crude fat methods (sometime called crude oil) are "defining" or "empirical" methods; the crude fat fraction is defined by the solvent and the extraction conditions (time, temperature, particle size, ratio of solvent to test portion, etc.) and is not specific for the extraction of lipid material. A number of studies have appeared in the literature over time, but a recent study specific to DDGS and corn reported on the effect of particle size, solvent type, drying time, and extraction time on the amount of crude oil extracted from these materials using a high temperature solvent extraction method (Liu, 2010). In addition to lipids, crude fat methods can coextract any other substances that are soluble under the conditions of the method. These might be substances such as residual moisture, residual ethanol, pigments, carotenes, urea, and others. Due to the nature of the methods, they are not specific to lipids, nor do the extraction conditions ensure that 100% of the lipid material will be extracted. The acid hydrolysis step is applicable for baked products such as pet food, and facilitates the extraction of fatty acids from glycerides, glycol- and phospholipids, and sterol esters that might otherwise be left unextracted. However, it can also facilitate coextraction of additional nonlipid materials. Any modification of any of the crude fat methods yields a slightly different fraction and produces different results and is, therefore, actually a different method. Therefore, care must be taken with any empirical method to follow the method exactly as directed and report the extraction conditions with the results.

Crude fat is commonly determined in feed laboratories by one of the following methods:

AOAC **2003.05** (AOAC, 2005ac), Crude Fat (Randall/Soxtec/Diethyl Ether—Submersion Method)

AOAC **945.16** (AOAC, 2005h), Oil in Cereal (Petroleum Ether)

AOAC **2003.06** (AOAC, 2005ad), Crude Fat (Randal/Soxtec/Hexanes–Submersion Method)

AOAC **954.02** (AOAC, 2005j), Fat (Acid Hydrolysis), and

AOCS Am-5 (Rapid Determination of Oil/Fat Utilizing High Temperature Solvent Extraction) (AOCS, 2004)

10.2.4 CRUDE FIBER ANALYSIS

Again, crude fiber methods are "defining" or "empirical" methods; the crude fiber fraction is defined by the method. Crude fiber is the residue remaining after digestion with 0.255 N sulfuric acid followed by digestion with 0.313 N HCl to solubilize proteins, sugars, and starches. The residue remaining, after drying and correcting for ash, is considered "crude fiber." Any modification of the method yields a different fraction and is, therefore, a different "crude fiber" method. Two crude fiber methods widely used in feed laboratories are AOAC **978.10** (AOAC, 2005p) and AOCS **Ba 6a-05** (AOCS, 2005). The chemistry for the methods is similar; but the equipment and extractions conditions are different. The traditional AOAC method specifies extractions in open glassware and filtration through fritted glass (F. G.) crucibles of a coarse porosity to separate the fiber portion. The AOCS method specifies extraction under pressure with the extraction occurring inside a synthetic filter bag where the pore size determines the fraction retained. A third technique new to the marketplace utilizes a filter bag (or "fiberbag") without pressure and provides for automated addition of reagents and filtering process.

10.2.5 INDUSTRY RECOMMENDED METHODS

With a goal of establishing industry standards for DDGS quality, a study was contracted by the American Feed Industry Association (AFIA), the Renewable Fuels Association (RFA) and the National Corn Grain Association (NCGA) to evaluate the efficacy, applicability, and the intralaboratory variation of a number of methods for moisture, crude protein, crude fat, and crude fiber in distillers dried grains with solubles (DDGS). Phase I of the study was a side-by-side comparison of the various analytical methods of interest in a single laboratory at South Dakota State University (SDSU). A second phase of the study assessed interlaboratory variation of the same methods in 23 laboratories. The materials used in Phase I of the study were 30 DDGS materials collected from six locations (five from each location). In Phase II, one material from each of five of the six locations was used, resulting in a total of five materials. Based on the results of these studies, the sponsoring associations established recommended reference methods for use in the commercial trade of DDGS (Thiex, 2009). The full text of the recommendations can be obtained from at least two websites: the AFIA website (AFIA, 2007a) and from the Distillers Grain Technology Council website (AFIA, 2007b).

10.2.5.1 Phase I Study Summary

Details of the Phase I of the Industry Study were published in 2009 (Thiex, 2009). A summary of the results are found in Table 10.1. Conclusions drawn from Phase 1 of the Industry Study regarding moisture testing were that DDGS is easily dried and contains volatile or degradable substances other than water; and that AOAC **2001.12** 2005z (Karl Fischer) should be used as the reference method to determine water in DDGS. Moisture results using the National Forage Testing Association (NFTA) Method **2.2.2.5** most closely approximated the results obtained by the Karl Fischer reference method, with an average recovery of 110% water and the most consistent recovery over the 30 materials in the study with a recovery range of 90%–131%. Therefore provided the best estimate of water loss on drying.

The two crude protein methods studied produced very similar results (Table 10.1). An average of 26.85% protein by AOAC **990.03** 2005t over all DDGS materials compared well to an average of 26.75% protein by AOAC **2001.11**, with both methods performing well. Either method was found to be suitable as a reference method.

Average crude fat on study materials by AOAC **2003.05** (2005ac), AOAC **945.16** (2005h), and AOAC **2003.06** (2005ad) was reported as 9.22%, 8.85%, and 9.00%, respectively, with similar method performance (Table 10.1). The fourth method, AOAC **954.02**, which uses an acid hydrolysis step, yielded higher results with an average crude fat of 13.03% (almost 150% of the nonhydrolysis

TABLE 10.1
Phase I of Industry Study Results by Method for 30 DDGS Materials

Method	Mean[b] (%)	Mean RSD	Average % Recovery (Compared to KF)[a]
Moisture results[c] by method			
AOAC 2001.12	9.03[d]		100
NFTA 2.2.2.5	9.87d,[e]		110
AOAC 935.29	10.17[e]		113
AOAC 930.15	12.69[f]		142
AOAC 934.01	10.67[e]		119
Crude protein results[c] by method			
AOAC 990.03	26.85[f]	0.64	
AOAC 2001.11	26.75[f]	0.58	
Crude fat results[c] by method			
AOAC 2003.05	9.22[g]	2.95	
AOAC 945.16	8.85[g]	2.63	
AOAC 2003.06	9.00[g]	2.09	
AOAC 954.02	13.03[h]	4.38	
Crude fiber results by method			
AOAC 978.10[c]	7.58[i]	4.02	
AOCS Ba 6a-05 with F58 filter bag[c]	7.64[i]	7.08	
AOCS Ba 6a-05 with F57 filter bag	6.99[j]	5.13	

Source: Adapted from Thiex, N. *Journal of AOAC International,* 92(1), 61–73, 2009.

[a] KF = Karl Fischer or AOAC 2001.12.

[b,d,e,f,g,h,i,j] Means with the same letter are not significantly different ($p < 0.0001$).

[c] Each value is the mean of triplicate measurements on 30 DDGS materials (90 measurements).

methods) and poorer precision. From Phase I, it was concluded that AOAC **2003.05**, AOAC **945.16**, and AOAC **2003.06** were all suitable for use on DDGS material.

Average crude fiber of the on study materials was reported as 7.58% and 7.64% by AOAC **978.10** and AOCS **Ba 6a-05**, respectively (Table 10.1), with better precision for AOAC **978.10**. Additional discussion around the porosity of the filter bag is included in the original publication. Since the F58 bag was not commercially available at the time of the study and only a limited number of them could be obtained, a modification of the AOCS method substituting the F57 filter bag, which has a coarser pore size, was also compared. The results using the F57 bag were about 10% lower than those obtained with the F58 bag. Due to the empirical nature of the method, there is a critical relationship between the fineness of the grind of the analytical sample and the pore size of the bags. It is possible that a coarser particle size could offset the pore size of the filter bag.

10.2.5.2 Phase II Study Summary

Details of Phase II were reported in 2009 (Theix, 2009). Results of Phase II of the Industry Study are summarized in Table 10.2, including methods studied, summaries of laboratories reporting data, number of laboratories that submitted data, number of laboratories remaining after removal of invalid data, total number of outliers, within laboratory repeatability (RSD_r), and among laboratory reproducibility (RSD_R) for each method.

The most widely used moisture method was also the poorest performing loss on drying method (AOAC 930.15, 135°C/2 h), while AOAC 2001.12 (Karl Fisher) was found to be the least widely used by laboratories.

Protein measurements in DDGS were found to be consistent among laboratories using either AOAC **990.03** or AOAC **2001.11**. Bias between protein methods was not significant in this study, even though bias with the combustion technique typically tends to be high, while bias with the Kjeldahl technique tends to be low due to the inherent nature of the methods.

TABLE 10.2

Summary of Repeatability (RSD_r) and Reproducibility (RSD_R) for Laboratories Reporting Data in Phase II of the Industry Study

Methods	Number of Labs	Number of Labs After Removal of Invalid Data	Additional Outlier Results	RSDr, % Range for all Materials	RSDR, % Range for all Materials
Moisture					
AOAC **934.01**	10	8		2.24–5.11	5.92–12.60
AOAC **935.29**	9	7		2.05–4.92	4.89–8.18
NFTA **2.2.2.5**	14	11	7	1.94–3.06	2.57–6.27
AOAC **930.15**	21	17	4	1.83–2.86	3.69–11.94
AOAC **2001.12**	1	1		0.68–5.55	Not applicable
Protein					
AOAC **990.03**	19	17		0.72–1.27	1.27–2.29
AOAC **2001.11**	8	8		0.61–1.05	0.96–1.35
Crude fat					
AOAC-**2003.05**	7	7	5	2.23–4.40	3.67–12.61
AOAC-**945.16**	10	8	2	1.76–2.07	2.72–3.35
AOAC-**2003.06**	5	5	3	1.24–5.37	3.59–6.90
AOAC-**954.02**	12	9	1	2.14–2.92	3.72–9.93
Crude fiber					
AOAC-**978.10**	10	6	2	2.34–4.47	15.33–19.73
AOCS-**Ba 6a-05**	7	6	1	1.70–5.28	6.01–8.48

Source: Adapted from Thiex, N. *Journal of AOAC International*, 92(1), 61–73, 2009.

The range of crude fat and crude fiber results from participating laboratories were both highly variable with all methods. Crude fat results were somewhat improved if acid hydrolysis results were omitted. For both crude fat and crude fiber, much of the variability was eliminated with the removal of the outlier laboratories. The variability of the crude fat and crude fiber results underscores the importance of following empirical methods as written to avoid introduction of bias.

10.2.5.3 AFIA Recommendations Based on the Industry Study

The full recommendations of the Committee, with reasoning, can be found on the AFIA and DGTC website (AFIA, 2007a, 2007b). The methods chosen are as follows:

NFTA **2.2.2.5** Lab Dry Matter (105°C/3 h), was selected as the recommended loss on drying method for the analysis of moisture in DDGS.

AOAC **990.03**, Protein (Crude) in Animal Feed—Combustion, and AOAC **2001.11**, Protein (Crude) in Animal Feed and Pet Food Copper Catalyst, were both selected to be used interchangeably for the analysis of protein in DDGS.

AOAC **945.16**, Oil in Cereal Adjuncts (Petroleum Ether) was selected as the recommended test method for the analysis of fat in DDGS.

The AOAC **978.10** Fiber (Crude) in Animal Feed and Pet Food (F.G. Crucible) was selected as the recommended method for crude fiber analysis in DDGS.

10.3 ANALYTICS FOR NUTRIENTS FOR INCLUSION IN LIVESTOCK FEEDS

In addition to those analytes described in Section 10.2, analytes of interest for inclusion of DDGS in livestock feeds include detergent fibers, ash, trace elements, amino acids, and starch.

10.3.1 DETERGENT FIBER

As with crude fiber, detergent fiber methods are "defining" or "empirical" methods; the respective fiber fraction is defined by the method used. Acid detergent fiber (ADF) and amylase-treated neutral detergent fiber (aNDF) are used to formulate diets for many animal species. A good resource for tips on the measurement of detergent fibers is "Critical Conditions in Determining Detergent Fibers" (Mertens, 1992), which has been reproduced in the NFTA method manual on the NFTA website (Undersander et al., 1993).

10.3.1.1 Acid Detergent Fiber and Acid Detergent Lignin

The method for measuring ADF utilizes strong acid to solubilize protein and hydrolyze all polysaccharides except cellulose and lignin, which are therefore the only components in ADF (Van Soest, 1963). The cellulose is further hydrolyzed with 72% sulfuric acid, leaving only the lignin. AOAC Official Method **973.18** Fiber, Acid Detergent, and Lignin, H_2SO_4 in Animal Feed (AOAC, 2005n) and ISO Standard (ISO, 2008; ISO) specify a method for the determination of ADF insoluble residue and acid detergent lignin (ADL) in all types of animal feeding stuffs. The AOAC and ISO methods are equivalent.

10.3.1.2 Amylase-Treated Neutral Detergent Fiber (aNDF)

The aNDF method provides an estimate of all insoluble cell wall material (lignin, cellulose, and hemicellulose (Van Soest and Wine, 1967). There are many versions in circulation; however the only procedures recognized as official methods are AOAC and ISO methods. The aNDF method is published as AOAC Official Method **2002.04** Amylase Treated Neutral Detergent Fiber in Feeds (AOAC, 2005aa) and as ISO Standard 16472:2006 (ISO, 2006), which "specifies methods for the determination of amylase-treated neutral detergent insoluble fibrous residue content in all types of animal feed." The AOAC and ISO methods are equivalent. The method also provides for determination of aNDF and amylase-treated NDF organic matter (aNDFom), which expresses the fiber result on an ash-free or organic matter basis.

10.3.2 ASH

Ash is the inorganic residue remaining after the water and organic matter have been removed by heating in a muffle furnace. Ash or "residue on ignition" provides a measure of the inorganic matter (or total mineral content). In addition, it can provide a quick estimate of dirt or soil contamination, which will be included in the ash residue due to the high silica content of soil. The AOAC Method **942.05** (AOAC, 2005f) specifies heating in a muffle furnace at 600°C for 2 h. ISO 5984:2002 (ISO, 2002) specifies 550°C for 3 h. With either method, the ash should be inspected, and if the ash contains unoxidized carbon, the test portion should be re-ashed as needed. Since these are "defining" or "empirical" methods, bias should be expected between the two methods.

10.3.3 TRACE ELEMENTS

Trace elements, for the purpose of this chapter, include those established as essential for optimum animal health and productivity. These include calcium, chlorine, chromium, cobalt, copper, fluorine, iodine, iron, magnesium, manganese, molybdenum, phosphorus, potassium, selenium, sodium, sulfur, and zinc (Mineral Tolerances of Animals, 2005). Generally mineral methods employ two basic steps: solubilization or digestion and detection. The solubilization step employs either dry ash followed by dissolving in acid or wet ash employing various acids, depending upon the element of interest and the detection technique. Detection techniques of current use in laboratories include gravimetric techniques, visible spectrophotometry, flame and graphite furnace atomic absorption spectroscopy, and inductively coupled plasma spectroscopy with atomic emission spectroscopic detection (ICP-AES) or with atomic mass spectroscopic detection (ICP-MS).

10.3.3.1 Sulfur (S)

Organic sulfur (supplied by sulfur amino acids) is essential in the diet of many animals. Some inorganic sulfur may be supplemented, but high dietary sulfur concentrations are associated with a risk for sulfur-associated polioencephalomalacia (PEM) in ruminants (Mineral Tolerances of Animals, 2005). Sulfur is of particular interest in distillers grains because of the use of sulfuric acid as a cleaning agent in ethanol plants. Excess sulfur in the form of sulfates has been associated with PEM in some batches of distillers grains due to this use. Methods for testing sulfur in distillers grains are analogous to those used for other feed stocks. They typically include three types of methods: gravimetric, ICP-AES, and combustion. One study recently presented comparability data for three methods on 9 DDGS materials (Van Erem et al., 2008). Two methods with lesser validation were compared to AOAC **923.01** (Sulfur in Plants) (AOAC, 2005b). The two methods compared were combustion and ICP-AES following dissolution by microwave digestion.

The gravimetric method was performed as described in the *AOAC Official Methods of Analysis*, 20th edition, Method **923.01**. A test portion was heated with a magnesium nitrate solution to oxidize the sulfur to sulfate and to prevent the volatilization of sulfur during ashing, and then ashed in a muffle furnace to destroy the remaining organic matter. The ash was dissolved in a hydrochloric acid solution, filtered, and barium chloride was added to the filtrate to precipitate barium sulfate. The barium sulfate precipitate was weighed and percent sulfur in the original sample was calculated from the amount of barium sulfate precipitate.

The ICP-AES method was performed as described and single lab validated in "Determination of Calcium, Cobalt, Copper, Iron, Potassium, Magnesium, Manganese, Molybdenum, Sodium, Phosphorus, Sulfur, and Zinc in Animal Feed and Pet Food by Microwave Digestion and ICP-AES" (Schilling et al., 2006) and was very similar to ISO 27085:2009 (ISO, 2009). A test portion was digested in a microwave oven with strong oxidizing agents (nitric acid and hydrogen peroxide) to destroy the organic matter. The diluted digest was then aspirated into an ICP-AES spectrometer where the sample signal was compared to a standard curve. The wavelengths used for sulfur were 180.676 nm and 181.978 nm.

The combustion method was performed using the LECO TruSpec Add-On Sulfur Module. A test portion was weighed into a combustion boat and mixed with approximately 1 g of a tungsten (VI) oxide-based accelerator. The sample was ignited in the pure oxygen environment of the combustion tube at 1350°C with additional oxygen supplied through a lance to accelerate the combustion of all materials. The gases produced as a result of the oxidation were swept through the remainder of the two tube combustion system, through a tube containing magnesium perchlorate to remove moisture, through a flow controller, and finally through the infrared detection cell located in the main TruSpec module where the concentration of sulfur dioxide was determined. The instrument converted the sulfur dioxide concentration using an equation preset in the software that took into account the test portion weight, the calibration, and a moisture value (if one is known).

The results of the study are presented in Table 10.3. The mean sulfur concentration for nine distillers grain materials was reported for gravimetric, ICP, and combustion methods as 0.74%, 0.77%, and 0.74%, respectively. None of the means were statistically different; the mean relative standard deviations (RSDs) were 1.87, 1.74, and 0.57, respectively. The percent recovery obtained on the NIST Certified Reference Material (0.96% S) was 103 for the gravimetric method, 103 for ICP, and 105 for the combustion method. It was concluded that the three methods were comparable.

10.3.3.2 Phosphorus (P)

Phosphorus is reliably measured using visible spectrophotometric methods following dry ashing of the feed matrix. AOAC **965.17** (AOAC, 2005k) *Phosphorus in Animal Feed, Photometric Method* is widely used. This method is the classic molybdovanadate reaction with phosphorus following dry ashing and solubilization in dilute HCl. The phosphomolybdate complex is read at 400 nm. ISO 6491:1998 *Determination of total phosphorus content—spectrophotometric method* is also available from ISO (ISO, 1998). Phosphorus can also be determined in the multianalyte ICP-AES method described in 2.3.3. (ISO, 2009).

10.3.3.3 Multianalyte Methods for Trace Minerals

Multianalyte methods for trace minerals (calcium, cobalt, copper, iron, sodium, magnesium, manganese, molybdenum, phosphorus, potassium, and zinc) have become very popular due to

---

TABLE 10.3
Comparability Data for Three Methods to Determine Sulfur (%) in DDGS

Material	%S, Gravimetric		%S, ICP-AES		%S, Combustion	
	Mean	RSD	Mean	RSD	Mean	RSD
DDG A	1.03	1.36	1.06	1.99	1.00	0.3
Distillers wet grains B	0.83	0.24	0.86	0.12	0.81	0.62
DDGS C	1.03	2.05	1.09	1.94	1.02	0.69
DDGS D	0.62	2.05	0.62	4.18	0.62	0.32
DDGS E	0.62	1.61	0.62	1.14	0.61	0.33
DDGS F	0.47	2.79	0.49	1.65	0.48	0.83
DDGS G	0.7	0.57	0.73	0.55	0.69	0.29
DDGS H	1.03	2.05	1.09	3.85	1.00	0.9
DDGS I	0.37	2.14	0.4	0.25	0.4	0.75
NIST 1573a tomato leaves	0.96	2.39	0.94	—	0.95	0.63
Mean of DDG materials (A-I)	0.74	1.87	0.76	1.74	0.73	0.57

Source: Adapted from Van Erem et al. Comparison of gravimetric, ICP-OES and combustion methods for determining sulfur in distillers grain, animal feed and feed ingredients. *AOAC International Midwest Section Official Meeting Program*, Bozeman, MT, 2008.

the ability to obtain results on multiple minerals with a single digest. The single Official AOAC multielement method is AOAC Method **968.08**, atomic absorption spectroscopy method for copper, iron, manganese, zinc, and calcium (AOAC, 2005l). This method allows for two types of digestion, either dry ash at 550°C with subsequent solubilization in 3 M HCl, or wet digestion with concentrated nitric acid. The detection is by atomic absorption spectroscopy with flame conditions and wavelengths specified in the method. ISO 6869:2000 provides for the determination of calcium, copper, iron, magnesium, manganese, potassium, sodium, and zinc using atomic absorption spectrometry (ISO, 2000).

ISO 27085:2009 (ISO, 2009) specifies an inductively coupled plasma atomic emission spectroscopic (ICP-AES) method for the determination of (1) the minerals calcium, sodium, phosphorus, magnesium, and potassium, and the elements iron, zinc, copper, manganese, cobalt, and molybdenum in animal feed stuffs; and (2) the elements arsenic, lead, and cadmium in minerals on their own, in premixes or mixtures for use in animal feed stuffs. The detection limit for each element is dependent on the sample matrix as well as on the instrument. The method is not applicable for determination of low concentrations of elements. The limit of quantification is 3 mg/kg or lower.

Additional official methods do not exist at this time. There are a number of in-house methods in use in laboratories, with or without published validations.

10.3.3.4 Selenium (Se)

Methods for the determination of selenium in animal feed have been validated and published as AOAC Official Method **996.16** Selenium in Feeds and Premixes, Fluorometric Method (AOAC, 2005w), and as AOAC Official Method **996.17** Selenium in Feeds and Premixes, Continuous Hydride Generation Atomic Absorption Method (AOAC, 2005x). Both methods are based on digestion with nitric and perchloric acids to destroy organic matter and convert selenium to inorganic forms (Se^{+4} and Se^{+6}). The presence of perchloric acid prevents loss of selenium. For both methods, all forms of selenium in the digestion mixture are reduced to Se^{+4} with HCl. With the fluorometric method, the selenite (Se^{+4}) is complexed with 2,3-diaminonaphthalene to form a fluorescent compound. With the hydride generation atomic absorption method, selenite (Se^{+4}) is converted to the hydride vapor and determined using atomic absorption. The methods are quantitative to 4 ng (in the test portion), which corresponds to about 20 ng/g (ppb) in distillers products.

10.3.3.5 Chromium (Cr)

Validated methods for the determination of chromium in animal feeds do not exist. Graphite furnace atomic absorption spectroscopy has been the method of choice for determining chromium in biological and/or agricultural matrices, with the limit of detection for this technology at 0.005 µg/L when an appropriate background correction method is used. Flame atomic absorption spectrometry and ICP-AES are not sensitive enough to detect chromium concentrations typically found in biological materials, but can be used to determine potentially toxic concentrations of chromium in feedstuffs (Mineral Tolerances of Animals, 2005). There is a need for fully validated methods for the determination of chromium in feedstuffs.

10.3.3.6 Chlorine (Cl)

Chlorine is generally measured in feedstuffs only in the form of chloride; other forms are not detectable. The most commonly used method is a potentiometric method, AOAC **969.10** (AOAC, 2005m). In this method, chloride is extracted into water and is titrated with silver nitrate using an electrode to determine the endpoint. It is a rugged and reproducible method. The next most commonly used method is a titrimetric method, AOAC **943.01**, commonly known as the Volhard method (AOAC, 2005g). In this method chloride is extracted with ferric sulfate under basic conditions, silver nitrate is added in excess to precipitate the chloride, and the excess silver nitrate is back titrated with potassium thiocyanate. Both methods are also detailed in ISO 6495:1999 (ISO, 1999).

10.3.3.7   Fluorine (F)

There is a long list of problems with the measurement of fluorine in any biological matrix. Appropriate sample preparation is the critical step, especially where only the free fluoride ion is measured. Fluorine is too reactive to be analyzed directly in biological/environmental materials, and its measurement is restricted to the detection of the free anion (F^-). The most widely used method is a potentiometric technique that measures the free anion using a fluoride ion-selective electrode. A number of other methods can be found in the literature, including gas chromatography, ion chromatography, capillary electrophoresis, atomic absorption, and photon activation. The most accurate method is a microdiffusion technique (Mineral Tolerances of Animals, 2005). Obviously, a fully validated method for fluorine feedstuffs is still needed.

10.3.3.8   Iodine (I)

Methods to determine iodine are only slightly better than those to determine fluorine. In nature, iodine exists as iodide, iodate, and molecular iodine. The first step, as with other elements, is a dissolution step to trap volatile forms, isolate iodine from the organic matrix, and convert the iodine into a form suitable for detection. Digestion methods include digestion with perchloric and nitric acids, alkali extraction with potassium hydroxide and tetramethylammonium hydroxide, and tetramethylammonium hydroxide alone. Ion chromatography, gas chromatography, and neutron activation have been used to determine iodine in feed digests. Due to the high sensitivity and selectivity of ICP-MS, it is more commonly used (Mineral Tolerances of Animals, 2005). ICP-MS is a potential technique for a fully validated method for iodine in feedstuffs.

10.3.4   Amino Acids

Amino acids are determined by hydrolyzing protein to liberate amino acids, followed by separation and determination of individual amino acids.

10.3.4.1   Standard Amino Acids

Alanine, arginine, aspartic acid, cystine, glutamic acid, glycine, histidine, isoleucine, leucine, lysine, methionine, phenylalanine, proline, serine, threonine, and valine are determined with AOAC Official Method **994.12** (AOAC, 2005u). This method utilizes performic acid oxidation prior to hydrolysis to oxidize cystine and methionine to cysteic acid and methionine sulfone, respectively. Sodium metabisulfite is added to decompose performic acid. Amino acids are liberated from protein by hydrolysis with 6 N HCl. Hydrolysates are evaporated or neutralized, pH is adjusted to 2.20 with sodium citrate buffer, and individual amino acid components are separated by ion exchange chromatography, followed by postcolumn derivatization with ninhydrin and measurement of the ninhydrin chromofore with visible detection. Tyrosine is partially destroyed by oxidation, and tryptophan is destroyed by acid hydrolysis, so those amino acids cannot be determined by this method. As an alternative to sodium metabisulfite, hydrobromic acid can be used to destroy the excess performic acid. With this alternative, tryptophan is still destroyed by hydrolysis and tyrosine, phenylalanine and histidine are destroyed during the oxidation process and by reaction with bromine, and cannot therefore be accurately measured. A similar method is ISO 13903:2005, which provides for the determination of free (synthetic and natural) and total (peptide-bound and free) amino acids in feed stuffs, using an amino acid analyzer or High Performance Liquid Chromatography (HPLC) equipment (ISO, 2005). It is applicable to the following amino acids: sum of cystine and cysteine; methionine; lysine; threonine; alanine; arginine; aspartic acid; glutamic acid; glycine; histidine; isoleucine; leucine; phenylalanine; proline; serine; tyrosine; valine.

AOAC **994.12** (AOAC, 2005u) also provides for alternatives to performic acid oxidation. If the oxidation step is left out, tyrosine is accurately quantitated; however, the sulfur amino acids are partially oxidized and the tryptophan is completely destroyed, so neither the sulfur amino acids nor tryptophan can be determined.

None of these methods distinguishes between the salts of amino acids, nor do they differentiate between D and L forms of amino acids.

10.3.4.2 Tryptophan

AOAC Method **988.15** (AOAC, 2005s) is used to determine tryptophan. Protein is hydrolyzed under vacuum with 4.2 M NaOH. After pH adjustment and clarification, tryptophan is separated by ion exchange chromatography with measurement of the ninhydrin chromophore or by reverse phase liquid chromatography with UV detection.

10.3.4.3 Alternative Methods

Alternative separation and detection systems are in practice, but at this point in time, none are collaboratively validated for distillers grains or other feedstuffs. These variations include various precolumn and postcolumn techniques. Studies currently underway by the National Corn-to-Ethanol Research Center, Edwardsville, Illinois are evaluating amino acid methods for distillers grain materials. Some of the chromatographic methods in the comparison include AOAC Official Method **994.12** Liquid Chromatographic (LC) Postcolumn Ninhydrin Derivatization (AOAC, 2005w) discussed above, LC precolumn OPA derivatization (Schuster, 1988), LC OPA postcolunn derivatization (Roth and Hampai, 1973), ultra performance LC (UPLC) precolumn derivatization (Boogers et al., 2008) and LC tandem mass spectroscopy (LC/MS/MS) precolumn derivatization (Zhang et al., 2006). Results should be forthcoming soon.

10.3.5 STARCH

Starch methods are typically enzymatic assays involving a gelatinization step followed by the enzymatic digestion and colorimetric detection of glucose. AOAC Method **920.40** (AOAC, 2005a) for starch in animal feeds is no longer valid because of discontinued production of the enzyme Rhozyme-S (Rohm and Haas, Philadelphia, PA) specified in the procedure. A comparison of existing methods was recently published in the *Journal of AOAC International* as an effort to find a potential replacement and a method for AOAC collaborative study (Hall, 2009; Hall and Keuler, 2009). Methods with known methodological defects were excluded and the three remaining enzymatic-colorimetric assays were studied and compared by Hall for applicability to animal feeds. The assays all used two-stage, heat-stable, α-amylase and amyloglucosidase hydrolyses, but they differed in the gelatinization solution. One method employed hot water, modified from the method of Holm (Holm et al., 1986); the second employed 3-(*N*-morpholino) propanesulfonic acid buffer, an extension of **AOAC 996.11** method for starch in cereal grains (AOAC, 2005v); and the third employed acetate buffer, modified from the method of Bach Knudsen (Knudsen, 1997). The method chosen to advance for collaborative study was that utilizing acetate buffer.

It should be noted that a commercialized version of AOAC **996.11** is available from Megazyme International Ireland Ltd as Total Starch (AA/AMG) Part No. K-TSTA, and this method has been reported in the literature for use with DDGS by some authors. The method was considered by Hall for use in feed labeling, but not chosen due to defects (Hall, 2009).

10.3.6 LABORATORY PROFICIENCY PROGRAMS FOR TESTING DISTILLERS GRAINS

The *Association of American Feed Control Officials*, Inc. (AAFCO) Feed Check Sample Program (www.aafco.org/NewsInfo/CheckSampleProgram/tabid/74/Default.aspx) is a laboratory proficiency program for analyzing animal feeds and pet foods. A subscribing laboratory receives specially ground and mixed samples on a monthly basis. The series, which begins in January of each year, includes a variety of feeds and supplements and is an excellent tool for laboratories to monitor performance on various methods. Because data is reported on individual methods, it is also a tool for assessing the performance of analytical methods in laboratories around the world. Distillers grains was offered as proficiency testing materials in the AAFCO Check Sample Program in 2006, in 2008, and is scheduled in the future.

10.3.7 DATABASES OF ANALYTICAL DATA FOR DDGS

Two databases are in the public domain at the time of this publication that house representative data on distillers grain. They are http://www.ddgs.umn.edu/profiles.htm and www.value-added.org/renewableEnergy/ethanol/ddgs/.

10.4 ANALYTICS FOR CONTAMINANTS

10.4.1 Mycotoxins

Mycotoxins are naturally occurring, secondary metabolites of molds and fungi and are common contaminants of corn. Each type thrives in specific growing, harvesting, or storage conditions; however, more than one type of mycotoxin is often found in a contaminated product. The general toxicity of the various mycotoxins is described in the 2003 Task Force Report No. 139 of the Council for Agricultural Science and Technology (CAST, 2003). For some of the "families" of mycotoxins there are good reports from the World Health Organization (WHO) by the Joint Committee on Food Additives (JECFA). These are listed at www.inchem.org/pages/jecfa.html, are very comprehensive, and are free (Maragos, 2010).

Mycotoxin concentrations in distillers grains are of concern when climatic or storage conditions have encouraged their growth in corn because they have varying adverse affects on human and animal health. Due to the removal of starch from the corn, mycotoxins are concentrated three to four times in the distillers grains. Mycotoxins of the most concern in distillers products are aflatoxin, fumonisins, trichothecenes (deoxynivalenol, T-2 toxin), zearalenone, and ochratoxin. While mycotoxin growth in the United States is not a widespread problem every year, cool damp climatic conditions in 2009 favored the growth of mycotoxins in the corn crop in many areas of the United States; therefore problems with mycotoxins in DDGS escalated in 2009 and 2010. Additional information on mycotoxin occurrence in DDGS can be found in Chapter 11.

The analysis of mycotoxins begins with sampling; mycotoxins are the classic example of an analyte difficult to sample. The variability for each step of the mycotoxin test procedure, as measured by the variance statistic, is shown to increase with mycotoxin concentration. Sampling is usually the largest source of variability associated with the mycotoxin test procedure. Sampling variability is large because a small percentage of kernels are contaminated and the level of contamination on a single seed can be very large (Whitaker, 2004). It is therefore necessary to collect and composite more and larger subsamples for the laboratory and to adhere to established sampling protocols, where they are established. Sampling for mycotoxins is discussed in greater detail in Chapter 11.

The immunoassay-based methods are useful as screening tools. They offer high specificity with rapidity. These semiquantitative or quantitative tests are of three types depending on the observed endpoint: radioimmunoassays (RIA), fluorogenic or fluorescent immunoassays (FIA), or chromogenic or enzyme immunoassays (ELISA or EIA). ELISA rapid test methods are often used for screening mycotoxin samples due to their speed and the throughput such methods allow. Other rapid methods in use include lateral flow devices, such as immunodipsticks, immunostrips, and immunofiltration. These methods are recommended to be used with caution due to cross-reactivity and false positives. Anything that interferes with antibody binding in the assay can be misconstrued as toxin. Therefore factors like pH, solvents, and matrix components can have dramatic effect on assay results.

Third party validations for rapid test kits are performed by USDA's Grain Inspection, Packers and Stockyards Administration (GIPSA), and by AOAC International Research Institute. The validation of rapid test kits is mostly, if not wholly, based on spiked samples, a procedure that can lead to false confidence in a test method to successfully measure a naturally occurring analyte. In the experience of some laboratories, a mycotoxin test kit that is validated only by using spiked samples has not been sufficiently validated. A good tool to validate a rapid test kit for mycotoxins is by comparison to a fully validated liquid chromatographic (LC) method using naturally contaminated product (Van Hulzen). Increasingly, certified reference materials are being made available for mycotoxins in commodities and foods. The use of such materials is another way to help ensure the validity of the protocols.

Readers should be aware of these issues, request validation data for distillers grain from the kit manufacturers, and perform their own in-house validation when choosing a kit to use for screening purposes. Suspect values should be verified by a second validated method of a different technology. A list of all methods approved by GIPSA is posted on the GIPSA web site at archive.gipsa.usda. gov/pubs/mycobook.pdf (GIPSA, 2009) and the methods approved by AOAC RI are posted on the AOAC website at www.aoac.org/testkits/testedmethods.html (AOACRI, 2010). Rapid test kit methods that have received some form of validation (GIPSA, manufacturer or AOAC RI) for mycotoxins in distillers grain are listed in Table 10.4.

Although LC methods are considered the "gold standard" in the detection of mycotoxins because of the reliability and low limits of quantitation, it is important to make sure that these methods have been validated as well. Due to the complexity, LC methods are more costly to run, both in terms of equipment and operator training. LC mass spectroscopy (LC/MS) and LC tandem mass spectroscopy (LC/MS/MS) methods are also becoming analytical methods of choice for many for mycotoxin analysis, particularly in cases where the identity of the toxin must be confirmed (Krska et al., 2007, 2008; Ren et al., 2007; Songsermsakul and Razzazi-Fazeli, 2008; Sulyok et al., 2007).

Mycotoxins are often isolated from DDGS using clean-up columns designed to remove matrix interferences prior to LC analysis. There are two different types of clean-up columns that are commonly used for chromatographic mycotoxin analysis: (1) solid phase extraction columns (SPE) and (2) immunoaffinity columns. The SPE columns are of two major types. The most common type of SPE column works by trapping the analyte and allowing the impurities to pass through; the toxin is then eluted from the column for detection. The second type of SPE column traps impurities and allows the analyte to pass through (one example is the Romer MycoSep column). The immunoaffinity columns function like the first type of SPE column in that they trap the analyte, but they use antibodies specific to the toxin being measured rather than a less specific adsorbent to retain the analyte. Users should be aware of the limitations of immunoaffinity columns. If the specificity of the antibody chosen is too high, it may not retain all the related toxins, allowing some to flow through with the matrix producing erroneously low results. An example would be with the aflatoxins: antibody with high specificity for aflatoxin B1 and low cross-reactivity to aflatoxin G1 would retain all of the aflatoxin B1, but might not completely retain all of the aflatoxin G1. Conversely, if the specificity of the antibody is too low (not specific enough), it can retain molecules other than the toxin being measured and, depending on the detection system, potentially lead to erroneously high results. This would most commonly occur with some of the metabolites of the toxin, which typically have a similar chemical structure and may not be distinguished by the antibody used in the column. This issue should be able to be resolved with good chromatography; however, if the chromatographic method is unable to separate the analyte from its metabolites and if the resulting co-elutant is coquantified, it could lead to erroneously high results (Van Hulzen). If either the SPE or immunoaffinity columns are overloaded with analyte, the amount of adsorbent or antibody may be insufficient to retain all the toxin molecules and this will produce erroneously low results. Generally speaking, SPE columns are cheaper and easier to use. However, when the specificity of antibody is appropriate, analysts may prefer the immunoaffinity columns because they usually give a much cleaner extract, meaning that the chromatographic conditions during the detection step can be less demanding (Maragos, 2010).

Excellent reviews on the general state of the art of mycotoxin analysis are available in the past annual report of the AOAC International Committee on Natural Toxins and Food Allergens (Fremy et al., 2002; Trucksess, 2003; Hungerford and Trucksess, 2004; Trucksess, 2005, 2006;

TABLE 10.4
Rapid Test Kit Methods for Mycotoxins in Distillers Grains

Toxin	Type		Company	Test Kit Name (Part Number)	Concentration Range or Cut Off	Validation Source				Test Format
						For DDGS		For Other Products		
	Quantitative	Qualitative				GIPSA	Manufacturer	GIPSA	AOAC RI	
Aflatoxin	X		Charm Sci	ROSA®Alfatoxin (LF-AFQ)	5–150 ppb	X				Lateral flow strip
Aflatoxin	X		Neogen	Veratox Alfatoxin (8030)	5–50 ppb		X	X	X	ELISA
Aflatoxin	X		R-Biopharm	Ridascreen FAST Aflatoxin SC (R9002)	3–100 ppb	X		X		Microtiter well plate
Aflatoxin	X		Romer	FluoroQuant Afla IAC (COKFA 4010)	5–100 ppb	X				Fluorometric immunoaffinity column
Aflatoxin	X		Romer	AgraQuant Afla (COKAQ1100)	2–20 ppb		X			ELISA
Aflatoxin	X		Romer	AgraQuant Afla (COKAQ1000)	4–40 ppb		X			ELISA
Aflatoxin	X		Vicam	Alfatest (12022)	5–100 ppb	X				Immunoaffinity column
Aflatoxin		X	Neogen	Agri-Screen for Aflatoxin (8010)	20 ppb		X			ELISA
Aflatoxin		X	Romer	AgraStrip Alfa (COKAS1000DSU)	20 ppb		X			Lateral flow strip
Fumonisins	X		Charm Sci	ROSA Fumonisin (LF-FUMQ-ETOH)	0.25–6.0 ppm		X	X		Lateral flow strip
Fumonisins	X		Neogen	Veratox Fumonisin (8830)	1.0–6.0 ppm		X	X		ELISA
Fumonisins	X		R-Biopharm	Ridascreen FAST Fumonisin (R5602)	0.22–6 ppm		X		X	Microtiter well plate
Fumonisins	X		Romer	AgraQuant Total Fumonisin (COKAQ3000)	0.5–5.0 ppm	X				ELISA
Fumonisins	X		Vicam	FumoniTest (G1029)	0.5–5.0 ppm		X	X		Fluorometric, solid phase cleanup
Fumonisins		X	Neogen	Agri-Screen for Fumonisin (8810)	5 ppm		X			ELISA
DON*	X		Charm Sci	ROSA® DON (LF-DONQ)	0.25–6 ppm		X	X		Lateral flow strip
DON*	X		Neogen	Veratox DON (8335)	0.5–5.0 ppm		X	X		ELISA
DON*	X		R-Biopharm	Ridascreen FAST DON (R5901)	0.22–6 ppm		X		X	Microtiter well plate
DON*	X		Romer	AgraQuantDON (COKAQ 4000)	0.25–5.0 ppm		X	X		ELISA
DON*		X	Neogen	Agri-Screen DON (8310)	1 ppm		X	X		ELISA
Zearalenone	X		Charm Sci	ROSA® Zeralenone (LF-ZEARQ)	25–1400 ppb	X				Lateral flow strip
Zearalenone	X		Neogen	Veratox Zearalenone (8110)	25–500 ppb		X			ELISA
Zearalenone	X		Romer	AgraQuant ZON (COKAQ5000)	40–1000 ppb		X			ELISA
T-2/H-2	X		Neogen	Veratox T-2/H-2	25–250 ppb		X			ELISA
T2	X		Romer	AgraQuant T2	75–500 ppb		X			ELISA

Note: Many of these test DON kits require a pH adjustment for use with distillers grains.

Goto et al., 2007; Shepard, 2008). These summaries are now published elsewhere, with the 2010 summary available from *World Mycotoxin Journal* (Shephard, 2010).

10.4.1.1 Aflatoxin

Aflatoxins are a group of mycotoxins produced by some *Aspergillus* species like *A. flavus* or *A. parasiticus*. The primary aflatoxins are B1, B2, G1, and G2 with aflatoxin B1 the most frequently occurring. Nursing animals may be severely affected by a toxic derivative of aflatoxin (aflatoxin M1) that can be passed through milk. High temperatures, drought stress, and insect injury may contribute to increased aflatoxin contamination in maize (Schmale and Munkvold, 2009).

Due to the carcinogenetic properties of aflatoxins, in 1969, the U.S. Food and Drug Administration (FDA) set an action level for aflatoxins at 20 ppb for all foods, including animal feeds, based on FDA's analytical capability and the agency's aim of limiting aflatoxin exposure to the lowest possible level. Animal feeding studies conducted in the 1970s and 1980s, however, demonstrated that levels of aflatoxins above 20 ppb could be fed to certain food-producing animals without presenting a danger to the health of these animals or posing a risk to consumers of food derived from the exposed animals. On the basis of these scientific studies, the agency revised its action level in 1982 to 300 ppb for aflatoxins in cottonseed meal intended for use as a feed ingredient for beef cattle, swine, and poultry; and in 1989 to varying levels for corn intended for use as a feed ingredient for subgroups of the same animals. In 1990, FDA issued guidance that aflatoxins in peanut products (i.e., peanuts, peanut meal, peanut hulls, peanut skins, and ground peanut hay) intended for use as a feed ingredient are no more toxic to these same subgroups of animals than is aflatoxin in corn. FDA action levels for total aflatoxins in livestock feed (USFDA, 1994; Henry, 2006) and FDA action levels for aflatoxins in food (CFSAN, 2000) are provided in Table 10.5.

Rapid test kit methods that have that received some form of validation (GIPSA, manufacturer or AOAC RI) for aflatoxin in distillers grain are listed in Table 10.4. Other nonkit validated methods include liquid chromatographic methods validated by AOAC and ISO. AOAC Official Method 2003.02 Aflatoxin B1 in Cattle Feed (AOAC 2005ac), is an immunoaffinity column liquid chromatography method for Aflatoxin B1 in cattle feed. ISO 17375:2006 specifies a method for the determination of aflatoxin B1 in animal feeding stuffs using high-performance liquid chromatography with postcolumn derivatization; it is applicable to animal feeds with a fat content of up to 50%.

TABLE 10.5
FDA Action Levels for Total Alfatoxin in Animal Feeds and Human Foods

Class of Animals/Human food	Feed Type or Commodity or Food	Action Level, Aflatoxin Concentration (ppb)
Finishing beef cattle	Corn and peanut products	300
Beef cattle, swine or poultry	Cottonseed meal	300
Finishing swine over 100 lb	Corn and peanut products	200
Breeding cattle, breeding swine and mature poultry	Corn and peanut products	100
Immature animals	Animal feeds and ingredients, excluding cottonseed meal	20
Dairy animals, animals not listed above, or unknown use	Animal feeds and ingredients	20
Human food	Brazil nuts	20
Human food	Foods	20
Human food	Milk	0.5 (aflatoxin M1)
Human food	Peanuts and peanut products	20
Human food	Pistachio nuts	20

Sources: Adapted from USFDA, 1994; Henry, 2006; CFSAN, 2000.

A multi-analyte liquid chromatographic method validated for aflatoxins, ochratoxin A, zearale-neone, and fumonisins in distillers grains is offered by Pickering Laboratories. In this method aflatoxins B1, B2, G1, and G2 are separated on reversed phase C18 column, derivatized with a postcolumn photochemical reactor, and detected by fluorescence (Ofitserova et al., 2007, 2009; Multi-Residue Analysis of Mycotoxins in Dry Distillers Grains, 2007).

10.4.1.2 Fumonsins

The fumonisins are a group of mycotoxins produced primarily by *Fusarium verticillioides* and *Fusarium proliferatum*, although a few other *Fusarium* species also may produce them. *Fusarium verticillioides* and fumonisins are distributed worldwide. Infection is increased if kernels are physically damaged, especially by insect feeding, or if drought stress is followed by warm, wet weather. Fungal growth and fumonisin production cease when grain is dried below about 19% moisture content, but the fumonisins remain intact, and the fungus can grow and produce additional fumonisins in storage if proper conditions are not maintained. There are at least 28 different forms of fumonisins, most designated as A-series, B-series, C-series, and P-series. Fumonisin B_1 is the most common and economically important form, followed by B_2 and B_3. Maize is the most commonly contaminated crop, and fumonisins are the most common mycotoxins in maize, although these toxins can occur in a few other crops as well. The primary health concerns associated with fumonisins are acute toxic effects in horses and swine, and carcinogenic properties (Schmale and Munkvold, 2009).

Based on the wealth of available information on the adverse animal health effects associated with fumonisins, FDA determined that human health risks associated with exposure to fumonisins are possible. In 2001, FDA published guidance for the maximum levels for fumonisins in human foods and in animal feeds. The levels were based on what FDA considered adequate to protect human and animal health and that FDA considered achievable in human foods and animal feeds with the use of good agricultural and good manufacturing practices. FDA guidance levels for total fumonisins in animal feeds and human foods are provided in Table 10.6 (USFDA, 2001; Henry, 2006).

Rapid test kit methods that are known to have been validated for fumonisins in distillers grain are listed in Table 10.4. A multi-analyte liquid chromatographic method validated for aflatoxins, ochratoxin A, zearaleneone, and fumonisins in distillers grains is offered by Pickering Laboratories. In this method fumonisin B1 and B2 are separated on reversed phase C18 column, derivatized post-column with *o*-phthalaldehyde (OPA), and detected by fluorescence (Ofitserova et al., 2007, 2009; Multi-Residue Analysis of Mycotoxins in Dry Distillers Grains, 2007).

10.4.1.3 Trichothecenes

The trichothecenes are the largest group of mycotoxins known to date, consisting of more than 200 chemically related toxic compounds that have a strong impact on the health of animals and humans due to their immunosuppressive effects. These mycotoxins are produced by several species of *Fusarium*, *Stachybotrys*, *Trichoderma*, and *Trichothecium* fungi, most notably *Fusarium graminearum* and *Fusarium sporotrichioides*. The most important structural features causing the biological activities of the trichothecences are the 12, 13-epoxy ring, the presence of hydroxyl or acetyl groups, and the structure and position of the side chains (Schmale and Munkvold, 2009).

Fusarium trichothecenes have been classified based on structural properties. The trichothecences are divided into two groups: macrocyclic and nonmacrocyclic. The nonmacrocyclic group is also divided into the Type A and Type B trichothecences. Type A and Type B trichothecenes differ in their A ring oxygenation pattern. Type A trichothecenes have a C-8 hydroxyl (neosolaniol), C-8 ester (T-2 toxin), or no C-8 substitution (diacetoxyscirpenol or DAS) and Type B trichothecenes have a C-8 keto group (deoxynivalenol, nivalenol, 3- and 15-acetyldeoxynivalenol, and 4-acetylnivalenol).

TABLE 10.6
FDA Guidance Levels for Total Fumonisins in Animal Feeds and Human Foods

Class of Animal/Human food	Feed Ingredient & Portion of Diet or Type of Food	Levels in Corn & Corn Byproducts (ppm)	Levels in Finished Feed/Human Food (ppm)
Equids and rabbits	Corn and corn byproducts not to exceed 20% of the diet[b]	5	1
Swine and catfish	Corn and corn byproducts not to exceed 50% of the diet[b]	20	10
Breeding ruminants, breeding poultry and breeding mink[a]	Corn and corn byproducts not to exceed 50% of the diet[b]	30	15
Ruminants ≥3 months old being raised for slaughter and mink being raised for pelt Production	Corn and corn byproducts not to exceed 50% of the diet[b]	60	30
Poultry being raised for slaughter	Corn and corn byproducts not to exceed 50% of the diet[b]	100	50
All other species or classes of livestock and pet animals	Corn and corn byproducts not to exceed 50% of the diet[b]	10	5
Human food	Degermed dry milled corn products (e.g., flaking grits, corn grits, corn meal, corn flour with fat content of <2.25%, dry weight basis)		2
Human food	Whole or partially degermed dry milled corn products (e.g., flaking grits, corn grits, corn meal, corn flour with fat content of ≥2.25%, dry weight basis)		4
Human food	Dry milled corn bran		4
Human food	Cleaned corn intended for mass production		4
Human food	Cleaned corn intended for popcorn		3

Sources: Adapted from USFDA, 2001; Henry, 2006.

[a] Includes lactating dairy cattle and hens laying eggs for human consumption.

[b] Dry weight basis.

10.4.1.3.1 Deoxynivalenol (Vomitoxin)

The most important trichothecene mycotoxin in the United States is deoxynivalenol (DON), a common contaminant of wheat, barley, and maize. DON is commonly called vomitoxin because of its deleterious effects on the digestive system of monogastric animals. This mycotoxin is produced by *Fusarium graminearum* that often occurs on corn (Gibberella ear rot). The mold usually develops during cool damp weather, resulting in a white or reddish fungus (Schmale and Munkvold, 2009).

In 1993, due to an unusually wet summer in the Midwest, the FDA issued an advisory for vomitoxin. Based upon the data and information available to them, FDA advised the levels of DON in wheat and wheat derived products that would not appear to present a public health hazard. The levels were also applied to other grains used as animal feeds. These advisory levels are provided in Table 10.7 (USFDA, 1993; Henry, 2006). The FDA advisory levels for DON in human food is 1 ppm DON for finished wheat products, for example, flour, bran, and germ, that may potentially be consumed by humans (USFDA, 1993) (Table 10.7).

Rapid test kit methods that have undergone some form of validation for DON in distillers grain are listed in Table 10.4. Gas chromatography–mass spectrometry (GS–MS) is used for the determination

TABLE 10.7

Advisory Levels for Vomitoxin (DON) in Livestock Feed and Human Foods

Class of Animal/Human Food	Feed Ingredient & Portion of Diet or Food	DON Levels (ppm) in		
		Grains & Grain Byproducts	Finished Feed	Food
Ruminating beef and feedlot cattle older than 4 months	Grain and grain byproducts not to exceed 50% of the diet	10	5	
Chickens	Grain and grain byproducts not to exceed 50% of the diet	10	5	
Swine	Grain and grain byproducts not to exceed 20% of the diet	5	1	
All other animals	Grain and grain byproducts not to exceed 40% of the diet	5	2	
Human food	Finished wheat products, for example, flour, bran, and germ			1

Sources: Adapted from USFDA, (1993); Henry, (2006).

of DON, 3-acetyldeoxynivalenol, 15-acetyldeoxynivalenol, nivalenol, T-2 toxin, iso T-2 toxin, acetyl-T-2 toxin, HT-2 toxin, T-2 triol, T-2 tetraol, fusarenon-X, diacetoxyscirpenol, scirpentriol, 15-acetoxyscirpentriol, neosolaniol, zearalenone, and zearalenol. The mycotoxins are extracted with acetonitrile:water (84:16). An aliquot of the supernatant is filtered through 1.5 g of C18:alumina (1:1) and an aliquot of the eluent is evaporated at 65°C. The residue is derivatized (*N*-trimethylsilyimidazole + trimethylchlorosilane + *n,o*-bis[trimethylsily]triflouroacetamide + pyridine) to form trimethylsily (TMS) ester derivatives of tricothenes and estrogens. The TMS-mycotoxin derivatives are then separated on a gas chromatograph using capillary column, and assayed by select ion monitoring and electron ionization in a mass spectrometer, using three or four ion fragments for identification and quantification of each mycotoxin (Groves et al., 1999; Raymond et al., 2003).

10.4.1.3.2 T-2 Toxin

T-2 toxin production is greatest with increased humidity and cool temperatures (6°C–24°C) (Romer Labs). T-2 is believed to have been associated with outbreaks of a human toxicosis known as alimentary toxic aleukia that resulted from consumption of cereal grains that had overwintered in fields. It also has been considered as a potential bioweapon agent.

There are no current levels of FDA action, advisory, or guidance established for T-2 toxin in animal feed. Rapid test kit methods that are known to have been validated for T-2 in distillers grain are listed in Table 10.4. A gas chromatography–mass spectroscopy (GC–MS) method previously described in 3.1.3.1. is also used to determine T-2 toxin, iso T-2 toxin, acetyl-T-2 toxin, HT-2 toxin, T-2 triol, T-2 tetraol, in addition to other trichothecenes.

10.4.1.4 Zearalenone

Zearalenone is a mycotoxin that mimics the reproductive hormone estrogen and hence affects reproduction. This mycotoxin is produced primarily by the fungus *Fusarium graminearum*, the same fungus that produces deoxynivalenol in maize and small grains. Zearalenone contamination is economically important in maize and high humidity and low temperatures favor the production of zearalenone by *F. graminearum* in maize (Schmale and Munkvold, 2009).

There is no current FDA action, advisory, or guidance levels established for zearalenone in U.S. feed. Rapid test kit methods that are known to have been validated for zearalenone in distillers grain are listed in Table 10.4. ISO 17372:2008, determination of zearalenone by immunoaffinity column chromatography and high-performance LC, is applicable to the analysis of zearalenone in animal feed and feed ingredients, including barley, corn, oats, rye, wheat, soybean meal, canola (rapeseed) meal, corn gluten, dried distillers' grains, lentils, and sugar beet pulp. The limit of quantification is 0.05 mg/kg (50 µg/kg). A lower limit of quantification may be achievable subject to appropriate validation being conducted by the user laboratory. Another liquid chromatographic multi-analyte method validated for aflatoxins, ochratoxin A, zearaleneone, and fumonisins in distillers grains is offered by Pickering Laboratories. In this method zearalenone is separated on reversed phase C18 column and detected by fluorescence (Ofitserova et al., 2007, 2009; Multi-Residue Analysis of Mycotoxins in Dry Distillers Grains, 2007). A gas chromatography–mass spectrometry (GC–MS) method used for the determination of zearalenone, and zearalenol, along with trichothecences is described in some detail in Section 3.1.3.1 (Groves et al., 1999; Raymond et al., 2003).

10.4.1.5 Ochratoxin A

Ochratoxin A, B, and C are mycotoxins produced by *Aspergillus* sp. (*A. ochraceus*) and *Penicillium* sp. (*P. viridicatum*). Ochratoxin contamination is economically important in cereal grains, grapes, coffee, tree nuts, and figs. At least nine ochratoxins have been identified, but ochratoxin A is the most common and has the greatest toxicological significance, being nephrotoxic and a suspected carcinogen. Citrinine and oxalic acid are also produced by those molds. Ochratoxins are ubiquitous in both tropical and temperate climates. Ochratoxins may be transferred through milk, blood, and meat (Schmale and Munkvold, 2009).

There is no FDA action, advisory, or guidance levels established for ochratoxin A in U.S. feed. There are no known rapid test kits that have been validated for ochratoxin in distillers grains. A liquid chromatographic multi-analyte method validated for aflatoxins, ochratoxin A, zearaleneone, and fumonisins in distillers grains is offered by Pickering Laboratories. In this method ochratoxin A is separated on reversed phase C18 column and detected by fluorescence (Ofitserova et al., 2007, 2009; Multi-Residue Analysis of Mycotoxins in Dry Distillers Grains, 2007).

10.4.2 Antibiotic Residues

The FDA Center for Veterinary Medicine (CVM) recently announced an assignment for a nation-wide survey in 2010 to determine the extent and level of antibiotic residues in a limited number of domestic and imported samples of distillers grains (McChesney, 2009). FDA CVM expressed concern that if antibiotic residues were present in distillers grains, there would be potential for subsequent animal and human health hazards from the use of distillers products used as animal feed ingredients. A provision of the announcement was that FDA laboratories were encouraged, but are not required, to analyze the distillers grains for mycotoxins (aflatoxins, fumonisins, vomitoxins, zearalenone, and/or ochratoxin A) and elements (especially sulfur). Distillers grains were defined as feed ingredients that meet the Association of American Feed Control Officials Incorporated (AAFCO) definitions for distillers dried grains (DDG), distillers wet grains (DWG), DDGS, distillers dried solubles (DDS), and condensed distillers solubles (CDS) and are offered for sale or for use in animal feed (2010 Official Publication of the Association of American Feed Control Officials, 2010).

In the announcement, FDA CVM provided justification for the survey as follows (McChesney, 2009):

> Currently, antibiotics such as virginiamycin (VIR), penicillin (PEN) and erythromycin, etc. are believed to be used in the U.S. as antimicrobials during the fermentation process to control bacterial contamination. Bacterial contamination of the fermenter is an ongoing problem for commercial fuel ethanol production facilities. Both chronic and acute contamination is of concern since bacteria compete with the ethanol-producing yeast for sugar substrates and micronutrients. Lactic acid levels often rise during bouts of contamination, suggesting that the most common contaminants are lactic acid producing bacteria. FDA CVM is concerned because the available literature has shown that not all antibiotics are markedly degraded or metabolized to inactive compounds during ethanol production. In fact, some antibiotics may be concentrated in the distillers grains where they may be present at levels as high as three times the level added to the fermenter on a dry weight basis.

The analytical method specified by FDA CVM for antibiotic residues in distillers grains is LIB 4438 "Analysis of Antibiotic in Distillers Grains Using Liquid Chromatography and Ion Trap Tandem Mass Spectrometry" (Heller, 2009). The method was developed by FDA CVM for the screening, confirmation, and determination of 13 antibiotics in distillers grain: streptomycin, ampicillin, oxytetracycline, chlortetracycline, bacitracin A, erythromycin, tylosin, chloramphenicol, clarithromycin, penicillin G, virginiamycin M1, and monensin. Extraction efficiency ranged from 65% to 97%, and limits of quantitation ranged from 0.1 to 1.0 µg/g. Method accuracy ranged from 88% to 111% with CVs ranging from 4% to 30%, depending on the compound. The method was also tested in DWG and corn syrup residue from ethanol production and method performance in those matrices was comparable to DDG.

It was reported at the January 2010 meeting of the AAFCO Laboratory Method and Services Committee that Phibro's Ethanol Process laboratory in St. Paul, MN, in cooperation with a commercial laboratory, has validated a microbiological method for the determination of virginiamycin in distillers grain and animal feed that provides good recoveries in the range of 0.5–1.5 ppm (Reimann, 2010). Virginiamycin consists of two major factors and several minor factors that interact in a synergistic fashion and therefore should be assayed using a biological activity-based assay. The method may be obtained from Phibro (QA@Phibro.com).

10.5 SUMMARY

This chapter has provided a discussion on analytical methods for distillers grain for three primary purposes: for trading (or quality standards); to determine nutrients for inclusion in livestock feeds; and to determine contaminants that may limit inclusion in livestock feeds. Perhaps more importantly, it has tried to provide the reader with a sense of the importance of method validation, method types, and an introductory sense of fit for purpose methods.

In summary, methods for trading purposes include moisture, crude protein, crude fat, and crude fiber; nutrient methods for incorporation of DDGS into animal feeds include detergent fibers, ash, trace elements, amino acids, and starch in addition to the four listed for trading purposes; and known contaminants of DDGS that warrant analysis includes various mycotoxins and antibiotics.

REFERENCES

Official Publication of the Association of American Feed Control Officials. 2010. *Association of American Feed Control Officials, Inc.* Oxford, IN.

AFIA. 2007. *Evaluation of Analytical Methods for Analysis of Distillers Grain with Solubles: AFIA Sub Working Group Final Report.* American Feed Industry Association [cited March 2010a]. Available online: www.afia.org/Afia/Files/BAMN-%20BSE-%20DDGS-%20Biosecurity%20Awareness/DDGS%20 FINAL%20Report%20and%20Recommendations2-07.pdf. Accessed February 15, 2011.

AFIA. 2007. *Evaluation of Analytical Methods for Analysis of Distillers Grain with Solubles: AFIA Sub Working Group Final Report.* Distillers Grain Technology Council [cited February 2010b]. Available online: www.distillersgrains.org/files/grains/Typical%20Analysis.pdf. Accessed February 15, 2011.

AOAC. 2005a. AOAC Official Method 920.40 Starch in Animal Feed. In *Official Methods of Analysis of AOAC International (OMA).* AOAC International, Gaithersburg, MD.

AOAC. 2005b. AOAC Official Method 923.01 Sulfur in Plants. In *Official Methods of Analysis of AOAC International (OMA).* AOAC International, Gaithersburg, MD.

AOAC. 2005c. AOAC Official Method 930.15 Loss on Drying (Moisture) for Feed (at 135C for 2 Hours); Dry Matter on Oven Drying for Feed (at 135 C for 2 Hours) In *Official Methods of Analysis of AOAC International (OMA).* AOAC International, Gaithersburg, MD.

AOAC. 2005d. AOAC Official Method 934.01 Loss on Drying (Moisture) at 95–100 C for Feed; Dry Matter on Oven Drying at 95–100 C for Feed. In *Official Methods of Analysis of AOAC International (OMA).* AOAC International, Gaithersburg, MD.

AOAC. 2005e. AOAC Official Method 935.29 Moisture in Malt. In *Official Methods of Analysis of AOAC International (OMA).* AOAC International, Gaithersburg, MD.

AOAC. 2005f. AOAC Official Method 942.05 Ash in Animal Feed. In *Official Methods of Analysis of AOAC International (OMA)*. AOAC International, Gaithersburg, MD.

AOAC. 2005g. AOAC Official Method 943.01 Chlorine (Soluble) in Animal Feed. In *Official Methods of Analysis of AOAC International (OMA)*. AOAC International, Gaithersburg, MD.

AOAC. 2005h. AOAC Official Method 945.16 Oil in Cereal Adjuncts. In *Official Methods of Analysis of AOAC International (OMA)*. AOAC International, Gaithersburg, MD.

AOAC. 2005i. AOAC Official Method 954.01 Protein (Crude) in Animal Feed and Pet Food. In *Official Methods of Analysis of AOAC International (OMA)*. AOAC International, Gaithersburg, MD.

AOAC. 2005j. AOAC Official Method 954.02 Fat (Crude) or Ether Extract in Pet Food. In *Official Methods of Analysis of AOAC International (OMA)*. AOAC International, Gaithersburg, MD.

AOAC. 2005k. AOAC Official Method 965.17 Phosphorus in Animal Feed and Pet Food. In *Official Methods of Analysis of AOAC International (OMA)*. AOAC International, Gaithersburg, MD.

AOAC. 2005l. AOAC Official Method 968.08 Minerals in Animal Feed and Pet Food. In *Official Methods of Analysis of AOAC International (OMA)*. AOAC International, Gaithersburg, MD.

AOAC. 2005m. AOAC Official Method 969.10 Chlorine (Soluble) in Animal Feed. In *Official Methods of Analysis of AOAC International (OMA)*. AOAC International, Gaithersburg, MD.

AOAC. 2005n. AOAC Official Method 973.18 Fiber (Acid Detergent) and Lignin (H_2SO_4) in Animal Feed. In *Official Methods of Analysis of AOAC International (OMA)*. AOAC International, Gaithersburg, MD.

AOAC. 2005o. AOAC Official Method 976.06 Protein (Crude) in Animal Feed and Pet Food. In *Official Methods of Analysis of AOAC International (OMA)*. AOAC International, Gaithersburg, MD.

AOAC. 2005p. AOAC Official Method 978.10 Fiber (Crude) in Animal Feed and Pet Food. In *Official Methods of Analysis of AOAC International (OMA)*. AOAC International, Gaithersburg, MD.

AOAC. 2005q. AOAC Official Method 984.13 Protein (Crude) in Animal Feed and Pet Food. In *Official Methods of Analysis of AOAC International (OMA)*. AOAC International, Gaithersburg, MD.

AOAC. 2005r. AOAC Official Method 988.05 Protein (Crude) in Animal Feed and Pet Food. In *Official Methods of Analysis of AOAC International (OMA)*. AOAC International, Gaithersburg, MD.

AOAC. 2005s. AOAC Official Method 988.15 Tryptophan in Foods and Food and Feed Ingredients. In *Official Methods of Analysis of AOAC International (OMA)*. AOAC International, Gaithersburg, MD.

AOAC. 2005t. AOAC Official Method 990.03 Protein (Crude) in Animal Feed. In *Official Methods of Analysis of AOAC International (OMA)*. AOAC International, Gaithersburg, MD.

AOAC. 2005u. AOAC Official Method 994.12 Amino Acids in Feeds. In *Official Methods of Analysis of AOAC International (OMA)*. AOAC International, Gaithersburg, MD.

AOAC. 2005v. AOAC Official Method 996.11 Starch (Total) in Cereal Products. In *Official Methods of Analysis of AOAC International (OMA)*. AOAC International, Gaithersburg, MD.

AOAC. 2005w. AOAC Official Method 996.16 Selenium in Feeds and Premixes. In *Official Methods of Analysis of AOAC International (OMA)*. AOAC International, Gaithersburg, MD.

AOAC. 2005x. AOAC Official Method 996.17 Selenium in Feeds and Premixes. In *Official Methods of Analysis of AOAC International (OMA)*. AOAC International, Gaithersburg, MD.

AOAC. 2005y. AOAC Official Method 2001.11 Protein (Crude) in Animal Feed, Forage (Plant Tissue), Grain and Oilseeds. In *Official Methods of Analysis of AOAC International (OMA)*. AOAC International, Gaithersburg, MD.

AOAC. 2005z. AOAC Official Method 2001.12 Water/Dry Matter (Moisture) in Animal Feed, Grain and Forage (Plant Tissue). In *Official Methods of Analysis of AOAC International (OMA)*. AOAC International, Gaithersburg, MD.

AOAC. 2005aa. AOAC Official Method 2002.04 Amylase-Treated Neutral Detergent Fiber in Feeds. In *Official Methods of Analysis of AOAC International (OMA)*. AOAC International, Gaithersburg, MD.

AOAC. 2005ab. AOAC Official Method 2003.05 Crude Fat in Feeds, Cereal Grains, and Forages. In *Official Methods of Analysis of AOAC International (OMA)*. AOAC International, Gaithersburg, MD.

AOAC. 2005ac. AOAC Official Method 2003.02 Aflatoxin BI in Cattle Feed. In *Official Methods of Analysis of AOAC International (OMA)*. AOAC International, Gaithersburg, MD.

AOAC. 2005ad. AOAC Official Method 2003.06 Crude Fat in Feeds, Cereal Grains, and Forages. In *Official Methods of Analysis of AOAC International (OMA)*. AOAC International, Gaithersburg, MD.

AOACRI. 2010. *Performance Tested Methods Validated Methods*. AOAC Research Institute, AOAC International, Gaithersburg, MD [cited March 2010]. Available online: http://www.aoac.org/testkits/test-edmethods.html. Accessed February 15, 2011.

AOCS. 2004. AOCS Official Method AOCS Am-5 (Rapid Determination of Oil/Fat Utilizing High. Temperature Solvent Extraction). In *Official Methods and Recommended Practices of the AOCS*. Association of Oil Chemists Society, Champaign, IL.

AOCS. 2005. AOCS Official Method Ba 6a-05 Crude Fiber Analysis in Feeds by Filter Bag Technique. In *Official Methods and Recommended Practices of the AOCS*. Association of Oil Chemists Society, Champaign, IL.

Boogers, I., W. Plugge, Y. Q. Stokkermans, and A. L. Duchateau. 2008. Ultra-performance liquid chromatographic analysis of amino acids in protein hydrolysates using an automated pre-column derivatisation method. *Journal of Chromatography* 1189(1–2): 406–9.

CAST. 2003. *Mycotoxins: Risks in Plant, Animal, and Human Systems*, Task Force Report No. 139. Council for Agricultural Science and Technology. Ames, IA.

CFSAN. 2000. *Guidance for Industry: Action Levels for Poisonous or Deleterious Substances in Human Food and Animal Feed*. Center for Food Safety and Applied Nutrition (CFSAN), Food and Drug Administration. Available online: www.fda.gov/Food/GuidanceComplianceRegulatoryInformation/GuidanceDocuments/ChemicalContaminantsandPesticides/ucm077969.htm#afla. Accessed February 15, 2011.

Codex Alimentarius Procedural Manual. 2010. Joint WHO/FAO Food Standards Programme FAO, Rome, Italy. Available online: ftp://ftp.fao.org/codex/Publications/ProcManuals/Manual_19e.pdf. Accessed July 2010.

Fremy, J. M., E. Usleber, T. Goto, W. M. Hagler, D. Hite, T. Nowicki, G. S. Shephard, H. P. van Egmond, A. M. Champenari, and A. Vindiola. 2002. Committee on Natural Toxins and Food Allergens. *Journal of AOAC International* 85(1): 281–284.

GIPSA. 2009. *GIPSA Performance Verified Rapid Test Kits for the Analysis of Mycotoxins*. GIPSA, USDA, Washington, DC [cited March 2010]. Available online: archive.gipsa.usda.gov/tech-servsup/metheqp/testkits.pdf. Accessed February 15, 2011.

Goto, T., C. M. Maragos, J. -M. Fremy, W. M. Hagler, S. Hefle, R. A. Labudde, T. W. Nowicki, S. H. Pincus, B. Popping, J. Stroka, M. W. Trucksess, H. P. van Egmond, A. G. Vindiola, and K. M. Williams. 2007. Committee on Natural Toxins and Food Allergens. *Journal of AOAC International* 90(1): 3.

Groves, F. D., L. A. Zhang, Y. S. Chang, P. F. Ross, H. Casper, W. P. Norred, W. C. You, and J. F. Fraumeni. 1999. Fusarium mycotoxins in corn and corn products in a high-risk area for gastric cancer in Shandong Province, China. *Journal of AOAC International* 82(3): 657–662.

Hall, M. B. 2009. Determination of starch, including maltooligosaccharides, in Animal Feeds: Comparison of Methods and a Method Recommended for AOAC Collaborative Study. *Journal of AOAC International* 92(1): 42–49.

Hall, M. B., and N. S. Keuler. 2009. Factors affecting accuracy and time requirements of a glucose oxidase-peroxidase assay for determination of glucose. *Journal of AOAC International* 92(1): 50–60.

Heller, D. N. 2009. *Analysis of Antibiotics in Distillers Grains Using Liquid Chromatography and Ion Trap Tandem Mass Spectroscopy*, ed. C. f. V. M. Food and Drug Administration, Office of Research. Department of Health and Human Services. Rockville, MD.

Henry, M. H. 2006. *Mycotoxins in Feeds: CVM's Perspective*. US Food and Drug Administration, Rockville, MD [cited March 2010]. Available online: www.fda.gov/AnimalVeterinary/Products/AnimalFoodFeeds/Contaminants/ucm050974.htm. Accessed February 15, 2011.

Holm, J., I. Björck, A. Drews, and N. G. Asp. 1986. A rapid method for the analysis of starch. *Starch—Stärke* 38: 224–226.

Horwitz, W. 1992. History of the IUPAC/ISO/AOAC Harmonization Program. *Journal of AOAC International* 75(2): 368–371.

Hungerford, J. M., and M. W. Trucksess. 2004. Committee on Natural Toxins and Food Allergens. *Journal of AOAC International* 87(1): 270–84.

ISO. 2011. International Organization for Standardization. Geneva, Switzerland. Available online: www.iso.org/iso/iso_catalogue.htm. Accessed February 15, 2011.

ISO. 1998. *ISO 6491:1998, Animal Feeding Stuffs—Determination of Phosphorus Content—Spectrometric Method*. International Organization for Standardization, Geneva, Switzerland.

ISO. 1999. *ISO 6495:1999, Animal Feeding Stuffs—Determination of Water-Soluble Chlorides Content*. Geneva, Switzerland: International Organization for Standardization.

ISO. 2000. *ISO 6869:2000 Animal Feeding Stuffs—Determination of the Contents of Calcium, Copper, Iron, Magnesium, Manganese, Potassium, Sodium and Zinc—Method Using Atomic Absorption Spectrometry*. International Organization for Standardization, Geneva, Switzerland.

ISO. 2002. *ISO 5984:2002, Animal Feeding Stuffs—Determination of Crude Ash*. International Organization for Standardization, Geneva, Switzerland.

ISO. 2005. *ISO 13903:2005 Animal Feeding Stuffs—Determination of Amino Acids Content*. International Organization for Standardization, Geneva, Switzerland.

ISO. 2006. *ISO 16472:2006, Animal Feeding Stuffs—Determination of Amylase-Treated Detergent Fiber Content (aNDF)*. International Organization for Standardization, Geneva, Switzerland.

ISO. 2008. *ISO 13906:2008, Animal Feeding Stuffs—Determination of Acid Detergent Fiber (ADF) and Acid Detergent Lignin (ADL) Contents*. International Organization for Standardization, Geneva, Switzerland.

ISO. 2009. *ISO 27085:2009, Animal Feeding Stuffs—Determination of Calcium, Sodium, Phosphorus, Magnesium, Potassium, Iron, Zinc, Copper, Manganese, Cobalt, Molybdenum, Arsenic, Lead and Cadmium by ICP-AES*. International Organization for Standardization, Geneva, Switzerland.

Knudsen, K. E. B. 1997. Carbohydrate and lignin contents of plant materials used in animal feeding. *Animal Feed Science and Technology* 67(4): 319–338.

Krska, R., P. Schubert-Ullrich, A. Molinelli, M. Sulyok, S. Macdonald, and C. Crews. 2008. Mycotoxin analysis: An update. *Food Additives and Contaminants* 25(2): 152–163.

Krska, R., E. Welzig, and H. Boudra. 2007. Analysis of Fusarium toxins in feed. *Animal Feed Science and Technology* 137(3–4): 241–264.

Liu, 2010. Selected factors affecting crude oil analysis of distiller dried grains with solubles (DDGS) as compared with milled corn. *Cereal Chemistry* 87(3): 243–249.

Maragos, Chris M. 2010. *Lead Scientist*, USDA-ARS-NCAUR. Peoria, IL. March 19, 2010.

McChesney, D. G. 2009. *FY 2010 Nationwide Survey of Distillers Grains for Antibiotic Residues*. US Department of Health and Human Services, Rockville, MD [cited March 2010]. Available online: http://www.fda.gov/AnimalVeterinary/Products/AnimalFoodFeeds/Contaminants/ucm190907.htm. Accessed February 15, 2011.

Mertens, D. R. 1992. Critical Conditions in Determining Detergent Fibers. Paper read at National Forage Testing Association Forage Analysis Workshop, Denver, CO.

Mineral Tolerances of Animals. 2005. The National Academies Press, Washington, DC.

Multi-Residue Analysis of Mycotoxins in Dry Distillers Grains, Single Run Analysis of Aflatoxins, Ochratoxin A, Zearaleneone and Fumonisins by HPLC and Post-Column Derivatization: Technical Note. 2007. Pickering Laboratories Incorporated, Mountain View, CA.

Ofitserova, M., S. Nerkar, M. Pickering, L. Torma, and N. Thiex. 2009. Multiresidue mycotoxin analysis in corn grain by column high-performance liquid chromatography with postcolumn photochemical and chemical derivatization: Single-laboratory validation. *Journal of AOAC International* 92(1): 15–25.

Ofitserova, M., S. Nerkar, L. Torma, and M. Pickering. 2007. Single laboratory validation of multi-residue mycotoxin analysis of grains and feeds. In *Final Program, 121st AOAC International Annual Meeting and Exposition*, Anaheim, CA: AOAC International, Gaithersburg, MD.

Raymond, S. L., T. K. Smith, and H. V. L. N. Swamy. 2003. Effects of feeding a blend of grains naturally contaminated with Fusarium mycotoxins on feed intake, serum chemistry, and hematology of horses, and the efficacy of a polymeric glucomannan mycotoxin adsorbent. *Journal of Animal Science* 81(9): 2123–2130.

Reimann, L. 2010. *Minutes of the Meeting of the AAFCO Laboratory Methods and Services Committee*. Redondo Beach, CA.

Ren, Y. P., Y. Zhang, S. L. Shao, Z. X. Cai, L. Feng, H. F. Pan, and Z. G. Wang. 2007. Simultaneous determination of multi-component mycotoxin contaminants in foods and feeds by ultra-performance liquid chromatography tandem mass spectrometry. *Journal of Chromatography A* 1143(1–2): 48–64.

Romer Labs, Inc. 2010. *T-2 Toxin*. Romer Labs, Inc., Union, MO [cited March 2010]. Available online: www.romerlabs.com/downloads/Mycotoxins/T2-Toxin.pdf. Accessed February 15, 2011.

Roth, M., and A. Hampai. 1973. Column chromatography of amino acids with fluorescence detection. *Journal of Chromatography* 83: 353–356.

Schilling, N., T. V. Erem, and N. J. Thiex. 2006. The Determination of Calcium, Cobalt, Copper, Iron, Potassium, Magnesium, Manganese, Molybdenum, Sodium, Phosphorus, Sulfur and Zinc in Animal Feed and Pet Food by Microwave Digestion and ICP-OES. Paper read at AOAC International Annual Meeting and Exposition Final Program, Minneapolis, MN.

Schmale, D. G., and G. P. Munkvold. 2009. *Mycotoxins in Crops: A Threat to Human and Domestic Animal Health*. The American Phytopathological Society (APS), St. Paul, Minnesota [cited March 2010]. Available online: www.apsnet.org/education/IntroPlantPath/Topics/mycotoxins/. Accessed February 15, 2011.

Schuster, R. 1988. Determination of amino acids in biological, pharmaceutical, plant and food samples by automated precolumn derivatization and high-performance liquid chromatography. *Journal of Chromatography* 431(2): 271–284.

Shepard, G. S. 2008. Committee on Natural Toxins and Food Allergens. *Journal of AOAC International* 91(1): 16.

Shephard, G. S. 2010. Developments in mycotoxin analysis: An update for 2008–2009. *World Mycotoxin Journal* 3(1): 3–23.

Songsermsakul, P., and E. Razzazi-Fazeli. 2008. A review of recent trends in applications of liquid chromatography-mass spectrometry for determination of mycotoxins. *Journal of Liquid Chromatography & Related Technologies* 31(11–12): 1641–1686.

Sulyok, M., R. Krska, and R. Schuhmacher. 2007. A liquid chromatography/tandem mass spectrometric multi-mycotoxin method for the quantification of 87 analytes and its application to semi-quantitative screening of moldy food samples. *Analytical and Bioanalytical Chemistry* 389(5): 1505–1523.

Thiex, N. 2009. Evaluation of analytical methods for the determination of moisture, crude protein, crude fat, and crude fiber in distillers dried grains with solubles. *Journal of AOAC International* 92(1): 61–73.

Thiex, N. 1994. The Do's and Don'ts of Kjeldahl Analysis. Paper read at National Forage Testing Association Forage Analysis Workshop, Columbia, MO.

Thompson, M., S. L. R. Ellison, and R. Wood. 2002. Harmonized guidelines for single-laboratory validation of methods of analysis. (IUPAC technical report). *Pure and Applied Chemistry* 74(5): 835–855.

Trucksess, M. W. 2003. Committee on natural toxins and food allergens. Mycotoxins. *Journal of AOAC International* 86(1): 129–138.

Trucksess, M. W. 2005. Committee on Natural Toxins and Food Allergens. Mycotoxins. *Journal of AOAC International* 88(1): 314–24.

Trucksess, M. W. 2006. Mycotoxins. *Journal of AOAC International* 89(1): 270–284.

Undersander, D., D. R. Mertens, and N. Thiex. 2002. Forage Analysis Test Procedures: National Forage Testing Association, Omaha, NE. Available online: www.foragetesting.org/lab_procedure/sectionB/2.2/part2.2.2.5.htm. Accessed March 2010.

Undersander, D., D. R. Mertens, and N. Thiex. 1993. Critical conditions in determining detergent fibers. In *Forage Analysis Procedures*. National Forage Testing Association, Omaha, NE.

USFDA. 1993. *Guidance for Industry and FDA: Letter to State Agricultural Directors, State Feed Control Officials, and Food, Feed, and Grain Trade Organizations*. US Food and Drug Administration, Rockville, MD [cited March 2010]. Available online: www.fda.gov/Food/GuidanceComplianceRegulatoryInformation/GuidanceDocuments/NaturalToxins/ucm120184.htm. Accessed February 15, 2011.

USFDA. 1994. *CPG Sec. 683.100 Action Levels for Aflatoxins in Animal Feeds*. US Food and Drug Administration, Rockville, MD [cited March 2010]. Available online: http://www.fda.gov/ICECI/ComplianceManuals/CompliancePolicyGuidanceManual/ucm074703.htm. Accessed February 15, 2011.

USFDA. 2001. *Guidance for Industry: Fumonisin Levels in Human Foods and Animal Feeds; Final Guidance*. US Food and Drug Administration, Rockville, MD [cited March 2010]. Available online: www.fda.gov/Food/GuidanceComplianceRegulatoryInformation/GuidanceDocuments/ChemicalContaminantsandPesticides/ucm109231.htm. Accessed February 15, 2011.

Van Erem, T., H. Manson, and N. J. Thiex. 2008. Comparison of Gravimetric, ICP-OES and Combustion Methods for Determining Sulfur in Distillers Grain, Animal Feed and Feed Ingredients. Paper read at AOAC International Midwest Section Official Meeting Program, Bozeman, MT.

Van Hulzen, S. Quality Control Director, POET Plant Management. Sioux Falls, SD 57104. March 9, 2010.

Van Soest, P., and R. H. Wine. 1967. Use of detergents in the analysis of fibrous feeds. IV. Determination of plant cell wall constituents. *Journal of the Association of Official Analytical Chemists* 50: 50–55.

Whitaker, T. B. 2004. Sampling feeds for mycotoxins. In *Mycotoxins in Food: Detection and Control*, eds. N. Magan, and M. Olsen. Woodhead Publishing Limited, Cambridge, UK.

Zhang, Y., L. Harken, S. Daniels, S. Nimkar, and B. Purkayastha. 2006. A Sensitive and Rapid LC/MS/MS based Amino Acid Analysis Method. Paper read at 120th AOAC International Annual Meeting, Minneapolis, MN.

11 Mycotoxin Occurrence in DDGS

*John Caupert, Yanhong Zhang, Paula Imerman,
John L. Richard, and Gerald C. Shurson*

CONTENTS

11.1 INTRODUCTION

The previous chapter (Chapter 10) has provided an overview of analytical and laboratory methods used for ethanol coproducts; in this chapter, we will focus on the study and testing of mycotoxins in ethanol coproducts.

Mycotoxins are unavoidable contaminants in crops (CAST, 2003) and therefore, they also occur in commodities entering the marketing chain including those grains to be used in ethanol production. Currently, corn (maize) is the primary commodity used for ethanol production in the United States.

However, depending on the geographical location of an ethanol plant and price relative to corn, sorghum and wheat alone, or blended with corn are sometimes used to produce ethanol and distiller dried grains with solubles (DDGS). Several mycotoxins can potentially be found in corn including aflatoxins, deoxynivalenol (DON), fumonisin, T-2 toxin, and zearalenone (ZON). Most of these toxins can occur in preharvested corn and are present in the grain at harvest. Such occurrence, however, is dependent upon the unique environmental conditions that are conducive to the growth of specific molds that produce these mycotoxins during crop development. Therefore, mycotoxin contamination of corn does not necessarily occur every year because the appropriate environmental conditions for the growth of the specific responsible fungi are often lacking. Among the toxins, T-2 toxin is not a major preharvest contaminant in grains, and is likely the result of inadequate grain storage allowing for its production by the responsible fungi.

During the corn-to-ethanol production process, approximately two-thirds of the grain, mainly starch, is fermented by yeast to produce ethanol and carbon dioxide, neither of the two products contains mycotoxins if contaminated corn was used. The remaining coproduct, DDGS, however, contains higher concentrations of mycotoxins than present in the grain original. The increased level of a given mycotoxin in DDGS was reported to be approximately three times higher than the level present in the grain (Bothast et al., 1992; Bennett and Richard, 1996). The tremendous growth in the fuel ethanol industry has been accompanied by concomitant growth in the production of DDGS, and the potential for increased use of DDGS as animal feed is great. As a result, more attention has been paid to the prevalence and levels of mycotoxins in DDGS.

In this chapter, the state-of-the-art testing methods available on the market for mycotoxin analysis in DDGS are discussed, and several independent studies and publications that have been conducted to determine mycotoxins in DDGS are reviewed. More importantly, detectable mycotoxin levels in DDGS from those studies are compared with the current action levels, advisory levels, and guidance levels of the FDA. This paper attempts to consolidate data from these studies and publications, and evaluate the meaning of the results relative to the toxicity potential in animals to which DDGS might likely be fed.

11.2 MYCOTOXINS IN CORN

Mycotoxins are fungal secondary metabolites that adversely affect the health, growth, or reproduction of animals, especially humans and domestic animals. Aflatoxins, including aflatoxin B1, B2, G1, and G2, are the most toxic and carcinogenic of the known mycotoxins, and they are produced by several *Aspergillus* species. Corn becomes susceptible to aflatoxin formation during growth under drought conditions, or in high moisture/humid storage (Richard, 2000).

Fusarium graminearum is the principal deoxynivalenol-producing fungus in grains in the United States (CAST, 2003). Deoxynivalenol may coexist with other toxins, like zearalenone. The organism survives on old infested residue left in the field from the previous growing season, where cold, moist conditions are favorable for the fungus to grow on corn. Generally, storage is not considered a potential source for contamination if the corn was mature and was stored at moisture level lower than 14% (Richard, 2000).

The major producer, *Fusarium verticillioides*, is capable of producing the fumonisins, mainly FB1, FB2, and FB3 (Gelderblom et al., 1988). Corn is the major commodity affected by the fungi that produce these toxins. The exact conditions for this fumonisin production are unknown, but it is suggested that drought stress followed by warm, wet weather during flowering seems to be important. It is reported that the organism is present virtually in every seed, is present in the corn plant throughout its growth, and sometimes there is considerable amount of fumonisins present in symptomless kernels of corn. Since the discovery of this toxin was fairly recent (1988), there is limited information available for this toxin (Richard, 2000).

T-2 is a member of fungal metabolites known as the trichothecenes. *Fusarium sporotrichioides* is the principal fungus responsible for the production of T-2. The production of T-2 is the greatest with increased humidity and temperatures of 6°C–24°C (CAST, 2003).

Zearalenone is an estrogenic fungal metabolite. The major fungus responsible for producing this toxin is *Fusarium graminearum*. Moist and cool growing conditions are favorable for this fungus to grow, which are the same conditions conducive for deoxynivalenol production. For storage, controlling moisture lower than 14% is important to avoid production of these mycotoxins.

11.3 REGULATIONS AND GUIDANCE

As of July, 2008, FDA established regulatory levels for mycotoxins in feed ingredients. Action levels for aflatoxins in animal feeds have been established for different animal species and at different production stages. The FDA "action level" represents the minimum limit at which the FDA can take legal action to remove feed ingredients from the market. Table 11.1 shows the action levels established for aflatoxins in animal feed in 2000 (FDA, 2011). Also included in Table 11.1 are the advisory levels for deoxynivalenol in animal feeds and the recommended maximum levels for fumonisins in animal feeds set by the FDA. No action levels, advisory levels, or guidance levels for T-2 toxin or zearalenone are available from the FDA at this time (FDA website).

11.4 ANALYTICAL TESTING METHODS FOR MYCOTOXINS

Testing for mycotoxins in DDGS involves obtaining an adequate sample, preparing it for analysis, and choosing a scientifically validated method for the specific mycotoxins of interest. Every step of

TABLE 11.1
FDA Action Levels for Mycotoxins in Feed Ingredients

Aflatoxins

Animals	Action Levels (ppb)
Finishing beef (i.e., feedlot) cattle	300
Finishing swine (>100 lb)	200
Breeding beef cattle, breeding swine, or mature poultry	100
Immature animals, dairy cattle, or intended use is not known	20

Deoxynivalenol

Animals	Advisory Levels (ppm)
Ruminating beef and feedlot cattle older than 4 months, and chickens with the added recommendation that these ingredients not exceed 50% of the diet of cattle and chickens	10
All other animals with the added recommendation that these ingredients not exceed 40% of the diet of cattle and chickens	5
Swine with the added recommendation that these ingredients not exceed 20% of the diet of cattle and chickens	5

Fumonisins

Animals	Advisory Levels (ppm)
Poultry being raised for slaughter, no more than 50% of the diet	100
Ruminants older than 3 months raised for slaughter and mink being raised for pelt production, no more than 50% of the diet	60
Breeding ruminants, poultry, and mink, no more than 50% of the diet	30
Swine and catfish, no more than 50% of the diet	20
All other species or classes of livestock and pet animals, no more than 50% of the diet	10
Equids and rabbits, no more than 20% of the diet	5

Source: Adapted from FDA. 2011. Aflatoxin in feeds and feed ingredients: Available online: www.cfsan.fda.gov/~lrd/fdaact.html#afla. Accessed February 16, 2011; Fumonisins in feeds and feed ingredients: Available online: http://www.fda.gov/Food/GuidanceComplianceRegulatoryInformation/GuidanceDocuments/ChemicalContaminantsandPesticides/ucm109231; Deoxynivalenol (DON) in feeds and feed ingredients: Available online: www.cfsan.fda.gov/~dms/graingui.html.

the process is important in order to obtain results that accurately reflect the mycotoxin concentration in the original lot (CAST, 2003). Refer to Chapter 10 for more information about analytical and testing methods.

Since the 1960s, many analytical methods have been developed to test mycotoxins in human foods and animal feeds due to human health concerns (Trucksess, 2000). Among the developed methods, thin-layer chromatography (TLC), enzyme-linked immunosorbent assay (ELISA), and immunosensor-based methods have been widely used for rapid screening, while high-performance liquid chromatography (HPLC) with fluorescence detection (FD) and mass spectrometry detection (MS) have been used as confirmatory and reference methods (Krska et al., 2008). In this paper, state-of-the-art mycotoxin analytical methods for DDGS will be discussed by focusing on testing kits approved by the Grain Inspection, Packers, and Stockyards Administration (GIPSA) of the United States Department of Agriculture (USDA GIPSA website), and confirmatory methods used by major DDGS testing labs in the United States (Zhang and Sido, 2008).

11.4.1 SAMPLE COLLECTION

The greatest variability or source of error in overall mycotoxin testing comes from sampling (CAST, 2003). To obtain accurate mycotoxin results, it is crucial to establish a well-enforced sampling program, which ensures an adequate number of sampling probes, sufficient sample size, and appropriate particle size. Once an adequate sample has been collected, it must be ground to reduce particle size, mixed thoroughly and then an aliquot is taken for analysis.

The goal of sampling is to remove an appropriate quantity for testing from a large bulk lot, in such a way that the proportion and distribution of the factors being tested are the same in both the whole (lot) and the part removed (sample) (Richard, 2006). The sampling process consists of taking a number of small samples from a lot and pooling them into a large composite sample. For a sample to be considered representative, it must be:

1. Obtained with appropriate equipment, such as probe (trier) for stationary grain, a diverter-type mechanical sampler or pelican sampler for moving grain.
2. Obtained using a sampling pattern and procedure designed to collect samples from all areas of the lot to make a composite sample.
3. Of adequate size. The GIPSA recommended minimum sample size for an aflatoxin test is 2 lb (about 1 kg) from a truck load, 3 lb (about 1.5 kg) from a railcar load, and 10 lb (about 4.5 kg) from a barge load.
4. Adequately identified and labeled.
5. Handled adequately, such as stored in cool and dry place, placed in double or triple lined paper bags or breathable cloth bags, and so forth.

Detailed information on the sampler and sampling pattern is listed at the USDA website (www. gipsa.usda.gov/GIPSA/webapp?area=home&subject=lr&topic=hb).

11.4.2 SAMPLE PREPARATION

One of the objectives of sample preparation is to obtain a small sample of grain to be used for analytical testing. Based on the recommendations from the Association of Analytical Chemists (AOAC) and the GIPSA, we suggest the following steps for sample preparation:

1. Grind the total sample using an appropriate mill to pass #14 mesh sieve. There are several grinders recommended by the GIPSA (USDA website).
2. Split the sample using a sample splitter until 1–2 kg is obtained.

3. Regrind 1–2 kg to completely pass a #20 mesh sieve.
4. Mix the reground portion thoroughly in a tumble blender or planetary mixer.
5. Take a 500 g subsample from the mix for further analytical testing.

11.4.3 SAMPLE ANALYSIS

The determination of mycotoxins in DDGS can be divided into four steps: extraction of the sample, cleanup of the extract, separation, and detection.

Mycotoxins are extracted from the sample to liberate the mycotoxins of interest from the sample matrix. The closer the recovery is to 100%, the more accurate the final analytical result will be. The most common extraction solution used is a mixture of water and other polar solvents.

Cleanup of extracts is usually necessary to eliminate interfering substances from the sample matrix to ensure better selectivity of the detection and allow for any low level of mycotoxins from the original sample to be detected. The current cleanup technology usually employs immunoaffinity columns (IACs) or multifunctional MycoSep® columns (Krska et al., 2008).

There has been a high demand for rapid and cost-effective, on-site determination of mycotoxins. These methods usually detect a single mycotoxin, allow for ease of operation, and sensitive quantitation with high sample throughput. As of September, 2008, there were six GIPSA approved methods for mycotoxin testing in DDGS: four of them for aflatoxins, one for fumonisins, and one for zearalenone (USDA GIPSA website; Table 11.2).

High-performance liquid chromatography (HPLC) has become the method of choice for confirmatory analysis of mycotoxin levels in animal feeds. This analysis can be performed with a variety of detectors that are easily coupled to an HPLC (Krska et al., 2008). The methods used by major DDGS testing labs in the United States are described in Table 11.3. These methods have been validated by individual labs and recently published in peer-reviewed scientific journals.

Liquid chromatography with mass spectrometry detection (LC/MS) has gained considerable attention recently because this technology can simultaneously detect and identify multiple mycotoxins in animal feed. This method provides definite confirmation of the molecular identity, uses simple extraction with no cleanup, and has high selectivity and sensitivity (Table 11.3).

TABLE 11.2
Rapid Mycotoxin Testing Kits for DDGS (Approved by GIPSA, 2011)

Brand Name	Manufacturer	Test Range	Test Format	Extraction	Cleanup
Aflatoxins					
Veratox Aflatoxin	Neogen Corporation	5–50 ppb	Microtiter Well Plate Assay	Methanol/water (70 + 30)	ELISA
Ridascreen FAST SC	R-Biopharm	5–100 ppb	Microtiter Well Plate Assay	Methanol/water (70 + 30)	ELISA
Aflatest	Vicam	5–100 ppb	Immunoaffinity Column	Methanol/water (80 + 20)	Affinity column
FluroQuant® Afla IAC	Romer	5–100 ppb	Fluorometry	Methanol/water (80 + 20)	Affinity column
Fumonisins					
AgraQuant Total Fumonisin 0.25/5.0	Romer	0.5–5 ppm	Direct Competitive ELISA	Methanol/water (70 + 30)	ELISA
Zearalenone					
ROSA® Zearalenone	Charm Sciences, Inc.	50–1000 ppb	Lateral Flow Strip	Methanol/water (70 + 30)	

TABLE 11.3
Instrumental Methods for Mycotoxin Testing in Animal Feed

Target	Testing	Detection Range	Extraction	Cleanup	Reference
Aflatoxins					
Corn, almonds, Brazil nuts, peanuts, and pistachio nuts	HPLC–FD	5–30 ppb	Acetonitrile–water (90 + 10)	MycoSep column	AOAC 994.08
Deoxynivalenol					
Cereals and cereal products	HPLC–UV	ppm (detection limit)	Water	Immunoaffinity column	MacDonald et al., 2005a
Fumonisins					
Corn and corn flakes	HPLC–FD	0.5–2 ppm	Methanol–acetonitrile–water (25 + 25 + 50)	Immunoaffinity column	AOAC 2001.04
Corn and corn-based feedstuffs	Thin-layer chromatography (TLC)	ppm (detection limit)	Acetonitrile–water (50 + 50)	C-18 column	Rottinghaus et al., 1992
T-2					
Food and feed	Thin-layer chromatography (TLC)	ppm (detection limit)	Acetonitrile–water (84 + 16)	Charcoal/alumina	Romer, 1986
Zearalenone					
Corn, wheat, and feed	Microtiter Well Plate Assay	0.8 ppm (detection limit)	Methanol–water (70 + 30)	MycoSep column	AOAC 994.01
Barley, maize and wheat flour, polenta, and maize-based baby foods	HPLC–FD	0.05 ppm (detection limit)	Acetonitrile–water	Immunoaffinity column	MacDonald et al., 2005b
Aflatoxins, Deoxynivalenol, Fumonisins, T-2, Zearalenone					
Food and feed	LC/MS/MS	Aflatoxins (1–100 ppb); Deoxynivalenol, (1, 1000 ppb) Fumonisin (16–3200 ppb) T-2, (2–1000 ppb) Zearalenone (20–1000 ppb)	Acetonitrile–water–acetic acid (79 + 20 + 1)		Sulyok et al., 2007

11.5 REVIEW AND ANALYSIS OF INDEPENDENT DATA AND RECENT STUDIES

11.5.1 Study Conducted by the National Corn-to-Ethanol Research Center

11.5.1.1 Sampling and Testing Methods

The National Corn-to-Ethanol Research Center (NCERC, Edwardsville, IL, USA) conducted a study of mycotoxin prevalence and levels in DDGS based on samples from their National DDGS Library (Zhang et al., 2008). As part of the study, 20 samples from the Library were selected. These samples were generated from 14 ethanol plants representing seven states in the Midwestern

United States. These samples were collected at the ethanol plants immediately after they were produced and shipped to the NCERC overnight. Immediately after they arrived, the samples were vacuum sealed and stored in a freezer at −20 °C (Zhang et al., 2009).

Mycotoxin testing was performed at Trilogy Analytical Laboratories (Washington, MO, USA). Samples were analyzed for aflatoxins B_1, B_2, G_1, G_2, deoxynivalenol, fumonisins B_1, B_2 and B_3, and zearalenone by *high-performance liquid chromatography* (HPLC) and for T-2 toxin by *thin-layer chromatography* (TLC). Aflatoxins B1, B2, G1, and G2 were detected after extraction with acetonitrile/water (84/16), isolation using a solid phase cleanup column (Trilogy TC-M160), and detection by fluorescence after postcolumn derivatizatoin with a Kobra cell (AOAC 994.08). Fumonisin B1, B2, and B3 were detected after extraction with methanol/water (3/1), isolation using an immunoaffinity cleanup column, and detection by fluorescence after precolumn derivatization with naphthalene dicarboxaldehyde (NDA) (AOAC 2001.04). Deoxynivalenol was detected after extraction with acetonitrile/water (84/16), isolation using a combination of solid phase (Trilogy TC-M160 and TC-C210) and immunoaffinity cleanup columns, and detection with a UV detector (MacDonald et al., 2005a). The extraction of T-2 toxin was performed with acetonitrile/water (84/16), and this was followed by isolation using a combination of solid phase cleanup columns (Trilogy TC-M160 and TC-C210) and detection using TLC (Romer, 1986). Zearalenone was detected after extraction with acetonitrile/water (84/16), isolation using a combination of solid phase (Trilogy TC-M160) and immunoaffinity cleanup columns, detection with a fluorescence detector (MacDonald et al., 2005b). The detection limits for the tests were 1 ppb for each aflatoxin, 0.1 ppm for deoxynivalenol, 0.1 ppm for each fumonisin, 0.1 ppm for T-2 toxin, and 0.05 ppm for zearalenone.

11.5.1.2 Results

Almost none of the 20 samples contained mycotoxin levels higher than the recommended levels established by the FDA for animal feeds. The exceptions were two DDGS samples that contained fumonisins higher than 5 ppm, but lower than 10 ppm. The results of this study are shown in Table 11.4. The average level of certain mycotoxin in DDGS from a data set was calculated assuming zero level of that mycotoxin if it was not detected in DDGS.

Aflatoxins: 70% of the samples were below the limit of detection of 1 ppb, thus they can be described as having nondetectable levels. The maximum level detected was 3.7 ppb, which is well below all FDA action levels. The average level of aflatoxin across all 20 samples was 0.7 ppb. Therefore, all 20 samples were below the lowest action level established by the FDA for all animal species.

Deoxynivalenol: 25% of the samples were below the limit of detection of 0.1 ppm, thus they can be described as having nondetectable levels. The maximum level detected was 1.2 ppm, which is

TABLE 11.4
Mycotoxin Concentrations in DDGS from the NCERC Study

Mycotoxins	Number of Samples Submitted	Minimum Level	Maximum Level	Average Level (of all Samples)	Percentage of Samples above the Lowest FDA Level
Aflatoxin (ppb)	20	<1	3.7	0.7	0%
Deoxynivalenol (ppm)	20	<0.1	1.2	0.3	0%
Fumonisin (ppm)	20	<0.1	8.6	1.9	10%
T-2 toxin (ppm)	20	<0.1	<0.1	0	N.A.[a]
Zearalenone (ppm)	20	<0.05	0.143	0.038	N.A.

Source: Zhang et al. National DDGS Library nutritional profile. *International Fuel Ethanol Workshop & Expo,* Nashville, TN, 2008.
[a] Data not available.

lower than all FDA advisory levels. The average level of deoxynivalenol across all 20 samples was 0.3 ppm. All 20 samples were below the lowest advisory level established by the FDA for all animal species.

Fumonisins: only two DDGS samples contained total fumonisin levels in excess of the lowest recommended guidance levels established by the FDA. The average level of fumonisins across all 20 samples was 1.9 ppm. The average concentration of fumonisins, across all 20 samples, was below the lowest recommended guidance levels established by the FDA for all animal species.

T-2 toxin: there are no action levels, advisory levels, or guidance levels for T-2 by the FDA. None of the 20 samples had a level of T-2 that exceeded the limit of detection of 0.1 ppm. All 20 samples can be described as having nondetectable levels for T-2.

Zearalenone: there are no action levels, advisory levels, or guidance levels for zearalenone by the FDA. 55% of the samples were below the detection limit of 0.05 ppm. The average concentration of zearalenone across all 20 samples was 0.038 ppm. The average concentration of zearalenone across all 20 samples was below the detection limit of 0.05 ppm.

These results confirmed the detectable presence of aflatoxins, deoxynivalenol, zearalenone, and fumonisins, but not T-2 toxin, in samples from a large geographic region of the United States (seven states). However, even when a sample had detectable levels of mycotoxins, their concentrations were well below the maximum tolerable guidelines for use in animal feeds. The exception was two samples with fumonisin levels that could be a concern only if these DDGS sources were fed to equids or rabbits.

11.5.2 Study Conducted by an Ethanol Company with Multiple Ethanol Plants

11.5.2.1 Sampling and Testing Methods

From February of 2006 through November of 2007, DDGS samples were collected from two ethanol plants (Plants A and B, Midwest area, USA) owned by an ethanol producer. More than one DDGS sample was collected on a monthly basis from each ethanol plant and sent to the Midwest Laboratories (Omaha, NE, USA) for mycotoxin testing. Between February of 2008 and July of 2008, combined DDGS samples from four ethanol plants owned by the ethanol producer were collected weekly and sent to the MVTL Laboratories (New Ulm, MN, USA) for mycotoxin testing (Zhang et al., 2009).

Samples were analyzed for aflatoxins B_1, B_2, G_1, G_2, deoxynivalenol, fumonisins B_1, B_2 and B_3, T-2 toxin, and zearalenone. The methodology utilized by the Midwest Laboratories was LC/MS. Based on numerous methods from the literature, a proprietary method was developed to analyze aflatoxins B1, B2, G1, and G2, fumonisins B1, B2, B3, T-2, and zearalenone using a methanol/water extraction solution to dissolve the potential mycotoxins in samples. After vortexing and centrifuging, the extracts were passed through an affinity column. The affinity column was washed with a phosphate buffer and eluted with methanol. The extracts were analyzed by *liquid chromatography/mass spectrometry.* For deoxynivalenol, deionized water was used for extraction, and the remainder of the procedure was the same as for the other mycotoxins. Detection limits for the tests were 1 ppb for each aflatoxin, 0.1 ppm for deoxynivalenol, each fumonisin, T-2, and 0.05 ppm for zearalenone.

The methodology utilized by MVTL Laboratories was HPLC. For aflatoxin B1, B2, G1, G2, and zearalenone, the sample was extracted with 70/30 methanol/water, the extract was run through a Vicam AOZ immunoaffinity column and detected by a fluorescence detector. Detection limits for the tests were 3 ppb for aflatoxin B_1, 1 ppb for aflatoxin B_2, 15 ppb for aflatoxin G_1, 5 ppb for aflatoxin G_2, and 0.2 ppm for zearalenone (AOAC 985.18; 991.31). For fumonisin B1, B2, B3 measurement, the sample was extracted with 25/25/50 methanol/acetonitrile/water. After precolumn derivatization with o-phthalaldehyde, the extract was passed through a Vicam fumonitest immunoaffinity column and detected by a fluorescence detector. Detection limits for the tests were 0.2 ppm for fumonisin B1 and B2 (AOAC 2001.04). For deoxynivalenol, the sample was extracted with deionized water, and the extract was passed through a Vicam DONtest immunoaffinity column and detected by a UV/VIS detector. The detection limit was 0.2 ppm for deoxynivalenol.

11.5.2.2 Results

Almost none of the 162 samples contained mycotoxin concentrations higher than the recommendation levels established by the FDA for animal feeds. Only two DDGS samples from Plant A and eight DDGS samples from the four plants combined contained fumonisins higher than 5 ppm, but they were all below 10 ppm. The results of this study are shown in Table 11.5.

Aflatoxins: 96% of the samples from Plant A, 88% of the samples from Plant B, and 99% of the samples from the four plants combined were below the limit of detection of 1 ppb, thus they can be described as having nondetectable levels. The maximum level of aflatoxins detected for Plant A was 2.56 ppb, for Plant B 1.21 ppb, and for the four plants combined 1.12 ppb. These levels fall well below all FDA action levels. Overall, the 162 samples were below the lowest action level established by the FDA for all animal species.

Deoxynivalenol: the maximum level of deoxynivalenol detected in a sample from Plant A was 1.42 ppm, for Plant B was 1.68 ppm, and for the four plants combined 1.9 ppm. These levels fall below all FDA advisory levels. The average level of deoxynivalenol detected in the samples from Plant A was 0.64 ppm, for Plant B was 1.02 ppm, and for the four plants combined 0.5 ppm. All 162 samples were below the lowest advisory level established by the FDA for all animal species.

TABLE 11.5
Mycotoxin Concentrations in DDGS from the Ethanol Producer Study

Mycotoxins	Number of Samples Submitted	Minimum Level	Maximum Level	Average Level (of All Samples)	Percentage of Samples above the Lowest FDA Level
Plant A, 2/2006–11/2007					
Aflatoxins (ppb)	69	<1	2.56	0.08	0%
Deoxynivalenol (ppm)	69	<0.1	1.42	0.64	0%
Fumonisins (ppm)	69	0.12	5.88	2.33	3%
T-2 toxin (ppm)	69	<0.1	<0.1	0	N.A.[a]
Zearalenone (ppm)	69	<0.05	0.123	0.025	N.A.
Plant B, 7/2006–11/2007					
Aflatoxins (ppb)	16	<1	1.21	0.15	0%
Deoxynivalenol (ppm)	16	0.13	1.68	1.02	0%
Fumonisins (ppm)	16	0.28	2.77	1.47	0%
T-2 toxin (ppm)	16	<0.1	<0.1	0	N.A.
Zearalenone (ppm)	16	<0.05	0.113	0.042	N.A.
Four Plants Combined, 2/2008–7/2008					
Aflatoxins (ppb)	77	<1	1.12	0.01	0%
Deoxynivalenol (ppm)	77	0.2	1.9	0.5	0%
Fumonisins (ppm)	77	<0.2	7.2	2.7	10%
T-2 toxin (ppm)	N.A.	N.A.	N.A.	N.A.	N.A.
Zearalenone (ppm)	77	<0.2	<0.2	0	N.A.

Source: Zhang et al. *Journal of Agricultural and Food Chemistry*, 57(20), 9828–9837, 2009.

[a] Data not available.

Fumonisins: the maximum level of fumonisins detected in a sample from Plant A was 5.88 ppm, for Plant B was 2.77 ppm, and for the four plants combined 7.2 ppm. The average level of fumonisin detected in the samples from Plant A was 2.33 ppm, for Plant B was 1.47 ppm, and for the four plants combined 2.7 ppm. While ten samples did have fumonisin levels above 5 ppm, those ten samples were all less than 10 ppm. The average concentration of fumonisin across all 162 samples was below the lowest recommended guidance levels established by the FDA for all animal species.

T-2 toxin: there are no FDA action levels, advisory levels, or guidance levels for T-2. None of the 69 samples submitted by Plant A or the 16 samples submitted by Plant B contained a concentration of T-2 above the 0.1 ppm level of detection. The 77 samples submitted by the four combined plants were not tested for T-2. All 85 samples that were tested for T-2 can be described as having nondetectable levels.

Zearalenone: there are no FDA action levels, advisory levels, or guidance levels for zearalenone. 68% of the samples from Plant A and 44% of the samples from Plant B fell below the limit of detection of 0.05 ppm. All samples from the four combined plants fell below 0.2 ppm. The maximum level of zearalenone for samples submitted from Plant A was 0.123 ppm, for Plant B was 0.113 ppm, and for the four combined plants was lower than 0.2 ppm. The average concentration of zearalenone across all 162 samples was below the 0.2 ppm detection limit.

Similar to the results from the NCERC study (Zhang et al., 2008), aflatoxins, deoxynivalenol, fumonisins, and zearalenone, but not T-2 toxin, were detected at very low levels in only some of the samples over a two-year time period. The levels detected in the majority of the samples were well below the FDA maximum tolerable guidelines for use in animal feed.

11.5.3 STUDY CONDUCTED BY IOWA STATE UNIVERSITY, VETERINARY DIAGNOSTIC LAB AND NOVECTA LLC

In the Asian Pacific market, there is a concern about the time, environmental conditions and shipping procedures of DDGS from the United States. The concern is that these factors support or enhance mold growth of DDGS. This study investigated the mycotoxin content in DDGS before and after shipment from a port in the United States to a port in Taiwan (Zhang et al., 2009).

The project was conducted in two phases: Phase I was conducted in the Taiwan during the winter season of 2006. The study included 7 DDGS samples coming directly from different ethanol plants and 11 samples coming from U.S. port shipping containers resulting from those ethanol plants. All samples were acquired from different sources in the Midwestern United States, including Iowa, Illinois, Wisconsin, and Minnesota. Samples were collected over a period of 3 months. The same 11 containers were sampled again upon arriving in Taiwan. The samples were then shipped back to the Veterinary Diagnostic Lab at Iowa State University for analysis. Phase II was conducted in Taiwan during the summer season of 2007. The study included samples from 12 shipping containers at a U.S. port. The DDGS samples came from several ethanol plants in Midwest area, USA. The 12 shipping containers were sampled after arriving in Taiwan (summer season), and selected samples were then shipped back to the Veterinary Diagnostic Lab at Iowa State University for analysis.

11.5.3.1 Sampling and Testing Methods

Sampling of the shipping containers was to be done using the Kansas State University probe technique (Herrman, 2001). However, due to safety concerns, sampling was performed using pelican-style sampling at the loading area by taking 10 samples from the stream at varying intervals. This sample was mixed well and subsampled into 400–500 g samples before shipment. All sampling in the United States was overseen by USDA officials. Sampling in Taiwan, done similar to that in the United States, was overseen and performed by an independent sampler at the loading port in Taiwan. Samples from the shipping containers were assigned the container number for either U.S. or Taiwan origin for data comparison.

All analyses were performed at the Veterinary Diagnostic Lab at Iowa State University (Ames, Iowa). Samples received at the lab were stored at −20°C until analysis could be performed. The samples were extracted using acetonitrile/water and cleaned up using solid phase extraction columns. The sample extract was screened for aflatoxins (B_1, B_2, G_1, G_2), deoxynivalenol, total fumonisins, T-2, and zearalenone/zearalenol by TLC. The detection limits of the TLC method were 5 ppb for each aflatoxin, 0.5 ppm for deoxynivalenol, fumonisins and zearalenone, and 1 ppm for T-2. For samples with mycotoxin levels below the detection limit, the extract was spiked every fifth sample with that specific mycotoxin and screened by TLC again to confirm test sensitivity. The spiking levels were 10 ppb for aflatoxins, 1 ppm for deoxynivalenol, fumonisin and zearalenone/zearalenol, and 2 ppm for T-2. For the samples with mycotoxin levels above the detection limit, a confirmatory test was performed using HPLC or GC (Stahr, 1991). The detection limits for the HPLC method were 0.5 ppb for each aflatoxin, 0.1 ppm for each fumonisin and zearalenone, and the detection limits for GC method were 0.1 ppm for deoxynivalenol and 0.3 ppm for T-2.

11.5.3.2 Results

None of the 53 samples, from either Phase I or II, contained mycotoxin levels higher than the FDA recommended levels for animal feed. The results of this study are shown in Table 11.6.

Aflatoxins: None of the 53 samples had a level of aflatoxin that exceeded the limit of detection of 5 ppb. All 53 samples were below the lowest action level established by the FDA for all animal species.

Deoxynivalenol: Only four samples from Phase I, two from the ethanol plants, and two from the Taiwan port, contained detectable deoxynivalenol levels. The maximum level detected was 3.4 ppm, which was lower than all FDA advisory levels (Table 11.6). None of the samples from Phase II had a level of deoxynivalenol that exceeded the limit of detection of 0.5 ppm (Tables 11.6). All 53 samples were below the lowest advisory level established by the FDA for all animal species. No increase in deoxynivalenol was observed in the shipment of DDGS from the United States to Taiwan.

Fumonisins: All 53 samples contained detectable fumonisins, but they were below the lowest recommended guidance levels established by the FDA for all animal species (Table 11.6). The maximum level of fumonisins detected in a sample from Phase I was 2.9 ppm, and from Phase II was 2.4 ppm. In Phase I study, the average levels of fumonisins were found to be 2.3 ppm for the samples from the ethanol plants, 1.9 ppm for the samples from the U.S. port, and 1.2 ppm for the samples from the Taiwan port. In Phase II study, the average levels of fumonisins were found to be 0.9 ppm for the samples from the U.S. port and 1.5 ppm for the samples from the Taiwan port. All 53 samples contained fumonisins lower than the lowest recommended guidance levels established by the FDA for all animal species. No increase in fumonisins was observed in the shipment of DDGS from the United States to Taiwan.

T-2 toxin: There are no action levels, advisory levels, or guidance levels for T-2 by the FDA. None of the 53 samples had a level of T-2 that exceeded the limit of detection of 1 ppm. All 53 samples can be described as having nondetectable levels of T-2.

Zearalenone: There are no action levels, advisory levels, or guidance levels for zearalenone by the FDA. None of the 53 samples had a level of zearalenone that exceeded the limit of detection of 0.5 ppm. All 53 samples can be described as having nondetectable levels of zearalenone.

The results from this study indicate that only deoxynivalenol and fumonisins were detected in a portion of the samples, but at very low levels, and well below the maximum tolerable for use in animal feed. No apparent increase in mycotoxins was observed in the shipment from the United States to Taiwan during winter and summer, which indicate that the port containers themselves did not contribute to mycotoxin production, and neither did environmental conditions.

TABLE 11.6
Mycotoxin Concentrations in DDGS from the Iowa State University Study

Mycotoxins	Number of Samples Submitted	Minimum Level	Maximum Level	Average Level (of All Samples)	Percentage of Samples above the Lowest FDA Level
DDGS Samples Directly from Ethanol Plants					
Aflatoxins (ppb)	7	<5	<5	0	0%
Deoxynivalenol (ppm)	7	<0.1	3.4	0.6	0%
Fumonisins (ppm)	7	1.8	2.9	2.3	0%
T-2 toxin (ppm)	7	<1	<1	0	N.A.[a]
Zearalenone/Zearalenol (ppm)	7	<0.5	<0.5	0	N.A.
Phase I (winter) DDGS Samples from U.S. Port Containers					
Aflatoxins (ppb)	11	<5	<5	0	0%
Deoxynivalenol (ppm)	11	<0.5	<0.5	0	0%
Fumonisins (ppm)	11	0.7	2.4	1.9	0%
T-2 toxin (ppm)	11	<1	<1	0	N.A.
Zearalenone/Zearalenol (ppm)	11	<0.5	<0.5	0	N.A.
Phase I (winter) DDGS Samples from Taiwan Port Containers					
Aflatoxins (ppb)	11	<5	<5	0	0%
Deoxynivalenol (ppm)	11	<0.1	1.0	0.1	0%
Fumonisins (ppm)	11	0.7	2.0	1.2	0%
T-2 toxin (ppm)	11	<1	<1	0	N.A.
Zearalenone/Zearalenol (ppm)	11	<0.5	<0.5	0	N.A.
Phase II (summer) DDGS Samples from U.S. Port Containers					
Aflatoxins (ppb)	12	<5	<5	0	0%
Deoxynivalenol (ppm)	12	<0.5	<0.5	0	0%
Fumonisins (ppm)	12	0.5	1.4	0.9	0%
T-2 toxin (ppm)	12	<1	<1	0	N.A.
Zearalenone/Zearalenol (ppm)	12	<0.5	<0.5	0	N.A.
Phase II (summer) DDGS Samples from Taiwan Port Containers					
Aflatoxins (ppb)	12	<5	<5	0	0%
Deoxynivalenol (ppm)	12	<0.5	<0.5	0	0%
Fumonisins (ppm)	12	0.4	2.4	1.5	0%
T-2 toxin (ppm)	12	<1	<1	0	N.A.
Zearalenone/Zearalenol (ppm)	12	<0.5	<0.5	0	N.A.

Source: Zhang et al. *Journal of Agricultural and Food Chemistry*, 57(20), 9828–9837, 2009.

[a] Data not available.

11.5.4   A Publication on Mycotoxins in DDGS by Asian Feed

11.5.4.1   Sampling and Testing Methods

There were 103 DDGS samples included in this study, 67% from the United States and 33% from Asia (Table 11.7). No information was provided on how representative the DDGS samples from the United States were, or how they were collected (Rodrigues, 2008).

In this study, aflatoxins, fumonisins, and zearalenone were analyzed by HPLC and deoxynivalenol by TLC. The analytical method for measuring T-2 toxin was not described. The detection limits for aflatoxins, deoxynivalenol, fumonisins, T-2, and zearalenone were 0.5 ppb, 0.15 ppm, 0.025 ppm, 0.03 ppm, and 0.01 ppm, respectively. The analyses were conducted at two commercial laboratories, but it is unclear which samples were analyzed by which laboratory. It is well documented that considerable lab-to-lab variation can exist when analyzing various nutrients and toxins present in feedstuffs.

TABLE 11.7
Mycotoxin Concentrations in DDGS from the Asian Feed Study

Mycotoxins	Number of Samples Submitted	Mean Conc. (of Detectable Levels)	Maximum Level	Percentage of Samples above the Lowest FDA Level
Aflatoxins (ppb)	103	24	89	N.A.[a]
Deoxynivalenol (ppm)	103	2.13	12.0	N.A.
Fumonisins (ppm)	103	0.596	9.042	N.A.
T-2 toxin (ppm)	103	0.113	0.218	N.A.
Zearalenone (ppm)	103	0.333	8.107	N.A.

Source: Rodrigues, I. Crucial to monitor mycotoxins in DDGS. *Feed Business Asia*, January/February, 36–39, 2008.
[a] Data not available.

11.5.4.2   Results

Rodrigues (2008) reported that 99% of the samples contained at least one detectable mycotoxin, with 8% containing detectable aflatoxins, 64% containing detectable deoxynivalenol, 87% containing detectable fumonisins, 26% containing T-2 toxin, and 92% containing detectable zearalenone. However, since all values for the mycotoxins were reported as maximums and averages, it was difficult to determine the percentage of these samples that contained concentrations of mycotoxins above the FDA action levels or recommended maximum tolerable levels for use in animal feed. Furthermore, it was not clear how many of the samples with detectable levels of mycotoxins were from DDGS produced in the United States.

Aflatoxins: Only 8% of the DDGS samples contained aflatoxin levels higher than 0.5 ppb. The average (24 ppb) and maximum (89 ppb) concentrations found were below the FDA action levels for all animal species except for lactating dairy cattle, where the maximum tolerable level is 20 ppb. The proportion of the 8% DDGS samples containing aflatoxin levels higher than 20 ppb that were produced in Asia compared to the United States is unknown.

Deoxynivalenol: Approximately 64% of the DDGS samples contained deoxynivalenol levels higher than 0.15 ppm. However, the average concentration of deoxynivalenol detected was approximately 2 ppm, indicating that a high proportion of these samples could be fed to all animal species. The maximum concentration detected was 12 ppm, higher than the FDA advisory level for any animal species. It is unknown what proportion of the DDGS samples with deoxynivalenol (64%) contained levels higher than 5 ppm, or how many of those samples were produced in Asia compared to the United States.

Fumonisins: A high proportion (87%) of the DDGS samples contained more than 0.025 ppm fumonisins. The average fumonisin concentration was around 0.6 ppm, and would be considered safe for use in feed for all animal species. The maximum fumonisin concentration was 9 ppm, which would be acceptable for use in feed for any animal species except equids and rabbits. It is unknown what proportion of the 87% DDGS samples contained more than 5 ppm fumonisin, or the number of those samples that were from Asia compared to the United States.

T-2 toxin: Only 26% of the samples contained more than 0.03 ppm T-2. It would be interesting to know if any of the U.S. samples analyzed in this study contained this mycotoxin.

Zearalenone: Is not regulated by the FDA and no recommendations have been suggested for industry guidance relative to incorporation in animal diets of feeds with this mycotoxin. Of the samples collected and analyzed in this study, 92% contained more than 0.01 ppm zearalenone. It is unknown how many of the samples that contained zearalenone had more than 1 ppm, or their origin.

The results of this study are difficult to interpret because the sampling methodology and geographic distribution of DDGS sources were not adequately described. Furthermore, the distribution of mycotoxin concentrations in contaminated samples was not reported. In addition, the author did not indicate what aflatoxins or fumonisins were measured. Although the maximum values for many of the mycotoxins were higher than those reported in previous U.S. studies, it appears that most of the samples contained concentrations below accepted FDA action levels.

11.5.5   A PUBLICATION ON MYCOTOXINS IN DDGS BY SOUTH DAKOTA COOPERATIVE EXTENSION SERVICE

11.5.5.1   Sampling and Testing Methods

This study was a review of data generated by Dairy One Forage Laboratory in Ithaca, New York (Garcia et al., 2008). The data were based on samples of DDGS and distiller wet grains (DWG) submitted to the lab from 2000 through 2007 for mycotoxin analysis. No information was provided regarding the geographic distribution of the samples or how the DDGS samples were collected (Dairy One Forage Laboratory website).

Neogen Veratox Quantitative Test Kits were used for the measurement of aflatoxins, deoxynivalenol, fumonisins, T-2, and zearalenone. The detection limits for each of them were 5 ppb, 0.5 ppm, 1 ppm, 0.025 ppm, and 0.025 ppm. Data reported for DWG were on a dry matter basis.

11.5.5.2   Results

All of the mycotoxins except deoxynivalenol, were analyzed in both coproducts, DDGS and distillers wet grains with solubles (DWGS), were well below the FDA recommendations or guidelines for each mycotoxin in animal feed. The results of this study are shown in Table 11.8.

Aflatoxins: Average aflatoxin concentration was 4.609 ppb, with a maximum concentration of 7.097 ppb. All DDGS samples tested for aflatoxin were below the FDA action levels for all animal species. The average (2.170 ppb) and maximum concentrations (6.785 ppb) of aflatoxin found in 28 DWG samples were also below the FDA action levels for all animal species.

TABLE 11.8
Mycotoxin Concentrations in DDGS from the South Dakota Study

Mycotoxins	Number of Samples Submitted	Average Level (of All Samples)	Maximum Level	Percentage of Samples above the Lowest FDA Level
DDGS				
Aflatoxin (ppb)	30	4.609	7.097	0%
Deoxynivalenol (ppm)	54	3.620	7.743	N.A.[a]
Fumonisin (ppm)	20	0.740	1.959	0%
T-2 toxin (ppm)	11	0.031	0.065	N.A.
Zearalenone (ppm)	16	0.239	0.510	N.A.
DWG				
Aflatoxin (ppb)	28	2.170	6.785	0%
Deoxynivalenol (ppm)	44	1.905	4.257	0%
Fumonisin (ppm)	27	0.688	1.729	0%
T-2 toxin (ppm)	14	0.122	0.240	N.A.
Zearalenone (ppm)	14	0.374	0.869	N.A.

Source: Garcia et al. Mycotoxins in corn distillers grains, a concern for ruminants? *South Dakota Cooperative Extension Service Extension Extra*, March 1–3, 2008.

[a]   Data not available.

Deoxynivalenol: The average concentration of deoxynivalenol in the 54 DDGS samples was 3.620 ppm, which was below the FDA advisory level for any animal diet. The maximum concentration of 7.743 ppm in DDGS was higher than the advisory level for swine, cattle, chickens, and other animals. At this level, the inclusion of the deoxynivalenol contaminated DDGS should not exceed 20% of the animal diet. It is unknown the proportion of DDGS samples that contained concentrations greater than 5 ppm. As for the 44 DWG samples analyzed for deoxynivalenol, the average was 1.905 ppm, and the maximum 4.257 ppm, with both levels below the FDA advisory level for any animal diet.

Fumonisins: Average fumonisin concentration was 0.740 ppm, with a maximum of 1.959 ppm among the 20 DDGS samples tested for fumonisins, and below the FDA recommendation levels for all animal species. The average (0.688 ppm) and maximum concentrations (1.729 ppm) of fumonisins found in 27 DWG samples were also below the FDA action levels for all animal species.

T-2 toxin: For DDGS samples, the T-2 average was 0.031 ppm and the maximum 0.065 ppm, while for DWG samples, the average was 0.122 ppm and the maximum 0.240 ppm.

Zearalenone: For DDGS samples, the average was 0.239 ppm zearalenone and the maximum 0.510 ppm, while for DWG samples, the average was 0.374 ppm and the maximum 0.869 ppm.

This study showed that aflatoxins, deoxynivalenol, fumonisins, T-2 toxin, and zearalenone were detected in DDGS and DWG samples submitted to Dairy One Forage Lab from 2000 to 2007. However, the levels of aflatoxins, fumonisins, T-2 toxin, and zearalenone were well below the maximum tolerable guidelines for use in animal feed, while certain samples contained more than 5 ppm deoxynivalenol but less than 10 ppm. The results of this study are difficult to interpret because the geographic distribution of the DDGS sources was not adequately described. In addition, the authors did not indicate what aflatoxins or fumonisins were measured.

11.6 CONCLUSIONS

We have examined the available data relative to concentrations of various mycotoxins, both regulated and unregulated by the FDA, in DDGS from two independent studies, one research study, and two publications. From these results all concentrations of mycotoxins in DDGS, and also in DWGS from one study, were generally below the FDA regulations for the specific mycotoxins. Only in a couple of exceptions were the concentrations of deoxynivalenol or fumonisins either at, or slightly above, the recommendations for selected sensitive animal species, and in those instances the occurrence was in less than 10% of the samples tested. These concentrations could fall well below any harmful concentration when the DDGS is blended with other ingredients to make up the overall animal diet.

The methodology for analysis of mycotoxins in grain and coproducts such as DDGS is quite sensitive, and the fact that mycotoxins are detectable has no relationship with their toxicity in a particular animal species. The dosage makes the toxins, and the animals fed DDGS in today's markets, not as sensitive as perhaps other animal species such as companion animals and humans. More discussions on feeding DDGS to animals will be presented in the next several chapters.

ACKNOWLEDGMENTS

We thank the following individuals for help with this paper: Gary Delong, Barb Randle, Ines Rodrigues. We thank U.S. Grains Council for providing the financial support for this study and permission to use the material from the study report for this chapter.

REFERENCES

AOAC Official Method 985.18, *Official Methods of Analysis of AOAC International*, 18th ed., 49.9.02.
AOAC Official Method 991.31, *Official Methods of Analysis of AOAC International*, 18th ed., 49.2.18
AOAC 994.01, 2000a. *Official Methods of Analysis of AOAC International*, 17th ed., Ch. 49, 56–59.

AOAC 994.08, 2000b. *Official Methods of Analysis of AOAC International*, 17th ed., Ch. 49, 26–27.

AOAC Official Method 2001.04. *Official Methods of Analysis of AOAC International*, 18th ed., 49.5.02

Bennett, G. A., and J. L. Richard. 1996. Influence of processing on *Fusarium* mycotoxins in contaminated grains. *Food Technology* 50: 235–238.

Bothast, R. J., G. A., Bennett, J. E.,Vancauwenberge, and J. L. Richard. 1992. Fate of fumonisin B_1 in naturally contaminated corn during ethanol fermentation. *Applied and Environmental Microbiology* 58: 233–236.

CAST. 2003. Mycotoxins: Risks in plant, animal, and human systems. Task Force Report No. 139. Ames, Iowa: Council for Agricultural Science and Technology

Dairy One Forage laboratory website: www.dairyone.com/Forage/FeedComp/disclaimer.asp.

FDA. 2011. Aflatoxin in feeds and feed ingredients: Available online: www.cfsan.fda.gov/~lrd/fdaact. html#afla. Accessed February 16, 2011; Fumonisins in feeds and feed ingredients: Available online: http://www.fda.gov/Food/GuidanceComplianceRegulatoryInformation/GuidanceDocuments/ ChemicalContaminantsandPesticides/ucm109231; Deoxynivalenol (DON) in feeds and feed ingredients: Available online: www.cfsan.fda.gov/~dms/graingui.html.

Garcia, A., K. Kalscheur, A. Hippen, D. Schingoethe, and K. Rosentrater. 2008. Mycotoxins in corn distillers grains, a concern for ruminants? South Dakota Cooperative Extension Service Extension Extra, March 2008: 1–3.

Gelderblom, W. C. A., A. K. Jaskiewicz, W. F. O. Marasas, P. G. Thiel, R. M. Horak, R. Vleggaar, and N. P. J. Kriek. 1988. Fumonisins: Novel mycotoxins with cancer-promoting activity produced by *Fusarium moniliforme*. *Applied and Environmental Microbiology* 54: 1806–1811.

Herrman, T. 2001. Sampling: Procedures for feed MF-2036. Kansas State University, Agricultural Experiment Station and Cooperative Extension Service. Available online: www.oznet.ksu.edu/library/grsci2/MF2036. pdf. Accessed February 15, 2011.

Imerman, P. 2006–2007. Study Conducted by Iowa State University and Novecta LLC Veterinary Diagnostic Lab, Iowa State University, Ames, IA 50011.

Krska, R., P. Schubert-Ulirich, A. Molinelli, M. Sulyok, S. MacDonald, and C. Crews. 2008. Mycotoxin analysis: An update. *Food Additives and Contaminants* 25(2): 152–163.

MacDonald, S. J., D. Chan, P. Brereton, A. Damant, and R. Wood. 2005a. Determination of deoxynivalenol in cereals and cereal products by immunoaffinity column cleanup with liquid chromatography: Interlaboratory study. *Journal of AOAC International* 88(4): 1197–1204.

MacDonald, S. J., S. Anderson, P. Brereton, R. Wood, and A. Damant. 2005b. Determination of zearalenone in barley, maize and wheat flour, polenta, and maize-based baby food by immunoaffinity column cleanup with liquid chromatography: Interlaboratory study. *Journal of AOAC International* 88(6): 1733–1740.

Richard, J. L. 2000. Mycotoxins—An overview. Romer Labs' Guide to Mycotoxins. *Romer Labs Guide to Mycotoxins* Vol. 1 Romer Labs, Inc.

Richard, J. L. 2006. Sampling and sample preparation for mycotoxin analysis. *Romer Labs Guide to Mycotoxins* Vol. 2. Romer Labs, Inc.

Rodrigues, I. 2008. Crucial to monitor mycotoxins in DDGS. *Feed Business Asia* January/February, 36–39.

Romer, T. R. 1986. Use of small charcoal/alumina cleanup columns in determination of trichothecene mycotoxins in foods and feeds. *Journal* of AOAC International 69(4): 699–703.

Rottinghaus, G. E., C. E. Coatney, and C. H. Minor. 1992. A rapid, sensitive, thin layer chromatography procedure for the detection of fumonisin B_1 and B_2. *Journal of Veterinary Diagnostic Investigation* 4: 326–329.

Stahr, H. M., ed. 1991. *Analytical Methods in Toxicology*, pp. 101–103. New York: John Wiley & Sons Inc.

Sulyok M., R. Krska, and R. Schuhmacher. 2007. A liquid chromatography/tandem mass spectrometric multimycotoxin method for the quantification of 87 analytes and its application to semi-quantitative screening of moldy food samples. *Analytical and Bioanalytical Chemistry* 389: 1505–1523.

Trucksess, M. W. 2000. Natural Toxins. *Official Methods of Analysis of AOAC International*, 17th ed., Ch. 49, 1–2.

USDA GIPSA. 2011. www.gipsa.usda.gov/GIPSA/webapp?area=home&subject=lr&topic=hb.

Zhang Y., and J. Sido. 2008. DDGS quality assessment survey of DDGS testing laboratories in the United States & their analytical methodologies, report for the United States Grains Council.

Zhang Y., J. Sido, C. Willaredt, and B. A. Wrenn. 2008. National DDGS Library Nutritional Profile. Presentation at the 2008 International Fuel Ethanol Workshop & Expo.

Zhang Y., J. Caupert, P. M. Imerman, J. L. Richard, and G. C. Shurson. 2009. The occurrence and concentration of mycotoxins in U.S. distiller's dried grains with solubles. *Journal of Agricultural and Food Chemistry* 57(20): 9828–9837.

Part III

Traditional Uses

12 Feeding Ethanol Coproducts to Beef Cattle

Alfredo Dicostanzo and Cody L. Wright

CONTENTS

12.1 INTRODUCTION

The previous chapter in this publication addressed issues related to the mycotoxin concentration in ethanol coproducts. While certain mycotoxins can be highly detrimental to livestock, particularly swine and dairy cattle, beef cattle are generally fairly tolerant of mycotoxins. This chapter will discuss issues related to use of distillers grains and other ethanol coproducts in the diets of various classes of beef cattle.

Rapid expansion of the fuel ethanol industry has resulted in a dramatic increase in the availability of distillers coproducts as livestock feed. Continued development of the biofuels industry has brought challenges and opportunities for beef producers and cattle feeders alike. A recent study (Shurson, 2009a) indicated that 38% of the national distillers grains production not destined for export is consumed by beef cattle. This value is likely to remain similar, relative to total production of distillers grains, as new coproduct streams become available (Shurson, 2009b). Between 2007 and 2008, three independent reports indicated that from 36% to 42% of the cattle feeding operations surveyed used distillers grains (USDA, 2007; Arora et al., 2008; RFA, 2008). Increased supply of distillers coproducts or acceptance of these coproducts as a partial substitute for corn and protein sources has led to increased use rates in cow–calf and cattle feeding operations.

Beef cattle operations interested in incorporating distillers coproducts into their nutrition programs must focus on three issues: (1) nutrient composition of the coproducts and the potential variation as influenced by production plant and season, (2) storage and handling, and (3) diet formulation with dry grind coproducts that enhances nutrient compatibility with other feed ingredients. This chapter reviews factors that affect the nutrient composition of distillers coproducts, feeding strategies for various classes of cattle, common methods to store and handle distillers coproducts, and associative effects between nutrients derived from distillers coproducts and other ingredients in diets of beef cattle.

12.2 NUTRIENT COMPOSITION

Distillers coproducts are derived from the dry grind ethanol production process (Chapter 5). The dry grind process begins with the enzymatic hydrolysis of starch to produce long-chain sugars. These sugars are further processed to produce a simple sugar called dextrose. *Saccharomyces cerevisiae*, a yeast species, is then added to convert dextrose into ethanol. Throughout this process, ground corn kernels are immersed in liquid. The combination of ground corn, water, enzymes, and yeast forms a mash frequently referred to as "beer." When the fermentation process is completed (approximately 40 to 60 h), ethanol is stripped from the "beer" and the remaining material is centrifuged to partially separate the liquid and solid fractions. The centrifugation process yields two main coproduct streams: thin stillage and distillers wet grains (DWG). Thin stillage is often further dried to yield condensed distillers solubles (CDS). DWG may be sold as is (approximately 35% dry matter) or dried (DDG; approximately 90% dry matter). In many new generation dry grind plants, CDS is added to the DWG to produce DWG plus solubles (DWGS). That product can be either partially dried to produce distillers modified wet grains plus solubles (DMWGS; approximately 50% dry matter) or completely dried to produce DDG plus solubles (DDGS). Some facilities produce DMWGS by adding CDS to DDGS until the final product reaches approximately 50% dry matter. This method allows for multiple additions of CDS to the coproduct. Table 12.1 summarizes ranges for nutrient compositions of distillers coproducts. Refer to Chapter 5 for discussion of the ethanol production process and Chapter 8 for more information on the chemical composition of DDGS.

The nutrient composition of distillers coproducts is generally very favorable for use in every segment of the beef cattle industry. Values listed in Table 12.1 indicate that these coproducts can be considered as both energy and protein sources. Differences in energy and protein concentrations between coproducts result largely from whether the coproduct was dried or not. DWG and DWGS contain more energy and slightly more protein than DDG or DDGS.

Most dry grind coproducts contain between 20% and 35% crude protein and approximately 50% of the crude protein is undegradable intake protein (UIP). This combination of crude protein (CP) and UIP is particularly advantageous in high producing animals such as growing calves and lactating cows. However, since a relatively high percentage of the protein in distillers coproducts is not available to the rumen microbial population, there was initially a concern as to whether or not there was enough degradable intake protein (DIP) in distillers grains to support fermentation. Research has clearly demonstrated that DDGS can function as the sole source of supplemental protein in forage-based diets (Doering-Resch et al., 2005). Furthermore, additional DIP, in the form of urea, provides no benefit to forage-fed cattle being supplemented with DDGS (Stalker et al., 2007).

While the exact energy concentration in distillers grains has been, and is currently, a subject of vigorous debate in the research community, most researchers and practicing nutritionists would agree that distillers grains has, at a minimum, equal energy values to corn and may have as much as 20% more energy than corn. The sources of the energy in distillers grains, largely highly digestible

TABLE 12.1

Nutrient Composition of Dry Grind Coproducts (Dry Matter Basis) Relevant to Feeding Beef Cattle

Nutrient	CDS[a] Min	CDS[a] Max	DWG[b] Min	DWG[b] Max	DMWGS[c] Min	DMWGS[c] Max	DDG[d] Min	DDG[d] Max	DDGS[e] Min	DDGS[e] Max
DM[f], %	30	50	25	35	45	50	88	90	85.6	91.9
CP[g], %	20	30	30	35	30	35	25	35	26.3	34.0
DIP[h], %	20	50	45	53	45	53	40	50	40	50
Fat, %	9	15	8	12	8	12	8	10	9.2	12.6
NDF[i], %	10	23	30	50	30	50	40	44	30.0	54.1
ADF[j], %	2	7	14	22			18	21	7.0	25.4
TDN[k], %	75	120	70	110	70	110	77	88	85	90
NE$_m$[l], Mcal/lb	1.00	1.15	0.90	1.10	0.90	1.10	0.89	1.00	0.98	1.00
NE$_g$[m], Mcal/lb	0.80	0.93	0.70	0.80	0.70	0.80	0.67	0.70	0.68	0.70
Ca, %	0.03	0.17	0.02	0.03	0.02	0.03	0.11	0.20	0.0	0.13
P, %	1.30	1.45	0.50	0.80	0.50	0.80	0.41	0.80	0.68	1.09
S, %	0.37	0.95	0.50	0.70			0.48		0.12	0.82

Source: Adapted from Spiehs, M. J. et al. *Journal of Animal Science*, 80, 2639–2645, 2002;, Loy, D. 2008. Ethanol coproducts for cattle the process and products. Available online: www.extension.iastate.edu/Publications/IBC18.pdf. Accessed October 28, 2009; and Tjardes, K. and C. Wright. 2002. Feeding corn distiller's coproducts to beef cattle. Extension Extra 2036. *South Dakota Cooperative Extension Service.* Brookings, SD.

[a] Condensed distillers solubles.
[b] Distillers wet grains.
[c] Distillers modified wet grains with solubles.
[d] Distillers dried grains.
[e] Distillers dried grains with solubles.
[f] Dry matter.
[g] Crude protein.
[h] Degradable intake protein (expressed as a percentage of CP).
[i] Neutral detergent fiber.
[j] Acid detergent fiber.
[k] Total digestible nutrients.
[l] Net energy of maintenance.
[m] Net energy for gain.

fiber and fat, are particularly beneficial for ruminants. This allows for inclusion in forage-based diets without eliciting negative associative effects on fiber digestion frequently observed with starch supplementation.

A review of the concentrations of each of the nutrients listed in Table 12.1 indicates that there is substantial variability in the nutrient composition of each of these coproducts. Tabular values from the Nutrient Requirements of Beef Cattle (NRC, 2000) list average CP and DIP concentrations of DDGS at 30.4% and 52% (as a percent of CP), respectively. However, these averages had coefficients of variation (CV) of 11.7% and 38.5%, respectively. The Nutrient Requirements of Dairy Cattle (NRC, 2001) and an analysis of DDGS procured from various modern dry grind plants in Minnesota and South Dakota (Spiehs et al., 2002) list average CP concentrations of 29.7% and 30.4% with coefficients of variation of 11.1% and 6.4%, respectively. Lower variation associated with the average CP concentration of DDGS samples obtained by Spiehs et al. (2002) is likely a reflection of the modern type of plants from whence the samples came. It is interesting to note that within modern plants overall variation in CP of DDGS was relatively small within plants. However, substantially more variation was observed between plants.

Variation in nutrient composition of dry grind coproducts has implications on diet formulation strategies. At a 6.4% CV for CP content of DDGS, CP concentration of a diet formulated to contain 12.5% CP derived from mixing 15% DDGS (30.4% CP), 75% corn grain (9.5% CP), and 15% corn silage (8.5% CP) would vary from 11.1% to 13.1%, assuming no variation in CP concentration of corn grain or corn silage. Similarly, for wet coproducts such as DWGS, fed at 31% of the diet (as fed basis) with 52% corn grain and 18% corn silage, a change in dry matter (DM) concentration from the minimum to the maximum concentrations listed in Table 12.1 results in a change in diet DM of three percentage units, and alters daily dry matter intake (DMI) by 1 lb (assuming no change in DM concentration of corn grain or corn silage). DMI variation is known to affect performance; thus, cattle feeders must be prepared to evaluate nutrient concentrations of coproducts and other feeds (especially high-moisture feeds) when mixing diets to avoid such variability in DM and nutrient intake. This is particularly true when feedlot inventory is large and source of coproduct is variable.

12.3 COWS

Since distillers coproducts contain 20% to 35% crude protein, highly digestible fiber, and phosphorus, they are an excellent alternative to expensive commercial feeds as a beef cow supplement. DDGS has been shown to be an effective replacement for oilseed meals as a protein supplement in the diets of beef cows consuming poor-quality forage (Doering-Resch et al., 2005). DDGS has also been effectively used as a supplement for cows grazing corn stalks. Gustad et al. (2005) fed growing steers DDGS at 0.29%, 0.49%, 0.69%, 0.88%, 1.08%, and 1.27% of body weight (BW). Average daily gain (ADG) increased linearly from 0.90 to 1.81 lb per day as the amount of DDGS fed increased.

Combinations of DWGS or other distillers coproducts such as CDS with poor-quality forages can provide beef cows with much, if not all, of their nutrient needs. Kovarik et al. (2008) compared limit-fed combinations (17.0 lb DM per head per day) of DWGS or CDS with corn stalks with a more traditional grass hay, corn stalks, alfalfa haylage diet fed *ad libitum*. Cows fed the combination of DWGS and corn stalks were heavier at the end of the experiment, but there were no differences in body condition score. As the cost of pasture increases, limit feeding beef cows has become an increasingly more attractive option for beef producers. Dried distillers grains with solubles can be included at up to 75% of the diet DM for limit-fed (23.2 lb DM per head per day) gestating beef cows without sacrificing body condition score or reproductive performance (Shike et al., 2009). However, careful attention to the total diet sulfur (S) concentration is essential in this scenario. This will be discussed in greater detail later in this chapter.

Frequency of supplementation does not appear to affect beef cows to the degree that it does young cattle. Supplementing DWGS to gestating beef cows either three or six days per week had no effect on cow BW or body condition score (Musgrave et al., 2009). The lack of an effect in mature cows likely results from the comparatively smaller amounts being supplemented and the lower percentage of fat being contributed to the total diet.

In certain instances, the fat contained in distillers coproducts may be beneficial for beef cows. Supplemental unsaturated fat has resulted in improved reproductive performance in beef cows (Bellows, 2001). Postpartum UIP (Wiley et al., 1991; Triplett et al., 1995) and fat (Webb et al., 2001) supplementation have been shown to reduce the postpartum interval in beef cows. Larson and Funston (2009) compared wet corn gluten feed (WCGF) with DDGS as a postpartum supplement for lactating beef cows. Their results suggest that both ingredients are acceptable supplements and elicit similar reproductive responses.

12.4 CREEP FEED

While the combination of energy and protein found in DDGS is highly beneficial to young, rapidly growing calves, there has been minimal research conducted to evaluate the use of DDGS in creep diets. Including DDGS at up to 50% of a creep supplement resulted in similar calf performance when compared with creep supplements based on soybean meal, soybean hulls, and wheat middlings (Reed et al., 2006). Dried distillers grains has been successfully used as a replacement for soybean meal as a protein source in corn and soybean hulls-based creep feeds without sacrificing animal performance (Lancaster et al., 2007).

One of the primary concerns when including distillers grains in creep feeds is the flow of the product in the complete feed. Creep feeds are frequently formulated with grains (whole or minimally processed) and pelleted feeds. Since DDGS is sold in a meal form formulating a diet with similar particle size and density can be challenging. If the particle size and densities of the feed ingredients are too different, substantial separation can occur as the feed flows through the feeder. Calves will also be more prone to sorting. Including DDGS in a pellet is a viable option to reduce the risk for separation. Unfortunately, DDGS alone does not pellet well and inclusion of other commodity feeds or commercial binders increases the cost and erodes the economic advantages of including DDGS.

12.5   HEIFER DEVELOPMENT

The nutrient requirements of bred heifers increase exponentially during the last trimester of gestation (NRC, 2000). For spring calving cowherds, the last third of gestation often falls during the winter when forage quality is at its nadir. Because of their nutrient profile, distillers coproducts are uniquely suited as a supplement for developing heifers. Supplementing DDGS resulted in similar weight gains and age at puberty, but greater conception rates to artificial insemination when compared to an isocaloric supplement comprised of dried corn gluten feed, whole corn germ, and urea (Martin et al., 2007). Heifers supplemented with UIP or fat in late gestation have experienced increased pregnancy rates relative to controls. Since DDGS contains high concentrations of both UIP and fat, they are excellent additions to prepartum diets for beef heifers. Engel et al. (2008) compared DDGS with soybean hulls in late gestation heifer diets. While final body condition scores were similar, DDGS-fed heifers had greater BW gain during the prepartum period. Furthermore, a greater percentage of DDGS-fed heifers became pregnant during the subsequent breeding season.

12.6   GRAZING CALVES

The combination of protein, highly digestible fiber, fat, and phosphorus underscores the utility of distillers coproducts in grazing applications. The lack of starch in distillers coproducts allows for inclusion at substantially higher dietary concentration than would be effective with grains. Leupp et al. (2009a) observed no adverse effects on fiber digestion or fermentation when feeding DDGS at 0.3% to 1.2% of BW to cattle consuming moderate-quality forage. Furthermore, since distillers coproducts contain at least as much energy as corn, cattle supplemented with distillers coproducts gain at increasingly faster rates as the inclusion rate increases. Griffin et al. (2008) conducted a meta-analysis of seven trials where grazing cattle were supplemented with DDGS. Results from their analysis indicate that ADG and final BW increase linearly as the level of supplementation is increased. Griffin et al. (2009) supplemented DDGS to yearling steers grazing subirrigated meadow pastures at 0%, 0.6%, or 1.2% of BW. The authors observed a linear increase in ADG over the 91-day grazing period as the level of DDGS supplementation increased. Furthermore, ADG in the subsequent feedlot period was increased linearly with the level of supplementation during the grazing period. Rolfe et al. (2009) supplemented long yearling steers with DMWGS at either 0% or 0.6% of BW while grazing native range pastures. During the grazing season, the ADG of supplemented steers was 61.8% greater than that of the control steers (2.2 vs. 1.4 lb, respectively). Research also suggests that the improvement in performance of grazing cattle supplemented with DDGS is not solely due to either UIP or fat, but the combination of the two (MacDonald et al., 2007).

Timing of supplementation for grazing cattle is of particular interest for beef producers. Less frequent supplementation may provide opportunities for reduced labor and fuel expenses. Stalker et al. (2009) evaluated the effect of timing of DDGS supplementation on diet digestibility and animal performance of forage-fed cattle. Reducing the frequency of supplementation, but providing a similar amount of DDGS over a 1-week period resulted in significant reductions in DMI and total diet and neutral detergent fiber (NDF) digestibility. Steers and heifers supplemented with a daily equivalent of 4.2 or 2.9 lb, respectively, of DDGS six times per week gained faster than their counterparts that were fed the same daily equivalent three times per week. Feeding DDGS on alternate days also tended to reduce hay DMI in beef heifers (Loy et al., 2007). When the heifers in the experiment conducted by Stalker et al. (2009) were supplemented with DDGS three times per week, the DDGS resulted in an addition of 5.4%. This level of supplemental fat was likely sufficient to depress fiber digestibility and subsequently animal performance. High levels of fat in the rumen are known to depress fiber digestion (Jenkins, 1993). Furthermore, unsaturated fats are more detrimental to fermentation than saturated fats (Jenkins, 1993). The fat contained in DDGS is derived from the parent corn. Coppock and Wilks (1991) determined that the fat in corn oil is largely comprised of unsaturated fats.

Supplementation of forage-based diets with distillers coproducts also reduces forage intake. When Loy et al. (2007) supplemented DDGS to heifers at 0.4% of BW total DMI increased, but forage intake decreased. For each 2.2 lb of DDGS fed, forage DMI decreased by 1.2 lb. Reduced forage intake may allow beef producers to harvest or purchase less forage or increase stocking rates in pastures and may be particularly important when confronted with forage availability issues resulting from drought or overstocking.

When feeding cattle on pasture, it is advisable to deliver wet coproduct feeds in feed bunks. In two experiments, Musgrave et al. (2009) determined that supplementing DWGS to either growing steers or gestating cows in a bunk compared to directly on the ground resulted in greater BW and body condition score gains. Using NRC (1996) equations, the authors calculated an estimated 13% to 20% feed waste for the steers fed DWGS on the ground.

12.7 BACKGROUNDING AND GROWING CALVES

Distillers coproducts are particularly well suited for utilization as protein and energy sources in growing diets. Furthermore the additional moisture content wet coproducts feeds (DWG, DWGS, and CDS) complements the dry forages frequently utilized in backgrounding and growing diet.

The efficacy of distillers coproduct in growing diets has been clearly demonstrated in the literature. Hay-based backgrounding and growing diets supplemented with DWGS and DDGS were comparable to diets supplemented with soybean or blood meal in one of 2 years of an experiment conducted by Mateo et al. (2004). In the second year of the study, calves fed DWGS had better feed conversion than those fed the control or DDGS diet. Using only DDGS and soybean hulls and DDGS, Mueller and Boggs (2005) effectively substituted corn and soybean in growing diets. Heifers fed low- or high-quality forages supplemented with 1.5 to 6 lb DDGS per head per day had increased total DM intake, albeit lower forage intake, and greater gain (Morris et al., 2004). Inclusion of either DDGS or corn bran (Dakota Bran) at 30% of the diet DM in hay and haylage-based diets enhanced DMI, ADG, and feed conversion in growing cattle (Buckner et al., 2006b). The authors determined that the energy value of DDGS greater than corn, and that of corn bran was equal to corn (Buckner et al., 2006b). This observation was corroborated in a subsequent study (Corrigan et al., 2006) using alfalfa hay and sorghum silage diets; however, the concentration of CDS in the DDGS appeared to have an impact on performance and energy values calculated from animal performance. At moderate CDS inclusion levels, ADG and feed efficiency responded positively with increasing concentrations of DDGS in the diet (Corrigan et al., 2006). Based on the aforementioned literature, the energy concentration of DDGS appears to be similar or slightly greater than that of corn.

Inclusion of wet coproducts in growing diets is generally more convenient when cattle are fed in confinement as compared to grazing applications. This convenience results from the large animal capacities of growing feedyards and the equipment they commonly have access to on site (bunkers, tractors, mixers, etc.) to effectively handle and store wet feeds. Recent research with wet corn coproducts has focused on mixing coproduct with forage for storage purposes has led to renewed interest in feeding distillers coproducts in growing diets. Storage of wet coproducts will be discussed later in this chapter.

Feeding DDGS, DMWGS, or DWGS to growing cattle fed high forage diets appears to make no difference in terms of performance. In the first of two sequential studies, numerical trends for differences were observed in performance when comparing to dry-rolled corn (Nuttelman et al., 2008). In the second study, feeding DWGS led to more rapid gains and improved feed efficiency (Nuttelman et al., 2009). When factoring in the moisture content of the base forage, feeding DMWGS or DDGS with wet forages resulted in better performance than when feeding these coproducts with corn and hay-based diets (Wilken et al., 2008b). However, when comparing CDS to DWGS in growing diets, steers fed DWGS had better gain and feed conversion (Wilken et al., 2008a). Gilbery et al. (2006) supplemented steers fed low-quality switchgrass hay with 5%, 10%, or 15% CDS on a dry matter basis. Their results suggest that CDS improves nutrient availability and utilization of poor-quality forges. Furthermore, the authors estimated that 86.7% of the crude protein in CDS is DIP.

Corn silage and alfalfa hay diets supplemented with DDGS, DWGS, or a mixture of DWGS and wheat straw fed at 2 to 6 lb DM per head to growing steer calves supported greater gain and gain efficiency (Nuttelman et al., 2007). In this study, there were no differences in ADG or feed efficiency between DWGS and DDGS, but steers fed the mixture of DWGS and wheat straw had lower DMI. Mixing DWGS with hay either daily prior to feeding or in a bunker upon storage resulted similar ADG and feed conversion as a corn and hay diet (Loy et al., 2009). Feeding ensiled DWGS only with straw to growing cattle yielded better performance than mixing it fresh in two studies (Wilken et al., 2008a; Buckner et al., 2009). In contrast, results from a similar study design (Loy et al., 2008) using CDS instead of DWGS demonstrated that mixing CDS and hay daily led to similar performance as the control diet (corn and hay). However, mixing CDS and hay at the time of storage resulted in poorer gains and feed conversion. The authors hypothesized that mixing CDS and hay ahead of time permitted sufficient time for undesirable aerobic fermentation to occur, leading to accumulation of mold throughout the stored material.

12.8   FINISHING DIETS

12.8.1   Use of Distillers Coproducts in Finishing Diets

A summary of university studies published in 1984 (NCR, 1984) projected that if supplies of distillers grains would increase, their application in feeding cattle would extend to supply energy as well as protein. In studies reported therein, daily gain and feed efficiency were improved by including distillers wet grains at up to 63.8% of the diet dry matter. More recent studies suggest that there are diminishing returns to increased concentration of dry grind coproducts in diets of feedlot cattle (Larson et al., 1993). Similarly, it was determined that DWGS and DDGS were both of higher value than corn, but DWGS contained greater concentrations of net energy for gain (NE_g) (0.98 Mcal/lb) than DDGS (0.85 Mcal/lb; Ham et al., 1994). Acid detergent insoluble CP concentration (up to 15% of CP) did not reduce the energy value of DDGS (Ham et al., 1994).

The difficulty with interpreting the combined results of various studies published since 1994 is compounded by the fact that distillers grains contains significant amounts of rumen undegradable protein (38% to 66%; Mustafa et al., 2000), and supplements formulated to balance nitrogen concentration vary in the amount and source of rumen undegradable protein. Thus, data reported in various university studies published since 1993 in refereed journals and university research reports were compiled and analyzed to determine the effects of type and concentration of corn dry grind coproducts on gain, intake, and feed efficiency of feedlot cattle. Data were adjusted for differences in location, sex, age (yearling vs. calf), dietary CP, and supplemental protein source (urea or soybean meal). However, there were not sufficient data for evaluation of CDS inclusion in feedlot cattle diets; thus, only trends observed in the original studies will be discussed. Similarly, results using modified distillers grains are just coming in from university reports. These will be discussed towards the end of this section.

In a meta-analysis, inclusion of distillers dried grains had no effect on feed conversion, but had a positive effect on ADG up to 15% inclusion (Reinhardt et al., 2007). At approximately 30% inclusion, DDGS led to similar ADG response as cattle fed no DDGS. This gain response appears to be supported by a DMI response up to 23% inclusion. Intake of DM at 0% and 50% DDGS was similar. Feed conversion was not affected by inclusion of DDGS in the diet indicating that the energy value of DDGS and that of corn are similar.

Similarly, optimum inclusion concentration for DDGS in dry-rolled corn-based diets using gain response was between 10% and 20% of diet DM (Klopfenstein et al., 2008a). At 30% inclusion, there were no effects of DDGS on gain, intake, or feed conversion (Leupp et al., 2009) in dry-rolled corn-based diets. In contrast, feeding distillers dried grains at 13% or 15% of the diet DM (a value similar to the optimized ADG value observed in our analysis) yielded no gain or intake response in heifers fed a steam-flaked corn-based diet (Depenbusch et al., 2008a, 2009a). Increasing concentrations of DDGS in steam-flaked corn-based diets reduced gain and feed conversion (Depenbusch et al., 2009a).

Meta-analysis results yielded maximum gain at 21% inclusion of DWGS, while optimized feed conversion occurred at 31% inclusion of DWGS. A recent meta-analysis of results from studies conducted at Nebraska demonstrated improvements in gain and feed efficiency with increasing DWGS concentrations up to 30% and 50%, respectively (Klopfenstein et al., 2008a). Similarly, using dry-rolled corn diets, increasing concentrations of DWGS increased gain and feed conversion up to 40% of diet DM (Corrigan et al., 2009). Gain was optimized within 15% and 27.5% DWGS inclusion in high-moisture corn-based diets, while gain appeared to decrease with DWGS inclusion in steam-flaked corn-based diets (Corrigan et al., 2009). Feed conversion improved with inclusion of DWGS up to 40% of diet DM in high-moisture corn-based diets, but not with steam-flaked corn-based diets. Similarly, at 26% inclusion DWGS in steam-flaked corn-based diets, gain and feed conversion were poorer than for diets without DWGS (Depenbusch et al., 2008b). At low inclusion (13%) DWGS had no negative impact on ADG or feed conversion in steam-flaked corn-based diets (Depenbusch et al., 2009b). This difference in response by feedlot cattle fed steam-flaked corn diets relative to those fed corn processed as dry rolled or high moisture has been the focus of several recent studies.

Several studies demonstrated a lack of response or negative response to DWGS in steam-flaked corn-based diets. A recent report on the interaction of corn processing and feeding DWGS led to similar conclusions, 30% DWGS inclusion in steam-flaked corn-based diets resulted in similar feed conversion, but poorer gain (Vander Pol et al., 2008). In the study of Corrigan et al. (2006), one would conclude that the energy value of DWGS is similar to that of steam-flaked corn as feed efficiency was not changed when DWGS substituted steam-flaked corn. This observation would be reconcilable if it were not for the numerous studies where inclusion of DWGS in steam-flaked corn-based diets at concentrations beyond 20% resulted in poorer feed conversion (May et al., 2007; Depenbusch et al., 2008b).

An evaluation of 37 trials involving utilization of distillers grains (without regard to whether corn and sorghum or corn and sorghum distillers grains were used) resulted in estimation of the net energy (NE) value for maintenance and gain of distillers grains of 114.5 ± 15 and 80.9 ± 14 Mcal/cwt in dry-rolled corn diets, while NE values for maintenance and gain of distillers grains were predicted to be 92.3 ± 29 and 61.8 ± 25 Mcal/cwt in steam-flaked corn diets (Cole et al., 2006). The authors concluded that this difference may be the result of differences in dietary fat, ethanol contamination, yeast effects, errors in DM determination, lower methane production, and a host of other factors. Results of our meta-analysis demonstrated that feeding steam-flaked corn with no DWGS reflected a 4% improvement in feed efficiency, and as DWGS inclusion was modeled, feed efficiency was reduced. Dry-rolled and high-moisture corn-based diets containing 19% DWGS yielded similar feed conversion as control steam-flaked corn diets containing no DWGS. This value reflects the inclusion level at which several studies with DWGS in steam-flaked corn diets demonstrated little or no change in feed efficiency. Our estimates of NE for maintenance (NE_m) and gain (NE_g) for DWGS when fed in dry-rolled or high-moisture corn and steam flaked corn diets were 1.21 and 0.86 Mcal NE_m and NE_g/lb DM and 0.78 and 0.50 NE_m and NE_g/lb DM, respectively. When determining NE values for DDGS from performance, two things are evident: NE value of DDGS is similar to that of corn and the depression in energy due to the associative effect of feeding it with steam flaked corn is not as severe as that observed when using DWGS. Estimates of NE_m and NE_g for DDGS when fed in dry-rolled or high-moisture corn and steam flaked corn diets were 1.06 and 0.74 Mcal NE_m and NE_g/lb DM and 1.34 and 0.61 NE_m and NE_g/lb DM, respectively. These values are similar to those described by Cole et al. (2006).

It is important to emphasize that although performance-derived energy values from these observations appear to show a negative associative effect when feeding DWG and steam flaked corn, production processes and individual feeding conditions have changed sufficiently since those studies were published so that these values are mere approximations and guidelines, and may not fully reflect current or future feeding conditions.

12.8.2 ROLE OF DEGRADABLE INTAKE PROTEIN

Perhaps a factor in the negative interaction between steam-flaked corn and distillers grains results from inclusion of sufficient amounts of DIP in the diet, particularly at distillers grains inclusion under 20% of the diet DM. Results from a meta-analysis indicated that appropriate DIP supplementation in steam-flaked corn diets is around 8.5% of diet DM (Reinhardt et al., 2007). Inclusion of protein supplements containing 30% crude protein, mostly from nonprotein nitrogen, is common when feeding 15% to 20% distillers grains. Protein supplements fed with 20% to 30% distillers grains diets typically contain 10% to 20% protein. Using the NRC (2000) model to determine DIP balance, researchers have discovered that DIP balance for supplements in low distillers grains inclusion diets containing 10% to 30% crude protein is below 50 g supply. Data indicate that in order for distillers grains diets to be balanced for DIP, a supply of 700 g DIP daily is required (170 g over that modeled by the NRC, 2000). Although feeding distillers grains at concentrations greater than 20% of diet DM should supply sufficient DIP, intake and actual degradability of protein play a significant role in determining DIP supply. This is particularly critical when feeding steam-flaked corn diets that lead to lower intakes, when feeding heifers or cattle of lower intake drive, and when feeding "natural" cattle. Also, two surveys of DIP concentration in DDGS; one from Dairy One, Ithaca, NY (accessible at www.dairyone.com/Forage/FeedComp/MainLibrary.asp; summarized from 2001 to 2007), and one from Kleinschmit et al. (2007) demonstrated that DIP concentration in DDGS may be in the range of 20% to 40% of dietary CP.

Research investigating the effect of DIP inclusion in distillers grains diets has produced conflicting results. Vander Pol et al. (2004) determined that additional DIP, in the form of urea, provided no benefit to cattle fed dry-rolled corn-based diets. In contrast, a recent study confirmed the need for DIP supplementation when using low concentrations of distillers grains steam-flaked corn-based diets (Ponce et al., 2009). Daily gain was greater for steers fed 15% DWGS and steam-flaked corn diets when diets contained 1.5% and 3.0% nonprotein nitrogen. Similarly, feed efficiency was greater for steers fed 15% DWGS and steam-flaked corn diets when diets contained 1.5% nonprotein nitrogen. Differences in ruminal starch degradation rates likely account for differences in the response to supplemental DIP.

12.8.3 DISTILLERS MODIFIED WET GRAINS WITH SOLUBLES

The evolution of processes at ethanol plants continues as plants strive to derive as much ethanol and high-value coproducts. This has led to the development of alternative coproduct streams. One such product, DMWGS, is generated either by partially drying DWG or fully drying DWG and adding

CDS. A recent survey of 169 feedlots in Iowa (Lain et al., 2008) reported that almost the same number of feedlots used DMWGS and DWGS (25% vs. 31%, respectively). Similarly, a recent survey of feedlot consultants in Minnesota (G. I. Crawford, personal communication) revealed that use of DMWGS is almost as high as that of DWGS, and in some cases, DMWGS has replaced DWGS. There are several reasons for this: price, availability, shelf life, and consistency. However, published results of studies conducted with DMWGS are limited. At 25% or 47% inclusion, Trenkle (2007) reported that gain and feed conversion were similar to the dry-rolled corn control. Huls et al. (2008) reported maximal ADG with 20% DMWGS inclusion, and feed conversion improved linearly up to 50% DMWGS inclusion. Similarly, increasing concentrations of DMWGS in dry-rolled corn diets led to similar gain and feed efficiency as the control up to 40% inclusion, at which point both gain and efficiency declined (Trenkle, 2008). Thus, it appears that the energy concentration of DMWGS is likely intermediate between DWGS and DDGS. Other authors concur with this observation (Klopfenstein et al., 2008b).

12.8.4 Coproduct Combinations

In regions of the country where coproduct streams from both dry and wet grind processes are available, feedlot operators have considered use of combinations of corn coproducts. WCGF or CDS in combination with DDGS often presents interesting price options for astute cattle feeders with knowledge and experience feeding one or the other. WCGF contains approximately 90% of the energy value of corn, and it ranges in crude protein content from the 15% to 24%. Its protein fraction is more rumen degradable than that of DDGS, and its fat content is often under 3.5%. In addition, effectiveness of NDF fiber from CGF appears to be greater than that of DDGS. Cattle fed diets with no other roughage source than the NDF fraction from WCGF performed similarly to those fed a dry-rolled corn control diet (Dahlen et al., 2003). Thus, DWGS and WCGF may complement each other quite well. Indeed, cattle fed DWGS with no added roughage fiber consumed less DM and gained less than those fed DWGS, WCGF, and corn silage roughage (Rich et al., 2009). When diets containing 35% WCGF were supplemented with increasing (13.3% to 40% of diet DM) concentrations of DWGS, gain decreased linearly but feed efficiency was not affected by treatment (Bremer et al., 2008). Similarly, when diets containing 30% DMWGS were supplemented with increasing (15% to 30%) concentrations of WCGF, gain and feed conversion decreased linearly (Benton et al., 2008). In contrast, 50:50 combinations of DWGS and WCGF (DM basis) fed at 30% or 60% of diet DM improved feed conversion in dry-rolled corn:high-moisture corn diets (Buckner et al., 2006a). Using similar diets, Loza et al. (2006) determined that inclusion of DWGS in 30% WCGF did not result in significantly improved feed conversion. These observations reveal two important items: DMWGS definitely has a lower energy value than DWGS, particularly in combination with WCGF. Secondly, inclusion of DWGS and WCGF may not necessarily enhance feed conversion over that of DMWGS inclusion alone, but under high grain prices (and relatively lower coproduct prices), combinations of DWGS and WCGF, 50:50 DM blends, at inclusion rates under 60% of diet DM will likely result in lower feed costs without detriment to performance.

12.8.5 Condensed Distillers Solubles

Results from a meta-analysis indicated that ADG increased with increasing concentrations of CDS in the diet, but it appeared to decrease once concentrations of CDS exceeded 10% of the diet DM (Reinhardt et al., 2007). Although DMI increased in a quadratic response to increasing CDS concentration, feed efficiency was optimized when CDS was fed at 15% of the diet DM. These observations are corroborated by those of Trenkle and Pingel (2004). In their experiment, yearling steers fed up to 12% CDS in dry-rolled corn diets had similar performance as the corn control. Similarly, in steers fed diets containing 35% WCGF, feeding from 6.7% to 20% CDS had no effect on gain or feed efficiency (Bremer et al., 2008). Although a formal evaluation of the energy value of CDS has not been conducted, it would appear from the aforementioned that its energy value is similar to that of corn. Furthermore, it appears that CDS complements other corn coproducts well, and it may play a significant role in reducing diet costs for feedlot operators who have access to this coproduct, and have established facilities and handling procedures for it.

12.9 NEW COPRODUCTS

Corn kernels are comprised of four major components, each of which is associated with a nutrient fraction of economic interest to the ethanol and food processing industry. The endosperm and germ are contained within the pericarp and tip cap. Both pericarp and tip cap represent the bran fraction, while the germ and endosperm each contain protein and oil, and starch, respectively. The endosperm is the fraction of greatest interest to ethanol production as it is the fraction that is fermented to generate ethanol. Currently, ethanol production technology ferments the entire corn kernel, which yields distillers coproducts containing a substantial amount of oil, protein, and fiber. Refer to Chapter 4 for more information on grain structures and composition.

As the ethanol industry moves forward to become more efficient, and to position itself to maximize revenue streams associated with each of the corn fractions, kernel fractioning prior to processing and oil extraction from coproducts appear to be the most viable processes currently being evaluated. Several plants are currently extracting corn oil from ethanol coproduct streams and pilot plants, mostly under proprietary processes, are separating bran, endosperm, and germ prior to fermenting the endosperm. Refer to Chapters 5, 23, and 25 for more information on fractionation systems and resulting products.

Oil extraction from DDGS yields 15 lb de-oiled DGS per bu corn processed containing 36% CP and 2.5% fat (Shurson, 2009b). Coproduct streams from fractionation can be derived either from the fermentable or germ fraction. Fractionation of 1 bu corn yields 10 lb high-protein DDGS, 3 lb corn bran, and 4 lb corn germ. High-protein DDGS contains 45% CP and 4% fat (Shurson 2009b). Depending on the proprietary process, corn bran enhanced with CDS (Dakota Bran; Poet Nutrition, Sioux Falls, SD) or a meal derived after extracting oil from germ, and bran (ECorn; Renessen LLC, Wayzata, MN), and prefermentation-derived, high-protein, low-oil DDGS (EPro; Renessen LLC, Wayzata, MN) are examples of two products already available to the dairy and livestock industries. Current projections of the proportion of total dry grind that will rely on fractionation, oil extraction, or conventional grind are conservative. Prevalence of each process was projected by Shurson (2009b) to be 20%, 22%, and 58% for fractionation, oil extraction, and conventional dry grinding, respectively for 2022. These estimates would yield 4.4, 7.2, and 22 million metric tonnes of high-protein DDGS, de-oiled DDGS, and conventional DDGS, respectively, in 2022.

Fractionation or oil extraction coproducts are new, and have not been tested extensively. A recent report from Nebraska indicated that ECorn fed at concentrations from 20% to 40% of the diet DM led to similar gains and feed efficiency in yearling steers (Godsey et al., 2009). Interestingly, increasing concentrations of ECorn in the diet DM led to reductions in marbling score and fat depth while increasing ribeye area. Authors reported that it was unclear why inclusion of ECorn had such profound effects on carcass characteristics.

Corn bran coproducts were tested in each of two experiments with growing and finishing cattle. When fed to growing cattle as a dry coproduct (Dakota Bran) at 15% or 30% inclusion, gain and feed efficiency were not different when compared to similar inclusion concentrations using DDGS (Buckner et al., 2006b). As a wet coproduct (Dakota Cake) fed to finishing yearling cattle at up to 45% of diet DM improved gain and feed efficiency (Bremer et al., 2005). The authors concluded that the energy value of this coproduct was 108% that of dry-rolled corn at the 45% inclusion concentration.

Another study contrasted finishing heifer response to partial fractionation distillers grains (Dakota Gold HP) containing 43% CP and 4% fat to conventional DDGS containing 26% CP and 12% fat (Depenbusch et al., 2008a) in steam-flaked corn-based diets at 13% inclusion. In spite of statistically lower DMI, heifers fed partial fractionation DDG gained and converted at the same rate as those fed conventional DDGS or steam-flaked corn.

Our current state of knowledge on use and applications of new corn coproducts in feedlot diets demonstrates clear roles for bran-based and de-oiled or high-protein distillers grains. In a recent study evaluating displacement values for each of these ingredients, Shurson (2009b) concluded that high-protein distillers grains and conventional distillers grains have corn displacement ratios greater than 1. Displacement ratios for de-oiled distillers grains and corn bran were 0.55 and 0.63, respectively. Indeed, Shurson (2009b) projected consumption of corn bran total production in 2022 to be greatest for beef cattle and secondly for dairy cattle.

12.10 EFFECT OF DISTILLERS COPRODUCTS ON CARCASS AND PRODUCT QUALITY

A white paper review of factors reducing marbling deposition in cattle (Corah and McCully, 2006) led the authors to conclude that marbling deposition declined because feeding distillers grains leads to reduced starch supply. This observation drove an increase in investigations on existing data (meta-analysis) and original research. Results from a meta-analysis (Reinhardt et al., 2007) demonstrated that at low and moderate inclusion of distillers grains, marbling increased albeit at the expense of greater yield grade (Koger et al., 2010). At concentrations beyond 30%, distillers grains decreased marbling score. The authors hypothesized that several factors may be contributing to this decline, unrelated to distillers grains per se. For instance, at moderate inclusion, dietary crude fat ranged from 4% to 6% and marbling score was actually enhanced. At greater distillers grains inclusion, fat content increased beyond 8%, which may affect intake and/or marbling deposition. Similarly, at greater distillers grains inclusion, total metabolizable energy intake is diminished due to reduced intake. The authors further hypothesized that low ME intake may lead to reduced marbling, particularly in studies when cattle are harvested after the same number of days on feed.

In support of these observations, Klopfenstein et al. (2008a) reported that cattle fed DWGS at greater than 30% of the diet DM gained less and marbled less. Similar observations were made when effects of DDGS on marbling were studied (Buckner et al., 2008a). Several original research projects have corroborated these observations. Inclusion of DDGS up to 45% of the diet DM increased marbling score, which then declined to 75% DDGS inclusion (Depenbusch et al., 2009a). In that study, percentage of carcasses reaching USDA Choice declined linearly with increasing DDGS inclusion. Effects of fat content or reduced energy intake on marbling could not be ruled out in that study as fat content increased from 3.7% to 8.0% and DMI decreased from 16.5 to 15.4 lb. A recent summary of three studies confirmed that DWGS actually may be enhancing marbling (Calkins et al., 2009).

When balanced for fat content of the diet using tallow, there were no differences in marbling score relative for diets containing up to 40% DDGS (Vander Pol et al., 2008). However, in that study, when balanced for fat content of the diet using corn oil, DDGS diets led to greater marbling deposition than fat-balanced diets using corn oil. Indeed, their metabolism study indicated that feeding DWGS resulted in greater concentrations of propionic acid in the rumen, greater fat digestibility, and more unsaturated fatty acids reaching the duodenum. The observation that greater propionic acid concentration results from feeding DWGS actually contradicted the initial perception that feeding distillers grains led to a reduction in glucose precursors as an explanation for reduced marbling at greater distillers grains inclusion.

Results from a recent study confirm that greater fat content, lower DMI, or even greater protein content may be contributing to lower marbling deposition in high inclusion distillers grains diets. Gunn et al. (2009) fed diets isonitrogenous and(or) isolipidic to 50% DDGS diets, and compared performance and carcass characteristics to a 25% DDGS control diet. All diets contained dry-rolled corn. Their results demonstrated that both greater fat or protein, and both fat and protein content of the 50% distillers grains diet or its isonitrogenous, isolipidic, or isonitrogenous and isolipidic counterparts led to reductions in marbling score and quality grade.

Distillers coproducts may also affect case life and fatty acid composition of beef from cattle fed the coproducts. A greater proportion of steaks from Holstein steers fed either DWGS or DDGS at 40% of their diet DM were considered moderately unacceptable during retail display when compared to cattle fed lower concentrations of distillers grains (Roeber et al., 2005). Furthermore, the authors determined that steaks from cattle fed distillers grains had greater a* values (more red) than steaks from steers fed 0% or 50% distillers grains (Roeber et al., 2005).

Research evaluating the effect of distillers grains on tenderness has produced inconsistent results. In cattle fed distillers grains at dietary inclusion rates ranging from 0% to 75%, Depenbusch et al. (2009a) reported that myofibrillar and overall tenderness increased linearly as the dietary concentration of distillers grains increased. This research is in contrast to previous research where no differences were observed between cattle fed diets with or without distillers grains (Koger et al., 2010; Roeber et al., 2005; Gill et al., 2008).

Feeding distillers grains has also altered the fatty acid composition of beef. Most research has reported no differences in the concentrations of saturated fatty acids and monounsaturated fatty acids (Gill et al., 2004; Koger et al., 2010; Depenbusch et al., 2009a). However, increased concentrations of the *trans*-10, *cis*-12 isomer have been reported in cattle fed DDGS. (Depenbusch et al., 2009a) also reported that greater concentrations of linoleic acid (18:2n-6cis), total n-6 fatty acids, and total PUFA in cooked ribeye steaks increased as dietary distillers grains concentrations increased.

12.11 DISTILLERS COPRODUCTS AND *E. COLI*

Results from recent studies indicate that there is a relationship between feeding distillers grains and prevalence of *E. coli* O157:H7. This pathogen is known to occur naturally in cattle, and feeding steam-flaked grains appears to increase its prevalence (Fox et al., 2007; Depenbusch et al., 2008c). However, it is not clear what hindgut conditions lead to increased prevalence. One hypothesis is that low starch flow postruminally, such as is observed with steam-flaked grains, leads to higher pH and lower organic acid concentrations, which may be conducive to growth of *E. coli* O157:H7. Similarly, due to the lower overall supply of starch in distillers grains-based diets, ruminal pH is expected to be lower, thereby providing an environment conducive to growth of *E. coli* O157:H7. Incubations of rumen fluid or fecal material and DDGS led to increases in nalidixic acid-resistant *E. coli* O157:H7 (Jacob et al., 2008b). Also, greater prevalence of *E. coli* O157:H7 occurred in cattle fed DWGS than those fed steam-flaked corn (Jacob et al., 2008c). The authors speculated that decreased starch content or increased fiber content in the hindgut was stimulating to growth of *E. coli* O157:H7. However, fecal pH and starch concentration were not different for heifers shedding or not *E. coli* O157:H7 (Depenbusch et al., 2008c). Attempts to explain the reason for increased prevalence of *E. coli* O157:H7 in calves fed DDG include involvement by yeast fermentation intermediates and organic acids (Jacob et al., 2008a).

Recent work conducted at the USDA Meat Animal Research Center has supported the observation that cattle fed DWGS have greater prevalence of *E. coli* O157:H7 in both fecal and hide samples (Wells et al., 2009). In support to previous observations and speculation, fecal pH was greater for cattle fed DWGS. Fecal L-lactate and total VFA were greater for cattle fed no DWGS, but perhaps not sufficiently different to be inhibitory to *E. coli* O157:H7. In another study, manure slurries collected from cattle fed from 0% to 60% DWGS, L-lactate was at concentrations beyond inhibitory levels to *E. coli* O157:H7 (Varel et al., 2008). In that study, inoculations of *E. coli* O157:H7 in manure slurries from cattle fed DWGS supported greater persistence of this pathogen. The authors concluded that both L-lactate concentrations (below 15 umol/g) and pH ranging from 6.0 to 8.0 may have been supportive of *E. coli* O157:H7 growth.

Although these observations lead to inconclusive reasons for the association between feeding distillers grains and prevalence of *E. coli* O157:H7, it seems clear that certain conditions are created in the rumen and hindgut of cattle fed distillers grains that lead to greater colonization by this pathogen. Understanding the rumen and hindgut conditions generated by feeding distillers grains is paramount to devise interventions and/or management strategies to reduce prevalence of *E. coli* O157:H7 in cattle fed distillers grains.

12.12   NUTRITIONAL LIMITATIONS

From a nutritional perspective, the utilization of distillers coproducts in ruminant diets is largely limited by one of two factors: fat or S. When lipids are included in ruminant diets, fermentation can be disrupted and the digestibility of nonlipid energy sources can be reduced (Jenkins, 1993). This is of particular importance in forage-based diets. Dietary fat is more detrimental to the digestibility of structural carbohydrates than it is to nonstructural carbohydrates (Jenkins, 1993). The lipid content of distillers coproducts is largely comprised of corn oil. Researchers have previously demonstrated that corn oil is high in unsaturated fatty acids. Unfortunately, unsaturated fats inhibit microbial fermentation to a greater extent than saturated fatty acids (Palmquist and Jenkins, 1980). The total lipid content and the composition of the lipids in distillers coproducts underscore the necessity of proper diet formulation to avoid reductions in diet digestibility and consequently animal performance.

The second factor that limits the inclusion of DDGS in beef cattle diets is S. Regardless of the dietary source, S can be problematic in ruminant diets. When S enters the rumen, it is reduced, in a series of steps, to sulfide. The sulfide produced then forms hydrogen sulfide gas (H_2S). Hydrogen sulfide is a neurotoxic gas that accumulates in the ruminal gas cap and is eructated by the animal. It is believed that a portion of the H_2S is then re-inhaled by the animal. Upon re-inhalation, the H_2S gas is absorbed into the blood stream where it elicits a number of neurological effects. The mechanism underlying the neurological effects of H_2S is believed to be interference with the cytochrome oxidase system in brain mitochondria. Reduced cytochrome oxidase function leads to the classical signs of S-induced polioencephalomalacia (sPEM), the condition commonly associated with overconsumption of S.

Traditionally, polioencephalomalacia (PEM) has been associated with reductions in circulating thiamine concentrations. However, researchers have demonstrated that blood thiamine concentrations of steers that had developed signs of PEM from high-S diet (sPEM) were normal (Sager et al., 1990; Gould et al., 1991). Gould (1998) concluded that evidence of thiamine deficiency was not demonstrable in sPEM. Thiamin supplementation to livestock fed high-S diets has produced mixed results. In cattle consuming high-S water (3790 mg SO_4/L), Ward and Patterson (2004) observed improved ADG when cattle were supplemented with 1 g thiamin per animal per day. However, in lambs fed diets containing 0.7% S from DDGS, up to 150 mg thiamin per animal per day had minimal effect on animal health and performance (Neville et al., 2009).

When coupled with potentially high-S water concentrations in some major cow–calf and cattle feeding areas, high-S content of distillers coproducts can lead to detrimental effects on animal performance (Loneragan et al., 2001). To avoid sPEM, the National Research Council (2005) has recommended maximum tolerable S concentrations of 0.3% of the diet dry matter for cattle consuming 85% concentrate or more and 0.5% of the diet dry matter for cattle consuming 45% roughage or more. These guidelines are conservative estimates intended to prevent disease. However, the research community has a great deal to learn regarding S nutrition in ruminant diets. Several research experiments have exposed cattle to S concentrations in excess of these guidelines with no negative consequences. Careful consideration of the total dietary S concentration is essential to safe and effective diet formulation using DDGS.

Corn is not inherently high in S; however, sulfuric acid is frequently utilized in the ethanol production process, which substantially increases the S content of the DDGS. While the S concentration of DDGS has been reported to be 0.47 (Spiehs et al., 2002), a recent review of mineral concentrations in corn coproducts revealed that some samples may reach concentrations of S as high as 1.5%, particularly in DWG, DWGS and/or CDS (Buckner et al., 2008b). A similar survey in Iowa demonstrated that concentrations of S in corn coproducts were 0.50% to 0.57% in DMWGS and DWGS, but with substantial variation (32% and 22% CV) around the mean (DeWitt et al., 2008). In that survey, concentrations of S in DDGS and CDS were 0.76% and 1.35%, respectively, with CV of 24% and 29%. Samples from total mixed rations containing less than 50 Mcal NEg/cwt, and those containing more than 50 Mcal NEg/cwt averaged 0.37% and 0.33% S, respectively (DeWitt et al., 2008). Vanness et al. (2009) evaluated the incidence of PEM and H_2S concentrations in ruminal fluid of cattle fed distillers grains and supplemented or not with roughage and concluded that diets containing greater than 0.30% S may be safely fed to cattle if the S source is the coproduct. There is some evidence that cattle can tolerate organic S better than inorganic S. A summary of several studies with a total of over 4000 animal observations indicated that incidence of PEM was low (0.14%) in diets containing 0.46% S or less (Vanness et al., 2008a). Incidence of PEM increased as dietary S reached past 0.56%. Feedlot cattle can tolerate inclusion of ethanol coproducts in dietary concentrations, taking into account water S, that yield S intake concentrations between 0.30% and 0.50%. Whether lower DMI response usually observed in cattle fed DWGS, DMWGS, DDGS, or CDS, or their combinations, at concentrations beyond 40% is due to higher concentrations of dietary S is not yet known.

Although sPEM is the greatest threat and perhaps the single most visible indicator of S toxicity, performance effects of high S inclusion occur well before clinical signs of sPEM occur. At 0.25% S derived from ammonium sulfate, steers consumed less DMI, gained slower, and had poorer feed conversions (Zinn et al., 1997). Similarly, 30% DDGS diets were fed at 0.42% or 0.65% of the diet DM to finishing steers (Uwituze et al., 2009). Sulfuric acid was added to the diet to increase S concentration. At the higher S concentration, steers had lower DMI, poorer ADG, and tended to have lower feed efficiency. Similar responses have been observed in cattle consuming high-sulfate water. Sexson et al. (2009) compared the performance of feedlot cattle exposed to well water containing 1933 mg SO_4/L to those exposed to well water blended with water treated by reverse osmosis containing 604 mg SO_4/L. High-SO_4 water reduced ADG and tended to reduce both water and DMI.

Recent research has attempted to identify feed ingredients and feeding strategies that can effectively ameliorate the deleterious effects of high dietary S concentrations. Manganese oxide is a powerful oxidizing agent that is also used as a Mn source in mineral supplements fed to feedlot and range cattle. In aqueous solution, manganese oxide oxidized hydrogen sulfide to sulfate and elemental S at pH of 5 (most cattle on feedlot diets have pH in the range of 5.25 to 6) at the rate of 1.49 g of manganese oxide to 1 g of hydrogen sulfide (Herszage and Afonso, 2003). In vitro incubations of 2000 ppm Mn as MnO in batch cultures containing 0.65% S reduced total H_2S (Kelzer et al., 2010). Incubations with 1000 and 2000 ppm Mn as MnO in batch cultures containing 0.65% S tended to have lower H_2S/mL gas produced. In rumen cannulated cattle consuming a diet containing 0.50% dietary S, 1000 ppm Mn as MnO reduced average and cumulative rumen H_2S concentrations (Kelzer et al., unpublished data). This response to MnO supplementation appears to be partially mediated by increased ruminal pH.

Management of high S concentrations in distillers coproducts may also be achieved through increased forage content of diets. At dietary S concentrations ranging from 0.41 to 0.47%, derived from feeding WCGF and(or) DWGS, increasing dietary forage concentrations from 0 to 15% led to reductions in H_2S concentrations in the rumen (Vanness et al., 2008b).

In addition to the potential for sPEM, high dietary S concentrations may also reduce the bioavailability of copper (Cu). Dietary S has been shown to reduce Cu absorption and status either alone (Suttle, 1974) or in combination with Mo (Suttle, 1975; Ward et al., 1993). While the effect of high-S from distillers coproducts on Cu availability has not been evaluated, it is highly likely that responses will be similar to those identified in previous research.

It appears that in spite of potential challenges with S concentrations in distillers coproducts, beef producers, feedlot managers, and consultants already have some tools at their disposal to address these concerns. On the one hand, higher concentrations of dietary S derived from coproducts may behave differently than that derived from inorganic sources. Also, inclusion of even low concentrations of roughage in the diet has resulted in reduced H_2S concentrations in the rumen.

Further enhancements in managing S concentrations in corn coproduct diets will come from addressing interactions of S, Mo, Fe, and Cu, and use of oxidizing agents such as Mn and other elements.

Although not a limitation per se, phosphorus (P) nutrition should be considered when formulating beef cattle diets. Given the elevated P concentrations in distillers coproducts relative to the parent corn, their incorporation into beef cattle diets can substantially increase P intake of the animals. The relative benefit or detriment of the additional P is highly dependent upon the diet and type of animal being fed. Early in the growing season, forages grazed by beef cattle may contain enough P to meet the needs of the livestock. However, as the plants mature, P concentrations decrease and consequently the potential for a P deficiency increases as the grazing season progresses. Furthermore, many beef production operations graze crop residues to capitalize on inexpensive forage once the crops have been harvested. These crop residues frequently contain very low P concentrations. In either of these scenarios, P is commonly supplemented via free choice mineral mixes that contain either monocalcium phosphate, dicalcium phosphate, or some combination thereof. Unfortunately, because of demand for these products from other livestock species and agronomic enterprises, the price can be highly variable, and may become cost prohibitive. The additional P contained in distillers coproducts can be used to offset any deficiency and reduce, or eliminate, the need for P in mineral supplements, resulting in cost savings for beef producers.

On the other hand, in feedlot diets, distillers coproducts are commonly included as a partial replacement for corn, thereby increasing the P concentration of the diet. However, research suggests that the P requirement of feedlot cattle may be lower than previously thought (Erickson et al., 1999, 2002)—low enough that the P contained in commonly used feed ingredients is sufficient. As such, additional dietary P from distillers coproducts generally results in increased P excretion and greater P concentrations in manure. While not directly problematic, P is of great concern relative to nutrient management. Control of nutrient runoff generally receives the most attention during discussions regarding nutrient management. However, even in feedlots with state-of-the art nutrient management and containment systems, additional P can create concern. In many states, the amount of manure that can be spread on a given location is dependent upon the P concentration of the soil at that location and the P concentration of the manure. Consequently, if the P concentration of the manure increases, feedlot operators may be forced to haul manure greater distances, which will increase cost. With that said, the nutrients in the manure have value to the agronomic enterprise, so it may be possible for the feedlot operator to recover at least a portion of the cost associated with distribution.

12.13 PURCHASING CONSIDERATIONS

12.13.1 Availability

Most dry grind coproducts procured in amounts typically required by feedlots are available from ethanol plants, brokers, and local feed elevators, especially those coproducts that contain little moisture. Although conditions vary, feeders operating small yards can access dried coproducts from most local elevators and dealers, who are willing to sell dried coproducts in amounts as little as 1 ton/delivery. However, feeders are encouraged to contact the plant of their choice, broker, or local elevators to determine specific purchasing considerations beyond those listed herein.

When wet coproducts are sought, certain purchasing restrictions may apply. For instance, plants may require that DWGS be purchased in semiloads (25 ton). Feeders operating small yards, wishing to utilize this coproduct must either group together to purchase a semiload at regular intervals of two weeks or less, or purchase singly and be prepared to preserve a semiload at less regular intervals. Some ethanol plants have more liberal purchase amount minimums (less than a semiload). Approximate days to consume a full semiload are provided in Table 12.2 for various concentrations of dry matter, inclusion rates, and cattle on inventory.

TABLE 12.2

Approximate Days to Consume a Full Semiload (25 ton) at Various Feedlot Inventories and Inclusion Rates for Three Moisture Levels of Ethanol Coproducts

Inclusion, % of Diet Dry Matter	Feedlot Inventory[a]					
	250	500	1000	1500	2000	2500
Distillers wet grains[b]						
5	58	29	15	10	7	6
10	29	15	7	5	4	3
15	19	10	5	3	2	2
20	15	7	4	2	2	1
25	12	6	3	2	1	1
30	10	5	2	2	1	1
35	8	4	2	1	1	1
40	7	4	2	1	1	1
45	6	3	2	1	1	1
50	6	3	1	1	1	1
Distillers modified wet grains[b]						
5	82	41	20	14	10	8
10	41	20	10	7	5	4
15	27	14	7	5	3	3
20	20	10	5	3	3	2
25	16	8	4	3	2	2
30	14	7	3	2	2	1
35	12	6	3	2	1	1
40	10	5	3	2	1	1
45	9	5	2	2	1	1
50	8	4	2	1	1	1
Distillers dried grains[b]						
5	164	82	41	27	20	16
10	82	41	20	14	10	8
15	55	27	14	9	7	5
20	41	20	10	7	5	4
25	33	16	8	5	4	3
30	27	14	7	5	3	3
35	23	12	6	4	3	2
40	20	10	5	3	3	2
45	18	9	5	3	2	2
50	16	8	4	3	2	2

[a] Assuming 22 lb dry matter intake.

[b] Distillers wet, modified wet, and dried grains assumed to be 32%, 45%, and 90% dry matter, respectively.

12.13.2 ENERGY VALUES AND OPPORTUNITY PRICE RELATIVE TO CORN

Opportunity prices considering only the energy contribution of distillers grains were generated for a range in corn price from $3.00 to $7.00/bu. Opportunity price calculations took into consideration dry matter and energy content of distillers grains. Energy content was derived from no-intercept regressions of ME intake on the amounts of grain, coproduct, roughage, and energy-containing ingredients in supplements from meta-analysis dataset of Reinhardt et al. (2007). Metabolizable energy of DDGS and DMWGS derived by this method averaged 1.47 ± 0.052 Mcal/lb DM and 1.63 ± 0.086 Mcal/lb DM, respectively, which correspond to 100% and 110% of the energy value of corn. Because there were not sufficient data to model performance response using DMWGS, an intermediate value was selected. Resulting opportunity prices for various corn prices were regressed (no intercept) to determine a simple price relationship between distillers grains and corn representative of the opportunity price. Opportunity price relationships between distillers grains and corn price were 37.34, 21.17, and 13.48 for DDGS, DMWGS, and DWGS, respectively. These values multiplied by the price of corn ($/bushel) provide a simple approach to the opportunity price of each distillers grains.

The author has maintained a dataset obtained from the Iowa Ethanol Plant Report (NW_GR111; USDA Agricultural Marketing Service available at www.ams.usda.gov/AMSv1.0/ams.fetchTemplateData.do?template=TemplateN&navID=MarketNewsAndTransportationData&leftNav=MarketNewsAndTransportationData&page=Bioenergy). High and low values derived from the report were averaged for each Monday report (Tuesdays were used when Mondays were holidays) and plotted since the report became available (10/2/2006). Figure 12.1 represents raw averages for corn ($/100 bushel), ethanol ($/100 gallons), and distillers grains ($/ton). During this time period, distillers grains price was regressed on corn price (Figure 12.2) to determine the distillers grains:corn price relationship over time. Data from Figure 12.2 revealed that distillers grains sold at values lower than the opportunity price relationship between distillers grains and corn. Indeed, DDGS, DMWGS, and DWGS sold on average at price relationships of 32.25, 14.98, and 10.14, respectively (Figure 12.3). This means that DDGS, DMWGS, and DWGS traded at 86%, 71%, and 75% of their opportunity price during a time period when corn grain traded from $2.27 to $7.02/bu. During this period, only DDGS traded at a relationship greater than its opportunity price to corn ratio; both DMWGS and DWGS always traded at lower than their opportunity price to corn price ratio. Time periods when DDGS traded at a ratio greater than its opportunity price to corn price ratio were in late 2006, 2007, and 2008. During these times, corn grain prices were rallying. Corn price rallies observed in the

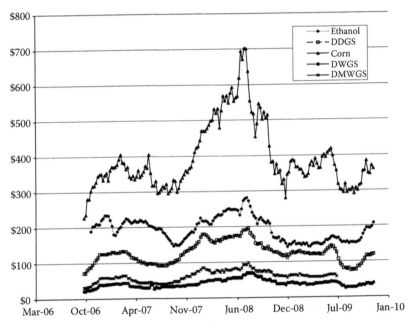

FIGURE 12.1 Raw averages for weekly (Mondays or Monday holiday on Tuesdays) spot prices for ethanol, yellow corn, distillers dried grains with solubles, distillers modified wet grains with solubles, and distillers wet grains with solubles in northwest Iowa (USDA AMS).

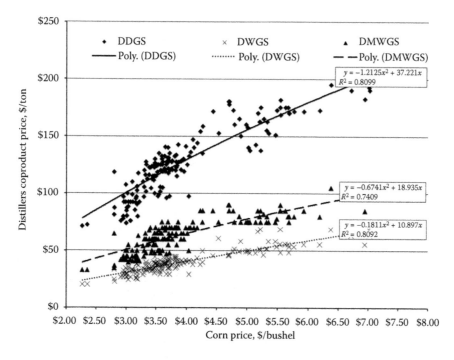

FIGURE 12.2 Relationship between northwest Iowa spot distillers coproduct prices ($/ton, as-is basis) and yellow corn price ($/bushel). Solving for the equations represents the modeled coproduct price from corn price.

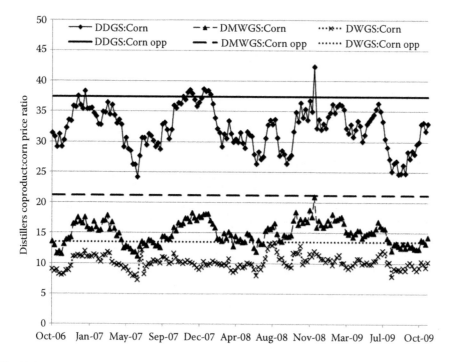

FIGURE 12.3 Distillers coproduct:corn price ratios for the time period between October 2, 2006 and November 23, 2009. Opportunity price ratios for distillers coproduct:corn prices are represented by the straight lines.

summer do not seem to precipitate great increases in distillers grains, particularly DDGS, likely because there are fewer cattle on feed in areas where distillers grains are produced during that time period. Refer to Chapter 2 for a discussion about the relationship between DDGS and corn price in more depth.

12.14 STORAGE AND HANDLING

12.14.1 Storage

Dry coproducts require relatively minimal storage facilities. Selecting and preparing a site that is high, dry, and protected from the elements are important to prevent losses from wind or moisture. Storage areas (old sheds or high-ground sites walled off with large bales) can be surfaced with hardened clay, gravel, blacktop, or cement and covered with tin sheets and(or) heavy tarps. Delivery and loading areas must face away from prevailing winds. Also, protecting delivery and loading areas (using heavy rubber to protect elevator chutes or loading wagons or trucks) within a protected area will reduce wind losses. Caution should be exercised when storing dry coproducts in upright bins. Dry coproducts tend to settle and bridge in these types of storage facilities. Anecdotal reports indicate that allowing the products to cool completely upon arrival may help alleviate some of these concerns.

Wet coproducts require more elaborate storage because heavy losses can occur from spoilage. Typically, DWG and DWGS have a shelf life of fewer than 5 days (Durham, 2001) in the summer. During the winter, DWG and DWGS may last up to 10 days. Ensiling wet coproducts will help preserve them. DWG and DWGS can be effectively stored in silage bags alone or in combinations with either high fiber concentrate feeds or forages (Garcia and Kalscheur, 2004; Erickson et al., 2008). However, investments in silo structures or bags need to be evaluated carefully to prevent excessive storage costs that will offset advantages of using coproducts.

Silage wedges made within walls of large bales may reduce investment; however, care must be taken to ensure that the wedge is firmly packed. Wedges made of DWG against a concrete bunker, covered with a tarp, and tapped with a front-end loader bucket have lasted at least 9 months without deteriorating. DWG and DWGS can also be preserved by spreading two 50-lb bags of livestock salt on top of the windrow left behind by the semitrailer as it delivers the load (Durham, 2001). Similarly, tapping a windrow left behind by the semitrailer as it unloaded, and then covering with a tarp secured at the base where the tarp meets the ground using sand or gravel permit storage of DWG for at least 6 months.

Storing DWGS alone in silage bags is also a viable storage option; however caution when loading the bags is required. The product causes excessive pressure on the bag and may result in tearing or failure of the bags. When loading silage bags with DWGS alone, it is recommended to apply only enough pressure to ensure air exclusion in the bag. Blending DWGS with other feeds can minimize the risk of bag failure. Researchers at South Dakota State University recommend mixing blends of 70% DWGS and 30% soybean hulls when filling bags or silos (Garcia and Kalscheur, 2004).

Wet coproducts can also be effectively stored in bunker silos in combination with poor-quality forages. An extensive evaluation of use of bags and bunkers to store DWG was conducted at the University of Nebraska (Erickson et al., 2008). Mixing DWG with (dry matter basis) 15% grass hay, 22.5% alfalfa hay, 12.5% wheat straw, 50% distillers dried grains, or 60% WCGF was optimal for storage in bags. At lower concentrations of grass hay or corn gluten feed, the bags failed. Wet coproducts can be blended with numerous forages; however, to facilitate effective storage in both bag and bunker applications, feeds should be blended to achieve a maximum of 50% dry matter.

Ensiling DWGS with poor-quality forages results in similar performance may improve performance relative to diets formulated using the same ingredients, but not ensiled together. Feeding growing steers DWGS ensiled with wheat straw resulted in similar ADG and feed efficiency compared to those fed nonensiled DWGS and wheat straw (Peterson et al., 2008). In contrast, Buckner et al. (2009) compared the performance of growing steers fed two combinations of wheat straw and DWGS (30:70 wheat straw:DWGS and 45:55 wheat straw:DWGS) that were either blended and ensiled prior to feeding or mixed fresh every other day. Steers fed the ensiled combinations tended to have greater ADG and better feed efficiency ($p = 0.16$ and 0.10, respectively) than those fed the freshly mixed combinations.

Due to their lower moisture, DMWGS are more difficult to store by compaction and covering as described above. However, modified distillers grains bagged at normal pressure without any forage additive are stored well (Erickson et al., 2008). Similarly, a tarp covering on the windrow left behind after unloading a semitrailer permitted storage of DMWGS during cool fall temperatures (author; personal observation).

Storing CDS can prove significantly more challenging than storing other distillers coproducts. Challenges associated with storing CDS result primarily from two characteristics of the product—CDS will freeze and separate. Consequently, in winter months, storing CDS either in underground tanks or in heated facilities is essential. Regardless of season, CDS needs to be stored in a tank equipped with a recirculating pump or some other aeration device to prevent settling and spoilage.

12.14.2 MIXING AND DELIVERY CONSIDERATIONS

Combinations of dry feeds with largely different particle sizes and densities (e.g., DDGS and corn grain) are difficult because insufficient mixing (shorter or longer than required) will permit one of the ingredients to fall out of the mix. Mixing wet forages and wet coproducts also presents a challenge, because particle size, density, and moisture content differ greatly.

Loading sequence and mixing time are dependent on moisture content of various feeds used, type of mixer, and delivery output and speed. These variables are difficult to establish for all combinations of mixer models and types of diets for reasons already stated. However, loading corn, liquid supplement, soybean meal, and hay, in that order, to a reel type wagon to mix a finishing diet (79% corn on a DM basis) led to a consistent ration within 4 min of mixing (Wagner, 1995). However, just altering the sequence so that liquid supplement was added ahead of corn resulted in an inconsistent ration, unless mixing time was increased to 8 min (Wagner, 1995). In that study, adding hay ahead of corn, soybean meal, and liquid supplement, or after adding corn, liquid supplement, and soybean meal did not appear to affect ration adequacy in a three-auger mixer as long as the mixer was permitted to run for 8 min. In a similar study (Johnson et al., 2000) conducted using a grower diet (66.6% hay and soyhulls) where corn, soyhulls, and liquid supplement were added as the first three ingredients, permitted to mix for 30 s, and followed by WCGF and grass hay, the observation was made that mixing times for reel type and three-auger mixers should be 4 and 8 min, respectively. Mixing time for a four-auger mixer tested in this study was optimized at 2 min. Recent observations reported by Bierman (2008), using a four-auger mixer indicate that adding modified distillers or WCGF as the last ingredient yielded the lowest CV for monensin concentration of the diet. It is important to emphasize that these mixing consistency results change depending on mixer type. Lowest CV for DM, CP, ADF, S, monensin, and chloride mixed ration concentrations were obtained in a study where DMWGS was added first (Loe et al., 2007).

Other recommendations include preventing overloading of mixers and setting mixing times using timers to prevent undermixing, especially when using wet feeds or liquid supplements. Also, it is recommended that feeders routinely evaluate the efficiency of mixing to avoid over- or undermixing. It is clear that at moisture concentrations of most corn coproducts, overloading and undermixing are situations that must be avoided.

12.15 CONCLUSIONS

Research has clearly demonstrated the utility of distillers coproducts in beef cattle diets. The nutrient composition of distillers coproducts is generally complimentary to both grain and forage-based ruminant diets. However, some challenges do exist when using distillers coproducts in beef cattle diets. Excess fat and S can result in reduced animal performance and, in the case of S, can be fatal. As such, careful analysis and formulation are essential to capture the economic advantages commonly associated with distillers coproducts. Nevertheless, these feeds present beef producers and cattle feeders with a tremendous opportunity to reduce input costs and enhance profitability of their operations.

Many of the nutritional attributes, positive and negative, of including distillers coproducts in beef cattle diets are applicable to dairy cattle diets as well. However, given the tremendous nutrient demand for milk production and the importance of milk composition, some unique nuances exist in dairy cattle diets. Chapter 13 will provide a thorough discussion regarding the incorporation of distillers coproducts into the diets of dairy cattle.

REFERENCES

Arora, S., M. Wu, and M. Wang. 2008. Update of distillers grains displacement ratios for corn ethanol life cycle analysis. Res. Rep. Center for Transportation Research. Energy System Division. Argonne National Laboratory. Available online: www.transportation.anl.gov/pdfs/AF/527.pdf. Accessed November 23, 2009.

Bellows, R. A., E. E. Grings, D. D. Simms, T. W. Geary, and J. W. Bergman. 2001. Effects of feeding supplemental fat during gestation to first-calf heifers. *The Professional Animal Scientist* 17: 81–89.

Benton, J. R., G. E. Erickson, T. J. Klopfenstein, M. K. Luebbe, and R. U. Lundquist. 2008. Effects of wet corn gluten feed and roughage inclusion levels in finishing diet containing modified distillers grains plus soluble. MP 92. *Nebraska Beef Cattle Rep.*, pp. 70–72. Lincoln, NE.

Bierman, S. 2008. Mixing integrity for ruminant diets containing by-products. *Proc. 69th Minnesota Nutrition Conference*, pp. 145–159. Owatonna, MN: University of Minnesota.

Bremer, V. R., J. R. Benton, M. K. Luebbe, K. J. Hanford, G. E. Erickson, T. J. Klopfenstein, and R. A. Stock. 2008. Wet distillers grains plus solubles or soluble in feedlot diets containing wet corn gluten feed. MP 92. *Nebraska Beef Cattle Rep.*, pp. 64–65. Lincoln, NE.

Bremer, V. R., G. E. Erickson, T. J. Klopfenstein, M. L. Gibson, K. J. Vander Pol, and M. A. Greenquist. 2005. Evaluation of a low protein distillers by-product for finishing cattle. MP 88. *Nebraska Beef Cattle Rep.*, pp. 57–58. Lincoln, NE.

Buckner, C. D., G. E. Erickson, T. J. Klopfenstein, R. A. Stock, and K. J. Vander Pol. 2006a. Effect of feeding a by-product combination at two levels or by-product alone in feedlot diets. MP 90. *Nebraska Beef Cattle Rep.*, pp. 25–26. Lincoln, NE.

Buckner, C. D., T. J. Klopfenstein, G. E. Erickson, K. J. Vander Pol, K. K. Karges, and M. L. Gibson. 2006b. Comparing a modified dry by-product to dry distillers grains with soluble in growing calf diets. MP 90. *Nebraska Beef Cattle Rep.*, pp. 15–16. Lincoln, NE.

Buckner, C. D., T. L. Mader, G. E. Erickson, S. L. Colgan, D. R. Mark, V. R. Bremer, K. K. Karges, and M. L. Gibson. 2008. Evaluation of dry distillers grains plus solubles inclusion on performance and economics of finishing beef steers. *The Professional Animal Scientist* 24: 404–410.

Buckner, C. D., S. J. Vanness, G. E. Erickson, T. J. Klopfenstein, and J. R. Benton. 2007. Sampling wet distillers grains plus solubles to determine nutrient variability. MP 91. *Nebraska Beef Cattle Rep.*, pp. 126–127. Lincoln, NE.

Buckner, C. D., T. J. Klopfenstein, G. E. Erickson, W. A. Griffin, and J. R. Benton. 2009. Comparing ensiled or fresh mixed wet distillers grains with solubles with straw at two inclusions in growing calf diets. MP 93. *Nebraska Beef Rep.*, pp. 42–43. Lincoln, NE.

Calkins, C. R., A. S. de Mello, and L. S. Senaratne. 2009. Effects of distillers grains on beef carcass quality and palatability. *Journal of Animal Science* 87(Suppl. 2): 328–329.

Cole, N. A., M. L. Galyean, J. MacDonald, and M. S. Brown. 2006. Interaction of grain coproducts with grain processing: Associative effects and management. MP-17. *Proc. Cattle Grain Processing Symposium*, pp. 193–204. Tulsa, OK: Oklahoma State University.

Coppock, C. E., and D. L. Wilks. 1991. Supplemental fat in high-energy rations for lactating cows: Effects on intake, digestion, milk yield, and composition. *Journal of Animal Science* 69: 3826–3837.

Corah, L. R., and M. A. McCully. 2006. Declining quality grades: A review of factors reducing marbling deposition in beef cattle. CAB Res. Doc. Available online: www.cabpartners.com/news/research. Accessed October 19, 2009.

Corrigan, M. E., G. E. Erickson, T. J. Klopfenstein, K. J. Vander Pol, M. A. Greenquist, M. K. Luebbe, K. Karges, and M. L. Gibson. 2006. Effect of distillers grains composition and level on steers consuming high-quality forage. MP 90. *Nebraska Beef Cattle Rep.*, pp. 17–18. Lincoln, NE.

Corrigan, M. E., G. E. Erickson, T. J. Klopfenstein, M. K. Luebbe, K. J. Vander Pol, N. F. Meyer, C. D. Buckner, S. J. Vanness, and K. J. Hanford. 2009. Effect of corn processing method and wet corn distillers grains plus solubles inclusion level in finishing steers. *Journal of Animal Science* 87: 3351–3362.

Dahlen, C. R., A. DiCostanzo, R. T. Ethington, T. L. Durham, J. E. Larson, and G. C. Lamb. 2003. Evaluation of forage sources for finishing diets containing wet corn gluten feed. *Journal of Animal Science* 81(Suppl. 1): 111–112.

Depenbusch, B. E., C. M. Coleman, J. J. Higgins, and J. S. Drouillard. 2009a. Effects of increasing levels of dried corn distillers grains with solubles on growth performance, carcass characteristics, and meat quality of yearling heifers. *Journal of Animal Science* 87: 2653–2663.

Depenbusch, B. E., E. R. Loe, J. J. Sindt, N. A. Cole, J. J. Higgins, and J. S. Drouillard. 2009b. Optimizing use of distillers grains in finishing diets containing steam-flaked corn. *Journal of Animal Science* 87: 2644–2652.

Depenbusch, B. E., E. R. Loe, M. J. Quinn, M. E. Corrigan, M. L. Gibson, K. K. Karges, and J. S. Drouillard. 2008a. Corn distillers grains with solubles derived from a traditional or partial fractionation process: Growth performance and carcass characteristics of finishing feedlot heifers. *Journal of Animal Science* 86: 2338–2343.

Depenbusch, B. E., J. S. Drouillard, E. R. Loe, J. J. Higgins, M. E. Corrigan, and M. J. Quinn. 2008b. Efficacy of monensin and tylosin in finishing diets based on steam-flaked corn with and without corn wet distillers grains with solubles. *Journal of Animal Science* 86: 2270–2276.

Depenbusch, B. E., T. G. Nagaraja, J. M. Sargeant, J. S. Drouillard, E. R. Loe, and M. E. Corrigan. 2008c. Influence of processed grains on fecal pH, starch concentration, and shedding of Escherichia coli O157 in feedlot cattle. *Journal of Animal Science* 86: 632–639.

DeWitt, D., S. Ensley, P. Imerman, B. Doran, K. Kohl, and P. Summer. 2008. Evaluation and observations of total sulfur intake with corn coproduct diets for feedlot cattle. R-2295. *Iowa State University Animal Ind. Rep.* Ames, IA.

Doering-Resch, H., C. Wright, K. Tjardes, G. Perry, K. Bruns, and B. Rops. 2005. Effectiveness of dried distillers grains with solubles as a replacement for oilseed meal in supplements for cattle consuming poor-quality forage. BEEF 2005-14. *South Dakota State University Beef Rep.*, pp. 68–72. Brookings, SD.

Durham, T. L. 2001. Corn milling and the use of the coproducts. B-476. Minnesota Cattle Feeder Day Rep.

Engel, C. L., H. H. Patterson, and G. A. Perry. 2008. Effect of dried corn distillers grains plus solubles compared with soybean hulls, in late gestation heifer diets, on animal and reproductive performance. *Journal of Animal Science* 86: 1697–1708.

Fox, J. T., B. E. Depenbusch, J. S. Drouillard, and T. G. Nagaraja. 2007. Dry-rolled or steam-flaked grain-based diets and fecal shedding of Escherichia coli O157 in feedlot cattle. *Journal of Animal Science* 85: 1207–1212.

Erickson, G., T. Klopfenstein, R. Rasby, A. Stalker, B. Plugge, D. Bauer, D. Mark, D. Adams, J. Benton, M. Greenquist, B. Nuttleman, L. Kovarik, M. Peterson, J. Waterbury, and M. Wilken. 2008. *Storage of Wet Corn Coproducts*, 1st ed. Nebraska Corn Board. Lincoln: University of Nebraska.

Erickson, G. E., T. J. Klopfenstein, C. T. Milton, D. Hanson, and C. Calkins. 1999. Effect of dietary phosphorus on finishing steer performance, bone status, and carcass maturity. *Journal of Animal Science* 77: 2832–2836.

Erickson, G. E., T. J. Klopfenstein, C. T. Milton, D. Brink, M. W. Orth, and K. M. Whittet. 2002. Phosphorus requirement of finishing feedlot calves. *Journal of Animal Science* 80: 1690–1695.

Garcia, A. D., and K. F. Kalscheur. 2004. Ensiling wet distillers grains with other feeds. Extension Extra 4029. *South Dakota Cooperative Extension Service.* Brookings, SD.

Gilbery, T. C., G. P. Lardy, S. A. Soto-Navarro, M. L. Bauer, and J. S. Caton. 2006. Effects of corn condensed distillers solubles supplementation on ruminal fermentation, digestion, and in situ disappearance in steers consuming low-quality hay. *Journal of Animal Science* 84: 1468–1480.

Gill, R. K., D. L. VanOverbeke, B. Depenbusch, J. S. Drouillard, and A. DiCostanzo. 2008. Impact of beef cattle diets containing corn or sorghum distiller's grains on beef color, fatty acid profiles, and sensory attributes. *Journal of Animal Science* 86: 923–935.

Godsey, C. M., M. K. Luebbe, J. R. Benton, G. E. Erickson, T. J. Klopfenstein, C. Ibanez, P. Guiroy, M. Greenquist, and J. Kazin. 2009. Evaluation of feedlot and carcass performance of steers fed different levels of E-corn, a potential new feed product from ethanol plants. MP 93. *Nebraska Beef Cattle Rep.*, pp. 69–71. Lincoln, NE.

Gould, D. H. 1998. Polioencephalomalacia. *Journal of Animal Science* 76: 309–314.

Gould, D. H., M. M. McAllister, J. C. Savage, and D. W. Hamar. 1991. High sulfide concentrations in rumen fluid associated with nutritionally induced polioencephalomalacia in calves. *American Journal of Veterinary Research* 52: 1164–1169.

Griffin, W. A., B. L. Nuttleman, T. J. Klopfenstein, L. A. Stalker, R. N. Funston, and J. A. Musgrave. 2009. Supplementing dried distillers grains to steers grazing cool season meadow. MP 93. *Nebraska Beef Rep.,* pp. 36–38. Lincoln, NE.

Griffin, W. A., V. R. Bremer, T. J. Klopfenstein, L. A. Stalker, L. W. Lomas, J. L. Moyer, and G. E. Erickson. 2008. Summary of grazing trials using dried distillers grains supplementation. MP 92. *Nebraska Beef Rep.,* pp. 37–39. Lincoln, NE.

Gunn, P. J., A. D. Weaver, R. P. Lemenager, D. E. Gerrard, M. C. Claeys, and S. L. Lake. 2009. Effects of dietary fat and crude protein on feedlot performance, carcass characteristics, and meat quality in finishing steers fed differing levels of dried distillers grains with solubles. *Journal of Animal Science* 87: 2882–2890.

Gustad, K. H., T. J. Klopfenstein, G. E. Erickson, K. J. Vander Pol, J. C. MacDonald, and M. A. Greenquist. 2005. Dried distillers grains supplementation of calves grazing corn residue. MP 88-A. *Nebraska Beef Rep.,* pp. 36–37. Lincoln, NE.

Ham, G. A., R. A. Stock, T. J. Klopfenstein, E. M. Larson, D. H. Shain, and R. P. Huffman. 1994. Wet corn distillers byproducts compared with dried corn distillers grains with solubles as a source of protein and energy for ruminants. *Journal of Animal Science* 72: 3246–3257.

Herszage, J., and M. dos Santos Afonso. 2003. Mechanism of hydrogen sulfide oxidation by manganese (IV) oxide in aqueous solutions. *Langmuir* 19: 9684–9692.

Huls, T. J., M. K. Luebbe, G. E. Erickson, and T. J. Klopfenstein. 2008. Effect of inclusion level of modified distillers grains plus soluble in finishing steers. MP 91. *Nebraska Beef Cattle Rep.,* pp. 41–42. Lincoln, NE.

Jacob, M. E., G. L. Parsons, M. K. Shelor, J. T. Fox, J. S. Drouillard, D. U. Thomson, D. G. Renter, and T. G. Nagaraja. 2008a. Feeding supplemental dried distiller's grains increases faecal shedding of Escherichia coli O157 in experimentally inoculated calves. *Zoones Public Health* 55: 125–132.

Jacob, M. E., J. T. Fox, J. S. Drouillard, D. G. Renter, and T. G. Nagaraja. 2008b. Effects of dried distillers' grain on fecal prevalence and growth of Escherichia coli O157 in batch culture fermentations from cattle. *Applied and Environmental Microbiology* 74: 38–43.

Jacob, M. E., J. T. Fox, S. K. Narayanan, J. S. Drouillard, D. G. Renter, and T. G. Nagaraja. 2008c. Effects of feeding wet corn distillers grain with solubles with or without monensin and tylosin on the prevalence of antimicrobial susceptibilities of fecal foodborne pathogenic and commensal bacteria in feedlot cattle. *Journal of Animal Science* 86: 1182–1190.

Jenkins, T. C. 1993. Lipid metabolism in the rumen. *Journal of Dairy Science* 76: 3851–3863.

Jenkins, K. H., J. C. MacDonald, F. T. McCollum, and S. H. Amosson. 2009. Effects of level of dried distillers grain supplementation on native pasture and subsequent effects on wheat pasture gains. *The Professional Animal Scientist* 25: 596–604.

Johnson, B. J., R. H. Pritchard, S. L. Bjornson, and W. M. Cerkoney. 2000. An evaluation of three TMR feed-mixing wagons. CATTLE 00-9. *South Dakota Beef Rep.,* pp. 38–43. Brookings, SD.

Kelzer, J. M., M. Ruiz-Moreno, A. DiCostanzo, and G. I. Crawford. 2010. Effects of manganese oxide on total in vitro gas and hydrogen sulfide production in cultures using high-sulfur distillers grains-based substrate. *Journal of Animal Science* 88(E-suppl.3): 125(Abstr.).

Kleinschmit, D. H., J. L. Anderson, D. J. Schingoethe, K. F. Kalscheur, and A. R. Hippen. 2007. Ruminal and intestinal degradability of distillers grains plus solubles varies by source. *Journal of Dairy Science* 90: 2909–2918.

Klopfenstein, T. J., G. E. Erickson, and V. R. Bremer. 2008a. Use of distillers by-products in the beef cattle industry. *Journal of Animal Science* 86: 1223–1231.

Klopfenstein, T. J., G. E. Erickson, and V. R. Bremer. 2008b. Use of distillers coproducts in diets fed to beef cattle. *Using distillers grains in the U.S. and international livestock and poultry industries* (electronic edition), eds. B. A. Babcock, D. J. Hayes, and J. D. Lawrence, pp. 5–55. Ames: Midwest Agribusiness Trade Research and Information Center.

Koger, T. J., D. M. Wulf, A. D. Weaver, C. L. Wright, K. E. Tjardes, K. S. Mateo, T. E. Engle, R. J. Maddock, and A. J. Smart. 2010. Influence of feeding various quantities of wet and dry distillers grains to finishing steers on carcass characteristics, meat quality, retail-case life of ground beef, and fatty acid profile of longissimus muscle. *Journal of Animal Science* 88: 3399–3408.

Kovarik, L. M., M. K. Leubbe, R. J. Rasby, and G. E. Erickson. 2008. Limit feeding beef cows with bunkered wet distillers grains plus solubles or distillers solubles. MP 92. *Nebraska Beef Rep.,* pp. 11–12. Lincoln, NE.

Lain, J., D. DeWitt, R. Euken, D. Schwab, and D. Loy. 2008. A producer survey of feeding corn coproducts in Iowa. R-2293. *Iowa State University Animal Ind. Rep.* Ames, IA.

Lancaster, P. A., J. B. Corners, L. N. Thompson, M. R. Ellersieck, and J. E. Williams. 2007. Effects of distillers dried grains with solubles as a protein source in a creep feed. 1. Suckling calf and dam performance. *The Professional Animal Scientist* 23: 83–90.

Lancaster, P. A., J. B. Corners, L. N. Thompson, M. R. Ellersieck, C. D. Buckner, and J. E. Williams. 2007. Effects of distillers dried grains with solubles as a protein source in a creep feed. 2. Subsequent feedlot performance, carcass measurements, and plasma parameters. *The Professional Animal Scientist* 23: 83–90.

Larson, E. M., R. A. Stock, T. J. Klopfenstein, M. H. Sindt, and R. P. Huffman. 1993. Feeding value of wet distillers byproducts for finishing ruminants. *Journal of Animal Science* 71: 2228–2236.

Larson, D. M., and R. N. Funston. 2009. The effects of corn coproduct supplementation on primiparous cow reproduction and calf performance. *The Professional Animal Scientist* 25: 536–540.

Leupp, J. L., G. P. Lardy, K. K. Karges, M. L. Gibson, and J. S. Caton. 2009a. Effects of increasing levels of corn distillers grains with solubles to steers offered moderate-quality forage. *Journal of Animal Science* 87: 4064–4072.

Leupp, J. L., G. P. Lardy, M. L. Bauer, K. K. Karges, M. L. Gibson, J. S. Caton, and R. J. Maddock. 2009b. Effects of distillers dried grains with solubles on growing and finishing steer intake, performance, carcass characteristics, and steak color and sensory attributes. *Journal of Animal Science* 87: 4118–4124.

Loe, E. R., B. D. Rops, and J. T. Keimig. 2007. Evaluation of mixing characteristics of diets containing modified distillers grains. BEEF 2007-07. *South Dakota Beef Rep.,* pp. 35–38. Brookings, SD.

Lonergan, G. H., J. J. Wagner, D. H. Gould, F. B. Garry, and M. A. Thoren. 2001. Effects of water sulfate concentration on performance, water intake, and carcass characteristics of feedlot steers. *Journal of Animal Science* 79: 2941–2948.

Loy, D. 2008. Ethanol coproducts for cattle the process and products. Available online: www.extension.iastate.edu/publications/ibc18.pdf. Accessed October 28, 2009.

Loy, D., D. Strohbehn, R. Berryman, and D. Morrical. 2008. Animal performance, storage losses and feasibility of ensiling a mixture of tub ground low quality hay and condensed distillers' solubles for growing cattle. R-2290. *Iowa State University Animal Ind. Rep.* Ames, IA.

Loy, D., D. Strohbehn, R. Berryman, D. Morrical, J. Sellers, and A. Trenkle. 2009. Animal performance, storage losses and feasibility of ensiling a mixture of tub ground low quality hay and wet distillers' grains for growing cattle. R-2412. *Iowa State University Animal Ind. Rep.* Ames, IA.

Loy, T. W., J. C. MacDonald, T. J. Klopfenstein, and G. E. Erickson. 2007. Effect of distillers grains or corn supplementation frequency on forage intake and digestibility. *Journal of Animal Science* 85: 2625–2630.

Loza, P. L., K. J. Vander Pol, M. A. Greenquist, G. E. Erickson, T. J. Klopfenstein, and R. A. Stock. 2006. Effects of different inclusion levels of wet distillers grains in feedlot diets containing wet corn gluten feed. MP 90. *Nebraska Beef Cattle Rep.,* pp. 28–28. Lincoln, NE.

MacDonald, J. C., T. J. Klopfenstein, G. E. Erickson, and W. A. Griffin. 2007. Effects of dried distillers grain and equivalent undegradable intake protein or ether extract on performance and forage intake of heifers grazing smooth brome grass pastures. *Journal of Animal Science* 85: 2614–2624.

Martin, J. L., A. S. Cupp, R. J. Rasby, Z. C. Hall, and R. N. Funston. 2007. Utilization of dried distillers grains for developing beef heifers. *Journal of Animal Science* 85: 2298–2303.

Mateo, K. S., K. E. Tjardes, C. L. Wright, and B. D. Rops. 2004. Evaluation of feeding wet distillers grains with soluble, dry distillers grains with soluble and blood meal to growing steers. BEEF 2004-02. *South Dakota Beef Rep.,* pp. 8–13. Brookings, SD.

May, M. L., J. S. Drouillard, M. J. Quinn, and C. E. Walker. 2007. Wet distiller's grains with solubles in beef finishing diets comprised of steam-flaked or dry-rolled corn. *Kansas State University Beef Cattle Res. Rep.,* pp. 57–59. Manhattan, KS.

Morris, S. E., T. J. Klopfenstein, D. C. Adams, G. E. Erickson, and K. J. Vander Pol. 2004. The effects of dried distillers grains on heifers consuming low or high quality forage. MP 83-A. *Nebraska Beef Cattle Rep.,* pp. 18–20. Lincoln, NE.

Mueller, C. J., and D. L. Boggs. 2005. Use of corn coproducts in soybean hull-based feedlot receiving diets. South Dakota State University. BEEF 2005-08. *South Dakota Beef Rep.,* pp. 33–39. Brookings, SD.

Musgrave, J. A., L. A. Stalker, M. C. Stockton, and T. J. Klopfenstein. 2009. Comparison of feeding wet distillers grains in a bunk or on the ground to cattle grazing native Sandhills winter range. MP 93. *Nebraska Beef Rep.,* pp. 17–18. Lincoln, NE.

Mustafa, A. F., J. J. McKinnon, and D. A. Christensen. 2000. The nutritive value of thin stillage and wet distillers' grains for ruminants—A review. *Asian–Aus. Journal of Animal Science* 13: 1609–1618.

NCR 1984. The Nutritional Value of Grain Alcohol Fermentation By-Products for Beef Cattle. *North Central Res. Pub. No. 297.*

Neville, B. W., C. S. Schauer, K. Karges, M. L. Gibson, M. M. Thompson, P. T. Berg, and G. P. Lardy. 2009. Influence of thiamin supplementation on feedlot performance, carcass quality, and incidence of polioencephalomalacia in lambs fed a 60% distillers dried grains with solubles finishing ration. *Proceedings, Western Section, American Society of Animal Science* 60: 135–139.

NRC 1996. *Nutrient Requirements of Beef Cattle,* 7th rev. ed. Nutrient Requirements of Domestic Animals. Washington, DC: National Academy Press.

NRC 2000. *Nutrient Requirements of Beef Cattle—Update 2000,* 7th rev. ed. Nutrient Requirements of Domestic Animals. Washington, DC: National Academy Press.

NRC 2001. *Nutrient Requirements of Dairy Cattle,* 7th rev. ed. Nutrient Requirements of Domestic Animals. Washington, DC: National Academy Press.

NRC 2005. *Mineral Tolerance of Animals,* 2nd rev. ed. Washington, DC: National Academy Press.

Nuttelman, B. L., T. J. Klopfenstein, G. E. Erickson, W. A. Griffin, and M. K. Luebbe. 2007. The effects of supplementing wet distillers grains mixed with wheat straw to growing steers. MP 91. *Nebraska Beef Cattle Rep.,* pp. 33–34. Lincoln, NE.

Nuttelman, B. L., M. K. Luebbe, J. R. Benton, T. J. Klopfenstein, L. A. Stalker, and G. E. Erickson. 2008. Energy value of wet distillers grains in high forage diets. MP 92. *Nebraska Beef Cattle Rep.,* pp. 28–29. Lincoln, NE.

Nuttelman, B. L., M. K. Luebbe, T. J. Klopfenstein, J. R. Benton, and G. E. Erickson. 2009. Comparing the energy value of wet distillers grains to dry rolled corn in high forage diets. MP 93. *Nebraska Beef Cattle Rep.,* pp. 43–45. Lincoln, NE.

Palmquist, D. L., and T. C. Jenkins. 1980. Fat in lactation rations: A review. *Journal of Dairy Science* 63: 1–14.

Peterson, M. M., M. K. Luebbe, R. J. Rasby, T. J. Klopfenstein, G. E. Erickson, and L. M. Kovarik. 2008. Level of wet distillers grains plus solubles and solubles ensiled with wheat straw for growing steers. MP 92. *Nebraska Beef Cattle Rep.,* pp. 35–36. Lincoln, NE.

Ponce, C. H., M. S. Brown, N. A. Cole, C. L. Maxwell, and J. C. Silva. 2009. Effects of ruminally degradable N in diets containing wet corn distillers grains and steam-flaked corn on feedlot cattle performance and carcass characteristics. *Journal of Animal Science* 87(Suppl. 2): 190.

Reed, J. J., G. P. Lardy, M. L. Bauer, M. Gibson, and J. S. Caton. 2006. Effects of season and inclusion of corn distillers dried grains with solubles in creep feed on intake, microbial protein synthesis and efficiency, ruminal fermentation, digestion, and performance of nursing calves grazing native range in southeastern North Dakota. *Journal of Animal Science* 84: 2200–2212.

Reinhardt, C. D., A. DiCostanzo, and G. Milliken. 2007. Distillers byproducts alters carcass fat distribution of feedlot cattle. *Journal of Animal Science* 85(Suppl. 2): 132.

RFA. 2008. Ethanol industry outlook 2008. Renewable Fuels Association. Available online: www.ethanolrfa. org/industry/outlook/. Accessed October 28, 2009.

Rich, A. R., M. K. Luebbe, G. E. Erickson, T. J. Klopfenstein, and J. R. Benton. 2009. Feeding fiber from wet corn gluten feed and corn silage in feedlot diets containing wet distillers grains plus soluble. MP 93. *Nebraska Beef Cattle Rep.,* pp. 76–77. Lincoln, NE.

Roeber, D. L., R. K. Gill, and A. DiCostanzo. 2005. Meat quality responses to feeding distiller's grains to finishing Holstein steers. *Journal of Animal Science* 83: 2455–2460.

Rolfe, K. M., M. K. Luebbe, W. A. Griffin, T. J. Klopfenstein, G. E. Erickson, and D. E. Bauer. 2009. Supplementing modified wet distillers grains with solubles to long yearling steers grazing native range. MP 93. *Nebraska Beef Rep.,* pp. 34–35. Lincoln, NE.

Sager, R. L., D. W. Hamar, and D. H. Gould. 1990. Clinical and biochemical alterations in calves with nutritionally induced polioencephalomalacia. *American Journal of Veterinary Research* 51: 1969–1974.

Sexson, J. L., J. J. Wagner, T. E. Engle, and J. W. Spears. 2010. Effects of water quality and dietary potassium on performance and carcass characteristics of yearling steers. *Journal of Animal Science* 88: 296–305.

Shike, D. W., D. B. Faulkner, D. F. Parrett, and W. J. Sexten. 2009. Influences of corn coproducts in limit-fed rations on cow performance, lactation, nutrient output, and subsequent reproduction. *The Professional Animal Scientist* 25: 132–138.

Shurson, J. 2009a. Analysis of current feeding practices of distiller's grains with solubles in livestock and poultry relative to land use credits associated with determining the low carbon fuel standard for ethanol. Res. Rep. Renewable Fuels Association. Available online: www.ethanolrfa.org/resource/reports/index. php. Accessed October 29, 2009.

Shurson, J. 2009b. A scientific assessment of the role of distiller's grains (DGS) and predictions of the impact of corn coproducts produced by front-end fractionation and back-end oil extraction technologies on indirect land use change. Res. Rep. Renewable Fuels Association. Available online: www.ethanolrfa.org/resource/reports/index.php. Accessed October 29, 2009.

Spiehs, M. J., M. H. Whitney, and G. C. Shurson. 2002. Nutrient database for distiller's dried grains with solubles produced from new ethanol plants in Minnesota and South Dakota. *Journal of Animal Science* 80: 2639–2645.

Stalker, L. A., D. C. Adams, and T. J. Klopfenstein. 2007. Urea inclusion in distillers dried grains supplements. *The Professional Animal Scientist* 23: 390–394.

Stalker, L. A., D. C. Adams, and T. J. Klopfenstein. 2009. Influence of distillers dried grain supplementation frequency on forage digestibility and growth performance of beef cattle. *The Professional Animal Scientist* 25: 289–295.

Suttle, N. F. 1974. Effects of organic and inorganic sulfur on the availability of dietary copper to sheep. *British Journal of Nutrition* 32: 559–568.

Suttle, N. F. 1975. The role of organic sulfur in the copper-molybdenum-S interrelationship in ruminant nutrition. *British Journal of Nutrition* 34: 411–420.

Tjardes, K. and C. Wright. 2002. Feeding corn distiller's coproducts to beef cattle. Extension Extra 2036. *South Dakota Cooperative Extension Service.* Brookings, SD.

Trenkle A. 2007. Performance of finishing steers fed modified wet distillers grains. R-2183. *Iowa State University Animal Ind. Rep.* Ames, IA.

Trenkle, A. 2008. Performance of finishing steers fed low, moderate and high levels of wet distillers grains. R-2286. *Iowa State University Animal Ind. Rep.* Ames, IA.

Trenkle, A., and D. Pingel. 2004. Effects of replacing corn grain and urea with condensed distillers solubles on performance and carcass value of finishing steers. R-1884. *Iowa State University Rep.* Ames, IA.

Triplett, B. L., D. A. Neuendorff, and R. D. Randel. 1995. Influence of undegraded intake protein supplementation on milk production, weight gain, and reproductive performance in postpartum Brahman cows. *Journal of Animal Science* 73: 3223–3229.

USDA. 2007. Ethanol coproducts used for livestock feed. National Agricultural Statistics Service, Agricultural Statistics. Available online: usda.mannlib.cornell.edu/MannUsda/viewDocumentInfo.do?documentID=1756. Accessed October 19, 2009.

Uwituze, S., M. K. Shelor, G. L. Parsons, K. K. Karges, M. L. Gibson, L. C. Hollis, and J. S. Drouillard. 2009. High sulfur content in distillers grains with solubles may be deleterious to performance and carcass quality of finishing steers. *Journal of Animal Science* 87(Suppl.2): 193.

Vander Pol, K. J., G. E. Erickson, and T. J. Klopfenstein. 2004. Degradable intake protein in finishing diets containing dried distillers grains. MP 83-A. *Nebraska Beef Rep.,* pp. 42–44. Lincoln, NE.

Vander Pol, K. J., M. A. Greenquist, G. E. Erickson, T. J. Klopfenstein, and T. Robb. 2008. Effect of corn processing in finishing diets containing wet distillers grains on feedlot performance and carcass characteristics of finishing steers. *The Professional Animal Scientist* 24: 439–444.

Vander Pol, K. J., M. K. Luebbe, G. I. Crawford, G. E. Erickson, and T. J. Klopfenstein. 2009. Performance and digestibility characteristics of finishing diets containing distillers grains, composites of corn processing coproducts, or supplemental corn oil. *Journal of Animal Science* 87: 639–652.

Vanness, S. J., W. A. Griffin, V. R. Bremer, G. E. Erickson, and T. J. Klopfenstein. 2009. Relating hydrogen sulfide levels to polioencephalomalacia. MP 93. *Nebraska Beef Cattle Rep.,* pp. 74–75. Lincoln, NE.

Vanness, S. J., T. J. Klopfenstein, G. E. Erickson, and K. K. Karges. 2008a. Sulfur in distillers grains. MP 92. *Nebraska Beef Cattle Rep.,* pp.79–80. Lincoln, NE.

Vanness, S. J., N. F. Meyer, T. J. Klopfenstein, and G. E. Erickson. 2008b. Rumen sulfide levels in corn byproduct diets with varying roughage levels. MP 92. *Nebraska Beef Cattle Rep.,* pp. 81–83. Lincoln, NE.

Varel, V. H., J. E. Wells, E. D. Berry, M. J. Spiehs, D. N. Miller, C. L. Ferrell, S. D. Shackelford, and M. K. Koohmaraie. 2008. Odorant production and persistence of Escherichia coli in manure slurries from cattle fed zero, twenty, forty, or sixty percent wet distillers grains with solubles. *Journal of Animal Science* 86: 3617–3627.

Wagner, J. J. 1995. Sequencing of feed ingredients for ration mixing. CATTLE 95-14. *South Dakota Beef Rep.,* pp. 52–52. Brookings, SD.

Ward, E. H., and H. H. Patterson. 2004. Effects of thiamin supplementation on performance and health of growing steers consuming high sulfate water. *Proceedings, Western Section, American Society of Animal Science* 55: 375–378.

Ward, J. D., J. W. Spears, and E. B. Kegley. 1993. Effect of copper level and source (copper lysine vs copper sulfate) on copper status, performance, and immune response in growing steers fed diets with or without supplemental molybdenum and sulfur. *Journal of Animal Science* 71: 2748–2755.

Webb, S. M., A. W. Lewis, D. A. Neuendorff, and R. A. Randel. 2001. Effects of dietary rice bran, lasalocid and sex of calf on postpartum reproduction in Brahman cows. *Journal of Animal Science* 79: 2968–2974.

Wells, J. E., S. D. Shackelford, E. D. Berry, N. Kalchayand, M. N. Guerini, V. H. Varel, T. M. Arthur, J. M. Bosilevac, H. C. Freetly, T. L. Wheeler, C. L. Ferrel, and M. Koohmaraie. 2009. Prevalence and level of Escherichia coli O157:H7 in feces and on hides of feedlot steers fed diets with or without wet distillers grains with solubles. *Journal of Food Protection* 72: 1624–1633.

Wiley, J. S., M. K. Peterson, R. P. Ansotegui, and R. A. Bellows. 1991. Production from first-calf heifers fed a maintenance or low level of prepartum nutrition and ruminally undegradable or degradable protein postpartum. *Journal of Animal Science* 69: 4279–4293.

Wilken, M. E., M. K. Luebbe, G. E. Erickson, T. J. Klopfenstein, and J. R. Benton. 2008a. Feeding corn distillers soluble or wet distillers grains plus soluble and cornstalks to growing calves. MP 92. *Nebraska Beef Cattle Rep.*, pp. 30–32. Lincoln, NE.

Wilken, M. E., T. L. Mader, G. E. Erickson, and L. L. Johnson. 2008b. Comparison of dry distillers or modified wet distillers grains plus soluble in wet or dry forage-based diets. MP 92. *Nebraska Beef Cattle Rep.*, pp. 33–34. Lincoln, NE.

Zinn, R. A., E. Alvarez, M. Mendez, M. Montano, E. Ramirez, and Y. Shen. 1997. Influence of dietary sulfur level on growth performance and digestive function in feedlot cattle. *Journal of Animal Science* 75: 1723–1728.

13 Feeding Ethanol Coproducts to Dairy Cattle

Kenneth F. Kalscheur, Arnold R. Hippen, and Alvaro D. Garcia

CONTENTS

13.1 INTRODUCTION

Ethanol coproducts are often added to beef cattle diets as a source of protein and energy, as discussed in the previous chapter (Chapter 12). They are also frequently used in dairy cattle diets. Protein can be degraded in the rumen to a variable extent, depending on the type of ethanol coproduct, while energy is supplied by fat and fermentable fiber. Mineral concentrations in ethanol coproducts can vary greatly, and this needs to be considered when formulating diets for livestock to prevent excessive or deficient dietary mineral concentrations. As new or altered ethanol coproducts appear in the marketplace with further development of the ethanol industry, there is a need to characterize these products from a nutritional standpoint to determine inclusion rates that will optimize their use in dairy cattle diets.

13.2 NUTRIENT COMPOSITION OF ETHANOL COPRODUCTS

The process by which ethanol and its coproducts are produced is described in Chapter 5. Ethanol coproducts commonly fed to dairy cattle include distillers dried grains with solubles (DDGS), distillers wet grains with solubles (DWGS), distillers modified wet grains with solubles (DMWGS), and condensed distillers solubles (CDS). When formulating diets for dairy cattle, accurate chemical composition analysis of ethanol coproducts is critical. Laboratory testing of purchased distillers grains is highly recommended because nutrient profiles of distillers grains can vary considerably from different ethanol plants. Refer to Chapters 2 and 8 for more information about composition of DDGS and other ethanol coproducts.

This chapter discussion will focus entirely on the nutritional composition of corn-ethanol coproducts as it pertains to their use in dairy cattle diets. Typical ranges of chemical composition of DDGS, DWGS, DMWGS, and CDS are presented in Table 13.1. The values presented for DDGS are those published in the Nutrient Requirements of Dairy Cattle (1989, 2001) and from a commercial laboratory. Chemical composition of ethanol coproducts can be influenced by factors such as grain type and quality, milling process, fermentation process, drying temperature, and amount of solubles blended back into distillers wet grains at drying time. As shown in Table 13.1, the chemical composition of DDGS and CDS varies considerably. Therefore, depending on the ratio of distillers grains to CDS in the final product, the nutrient profiles of DDGS, DWGS, and DMWGS can also vary considerably (Cao et al., 2009). In addition, ethanol can be produced from many types of grain (corn, barley, wheat, triticale, and sorghum) and this can significantly alter the nutrient profile of the distillers grains produced accordingly with the nutrient profile of the original feedstock (Chapter 6).

13.2.1 PROTEIN

Currently, distillers grains being fed in dairy cow diets contain more protein than older reference values (NRC, 1989). The latest edition of the Dairy NRC (2001) lists crude protein (CP) at 29.7% for DDGS, a number similar to values reported by commercial laboratories. According to data reported by Dairy One Forage Labs (Table 13.1), the average CP for DDGS is around 31%, but ranges from 27% to 35%.

Of particular interest to dairy nutritionists is that DDGS is a good source of rumen-undegraded protein (RUP). RUP values can vary depending on the method used to evaluate degradability, which needs to be considered when comparing RUP. In situ reported RUP values for distillers grains ranged from 40% to 67% (Kleinschmit et al. 2007; Cao et al., 2009; Mjoun et al. 2010b). In these trials, DDGS had greater RUP than did DWGS (62.0% vs. 46.9%), and RUP decreased as solubles inclusion in the final product increased (Cao et al., 2009). Kleinschmit et al. (2007) evaluated five different sources of DDGS and found that RUP varied from 59.1% to 71.7%. Mjoun et al. (2010b) evaluated three types of DDGS and found RUP varied from 52.3% to 60.4%. Both studies (Kleinschmit et al., 2007; Mjoun et al., 2010b) included DDGS and DWGS samples and both confirmed that DWGS had greater protein degradability. Some of the rumen degradable protein (RDP)

TABLE 13.1

Chemical Composition (% of DM Unless Otherwise Noted) of Distillers Dried Grains with Solubles (DDGS), Distillers Wet Grains with Solubles (DWGS), Distillers Modified Wet Grains with Solubles (DMWGS), and Condensed Distillers Solubles (CDS)

Item[a] (%)	DDGS (1989)[b]	DDGS (2001)[c]	DDGS[d]	DWGS[e]	DMWGS[f]	CDS[g]
DM, % as is	92	90.2	88.1 ± 7.0[h]	31.4 ± 7.0[h]	48.3	31.4
CP	25	29.7	30.9 ± 4.0	30.0 ± 9.9	28.2	20.3
SP, % of CP	—	—	16.7 ± 7.2	22.9 ± 15.2	16.1	63.9
ADICP	—	5.0	4.6 ± 2.0	3.6 ± 2.1	1.3	0.6
NDICP	—	8.6	9.5 ± 2.9	8.1 ± 3.7	1.9	1.9
NDF	44	38.8	33.8 ± 4.6	30.6 ± 9.1	24.4	4.0
ADF	18	19.7	16.9 ± 3.4	15.1 ± 5.3	8.6	1.9
Lignin	4	4.3	5.1 ± 1.6	4.7 ± 1.7	5.3	0.4
Starch	—	—	5.4 ± 4.4	5.6 ± 9.0	7.3	5.4
Crude Fat	10.3	10.0	13.0 ± 3.0	12.8 ± 4.0	12.0	18.5
Ash	4.8	5.2	5.9 ± 1.0	5.6 ± 1.6	5.9	9.6
Ca	0.15	0.22	0.08 ± 0.15	0.07 ± 0.11	0.06	0.07
P	0.71	0.83	0.88 ± 0.16	0.85 ± 0.17	0.88	1.57
Mg	0.18	0.33	0.32 ± 0.07	0.32 ± 0.08	0.41	0.67
K	0.44	1.10	1.05 ± 0.24	1.00 ± 0.29	1.25	2.26
Na	0.57	0.30	0.19 ± 0.17	0.17 ± 0.13	0.36	0.33
S	0.33	0.44	0.64 ± 0.18	0.58 ± 0.15	0.79	1.10
TDN	88	79.5	83.3 ± 4.9	85.1 ± 5.1	—	103.2
NE_L, Mcal/kg	2.04	1.97	2.07	2.11	—	1.19
NE_M, Mcal/kg	2.18	2.07	2.18	2.23	—	1.29
NE_G, Mcal/kg	1.50	1.41	1.50	1.54	—	0.92

[a] Nutrients: DM = dry matter, CP = crude protein, SP = soluble protein, ADICP = acid detergent insoluble CP, NDICP = neutral detergent insoluble CP, NDF = neutral detergent fiber, ADF = acid detergent fiber, TDN = total digestible nutrient, NE_L = net energy for lactation, NE_M = net energy for maintenance, and NE_G = net energy for gain.

[b] NRC Nutrient Requirements of Dairy Cattle, 6th rev. ed. (1989).

[c] NRC Nutrient Requirements of Dairy Cattle, 7th rev. ed. (2001).

[d] Analyzed by Dairy One Forage Lab (www.dairyone.com) from May 2000 to April 2009. Number of samples—1875 to 4948 depending on nutrient analyzed.

[e] Analyzed by Dairy One Forage Lab (www.dairyone.com) from May 2000 to April 2009. Number of samples of DWGS—863 to 1923 depending on nutrient analyzed.

[f] DMWGS analysis is from two samples evaluated at South Dakota State University.

[g] Analyzed by Dairy One Forage Lab (www.dairyone.com) from May 2000 to April 2009. Number of samples of CDS—103 to 757 depending on nutrient analyzed.

[h] Data is reported as means ± the standard deviation.

in corn is altered in the fermentation process to produce ethanol; therefore the protein remaining in DDGS is expected to have greater RUP than the original corn. The lower RUP values observed for DWGS were likely due to the absence of drying and possibly greater quantities of solubles returned to the DWGS compared to the DDGS.

Protein quality in DDGS can be good, although as with most corn products, lysine is the first limiting amino acid. Very high RUP (e.g. >80% of CP) in DDGS, usually results from heat damaged, indigestible protein. Heat damaged protein may be indicated by a high acid detergent insoluble crude protein (ADICP), although in DDGS there is no clear relationship between ADICP and protein digestibility as in some other feeds. Extensive heating creates darker DDGS and is believed to decrease the concentration of digestible lysine as this amino acid is very sensitive to high temperatures. Boucher et al. (2009) demonstrated that heating substantially reduced intestinally available lysine; however, it should be noted that the type of grain and the amount of solubles added back to distillers grains can create darker products without necessarily reducing amino acid quality. Recently, Mjoun et al. (2010b) evaluated the intestinal digestibility of protein of four distillers grains products (conventional DDGS, reduced-fat DDGS, high-protein distillers dried grains (DDG), and DMWGS) and found that while these products were slightly less digestible than soybean products, 92.4% and 97.7%, respectively, their values were greater than the 80% RUP digestibility used in feed formulation models such as the NRC (2001). Intestinal digestibility of the essential amino acids exceeded 92% across all feedstuffs, with the exception of lysine, where distiller grains was less (84.6%) compared with soybean feedstuffs (97.3%) (Mjoun et al., 2010b).

13.2.2 FIBER

Neutral detergent fiber (NDF) values are often between 30% and 40%, but can vary considerably between individual ethanol plants. Some newer DDGS have been reported to have concentrations of NDF considerably lower than NRC values (NRC, 2001; Robinson et al., 2008). Although DDGS contains a considerable amount of NDF, this fiber should not be considered a source of physically effective fiber in diets. Because the corn is ground prior to fermentation to produce ethanol, the resulting DDGS has very small particle size (Kleinschmit et al., 2007). Replacing forage fiber with nonforage fiber provided by DDGS can create unfavorable fermentation in the rumen and potentially result in milk fat depression (Cyriac et al., 2005). While fiber provided by DDGS is a good source of energy, it should not replace forage fiber in diets of high-producing dairy cows.

13.2.3 STARCH

Maximizing the fermentation of starch to ethanol is always the goal of ethanol production; however, there is usually some starch remaining in distillers grains. During the 1980s and 1990s, starch in DDGS was determined to be 10% to 15% (Belyea et al., 1989; Batajoo and Shaver, 1998). Samples from newer fuel ethanol plants have resulted in values of 5%–6% starch, with some samples greater than 8% (Mjoun et al., 2010b). Improved processes to ferment starch to ethanol is most likely the reason for decreased starch concentrations in DDGS.

13.2.4 FAT

One concern of nutritionists is that the concentration of fat in distillers grains can vary greatly, and potentially exceed 12%, much greater than values reported in the NRC (2001). Distillers grains contain high levels of unsaturated oil, predominantly linoleic acid (C18:2), reflecting the composition of corn oil (Elliot et al., 1993). Dried or wet distillers grains that contains greater proportions of CDS results in greater concentrations of fat in the final product (Cao et al., 2009). It should be noted that the method of analysis can significantly affect the crude fat value (Cao et al., 2009). Recently, a study evaluating methods for crude fat analysis recommended the use of petroleum ether when analyzing distillers grains samples (Thiex, 2009).

High concentrations of unsaturated fatty acids are a concern when including distillers grains in diets for lactating dairy cows because the presence of unsaturated fatty acids can increase incomplete biohydrogenation in the rumen, which has been related to observed milk fat depression. However, if diets are formulated to provide sufficient levels of physically effective fiber, increasing the concentration of polyunsaturated fat will most often not result in milk fat depression (Ranathunga et al., 2010).

13.2.5 MINERALS

Environmental concerns regarding excessive phosphorus (P) has increased the awareness of phosphorus concentrations in DDGS. Most DDGS contains between 0.65% and 0.95% P and this value increases with the amount of CDS added to the distillers grains (Table 13.1). Even though DDGS protein is relatively undegraded in the rumen, phosphorus has been shown to be highly available (Mjoun et al., 2008). Fortunately, high-producing dairy cows often need some supplemental P; therefore inclusion of DDGS can replace more expensive inorganic sources. The greatest concern of feeding DDGS will be in regions of the United States where soils are already high in P. In order to minimize excess P in manure, diets should be formulated close to the animal's requirement (NRC, 2001). The other mineral that can be highly variable is sulfur (S). Although an average S concentration in DDGS is about 0.64% (Table 13.1), it has been reported to exceed 1.0% in some samples. Distillers grains products with greater concentrations of CDS often contain greater S concentrations (Cao et al., 2009). Though rarely reported in dairy cattle, excessive S concentrations in feed and water can result in central nervous system disorders, which can lead to poor performance or death. More about this is discussed in Chapter 12.

13.2.6 ENERGY

Some researchers speculate that the distillers grains available today may contain more energy than indicated by the NRC reference values. Birkelo et al. (2004) determined the energy value of DWG for lactating cows. In this study, digestible energy (DE), metabolizable energy (ME), and NEL of distillers wet grains were 4.09, 3.36, and 2.27 Mcal/kg, respectively. These energy values are 7% to 11% and 10% to 15% higher than previously published values reported in the 1989 and 2001 NRC (Table 13.1). These greater values are likely attributed to increased fat concentration as well as more highly digestible fiber measured in distillers grains products than assumed by the NRC (2001).

13.3 FEEDING DDGS TO DAIRY CATTLE

13.3.1 FEEDING DDGS TO LACTATING COWS

Numerous studies evaluating the addition of distillers grains products to dairy cow diets have been reported in the United States since 1982. Prior to the late 1990s, many of the experiments evaluating DDGS in livestock diets used DDGS originating from production of drinking alcohol. Since that time, most DDGS evaluated in dairy cow studies have used DDGS originating from modern fuel ethanol plants. This discussion on feeding DDGS to lactating dairy cows will consider all studies, both old and new, and how inclusion of DDGS affects milk production.

13.3.1.1 DDGS as a Protein Source

Since DDGS was first recognized as an inexpensive source of protein, research has been conducted to test DDGS as an alternative source of CP and RUP for dairy cow diets. Most previous studies added DDGS in substitution for soybean meal alone or together with corn. Palmquist and Conrad (1982) found different milk production responses in Holsteins versus Jersey cows fed isonitrogenous diets containing DDGS. Jersey cows responded positively to the addition of 24% dried DDGS; however, Holstein cows were unaffected. Van Horn et al. (1985) reported decreased milk yield and protein percentage when DDGS was fed at 22.5% and 41.6% of the diet to provide 14% and 18% dietary CP. The authors attributed the loss of production to the high acid detergent insoluble nitrogen (ADIN) concentration in the DDGS fed. Weiss et al. (1989) fed a DDGS produced from a blend of 65% barley and 35% corn. In this experiment, DDGS was fed up to 12.9% of the diet dry matter (DM) with no effects on dry matter intake (DMI) or on milk production and milk composition.

Owen and Larson (1991) reported the results of a study comparing DDGS and soybean meal in diets for early lactation cows. Milk production was increased when DDGS was fed at 18.8% of diet DM, most likely because of increased dietary RUP, but production decreased when DDGS was included at 35.8% of the diet DM. The authors concluded that lower protein digestibility and low lysine concentrations contributed to the decreased performance of cows fed 35.8% DDGS. The decrease in milk protein percentage for cows fed the DDGS corn silage-based diets in this study indicates that available lysine may have been deficient.

Powers et al. (1995) compared three different sources of DDGS with soybean meal at two dietary CP concentrations (14% and 18%). At 14% CP, DDGS was included at 13% of the diet, whereas at 18% CP, DDGS was included at 26% of the diet (DM basis). Designated by source, DDGS1 and DDGS2 were light in color and were lower in ADIN, whereas DDGS3 was dark in color with a higher ADIN. Dry matter intake was not affected by source of DDGS or by CP concentration. Cows fed DDGS1 and DDGS2 produced more milk than cows fed 0% DDGS (soybean meal diet) and the DDGS3 diet. In addition, cows fed DDGS incorporated into the diet at 26% of the diet DM produced more milk than those fed 13%.

More recently, Kleinschmit et al. (2006) compared three different sources of DDGS from modern fuel ethanol plants. The three DDGS were processed differently, each with the intent to reduce protein damage. The source of DDGS had very little impact on overall lactation performance. Dry matter intake did not change, but cows fed diets containing DDGS (20% of diet DM) produced more milk than cows fed the control diet with no DDGS. Milk fat percentage did not change with the addition of DDGS, but milk protein percentage declined from 3.28 to 3.16 with the addition of DDGS.

Since DDGS is often used as a source of RUP, Pamp et al. (2006) compared DDGS to soybean protein to investigate the effect of source and level of RUP in lactation diets. Soybean protein included a combination of soybean meal, extruded soybeans, soybean hulls, and expeller soybean meal. These feeds were combined to create similar nutrient content to DDGS. Diets were formulated to contain 5.3%, 6.8%, and 8.3% RUP (DM basis) and DDGS was included at 0%, 11%, or 22% as the source of RUP. Milk production increased with the addition of RUP and was greater for cows fed diets containing DDGS versus soybean protein. Milk fat percentage was not affected by added RUP or by its source, but milk protein percentage was greater for cows fed DDGS compared to cows fed soybean protein (2.85% vs. 2.80%). This research supports the conclusion that DDGS can replace soybean protein as a source of RUP without reducing milk production and milk protein percentage if diets are formulated properly.

Distillers dried grains with solubles are a good source of methionine, but are low in lysine, therefore supplementation of rumen-protected lysine would theoretically allow maximization of DDGS in dairy cow diets. Nichols et al. (1998) reported an increase in milk production when cows were fed a ruminally protected lysine and methionine product in diets where DDGS were included at 20.25% of the diet DM; however, in a following experiment, Liu et al. (2000) did not observe increased milk production when ruminally protected lysine and methionine were fed to cows with 18.85% DDGS in the diet. Overall, results of experiments feeding DDGS with supplemental ruminally protected amino acids or with feedstuffs such as blood meal to provide lysine have been mixed. In general, DDGS is a good quality protein source that can often be used as the only source of supplemental protein in properly formulated diets.

Recently, Mjoun et al. (2010c) evaluated the inclusion of a reduced-fat DDGS (3.5% crude fat) on milk production and amino acid utilization of midlactation dairy cows. Cows were fed 0%, 10%, 20%, or 30% reduced-fat DDGS on a DM basis. As expected, lysine decreased as diets increased in DDGS. Increasing DDGS had no effect on DMI or milk production, but milk protein percentage responded quadradically with an increase at 10% and 20% inclusion, but a decrease with 30% DDGS. Although arterial lysine concentrations decreased linearly with the inclusion of DDGS, extraction of lysine by the mammary gland increased. In this experiment, even though lysine concentrations were less than recommended when DDGS was added, milk protein percentage did not decrease until DDGS was at 30% of the diet. This suggests that lysine was not limiting milk protein synthesis until cows were fed at 30% of diet.

13.3.1.2 DDGS as a Carbohydrate Source

Research has also been conducted to evaluate the ability of DDGS to replace a portion of the forages found in dairy cow diets. Since DDGS is high in NDF, it is tempting to replace some of the NDF supplied by forages with NDF from DDGS. Clark and Armentano (1993) tested the effects on milk production and composition of replacing alfalfa NDF with NDF from DDGS. Replacing 12.7% of the alfalfa DM with DDGS resulted in both milk production and milk protein percentage increases. Milk fat production was not altered with the inclusion of DDGS even though the proportion of alfalfa in the diet dropped from 44% to 31% of the diet DM.

One concern about replacing forage NDF in dairy diets is that removal of physically effective fiber may alter normal rumen function causing milk fat depression. Cyriac et al. (2005) evaluated the replacement of forage fiber from corn silage with nonforage fiber from DDGS. Diets were formulated to include 0%, 7%, 14%, and 21% DDGS replacing an equal portion of corn silage. The basal diet included 40% corn silage and 45% concentrate. All diets contained 15% alfalfa hay. Dry matter intake increased as DDGS increased in the diets, but milk production remained the same. Milk fat percentage, however, decreased (3.34%, 3.25%, 3.04%, and 2.85%) and milk true protein increased (2.82%, 2.90%, 2.93%, and 3.04%) as the percentage of DDGS increased in the diet. When formulating diets with DDGS, sufficient physically effective fiber from forage must be included to prevent changes in milk fat composition.

As DDGS became competitively priced with corn, there was greater interest in replacing energy provided by starch in corn grain with nonforage fiber provided by DDGS. Ranathunga et al. (2010) evaluated the replacement of corn with DDGS and soy hulls (SH) such that dietary starch concentrations decreased from 29% to 20% of the diet (DM basis). Replacing starch from corn with nonforage fiber from DDGS and SH did not affect milk production or composition. It did, however, decrease DMI, resulting in a trend for increased feed efficiency. Greater conversion of feed to milk combined with lower feed costs was shown to improve profitability when nonforage fiber sources replaced corn in the diets of lactating cows.

13.3.1.3 DDGS as a Fat Source

In addition to providing fermentable fiber and protein in dairy cow diets, DDGS has long been recognized as a high-energy supplement for lactating dairy cows. As early as 1952, Loosli et al. documented increases in milk yield and milk fat production when cows were fed DDGS. Loosli attributed these responses to the fat contained within DDGS. With DDGS typically containing 9% to 11% fat, 25% to 30% of the energy content of the DDGS is contributed by this nutrient. Technically, the fat in DDGS is more appropriately referred to as oil as it remains liquid at room temperature because of its high degree of unsaturation. During the wet milling process, the oil is normally extracted from the byproduct residues to be marketed as corn oil. However, for dry grind ethanol production, the oil is not usually extracted from the distiller grains or solubles. Reported values for ether extract (oil) content in solubles range from 18.5% to 21.5% of DM (DaCruz et al., 2005; Kalscheur et al., 2008, Sasikala-Appukuttan et al., 2008). Therefore, the contribution of oil from corn solubles added to the distillers grains increases the ether extract of the latter from 8.9% to 11.7% for most DDGS depending upon the amount of solubles added to the distillers grains by the processor (Noll et al., 2007). For DMWGS, described elsewhere, the amount of solubles added to the distillers grains may increase the ether extract to 18% or more.

The oil in distillers grains and solubles contains a fatty acid profile similar to corn oil with approximately 53.3% linoleic (C18:2), 25.7% oleic (C18:1), 14.6% palmitic (C16:0), and 1.6% linolenic (C18:3) acids (Leonardi et al., 2005). Because of the large proportion of unsaturated fatty acids, there is concern of the negative effects of the oil in distillers grains on milk fat synthesis creating milk fat depression. Milk fat depression is a reduction in milk fat yield that has been associated with the presence of certain biohydrogenation intermediates of 18 carbon unsaturated fatty acids arising from microbial alteration of those present in the rumen (Bauman and Griinari, 2003). Bauman and Griinari (2003) have described the regulatory role that these intermediates have in milk fat synthesis.

If sound diet formulation guidelines are followed when feeding distillers grains, or other corn coproducts, milk fat depression is not likely to occur. Among recommendations for preventing milk fat depression is the inclusion of adequate effective fiber. A meta-analysis of research on feeding distillers grains to dairy cows by Kalscheur (2005), discussed later in this chapter, found that milk fat percentages were decreased slightly when distillers grains was included in diets at 20% to 30% of diet DM but not at 40%. Further analysis revealed that these decreases in milk fat were more highly related to forage content of the diets. When diets contained more than 50% DM as forage, milk fat production was not changed compared to control diets.

Mpapho et al. (2006) fed Holstein and Brown Swiss cows diets with 50% forage and DWGS at 15% of DM for an entire lactation. Milk fat percentages averaged 4.1% for cows fed DWGS compared with 3.7% for diets not containing DWGS. In a study comparing 10% and 20% dietary distillers grains and DWGS versus DDGS, Anderson et al. (2006) obtained the greatest yields of milk fat with diets that contained distillers grains in both forms compared with the control. Likewise, Kleinschmit et al. (2006) observed increased milk fat yields when different sources of DDGS were added to diets at 20% of DM. Cows fed diets containing the DDGS produced 1.26 kg/d of milk fat compared with 1.14 kg/d for cows fed the control diet. For both of these studies, the treatment and control diets were not normalized for total lipid content with fat concentration 2% to 3% greater in diets containing distillers grains. Thus results obtained in these experiments reflect the observations by Loosli (1952) of increased yields of milk fat when distillers grains was added to dairy cow diets.

Because distillers grains contains high amounts of NDF, nutritionists may replace forage NDF with NDF from distillers grains when formulating diets. The NDF in distillers grains is not "effective" fiber, however, because of small particle size and high fermentability. In a classic experiment, Griinari et al. (1998) demonstrated severe milk fat depression when unsaturated fat from corn oil was added to diets containing decreased forage. This provided indication of the potential effects of fat from distillers grains on milk fat production when forage concentrations are marginal. More recently, Cyriac et al. (2005) showed a linear decrease in milk fat concentrations when DDGS and soy hulls replaced corn silage on a one-to-one ratio for NDF content. This experiment demonstrated the need to maintain effective fiber from forage when feeding diets containing DDGS. In diets containing less forage (45% of diet DM), Leonardi (2005) demonstrated a linear decrease in milk fat percentage as DDGS increased from 0% to 15% of diet DM; however, yields of milk fat were not decreased. Hippen et al. (2010) demonstrated both decreased concentrations and yields of milk fat feeding 20% DDGS in diets that contained 43% forage and less than 20% of diet DM as forage NDF.

To test the effects of starch fermentability and its interaction with DDGS, Owens et al. (2009) fed diets containing either high-moisture corn or dry ground shelled corn and 0% or 18% DDGS in a 2×2 factorial design. All diets contained 50% forage but only 20.9% forage NDF. This research demonstrated decreases in both concentrations (3.42 vs. 2.95%) and yields (1.20 vs. 1.02 kg/d) of milk fat when DDGS was included in the diet. The least concentration and yield of fat (2.76% and 0.90 kg/d) were observed when the highly fermentable high-moisture corn was fed along with the DDGS. This interaction of increased starch fermentability and unsaturated dietary fatty acids was described by Bauman and Griinari (2001) as a prerequisite for milk fat depression to occur when feeding low-fiber or high-starch diets.

The relative physical availability of the oil contained in the distillers grains also impacts their likelihood of creating milk fat depression. Abdelqader et al. (2009b) fed diets containing 2.5% supplemental oil from corn germ, DDGS, or corn oil. Milk fat concentrations averaged 3.88%, 3.80%, 3.59%, and 3.50% for the control, corn germ, DDGS, and corn oil treatments, respectively. Diets in this experiment contained 55% forage with 22.7% NDF from forage. Differences in milk fat concentrations were attributable to the relative availability of fatty acids in the products and their susceptibility to ruminal biohydrogenation. The oil contained in germ is more highly associated with the grain than is the oil contained in DDGS or DWGS that have gone through steeping and fermentation. Similar results were demonstrated when Leonardi et al. (2005) compared feeding DDGS to corn oil. The authors observed a linear decrease in milk fat concentrations as DDGS increased up to the greatest increment (15% of diet DM), where milk fat concentrations were similar to when cows were fed corn oil at 1.5%.

Milk fat fatty acid profiles are not changed significantly when distillers grains is fed and milk fat production is normal. When milk fat depression occurs from feeding distillers grains, milk fatty acid profiles are altered significantly, with increasing concentrations of long-chain fatty acids and decreased concentrations of short-chain fatty acids from de novo synthesis. Small increases in trans- and conjugated fatty acids have been observed (Leonardi et al., 2005; Anderson et al., 2006; Abdelqader et al., 2009b; Owens et al., 2009), which suggest that the mechanism of milk fat depression from feeding distillers grains appears to be attributable to altered biohydrogenation pathways of the unsaturated fatty acids.

In summary, feeding distillers grains will have minimal effects on, and may even increase, milk fat production when dietary guidelines are followed. To avoid milk fat depression when feeding distillers grains, adequate effective fiber must be provided, and fermentability of starch and the diet as a whole must be considered. Diets that provide healthy ruminal fermentation patterns will likely be successful candidates for inclusion of distillers grains with minimal effects on milk fat.

13.3.1.4 DDGS as a Mineral Source

Distillers grains products are relatively high sources of P. Although corn is high in phytate P, it has been hypothesized that this P undergoes hydrolysis during the fermentation process in the ethanol plant resulting in DDGS that is highly soluble. This was corroborated by Mjoun et al. (2008) who found that ruminal P disappearance was very high for DDGS. In another study, when DDGS substituted for soybean and inorganic P in diets of lactating cows, the addition of DDGS did not affect P utilization or increase P excretion by the cow compared to traditional sources of dietary P (Mjoun et al., 2007). This study demonstrated that feeding DDGS at concentrations needed to meet, but not exceed, the cow's P requirement is key to formulating diets that do not increase P excretion.

Other minerals have not been extensively studied in lactating dairy cows. There can be considerable variation in S concentration in DDGS. Therefore, S levels need to be considered to prevent toxicity.

13.3.1.5 Rates of DDGS Feeding

Several studies have been conducted evaluating inclusion rates of DDGS in dairy cow diets. Anderson et al. (2006) fed isonitrogenous diets formulated to include 0%, 10%, and 20% DDGS in diets containing 50% forage of which one-half was corn silage and the other half was alfalfa hay. Milk production was greatest for cows fed the 20% DDGS diets (39.8, 40.9, and 42.5 kg/d for 0%, 10%, 20% DDGS, respectively). Milk fat and protein composition were not affected by DDGS.

Hippen et al. (2004) conducted a study that included DDGS up to 40% of the diet on a DM basis. Diets were formulated to include DDGS at 0%, 13%, 27%, and 40% of the diet. These diets were not isonitrogenous, and CP increased from 16.5% to 18.9%. Corn silage was the predominant forage used in this study. Milk production was the greatest at 13% inclusion and decreased as DDGS was further added to the diet (40.7, 41.7, 39.1, 36.3 kg/d). Milk fat percentage declined from 3.40% to 3.16% for the cows fed the control diet versus the DDGS diets.

Janicek et al. (2008) evaluated the inclusion of DDGS up to 30% of the diet on a DM basis. Diets were formulated to include DDGS at 0%, 10%, 20%, and 30% of the diet replacing both forage and concentrate ingredients. Dry matter intake increased linearly with the inclusion of DDGS in the diet (21.4, 22.4, 23.0, and 24.0 kg/d), as did milk production (27.4, 28.5, 29.3, and 30.6 kg/d). Milk fat and protein composition were similar for all diets.

Grings et al. (1993) added DDGS at 0%, 10.1%, 20.8%, and 31.6% of the diet DM to alfalfa-based diets. These diets were formulated to provide 13.9%, 16.0%, 18.1%, and 20.0% CP with increasing concentration of DDGS. Although DMI did not vary across diets, milk yield and milk protein percentage increased linearly with increasing DDGS in the diet. It should be noted that the CP percentage of these diets was not kept constant; therefore as DDGS increased in the diet, so did CP. Cows fed the 31.6% DDGS diet produced the same amount of milk as those fed the 20.8% DDGS diet; therefore cows fed the 31.6% DDGS diet were fed in excess of their protein requirement and, consequently, no production response occurred.

13.3.1.6 Meta-Analysis of Distillers Grains Studies

In order to assess the impact of inclusion of distillers grains on lactation performance, a meta-analysis of previous feeding studies was conducted (Kalscheur, 2005). Twenty-three studies investigating the inclusion of distillers grains in dairy cow diets were compiled into a database with 96 treatment comparisons. An abbreviated summary of this extensive survey is listed in Table 13.2. Overall, DMI increased when distillers grains was included up to 20% of the DM in dairy cow diets; however, DMI was also affected by form of distillers grains. Whereas DMI of cows fed DWGS was highest at lower inclusion levels (<20% of the DM), cows fed DDGS increased DMI up to 30% inclusion. Milk production was not affected by form, but there was a curvilinear response of increasing distillers grains in dairy cow diets. Cows that were fed diets containing 4% to 30% distillers grains produced similar amounts of milk, approximately 0.4 kg/d more than cows fed diets containing no distillers grains. When cows were fed the highest inclusion rate (>30%), milk yield decreased. Cows fed DWGS decreased milk production when DWGS was included in the diet at greater than 20%. This was most likely related to decreased DMI.

Milk fat percentage varied among inclusion levels but was not significantly affected by inclusion level or form of distillers grains (Table 13.2). With the current dataset, the inclusion of distillers grains does not support the theory that feeding distillers grains results in milk fat depression. Many factors play an important role in causing milk fat depression. When formulating diets, it is important to include sufficient fiber from forages in order to maintain rumen function. Distillers grains provide 28% to 44% NDF, but this fiber is finely processed and rapidly digested in the rumen. High levels of unsaturated fat as provided from distillers grains may impact rumen function leading to milk fat depression, but it is often a combination of dietary factors that lead to significant reduction in milk fat percentage.

TABLE 13.2
Lactation Performance of Dairy Cows Fed Increasing Levels of Distillers Dried and Wet Grains

Inclusion Level (DM Basis)	DMI (kg/d)			Milk (kg/d)			Fat %	Protein %	FE[a]
	Dried	Wet	All	Dried	Wet	All			
0%	23.5[d]	20.9[c]	22.2[c]	33.2	31.4	33.0	3.39	2.95[b]	1.41
0%–10%	23.6[c,d]	23.7[b]	23.7[b]	33.5	34.0	33.4	3.43	2.96[b]	1.44
10%–20%	23.9[b,c]	22.9[b,c]	23.4[b,c]	33.3	34.1	33.2	3.41	2.94[b]	1.44
20%–30%	24.2[b]	21.3[b,c]	22.8[b,c]	33.6	31.6	33.5	3.33	2.97[b]	1.42
>30%	23.3[c,d]	18.6[d]	20.9[d]	32.2	31.6	32.2	3.47	2.82[c]	1.48
SEM	0.8	1.3	0.8	1.5	2.6	1.4	0.08	0.06	0.06

Source: Adapted from Kalscheur, K. F. Impact of feeding distillers grains on milk fat, protein, and yield. Proc. Distillers Grains Technology Council, 9th Annual Symposium, Louisville, KY, 2005.

[a] Feed efficiency (FE) = energy-corrected milk/DMI.

[b,c,d] Values within column followed by a different superscript letters differ significantly ($p < 0.05$). No superscript indicate no significant difference between treatments.

Milk protein percentage was not different for cows fed diets with 0% to 30% distillers grains inclusion and the form of the distillers grains did not alter composition (Table 13.2). Milk protein percentage did decrease by 0.13 percentage units, however, when distillers grains was included at concentrations greater than 30% of the diet compared to cows fed control diets. At inclusion levels greater than 30% of the diet, distillers grains most likely replaced all other sources of protein supplementation. Consequently, decreased intestinal protein digestibility, decreased lysine concentrations, and an unbalanced amino acid profile may all contribute to a lower milk protein percentage. It should be noted that the lower milk protein percentages were most evident in studies conducted in the 1980s and 1990s. More recent studies are not as consistent in showing this effect. Lysine is very heat sensitive, and can be negatively affected by processing and drying. Improved processing and drying procedures in the newer fuel ethanol plants may have improved amino acid quality of the product.

Feed efficiency defined as energy-corrected milk (kg/d)/DMI (kg/d) tended ($P = 0.06$) to increase with the inclusion of distillers grains in the diets (Table 13.2). This indicates that diets formulated with distillers grains improved the conversion of nutrients from feed to milk compared to traditional feeds 2% to 5% depending on the inclusion level.

This review found that forage type had no impact on DMI, milk production, or milk fat composition for diets containing distillers grains. Forage type did, however, affect milk protein composition. Cows fed diets containing 55% to 75% corn silage produced milk with the highest concentration of protein at 3.04%. Cows fed 100% alfalfa/grass with 0% corn silage resulted in the lowest concentration of protein in milk at 2.72%. Cows fed 45% to 54% corn silage and 100% corn silage produced milk with intermediate levels of protein at 2.98% and 2.82%, respectively. Cows fed diets with a blend of corn silage and alfalfa produced milk with greater milk protein percentage suggesting that diets formulated with one forage source are more likely to be insufficient in amino acids needed to maximize milk protein percentage.

Forage to concentrate ratio is another dietary factor that may affect lactation performance of the dairy cow when distillers grains are included in the diet. To evaluate the effect of forage to concentrate ratios, treatments were classified into one of three categories: diets containing <50% forage, diets containing 50% forage and 50% concentrate, and diets containing >50% forage. Dry matter intake, milk production, and milk protein percentage were not affected by forage to concentrate ratio (Table 13.3). The percentage of milk fat, however, was decreased by 0.36 percentage units in diets containing <50% forage.

TABLE 13.3
Effect of Forage to Concentrate Ratio in the Diet on Lactation Performance

% Forage in Diet	DMI (kg/d)	Milk (kg/d)	Fat (%)	Protein (%)
<50% forage	23.1	34.1	3.21[b]	2.95
50% forage	23.7	32.3	3.57[a]	2.90
>50% forage	22.4	32.5	3.46[a]	2.97
SEM	1.10	2.18	0.072	0.078

Source: Adapted from Kalscheur, K. F. Impact of feeding distillers grains on milk fat, protein, and yield. Proc. Distillers Grains Technology Council, 9th Annual Symposium, Louisville, KY, 2005.

[a,b] Values within a column followed by a different superscript letters differ significantly ($p < 0.05$). No superscript within a column would indicate that there is no significant difference.

This supports the hypothesis that lack of forage in the diet, likely attributed to insufficient effective fiber, is a major contributing factor in decreased milk fat percentages rather than inclusion of distillers grains in the diet. On paper, NDF levels may appear adequate because of the fiber provided by distillers grains, but this fiber has a small particle size and does not provide effective fiber needed for normal rumen function. The experiment of Cyriac et al. (2005) tested this hypothesis directly demonstrating that as forage decreased in the diet from 55% to 34% of diet DM, milk fat percentage decreased linearly from 3.34% to 2.85% even though total dietary NDF remained similar across diets. This review did not find a strong relationship between distillers grains inclusion and milk fat depression; however there could be interactions between the concentration of oil in distillers grains and a lack of effective fiber in diets, which can result in milk fat depression. More research needs to look into how different nutrients interact to cause negative effects on ruminal fermentation.

The last factor that was evaluated was formulation for amino acids. The review included experiments where rumen-protected lysine and methionine or a source of lysine, such as blood meal, was added to diets. Lysine may be deficient in diets when corn feedstuffs are the predominant ingredients. Milk protein percentage tended to increase when diets included a source of lysine. Additional research is needed to determine if supplemental lysine would allow for additional distillers grains to be included in dairy cow diets.

There are additional dietary factors to consider when deciding how much distillers grains to include in a lactating cow diet. Diets must be properly balanced to prevent disruptions in normal rumen function. The fiber provided from distillers grains can enhance rumen fermentation as long as it does not replace forage fiber required for rumen function. Disruptions in normal rumen function can lead to decreased DMI, lower milk productions, and milk fat depression. More research is needed to determine how specific dietary factors interact with the distillers grains in order to maximize inclusion of this coproduct in dairy cow diets.

13.3.1.7 Feeding DDGS to Grazing Dairy Cows

Little research has been conducted in feeding DDGS to grazing dairy cattle. Because DDGS represents a good source of RUP and has high energy content, it provides an ideal dietary complement to pasture grasses, which typically have high concentrations of RDP and low energy values as grasses mature. Shaver et al. (2009) examined DDGS as a supplement to dairy cows grazing ryegrass on Chilean dairy farms. On one farm, DDGS were fed at a rate of 2 kg/d in 5 kg of concentrate, replacing corn and soybean meal. Effects of DDGS varied by season, but feeding DDGS tended to increase milk by 1.8 to 1.9 kg/d regardless of season. Milk fat percentage decreased during the spring when grass quality was greater and contained more highly digestible fiber. On five other farms where DDGS was mixed in corn silage to provide 2.5 kg/d replacing a variety of concentrate sources, feeding DDGS in the winter increased milk production by 0.9 kg/d, but did not affect milk yield in the spring.

Research conducted by Nyoka et al. (2007) with dairy cows grazing alfalfa-based pastures examined the effects of supplementing cows with stored forages and concentrate to supply one-half of their daily requirement. The supplements varied by inclusion of (1) DDGS at 15% of estimated daily DMI or DDGS replaced by (2) soybean meal and extruded soybeans or (3) fishmeal and soy oil. There were no dietary effects on milk yield averaging 31.5 kg/d. The researchers observed no effect of protein supplements on milk yield. Concentrations of fat and yield of protein in milk were greatest for cows fed supplemental fish meal and soy oil, with DDGS-supplemented cows having the next greatest milk fat concentrations and protein yields. The results indicate that though protein quality of DDGS may limit production responses compared with fish meal when fed to grazing cattle, DDGS appears preferable as a protein supplement for grazing dairy cows compared with soybean meal.

13.3.2 Feeding DDGS to Dry Cows

Distillers grains has not been widely examined as supplements for nonlactating (dry) dairy cows, although their nutrient content and relative cost when compared with more traditional protein supplements suggest opportunities for formulating diets for dry cows. As described in the section on feeding distillers grains to dairy heifers, the nutritional profile of distillers grains is complementary to low energy, low protein forages commonly fed in dry cow diets. The nutritional goals of diets for nonlactating dairy cows are to provide adequate energy for maintenance of body weight (BW) and fetal growth, while avoiding overfeeding energy, which contributes to the subsequent postcalving fat-cow syndrome. Supplying adequate rumen fill in the form of long-stem forage during this time of the production cycle is important for regulation of energy intake. Frequently though, forages fed not only have low energy content, but are also deficient in protein and minerals required for the cow to prepare for the ensuing lactation. Specific potential nutritional advantages to inclusion of distillers grains in diets of these cattle are as follow.

Dry cow diets are commonly formulated with low content of highly fermentable soluble carbohydrate in the form of starch or sugars and high concentrations of forage NDF. Research summarized by Overton and Waldron (2008) indicate benefits to postpartum health of cows that have supplemental starch or sugars added to prepartum diets. As the authors pointed out, the inclusion of these ingredients, however, confounded the effects of the nonfiber carbohydrate source with energy content of the diets. Less equivocal, however, is research by Pickett et al. (2003) that demonstrated an improved energy status and milk production after being fed diets that replaced forage NDF with more highly fermentable NDF provided from byproduct feeds, such as distillers grains, during the dry period. Lastly, VandeHaar et al. (1999) demonstrated improved nutrient balance and decreased hepatic lipid content at calving of cows fed increased energy and protein prepartum. Feeding distillers grains as a supplement to low-quality forage during the prepartum period would provide the additional energy indicated as being beneficial as well as an additional source of RUP.

Research conducted during the 1990s (Van Saun et al., 1993; Moorby et al., 1996; Huyler et al., 1999) demonstrated increasing CP supplementation above NRC (1989) recommendations of 12% with RUP to close-up cows resulted in increased milk and(or) protein yield in the subsequent lactation. Robinson et al. (2001) demonstrated similar benefits particularly in primiparous cows fed high RUP diets prepartum when data were subjected to parallel curve analysis. Diets fed by Robinson et al. (2001) increased total dietary protein from 11.7% to 14% using 0.16 kg/d of DDGS along with expeller soybean meal, blood meal, and corn gluten feed as high-RUP protein sources. The authors did not provide an estimate of diet RUP, but it would be similar to that supplied entirely by supplementing with DDGS alone. Though not examined in dairy cattle, research with beef heifers supplemented with RUP during late gestation increased pregnancy rates relative to controls (Patterson et al., 2003).

Much of the energy content in distillers grains is in the form of fat, of which the major fatty acid is linoleic acid. Data on feeding fat to dairy cows prepartum has demonstrated mixed results as summarized by Overton and Waldron (2004). Kronfeld's (1982) glucogenic/lipogenic theory of milk synthetic requirements indicates beneficial effects of supplemental dietary fat along with glucogenic precursors to prevent ketosis postpartum. Accordingly, when fat has been added to low-energy diets, beneficial results for postpartum health and production have been observed (Overton and Waldron, 2004). Further, the addition of linoleic acid in whole oilseeds to diets of periparturient dairy cattle has demonstrated beneficial effects on reproduction (Colazo et al., 2009). Research by Abdelqader et al. (2009b) indicates that fat contained in DDGS may have ruminal characteristics similar to that of oilseeds. Similar results to that observed by Colazo et al. (2009) may be attainable by feeding DDGS.

Of the metabolic disorders in dairy cattle, milk fever (hypocalcemia) ranks as the number one preventable disorder by prepartum dietary manipulation. Feeding diets low in potassium and calcium along with an acidifying agent have been demonstrated to prevent this disorder. Typical acidifying agents are sources of chlorine or sulfur. The goal of dietary manipulation using this strategy is to obtain a low or negative dietary cation–anion difference (DCAD), most commonly measured as molar proportions of (sodium plus potassium) minus (chloride plus sulfur). Distillers grains, compared with oilseed meals, typically contains lesser concentrations of sodium, calcium, and potassium and greater concentrations of sulfur. The DCAD of DDGS ranges from −15 to 0 MEq/cwt, and depends primarily on sulfur content. In contrast, the DCAD of soybean meal is approximately +30 MEq/cwt. Feeding DDGS prepartum, may provide a means of accomplishing a negative DCAD without the use of unpalatable anionic salts as acidifying agents.

Research describing the effects of feeding DDGS to nonlactating dairy cows is limited. Engle et al. (2008) examined feeding diets containing 40% DDGS (DM basis) during late gestation to beef cattle, hypothesizing that the combined effects of fat and RUP would provide reproductive benefits after calving. During the 2 years of the experiment, heifers fed DDGS had a greater positive BW change than heifers fed the control diets. Furthermore, a greater percentage of cows fed DDGS prepartum became pregnant compared with those fed the control diet (94% vs. 84%). These results led the authors to conclude that "prepartum diets containing DDGS, a source of fat and RUP, benefited pregnancy rates in well-maintained, primiparous beef heifers."

The only research that documents the inclusion of significant amounts (15% of diet DM) of distillers grains in dry dairy cow diets is work by Mpapho et al. (2007a, 2007b). Cows were fed DWGS from 28 days before calving until 70 DIM replacing corn grain, soybean meal, and extruded and expeller soybean meal from the control diet. Diets were balanced to be isonitrogenous and isocaloric. Dry matter intake, both pre- and postpartum did not differ for cows fed DWGS compared with the control diet. During the subsequent lactation, yields of milk, fat-corrected milk, milk components, and feed efficiency were similar for the two diets. The percentage of protein in milk was, however, increased for cows fed DWGS. During the prepartum period, concentrations of glucose, urea nitrogen, cholesterol, β-hydroxybutyrate, and nonesterified fatty acids in blood were not affected by treatments. Postpartum, however, cows fed DWGS had decreased concentrations of urea nitrogen, NEFA, and BHBA and tended to have greater concentration of glucose in blood than did cows fed the control diets. The authors concluded that feeding DWGS at 15% of the ration DM improved energy balance and resistance to ketosis and metabolic disorders postpartum as indicated by primary metabolic indicators in blood.

Though limited, existing research indicates beneficial effects of feeding DDGS during the dry period. Dairy cows fed DDGS prepartum exhibited a more positive energy balance and decreased metabolic indicators of stress and susceptibility to ketosis. Beef brood cows fed DDGS during the same period demonstrated improved reproductive performance. Clearly, there is great potential for utilization of distillers grains in diets of nonlactating dairy cows.

13.3.3 Feeding DDGS to Dairy Calves

Though the nutritional profile of DDGS indicates that its inclusion in starter diets would support favorable growth and development of calves, DDGS has not been widely examined as a feedstuff for pre- and postweaned calves. Traditional concentrates contain easily digestible carbohydrates that promote rumen development compared with low soluble carbohydrate and high fiber content in DDGS. The low amount of soluble carbohydrate in DDGS, however, may increase rumen pH compared with that observed by feeding traditional grains. Lesmeister and Heinrichs (2004) observed low rumen pH in calves fed concentrates alone. Furthermore, Ham et al. (1994) suggested that distillers grains can reduce the incidence of ruminal acidosis because it has (1) low starch concentration; (2) greater fiber content, which is digested more slowly; and (3) more fat, which increases energy. The impact of potential changes in rumen pH from feeding distillers grains in calf starters has not been fully evaluated, but would probably benefit calves fed limited amounts of forages.

The protein content of DDGS at 28% to 32% provides an ideal replacement for the traditional corn/soybean meal combination in calf starters. There may be concern, however, of lesser protein quality in DDGS for growing calves than with soybean meal-based diets. In calves fed corn and corn gluten meal up to 3 months of age, lysine has been shown to be the first limiting amino acid (Abe et al., 1997). Because DDGS has a similar amino acid profile as other corn products, low lysine concentrations may be a concern. No amino acid was defined as limiting for calves older than

12 weeks, as rumen microbial protein synthesis begins by that age. To the contrary, in calves less than 11 week old fed a corn, soybean meal-based diet, Abe et al. (1998) found methionine as the first limiting amino acid followed by lysine. These results suggest that DDGS combined with soybean meal may provide an ideal protein supplement for calves.

Most experiments using distillers grains solubles in calf starters occurred during the 1940s and 1950s. Research has validated the inclusion of DDGS up to 25% of DM in starter diets without negatively affecting calf growth (Lassiter et al., 1955). An experiment using 44 Jersey calves, in which distillers grains solubles from corn, rye, and sorghum replaced equal amounts of skim milk powder in starter feeds resulted in similar growth.

Data on feeding growing beef calves is more readily available and provides indications of possible effects of DDGS in dairy calves. Loy et al. (2008) fed DDGS at 0.21% or 0.81% of body weight (BW) to crossbred heifers ($n = 120$; average BW 265 kg) fed high-forage diets. At the lesser DDGS concentration, heifers gained more and were more efficient than those supplemented with dried rolled corn or dried rolled corn with corn gluten meal at the same amount. Feed efficiency (gain to feed) was improved when feeding the greater amount of DDGS. The authors calculated the TDN content of DDGS to be 18% to 30% greater than corn, which led them to conclude that in high-forage diets for growing calves, DDGS has greater energy value than corn.

Somewhat more relevant to feeding DDGS in dairy calf starter feeds is work by Reed et al. (2006) evaluating inclusion of DDGS as protein supplement in creep feeds for nursing Angus steer calves. Calves were fed supplements containing either 41% soybean meal, 26.25% wheat middlings, and 26.25% soybean hulls or 50% corn DDGS, 14% soybean meal, 14.25% wheat middlings, and 14.25% soybean hulls in two experiments. In the first experiment, there were no differences in DMI or nutrient digestibilities. Calves consuming DDGS had decreased acetate to propionate ratio, increased molar proportion of butyrate, and decreased molar proportions of isobutyrate and isovalerate in the rumen. In the second experiment, intake of organic matter as percent of BW was less for calves fed DDGS compared with control, but there were no differences in performance or subsequent carcass composition. The authors concluded that inclusion of 50% corn DDGS in a creep supplement for nursing calves produced similar results compared with a control creep feed based on soybean meal. Similarly, a 2-year study by Lancaster et al. (2007) that replaced soybean meal in creep feeds as a protein source with DDGS found no differences in growth or performance of steer calves, but did demonstrate a considerable cost advantage for feeding DDGS.

To this point, the most recent evaluation of corn DDGS in diets of nursing dairy calves evaluated inclusion of DDGS at 0%, 28%, or 56% of diet DM in an intensified growth feeding program (Thomas et al., 2006a, 2006b). From 2 days of age, calves were offered the experimental starter feeds for ad libitum consumption in addition to the daily milk replacer. The experimental starter feeds contained on average 22.4% CP, formulated by replacing cracked corn, rolled oats, and soybean meal with DDGS. As DDGS increased in starters, concentrations of NDF, ADF, and fat increased and concentration of starch decreased, allowing diets to remain isocaloric. Concentrations of phosphorus and sulfur increased to 0.61% and 0.38% of starter DM, respectively. As the percentage of inclusion of DDGS in the starters increased, lysine concentration decreased from 1.0% to 0.50% and methionine increased from 0.27% to 0.36% of DM. Calves were weaned at 6 weeks and fed starter feeds for ad libitum consumption until 12 week of age.

Although calves fed starter containing 56% DDGS had greater DMI after weaning at week 6 when compared with the 0% or 28% DDGS, BW did not differ throughout the experiment (Thomas et al., 2006a). Feed efficiencies of calves fed the 56% DDGS diet decreased slightly compared to calves fed 0% DDGS. The decreased feed efficiency may have been attributable to decreased concentrations of lysine in the starter as first limiting amino acid for support of lean body mass accretion. Development of the digestive tract and the rumen, specifically, was not different when comparing empty organ weights (Thomas et al., 2006b). Calves fed DDGS, however, exhibited shorter, wider, and denser rumen papillae with lesser total surface area indicating a shift in ruminal VFA patterns for calves fed DDGS. Though differences in pH were not observed, short papillae and mucosal proliferations are adaptive changes to low pH (Zitnan et al., 2005). Darker colors of the papillae, greater papillae density, and shorter papillae length of calves fed 56% DDGS may be indicative of parakeratosis, though health status of calves did not differ between treatments.

The effects on growth observed in the experiment described previously may be partially explained by work conducted by Vazquez-Anon et al. (1993). In their experiment rumen-cannulated calves were used to evaluate in situ rates of protein disappearance from the rumen of DDGS compared with soybean meal, heat-treated soybean meal, and corn gluten feed at 2, 4, and 8 weeks after weaning. During the early growth, rates of rumen disappearance of CP were not different for the feedstuffs tested. For soybean meal and corn gluten feed, however, rates of CP disappearance increased as calves aged, but not for heat-treated soybean meal or DDGS. In fact, rates of CP disappearance between cows and calves were similar for both DDGS and corn gluten feed. During the first 4 weeks after weaning, feedstuffs high in degradable protein have rumen degradation rates similar to soybean meal, but at 8 weeks postweaning rumen, degradation rates of DDGS remain low and do not change. Thus prior to rumen development, the relative degradability of DDGS has little impact on calf nutrition, and amino acid balance may be the primary determinant of protein quality of DDGS in dairy calf diets.

In conclusion, DDGS provides an excellent protein supplement in starter feeds for dairy calves. Feeding greater than 25% to 30% of DM as DDGS should be approached with caution. Data on feeding at greater concentrations is limited and there are indications of potential ruminal acidosis effects because of the highly fermentable fiber content of DDGS replacing forage fiber. Furthermore, the amino acid profile of DDGS may well serve to balance the low methionine content typical of soybean meal-based supplement while decreasing cost of gain.

13.3.4 Feeding DDGS to Dairy Heifers

The NRC (2001) suggests that BW gains for large breed dairy heifers should be around 0.8 kg/d as excessive weight gains caused by unbalanced rations may compromise the development of mammary tissue. Recent research by Daniels et al. (2009) suggests that greater weight gains can be achieved without compromising the development of mammary tissue if diets are properly balanced. In order to accomplish this with the inclusion of distillers grains, lower quality forages can be utilized to balance the diet. In this feeding scheme, distillers grains products complement high fiber forages because of the high concentration in energy in distillers grains products. Maintaining homogeneous mixes between dry forages and other dry feedstuffs is often challenging as smaller particles tend to separate and settle towards the bottom of the mixed ration. This leads to uneven intake of nutrients by growing heifers with resultant differences in growth. Inclusion of DWGS, due to their stickiness, reduces this problem and results in diets that maintain the intended nutrient concentration as consumed. Since DWGS provides more protein, fat, and P than is required by growing dairy heifers, a good match to feed with DWGS is low-quality, high fiber feeds such as crop residues. Corn stalks or small grain straws make excellent alternatives to high protein and high energy containing forages such as corn silage and alfalfa hay. When low-quality forages and DWGS are blended together at appropriate concentrations, the resultant feed can supply recommended nutrient concentrations. Research performed at SDSU with distillers grains in growing dairy heifers has been mostly done with DWGS for this reason. Its use has been evaluated in combination with other agricultural byproducts such as soybean hulls.

Anderson et al. (2009) conducted a study to determine if the dietary energy supplied as fermentable fiber and fat versus starch would result in similar performance in growing heifers. Distillers wet grains with solubles (DWGS) was ensiled with soybean hulls (SH) at a ratio of 70% DWGS to 30% SH on an as-fed basis. This blend (DWGSH) was used in heifer diets to replace corn and soybean meal. The DWGSH blend was stored for approximately 3.5 months in a sealed silage bag until using for the experiment. Heifers weighed 184 kg on average and were approximately 150 days old at the start of this experiment. Diets were (1) control, (2) low inclusion of DWGSH, and

(3) high inclusion of DWGSH. All treatment diets contained 50% brome grass hay on a dry basis. The control diet had 50% of the diet (dry basis) as grain mix, which was composed of corn, soybean meal, and minerals. The low inclusion diet contained 24.4% DWGSH blend and 25.6% grain mix whereas the high inclusion diet contained 48.7% DWGSH blend and 1.3% mineral mix. Diets were formulated to be isoenergetic, but based upon calculations using the analytically determined nutrient composition of feeds and the NRC (2001), the addition of the DWGSH blend slightly decreased the NE_g of the diet.

Dry matter intake decreased as the amount of the DWGSH blend increased in the diets, while average daily gain did not differ among diets (Anderson et al., 2009). As a result of lesser DMI and similar gains, the gain to feed ratio increased linearly as heifers were fed the diets containing the DWGSH blend. Initial, final, and calculated daily changes in growth were similar across diets for wither height, hip height, body length, and heart girth. Initial and final body condition score (BCS) were similar across diets, but when overall BCS changes were calculated there was a linear decrease as the DWGSH blend increased in the diet. The inclusion of the blend resulted in greater concentrations of NDF, ADF, and EE and lesser concentrations of nonfibrous carbohydrates and starch in diets.

In this study, DMI decreased linearly as the DWGSH blend replaced corn and soybeans in heifer diets (Anderson et al., 2009). One possible cause is the increase in dietary fiber content as the DWGSH blend increased in the diets. In a study by Quigley et al. (1986), DMI and the proportion of dietary NDF were related quadradically, with maximum DMI occurring at NDF concentrations between 40% and 44% of the diet DM. In the study by Anderson et al. (2009), NDF concentration in the high DWGSH diet was 54.8% on a dry matter basis, which is much greater than the concentration found for maximizing DMI, but the control and low DWGSH diets were 34.8% and 45.4% on a dry basis, respectively, which are closer to the NDF concentration observed to maximize DMI by Quigley et al. (1986).

The lower DMIs observed in Anderson et al. (2009), however, may not be as affected by the NDF concentrations as in the results of Quigley et al. (1986). In Anderson's study, the additional NDF was supplied by nonforage fiber sources, which do not contribute to rumen fill as much as the forage used to increase NDF in the study by Quigley et al. (1986). Another possible reason for the decrease in DMI is that dietary fat increased as DWGSH was added to the diets. In Anderson's study more than 50% of the fatty acid profile in the DWGSH blend was constituted by C18:2n-6. An increase in unsaturated dietary fat may have decreased DMI because of stimulation of satiety receptors by the fat (Allen et al., 2005).

Average daily gains in Anderson et al. (2009) were similar across diets at 1.25 kg/d, greater than recommended by the NRC (2001) for growing dairy heifers. This was more than likely a combination of underestimating both dietary energy concentration and DMI in the original diet formulations. Similar BW gains and decreased DMI, as the DWGSH blend increased in the diets, resulted in improved feed to gain ratios. Studies reviewed by Klopfenstein et al. (2008) have also found improved feed conversion in growing beef cattle as DWGS increased in the diet. There is speculation that more fat and protein in the DWGS bypass the rumen and are used to a greater extent in the small intestine. Conversely, corn and soybean particles are subject to greater degradation and fermentation in the rumen, resulting in a less efficient conversion of feed for growth (Klopfenstein et al. 2008).

Anderson et al. (2009) concluded that a 70:30 (as fed) blend of DWGS and SH when fed in replacement of one-half or all of a traditional concentrate mix to growing dairy heifers, maintained performance and improved feed efficiency.

13.4 FEEDING OTHER ETHANOL COPRODUCTS

In addition to DDGS, other ethanol coproducts have been evaluated in dairy cattle diets. Descriptions of each are described in the following sections.

13.4.1 Distillers Wet Grains or Distillers Modified Wet Grains

Distillers wet grains or "wet cake" are terms sometimes applied indiscriminately in the field to ethanol coproducts dried to a variable extent. When formulating diets for dairy cattle, interchanging coproducts with different moisture contents can lead to costly mistakes. Traditional DWGS contains 30% to 35% DM and is similar in nutrient composition to DDGS on a DM basis (Table 13.1). Distillers modified wet grains with solubles is distillers grains that has either undergone partial drying or has been completely dried to DDGS and has had CDS added to achieve a higher-moisture product. Dry matter of DMWGS is typically between 45% and 55% (Table 13.1). Nutrient composition on a dry basis is similar to that reported for DWGS and DDGS, but again it can vary depending on processing factors, especially the amount of solubles added back to make the final product (Cao et al., 2009). Nutrient composition of DMWGS can vary significantly from plant to plant; therefore, nutrient analysis is highly recommended prior to use in specific diets.

One of the advantages of DWGS and DMWGS over DDGS is that it helps condition overly dry rations. This oftentimes results in a more homogeneous DMI of nutrients and less feed sorting by cattle. Both forms of DWGS can be fed to growing and lactating dairy cattle. There needs to be caution though with potentially excessive weight gains and overconditioning when fed to growing heifers. Excessive weight gains not accompanied by corresponding frame development may result in fat deposits in mammary tissue. It has been suggested that fat infiltration of the udder may compromise milk production in the subsequent lactation and increase the incidence of postcalving metabolic disorders. It is important to limit the dietary energy DMI by restricting the amount of feed offered (limit feeding), by diluting it with low-energy feeds such as crop residues (i.e., straw), or both. High fibrous forages limit DMI by a rumen fill effect. On a dry basis, distillers grains provides more protein, fat, sulfur, and P than required by growing dairy heifers. Combining them with high fiber feeds such as corn stalks or straw balances the excesses of one with the deficits of the other.

A constraint with feeding DWGS and DMWGS is the variability in nutrient content that exists between and sometimes even within ethanol plants. Not only can there be variability in nutrient concentration but also differences in digestibility of individual nutrients. When heated in excess during drying, protein binds to carbohydrates (rendering them indigestible) through the Maillard reaction. This reaction is dependent on the temperature, length of time of the drying process, and moisture content of the feed. All three parameters differ to a variable extent between ethanol plants.

Hippen et al. (2003) evaluated the effects of increasing DWGS to dairy cow diets. Inclusions rates were 10%, 20%, 30%, or 40% of the diet DM replacing soybean meal, soybean hulls, and animal fat. Diet DM, DMI, and milk production decreased as DWGS in the diets increased. Milk component concentrations did not change with diets; however, milk fat yield, lactose, and urea nitrogen decreased when DWGS increased in the diets. In this experiment feeding DWGS at more than 20% of the diet DM depressed DMI and milk production. Feed efficiency, on the other hand, increased with the inclusion of DWGS. Milk protein and fat concentration were not affected by diets; however, they were inverted on all treatments. It is speculated that high dietary moisture from fermented feedstuffs could have negatively impacted rumen fermentation and milk fat production.

Anderson et al. (2006) compared the inclusion of DDGS or DWGS within the same lactation study. Both forms of distillers grains (from the same plant) were included at 0% (control), 10%, or 20% of the DM. Feed efficiency was calculated as energy-corrected milk/DMI and it was 1.70, 1.79, 1.87, 1.84, and 1.92 for control, 10% DDGS, 20% DDGS, 10% DWGS, and 20% DWGS, respectively. Cows fed both forms of distillers grains had better lactation performance than those fed the control diet, with a tendency of greater production from cows fed DWGS compared to DDGS. There were trends for the cows fed diets with 20% distillers grains to have higher milk protein yields as well as higher feed efficiency when compared to cows fed diets with 10% distillers grains. This again suggests that the optimum supplementation level with either DDGS or DWGS is somewhere between 10% and 20% of the diet DM. This optimum might be modified depending on the feeds that constitute the rest of the diet.

Al-Suwaiegh et al. (2002) analyzed the nutritional value of distillers grains from sorghum or corn fermentation, in both wet (35.4% DM) and dried (92.2% DM) form when fed to lactating dairy cows. Both types of distillers grains were fed at 15% of the ration DM in either wet or dried forms. The authors found no effect of the source or form of distillers grains on DMI, rumen pH, and VFA, or in situ digestion kinetics of NDF from distillers grains. Milk production efficiency was not affected by diets, and both forms of distillers grains resulted in similar performance.

Much of the research with distillers grains products in lactation diets are for limited periods of time, 4 to 10 weeks. Mpapho et al. (2005, 2006) fed DWGS at 15% of the diet DM of dairy cows for an entire lactation to evaluate production, DMI, and milk composition. The DWGS replaced a combination of corn grain, SBM, and extruded and expeller soybean meals in the control diet. In early lactation (22 to 105 DIM), diets were balanced for 17% CP and 1.63 Mcal/kg of NEL. During this time, there were no differences in DMI or milk production and components of cows fed DWGS or control indicating DWGS can be fed at 15% of the diet DM without affecting early lactation performance. During later lactation (106 DIM to dry off) diets were adjusted to 16% CP while maintaining 15% DWGS in the treatment diet. When total lactation performance was considered, cows fed DWGS had improved feed efficiency (1.57 vs. 1.30 kg fat-corrected milk/kg DMI for DWGS and control, respectively), increased milk fat (4.07% vs. 3.75% and 1.38 vs. 1.20 kg/d, percentage and yield, respectively), and increased milk protein (3.22% vs. 3.05% and 1.07 vs. 0.96 kg/d). Cows fed DWGS also had greater body weight gain (43.9 vs. 34.6 kg) and BCS (3.41 vs. 3.29) at the end of lactation. The results indicated that published energy values for DWGS may be underestimated and that lactation performance can be maintained or improved with feeding DWGS for an entire lactation.

13.4.2 Condensed Distillers Solubles

Condensed distillers solubles (CDS) is usually blended with distillers grains to make distillers grains with solubles, which is marketed as DWGS or DDGS. Condensed distillers solubles is a good source of protein and fat (Rust et al., 1990; Larson et al., 1993), and thus energy on a dry basis. Very few studies have been conducted evaluating the use of CDS in dairy cow diets. Milk production increased slightly when cows were fed CDS processed from a mash blend of 60% to 70% corn, 16% to 18% rye, and 12% to 14% barley (Udedibie and Chase, 1988). Huhtanen and Miettinen (1992) reported more protein but less fat in Finnish CDS than routinely observed in the United States, and they also reported slightly increased milk production.

Da Cruz et al. (2005) observed a tendency toward decreased DMI when cows were fed CDS at 0%, 5%, or 10% of total diet DM in substitution for a portion of rolled corn and soybean meal. Milk production was higher (34.1 vs. 35.5 kg/d) when CDS was fed at 5% of the diet compared to the control but there was no advantage when they were further increased to 10%. Huhtanen and Miettinen (1992) also observed increased production when cows were fed 5.9% of their diet DM as CDS, but no difference when CDS was raised to 17.5% of the ration. It is likely that the increased milk production was a result of the added fat as has been observed in the past (Palmquist and Jenkins, 1980). Milk fat profile was changed by the inclusion of CDS resulting in milk with higher concentrations of stearic and oleic acids (DaCruz et al., 2005). There were more unsaturated fatty acids in the milk of cows supplemented with 10% compared to 5% CDS, with a trend for more unsaturated fatty acids in CDS-supplemented diets compared with the control. Rumen acetate decreased in diets that contained CDS, and tended to be less for cows fed 10% CDS compared to 5% CDS. Butyrate concentration increased with increased CDS concentration in the diet. Lower acetate concentration in the rumen fluid in CDS-supplemented diets may be the result of long-chain unsaturated fatty acid inhibition of fiber digestion.

Da Cruz et al. (2005) concluded that CDS may be an economical source of energy and protein for lactating dairy cattle that can increase production, milk protein, and lactose. Although milk fat percentage was slightly decreased this was offset by the greater fat yield due to increased milk production. The authors further suggested a maximum concentration in the diet of 5% CDS as there were no added benefits of greater concentrations other than changes in the fatty acid composition.

Sasikala-Appukuttan et al. (2008) fed CDS and DDGS in total mixed rations of lactating dairy cows to evaluate the optimal amount to include in diets, and determine the feasibility of feeding CDS with or without DDGS. Their experimental diets were (1) 0% distillers grains products (control), (2) 18.5% DDGS, (3) 10% CCDS, (4) 20% CCDS, and (5) a combination diet of 18.5% DDGS with 10% CCDS. In diets 2 and 3 there was 2% fat from DDGS or CCDS, whereas diet 4 contained 4% fat from CDS and diet 5 contained 4% fat from the blend of DDGS and CCDS. Although there were no differences in DMI across diets, milk production tended to be greater for the diets that contained corn coproducts compared to control. Milk urea nitrogen and blood urea nitrogen were decreased in diets 2 to 5 compared with diet 1. Concentrations of long-chain fatty acids as well as polyunsaturated fatty acids in milk were greater and medium-chain fatty acid concentrations were lower for diets 2 to 5 compared with diet 1. Concentrations of *cis*-9, *trans*-11 conjugated linoleic acid as well as *trans*-10, *cis*-12 CLA were greater for diets 2 to 5 compared with diet 1. Results of rumen VFA molar proportions were similar to Da Cruz et al. (2005) for all diets that contained corn coproducts.

Sasikala-Appukuttan et al. (2008) concluded that CDS can replace soybean meal and corn grain up to 20% of the diet DM in the total mixed ration without adversely affecting milk production or DMI provided that the overall diet does not contain more than 7% total fat. One other concern that has not been addressed by research with CDS is that the P content of diets containing CDS would be high as CDS contains about 1.5% P (Table 13.1).

13.5 FEEDING FRACTIONATED ETHANOL COPRODUCTS

Recently, new ethanol coproducts have been developed by modifying the prefermentation or postfermentation processes in ethanol production. Refer to Chapters 5, and 25 for more information on fractionation processes in ethanol production. Use of these ethanol coproducts is described in the following sections.

13.5.1 REDUCED-FAT DDGS

There has been interest in removing fat from DDGS for use in biodiesel production or as feed grade fat source. One such strategy is solvent extraction of DDGS. The resulting coproduct, reduced-fat DDGS (RFDDGS), has a much lower crude fat concentration (3.5% of DM), but slightly greater concentrations of the remaining nutrients compared to conventional DDGS (Tables 13.1 and 13.4). Mjoun et al. (2010b) evaluated RUP with in situ methods and found that RUP was higher in RFDDGS compared to DDGS (60.4% vs. 52.3%). In feeding studies, Mjoun et al. (2010c) concluded that RFDDGS could successfully replace soy-based ingredients at inclusions of 10%, 20%, or 30% of diet DM. Cows had similar DMI and milk production across inclusion levels. Milk from cows fed 30% RFDDGS had the highest fat percentages, whereas milk from cows fed 10% and 20% had the greatest milk protein percentages. Mjoun et al. (2010a) also evaluated the inclusion of 20% RFDDGS and 22% DDGS in early lactation diets. In this experiment, cows fed either DDGS diet had similar DMI and milk production to cows fed soybean meal diets. Cows fed the DDGS diets produced milk higher in protein percentage and yield even though lysine was determined to be limiting. These studies concluded that RFDDGS are a good source of metabolizable amino acids and that, at 20% of the diet, RFDDGS did not limit milk or milk protein production.

13.5.2 FRACTIONATED DISTILLERS GRAINS PRODUCTS

With the development of new processes to increase efficiency of extraction of starch from grains for ethanol production, new coproducts are available as feedstuffs to dairy producers. In one such

TABLE 13.4
Chemical Composition (% of DM Unless Otherwise Noted) of Feed Products from Fractionation Technologies for Production of Ethanol and Coproducts

Nutrient	Product			
	Bran[a]	Germ[b]	HPDDG[c]	RFDDGS[d]
DM, % as is	90.3	94.1 ± 1.2[e]	92.1 ± 1.3[e]	86.9
CP	15.3	16.1 ± 1.0	43.4 ± 2.2	34.3
SP, % of CP	—	53.4 ± 1.5	7.63 ± 2.67	10.9
ADICP	0.30	0.33 ± 0.05	2.75 ± 0.95	4.5
NDF	21.4	26.2 ± 3.2	26.5 ± 2.6	43.8
ADF	7.36	9.26 ± 3.63	12.5 ± 4.4	12.7
Lignin	2.63	2.23 ± 0.83	2.99 ± 1.55	—
Starch	—	23.8 ± 2.48	9.60 ± 1.61	4.7
Crude fat	9.49	19.0 ± 1.1	4.00 ± 0.77	3.5
Ash	3.84	5.90 ± 0.24	2.13 ± 0.28	5.2
Ca	—	0.02 ± 0.01	0.02 ± 0.01	0.12
P	—	1.21 ± 0.10	0.44 ± 0.05	0.81
Mg	—	0.50 ± 0.02	0.12 ± 0.02	0.36
K	—	1.49 ± 0.06	0.42 ± 0.06	0.98
Na	—	0.01 ± 0.001	0.13 ± 0.04	—
S	—	0.17 ± 0.01	0.80 ± 0.05	0.78
NE_L, Mcal/kg[f]	1.89	2.27	1.98	1.58

[a] Dakota Bran, Poet Nutrition, Sioux Falls, SD. Compilation of values from Janicek et al. (2007), Tedeschi et al. (2009), and Dakota Bran, Poet Nutrition, Sioux Falls, SD.

[b] Dakota Germ Corn Germ Dehydrated, Poet Nutrition, Sioux Falls, SD. Compilation of values reported by Robinson et al. (2008), Abdelqader et al. (2009a), Abdelqader et al. (2009b), Kelzer et al. (2009), and Tedeschi et al. (2009).

[c] HPDDG = high-protein dried distillers grains. Dakota Gold HP Distillers Dried Grains. Poet Nutrition, Sioux Falls, SD. Compilation of values reported by Robinson et al. (2008), Abdelqader et al. (2009b), Kelzer et al. (2009), Mjoun et al. (2010b), Tedeschi et al. (2009), and Christen et al. (2010).

[d] RFDDGS = reduced-fat distillers dried grains. Verasun Energy, Aurora, SD. Compilation of values reported by Mjoun et al. (2010b, 2010c).

[e] Data is reported as means ± the standard deviation.

[f] Calculated from NRC (2001) at 3X maintenance.

example, corn is milled into several fractions prior to fermentation so that the resulting products can be redirected into different processing streams (Gibson and Karges, 2006). This fractionation results in products such as dehydrated corn germ, corn bran, and distillers dried grains with increased protein and decreased fat and fiber content (HPDDG, Table 13.4). These products may be distributed "as is" or further modified, such as by addition of syrup to the bran resulting in a product being marketed as bran cake. These are generally proprietary products and are specific to individual companies. Because of the technology used and specific characteristics in the nutrient content of each fraction, the composition of these streams may vary considerably and be quite different from that of traditional DDGS. For example, the bran is the seed coat of the corn kernel; therefore it will contain more fiber, but less protein. Germ is removed because it has low starch concentration but is rich in lipids and P. The resulting product from fermenting the remaining endosperm (where most of the starch is located) yields high-protein DDG that is low in fat, fiber, and P.

Other strategies resulting in coproducts with unique nutritional characteristics would include postfermentation modifications of DDGS. As already mentioned, one example of this strategy is solvent or mechanical extraction of oil from traditional DDGS for further biofuel production. Compared with traditional DDGS the resulting product, labeled "de-oiled" or "reduced-fat" DDGS, has a low fat content (3% to 4%) and correspondingly greater concentrations of all other nutrients.

13.5.2.1 Corn Germ

With fractionation technology, the germ-enriched fraction removed from the kernel during predistillation milling is available as a feedstuff, and referred to as maize germ dehydrated (AAFCO, 2006). During wet milling, and after steeping, germ and fiber fractions are removed by differences in density and particle size, respectively (Rausch and Belyea, 2006). During the dry milling process, however, the germ is not subjected to the steeping process and thus retains more soluble proteins, P, and a greater proportion of the starch as well as most of the fat from the kernel.

Corn germ from dry milling contains 22% to 30% NDF (Table 13.4) and 21% to 26% residual starch, making it a highly fermentable feedstuff. Accordingly, corn germ has the fastest rate of fermentation for intact substrate compared with bran or DDGS (Tedeschi et al., 2009). Abdelqader et al. (2009a) demonstrated greater DM degradation rates for germ compared with SBM and two types of DDGS. When feeding lactating cows increasing amounts of corn germ, Abdelqader et al. (2009c) predicted the energy content of germ to be 2.39 Mcal/kg of NEL compared with an NRC (2001) prediction of 2.27 Mcal/kg. The greater energy prediction in this feeding study compared with TDN estimations by the NRC likely resulted from greater digestibility of the fiber fraction compared with values predicted by NRC.

There are considerable losses of water soluble proteins during the steeping process when corn germ is produced during wet milling (Parris et al., 2006). Because germ produced from dry milling does not undergo steeping these soluble proteins are not lost. Therefore, although protein content of germ from wet or dry milling is similar at 14% to 16% DM, germ from dry milling contains greater amounts of trichloroacetic acid precipitable proteins in the form of albumin, globulin, and zein while germ from wet milling contains greater proportions of glutelin and NPN. The increased NPN presumably arises from protein degradation during steeping (Parris et al., 2006). Accordingly, the soluble protein content of corn germ from dry milling has been determined to be 52% to 55% (Robinson et al., 2008; Kelzer et al., 2009). With in situ methods, Abdelqader et al. (2009a) determined the RDP fraction of germ to be 71.8% compared with RDP of 52.0% to 55.6% for DDGS.

The fat content of dry corn kernels is mostly contained in their embryo or germ portion (Moreau et al., 2005). Thus the corn germ produced during wet milling contains 40% to 50% fat. The fat content of germ produced through the dry grinding process however is somewhat less at about 17% to 20%. The separation of corn components in the dry milling process is not as complete as in wet milling. Small portions of the pericarp and endosperm remain attached to the germ in dry milling, which decreases the fat concentration (Rausch and Belyea, 2006). The amount of fat in the germ is 5 to 7 times greater than in corn grain and approximately double that of corn DDGS regardless of the source. The major fatty acids in corn germ are similar to that of other coproducts and include C18:2, C18:1, and C16:0 with approximate concentrations of 56%, 28%, and 11% of total fatty acids, respectively (Abdelqader et al., 2009b, 2009c).

Milk production and fat yields increased when corn germ from dry milling was fed at 7% and 14% of diet DM (Abdelqader et al., 2009c); however, feeding at 21% decreased the concentration and yield of milk fat and tended to decrease DMI. In this experiment, the diet that contained 21% germ had a total fat concentration of 8% because of inclusion of a basal amount of fat to the diet in addition to the germ. The negative effects of feeding 21% germ more likely resulted from total dietary fat rather than excessive contribution of fat from corn germ alone (NRC, 2001). To more directly assess the effects of the fat contribution from germ on milk fat, Abdelqader et al. (2009b) fed cows diets that were isolipidic at 6% ether extract, with 2.5% supplemental fat from either ruminally inert fat (control), 14% corn germ, 30% DDGS, or 2.5% corn oil. The DMI was greater for diets containing germ (27.2 kg/d) than for the control diet (24.8 kg/d) but similar to those that contained DDGS or corn oil (26.2 kg/d). In this experiment, milk fat concentration was not decreased when corn germ was fed but was decreased with corn oil and tended to decrease with DDGS. Concentrations of trans-fatty acids and conjugated linoleic acids (CLA), in particular *cis*-9, *trans*-11 CLA, in milk fat were slightly, though not significantly, increased by feeding corn germ compared with the control diet though were significantly increased by feeding the DDGS or the corn oil diet. These results indicate that the fat in the corn germ from dry milling has a degree of ruminal "inertness" compared with that in traditional DDGS or free corn oil.

Because the germ from dry milling is not steeped, its mineral composition also differs from that of traditional DDGS (Table 13.4). Notably, concentrations of phosphorus and potassium are greater, and because the germ is not combined with solubles from the distillation process, concentrations of sulfur are much less. Mjoun et al. (2008), investigated ruminal phosphorus disappearance from corn germ, several soybean products, and DDGS, and determined that phosphorus in germ is slightly less water soluble than in DDGS but more so than for soy products (82.1%, 77.0%, and 45% for DDGS, germ, and soy products, respectively). Total ruminal disappearance of phosphorus however, was similar to that of DDGS at 93.5% and 93.3% for DDGS and germ, respectively. Corn germ from the dry milling process is a good source of phosphorus and potassium for dairy cows while providing lesser amounts of sulfur, which is frequently a concern when ethanol coproducts are fed.

An experiment comparing feeding corn germ to DDGS, and a high-protein DDG from BFRAC technology, all at 15% of diet DM, with a control diet containing soybean meal, demonstrated the greatest DMI and milk yield when cows were fed the diet containing corn germ (Kelzer et al., 2009). Rumen fermentation parameters did not differ between corn coproduct treatments; however, cows fed all corn coproducts had lower concentrations of acetate in rumen fluid than those fed the control diet. Data to date indicate that corn germ from dry milling may be fed to lactating dairy cattle at concentrations of at least 15% of DM. Furthermore, Tedeschi et al. (2009) concluded that when energy is limiting, germ would be a preferable supplement to DDGS in dairy cattle diets.

13.5.2.2 Corn Bran

In the traditional cornethanol production system, corn kernels are ground, cooked, and then fermented to produce ethanol. Corn bran is a coproduct of this technology, and is currently produced by adding CDS to the bran fraction of the kernel. Most of the fat and protein fractions are contributed by CDS whereas most of the fiber comes from the corn grain pericarp. Its high content of fibrous carbohydrates and very little starch makes corn bran a fit for ruminant diets. The chemical composition of corn bran described in recent research is presented in Table 13.4.

In vitro disappearance of the NDF fraction has been determined to be approximately 87% (DeHaan, 1983). This suggests that in spite of its high fiber content, the energy supplied by this carbohydrate fraction is high. One of the advantages of high fiber supplements such as corn bran is that, although highly digestible, their patterns of rumen fermentation shift towards more acetate rather than lactate and as a result do not acidify rumen contents as much, and are less conducive to negative associative effects. Lignin in corn bran has a range of values from 7.3% to 16.7% (Tedeschi et al., 2009), which might suggest significant variations in this energy content. Tedeschi et al. (2009) suggested that the most influential variables that affect the rate of degradation of NDF also affect the predicted TDN values.

The relatively low protein concentration of corn bran results in an advantage to nutritionists as the overall protein amino acid balance can be improved through the inclusion of other feeds with higher lysine concentration. Corn grain protein is composed of proteins soluble in alcohol (zein, albumins, and globulins) and proteins soluble in alkali (glutelin). The soluble protein zein comprises nearly 60%–70% of the protein in the endosperm and it has few to no lysine residues (Coleman and Larkins, 1999). During the ethanol fermentation process, most soluble proteins will be in the alcohol solution (Paulis et al., 1969) and thus provide a greater portion of the protein in CDS. A higher contribution to the total protein by CDS suggests that less glutelin (insoluble in alcohol) will be available, decreasing its contribution to lysine concentrations. Germ protein has higher lysine content (5.4%) than the whole grain (2.7%). The reason for this is that dilution by abundant endosperm proteins having only 1.9% lysine on average. This suggests that the protein in corn bran has lower lysine concentrations, which needs to be taken into account at higher inclusion levels and when milk production amounts require limiting amino acids to be considered.

The fat concentration in corn bran is probably the first limiting factor for inclusion in dairy cow diets. Tedeschi et al. (2009) found that ether extract was the variable with the highest influence on total digestible nutrients (TDN). For each standard deviation (SD) change in EE, TDN of corn bran varied by 0.52 SD units, respectively. Due to similar fat concentrations as DDGS the inclusion of bran should be similar to that recommended for DDGS. When both DDGS and corn bran are included in the diet their combination should probably not exceed 20% of the diet DM to avoid milk fat depression. This is supported by results from Janicek et al. (2007) who fed diets to lactating dairy cows that included corn bran at 10%, 17.5%, and 25% of DM replacing corn silage and alfalfa. Although milk fat decreased by 0.26% when corn bran was increased from 10% to 25% of the diet DM total fat yield was unaffected. Corn bran increased milk protein by 0.12 kg/d when its concentration in the diet DM was increased from 10% to 25%. Milk yield also tended to increase, but no differences were observed on 3.5% fat-corrected milk. One important aspect of their findings was that feed conversion improved with the inclusion of corn bran in the diet reaching 1.55 ± 0.05 kg of milk/kg of feed at 25% inclusion.

Inclusion of corn bran in dairy cattle diets will be limited by the total fat present in the diet. Its high fiber content together with the unfavorable amino acid profile suggests that it should be limited to diets for growing animals with functioning rumens. It is possible that because of its high fat content, this product might undergo lipid oxidation after prolonged storage periods and possibly develop some palatability issues.

13.5.2.3 High Protein DDG

As previously described, high-protein distillers grains (HPDDG) is a DDG product from the prefermentation fractionation process. One particular HPDDG product has been evaluated in several dairy feeding studies. The chemical composition is presented in Table 13.4. This HPDDG does not have solubles (CDS) added back into the final product. Consequently, it is considerably higher in CP, but lower in soluble protein, phosphorus, and sulfur compared to conventional DDGS. Phosphorus concentrations are typically 0.41% of DM in HPDDG compared to 0.88% of DM in conventional DDGS. Because the germ is removed prefermentation, crude fat of the remaining DDG product is considerably lower (4.35% of DM) compared to conventional DDGS. Two studies determined RUP using multiple timepoint in situ methods. Both Abdelqader et al. (2009a) and Mjoun et al. (2010b) calculated the RUP of HPDDG at 55.1% whereas RUP in conventional DDGS was calculated at 52.1%.

High-protein DDG has been evaluated in three lactating dairy cow feeding studies. Hubbard et al. (2009) evaluated the inclusion of 20% HPDDG as replacement for soybean meal and soybean expeller meal. Cows fed the HPDDG diet produced greater milk, fat, and protein yields than cows fed the control, soybean-based diet. In addition, cows fed HPDDG had greater feed efficiency (milk/DMI) compared to control-fed cows. Kelzer et al. (2009) compared the inclusion of 14.4% HPDDG to a control soybean-based diet, as well as a 15% conventional DDGS-based diet. In this study, cows fed HPDDG produced similar to cows fed the soy-based control or the DDGS-based diets. In a third study, Christen et al. (2010) compared HPDDG at 12% of diet DM to three other protein supplement diets: soybean meal, canola meal, or DDGS. Each supplement provided 38% of the protein fed in each diet. Diets were formulated to be deficient in CP (15.0% to 15.6% CP) to determine if amino acids provided by each supplement were limiting to milk production. Regardless of the supplement fed, cows had similar DMI and milk production. Fat and protein concentrations in milk of cows fed HPDDG were similar to that from cows fed soybean meal, but higher than for cows fed DDGS. Although lysine was determined to be the first limiting amino acid for HPDDG, as with DDGS, it was concluded that HPDDG can successfully replace soybean meal and canola meal without reducing performance of lactating dairy cows.

13.6 PRACTICAL APPLICATION OF DDGS USE IN DAIRY CATTLE DIETS

13.6.1 REPLACEMENT OF OTHER FEEDSTUFFS WITH DDGS

As described in the previous sections on feeding, DDGS has been included in dairy cow diets as a source of protein, energy, and minerals. Replacement of other ingredients is highly dependent on the ingredients included in the dairy cow's diet. Traditionally, DDGS has replaced soybean meals (both traditional and expeller types) in dairy cow diets because of its value as a source of RUP (Pamp et al., 2006). More recently, DDGS has been valued in the marketplace similar to corn as a replacement for energy. As a result, greater emphasis has been placed upon inclusion of DDGS as a replacement for corn (Ranathunga et al., 2010). When considered as a mineral source, DDGS is high in phosphorus. While this can be an environmental concern if dietary phosphorus levels exceed the animal's requirement, phosphorus in DDGS has been shown to be highly available and can serve as a replacement for more expensive inorganic sources (Mjoun et al., 2007).

Since there is no one ingredient equal to distillers grains in chemical composition, all ingredient replacements in diets need to take into account changes in dietary nutrient profile. Therefore, if DDGS is included into the diet primarily as protein source, DDGS can be a replacement for a combination of soybean meal, soybean expeller meal, soybean hulls, and a fat source. On the other hand, if DDGS is included primarily as an energy source, up to 80% of its inclusion level can replace corn, while the remaining 20% may be soybean products. In most cases, one part DDGS will be typically replaced by 0.50 to 0.70 parts soybean feedstuffs and 0.30 to 0.50 parts corn or other byproducts feeds. Inclusion of DDGS is dependent on price of feedstuffs and balancing for the nutrient requirements of the dairy animal. As noted in the meta-analysis (Table 13.2), the conversion of feed into milk tends to increase 2% to 5% as DDGS replaces traditional feeds, such as corn and soybean feedstuffs, in dairy cow diets. This indicates there is greater value in DDGS in dairy cow diets compared to the feed it replaces.

13.6.2 CHALLENGES WITH DDGS

Research has shown that distillers grains products can effectively be included in dairy cattle diets. For practical application in the field, however, nutritionists and producers may face some challenges when including distillers grains in dairy cattle diets. These challenges include:

- Nutrient composition variability: Nutrient composition can vary considerably among ethanol plants. These differences can be attributed to factors such as the grain type, grain quality, milling process, fermentation process, drying temperature, and amount of solubles blended back to distillers wet grains before drying. Lack of adjustment for changes in coproduct nutrient composition results in diets not formulated as intended. These changes in nutrient composition can result in reduced DMI, milk production losses, and altered milk composition. It is highly recommended to obtain a nutrient composition profile of the distillers grains products intended for use in dairy cattle diets.
- Protein and amino acid quality: As interest in amino acid formulation continues to increase, the quality of protein and amino acids provided by distillers grains products becomes a concern. With inclusion of distillers grains products, lysine is predicted to be the first limiting amino acid. In most cases, feeding up to 20% distillers grains products does not reduce milk and milk protein production; however lysine adequacy in the diet needs to be maintained.
- Physically effective fiber: Even though distillers grains products are sources of dietary NDF, they do not effectively replace forage NDF in lactating dairy cow diets. Reducing physically effective fiber in diets alters normal rumen fermentation potentially resulting in milk fat depression. While distillers grains products may replace starch and byproduct fiber in dairy cow diets, they do not replace forage NDF.

- Polyunsaturated fatty acids (PUFA): Distillers grains products are proportionally high in PUFA, which can potentially alter normal rumen function; however it has been shown that if physically effective fiber is maintained in lactating dairy cow diets, milk fat depression is typically alleviated.
- High mineral concentrations: Some distillers grains products are high in available phosphorus. These products can be an inexpensive source of phosphorus when supplementation is required. Concentrations of phosphorus in feedstuffs, however, may be a concern in regions where soil phosphorus concentrations are already high. Distillers grains products can effectively replace inorganic mineral supplements as a source of phosphorus in dairy cattle diets. Finally, because some distillers grains products are high in sulfur, variations in the concentration of this mineral should be monitored to prevent potential health risks by feeding excessive sulfur.

13.7 CONCLUSIONS

Distillers grains with or without solubles is an excellent source of protein and energy for dairy cattle. Research suggests that distillers grains can replace more expensive sources of protein, energy, and minerals in dairy cow diets. Because distillers grains can be highly variable, it is recommended that sources be tested to determine precise nutrient compositions to properly formulate diets. When balancing diets with distillers grains, care must be taken to provide sufficient physically effective fiber to maintain normal rumen function and prevent milk fat depression in lactating cows. Nitrogen and phosphorus concentrations in distillers grains-based diets also need to be monitored to prevent excessive losses to the environment.

Recommended maximum inclusion levels for DDGS and DWGS are summarized in Table 13.5. Maximum inclusion levels of distillers grains for preweaned calves, growing heifers and dry cows are 25%, 30%, and 15% of the diet on a DM basis, respectively. Current recommendations for feeding distillers grains to lactating dairy cows would be to include it up to 20% and 15% of the diet DM for DDGS and DWGS, respectively. Diets with greater than 10% of the diet as DDGS or DWGS should be formulated using sound nutritional principles for dairy cattle respective of nutrient requirements. As technology improves, new ethanol coproducts will be developed and become

TABLE 13.5
Recommended Maximum Concentration for DDGS and DWGS for Dairy Cattle

	DDGS	DWGS
Preweaned calves[a]	25	—
Growing heifers/steers[b]	30	30
Far-off/close-up dry cows[c]	15	15
Cows in lactation[e]	20	15

Note: Inclusion of distillers grains is dependent on proper nutrient formulation and high levels of inclusion are not recommended if nutrients are fed in excess of requirements.

[a] Critical nutrient to formulate for in calf diets: Lysine, fiber, and fat.

[b] Critical nutrients to formulate for in growing heifers/steer diets: Fat/energy and sulfur.

[c] Critical nutrients to formulate for in far-off/close-up dry cow diets: Fat/energy, sulfur, calcium/phosphorus, and overall moisture with inclusion of other wet feeds.

[d] Critical nutrients to formulate for in lactating cow diets: Polyunsaturated fatty acid/total fat, physically effective fiber, sulfur, calcium/phosphorus, RUP, lysine, and overall moisture with inclusion of other wet feeds.

available to livestock producers. These new distillers grains products need to be examined individually with consideration of their unique nutritional profiles for optimal economic inclusion in diets of dairy cattle. Distillers grains is increasingly being used in monogastric diets as well, as discussed in the following chapter on swine.

REFERENCES

AAFCO. 2006. *Official Publication*. West Lafayette, IN. Association of American Feed Control Officials.

Abdelqader, M. M., J. L. Anderson, A. R. Hippen, D. J. Schingoethe, and K. F. Kalscheur. 2009a. In situ ruminal degradability of dry matter and protein from corn germ, distillers grains, and soybean meal. *Journal of Dairy Science* 92(5): 2362. (Abstr.)

Abdelqader, M. M., A. R. Hippen, K. F. Kalscheur, D. J. Schingoethe, and A. D. Garcia. 2009b. Isolipidic additions of fat from corn germ, corn distillers grains, or corn oil in dairy cow diets. *Journal of Dairy Science* 92: 5523–5533.

Abdelqader, M. M., A. R. Hippen, K. F. Kalscheur, D. J. Schingoethe, K. Karges, and M. L. Gibson. 2009c. Evaluation of corn germ from ethanol production as an alternative fat source in dairy cow diets. *Journal of Dairy Science* 92: 1023–1037.

Abe, M., T. Iriki, and M. Funaba. 1997. Lysine deficiency in postweaned calves fed corn and corn gluten meal diets. *Journal of Animal Science.* 75: 1974–1982.

Abe, M., T. Iriki, M. Funaba, and S. Onda. 1998. Limiting amino acids for a corn and soybean meal diet in weaned calves less than three months of age. *Journal of Animal Science* 75: 628–636.

Allen, M. S., B. J. Bradford, and K. J. Harvatine. 2005. The cow as a model to study food intake regulation. *Annual Review of Nutrition* 25: 523–547.

Al-Suwaiegh, S., K. C. Fanning, R. J. Grant, C. T. Milton, and T. J. Klopfenstein. 2002. Utilization of distillers grains from the fermentation of sorghum or corn in diets for finishing beef and lactating dairy cattle. *Journal of Animal Science* 80: 1105–1111.

Anderson, J. L., K. F. Kalscheur, A. D. Garcia, D. J. Schingoethe, and A. R. Hippen. 2009. Ensiling characteristics of wet distillers grains mixed with soybean hulls and evaluation of the feeding value for growing Holstein heifers. *Journal of Animal Science* 87: 2113–2123.

Anderson, J. L., D. J. Schingoethe, K. F. Kalscheur, and A. R. Hippen. 2006. Evaluation of dried and wet distillers grains included at two concentrations in the diets of lactating dairy cows. *Journal of Dairy Science* 89: 3133–3142.

Batajoo, K. K., and R. D. Shaver. 1998. In situ dry matter, crude protein, and starch degradabilities of selected grains and by-product feeds. *Animal Feed Science and Technology* 71: 165–176.

Bauman, D. E., and J. M. Griinari. 2001. Regulation and nutritional manipulation of milk fat: Low-fat milk syndrome. *Livestock Production Science* 70: 15–29.

Bauman, D. E., and J. M. Griinari. 2003. Nutritional regulation of milk fat synthesis. *Annual Review of Nutrition* 23: 203–227.

Belyea, R. L., B. J. Steevens, R. J. Restrepo, and A. P. Clubb. 1989. Variation in composition of by-product feeds. *Journal of Dairy Science* 72: 2339–2345.

Birkelo, C. P., M. J. Brouk, and D. J. Schingoethe. 2004. The energy content of wet corn distillers grains for lactating dairy cattle. *Journal of Dairy Science* 87: 1815–1819.

Boucher, S. E., S. Calsamiglia, C. M. Parsons, H. H. Stein, M. D. Stern, P. S. Erickson, P. L. Utterback, and C. G. Schwab. 2009. Intestinal digestibility of amino acids in rumen-undegraded protein estimated using a precision-fed cecectomized rooster bioassay: II. Distillers dried grains with solubles and fish meal. *Journal of Dairy Science* 92: 6056–6067.

Cao, Z. J., J. L. Anderson, and K. F. Kalscheur. 2009. Ruminal degradation and intestinal digestibility of dried or wet distillers grains with increasing concentrations of condensed distillers solubles. *Journal of Animal Science* 87: 3013–3019.

Christen, K. A., D. J. Schingoethe, K. F. Kalscheur, A. R. Hippen, K. Karges, and M. L. Gibson. 2010. Response of lactating dairy cows to high protein distillers grains or three other protein supplements. *Journal of Dairy Science* 93: 2095–2104.

Clark, P. W., and L. E. Armentano. 1993. Effectiveness of neutral detergent fiber in whole cottonseed and dried distillers grains compared to alfalfa haylage. *Journal of Dairy Science* 76: 2644–2650.

Colazo, M. G., A. Hayirli, L. Doepel, and D. J. Ambrose. 2009. Reproductive performance of dairy cows is influenced by prepartum feed restriction and dietary fatty acid source. *Journal of Dairy Science* 92: 2562–2571.

Coleman, C. E., and B. A. Larkins. 1999. Prolamins of maize. In *Seed Proteins*, eds. P. R. Shewry, and R. Casey, pp. 109–139. The Netherlands: Kluwer Academic Publishers.

Cyriac, J., M. M. Abdelqader, K. F. Kalscheur, A. R. Hippen, and D. J. Schingoethe. 2005. Effect of replacing forage fiber with non-forage fiber in lactating dairy cow diets. *Journal of Dairy Science* 88(Suppl. 1): 252 (Abstr.).

Da Cruz, C. R., M. J. Brouk, and D. J. Schingoethe. 2005. Lactational response of cows fed condensed corn distillers solubles. *Journal of Dairy Science* 88: 4000–4006.

Daniels, K. M., M. L. McGilliard, M. J. Meyer, M. E. Van Amburgh, A. V. Capuco, and R. M. Akers. 2009. Effects of body weight and nutrition on histological mammary development in Holstein heifers. *Journal of Dairy Science* 92: 499–505.

DeHaan, K. A. 1983. Improving the utilization of fiber and energy through the use of corn gluten feed and alkali compounds. Ph.D. dissertation, University of Nebraska, Lincoln.

Engel, C. L., H. H. Patterson, and G. A. Perry. 2008. Effect of dried corn distillers grains plus solubles compared with soybean hulls, in late gestation heifer diets, on animal and reproductive performance. *Journal of Animal Science* 86: 1697–1708.

Elliot, J. P., J. K. Drackley, D. J. Schauff, and E. H. Jaster. 1993. Diets containing high oil corn and tallow for dairy cows during early lactation. *Journal of Dairy Science* 76: 775–789.

Gibson, M. L., and K. Karges. 2006. By-products from non-food agriculture: Technicalities of nutrition and quality. In *Recent Advances in Animal Nutrition*, eds. P. C. Garnsworthy, and J. Wiseman, pp. 209–227. Nottingham, UK: Nottingham University Press.

Griinari, J. M., D. A. Dwyer, M. A. McGuire, D. E. Bauman, D. L. Palmquist, and K. V. Nurmela. 1998. Trans-octadecenoic acids and milk depression in lactating dairy cows. *Journal of Dairy Science* 81: 1251–1261.

Grings, E. E., R. E. Roffler, and D. P. Deitelhoff. 1992. Responses of dairy cows to additions of distillers dried grains with solubles in alfalfa-based diets. *Journal of Dairy Science* 75: 1946–1952.

Ham, G. A., R. A. Stock, T. J. Klopfenstein, E. M. Larson, D. H. Shain, and R. P. Huffman. 1994. Wet corn distillers by-products compared with dried corn distillers grains with solubles as a source of protein and energy for ruminants. *Journal of Animal Science* 72: 3246–3257.

Hippen, A. R., K. N. Linke, K. F. Kalscheur, D. J. Schingoethe, and A. D. Garcia. 2003. Increased concentrations of wet corn distillers grains in dairy cow diets. *Journal of Dairy Science* 86(Suppl. 1): 340. (Abstr.)

Hippen, A. R., K. F. Kalscheur, D. J. Schingoethe, and A. D. Garcia. 2004. Increasing inclusion of dried corn distillers grains in dairy cow diets. *Journal of Dairy Science* 87(6): 1965. (Abstr.)

Hippen, A. R., D. J. Schingoethe, K. F. Kalscheur, P. Linke, D. R. Rennich, M. M. Abdelqader, and I. Yoon. 2010. *Saccharomyces cerevisiae* fermentation product in dairy cow diets containing dried distillers grains plus solubles. *Journal of Dairy Science* 93: 2661–2669.

Hubbard, K. J., P. J. Kononoff, A. M. Gehman, J. M. Kelzer, K. Karges, and M. L. Gibson. 2009. The effect of feeding high protein dried distillers grains on milk production. *Journal of Dairy Science* 92: 2911–2914.

Huhtanen, P., and H. Miettinen. 1992. Milk production and concentrations of blood metabolites as influenced by the level of wet distillers solubles in dairy cows receiving grass silage-based diet. *Agricultural Science in Finland* 1: 279–290.

Huyler, M. T., R. L. Kincaid, and D. F. Dostal. 1999. Metabolic and yield responses of multiparous Holstein cows to prepartum rumen-undegradable protein. *Journal of Dairy Science* 82: 527–536.

Janicek, B. N., P. J. Kononoff, A. M. Gehman, and P. H. Doane. 2008. The effect of feeding dried distillers grains plus solubles on milk production and excretion of urinary purine derivatives. *Journal of Dairy Science* 91: 3544–3553.

Janicek, B. N., P. J. Kononoff, A. M. Gehman, K. Karges, and M. L. Gibson. 2007. *Short Communication:* Effect of increasing levels of corn bran on milk yield and composition. *Journal of Dairy Science* 90: 4313–4316.

Kalscheur, K. F. 2005. Impact of feeding distillers grains on milk fat, protein, and yield. Proceedings Distillers Grains Technology Council, 9th Annual Symposium, Louisville, KY.

Kalscheur, K., A. Garcia, K. A. Rosentrater, and C. Wright. 2008. Ethanol coproducts for ruminant livestock diets. *Fact Sheet 947*. Brookings, SD: South Dakota State University, Cooperative Extension Service, Available online: agbiopubs.sdstate.edu/articles/FS947.pdf. Accessed February 10, 2010.

Kelzer, J., M. P. J. Kononoff, A. M. Gehman, L. O. Tedeschi, K. Karges, and M. L. Gibson. 2009. Effects of feeding three types of corn-milling co-products on milk production and ruminal fermentation of lactating Holstein cattle. *Journal of Dairy Science* 92: 5120–5132.

Kleinschmit, D. H., D. J. Schingoethe, K. F. Kalscheur, and A. R. Hippen 2006. Evaluation of various sources of corn distillers dried grains plus solubles for lactating dairy cattle. *Journal of Dairy Science* 89: 4784–4794.

Kleinschmit, D. H., J. L. Anderson, D. J. Schingoethe, K. F. Kalscheur, and A. R. Hippen. 2007. Ruminal and intestinal degradability of distillers grains plus solubles varies by source. *Journal of Dairy Science* 90: 2909–2918.

Klopfenstein, T. J., G. E. Erickson, and V. R. Bremer. 2008. Board invited review: Use of distillers by-products in the beef cattle feeding industry. *Journal of Animal Science* 86: 1223–1231.

Kronfeld, D. S. 1982. Major metabolic determinants of milk volume, mammary efficiency, and spontaneous ketosis in dairy cows. *Journal of Dairy Science* 65: 2204–2212.

Lancaster, P. A., J. B. Corners, L. N. Thompson, M. R. Ellersieck, and J. E. Williams. 2007. Effects of distillers dried grains with solubles as a protein source in a creep feed. 1. Suckling calf and dam performance. *Professional Animal Scientist* 23: 83–90.

Larson, E. M., R. A. Stock, T. J. Klopfenstein, M. H. Sindt, and R. P. Huffman. 1993. Feeding value of wet distillers by-products for finishing ruminants. *Journal of Animal Science* 71: 2228–2236.

Lassiter, C. A., D. M. Seath, R. F. Elliott, and G. M. Bastin. 1955. Use of distillers' grain solubles in calf starters. Bulletin no. 623, Kentucky, Agricultural Experiment Station.

Leonardi, C., S. Bertics, and L. E. Armentano. 2005. Effect of increasing oil from distillers grains or corn oil on lactation performance. *Journal of Dairy Science* 88: 2820–2827.

Lesmeister, K. E., and A. J. Heinrichs. 2004. Effects of corn processing on growth characteristics, rumen development, and rumen parameters in neonatal dairy calves. *Journal of Dairy Science* 87: 3439–3450.

Liu, C., D. J. Schingoethe, and G. A. Stegeman. 2000. Corn distillers grains versus a blend of protein supplements with or without ruminally protected amino acids for lactating cows. *Journal of Dairy Science* 83: 2075–2084.

Loosli, J. K., K. L. Turk, and F. B. Morrison. 1952. The value of distillers feed for milk production. *Journal of Dairy Science* 35: 868–873.

Loy, T. W., T. J. Klopfenstein, G. E. Erickson, C. N. Macken, and J. C. MacDonald. 2008. Effect of supplemental energy source and frequency on growing calf performance. *Journal of Animal Science* 86: 3504–3510.

Mjoun, K., K. F. Kalscheur, A. R. Hippen, and D. J. Schingoethe. 2008. Ruminal phosphorus disappearance from corn and soybean feedstuffs. *Journal of Dairy Science* 91: 3938–3946.

Mjoun, K., K. F. Kalscheur, A. R. Hippen, and D. J. Schingoethe. 2010a. Performance and amino acid utilization of early lactating cows fed regular or reduced-fat dried distillers grains with solubles. *Journal of Dairy Science* 93: 3176–3191.

Mjoun, K., K. F. Kalscheur, A. R. Hippen, and D. J. Schingoethe. 2010b. Ruminal degradability and intestinal digestibility of protein and amino acids in soybean and corn distillers grains products. *Journal of Dairy Science* 93: 4144–4154.

Mjoun, K., K. F. Kalscheur, A. R. Hippen, D. J. Schingoethe, and D. E. Little. 2010c. Lactation performance and amino acid utilization of cows fed increasing amounts of reduced-fat dried distillers grains with solubles. *Journal of Dairy Science* 93: 288–303.

Mjoun, K., K. F. Kalscheur, B. W. Pamp, D. J. Schingoethe, and A. R. Hippen. 2007. Phosphorus utilization in dairy cows fed increasing amounts of dried distillers grains with solubles. *Journal of Dairy Science* 90(Suppl. 1): 451. (Abstr.)

Moorby, J. M., J. Dewhurst, and S. Marsden. 1996. Effect of increasing digestible undegraded protein supply to dairy cows in late gestation on the yield and composition of milk during the subsequent lactation. *Animal Science* 63: 201–213.

Moreau, R. A., D. B. Johnston, and K. B. Hicks. 2005. The influence of moisture content and cooking on the screw pressing and prepressing of corn oil from corn germ. *Journal of the American Oil Chemists Society* 82: 851–854.

Mpapho, G. S., A. R. Hippen, K. F. Kalscheur, and D. J. Schingoethe. 2005. Long term feeding of wet corn distillers grains and lactation performance of dairy cows. *Journal of Dairy Science* 88 (Suppl. 1): 394. (Abstr.)

Mpapho, G. S., A. R. Hippen, K. F. Kalscheur, and D. J. Schingoethe. 2006. Lactational performance of dairy cows fed wet corn distillers grains for the entire lactation. *Journal of Dairy Science* 89: 1811 (Abstr.)

Mpapho, G. S., A. R. Hippen, K. F. Kalscheur, and D. J. Schingoethe. 2007a. Production responses of dairy cows fed wet distillers grains during the transition period and early lactation. *Journal of Dairy Science* 90: 3080. (Abstr.)

Mpapho, G. S., A. R. Hippen, K. F. Kalscheur, and D. J. Schingoethe. 2007b. Blood metabolites profiles of dairy cows fed wet corn distillers grains during early lactation. *Journal of Dairy Science* 90(Suppl. 1): 351. (Abstr.)

National Research Council. 1989. *Nutrient Requirements of Dairy Cattle*, 6th rev. ed. Washington, DC: Natl. Acad. Sci.

National Research Council. 2001. *Nutrient Requirements of Dairy Cattle*, 7th rev. ed. Washington, DC: Natl. Acad. Sci.

Nichols, J. R., D. J. Schingoethe, H. A. Maiga, M. J. Brouk, and M. S. Piepenbrink. 1998. Evaluation of corn distillers grains and ruminally protected lysine and methionine for lactating dairy cows. *Journal of Dairy Science* 81: 482–491.

Noll, S. L., J. Brannon, and C. Parsons. 2007. Nutritional value of corn distiller dried grains with solubles (DDGs): Influence of solubles addition. *Poultry Science* 86(Suppl. 1): 68. (Abstr.)

Nyoka, R., A. R. Hippen, and K. F. Kalscheur. 2007. Supplementation of grazing dairy cows with high-fat dietary protein sources. *Journal of Dairy Science* 90(Suppl. 1): 103. (Abstr.)

Overton, T. R., and M. R. Waldron. 2004. Nutritional management of transition dairy cows: Strategies to optimize metabolic health. *Journal of Dairy Science* 87(E. Suppl.): E105–E119.

Owen, F. G., and L. L. Larson. 1991. Corn distillers dried grains versus soybean meal in lactation diets. *Journal of Dairy Science* 74: 972–979.

Owens, T. M., A. R. Hippen, K. F. Kalscheur, D. J. Schingoethe, D. L. Prentice, and H. B. Green. 2009. High-fat or low-fat distillers grains with dry or high-moisture corn in diets containing monensin for dairy cows. *Journal of Dairy Science* 92(E-Suppl. 1): 377. (Abstr.)

Palmquist, D. L., and H. R. Conrad. 1982. Utilization of distillers dried grains plus solubles by dairy cows in early lactation. *Journal of Dairy Science* 65: 1729–1741.

Palmquist, D. L., and T. C. Jenkins. 1980. Fat in lactation rations: Review. *Journal of Dairy Science* 63: 1–14.

Pamp, B. P., K. F. Kalscheur, A. R. Hippen, and D. J. Schingoethe. 2006. Evaluation of dried distillers grains versus soybean protein as a source of rumen-undegraded protein for lactating dairy cows. *Journal of Dairy Science* 89(Suppl. 1): 403. (Abstr.)

Parris, N., R. A. Moreau, D. B. Johnston, V. Singh, and L. C. Dickey. 2006. Protein distribution in commercial wet- and dry-milled corn germ. *Journal of Agricultural and Food Chemistry* 54: 4868–4872.

Patterson, H. H., D. C. Adams, T. J. Klopfenstein, R. T. Clark, and B. Teichert. 2003. Supplementation to meet the metabolizable protein requirements of primiparous beef heifers: II. Pregnancy and economics. *Journal of Animal Science* 81: 563–570.

Paulis, J. W., C. James, and J. S. Wall. 1969. Comparison of glutelin proteins in normal and high-lysine corn endosperms. *Journal of Agricultural and Food Chemistry* 17: 1301–1305.

Pickett, M. M., T. W. Cassidy, P. R. Tozer, and G. A. Varga. 2003a. Effect of prepartum dietary carbohydrate source and monensin on dry matter intake, milk production and blood metabolites of transition dairy cows. *Journal of Dairy Science* 86(Suppl. 1): 10. (Abstr.)

Poet Nutrition. 2010. Available online: www.dakotagold.com/files/nutrientprofiles/DakotaBran.pdf. Accessed January 27, 2010.

Powers, W. J., H. H. VanHorn, B. Harris, Jr., and C. J. Wilcox. 1995. Effects of variable sources of distillers dried grains plus solubles or milk yield and composition. *Journal of Dairy Science* 78: 388–396.

Quigley, J. D. III, R. E. James, and M. L. McGillard. 1986. Dry matter intake in dairy heifers. 1. Factors affecting intake of heifers under intensive management. *Journal of Dairy Science* 69: 2855–2862.

Ranathunga, S. D., K. F. Kalscheur, A. R. Hippen, and D. J. Schingoethe. 2010. Replacement of starch from corn with non-forage fiber from distillers grains and soy hulls in diets of lactating dairy cows. *Journal of Dairy Science* 93: 1086–1097.

Rausch, K. D., and R. L. Belyea. 2006. The future of co-products from corn processing. *Applied Biochemistry and Biotechnology* 128: 47–86.

Reed, J. J., G. P. Lardy, M. L. Bauer, M. Gibson, and J. S. Caton. 2006. Effects of season and inclusion of corn distillers dried grains with solubles in creep feed on intake, microbial protein synthesis and efficiency, ruminal fermentation, digestion, and performance of nursing calves grazing native range in southeastern North Dakota. *Journal of Animal Science* 84: 2200–2212.

Robinson, P. H., K. Karges, and M. L. Gibson. 2008. Nutritional evaluation of four co-product feedstuffs from the motor fuel ethanol distillation industry in the Midwestern USA. *Animal Feed Science and Technology* 146: 345–352.

Robinson, P. H., J. M. Moorby, M. Arana, R. Hinders, T. Graham, L. Castelanelli, and N. Barney. 2001. Influence of close-up dry period protein supplementation on productive and reproductive performance of Holstein cows in their subsequent lactation. *Journal of Dairy Science* 84: 2273–2283.

Rust, S. R., J. R. Newbold, and K. W. Metz. 1990. Evaluation of condensed distillers' solubles as an energy source for finishing cattle. *Journal of Animal Science* 68: 186–192.

Sasikala-Appukuttan, A. K., D. J. Schingoethe, A. R. Hippen, K. F. Kalscheur, K. Karges, and M. L. Gibson. 2008. The feeding value of corn distillers solubles for lactating dairy cows. *Journal of Dairy Science* 91: 279–287.

Shaver, R., R. Ehrenfeld, M. Olivares, and J. Cuellar. 2009. Feeding distiller's dried grains to lactating dairy cows in the pasture-region of Chile: Field trial results. Available online: www.uwex.edu/ces/dairynutrition. Accessed February 10, 2010.

Tedeschi, L. O., P. J. Kononoff, K. Karges, and M. L. Gibson. 2009. Effects of chemical composition variation on the dynamics of ruminal fermentation and biological value of corn milling (co)products. *Journal of Dairy Science* 92: 401–413.

Thomas, M., A. R. Hippen, K. F. Kalscheur, and D. J. Schingoethe. 2006a. Growth and performance of Holstein dairy calves fed distillers grains. *Journal of Dairy Science* 89(Suppl.1): 1864 (Abstr).

Thomas, M., A. R. Hippen, K. F. Kalscheur, and D. J. Schingoethe. 2006b. Ruminal development in Holstein dairy calves fed distillers grains. *Journal of Dairy Science* 89(Suppl. 1): 437 (Abstr).

Thiex, N. 2009. Evaluation of analytical methods for the determination of moisture, crude protein, crude fat, and crude fiber in distillers dried grains with solubles. *Journal of AOAC International* 92: 61–73.

Udedibie, A. B. I., and L. E. Chase. 1988. The value of corn condensed distillers solubles (CCDS) for milk production in dairy cows. *Nigeria Agricultural Journal* 23: 118–129.

VandeHaar, M. J., G. Yousif, B. K. Sharma, T. H. Herdt, R. S. Emery, M. S. Allen, and J. S. Liesman. 1999. Effect of energy and protein density of prepartum diets on fat and protein metabolism of dairy cattle in the periparturient period. *Journal of Dairy Science* 82:1282–1295.

Van Horn, H. H., O. Blanco, B. Harris, Jr., and D. K. Beede. 1985. Interaction of protein percent with caloric density and protein source for lactating cows. *Journal of Dairy Science* 68: 1682–1695.

Van Saun, R. J., S. C. Idleman, and C. J. Sniffen. 1993. Effect of undegradable protein fed prepartum on postpartum production in first lactation Holstein cows. *Journal of Dairy Science* 76: 236–244.

Vazquez-Anon, M., A. J. Heinrichs, J. M. Aldrich, and G. A. Varga. 1993. Effect of postweaning age on rate of in situ protein disappearance in calves weaned at 5 weeks of age. *Journal of Dairy Science* 76: 2749–2757.

Weiss, W. P., D. O. Erickson, G. M. Erickson, and G. R. Fisher. 1989. Barley distillers grains as a protein supplement for dairy cows. *Journal of Dairy Science* 72: 980–987.

Zitnan, R., S. Kuhla, P. Sanftleben, A. Bilska, F. Schneider, M. Zupcanova, and J. Voigt. 2005. Diet induced ruminal papillae development in neonatal calves not correlating with rumen butyrate. *Veterinary Medicine* 50(11): 472–479.

14 Feeding Ethanol Coproducts to Swine

Hans H. Stein

CONTENTS

14.1 INTRODUCTION

Utilization of ethanol coproducts to reduce diet costs has become important as more U.S.-grown corn is used in the fuel ethanol industry. While the majority of the distillers dried grains with solubles (DDGS) that is produced is used in diets fed to ruminants such as beef cattle (Chapter 12) and dairy cows (Chapter 13), DDGS has also been used in the feeding of swine for more than 50 years. However, new production technologies and new corn varieties have changed the gross composition of the grain, which now contains less protein and more fat than in the past. Please refer to Chapters 5 and 8 for details on ethanol production and DDGS composition. The changes in the composition of corn that have been observed during the past decades influence DDGS composition and may also change the response of pigs to diets containing DDGS. New production technologies in ethanol plants such as front-end fractionation or back-end separation have resulted in the production of a number of novel coproducts from the ethanol industry. Research to measure the nutritional value of these products has been conducted in recent years. Considerable new information about feeding ethanol coproducts to swine has, therefore, been generated during the last decade and currently much of this information is being implemented in the swine industry. The objective of this chapter is to summarize current knowledge about feeding ethanol coproducts to swine.

14.2   ETHANOL COPRODUCTS USED IN DIETS FED TO SWINE

Corn and sorghum are two grains that are most often used in the production of ethanol in the United States, with corn being predominant. Blends of sorghum and corn are sometimes also used (Urriola et al., 2009). In Europe and Canada, however, wheat and sometimes barley may be used in ethanol production. The composition of the DDGS obtained from ethanol production depends on the grain that was used and there are, therefore, differences among sources of DDGS depending on grain type. DDGS may also be produced as a result of beverage distillation, but the nutritional quality of DDGS originating from beverage distillation is not different from the quality of DDGS produced from ethanol distillation (Pahm et al., 2008a). Other factors that have been suggested to influence the composition and nutritional value of DDGS include the region in which the grain is produced (Fastinger and Mahan, 2006), but it has been demonstrated that in the United States, the variation in nutritional value of DDGS among regions is no greater than within regions (Pahm et al., 2008a; Stein et al., 2009).

While DDGS is the most common and best known coproduct from the ethanol industry, other coproducts are also produced. If the solubles are not added back to the fermented grain, a product called distillers dried grains (DDG) is produced. This product may have a greater nutritional value than DDGS, because addition of solubles to the fermented grain may increase the heat damage of the amino acids (AA) in the product (Pahm et al., 2008b).

When front-end grain fractionation is used before ethanol is produced, corn is usually degermed and dehulled and only the endosperm is used in ethanol production. This process results in production of either high-protein distillers dried grains (HP-DDG) or high-protein distillers dried grains with solubles (HP-DDGS). However, at this point, there is no commercial production of HP-DDGS, whereas there is a considerable production of HP-DDG (Widmer et al., 2007, 2008; Jacela et al., 2009; Kim et al., 2009b).

The hulls that are removed from grain in the fractionation process are usually not used in swine feeding, but the corn germ that is extracted may be fed to pigs. Corn germ has a relatively high concentration of nonstarch polysaccharides with a total fat concentration of approximately 18% (Widmer et al., 2007).

If oil is extracted from the DDGS, a de-oiled DDGS that contains between 2% and 4% fat is produced (Jacela et al., 2007). Because of the low fat level in de-oiled DDGS, this product contains less energy than normal DDGS. Oil may also be removed by centrifugation, which is less efficient than removal by extraction, and a low-fat DDGS containing 7%–8% fat result from this process.

If fiber is removed from the DDGS, enhanced DDGS is obtained (Soares et al., 2008). This product contains approximately 10% less nonstarch polysaccharides than conventional DDGS and is expected to have a greater concentration of DE and ME than conventional DDGS.

14.3   CONCENTRATION AND DIGESTIBILITY OF NUTRIENTS IN ETHANOL COPRODUCTS

14.3.1   Carbohydrates

Production of ethanol or alcohol from grain is accomplished by fermentation of the starch fraction of the grain. All coproducts from the ethanol industry are, therefore, characterized by containing very little starch. The concentration of starch in DDGS produced from corn is approximately 7% (Stein and Shurson, 2009), whereas corn grain contains approximately 65% starch (Table 14.1).

Dietary fiber is the sum of carbohydrates and lignin that are resistant to digestion by mammalian enzymes in the small intestine, but that may be partially or completely fermented in the hindgut (AACC, 2001; IOM, 2006). Methods to measure dietary fiber include the crude fiber analysis (Mertens, 2003), the ADF and NDF procedures (Van Soest, 1963), and the total dietary fiber (TDF) procedure, which may separate dietary fiber into insoluble (IDF) and soluble dietary fiber (SDF; Prosky et al., 1984). An alternative to analyzing samples for dietary fiber is to calculate the concentration of organic residue (OR) in a feed ingredient by subtracting CP, ash, moisture, ether extract, sugar, and starch from 100 (de Lange, 2008).

TABLE 14.1

Chemical Composition of Corn, Sorghum, and Wheat, and Ethanol Coproducts Produced from Corn, Sorghum, and Wheat and Fed to Swine (As-Fed Basis)

Item	Corn	Sorghum	Wheat	Corn DDGS	Sorghum DDGS	Wheat DDGS	Corn DDG	Corn HP-DDG	De-oiled Corn DDGS	Enhanced Corn DDGS	Corn Germ
N	4	1	1	34	3	4	1	1	1	2	1
Gross energy (kcal/kg)	3891	3848	3830	4776	4334	4817	—	4989	—	4742	4919
Crude protein (%)	8.0	9.8	12.44	27.5	31.0	38.2	28.8	41.1	31.2	29.1	14.0
Calcium (%)	0.01	0.01	0.04	0.03	—	0.15	—	0.01	0.05	0.27	0.03
Phosphorus (%)	0.22	0.24	0.38	0.61	0.64	1.04	—	0.37	0.76	0.86	1.09
Crude fat (%)	3.3	—	2.0	10.2	7.7	3.6	—	3.7	4.0	10.8	17.6
Crude fiber (%)	—	—	2.4	6.6	9.8	7.6	—	—	—	—	—
Starch (%)	—	—	—	7.3	—	—	3.83	11.2	—	—	23.6
Neutral detergent fiber (%)	7.3	7.3	14.2	37.6	40.7	32.4	37.3	16.4	34.6	29.7	20.4
Acid detergent fiber (%)	2.4	3.8	2.9	11.1	22.8	17.0	18.2	8.7	16.1	8.7	5.6
Total dietary fiber (%)	—	—	—	31.8	32.2	—	43.0	—	—	25.2	—
Ash	0.9	—	—	3.8	3.6	4.8	—	3.2	4.64	—	3.3
Indispensable amino acids (%)											
Arginine	0.39	0.32	0.57	1.16	1.10	1.53	1.15	1.54	1.31	1.34	1.08
Histidine	0.23	0.23	0.29	0.72	0.71	0.92	0.68	1.14	0.82	0.75	0.41
Isoleucine	0.28	0.37	0.43	1.01	1.36	1.35	1.08	1.75	1.21	1.04	0.45
Leucine	0.95	1.25	0.83	3.17	4.17	2.66	3.69	5.89	3.64	3.26	1.06
Lysine	0.24	0.20	0.36	0.78	0.68	0.65	0.81	1.23	0.87	0.93	0.79
Methionine	0.21	0.18	0.21	0.55	0.53	0.53	0.56	0.83	0.58	0.58	0.25
Phenylalanine	0.38	0.47	0.53	1.34	1.68	1.92	1.52	2.29	1.69	1.38	0.57
Threonine	0.26	0.29	0.33	1.06	1.07	1.21	1.10	1.52	1.10	1.03	0.51
Tryptophan	0.09	0.07	0.16	0.21	0.35	0.40	0.22	0.21	0.19	0.19	0.12
Valine	0.38	0.48	0.55	1.35	1.65	1.70	1.39	2.11	1.54	1.40	0.71
Dispensable amino acids (%)											
Alanine	0.58	0.86	0.44	1.94	2.90	1.48	2.16	3.17	2.13	1.99	0.91
Aspartic acid	0.55	0.60	0.62	1.83	2.17	1.92	1.86	2.54	1.84	1.80	1.05

continued

TABLE 14.1 (continued)
Chemical Composition of Corn, Sorghum, and Wheat, and Ethanol Coproducts Produced from Corn, Sorghum, and Wheat and Fed to Swine (As-Fed Basis)

	Grain or Coproduct										
Item	Corn	Sorghum	Wheat	Corn DDGS	Sorghum DDGS	Wheat DDGS	Corn DDG	Corn HP-DDG	De-oiled Corn DDGS	Enhanced Corn DDGS	Corn Germ
N	4	1	1	34	3	4	1	1	1	2	1
Cystine	0.16	0.18	0.27	0.53	0.49	0.73	0.54	0.78	0.54	0.52	0.29
Glutamic acid	1.48	1.92	3.57	4.37	6.31	9.81	5.06	7.11	4.26	4.06	1.83
Glycine	0.31	0.29	0.50	1.02	1.03	1.62	1.00	1.38	1.18	1.11	0.76
Proline	0.70	0.77	1.14	2.09	1.40	4.11	2.50	3.68	2.11	1.99	0.92
Serine	0.38	0.37	0.48	1.18	2.50	1.88	1.45	1.85	1.30	1.25	0.56
Tyrosine	0.27	0.25	0.27	1.01	—	—	—	1.91	1.13	1.04	0.41

Source: Data from Bohlke, R. A. et al. *Journal of Animal Science* 83, 2396–2403, 2005; Stein, H. H. et al. *Journal of Animal Science* 84, 853–860, 2006; Stein, H. H. et al. *Asian Australian Journal of Animal Science* 22, 1016–1025, 2009; Feoli, C. et al. Digestible energy content of corn- vs. sorghum-based dried distillers grains with solubles and their effects on growth performance and carcass characteristics in finishing pigs. *Kansas State University Swine Day Report 2007*, pp. 131–136. Manhattan, KS: Kansas State University, 2007; Jacela, J. Y. et al. Amino acid digestibility and energy content of corn distillers meal for swine. *Kansas State University Swine Day Report 2007*, pp. 137–141. Manhattan, KS: Kansas State University, 2007; Jacela, J. Y. et al. *Journal of Animal Science* 87(E-Suppl. 3), 105, 2009; Pedersen, C. et al. *Journal of Animal Science* 85, 2473–2483, 2007b; Widmer, M. R. et al. *Journal of Animal Science* 85, 2994–3003, 2007; Widyaratne, G. P., and R. T. Zijlstra. *Canadian Journal of Animal Science* 87, 103–114, 2007; Lan, Y. et al. *Animal Feed Science and Technology* 140, 155–163, 2008; Pahm, A. et al. *Journal of Animal Science* 86, 2180–2189, 2008a; Soares, J. A. et al. *Journal of Animal Science* 86(Suppl. 1), 522, 2008; Kim, B. G. et al. *Journal of Animal Science* 87, 4013–4021, 2009b; Urriola, P. E. et al. *Journal of Animal Science* 87, 2574–2580, 2009; Urriola, P. E. et al. *Journal of Animal Science* 88, 2373–2381, 2010; and Widyaratne, G. P. et al. *Canadian Journal of Animal Science* 89, 91–95, 2009.

Note: DDGS = distillers dried grains with solubles; DDG = distillers dried grains; and HP-DDG = high-protein distillers dried grain.

The concentration of the different fiber fractions is approximately three times greater in DDGS and DDG than in corn, but HP-DDG and HP-DDGS contain less fiber than DDG and DDGS because the corn was dehulled before fermentation (Guo et al., 2004). In contrast, the concentration of fiber in corn germ is greater than in DDGS because some of the fiber in corn is removed along with the mechanical removal of the germ. The digestibility of TDF in DDGS and in DDG is less than 35% in the small intestine and less than 50% over the entire gastrointestinal tract (Table 14.2). However, the digestibility of SDF in DDGS is greater than 90% (Urriola et al., 2010), which indicates that this fraction of fiber is easily digested by pigs. However, the digestibility of IDF is much less and as there is much more IDF than SDF in DDGS, this results in lower overall digestibility (Table 14.2). The fiber fraction, therefore, contributes relatively little to the energy value of these feed ingredients (Urriola et al., 2010). It is expected that the digestibility of fiber in other ethanol coproducts is equally low although this has not yet been confirmed.

The low digestibility of fiber in ethanol coproducts results in increased quantities of manure being excreted from pigs fed these products (Stein and Shurson, 2009). The overall digestibility of DM and fiber in diets containing ethanol coproducts is, therefore, reduced compared with diets containing only corn and soybean meal (Urriola and Stein, 2010). The digestibility of energy is also less in diets containing DDGS than in diets containing no DDGS (Urriola and Stein, 2010). Currently, much effort is directed towards developing feed additives such as enzymes or yeast products that can improve the digestibility of fiber in ethanol coproducts, but at this point, results have been disappointing (Jones et al., 2009). It is also possible that chemical treatment or processing of DDGS may result in improvement of the fiber digestibility. If the digestibility of the fiber in ethanol coproducts is improved, their energy value will also improve. It was recently reported that the apparent total tract digestibility (ATTD) of energy in DDGS increases after extrusion (Beltranena et al., 2009), which may be a result of increased fiber digestibility. Likewise, the ATTD of DM, OM, and energy is increased in diets containing DDGS if these diets are pelleted (Zhu et al., 2009), but it is not known if this is an effect of pelleting the DDGS or if it comes from the non-DDGS components of the diet. More research in this area is, therefore, needed.

TABLE 14.2
Concentration, Apparent Ileal Digestibility (AID), Apparent Total Tract Digestibility (ATTD), and Hindgut Fermentation of Different Fiber Fractions in Distillers Dried Grains with Solubles Produced from Corn (C), Sorghum (S), or a Blend of Corn and Sorghum (CS) Fed to Growing–Finishing Pigs

Item	Concentration (%)			AID (%)			ATTD (%)			Hindgut Fermentation (%)		
	C	S	CS	C	S	CS	C	S	CS	C	S	CS
N	8	1	1	8	1	1	8	1	1	8	1	1
Crude fiber	6.6	9.8	8.1	31.0	38.6	30.7	44.3	41.6	39.9	13.3	3.0	9.2
ADF	11.1	22.8	16.5	36.8	57.4	41.4	58.5	60.7	53.7	21.7	3.3	12.3
NDF	37.6	40.7	39.5	45.9	49.9	37.9	59.3	59.3	51.5	13.4	9.4	13.6
IDF[a]	30.7	34.1	35.4	20.0	27.7	4.8	40.3	41.3	28.6	20.3	13.6	23.8
SDF[a]	1.1	1.2	0.4	64.4	65.9	63.4	92.0	90.9	90.6	27.6	25.0	27.2
TDF[a]	31.8	32.2	35.8	28.9	33.4	15.9	49.5	48.8	39.2	20.6	15.4	23.3
OR[a]	46.5	45.6	46.9	58.6	41.6	32.9	77.1	72.5	68.4	18.5	30.9	35.5

Source: Data from Urriola, P. E. et al. Journal of Animal Science 88, 2373–2381, 2010.
[a] IDF = insoluble dietary fiber; SDF = soluble dietary fiber; TDF = total dietary fiber; and OR = organic residue.

14.3.2 AMINO ACIDS

The concentration and standardized ileal digestibility of AA have been measured in 34 sources of corn DDGS, in one source of sorghum DDGS, and in two sources of wheat DDGS (Table 14.3). The results showed that even when the DDGS are produced from the same type of grain, some variation among different samples of DDGS exists for AA digestibility (Stein et al., 2006, 2009; Pahm et al., 2008a; Urriola et al., 2009). This is particularly true for Lys, which is more variable in digestibility than all other indispensable AA (Stein et al., 2006; Pahm et al., 2008a). Variability in the digestibility of AA is not related to the region within the United States where the DDGS is produced (Pahm et al., 2008a) and even if the DDGS is produced within a relatively small geographical region, large

TABLE 14.3
Standardized Ileal Digestibility of AA in Corn, Sorghum, and Wheat, and in Ethanol Coproducts Produced from Corn, Sorghum, and Wheat and Fed to Growing–Finishing Pigs

						Grain or Coproduct				
Item	Corn	Sorghum	Wheat	Corn DDGS	Sorghum DDGS	Wheat DDGS	Corn DDG	Corn HP-DDG	Corn Germ	De-oiled Corn DDGS
N	2	1	1	34	1	2	1	3	1	1
Indispensable amino acids (%)										
Arginine	87	70	88	81	78	86	83	83	83	823
Histidine	83	65	86	78	71	77	84	81	69	75
Isoleucine	81	66	84	75	73	80	83	81	57	75
Leucine	87	70	86	84	76	83	86	91	68	84
Lysine	72	57	75	62	62	57	78	64	58	50
Methionine	85	69	86	82	75	81	89	88	68	80
Phenylalanine	84	68	86	81	76	86	87	87	64	81
Threonine	74	64	79	71	68	75	78	77	53	66
Tryptophan	70	57	86	70	70	86	72	81	67	78
Valine	79	64	81	75	72	82	81	80	62	74
Dispensable amino acids (%)										
Alanine	83	69	76	78	73	68	82	86	64	77
Aspartic acid	80	66	78	69	68	57	74	76	60	61
Cystine	82	64	86	73	66	75	81	82	64	64
Glutamic acid	80	52	84	80	76	86	87	88	72	78
Glycine	84	71	92	63	67	68	66	75	76	53
Proline	96	50	105	74	83	81	55	73	84	73
Serine	83	72	88	76	73	77	82	84	65	73
Tyrosine	82	67	81	81	—	—	—	88	59	81

Source: Data from Bohlke, R. A. et al. *Journal of Animal Science* 83, 2396–2403, 2005; Stein, H. H. et al. *Journal of Animal Science* 84, 853–860, 2006; Stein, H. H. et al. *Asian Australian Journal of Animal Science* 22, 1016–1025, 2009; Jacela, J. Y. et al. Amino acid digestibility and energy content of corn distillers meal for swine. *Kansas State University Swine Day Report 2007*, pp. 137–141. Manhattan, KS: Kansas State University, 2007; Jacela, J. Y. et al. *Journal of Animal Science* 87(E-Suppl. 3), 105, 2009; Pedersen, C. et al. *Journal of Animal Science* 85, 2473–2483, 2007b; Widmer, M. R. et al. *Journal of Animal Science* 85, 2994–3003, 2007; Widyaratne, G. P., and R. T. Zijlstra. *Canadian Journal of Animal Science* 87, 103–114, 2007; Lan, Y. et al. *Animal Feed Science and Technology* 140, 155–163, 2008; Pahm, A. et al. *Journal of Animal Science* 86, 2180–2189, 2008a; Kim, B. G. et al. *Journal of Animal Science* 87, 4013–4021, 2009b; and Urriola, P. E. et al. *Journal of Animal Science* 87, 2574–2580, 2009.

Note: DDGS = distillers dried grains with solubles; DDG = distillers dried grains; and HP-DDG = high-protein distillers dried grains; AA = anino acids.

variations in Lys digestibility among sources of DDGS may be observed (Stein et al., 2009). The main reason for this variability is that some production units overheat the DDGS during drying, which results in Maillard type destruction of Lys (Pahm et al., 2008a, 2008b). This will result in a reduction in the total concentration of Lys as well as in Lys digestibility, but the concentration of crude protein will not be changed. In undamaged corn DDGS, the concentration of Lys as a percentage of CP is between 3.1% and 3.3%, but in heat-damaged corn DDGS, this percentage can be as low as 2.10% (Stein, 2007; Stein et al., 2009). It is, therefore, recommended that the concentration of Lys is measured before corn DDGS is used in diets fed to swine and if the concentration of Lys expressed as a percentage of CP is less than 2.80%, the DDGS should not be used in diets fed to swine.

Some of the variability in AA digestibility, and Lys digestibility in particular, is caused by the addition of solubles to the fermented grain because the solubles contain some residual sugars that were not fermented. The presence of these sugars will increase the likelihood of Maillard reactions occurring when the fermented grain is dried. The digestibility of AA in corn DDG is, therefore, greater than in corn DDGS, because the solubles are not added to the fermented grain when DDG is produced (Pahm et al., 2008a).

Most AA in corn DDGS have a digestibility that is approximately 10 percentage units less than corn, which may be a result of the greater concentration of dietary fiber in DDGS than in corn, because the presence of fiber in a feed ingredient usually reduces the digestibility of AA (Stein et al., 2007). However, except for Lys, the variability among different sources of corn DDGS is within the normal range of variation observed in other feed ingredients.

The digestibility of AA in corn HP-DDG is within the range of values measured for corn DDGS (Widmer et al., 2007; Jacela et al., 2009; Kim et al., 2009b). However, the digestibility of AA in corn germ is less than in corn DDG and corn DDGS. The reason for this may be that the proteins in corn germ are different from other proteins in the grain kernel (Widmer et al., 2007). The relatively high concentration of fiber in corn germ may also contribute to a reduction in AA digestibility.

Although sorghum has a lower digestibility of AA than corn (Pedersen et al., 2007b), sorghum DDGS has AA digestibilities that are within the range of values observed in corn DDGS (Urriola et al., 2009). However, AA digestibility data for only one source of sorghum DDGS has been reported. Likewise, the digestibility of most AA in wheat DDGS is similar to the values obtained in corn DDGS although the digestibility of Lys in wheat DDGS is less than in corn DDGS (Table 14.3). The digestibility of AA was measured in one source of de-oiled corn DDGS and all values were reported to be within the range of values for conventional corn DDGS (Jacela et al., 2008b).

14.3.3 Phosphorus

The P concentration in corn and sorghum is usually between 0.22% and 0.26%, but wheat contains between 0.35% and 0.40% P (Table 14.4). When starch is removed during fermentation, the concentration of minerals in the resulting DDGS is increased and the P concentration in corn DDGS and in sorghum DDGS is; therefore, between 0.60% and 0.80%, whereas wheat DDGS contains more than 1% P (Stein and Shurson, 2009; Widyaratne et al., 2009). When corn grain is dehulled and degermed, the majority of the P ends up in the germ fraction. Corn germ, therefore, contains approximately 1.10% P, whereas HP-DDG contains only 0.40% P (Widmer et al., 2007).

The majority of the P in cereal grains is bound in the phytate complex (Eeckhout and de Paepe, 1994). Pigs do not have the ability to break down the phytate complex in the small intestine where P absorption occurs, and the phytate bound P is excreted in the feces. The digestibility of P in corn and sorghum is, therefore, very low and values between 15% and 30% are usually measured for P digestibility in corn and sorghum (NRC, 1998; Bohlke et al., 2005; Pedersen et al., 2007a). The digestibility of P in wheat is, however, somewhat greater than in corn and sorghum because wheat contains intrinsic phytase that partially degrades the phytate complex before the end of the

TABLE 14.4
Concentration (%) and Apparent Total Tract Digestibility (ATTD) of P (%) in Cereal Grain and Ethanol Coproducts Fed to Growing–Finishing Pigs

				Grain or Coproduct			
Item	Corn	Sorghum	Wheat	Corn DDGS	Wheat DDGS	Corn HP-DDG	Corn Germ
N	4	1	1	15	4	1	1
Phosphorus (%)	0.22	0.24	0.38	0.61	1.04	0.37	1.09
ATTD, P (%)	25.00	25.00	30.00	58.90	51.90	59.60	28.60
ATTD, P (%) with phytase	57.80	—	45.00	71.00	—	—	—

Source: Data from Sauvant, D. et al. *Tables of Composition and Nutritional Value of Feed Materials,* 2nd ed. Wageningen Academic Publishers, 2004; Bohlke, R. A. et al. *Journal of Animal Science* 83, 2396–2403, 2005; Nyachoti, C. M. et al. *Journal of the Science of Food and Agriculture* 85, 2581–2586, 2005; Pedersen, C. et al. *Journal of Animal Science* 85, 1168–1176, 2007a; Widmer, M. R. et al. *Journal of Animal Science* 85, 2994–3003, 2007; Widyaratne, G. P., and R. T. Zijlstra. *Canadian Journal of Animal Science* 87, 103–114, 2007; Lan, Y. et al. *Animal Feed Science and Technology* 140, 155–163, 2008; Stein, H. H. et al. *Asian Australian Journal of Animal Science* 22, 1016–1025, 2009; Widyaratne, G. P. et al. *Canadian Journal of Animal Science* 89, 91–95, 200; and Almeida, F. N., and H. H. Stein. *Journal of Animal Science* 88: 2968–2977, 2010.

Note: DDGS = distillers dried grains with solubles; and HP-DDG = high-protein distillers dried grain.

small intestine, which increases the absorption of P from wheat (NRC, 1998; Zimmermann et al., 2002).

When corn, sorghum, or wheat is fermented, some of the phytate-bound phosphorus is believed to be released, and coproducts that have gone through fermentation, therefore, have a greater digestibility of P than corn (Stein and Shurson, 2009). The ATTD of P has been measured in 15 different sources of corn DDGS with an average value of 59% (Pedersen et al., 2007a; Stein et al., 2009; Almeida and Stein, 2010). In corn HP-DDG, which is also a fermented product, the ATTD of P is 59.6%, very close to the value measured for corn DDGS (Widmer et al., 2007). It appears, therefore, that the ATTD of P in corn coproducts that have been fermented is close to 59%. In contrast, in corn germ, which is not a fermented coproduct, the ATTD of P is only 28.60%, which is close to the values for corn (Widmer et al., 2007).

The ATTD of P in wheat DDGS was measured in four sources and an average value of 51.9% was reported (Nyachoti et al., 2005; Widyaratne and Zijlstra, 2007; Widyaratne et al., 2009). It is, therefore, apparent that the increase in P digestibility in wheat compared with corn is not maintained during fermentation and in fermented DDGS products, this digestibility is actually slightly greater in corn DDGS than in wheat DDGS.

The effect of adding microbial phytase to corn DDGS was measured in one experiment (Almeida and Stein, 2010) and it was shown that the ATTD of P in corn DDGS is 71% when microbial phytase was included in the diet. At this point, there are no reports on the effects of adding microbial phytase to other ethanol coproducts, but work to address this question is currently being conducted at the University of Illinois.

Because of the greater digestibility of P in DDGS than in corn and soybean meal, the need for adding inorganic phosphorus sources is reduced when DDGS is included in the diet. The excretion of P from the pigs, will therefore, also be reduced without reducing performance (Almeida and Stein, 2010).

The availability of P in feed ingredients is sometimes measured as the relative bioavailability of phosphorus rather than the ATTD (Cromwell, 1992). While data for the relative availability of

P have been measured in most commonly used feed ingredients, they are not always useful in feed formulation because they are not always additive in mixed diets as different standards have been used to measure availability.

The relative availability of P in corn DDGS has, however, been measured and values between 70% and 90% have been reported in experiments where dicalcium phosphate was used as a standard (Burnell et al., 1989; Whitney and Shurson, 2001; Jenkin et al., 2007). These values correspond to ATTD between 56% and 72%, which are close to those of P measured in corn DDGS. There is, therefore, good agreement between the two different types of measurements, but for practical feed formulation, it is suggested that values for ATTD of P in feed ingredients should be used.

14.3.4 Ether Extract

The apparent ileal and the true ileal digestibility of ether extract in corn, DDGS, HP-DDG, and corn germ has been reported from one experiment (Kim et al., 2009a). Results of this experiment showed that the apparent ileal digestibility of ether extract is between 50% and 60% and the true ileal digestibility of ether extract is between 50% and 76% in DDGS, HP-DDG, and corn germ (Table 14.5). The ATTD of ether extract in DDGS was also measured by Stein et al. (2009) and reported at 70%. There is, however, a need for more information in this area because the relatively high concentrations of ether extract in DDGS and other ethanol coproducts may result in production of pork with soft fat. Soft and unsaturated fat is produced if pigs are fed diets containing relatively high amounts of unsaturated fat (Madsen et al., 1992), which may be the case if DDGS is included in the diets.

14.4 CONCENTRATION AND DIGESTIBILITY OF ENERGY IN ETHANOL COPRODUCTS

The concentration of energy is greater in DDGS and in most other corn-based ethanol coproducts than in corn (Table 14.6). However, the concentration of fiber in ethanol coproducts is greater than in corn and because of the low digestibility of energy in fiber (Urriola et al., 2010), the digestibility of energy is also less in DDGS and other ethanol coproducts than in corn. The concentration of digestible and metabolizable energy in corn DDGS is, therefore, not greater than in corn (Pedersen et al., 2007a; Stein et al., 2009). The same is the case for corn germ, which also has a similar concentration of digestible and metabolizable energy as corn (Widmer et al., 2007). However, HP-DDG has a reduced concentration of fiber compared with DDGS and corn germ, and the concentration of digestible and metabolizable energy in HP-DDG is, therefore, greater than in corn (Widmer et al., 2007; Kim et al., 2009b). In contrast, in de-oiled DDGS, the fiber is left and the oil is removed.

TABLE 14.5
True Digestibility of Ether Extract in Corn and Ethanol Coproducts Fed to Growing–Finishing Pigs

| Item | Grain or Coproduct | | | |
	Corn	Corn DDGS	Corn HP-DDG	Corn Germ
N	1	1	1	1
True ileal digestibility (%)	53.0	62.1	76.5	50.1
True total tract digestibility (%)	41.4	51.9	70.2	43.9

Source: Unpublished data from the University of Illinois.

Note: DDGS = distillers dried grains with solubles; and HP-DDG = high-protein distillers dried grain.

TABLE 14.6
Concentration of Energy in Corn, Sorghum, and Wheat and in Ethanol Coproducts Produced from Corn, Sorghum, and Wheat Fed to Growing–Finishing Pigs

									De-oiled
									Corn
				Corn	Sorghum	Wheat	Corn	Corn	Corn
Item Ingredient	Corn	Sorghum	Wheat	DDGS	DDGS	DDGS	HP-DDG	Germ	DDGS
N	2	—	—	10	2	—	2	1	1
Gross energy (kcal/kg DM)	4458	—	—	5434	4908	—	5553	5335	4090
Apparent total tract digestibility (%)	90.0	87.0	88.0	76.8	76.0	74.0	89.4	74.6	66.5
Digestible energy (kcal/kg DM)	4072	3798	3827	4140	3459	3581	4903	3979	2719
Metabolizable energy (kcal/kg DM)	3981	3753	3648	3897	—	3346	4583	3866	2506

Sources: Data from NRC. *Nutrient Requirements of Swine,* 10th rev. ed. National Academies Press, 1998; Sauvant, D. et al. *Tables of Composition and Nutritional Value of Feed Materials,* 2nd ed. Wageningen Academic Publishers, 2004; Feoli, C. et al. Digestible energy content of corn- vs. sorghum-based dried distillers grains with solubles and their effects on growth performance and carcass characteristics in finishing pigs. *Kansas State University Swine Day Report 2007,* pp. 131–136. Manhattan, KS: Kansas State University, 2007; Pedersen, C. et al. *Journal of Animal Science* 85, 1168–1176, 2007a; Widmer, M. R. et al. *Journal of Animal Science* 85, 2994–3003, 2007; Jacela, J. Y. et al. *Journal of Animal Science* 86(E-Suppl. 3), 87–88, 2008b; and Kim, B. G. et al. *Journal of Animal Science* 87, 4013–4021, 2009b.

Note: DDGS = distillers dried grains with solubles; and HP-DDG = high-protein distillers dried grain.

De-oiled corn DDGS, therefore, contains only 2719 and 2506 kcal digestible and metabolizable energy, respectively (Jacela et al., 2008b).

The concentration of digestible energy in sorghum DDGS has been measured in one experiment and it was reported that sorghum DDGS contained approximately 220 kcal/kg (as-is basis) less digestible energy than corn DDGS (Feoli et al., 2007). The reduced concentration of digestible energy in sorghum DDGS may be a result of a reduced concentration of ether extract when compared with corn DDGS.

14.5 INCLUSION OF ETHANOL COPRODUCTS IN DIETS FED TO SWINE

14.5.1 INCLUSION OF ETHANOL COPRODUCTS IN DIETS FED TO WEANLING PIGS

Effects of including corn DDGS in diets fed to weanling pigs have been investigated in 12 experiments (Table 14.7). In one experiment, only 10% DDGS was included in the diet and this level of inclusion did not significantly affect pig performance (Linneen et al., 2006), which is in agreement with data showing that 20% corn DDGS can be used in diets fed to weanling pigs (Senne et al., 1995; Almeida and Stein, 2010). Likewise, in two titration experiments, effects of adding 0%, 5%, 10%, 15%, 20%, or 25% DDGS to diets fed to weanling pigs from 4 days after weaning were investigated and no impact of DDGS level was observed (Whitney and Shurson, 2004). Up to 30% inclusion of DDGS in nursery diets were also reported to not influence performance if DDGS was introduced in the diets from 2 to 3 weeks postweaning (Gaines et al., 2006; Spencer et al., 2007; Barbosa et al., 2008; Burkey et al., 2008). Of the 12 experiments, which evaluated effects of adding

TABLE 14.7
Effects of Including Corn Distillers Dried Grains with Solubles (DDGS) in Diets Fed to Weanling Pigs

Item	N	Response to Dietary Corn DDGS		
		Increased	Reduced	Not Changed
ADG	12	0	0	12
ADFI	12	0	4	8
G:F	12	5	0	7
Mortality	2	0	0	2

Source: Data calculated from experiments by Senne, B. W. et al. Effects of distillers grains on growth performance in nursery and finishing pigs. *Kansas State University Swine Day Report,* pp. 68–71. Manhattan, KS: Kansas State University, 1995; Whitney, M. H., and G. C. Shurson. *Journal of Animal Science* 82, 122–128, 2004; Gaines, A. et al. *Journal of Animal Science* 84 (Suppl. 2), 120, 2006; Linneen, S. K. et al. Effects of dried distillers grain with solubles on nursery pig performance. *Kansas State University Swine Day Report,* pp. 100–102. Manhattan: Kansas State University, 2006; Spencer, J. D. et al. *Journal of Animal Science* 85(Suppl. 2), 96–97, 2007; Barbosa, F. F. et al. *Journal of Animal Science* 86(Suppl. 1): 446, 2008; Burkey, T. E. et al. *Journal of Animal Science* 86(Suppl. 2), 50, 2008; and Almeida, F. N., and H. H. Stein. *Journal of Animal Science* 88, 2968–2977, 2010.

DDGS to weanling pig diets, none reported a reduction in average daily gain (ADG), when DDGS was introduced no sooner than 2 to 3 weeks postweaning. The average daily feed intake (ADFI) was reduced in two experiments as DDGS was included in the diet (Gaines et al., 2006; Barbosa et al., 2008), but the gain to feed ratio (G:F) was improved when DDGS was added to the diet in five of the 12 experiments (Gaines et al., 2006; Spencer et al., 2007; Barbosa et al., 2008). Pig mortality was reported in two experiments and no differences between pigs fed DDGS containing diets and pigs fed control diets were observed. Based on these data it was concluded that weanling pigs may be fed diets containing up to 30% DDGS from 2 to 3 weeks postweaning. Inclusion of up to 30% de-oiled corn DDGS in diets fed to weanling pigs also results in no change in ADFI, ADG, and G:F (Jacela et al., 2008a).

Inclusion of sorghum DDGS in diets fed to weanling pigs has been investigated in three experiments, and results from one experiment suggest that it may be possible to include 30% sorghum DDGS in diets fed to weanling pigs without reducing pig performance (Senne et al. 1996). However, later results indicate that inclusion of 30% sorghum DDGS in diets fed to weanling pigs may reduce performance (Feoli et al., 2008). Differences in the quality of sorghum DDGS used in these experiments may be the reason for the different results observed among experiments and at this time, it is recommended that diets fed to weanling pigs contain no more than 20% sorghum DDGS.

There have been no experiments conducted to investigate the effects of including ethanol coproducts other than DDGS and de-oiled DDGS in diets fed to weanling pigs. It is, therefore, unknown if HP-DDG, corn germ, or other coproducts can be used in these diets.

14.5.2 INCLUSION OF ETHANOL COPRODUCTS IN DIETS FED TO GROWING-FINISHING PIGS

14.5.2.1 Effects of Ethanol Coproducts on Live Pig Performance

Data from a total of 32 experiments have been reported in which corn DDGS was fed to growing–finishing pigs (Table 14.8). In most of these experiments, no change in ADG, ADFI, and G:F has been reported. There are, however, some experiments that reported reduced performance as DDGS

TABLE 14.8
Effects of Including Corn Distillers Dried Grains with Solubles (DDGS) in Diets Fed to Growing–Finishing Pigs

Item	N	Response to Dietary Corn DDGS		
		Increased	Reduced	Not Changed
ADG	31	1	7	23
ADFI	29	2	7	20
G:F	31	4	6	21
Dressing percentage	21	0	10	11
Backfat (mm)	19	0	4	15
Lean meat (%)	18	0	1	17
Loin depth (cm)	17	0	3	14
Belly thickness (cm)	4	0	2	2
Belly firmness	3	0	3	0
Iodine value	9	8	0	1

Source: Data calculated from experiments by Gralapp et al. (2002), Fu et al. (2004), Cook et al. (2005), DeDecker et al. (2005), Whitney et al. (2006), McEwen (2006, 2008), Gaines et al. (2007a, 2007b), Gowans et al. (2007), Hinson et al. (2007), Jenkin et al. (2007), White et al. (2007), Widyaratne and Zijlstra (2007), Xu et al. (2007a, 2007b, 2008a, 2008b), Augspurger et al. (2008), Drescher et al. (2008, 2009), Duttlinger et al. (2008), Hill et al. (2008a), Linneen et al. (2008), Stender and Honeyman (2008), Weimer et al. (2008), Widmer et al. (2008), Amaral et al. (2009), Cromwell et al. (2009), Hilbrands et al. (2009), Lammers et al. (2009), and Stevens et al. (2009).

was included in the diets. The reason for this difference may be that if DDGS of poor quality is used, pig performance may be reduced, whereas good quality DDGS does not reduce performance (Drescher et al., 2009). The inclusion rate of DDGS in the experiments that were reported varied from 5% to 60%, but most experiments used 30% DDGS or less. In some of the experiments with 30% inclusion, no difference in performance was observed (Cook et al., 2005; DeDecker et al., 2005; Xu et al., 2007a). Lower inclusion rates have also been used without influencing pig performance (Gowans et al., 2007; Jenkin et al., 2007; Linneen et al., 2008; Amaral et al., 2009; Drescher et al., 2009; Hilbrand et al., 2009; Lammers et al., 2009; Stevens et al., 2009). However, data from other experiments in which 10%, 20%, or 30% DDGS was included in diets fed to growing–finishing pigs showed a linear reduction in live pig performance (Fu et al., 2004; Whitney et al., 2006; Linneen et al., 2008; Weimer et al., 2008). The reduced performance that is sometimes observed from pigs fed diets containing DDGS may be a result of reduced palatability, because pigs prefer to eat diets containing no DDGS (Hastad et al., 2005; Seabolt et al., 2010). It is also possible that in some of the experiments in which reduced performance was observed, diets were formulated based on the concentration of total AA rather than on digestible AA, which may lead to a reduced performance. Diets that contain DDGS may also be deficient in tryptophan (Trp) because DDGS has a low concentration of Trp. It may, therefore, sometimes be necessary to also include crystalline Trp in diets containing DDGS (Stein, 2007).

Based on the wide body of literature that is available on feeding corn DDGS to growing–finishing pigs, it is concluded that in most cases, at least 20% and probably 30% DDGS can be included in growing–finishing pig diets. At these inclusion levels, pig performance is usually not influenced by DDGS provided that a high-quality source of DDGS is used and diets are properly formulated.

Conclusions from eight experiments in which sorghum DDGS was included in diets fed to growing–finishing pigs also led to the conclusion that 30% sorghum DDGS may be used. At this inclusion rate, ADG, ADFI, and G:F are not reduced (Senne et al., 1996). However, if greater inclusion rates are used, pig performance will be reduced (Senne et al., 1996; Feoli et al., 2008).

Inclusion of corn HP-DDG in diets fed to growing–finishing pigs was reported in two experiments (Widmer et al., 2008; Kim et al., 2009b). In the experiment by Widmer et al. (2008), pigs were fed HP-DDG from 22 kg to market, whereas in the experiment by Kim et al. (2009b), pigs were fed diets containing HP-DDG only in the early and late finishing stages. In both experiments, no overall effects of HP-DDG on pig performance were observed. It was, therefore, concluded that corn HP-DDG may be included in corn-based diets fed to growing–finishing pigs at levels needed to replace all the soybean meal in the diets provided that crystalline AA are included in concentrations necessary to balance all indispensable AA.

Corn germ was also included in diets fed to growing–finishing pigs in the experiment by Widmer et al. (2008). Diets containing 5% or 10% corn germ, but no other ethanol coproducts, were used in all three stages of growth. A linear increase in the final weight of the pigs was observed as corn germ was included in the diets and a tendency for increased daily gain was observed with diets containing corn germ. It was, therefore, concluded that growing–finishing pigs will improve performance if they are fed diets containing 10% corn germ (Widmer et al., 2008). It is possible that greater inclusion rates for corn germ can be used, but research to investigate this possibility needs to be conducted. There have been no reports of experiments in which other ethanol coproducts were fed to growing–finishing pigs.

14.5.2.2 Effects of Ethanol Coproducts on Carcass Composition and Quality

The most consistent effect of including DDGS in diets fed to finishing pigs is that they will produce fat that is less saturated than if no DDGS is used. The reason for this is that DDGS contains approximately 10% fat, which consists of mainly unsaturated fatty acids. More information about specific lipids can be found in Chapter 8. When pigs are fed DDGS, they will, therefore, deposit more unsaturated fat, which results in softer body fat depots. This can be observed in the form of softer bellies and softer backfat when pigs are processed, which may reduce their value. Softness of fat in pigs is often assessed by measuring the so-called iodine values. Iodine values have been measured in nine experiments in which DDGS was included in the diets fed to finishing pigs, and in eight of these nine experiments, an increase in iodine value was reported. This observation indicates that the fat was softer in pigs fed DDGS than in pigs fed no DDGS. To correct this problem, DDGS may be withdrawn from the diets during the final 2 to 4 weeks before harvest, which can result in acceptable iodine values (Hill et al., 2008a; Xu et al., 2008b).

Reduced dressing percentage of pigs fed corn DDGS containing diets has also been reported in some of the experiments in which corn DDGS were fed to growing–finishing pigs (Cook et al., 2005; Whitney et al., 2006; Feoli et al., 2007; Gaines et al., 2007a, 2007b; Hinson et al., 2007; Xu et al., 2007a; Linneen et al., 2008; Weimer et al., 2008; Drescher et al., 2009; Stevens et al., 2009). However, in many other experiments, no effects of DDGS on dressing percentage were reported (Fu et al., 2004; McEwen, 2006, 2008; Xu et al., 2007b; Augspurger et al., 2008; Drescher et al., 2008; Hill et al., 2008a; Stender and Honeyman, 2008; Widmer et al., 2008; Hilbrands et al., 2009). Thus, a reduced dressing percentage was reported only in 10 of the 21 experiments conducted. Pigs that are fed diets containing sorghum DDGS also sometimes have no change in dressing percentage while at other times, differences may be observed (Senne et al., 1996, 1998; Feoli et al., 2007). At this point, it is not clear what the reason for these differences are, but increased concentrations of fiber and protein may be contributing factors (Kass et al., 1980; Ssu et al., 2004).

Other carcass quality measures such as backfat thickness, lean meat percentage, and loin depth are not influenced by the inclusion of corn DDGS in the diets. Belly thickness has been reported to decrease in some, but not all, experiments in which corn or sorghum DDGS was included in the diet. Pigs fed diets containing corn HP-DDG may also have softer bellies and increased iodine values compared with pigs fed corn–soybean meal diets (Widmer et al., 2008), but pigs fed diets containing corn germ have firmer bellies and reduced iodine values (Widmer et al., 2008).

The palatability of pork from pigs fed up to 20% corn DDGS, up to 40% corn HP-DDG, or up to 10% corn germ was measured in one experiment (Widmer et al., 2008). It was reported from this experiment that the overall acceptance of pork from pigs fed diets containing ethanol coproducts is not different from that of pigs fed corn–soybean meal diets. It is, therefore, unlikely that consumers will be able to tell whether or not the pork they are eating come from a pig that was fed ethanol coproducts or not.

14.5.3   INCLUSION OF ETHANOL COPRODUCTS IN DIETS FED TO SOWS

Gestating sows may be fed diets containing up to 50% corn DDGS without any negative impacts on performance measured as farrowing rate, litter weight, lactation performance, and return to estrus (Wilson et al., 2003). It is, however, possible that litter size is increased if DDGS is included in the diet for at least two parities (Wilson et al., 2003), which may be a result of the increased fiber concentration in DDGS containing diets compared with corn–soybean meal diets.

Diets fed to lactating sows may include up to at least 30% DDGS without negatively influencing sow or litter performance (Song et al., 2007a; Greiner et al., 2008), but experiments to measure the effects of using greater inclusion rates have not been conducted. The concentration of P in the manure from sows fed diets containing DDGS is also reduced compared with sows fed no DDGS (Hill et al., 2008b). There is no influence of adding corn DDGS to a corn–soybean meal-based diet on milk composition, apparent nitrogen digestibility, or nitrogen retention, but sows fed diets containing 20% or 30% corn DDGS have reduced values for blood urea nitrogen compared with sows fed a corn–soybean meal diet (Song et al., 2007b). It is, therefore, possible that sows fed DDGS containing diets receive a better balance of AA compared with sows fed the control diet. Sows fed diets containing DDGS also have a greater weight gain during lactation than those fed diets containing no DDGS (Greiner et al., 2008), which indicates that sows may digest the fiber in DDGS better than growing pigs, and therefore, are able to absorb more energy from DDGS.

There are no data on inclusion of HP-DDG, de-oiled DDGS, or corn germ in diets fed to sows. Likewise, the effects of adding wheat DDGS or sorghum DDGS have not been reported. It is, therefore, not known how much wheat DDGS or sorghum DDGS may be included in diets fed to sows, but it is expected that the responses to sorghum DDGS and wheat DDGS are similar to those observed for corn DDGS.

14.6   CONCLUSIONS

Many experiments have been conducted in recent years to elucidate the effects of adding DDGS to diets fed to swine. Although the results of these experiments are somewhat variable, it is concluded that with the exception of the initial 2 to 3 weeks postweaning, pigs may be fed diets containing up to 30% DDGS. Gestating sows may be fed up to 50% DDGS and lactating sows may also be fed at least 30% DDGS. At these inclusion levels, pig performance will likely not change provided that a source of DDGS of reasonable quality is used. It is, however, documented that pigs fed DDGS will have softer fat and greater iodine values and it may, therefore, be necessary to withdraw or reduce the inclusion rate of DDGS in diets during the late finishing period.

All the soybean meal in diets fed to growing and finishing pigs may be replaced by HP-DDG without any change in performance. Likewise, corn germ may be included at a concentration of at least 10% in growing–finishing diets.

Diets containing DDGS will usually contain the same amount of metabolizable energy as diets containing no DDGS. The inclusion of inorganic P in DDGS containing diets can, however, be reduced because DDGS has a much greater concentration of digestible P than corn and soybean meal. Diets that include DDGS need to be formulated on the basis of standardized ileal digestible AA and all indispensable AA need to be included at required levels. It is, therefore, necessary that diets containing DDGS or other biofuels coproducts are fortified with crystalline AA. Some sources of DDGS have been observed to contain less AA than average and DDGS of such qualities should not be used in diets fed to swine. In conclusion, DDGS and other coproducts from the ethanol indus-try are excellent feed ingredients that easily can be incorporated into diets fed to pigs if a few basic principles in diet formulations are observed. DDGS and other coproducts from the ethanol industry can also be used in poultry diets, as is discussed in Chapter 15.

REFERENCES

AACC . 2001. The definition of dietary fiber. AACC Report. *Cereal Foods World* 46: 112–126.

Almeida, F. N., and H. H. Stein. 2010. Performance and phosphorus balance of pigs fed diets formulated on the basis of values for standardized total tract digestibility of phosphorus. *Journal of Animal Science* 88: 2968–2977.

Amaral, R. B., L. J. Johnson, J. E. Anderson, S. K. Baidoo, and G. C. Shurson. 2009. Effects of low-solubles distillers dried grains (LS-DDGS) in diets for growing-finishing pigs: Energy content and pig growth performance. *Journal of Animal Science* 87(E-Suppl. 3): 80. (Abstr.)

Augspurger, N. R., G. I. Petersen, J. D. Spencer, and E. N. Parr. 2008. Alternating dietary inclusion of corn dis-tillers dried grains with solubles (DDGS) did not impact growth performance of finishing pigs. *Journal of Animal Science* 86(Suppl. 1): 523. (Abstr.)

Barbosa, F. F., S. S. Dritz, M. D. Tokach, J. M. DeRouchy, R. D. Goodband, and J. L. Nelsen. 2008. Use of distillers dried grains with solubles and soybean hulls in nursery pig diets. *Journal of Animal Science* 86(Suppl. 1): 446. (Abstr.)

Beltranena, E., J. Sánchez-Torres, L. Goonewardene, X. Meng, and R. T. Zijlstra. 2009. Effect of single- or twin-screw extrusion on energy and amino acid digestibility of wheat or corn distillers dried grains with solubles (DDGS) for growing pigs. *Journal of Animal Science* 87(Suppl. 3): 166. (Abstr.)

Bohlke, R. A., R. C. Thaler, and H. H. Stein. 2005. Calcium, phosphorus, and amino acid digestibility in low-phytate corn, normal corn, and soybean meal by growing pigs. *Journal of Animal Science* 83: 2396–2403.

Burkey, T. E., P. S. Miller, R. Moreno, S. S. Shepherd, and E. E. Carney. 2008. Effects of increasing levels of distillers dried grains with solubles (DDGS) on growth performance of weanling pigs. *Journal of Animal Science* 86(Suppl. 2): 50. (Abstr.)

Burnell, T. W., G. L. Cromwell, and T. S. Stahly. 1989. Bioavailability of phosphorus in dried whey, blood meal, and distillers grains for pigs. *Journal of Animal Science* 67(Suppl. 1): 262. (Abstr.)

Cook, D., N. Paton, and M. L. Gibson. 2005. Effect of dietary level of distillers dried grains with solubles (DDGS) on growth performance, mortality, and carcass characteristics of grow–finish barrows and gilts. *Journal of Animal Science* 83(Suppl. 1): 335. (Abstr.)

Cromwell, G. L. 1992. The biological availability of phosphorus in feedstuffs for pigs. *Pig News Info.* 13: 75N–78N.

Cromwell, G. L., M. J. Azain, O. Adeola, S. D. Carter, T. D. Crenshaw, S. W. Kim, D. C. Mahan, P. S. Miller, and M. C. Shannon. 2009. Corn distillers dried grains with soluble (DDGS) in diets for growing-finishing pigs—A cooperative study. *Journal of Animal Science* 87(E-Suppl. 3): 81. (Abstr.)

DeDecker, J. M., M. Ellis, B. F. Wolter, J. Spencer, D. M. Webel, C. R. Bertelsen, and B. A. Peterson. 2005. Effects of dietary level of distiller dried grains with solubles and fat on the growth performance of grow-ing pigs. *Journal of Animal Science* 83(Suppl. 2): 79. (Abstr.)

de Lange, C. F. M. 2008. Efficiency of utilization of energy from protein and fiber in the pig—A case for NE systems. *Proceeding of the Midwest Swine Nutrition Conference,* pp. 58–72. Indianapolis, IN.

Drescher, A. J., S. K. Baidoo, L. J. Johnston, and G. C. Shurson. 2009. Effect of DDGS source on growth performance and carcass characteristics of growing-finishing pigs. *Journal of Animal Science* 87(E-Suppl. 3): 80. (Abstr.)

Drescher, A. J., L. J. Johnston, G. C. Shurson, and J. Goihl. 2008. Use of 20% dried distillers grains with solubles (DDGS) and high amounts of synthetic amino acids to replace soybean meal in grower–finisher swine diets. *Journal of Animal Science* 86(Suppl. 2): 28. (Abstr.)

Duttlinger, A. W., M. D. Tokach, S. S. Dritz, J. M. DeRouchy, J. L. Goodband, R. D. Goodband, and H. J. Prusa. 2008. Effects of increasing dietary glycerol and dried distillers grains with solubles on growth performance of finishing pigs. *Journal of Animal Science* 86(Suppl. 1): 607. (Abstr.)

Eeckhout, W., and M. de Paepe. 1994. Total P, phytate P and phytase activity in plant feedstuffs. *Animal Feed Science and Technology* 47: 19–29.

Ewan, R. C., J. D. Crenshaw, T. D. Crenshaw, G. L. Cromwell, R. A. Easter, J. L. Nelssen, E. R. Miller, J. E. Pettigrew, and T. L. Veum. 1996. Effect of adding fiber to gestation diets on reproductive performance of sows. *Journal of Animal Science* 74(Suppl. 1): 190. (Abstr.)

Fastinger, N. D., and D. C. Mahan. 2006. Determination of the ileal amino acid and energy digestibilities of corn distillers dried grains with solubles using grower–finisher pigs. *Journal of Animal Science* 84: 1722–1728.

Feoli, C., J. D. Hancock, C. Monge, T. L. Gugle, S. D. Carter, and N. A. Cole. 2007. Digestible energy content of corn- vs. sorghum-based dried distillers grains with solubles and their effects on growth performance and carcass characteristics in finishing pigs. *Kansas State University Swine Day Report 2007*, pp. 131–136. Manhattan, KS: Kansas State University.

Feoli, C., J. D., Hancock, T. L. Gugle, and S. D. Carter. 2008. Effects of expander conditioning on the nutritional value of diets with corn- and sorghum-based distillers dried grains with solubles in nursery and finishing diets. *Journal of Animal Science* 86(Suppl. 2): 50. (Abstr.)

Fu, S. X., M. Johnston, R. W. Fent, D. C. Kendall, J. L. Usry, R. D. Boyd, and G. L. Allee. 2004. Effect of corn distiller's dried grains with solubles (DDGS) on growth, carcass characteristics, and fecal volume in growing finishing pigs. *Journal of Animal Science* 82 (Suppl. 2): 80. (Abstr.)

Gaines, A. M., G. I. Petersen, J. D. Spencer, and N. R. Augspurger. 2007a. Use of corn distillers dried grains with solubles (DDGS) in finishing pigs. *Journal of Animal Science* 85 (Suppl. 2): 96. (Abstr.)

Gaines, A. M., J. D. Spencer, G. I. Petersen, N. R. Augspurger, and S. J. Kitt. 2007b. Effect of corn distillers dried grains with solubles (DDGS) withdrawal program on growth performance and carcass yield in grow–finish pigs. *Journal of Animal Science* 85 (Suppl. 1): 438. (Abstr.)

Gaines, A., B. Ratliff, P. Srichana, and G. Allee. 2006. Use of corn distiller's dried grains and solubles in late nursery pig diets. *Journal of Animal Science* 84 (Suppl. 2): 120. (Abstr.)

Gowans, J., M. Callaahan, A. Yusupov, N. Campbell, and M. Young. 2007. Determination of the impact of feeding increasing levels of corn dried distillers grains on performance of growing–finishing pigs reared under commercial conditions. *Advances in Pork Production* 18: A-22. (Abstr.)

Gralapp, A. K., W. J. Powers, M. A. Faust, and D. S. Bundy. 2002. Effects of dietary ingredients on manure characteristics and odorous emissions from swine. *Journal of Animal Science* 80: 1512–1519.

Greiner, L. L., X. Wang, G. Allee, and J. Connor. 2008. The feeding of dry distillers grain with solubles to lactating sows. *Journal of Animal Science* 86(Suppl. 2): 63. (Abstr.)

Guo, L., X. Piao, D. Li, and S. Li. 2004. The apparent digestibility of corn by-products for growing–finishing pigs in vivo and in vitro. *Asian Australian Journal of Animal Science* 17: 379–385.

Hastad, C. W., J. L. Nelssen, R. D. Goodband, M. D. Tokach, S. S. Dritz, J. M. DeRouchey, and N. Z. Frantz. 2005. Effect of dried distillers grains with solubles on feed preference in growing pigs. *Journal of Animal Science* 83(Suppl. 2): 73. (Abstr.)

Hilbrands, A. M., L. J. Johnson, G. C. Shurson, and I. Kim. 2009. Influence of rapid introduction and removal of dietary corn distillers gains with soluble (DDGS) on pig performance and carcass characteristics. *Journal of Animal Science* 87(E-Suppl. 3): 82. (Abstr.)

Hill, G. M., J. E. Link, D. O. Liptrap, M. A. Giesemann, M. J. Dawes, J. A. Snedegar, N. M. Bello, and R. J. Tempelman. 2008a. Withdrawal of distillers dried grains with solubles (DDGS) prior to slaughter in finishing pigs. *Journal of Animal Science* 86(Suppl. 2): 52. (Abstr.)

Hill, G. M., J. E. Link, M. J. Rincker, D. L. Kirkpatrick, M. L. Gibson, and K. Karges. 2008b. Utilization of distillers dried grains with solubles and phytase in sow lactation diets to meet the phosphorus requirement of the sow and reduce fecal phosphorus concentrations. *Journal of Animal Science* 2008. 86: 112–118.

Hinson, R., G. Allee, G. Grinstead, B. Corrigan, and J. Less. 2007. Effect of amino acid program (low vs. high) and dried distillers grains with solubles (DDGS) on finishing pig performance and carcass characteristics. *Journal of Animal Science* 85(Suppl. 1): 437. (Abstr.)

IOM. 2006. Dietary, functional, and total dietary fiber. *Dietary Reference Intakes for Energy, Carbohydrates, Fiber, Fat, Fatty Acids, Cholesterol, Protein, and Amino Acids*, pp. 340–421. Washington, DC: National Academies Press.

Jacela, J. Y., L. Brandts, J. M. DeRouchey, S. S. Dritz, M. D. Tokach, R. D. Goodband, J. L. Nelssen, R. C. Thaler, D. Peters, and D. E. Little. 2008a. Effects of deoiled corn distillers dried grains with soluble, solvent extracted on nursery pig performance. *Journal of Animal Science* 86(E-Suppl. 2): 450. (Abstr.)

Jacela, J. Y., J. M. DeRouchey, S. S. Dritz, M. D. Tokach, R. D. Goodband, J. L. Nelssen, R. C. Sulabo, and R. C. Thaler. 2008b. Amino acid digestibility and energy content of corn distillers meal for swine. *Journal of Animal Science* 86(E-Suppl. 3): 87–88. (Abstr.)

Jacela, J. Y., J. M. DeRouchey, S. S. Dritz, M. D. Tokach, R. D. Goodband, J. L. Nelssen, R. C. Sulabo, and R. C. Thaler. 2007. Amino acid digestibility and energy content of corn distillers meal for swine. *Kansas State University Swine Day Report 2007*, pp. 137–141. Manhattan, KS: Kansas State University.

Jacela, J. Y., H. L. Frobose, J. M. DeRouchey, M. D. Tokach, S. S. Dritz, R. D. Goodband, and J. L. Nelssen. 2009. Amino acid and energy digestibility of high-protein corn distillers dried grains in pigs. *Journal of Animal Science* 87(E-Suppl. 3): 105. (Abstr.)

Jenkin, S., S. Carter, J. Bundy, M. Lachmann, J. Hancock, and N. Cole. 2007. Determination of P-bioavailability in corn and sorghum distillers dried grains with solubles for growing pigs. *Journal of Animal Science* 85(Suppl. 2): 113. (Abstr.)

Jones, C. K., J. R. Bergstrom, M. D. Tokach, J. M. DeRouchey, J. L. Nelssen, S. S. Dritz, and R. D. Goodband. 2009. Effects of commercial enzymes in diets containing dried distillers grains with soluble (DDGS) on nursery pig performance. *Journal of Animal Science* 87(E-Suppl. 3): 83. (Abstr.)

Kass, M. L., P. J. van Soest, and W. G. Pond. 1980. Utilization of dietary fiber from alfalfa by growing swine. I. Apparent digestibility of diet components in specific segments of the gastrointestinal tract. *Journal of Animal Science* 50: 175–191.

Kim, B. G., D. Y. Kil, and H. H. Stein. 2009a. Apparent and true ileal digestibility of acid hydrolyzed ether extract in various feed ingredients fed to growing pigs. *Journal of Animal Science* 87 (E-Suppl. 2): 579. (Abstr.)

Kim, B. G., G. I. Petersen, R. B. Hinson, G. L. Allee, and H. H. Stein. 2009b. Amino acid digestibility and energy concentration in a novel source of high-protein distillers dried grains and their effects on growth performance of pigs. *Journal of Animal Science* 87: 4013–4021.

Lammers, P. J., M. S. Honeymann, and B. J. Kerr. 2009. Biofuels co-products in swine diets: Combining DDGS and glycerol. *Journal of Animal Science* 87(E-Suppl. 3): 90. (Abstr.)

Lan, Y., F. O. Opapeju, and C. M. Nyachoti. 2008. True ileal protein and amino acid digestibilities in wheat dried distillers' grains with solubles fed to finishing pigs. *Animal Feed Science and Technology* 140: 155–163.

Linneen, S. K., M. U. Steidiger, M. D. Tokach, J. M. DeRouchy, R. D. Goodband, S. S. Dritz, and J. L. Nelssen. 2006. Effects of dried distillers grain with solubles on nursery pig performance. *Kansas State University Swine Day Report,* pp. 100–102. Manhattan: Kansas State University.

Linneen, S. K., J. M. DeRouchy, S. S. Dritz, R. D. Goodband, M. D. Tokach, and J. L. Nelssen. 2008. Effects of dried distillers grains with solubles on growing and finishing pig performance in a commercial environment. *Journal of Animal Science* 86: 1579–1587.

Madsen, A., K. Jacobsen, and H. P. Mortensen. 1992. Influence of dietary fat on carcass fat quality in pigs. A review. *Acta Agriculturae Scandinavica, Section A—Animal Science* 42: 220–225.

McEwen, P. 2008. Canadian experience with feeding DDGS. *Proceedings of the 8th London Swine Conference.* London, ON, pp. 115–120. April 1–2, 2008.

McEwen, P. L. 2006. The effects of distillers dried grains with solubles inclusion rate and gender on pig growth performance. *Canadian Journal of Animal Science* 86: 594. (Abstr.)

Mertens, D. R. 2003. Challenges in measuring insoluble dietary fiber. *Journal of Animal Science* 81: 3233–3249.

NRC. 1998. *Nutrient Requirements of Swine,* 10th rev. ed. Washington, DC: National Academies Press.

Nyachoti, C. M., J. D. House, B. A. Slominski, and I. R. Seddon. 2005. Energy and nutrient digestibilities in wheat dried distillers' grains with solubles fed to growing pigs. *Journal of the Science of Food and Agriculture* 85: 2581–2586.

Pahm, A. A., C. Pedersen, D. Hoehler, and H. H. Stein. 2008a. Factors affecting the variability in ileal amino acid digestibility in corn distillers dried grains with solubles fed to growing pigs. *Journal of Animal Science* 86: 2180–2189.

Pahm, A. A., C. Pedersen, and H. H. Stein. 2008b. Application of the reactive lysine procedure to estimate lysine digestibility in distillers dried grains with solubles fed to growing pigs. *Journal of Agricultural and Food Chemistry* 56: 9441–9446.

Pedersen, C., M. G. Boersma, and H. H. Stein. 2007a. Digestibility of energy and phosphorus in 10 samples of distillers dried grains with solubles fed to growing pigs. *Journal of Animal Science* 85: 1168–1176.

Pedersen, C., M. G. Boersma, and H. H. Stein. 2007b. Energy and nutrient digestibility in NutriDense corn and other cereal grains fed to growing pigs. *Journal of Animal Science* 85: 2473–2483.

Prosky, L., G. N. Asp, T. F. Schweizer, J. W. de Vries, and I. Furda. 1992. Determination of insoluble and soluble dietary fiber in foods and food products: collaborative study. *Journal of the Association of Official Analytical Chemists* 75: 360–367.

Sauvant, D., J. Perez, and G. Tran. 2004. *Tables of Composition and Nutritional Value of Feed Materials,* 2nd ed. The Netherlands: Wageningen Academic Publishers.

Seabolt, B. S., E. van Heugten, S. W. Kim, K. D. Ange-van Heugten, and E. Roura. 2010. Feed preferences and performance of nursery pigs fed diets containing various inclusion levels and qualities of distillers coproducts and flavor. *Journal of Animal Science* 88: 3725–3738.

Senne, B. W., J. D. Hancock, R. H. Hines, D. W. Dean, I. Mavromichalis, and J. R. Froetschner. 1998. Effects of whole grain and distillers dried grains with solubles from normal and heterowaxy endosperm sorghums on growth performance, nutrient digestibility, and carcass characteristics of finishing pigs. *Kansas State University Swine Day Report,* pp. 148–152. Manhattan, KS: Kansas State University.

Senne, B. W., J. D. Hancock, I. Mavromichalis, S. L. Johnston, P. S. Sorrell, I. H. Kim, and R. H. Hines. 1996. Use of sorghum-based distillers dried grains in diets for nursery and finishing pigs. *Kansas State University Swine Day Report,* pp. 140–145. Manhattan, KS: Kansas State University.

Senne, B. W., J. D. Hancock, P. S. Sorrell, I. H. Kim, S. L. Traylor, R. H. Hines, and K. C. Behnke. 1995. Effects of distillers grains on growth performance in nursery and finishing pigs. *Kansas State University Swine Day Report,* pp. 68–71. Manhattan, KS: Kansas State University.

Spencer, J. D., G. I. Petersen, A. M. Gaines, and N. R. Augsburger. 2007. Evaluation of different strategies for supplementing distillers dried grains with solubles (DDGS) to nursery pig diets. *Journal of Animal Science* 85(Suppl. 2): 96–97. (Abstr.)

Spiehs, M. J., M. H. Whitney, and G. C. Shurson. 2002. Nutrient database for distiller's dried grains with solubles produced from new ethanol plants in Minnesota and South Dakota. *Journal of Animal Science* 80: 2639–2645.

Soares, J. A., H. H. Stein, J. V. Singh, and J. E. Pettigrew. 2008. Digestible and metabolizable energy in distillers dried grains with solubles (DDGS) and enhanced DDGS. *Journal of Animal Science* 86(Suppl. 1): 522. (Abstr.)

Song, M., S. K. Baidoo, G. C. Shurson, and L. J. Johnson. 2007a. Use of dried distillers grains with solubles in diets for lactating sows. *Journal of Animal Science* 85(Suppl. 2): 97. (Abstr.)

Song, M., S. K. Baidoo, M. H. Whitney, G. C. Shurson, and L. J. Johnson. 2007b. Effects of dried distillers grains with solubles on energy and nitrogen balance, and milk composition of lactating sows. *Journal of Animal Science* 85(Suppl. 2): 100–101. (Abstr.)

Ssu, K. W., M. C. Brumm, and P. S. Miller. 2004. Effect of feather meal on barrow performance. *Journal of Animal Science* 82: 2588–2595.

Stender, D., and M. S. Honeyman. 2008. Feeding pelleted DDGS-based diets for finishing pigs in deep-bedded hoop barns. *Journal of Animal Science* 86(Suppl. 2): 50. (Abstr.)

Stein, H. H. 2007. Distillers dried grains with solubles (DDGS) in diets fed to swine. *Swine Focus #001.* Champaign, IL: University of Illinois Urbana.

Stein H. H., and G. C. Shurson. 2009. Board invited review: The use and application of distillers dried grains with solubles (DDGS) in swine diets. *Journal of Animal Science* 87: 1292–1303.

Stein, H. H., S. P. Connot, and C. Pedersen. 2009. Energy and nutrient digestibility in four sources of distillers dried grains with solubles produced from corn grown within a narrow geographical area and fed to growing pigs. *Asian Australian Journal of Animal Science* 22: 1016–1025.

Stein, H. H., C. Pedersen, M. L. Gibson, and M. G. Boersma. 2006. Amino acid and energy digestibility in ten samples of dried distillers grain with solubles by growing pigs. *Journal of Animal Science* 84: 853–860.

Stein, H. H., B. Seve, M. F. Fuller, P. J. Moughan, and C. F. M. de Lange. 2007. Invited review: Amino acid bioavailability and digestibility in pig feed ingredients: Terminology and application. *Journal of Animal Science* 85: 172–180.

Stevens, J. G., A. P. Schinkel, M. A. Latour, D. Kelly, B. Legan, and B. T. Richert. 2009. Evaluation of distillers dried grains with soluble withdrawal programs on grow-finish pig growth performance and carcass quality. *Journal of Animal Science* 87(E-Suppl. 3): 82. (Abstr.)

Urriola, P. E., and H. H. Stein. 2010. Effects of distillers dried grains with solubles on AA, energy, and fiber digestibility and on intestinal transit time of a corn–soybean meal diet fed to growing pigs. *Journal of Animal Science* 88: 1454–1462.

Urriola, P. E., C. Pedersen, H. H. Stein, and G. C. Shurson. 2009. Amino acid digestibility of distillers dried grains with solubles, produced from sorghum, a sorghum–corn blend, and corn fed to growing pigs. *Journal of Animal Science* 87: 2574–2580.

Urriola, P. E., G. C. Shurson, and H. H. Stein. 2010. Digestibility of dietary fiber in distillers co-products fed to growing pigs. *Journal of Animal Science* 88: 2373–2381.

Van Soest, P. J. 1963. Use of detergents in the analysis of fibrous feeds. II. A rapid method for the determination of fiber and lignin. *Journal of the Association of Official Analytical Chemists* 46: 829–835.

Weimer, D., J. Stevens, A. Schinckel, M. Latour, and B. Richert. 2008. Effects of feeding increasing levels of distillers dried grains with solubles to grow-finish pigs on growth performance and carcass quality. *Journal of Animal Science* 86(Suppl. 2): 51. (Abstr.)

White, H., B. Richert, S. Radcliffe, A. Schinckel, and M. Latour. 2007. Distillers dried grains decreases bacon lean and increases fat iodine values (IV) and the ratio og n6:n3 but conjugated linoleic acids partially recovers fat quality. *Journal of Animal Science* 85(Suppl. 2): 78. (Abstr.)

Whitney, M. H., and G. C. Shurson. 2001. Availability of phosphorus in distillers dried grains with solubles for growing swine. *Journal of Animal Science* 79(Suppl. 1): 108. (Abstr.)

Whitney, M. H., and G. C. Shurson. 2004. Growth performance of nursery pigs fed diets containing increasing levels of corn distillers dried grains with solubles originating from a modern Midwestern ethanol plant. *Journal of Animal Science* 82: 122–128.

Whitney, M. H., G. C. Shurson, L. J. Johnson, D. M. Wulf, and B. C. Shanks. 2006. Growth performance and carcass characteristics of grower–finisher pigs fed high-quality corn distillers dried grain with solubles originating from a modern Midwestern ethanol plant. *Journal of Animal Science* 84: 3356–3363.

Widmer, M. R., L. M. McGinnis, and H. H. Stein. 2007. Energy, phosphorus, and amino acid digestibility of high-protein distillers dried grains and corn germ fed to growing pigs. *Journal of Animal Science* 85: 2994–3003.

Widmer, M. R., L. M. McGinnis, D. M. Wulf, and H. H. Stein. 2008. Effects of feeding distillers dried grains with solubles, high-protein distillers dried grains, and corn germ to growing–finishing pigs on pig performance, carcass quality, and the palatability of pork. *Journal of Animal Science* 86: 1819–1831.

Widyaratne, G. P., and R. T. Zijlstra. 2007. Nutritional value of wheat and corn distillers dried grain with solubles: Digestibility and digestible contents of energy, amino acids and phosphorus, nutrient excretion and growth performance of grower–finisher pigs. *Canadian Journal of Animal Science* 87: 103–114.

Widyaratne, G. P., J. F. Patience, and R. T. Zijlstra. 2009. Effect of xylanase supplementation f diets containing wheat distillers dried grains with soluble on energy, amino acid and phosphorus digestibility and growth performance of grower-finisher pigs. *Canadian Journal of Animal Science* 89: 91–95.

Wilson, J. A., M. H. Whitney, G. C. Shurson, and S. K. Baidoo. 2003. Effects of adding distillers dried grains with solubles (DDGS) to gestation and lactation diets on reproductive performance and nutrient balance in sows. *Journal of Animal Science* 81(Suppl. 2): 47–48. (Abstr.)

Xu, G., S. K. Baidoo, L. J. Johnston, J. E. Cannon, and G. C. Shurson. 2007a. Effects of adding increasing levels of corn dried distillers grains with solubles (DDGS) to corn–soybean meal diets on growth performance and pork quality of growing–finishing pigs. *Journal of Animal Science* 85(Suppl. 2): 76. (Abstr.)

Xu, G., G. C. Shurson, E. Hubby, B. Miller, and B. de Rodas. 2007b. Effects of feeding corn–soybean meal diets containing 10% distillers dried grains with solubles (DDGS) on pork fat quality of growing-finishing pigs under commercial production conditions. *Journal of Animal Science* 85(Suppl. 2): 113. (Abstr.)

Xu, G., S. K. Baidoo, L. J. Johnston, J. E. Cannon, D. Bibus, and G. C. Shurson. 2008a. Effects of adding increasing levels of corn dried distillers grains with solubles (DDGS) to corn–soybean meal diets on pork fat quality of growing–finishing pigs. *Journal of Animal Science* 86(Suppl. 2): 51. (Abstr.)

Xu, G., S. K. Baidoo, L. J. Johnston, J. E. Cannon, D. Bibus, and G. C. Shurson. 2008b. Effects of dietary corn dried distillers grains with solubles (DDGS) and DDGS withdrawal intervals, on pig growth performance, carcass traits, and fat quality. *Journal of Animal Science* 86(Suppl. 2): 52. (Abstr.)

Zhu, Z. P., R. B. Hinson, L. Ma, and G. L. Allee. 2009. Effects of pelleting diets containing distillers dried grain with soluble on growth performance and nutrient digestibility in nursery pigs. *Journal of Animal Science* 87(E-Suppl. 3): 89. (Abstr.)

Zimmermann, B., H. -J. Lantzsch, R. Mosenthin, F. -J. Schoner, H. K. Biesalski, and W. Drochner. 2002. Comparative evaluation of the efficacy of cereal and microbial phytases in growing pigs fed diets with marginal phosphorus supply. *Journal of the Science of Food and Agriculture* 82: 1298–1304.

15 Feeding Ethanol Coproducts to Poultry

Amy B. Batal and Kristjan Bregendahl

CONTENTS

15.1 INTRODUCTION

It is now well recognized by the feed industry (as mentioned in Chapter 14, "Feeding DDGS to Swine") that increasing quantities of distillers dried grains with solubles (DDGS) are entering the ingredient market. Coproducts from dry grind processing of cereal grains for alcohol production have been available to poultry and livestock producers for many years. For those interested in a history of this ingredient, a substantial scientific literature exists in both refereed journal and proceedings of conferences, particularly the Distillers Feed Conference. In 1970, Dr. M. L. Scott presented a paper entitled "Twenty-Five Years of Research on Distillers Feeds in Broiler" (Scott, 1970). Aside from Dr. Scott, many prominent poultry nutritionists of the day were active in studying the use of this ingredient, these including G. F. Combs, Sr. (Combs and Bossard, 1969), T. D. Runnels (Runnels, 1966, 1968), R. H. Harms (Harms et al., 1969), and L. S. Jensen (Jensen et al., 1974, 1978). Although the coproducts were considered better suited for ruminants because of their relatively high fiber content, Morrison (1954) suggested that chick diets could contain up to 8% distillers grains and that diets for laying hens could contain up to 10% distillers grains without affecting performance. No adverse effects on growth performance of broiler chickens or egg production of laying hens were detected when diets with up to 20% DDGS from beverage-alcohol production were fed (Matterson et al., 1966; Waldroup et al., 1981), although the feed utilization of broilers tended to decrease when 25% corn DDGS was included in the diet (Waldroup et al., 1981). In addition to being a source of protein and energy, distillers grains was especially useful as a source of the water-soluble vitamins before chemical synthesis and the commercialization of vitamins (Morrison, 1954; Matterson et al., 1966). Nevertheless, in recent decades DDGS has only rarely been a component of broiler and layer feeds, despite the dramatic growth of the poultry industry. The lack of interest in DDGS seems to stem partly from a relatively limited supply, competing use in ruminant feeds without having to dry the product, and concerns over an occasionally inconsistent composition. The DDGS available many decades ago was derived primarily from the beverage industry based on the fermentation of a variety of grains, in addition to commercial production of alcohol. However, in recent years, policies encouraging the production of ethanol have stimulated an enormous increase in the production of DDGS, through a fermentation process that is slightly different from those of beverage-alcohol production. The DDGS ingredient that is currently becoming available differs from that of previous decades in that it is derived almost entirely from corn and is dried under less severe conditions. As a result, over 98% of the fermentation coproducts available today are from fuel ethanol production using corn grain as a substrate (University of Minnesota, 2008a).

Coproducts of the ethanol industry will contain approximately three times the level of nutrients in the feedstock (except for the starch/sugars) used for the fermentation of ethanol. Corn grain is composed of approximately two-third starch, which is consumed during fermentation, which essentially triples the concentration of the oil, fiber, protein, and other minerals in the final corn DDGS. Corn DDGS typically contains about 27% crude protein, 10% oil, and 0.8% phosphorus (Table 15.1; refer to Chapters 2 and 8 for more information on the composition of ethanol coproducts). Until recently, the majority of the dry grind ethanol plants used unmodified corn (or other cereal grains) to produce ethanol and some type of distillers dried grains (DDG). However, there is growing interest by many ethanol producers to modify the technology to get better ethanol yield from the feedstock will continue to result in coproducts that have markedly different nutritional composition. Depending on the specific ethanol plant, there can be several variations in the ethanol production process: some remove the oil-rich germ and fiber-rich hulls prior to fermentation to improve ethanol yield, some omit the jet-cooking process, others remove the oil from the thin stillage, and so on. Currently, many plants are implementing a modified dry milling process as the first step in the ethanol facility

TABLE 15.1
Average Nutrient Composition of Distillers Wet Grains, Distillers Solubles, Distillers Dried Grains with Solubles (DDGS), High-Protein Distillers Dried Grains (HP-DDG), and Corn Germ for Poultry

Component (%)	Distillers Wet Grains (DWG)	Distillers Solubles[a]	DDGS	HP-DDG	Corn Germ
TME (kcal/kg)	87	87	2,800	2750	3500
Crude protein (%)	31	18	27	42	15
Crude fat (%)	7.0	17	10	3.0	16
Crude fiber (%)	9.0	1.5	7	7	5
Ash (%)	2.5	8.5			
Phosphorus (%) total	0.48	1.25	0.75	0.40	1.20
Phosphorus (%) available[b]		1.00	0.49	0.20	0.36

Nutrient composition values are just averages and should only be used as a guide. For each ingredient significant differences in composition can and do exist.

Source: Data from Martinez-Amezcua, C. et al. *Poultry Science* 86, 2624–2630, 2007 and unpublished data.

[a] Also known as condensed distillers solubles, or CDS.

[b] Calculated assuming an estimated bioavailability of 65% for DDGS, 50% for HP-DDG, and 30% for corn germ (Batal, 2007; Kim et al., 2008).

TABLE 15.2
Effect of Soluble Addition on TMEn and Nutrient Characteristics of DDGS

Nutrient	Soluble Addition (gal/min) (%)			
	0 (0)	12 (30)	25 (60)	42 (100)
Protein, %	32	33	32	32
Fat, %	8.0	9.1	9.2	10.5
Phosphorus, %	0.53	0.66	0.77	0.91
TME (kcal/kg)	2712	2897	3002	3743

Source: Noll et al. *Poultry Science*, 85(Suppl. 1), 204, 2006.
Note: Soluble are also known as condensed distillers solubles, or CDS.

in which they are recovering the nonfermentables (germ and fiber) prior to the dry grind process. The whole corn is milled into three fractions; corn germ, bran, and the endosperm that is used for ethanol fermentation. The products from the fermentation of the endosperm are ethanol, syrup, CO_2, high-protein distillers dried grains (HP-DDG), and corn germ. The HP-DDG, resulting from the prefermentation fractionation of the corn grain, contains approximately 40% crude protein, whereas the dehydrated corn germ contains 15% crude protein (Table 15.1; Refer to Chapter 5 for more information on the processing of corn grain into ethanol and distillers grain). The solubles stemming from the fermentation of fractionated corn grain can be used differently depending on the ethanol plant. Some plants combine the solubles (known as condensed distillers solubles, or CDS) with the bran and sell it for use in ruminant feed, and some plan to burn it for energy. By definition, corn DDGS (International Feed Number 5-02-843) consists of a dried mixture of at least 75% of the solids in the whole stillage (AAFCO, 2007) and therefore includes the wet grains and (most of) the solubles. Corn DDG (International Feed Number 5-02-842), however, includes only the wet grains (AAFCO, 2007). Hence, corn DDG and HP-DDG do not contain the germ and the nutrient rich solubles fraction, resulting in a markedly different nutrient profile than that of corn DDGS (Table 15.1). DDGS, HP-DDG, and dehydrated corn germ are suitable feed ingredients for poultry and can be included in poultry diets as long as the nutrient and energy contents are known and the diet is formulated accordingly. However, one of the biggest challenges of feeding DDGS to poultry is the variation in nutrient content and digestibility among U.S. ethanol plants. The variation in DDGS is most likely due to the grain input (corn varieties and geographic location), processing at plant, drying conditions, and the rate of the addition of solubles back to the grains (Tables 15.1 and 15.2). This chapter will cover the nutritional characteristics of DDGS and discuss how the variation in the nutritional composition of DDGS affects it feeding value for poultry. Recommendations on the environmental impact, potential issues, and benefits of feeding DDGS to poultry will also be discussed.

15.2   NUTRIENT COMPOSITION OF DDGS FOR POULTRY

15.2.1   Energy

Energy is often the main parameter in determining the value of most feed ingredients for poultry. Thus, it is important that the available energy in an ingredient is known before it can accurately be used in a poultry diet. In the United States, the nitrogen-corrected metabolizable energy (MEn) system is the most common method used to determine the energy in a feed ingredient. This measure represents the gross energy of the feed minus the gross energy of the feces and urine, corrected for nitrogen retained in the body (NRC, 1994). True MEn (TMEn) is determined by taking into account endogenous energy losses in the feces. Because of the correction for endogenous energy losses, values for TMEn are usually greater than the corresponding apparent MEn values (NRC, 1994). The energy value of DDGS (referring to corn DDGS throughout unless otherwise noted) has been evaluated using the precision-fed rooster assay, in which a small amount (30 to 35 g) of DDGS is fed to adult male birds after a 24-h fast, and the resultant excreta is collected over a 48-h period; endogenous energy losses are estimated from the gross energy of excreta from birds fasted for 48 h (Sibbald, 1976, 1986).

The TMEn content of DDGS has been reported to range between 2308 and 3434 kcal/kg. Lumpkins et al. (2004) reported the TMEn content of a single DDGS sample to be 2905 kcal/kg. Batal and Dale (2006) determined the TMEn content of 17 different DDGS samples, representing products from six different ethanol plants. The determined TMEn values ranged from 2490 to 3190 kcal/kg with a mean of 2820 kcal/kg and an associated coefficient of variation of 6.4%. From a smaller data set with five samples of DDGS from five different ethanol plants, Fastinger et al. (2006) concluded that the TMEn content of DDGS averaged 2871 kcal/kg, although there was considerable variation among the samples (range of 563 kcal/kg). A large variation in TMEn values of DDGS was also reported by Parsons et al. (2006), who determined the mean TMEn value of 20 DDGS samples to be 2863 kcal/kg with a range of 447 kcal/kg. Waldroup et al. (2007) suggested that nutritionists use a TMEn value of 2851 kcal/kg for DDGS, based on a survey of published TMEn values. Roberson et al. (2005) determined the apparent MEn of a single DDGS sample with laying hens to be 2770 kcal/kg. This value was about 4% lower than the TMEn value determined for the same DDGS sample using cockerels. Roberson (2003) also observed that an ME value of 2870 kcal/kg was too high for DDGS when it was fed to turkeys and instead used an ME value of 2805 kcal/kg. A TMEn value of 2800 kcal/kg has often been stated as the average value for corn DDGS. However, the MEn or TMEn value should be determined for each plant and/or supplier; due to the variation in DDGS it is difficult to recommend an average MEn or TMEn value.

To determine if gross energy could be used to estimate the TMEn of DDGS, Fastinger et al. (2006) tested five DDGS. Based on the average gross energy (4900 kcal/kg) and TMEn (2871 kcal/kg) it appeared that the TMEn of DDGS is close to 60% of its gross energy content, similar to the relationship between gross energy and TMEn in other protein-rich ingredients, such as soybean meal (Leske et al., 1991). However, the relationship was decidedly lower (51%) for one sample of DDGS (Fastinger et al., 2006), so predicting TMEn of DDGS from its gross energy content is not recommended. Rather, the TMEn content may be predicted from the chemical composition of DDGS. To test this theory Batal and Dale (2006) developed equations with which to estimate the TMEn on the basis of proximate composition. The best single indicator of TMEn was fat ($r^2 = 0.29$). The addition of fiber, protein, and ash improved the accuracy of the TMEn prediction equation ($r^2 = 0.45$). However, the relatively low r^2 suggests that these prediction equations should only serve as a guide (Batal and Dale, 2006). The National Research Council (NRC, 1994) included MEn prediction equations for various feed ingredients, including DDGS, based on the chemical composition. When the MEn content of DDGS is calculated by entering the proximate analyses reported by Batal and Dale into the NRC-suggested equation, it is evident that the NRC equation underestimates the MEn content of DDGS when comparing the corresponding TMEn determined by Batal and Dale, even taking into consideration that MEn values of DDGS are about 4% to 5% lower than their corresponding TMEn values (Roberson et al., 2005). The TMEn values calculated using the prediction equation reported by Batal and Dale (2006) correspond well with the determined TMEn values and better than the MEn values calculated using the NRC equation. Due to the low correlation (0.45) of the prediction equation by Batal and Dale (2006) the equation should be verified with an independent set of DDGS samples before it is widely used (Black, 1995).

The best single predictor (although not a good predictor) of TMEn content in DDGS was the oil content ($r^2 = 0.29$) (Batal and Dale, 2006). Because the solubles contain over three times as much oil as the wet grains (Table 15.1), the rate of solubles addition during the DDGS manufacturing process is directly related ($r^2 = 0.88$) to the DDGS TMEn content (Noll et al., 2007a, 2007b) (Table 15.2). The oil content of corn DDGS has been reported to vary from 2.5% to 16% in DDGS samples (Batal and Dale, 2006; Parsons et al., 2006; University of Minnesota, 2008b), with substantial potential for variation in TMEn content. The ratio of blending the solubles with the wet grains fractions varies between and within ethanol plants. There are substantial differences in the nutrient composition of the soluble and grain components (Table 15.1) making it easy to see how any variation in blending would affect the DDGS composition, specifically the crude fat and the TMEn value.

Noll and coauthors (Noll et al., 2007a; Noll et al., 2007b) reported a strong inverse correlation (correlation coefficient, r = −0.98) between the degree of lightness (L* values) of DDGS and the rate of solubles addition, suggesting that darker DDGS has a greater content of TMEn. However, Fastinger et al. (2006) reported only a moderate relationship ($r^2 = 0.52$) between the degree of lightness and the TMEn content of DDGS. Dale and Batal (2005) reported no correlation between color and TMEn values. The TMEn and L* values reported in the Noll studies were from DDGS obtained from a single ethanol plant in which the solubles addition rate was experimentally varied, whereas the values by Fastinger et al. (2006) and Dale and Batal (2005) were commercial DDGS samples from different ethanol plants. Moreover, the variation among samples within in the Noll studies is much smaller and may have allowed the scientist to observe differences. Nevertheless, color is not believed to be a reliable indicator of energy content in DDGS.

Limited studies have been conducted on the new coproducts from ethanol plants. Kim et al. (2008) determined the TMEn value of HP-DDG and corn germ. The TMEn content of HP-DDG and corn germ was 2694 and 4137 kcal/kg, respectively. The value reported for the HP-DDG is similar to the average (2851 kcal/kg) and falls within the range (2667 to 3282 kcal/kg) reported by Batal (2007) and Jung and Batal (2009). However, the TMEn value reported for the corn germ was much higher than the average (3204 kcal/kg) reported by Batal (2007) but it did fall within the range reported (2911 to 3681 kcal/kg). Although HP-DDG has less oil (due to the removal of the germ prior to fermentation and no addition of the solubles) the TMEn value appears to be similar, or perhaps slightly lower than the energy values reported for DDGS. Dehydrated corn germ, however, contains about 12% to 20% more TMEn than the DDGS, likely due to the differences in the oil and protein contents between the two coproducts.

There has been increased interest in using different feedstocks/grains (sources of starch) for ethanol production. It is important to know the feedstock used for the production of the ethanol as this will directly affect the nutritional composition and the TMEn of the subsequent DDGS. Thacker and Widyaratne (2007) determined the gross and metabolizable energy contents of a single sample of wheat DDGS from fuel ethanol production to be 4724 and 2387 kcal/kg, respectively. It has been reported that the TMEn value of sorghum DDGS ranges from 2600 to 2860 kcal/kg (personal communication, Dr. Hancock at Kansas State University). The TMEn of barley DDGS can range from 2745 to 3225 kcal/kg, which may be correlated to the range in fat (4.9 to 8.6) that has also been observed (Batal, unpublished). Refer to Chapter 6 for more information on processing of alternative grains into ethanol and distillers grains.

15.2.2 Amino Acids/Protein

The protein content of DDGS (Tables 15.1 and 15.3) has been reported to vary between 23% and 32% (Spiehs et al., 2002; Evonik Degussa, 2005; Batal and Dale, 2006; Fastinger et al., 2006). This wide range is likely because of differences in the protein content of the corn grain used to produce DDGS and because of differences in residual starch content (diluting the concentrations of protein and other nutrients) caused by differences in fermentation efficiency. Although some DDGS suppliers go to great lengths to minimize variation in nutrient contents (Stein et al., 2006), the amino acid content in DDGS in general can vary substantially. For instance, the content of the first-limiting amino acid for poultry, methionine, has been reported to range from 0.42% to 0.65% (Spiehs et al., 2002; Evonik Degussa, 2005; Fastinger et al., 2006). Nevertheless, the amino acid content of DDGS is among the main reasons for including this coproduct in poultry diets.

The true digestibility of amino acids in DDGS has been reported to vary substantially among ethanol plants (Batal and Dale, 2006; Fastinger et al., 2006) and it could potentially vary from batch to batch within the same ethanol plant. It has been suggested that the main reason for the variation is the drying process (Fontaine et al., 2007). Different drying techniques (e.g., rotary kiln drying, ring drying),

TABLE 15.3
Total and Digestibility Coefficients of Selected Amino Acids of Corn Coproducts from the Fuel Ethanol Industry

	DDGS		HP-DDG		Corn Germ	
	Total (%)	Digestibility Coefficient (%)	Total (%)	Digestibility Coefficient (%)	Total (%)	Digestibility Coefficient (%)
Arginine	1.10	84	1.50	87	1.15	91
Isoleucine	1.07	82	1.75	85	0.45	80
Leucine	3.10	90	6.10	93	1.05	84
Lysine	0.80	70	1.15	74	0.80	85
Methionine	0.55	85	0.95	92	0.27	86
Met + Cystine	1.00	79	1.08	88	0.60	83
Threonine	1.00	75	1.55	82	0.55	75
Tryptophan	0.18	82	0.20	85	0.13	87
Valine	1.35	82	2.15	87	0.73	82

Sources: Data from Batal, A., and N. Dale. *Journal of Applied Poultry Research* 12, 400–403, 2003; Batal, A. *Poultry Science* 86 (Suppl. 1), 206, 2007; Jung, B. Y., and A. B. Batal. *Journal of Applied Poultry Research* 18: 741–751, 2009.

drying temperatures, and drying times can cause inconsistent drying (e.g., "hot spots") or overdrying (Refer to Chapter 5 for more information on the drying ethanol coproduct streams). Precooking the corn grain to remove unwanted microbial contamination may also be responsible for some of the heat damage. These processes are also the reasons why the amino acid digestibility is lower in DDGS than in corn grain. In particular, the digestibility of lysine varies substantially because of its susceptibility to heat damage during the drying process (Stein et al., 2006; Fontaine et al., 2007). The epsilon amino group on lysine can react with reducing sugars in a Maillard reaction. Because poultry do not produce the enzymes needed to break the bond between lysine and the sugar residue, the lysine involved in the Maillard reaction becomes unavailable to the birds. It is important to note that although lysine bound to a sugar is unavailable to the animal, the lysine can still be measured and will be accounted for when measuring total lysine. Thus, just measuring the total lysine content of the DDGS will not provide information regarding the availability. Batal and Dale (2006) measured the true digestibility of lysine in eight DDGS samples using the cecectomized rooster assay; lysine digestibilities ranged from 46% to 78%, with a mean digestibility of 70%. In this case the analyzed content of total lysine also varied considerably, from 0.39% total lysine in the DDGS with the lowest digestibility to 0.86% total lysine in the DDGS with the highest digestibility, but the total lysine content was really only low for the samples that had significantly lower true lysine digestibility coefficients. The production of complete Maillard reactions possibly resulted in the samples that had low total lysine content along with low lysine digestibility because the extreme heating not only caused the lysine to bind with a sugar it may have destroyed the lysine to such a degree, making it unavailable to the animal as well as immeasurable on currently analytical equipment (Cromwell et al., 1993; Fontaine et al., 2007; Martinez-Amezcua et al., 2007). The true amino acid digestibility also varied among the five different DDGS samples tested by Fastinger et al. (2006). The lysine digestibility varied from 65% to 82%, appearing to be correlated with total lysine content in the DDGS. As noted previously, variation in lysine digestibility was the most common issue with overheating, although the true cystine digestibility varied substantially as well (Batal and Dale, 2006). While the digestibility varied for all amino acids, the variation in lysine digestibility among DDGS samples was the greatest, suggesting varying degrees of heat damage through differences in drying temperatures and time among and within ethanol plants.

Pahm et al. (2009) evaluated an in vitro reactive lysine method to predict the concentration of standardized digestible (SDD) lysine. The concentration of SDD lysine was correlated ($r^2 = 0.84$, p < 0.05) with the concentration of reactive lysine in DDGS. The authors concluded that the values for reactive lysine may be used to estimate the concentration of SDD lysine. Fontaine et al. (2007) measured the contents of reactive lysine, in 80 DDGS samples and suggested that 10% to 40% of the lysine in DDGS is heat damaged, and that some overheated batches of DDGS lost up to 59% of their lysine. The DDGS digestibility values reported by the NRC (1994) are mainly from experiments with DDGS originating from beverage-alcohol production and probably should not be used for DDGS from fuel ethanol production because of differences in the processes, most notably drying. Current research is being conducted to determine the correlation of furosine and digestible lysine (amino acid) levels in DDGS. The research that has been conducted to date has shown a very high correlation between furosine levels and digestible lysine, suggesting that measuring the level of furosine may be a quick way to estimate amino acid digestibility of DDGS samples.

The degree of heat produced from the drying (i.e., a combination of the drying temperature and time) affects the amino acid digestibility in DDGS because of Maillard reactions. These reactions between amino acids and sugars generate a characteristic dark color, which can be used as a rough guide for the extent of heat damage to amino acids and lower amino acid digestibility (Cromwell et al., 1993; Batal and Dale, 2006; Fastinger et al., 2006). In general, samples of DDGS with a lighter and more yellow color (i.e., with greater L* and b* values, respectively) tend to have greater amino acid digestibility values and greater contents of true digestible amino acids (see Table 15.4 and Figure 15.1 for an example). Ergul et al. (2003) noted a significant positive correlation between lysine, cystine, and threonine digestibility and lightness (L*) and yellowness (b*) values of DDGS. The rate of addition of solubles to DDGS affects the product's color; a greater addition rate was associated with a darker, less yellow color (i.e., lower L* and b* values), which tended to correlate with a decrease in the true lysine digestibility, in part because the greater addition rate of solubles warranted greater drying temperatures (Noll et al., 2007a, 2007b). Pahm et al. (2009) evaluated color score and bioavailable lysine. Similar to reports with digestible lysine the authors concluded that greater L* (lightness) scores were associated with a greater ($r^2 = 0.90$, p < 0.05) concentration of bioavailable lysine in DDGS. However, contrary to their expectations, Martinez-Amezcua et al. (2007) did not detect a relationship between the rate of addition of solubles and the true digestibility of amino acids in DDGS.

As the use of new feedstocks (other than corn) to produce ethanol will change the energy value of the corresponding coproducts it will also significantly affect the total and digestible amino acid concentrations. Jung and Batal (2009) reported that the average (1.23%) and range (1.13% to 1.38%) of total lysine in HP-DDG was approximately twice as high as compared to conventional DDGS (Table 15.3). The total level of all amino acids doubled as seen with lysine, which would have been expected as the average crude protein level in HP-DDG is 44% versus an average of 28% for conventional corn DDGS. The total amino acid levels reported by Jung and Batal (2009) were slightly higher than the value reported from Kim et al. (2008). The lysine amino acid digestibility coefficient for HP-DDG averaged 76.1% and ranged from 67.5% to 85.6% (Jung and Batal, 2009) and was similar to the value of 73% reported by Kim et al. (2008). The true amino acid digestibility of HP-DDG was generally greater than that of unfractionated, conventionally produced DDGS, but, the true

TABLE 15.4
Example of the Relationship between Color and Lysine Digestibility[a] (see Figure 1)

| | Color | | | | | |
Sample	L* (Lightness)	b* (Yellowness)	a* (Redness)	Total Lysine	Lysine Digestibility Coefficient[b]	Digestible Lysine
1	60.3	25.9	5.0	0.86	76.8	0.66
2	57.7	18.3	6.2	0.82	72.1	0.59
3	50.4	7.41	5.2	0.39	45.8	0.18

FIGURE 15.1 Example of color variation observed in distillers dried grains with solubles.

lysine digestibility was not different between the two products (Kim et al., 2008; Jung and Batal, 2009). Batal (2007) reported a 0.83% total lysine level for corn germ with a digestibility coefficient of 85.7%. This agrees with the 0.80% total lysine and a 91% digestibility coefficient reported by Kim et al. (2008). The total lysine in sorghum DDGS has been stated to be between 0.55% to 0.60%, slightly lower than the 0.7% to 0.8% total lysine in conventional corn DDGS. The lysine digestibility coefficient for sorghum DDGS is around 70%, very similar to that which has been reported for corn DDGS. The total lysine levels in barley DDGS ranged from 0.9% to 1.43% with an average of 1.15% and a digestibility coefficient of 80%. These values seem somewhat reasonable although slightly low; if you estimated that the total lysine level (0.53%) in barley would increase approximately threefold one would calculate a total lysine level around 1.59%. The composition of barley DDGS appears to be highly dependent on variety and fermentation method (Batal, unpublished). Change in feedstocks as well as different fractionation process can greatly influence the amino acid composition and digestibility, thus it is important to have accurate information on the nutrient composition and digestibility of all coproducts.

15.2.3 PHOSPHORUS

DDGS contains a substantial amount of total phosphorus (Table 15.1) (0.72%; Martinez-Amezcua et al., 2004), most of which is highly bioavailable. As with other nutrients, the phosphorus content in DDGS varies, with reports ranging widely, from 0.59% to 0.95% (Spiehs et al., 2002; Batal and Dale, 2003; Martinez-Amezcua et al., 2004; Stein et al., 2006). The large range in phosphorus content stems in part from variation in phosphorus content in the corn grain and starch residue in the DDGS, but the rate of the addition of solubles to the wet grains prior to drying affects the phosphorus content as well, because the solubles contain more than three times as much phosphorus as do the wet grains (Table 15.1) (Martinez-Amezcua et al., 2007; Noll et al., 2007a, 2007b). As with amino acid digestibility, the total phosphorus content of DDGS can be predicted to some extent by observing significant variations in the color of the DDGS. Noll et al. (2006) and Noll et al. (2007) observed that as the rate of solubles being added back to the grains made the DDGS darker in color (lower L* color values) it also increased the phosphorus content of the DDGS (r^2 = 0.96 and 0.98, respectively).

The relative P bioavailability of DDGS has been reported to be higher than the values reported in NRC (1994). Martinez-Amezcua et al. (2004) reported the phosphorus bioavailability (relative to that of phosphorus in dipotassium hydrogen phosphoric acid (K_2HPO_4), which was considered to be 100% bioavailable), estimated from tibia ash using the slope ratio method to range from 69%

to 102%, among DDGS samples with varying degrees of lysine digestibility. Lumpkins and Batal (2005) determined the phosphorus bioavailability of two DDGS samples (relative to K_2HPO_4) to be 68% and 54%. While the phosphorus in corn is only about 30% bioavailable (Lumpkins and Batal, 2005), the bioavailability of phosphorus in DDGS is much greater; this may be due to the heat destruction of phytate during drying and may also be due to some production of phytase during fermentation (Martinez-Amezcua et al., 2004; Martinez Amezcua and Parsons, 2007). It can be postulated that the fermentation vessel acts much like a rumen, synthesizing phytase and thus the phytase being active in the resulting DDGS improving the phosphorus bioavailability. It has also been stated that ethanol plants are adding phytase to the fermentation tanks.

The phosphorus bioavailability of DDGS appears to be inversely correlated with lysine digestibility, and researchers have suggested that the degree of heat damage (which reduces lysine digestibility) increases phosphorus bioavailability (Martinez Amezcua and Parsons, 2007). This hypothesis was further examined in a study in which heat damage of DDGS was controlled by autoclaving or oven drying at different temperatures and times (Martinez Amezcua and Parsons, 2007). Increased heating of the DDGS increased phosphorus bioavailability from 69% in the control DDGS to as much as 91% in the DDGS sample that was oven dried at 55°C for three days and then oven dried at 121°C for 60 min. As expected, the lysine digestibility decreased with increasing heat treatment. The inclusion of citric acid in DDGS-containing diets improved phosphorus bioavailability (Martinez-Amezcua et al., 2006), as did the addition of a commercially available phytase enzyme. Martinez-Amezcua et al. (2006) estimated that phytase released from 0.049% to 0.072% phosphorus from DDGS (which indicates that phytase released approximately 20% to 28% of the nonbioavailable phosphorus in DDGS). However, the authors noted that the efficacy of the phytase enzyme in improving phosphorus bioavailability depends on the phytate-phosphorus content of the DDGS, which is likely to be affected by processing (heat treatment).

The phosphorus content of HP-DDG is lower than that of DDGS (0.3% vs. 0.7% for HP-DDG and DDGS, respectively) (Table 15.1). However, the relative phosphorus bioavailability of HP-DDG (50%) does not appear to be different from that of DDGS (to date only a limited number of samples have been evaluated for HP-DDG and corn germ), but the bioavailability of phosphorus in dehydrated corn germ is only 25% to 30% relative to the bioavailability of phosphorus in K_2HPO_4 (Kim et al., 2008; Batal, 2007). The phosphorus content of coproducts from corn fractionated prior to fermentation depends critically on the fractionation method used (Batal, 2007; Martinez-Amezcua et al., 2007; Kim et al., 2008; Jung and Batal, 2009) and, presumably, so does the bioavailability.

15.2.4 MINERALS

DDGS is often considered to be high in heavy metals, which may impact the rancidity rate of the fat, which is highly unsaturated, which can be as high as 15%. The mineral concentration of DDGS is approximately threefold that found in corn grain but the contents of sulfur and sodium can be substantially greater than what could be expected from the inherent mineral content in corn grain (Table 15.5). The sources of the "extra" sulfur in DDGS include the sulfur in yeast, well water, and sulfuric acid (H_2SO_4) added during the ethanol production process. Sulfuric acid is added at several stages in the process to adjust the pH to different optimum levels for the carbohydrases and the yeast. Depending on well-water quality and the need for pH adjustments, the sulfur content of DDGS can vary substantially, from 0.3% to well over 1% (Spiehs et al., 2002; Batal and Dale, 2003; University of Minnesota, 2008b). Broiler chickens can tolerate dietary sulfur levels of up to 0.5%, and laying hens can tolerate even greater levels (Leeson and Summers, 2001, 2005), so there does not appear to be any issues with feeding high sulfur DDGS to poultry. However, sulfur may interfere with calcium and trace mineral absorption in the small intestines and thus bone and eggshell strength could be affected (Leeson and Summers, 2001, 2005).

TABLE 15.5
Mineral Composition of Distillers Dried Grains with Solubles (DDGS)

	Sodium	Potassium	Phosphorus	Calcium	Magnesium	Sulfur
			(%)			
Average ± SD	0.25 ± 0.15	0.91 ± 0.11	0.68 ± 0.07	0.29 ± 0.27	0.28 ± 0.04	0.84 ±0.21
NRC (1994)	0.48	0.65	0.72	0.17	0.19	0.30
Projected[a]	0.10	0.90	0.84	0.06	0.36	0.24
	Manganese	**Iron**	**Aluminum**	**Copper**	**Zinc**	
			(ppm)			
Average ± SD	22 ± 12	149 ± 86	56 ± 30	10 ± 4.3	61 ±13	
NRC (1994)	24	280	NL[b]	57	80	
Projected[a]	21	138	NL[b]	9	54	

Source: Data from Batal, A., and N. Dale. *Journal of Applied Poultry Research* 12, 400–403, 2003; Batal, A. *Poultry Science* 86 (Suppl. 1), 206, 2007.

[a] Projected assuming a threefold increase compared to yellow corn grain.
[b] Not listed.
SD denotes standard deviation.

The sodium content of DDGS varies significantly from what would be expected, ranging from about 0.09% to 0.52% (Spiehs et al., 2002; Batal and Dale, 2003; University of Minnesota, 2008b). The source of variation is believed to be due primarily to sodium based cleaning products used in the ethanol plants. Thus, the sodium level appears to not only vary between plants but can also vary within plants depending on the cleaning schedule. Poultry can tolerate high levels of sodium in the diet (Klasing and Austic, 2003), but the levels should be monitored and adjusted (through changes in the rate of salt inclusion in the diets) depending on the sodium level of the DDGS. High dietary sodium levels lead to increased water consumption, which may increase the incidences of wet litter, foot pad lesions, and dirty eggs (Leeson et al., 1995; Klasing and Austic, 2003). Therefore, the sodium level needs to be known and taken into account when formulating diets in order to avoid any potential production problems.

15.2.5 CAROTENOID PIGMENTS (XANTHOPHYLL)

Xanthophyll pigments are of special interest to poultry producers. These pigments have to be obtained from the diet and corn grain contains significant levels of xanthophylls. When consumed by poultry, xanthophylls are absorbed and deposited in the skin, adipose tissue, and egg yolks, changing their color to the more desirable yellow or red color (Ouart et al., 1988; Leeson and Caston, 2004). Corn grain contains about 20 ppm of xanthophylls (NRC, 1994; Leeson and Summers, 2005), and one would expect the content in DDGS to be concentrated threefold, thus containing approximately 60 ppm. Thus, one could assume that corn DDGS would be an excellent source of xanthophylls. However, the actual xanthophyll content may be lower because xanthophylls are very heat sensitive and may suffer heat destruction during the drying of the DDGS. There is very little reported data on the xanthophyll level in DDGS making it hard for one to know what xanthophyll level to use for DDGS. Roberson et al. (2005) analyzed two DDGS samples for xanthophylls and reported 30 ppm in one sample, but only 3 ppm in the other much darker sample. The lower level in the darker sample leaves one to speculate that the xanthophylls were likely heat damaged because of the greater drying time needed due to increased soluble addition (which would account for the darker color). The results reported by Roberson et al. (2005) suggest, that as with the other nutrients in DDGS, the carotenoid content can be a significant variable and needs to be monitored.

15.2.6 Mycotoxins

Just as the levels of nutrients are tripled in DDGS as compared to the original grain, this rule also applies to the concentration of mycotoxins because mycotoxins are not destroyed by fermentation. While mycotoxins may occur, this is considered by the alcohol industry to be relatively unlikely. The profit from fermentation is clearly in the efficient production of alcohol. Corn or any feedstock that has been improperly stored and has developed aflatoxin or other mycotoxins may not give the same efficiency of alcohol production as higher quality feedstocks. However, it is a real possibility that DDGS can be contaminated with mycotoxins, and it will be highly dependent on the yearly corn (feedstock) crop. Thus, it is important for poultry producers to follow the levels of mycotoxins in the yearly corn crops or feedstocks and to know if the ethanol plant is testing the incoming feedstock and the DDGS coproduct being produced. Refer to Chapter 11 for more information on mycotoxin occurrence in DDGS and other ethanol coproducts.

15.3 FEEDING DDGS TO POULTRY

15.3.1 Meat Production (Broiler Chickens and Turkeys)

In past decades, DDGS was used as a feed ingredient in poultry diets, partially due to its unidentified growth factors. Later, it was determined that these unidentified growth factors were usually vitamins synthesized during the fermentation that were limited in the diets. Thus, feeding DDGS resulted in improved overall performance. Researchers have observed positive results when DDGS was incorporated in broiler diets. Day et al. (1972) observed an increase in weight gain when broilers were fed diets containing low levels of DDGS (2.5% and 5%). Other researchers have concluded that up to 25% DDGS can be incorporated in broiler diets if dietary energy is held constant (Waldroup et al., 1981). Parsons et al. (1983) concluded that up to 40% of the soybean meal protein could be replaced with DDGS if formulated for total lysine. However, it is presumed that most of the DDGS used in these early studies came from the beverage industry and DDGS from modern ethanol plants may differ in nutrient composition. The majority of the research conducted today on the feeding of DDGS is conducted on DDGS from modern ethanol plants.

Lumpkins et al. (2004) fed 0% or 15% DDGS from modern ethanol plants to Cobb-500 straight-run broiler chickens from 1 to 18 days of age and found no adverse effects on body-weight gain or feed utilization. However, when the diets were formulated to contain lower energy and protein contents to increase the likelihood of detecting differences in growth performance due to DDGS, feed utilization was adversely affected in broilers fed the 15% DDGS diet. This effect was evident during the first 2 weeks of age, but there were no effects of DDGS at 18 days of age. In a subsequent experiment, Lumpkins et al. (2004) fed 0%, 6%, 12%, and 18% DDGS to 1-day-old Cobb-500 chicks until 42 days of age. In this experiment, body-weight gain and feed utilization were unaffected by feeding up to 12% DDGS, but gain and feed utilization were lowered when broilers were fed the 18% DDGS diet, which the authors attributed to an amino acid deficiency (likely lysine) in the starter diet. Based on their study, Lumpkins et al. (2004) recommended that no more than 12% DDGS be included in the starter diets but that grower and finisher diets could contain 12% to 15% DDGS. However, the diets were formulated on a total amino acid basis. Given the relatively low amino acid digestibilities of DDGS, it is possible that the reduced growth performance at high DDGS inclusion levels was caused by an amino acid deficiency (e.g., lysine or arginine) and could have been overcome if the diets were formulated on a digestible amino acid basis. Wang et al. (2007c) balanced broiler diets on a digestible amino acid basis using a standardized nutrient matrix for DDGS (Waldroup et al., 2007) and dietary digestible amino acid levels based on industry averages. In the study, graded levels of DDGS between 0% and 25% were fed to male Cobb-500 chickens from 1 to 49 days of age with no treatment effects on body-weight gain. However, cumulative feed consumption from 0 to 35 and from 0 to 49 days of age increased compared with the control diet when the diet contained 25%

DDGS. As a result, the cumulative feed efficiency was negatively affected during the same age periods. A careful examination of the analyzed (total) amino acid concentrations in the diets suggests that the 25% DDGS diets may have been marginally limiting in arginine with observed arginine-to-lysine ratios between 102% and 104%, depending on the specific diet. The ideal arginine-to-lysine ratio (expressed on a digestible basis) is 105% to 111% (Baker and Han, 1994; Mack et al., 1999; Baker, 2003), somewhat greater than the arginine-to-lysine ratios calculated from data presented by Wang et al. (2007c).

Young birds can be more sensitive to poor quality feed ingredients because their digestive tract is not fully developed until 10 to 14 days of age (Batal and Parsons, 2002a, 2002b). Thus, care should be taken when including high levels of DDGS during the first 2 weeks posthatch because of the high fiber content and low amino acid digestibility. Wang et al. (2007b) observed a trend toward decreased body weight during the initial 2 weeks after hatch in broilers fed diets containing 30% DDGS compared to 0% or 15% DDGS, and the broilers' body weights continued to lag behind those of control-fed broilers throughout the 42-day study. Feed utilization was significantly lower throughout the study for broilers fed the 30% DDGS diet compared to broilers fed 0% or 15% DDGS. When DDGS was omitted from starter diets (1 to 14 days of age), but introduced in grower diets (14 to 35 days of age) at 15% and subsequently kept constant or further increased in the finisher diets fed for the last 7 days before slaughter (i.e., 0%–15%–15% or 0%–15%–30% DDGS), body weight, feed consumption, and feed utilization at 42 days of age were similar to those of broilers fed no DDGS. However, when 30% DDGS content was included in the diets, either from 1 day of age or introduced in grower or finisher diets, growth performance was depressed. Similarly, in a third study, Wang et al. (2007a) fed diets containing 0%, 15%, or 30% DDGS to Cobb-500 broiler chicks throughout the growing period from 1 to 42 days of age and observed no effect of feeding 15% DDGS, but depressed growth performance was observed when the diet containing 30% DDGS was fed. In both studies, an arginine deficiency may be the culprit, as the arginine-to-lysine ratios, calculated from the reported dietary total amino acid values, were between 82% and 93% in the diets containing 30% DDGS (Wang et al., 2007a, 2007b). The diets in the studies by Wang et al. (2007a, 2007b) were pelleted, and the authors observed a decrease in pellet quality as the level of DDGS inclusion increased. Therefore, it is possible that the negative performance observed in broilers fed diets with 30% DDGS could partly be due to poor pellet quality.

Although it has been speculated that the inclusion of high levels of DDGS decreases dressing percentage and breast-meat yield the dressing percentage was not affected in broilers fed diets containing up to 30% DDGS (Lumpkins et al., 2004; Wang et al., 2007a, 2007b). In a study by Wang et al. (2007c), dressing percentage appeared to decrease linearly with increased DDGS content. Compared to the control diet, the dressing percentage was lower when broilers were fed diets containing 15% and 25% DDGS, but not in diets containing 5%, 10%, and 20% DDGS. Despite decreased growth performance in broilers fed 18% DDGS (Lumpkins et al., 2004), breast-meat yield and other cuts were unaffected by the dietary treatments whether they were measured on a gram-per-bird basis or a percentage-of-carcass-weight basis. Similarly, Wang et al. (2007a, 2007b) observed no effects on carcass quality when broilers were fed up to 15% DDGS. However, when fed 30% DDGS, broilers had lower breast-meat yield (Wang et al., 2007a, 2007b), likely attributable to an arginine deficiency (Corzo et al., 2003) as previously described. Corso et al. (2009) conducted a study to evaluate the effects of feeding DDGS (8%) on broiler breast and thigh meat quality. Overall, no differences were observed in the quality of breast or thigh meat due to the inclusion of 8% DDGS in the diet. However, feeding diets with 8% DDGS resulted in slight differences in consumer acceptability of the breast meat, the breast meat from the broiler fed no DDGS being slightly preferred. The meat from the broilers fed 8% DDGS has a greater percentage of linoleic and total polyunsaturated fatty acids, indicating that it may be slightly more susceptible to oxidation (Schilling et al., 2010).

Potter (1966) concluded that DDGS from the beverage industry could be included up to 20% in turkey diets if the lysine and energy levels are adjusted. Roberson (2003) conducted an experiment with turkey hens fed DDGS-containing diets from 56 to 105 days of age, at which time the hens were sent to a commercial processing plant. The diets contained up to 27% DDGS and were formulated on a digestible amino acid basis. Body-weight gain decreased linearly with increasing DDGS content, attributed to a deficiency in digestible lysine, likely caused by a lower-than-expected digestibility value. In a second experiment, DDGS was included in up to 10% of the diet with no effects on body-weight gain or feed conversion of the turkeys. Carcass quality was not investigated in the experiment by Roberson (2003), but Noll et al. (2002) reported no adverse effects on breast-meat yield of feeding DDGS to turkey toms as long as the essential amino acid levels were met. Noll and Brannon (2009) concluded that the inclusion of up to 20% DDGS in market tom diets had no negative effects on body weight or carcass yield. However, feed conversion was slightly increased when 20% DDGS was included in the diets. Similar results were reported by Noll and Brannon (2006, 2007) in which they reported that poultry performance was not negatively affected by the high (30%) inclusion levels of DDGS in starter diets (2 to 5 weeks of age). However the inclusion of 40% DDGS did decrease body-weight gain and negatively affected feed conversion.

Thacker and Widyaratne (2007) fed broiler diets containing 0%, 5%, 10%, 15%, and 20% wheat DDGS. No statistical differences were observed on performance; however performance tended to decrease when the inclusion of wheat DDGS was 20%. The author concluded that wheat DDGS could be used up to 15% in broiler diets.

15.3.2 EGG PRODUCTION (LAYING HENS)

Harms et al. (1969), Jensen et al. (1974), and Matterson et al. (1966) concluded that laying hens could be fed diets with up to 20% DDGS from the beverage industry without negatively influencing egg production and egg weight. It was also observed that distillers feed had a positive effect on interior egg quality (Jenson et al., 1974). Hughes and Hauge (1945) observed improvements in Haugh units, a measure of egg interior quality, when hens were fed diets consisting of 5% and 10% brewers dried grains. Lilburn and Jensen (1984) reported that the incorporation of 20% corn fermentation solubles into laying-hen diets resulted in a significant improvement of Haugh units. Jensen et al. (1978) also observed improved egg quality with the addition of corn fermentables, and Waldroup and Hazen (1979) reported a similar response in Haugh units with the use of corn-dried steep liquor concentrate.

Today the majority of the DDGS is from the ethanol industry and not the beverage-alcohol industry. Lumpkins et al. (2005) fed diets containing either 0% or 15% corn DDGS to Hy-Line W36 White Leghorn laying hens from 22 weeks of age (corresponding to about 4 weeks before peak egg production) to 43 weeks of age. The DDGS inclusion did not affect egg production, egg weight, feed consumption, or feed utilization. However the hens were fed a low-density diet in which energy, amino acids, and the nutrient-to-energy ratios were lowered to increase the likelihood that issues with feeding 15% DDGS, if any, could be detected. Compared to the control diet, the low-density 15% DDGS diet resulted in slightly lower egg production and poorer feed utilization. The diets were formulated on a total amino acid basis with equal contents of lysine and methionine, suggesting that the 15% DDGS diets were deficient in one or more amino acids due to the lower digestibility of amino acids in DDGS as discussed previously. Roberson et al. (2005) fed DDGS to laying hens at 0%, 5%, 10%, and 15% of the diet. The diets were fed to white leghorn-type hens from 48 weeks of age over a period of 8 weeks, during which there were inconsistent effects of DDGS on egg production, with a decrease during 2 of the experiment's 8 weeks. Egg weight was not affected, but egg mass (defined as percent egg production × grams of egg weight) decreased in the same weeks that egg production decreased. An additional experiment was conducted with the same hens from 58 to 67 weeks of age and using a different, darker-colored DDGS sample egg production and egg mass were not affected, although egg weights decreased linearly during 1 of the weeks. Neither feed consumption nor feed utilization was affected in either experiment. Roberson et al. concluded that DDGS could be fed to laying hens at levels as high as 15%, whereas Lumpkins et al. (2005)

recommended a DDGS inclusion level of no more than 10% to 12%. However, the experimental diets in both experiments were formulated using total amino acids, not digestible amino acids, the importance of which was discussed previously. Roberts et al. (2007b) fed diets containing 0% or 10% DDGS to white leghorn-type laying hens from 23 to 58 weeks of age and observed no effects on any egg production or egg quality parameters. The diets used in this study were formulated on a digestible amino acid basis and to contain similar amounts of apparent MEn.

Pineda et al. (2008) conducted an experiment to investigate whether egg production and egg quality would be affected by very high inclusion levels of DDGS. In their experiment, graded levels between 0% and 69% DDGS were fed to white leghorn-type laying hens 53 weeks of age for 8 weeks after a 4-week transition period, during which the dietary DDGS contents were gradually changed in steps of about 12 percentage points per week. Egg production decreased linearly during the 8-week experimental period, countered by an increase in egg weight. As a result, egg mass was unaffected by the dietary DDGS inclusion. Feed consumption increased with increasing dietary DDGS content, but feed utilization was unaffected. Egg quality measured as Haugh units, egg composition, and specific gravity was not affected by the DDGS inclusion. The experiment by Pineda et al. (2008) demonstrated that laying hens can be fed diets with high amounts of DDGS with no adverse effects on egg production and egg quality as long as the energy and nutrient contents of all feed ingredients (including DDGS) are considered and the diets are formulated on a digestible amino acid basis. The diets used by Pineda et al. (2008) may not have been practical in that they contained high levels of (expensive) supplemental vegetable oil to compensate for the relatively low energy content in DDGS. As a result, the flowability of the DDGS-containing diets was lower than would probably be acceptable on a commercial farm with automatic feeders. More practical inclusion levels were used in an experiment reported by Scheideler et al. (2008), in which white leghorn-type hens 24 weeks of age were fed diets containing graded levels of DDGS of between 0% and 25% for 22 weeks. Egg production, feed consumption, and body-weight gain were not affected by the dietary DDGS inclusion. Egg weights, however, were lower when the diets contained 20% and 25% DDGS, which the authors attributed to a dietary amino acid deficiency.

Reports from a number of recent studies have indicated that the inclusion of DDGS does not affect Haugh units or shell quality, as indicated by the shell breaking-strength or specific gravity of the eggs (Lumpkins et al., 2005; Roberson et al., 2005; Pineda et al., 2008). However, in past studies improvements in Haugh units were observed when laying hens were fed diets consisting of DDG from the beverage industry (Hughes and Hauge, 1945; Lilburn and Jensen, 1984). No differences in egg shell quality has been observed when DDGS is fed to laying hens; however, it has been reported that the consumption of a sulfur rich diet may interfere with absorption of dietary calcium (Leeson and Summers, 2001, 2005), which could negatively affect eggshell quality. The inclusion of DDGS in laying-hen diets has been show to affect yolk color (Roberson et al., 2005; Roberts et al., 2007b; Pineda et al., 2008). Increases in the L* and a* color scores (indicating darker and redder yolks, respectively) have been observed to increase with increasing dietary DDGS content.

Little if any recent research has been published on the use of DDGS in pullet, broiler breeders, or turkey hen diets; however, DDGS is being successfully incorporated into these diets. The current research conducted with broilers and laying hens and some older research (Couch et al., 1970; Manley et al., 1978) can be used as a rough guide for using DDGS in various poultry diets.

15.4 ENVIRONMENTAL ASPECTS OF FEEDING COPRODUCTS FROM THE ETHANOL INDUSTRY TO POULTRY

Protein levels in DDGS-containing diets are expected to be greater than if the diet is formulated with just corn and soybean meal because of the relatively low amino acid digestibility in DDGS and the poor amino acid profile of DDGS as compared to soybean meal. Thus, inclusion of DDGS in poultry diets is expected to increase both the nitrogen consumption and excretion from the birds, especially at higher levels of DDGS inclusions (Roberts et al., 2007b; Pineda et al., 2008). Leytem et al. (2008) reported that the inclusion of wheat DDGS increased the nitrogen and phosphorus output, as well as the solubility of P excreted. The authors concluded that care should be taken when including high levels of DDGS in poultry diets, because increases in N and P excretion are a concern from an environmental standpoint. Applegate et al. (2009) reported similar findings when HP-DDG was fed to broilers. The authors reported that due in a large part to the amino acid profile and digestibility of HP-DDG, high inclusion levels in broiler diets result in more manure and manure nitrogen. Although increased ammonia (NH_3) emission from the manure is associated with increased nitrogen excretion (Summers, 1993; Keshavarz and Austic, 2004), dietary DDGS appears to have an attenuating effect on ammonia emissions (Roberts et al., 2007a). Fiber is not digested by the birds, and some of it is instead fermented by microbes in the large intestines, producing short chain fatty acids, which in turn lower the manure pH. The lowered pH results in a shift in the NH_3 equilibrium toward the less-volatile ammonium ion ($NH_3 + H^+ \leftrightarrow NH_4^+$). Therefore, poultry fed DDGS may excrete more nitrogen, but because of the resultant lower manure pH, the nitrogen does not evaporate off. This effect of dietary fiber on manure acidification and NH_3 emission was first demonstrated in pigs by Canh et al. (1998a, 1998b) and later in laying hens by Roberts et al. (2007a) using DDGS-containing diets. Hence, at first glance, it appears that an increase in dietary crude protein content of DDGS-containing diets may have adverse effects on air quality and the environment because of increased nitrogen excretion. However, the nitrogen appears to remain in the manure, which, when applied correctly on fields, does not adversely affect the environment and may increase the fertilizer and therefore the economic value of the manure.

15.5   FORMULATING DIETS WITH DDGS FOR POULTRY

Many believe that all the feed problems created by the use of a substantial amount of the United States corn supply for ethanol production will be solved with the feeding of DDGS. However, DDGS does not directly replace corn or soybean meal in a poultry diet and has other constraints, such a price, handling, logistics, and nutrient variation that limit its use. When formulating feed for poultry, the two most critical nutritional components are protein and energy. Traditionally, corn grain has been the main source of energy while soybean meal has supplied the majority of amino acids in poultry diets in the United States (and much of the world). However, corn protein is relatively high in methionine and low in lysine, whereas the opposite is true in soybean protein. Hence, corn grain and soybean meal complement each other very well in meeting the amino acid needs of poultry. Because DDGS is a coproduct of corn the amino acid profile of corn grain and DDGS are similar. Therefore, the amino acids in corn grain and DDGS do not complement each other, as seen with corn and soybean meal. The large difference in the amino acid profile ("protein quality") between distillers coproducts and other protein supplements (specifically soybean meal) has been recognized for a long time (Morrison, 1954) and so, despite the relatively high protein content of DDGS it cannot be viewed as an adequate replacement for soybean meal or other protein supplements in poultry diets. DDGS also cannot be viewed as an adequate replacement for corn as a source of energy since it contains 17% less energy than corn. Since DDGS is not a superior source of energy or amino acids the inclusion of DDGS in poultry diets will replace both corn and soybean meal. The use of DDGS will also likely increase the need for supplemental lysine and fat.

DDGS can make up a substantial portion of the diets for broiler chickens, turkeys, and laying hens. When birds that are fed DDGS-containing diets failed to meet egg production, growth performance, or carcass quality expectations, it is almost always attributed to an amino acid deficiency, illustrating the important differences in amino acid profiles between DDGS and soybean meal and the differences in amino acid digestibility. Because of the relatively low amino acid digestibilities in DDGS, it is especially important to formulate poultry diets on a true digestible amino acid basis when including DDGS in diets. If the DDGS-containing diet is formulated on a total amino acid basis, only a relatively small amount (5% to 10%) can be included in poultry diets without affecting production. To illustrate the importance of formulating diets on a true digestible basis, three different sets of starter broiler diets with or without 15% DDGS were formulated using either a crude protein minimum of 22% (with a total methionine + cystientine minimum of 0.92%), on a total amino acid basis (with no crude protein minimum), or on a digestible amino acid basis (with no crude protein minimum). The lysine content was set to 1.22% total lysine in the diet formulated on total amino acids and to 1.18% true digestible amino acid basis (the levels used for all the amino acids can be seen in Table 15.6). The nutrient contents listed by Batal and Dale, (2009) were used for all ingredients. The diets were formulated on a "least-cost basis" and the diet costs were calculated using feed ingredient prices from the November 16th, 2009 edition of *Feedstuffs* magazine (Atlanta prices). If it is accepted that the best estimate of the broilers amino acid needs is the digestible amino acid requirements, then it is evident from Table 15.6 that a diet formulated to contain 22% crude protein from corn, soybean meal, and poultry byproduct meal will be marginally deficient in methionine + cystientine and lysine when amino acid digestibilities are considered. These deficiencies, especially lysine, were exacerbated when DDGS was included in the diet. Formulating on a total amino acid basis decreased the severity of the amino acid deficiencies, as based on the digestible amino acid requirements. However, deficiencies were still significant for lysine, methionine + cystine, threonine and arginine. The only scenario in which there were no deficiencies was when the diets were formulated on digestible amino acids, emphasizing the importance of formulating diets on a digestible amino acid basis when formulating with DDGS.

15.6 POTENTIAL PRACTICAL LIMITATIONS FOR USE OF DDGS IN POULTRY DIETS

The variation in the nutrient content of DDGS is not the only issue that affects the use of DDGS in poultry diets. DDGS may cause problems at nearly every phase of feed manufacturing. This includes railcars that simply won't unload, lack of bin space, problems with handling and flowability, and maybe of greatest concern is the decrease in pellet throughput and pellet quality. However, when economic restraints are removed and only the feed ingredients' content of energy and nutrients restrain its use in the diet formulation, as much as 70% DDGS can be included in a laying-hen and broiler diets (Pineda et al., 2008). While there may be no nutritional or production effects of such high dietary DDGS levels, other factors may limit the dietary inclusion rate of DDGS. For instance, the relatively low energy content of DDGS warrants a greater inclusion of supplemental oil or fat, which may increase the diet cost as well as decrease the flowability of the diet, thereby causing problems associated with bridging (Waldroup et al., 1981; Pineda et al., 2008). The bulk density of DDGS averages 570 g/L (35.7 lb/ft³), although with some variation among samples from different ethanol plants (U.S. Grains Council, 2008). In comparison, the bulk density of ground corn grain is approximately 580 g/L (36.2 lb/ft³), and that of soybean meal is around 630 g/L (39.4 lb/ft³) (Jurgens and Bregendahl, 2007), meaning that the density of DDGS-containing diets tends to decrease with increasing DDGS content (Wang et al., 2007a, 2007b, 2007c). The lower bulk density of DDGS-containing diets means that less feed (on a weight basis) can be transported in each truck from the feed mill to the poultry barn and that gut fill may limit feed consumption by the birds. As a result, the upper practical limit for DDGS inclusion in mash or meal diets is likely somewhere around 20% to 25% of the diet, with greater levels requiring pelleting, flow agents, or antioxidants. Clementson and Ileleji (2009) reported that there is a bulk density variation as DDGS is loaded and emptied at ethanol plants. Bulk density variation was shown to be primarily caused by particle segregation that takes place while filling the hopper and during discharge.

Pelleting of DDGS-containing diets is possible, but there may be difficulties if the diet contains more than 5% to 7% DDGS (Behnke, 2007). The pelleting difficulties stem in part from an increase in the dietary oil content (some of which comes from the DDGS) and in part because DDGS lacks starch, which otherwise helps bind the pellets together (Behnke, 2007). However, pelleting diets

TABLE 15.6
Examples of Broiler Diets Formulated Based on Crude Protein, Total Amino Acids, and Digestible Amino Acids with or without Corn Distillers Dried Grains with Solubles (DDGS)

Item	Minimum Recommendations	Formulated Using Crude Protein Minimum		Formulated Using Total Amino Acids		Formulated Using Digestible Amino Acids	
		0% DDGS	15% DDGS	0% DDGS	15% DDGS	0% DDGS	15% DDGS
Ingredient							
DDGS		0	15	0	15	0	15
Corn		60.73	52.53	67.04	57.54	66.28	52.88
Soybean meal		30.73	23.89	24.51	18.95	25.09	22.91
Poultry		5.00	5.00	5.00	5.00	5.00	5.00
Poultry Fat		1.23	1.37	0.68	0.83	0.70	1.51
Defluorinated		1.04	0.77	1.09	0.81	1.08	0.78
Limestone		0.44	0.68	0.51	0.70	0.51	0.68
Salt		0.35	0.28	0.32	0.27	0.32	0.28
DL-Methionine		0.18	0.16	0.24	0.21	0.29	0.24
L-Lysine HCL		0	0	0.26	0.36	0.35	0.35
L-Threonine		0	0	0.01	0.01	0.06	0.03
Vitamin and trace mineral premix		0.33	0.33	0.33	0.33	0.33	0.33
Calculated chemical composition							
MEn (kcal/kg)	3,030	3030	3030	3040	3030	3030	3030
Crude protein (%)	22.0	22.0	22.0	19.8	20.4	20.16	21.97
Crude fat (%)		4.40	5.52	4.00	5.10	4.00	5.67
Calcium	0.90	0.90	0.90	0.92	0.90	0.92	0.90
Phosphorus available	0.45	0.45	0.45	0.45	0.45	0.45	0.45
Total amino acids							
Arginine	1.26	1.54	1.42	1.34	1.26	1.35	1.39
Lysine	1.22	1.19	1.08	1.22	1.22	1.30	1.33
Methionine	0.50	0.55	0.54	0.58	0.56	0.63	0.61
Met + Cystine	0.92	0.92	0.92	0.92	0.92	0.97	0.99
Tryptophan	0.20	0.28	0.26	0.24	0.23	0.25	0.25
Threonine	0.79	0.88	0.86	0.79	0.79	0.85	0.88
Valine	0.92	1.20	1.18	1.06	1.07	1.07	1.16
Digestible amino acids							
Arginine	1.23	1.40	1.26	1.21	1.11	1.23	1.23
Lysine	1.18	1.05	0.93	1.10	1.08	1.18	1.18
Methionine	0.45	0.51	0.50	0.55	0.53	0.60	0.57
Met + Cystine	0.88	0.82	0.81	0.82	0.82	0.88	0.88
Threonine	0.18	0.24	0.21	0.20	0.19	0.21	0.21
Tryptophan	0.74	0.76	0.72	0.68	0.66	0.74	0.74
Valine	0.88	1.08	1.04	0.92	0.94	0.96	1.02
Diet cost $/ton		248.08	229.07	242.18	226.34	247.91	237.69

Source: Data from Dale, N. M., and A. B. Batal. *Poultry USA*, pp. 36–39. August 2005.

with high DDGS inclusion levels is manageable, as shown by Wang et al. (2007a, 2007b, 2007c). Although pellet durability was not specifically tested in these studies, Wang et al. (2007a, 2007b) reported that the pellet quality of the diet containing 15% DDGS was similar to that of the control diet, but the diet containing 30% DDGS pelleted poorly and contained numerous fines despite the addition of a pellet binder. However, Loar II et al. (2010) reported that the inclusion of 8% DDGS in broiler diets decreased pellet quality measured by pellet durability index and modified pellet durability index as well as a significant increase in total fines. The inclusion of 8% DDGS also leads to a decrease in energy usage at the pellet mill and a decrease in bulk density of the diets.

There have been many reports of problems with the flowability of DDGS (Ganesan et al., 2007a, 2007b; Bhadra et al., 2009, 2010). The handling problems with DDGS may be attributed to the flat, plate-like structure of the bran, which results in an arch formation when DDGS is stored (Shurson et al., 2007). Unfortunately, few studies have been conducted to find ways to overcome the negative issue of flowability. There are a few approaches that can help to improve the flow characteristics of DDGS, however most need to be implemented at the ethanol plant. The first is to simply hold the DDGS in the flat storage site until the moisture equilibrium has occurred, which will be longer during the summer months. Shurson et al. (2007) concluded that the increase in moisture content of DDGS from 9% to over 11.5% decreases the flowability of DDGS. Due to these practical issues the use of DDGS may add an additional cost at the feed mill.

15.7 CONCLUSIONS

Corn DDGS is a highly acceptable feed ingredient that can be used successfully in poultry diets. Corn DDGS and other distillers coproducts can be valuable sources of energy, amino acids, and phosphorus for poultry. However, care should be taken when formulating the diets with ethanol coproducts, because significant variation in the nutrient content and digestibility can be observed within and between ethanol plants. Preferably, the coproducts should be from a single source (single plant) to minimize variation, and it is important that chemical analyses are consistently monitored to verify the nutrient composition and to estimate nutrient availability. When the diets are formulated on a digestible amino acid basis, the coproducts can be included safely at higher levels (15% to 25%) for broilers, turkeys, and laying hens, although there are practical considerations that may limit the inclusion level. When working with a new coproduct nutritionists are advised to start at lower inclusion levels until they become comfortable with estimating the nutrient content and digestibility of the coproduct they are receiving and then gradually increase the inclusion level in the diets. Nutritionists are also advised to start at lower inclusion levels in the diets of young birds and slowly increase the level in the diets of older birds.

REFERENCES

Applegate, T. J., C. Troche, Z. Jiang, and T. Johnson. 2009. The nutritional value of high-protein corn distillers dried grains for broiler chickens and its effect on nutrient excretion. *Poultry Science* 88: 354–359.

Association of American Feed Control Officials (AAFCO). 2007. *AAFCO Official Publication. Reference Manual.* Oxford, IN: AAFCO.

Baker, D. H. 2003. Ideal Amino Acid Patterns for Broiler Chicks. *In Amino Acids in Animal Nutrition,* ed. J. F. P. D'Mello, pp. 223–235. Oxon, UK: CABI Publishing.

Baker, D. H., and Y. Han. 1994. Ideal amino acid profile for chicks during the first three weeks posthatching. *Poultry Science* 73: 1441–1447.

Batal A. B., and N. M. Dale. 2009. Ingredient analysis table: 2010 edition. *Feedstuffs* 80(38); 16.

Batal, A. 2007. Nutrient digestibility of high protein corn distillers dried grains with solubles, dehydrated corn germ and bran. *Poultry Science* 86 (Suppl. 1): 206.

Batal, A. B., and N. M. Dale. 2006. True metabolizable energy and amino acid digestibility of distillers dried grains with solubles. *Journal of Applied Poultry Research* 15: 89–93.

Batal, A., and N. Dale. 2003. Mineral composition of distillers dried grains with solubles. *Journal of Applied Poultry Research* 12: 400–403.

Batal, A. B., and C. M. Parsons. 2002a. Effects of age on development of digestive organs and performance of chicks fed a corn–soybean meal versus a crystalline amino acid diets. *Poultry Science* 81: 1338–1341.

Batal, A. B., and C. M. Parsons. 2002b. Effects of age on nutrient digestibility in chicks fed different diets. *Poultry Science* 81: 400–407.

Behnke, K. C. 2007. Feed manufacturing considerations for using DDGS in poultry and livestock diets. In *Proceedings of the 5th Mid-Atlantic Nutrition Conference,* ed. N. G. Zimmerman, pp. 77–81. College Park, MD: Maryland Feed Industry Council and Dept. of Animal and Avian Sciences, University of Maryland.

Bhadra, R., K. A. Rosentrater, and K. Muthukumarappan. 2009. Flowability properties of commercial distillers dried grains with solubles (DDGS). *Cereal Chemistry* 86(2): 170–180.

Bhadra, R., K. A. Rosentrater, and K. Muthukumarappan. 2010. Modeling of DDGS flowability with varying drying and storage parameters. *2010 ASABE Annual International Meeting.* Pittsburgh PA. June 20–23, 2010.

Black, J. L. 1995. The testing and valuation of models. In *Modeling Growth in the Pig,* eds. P. J. Moughan, M. W. A. Verstegen, and M. I. Visser-Reyneveld, pp. 23–31. EAAP Publication No. 78. Wageningen: Wageningen Press.

Canh, T. T., A. J. A. Aarnink, J. B. Schutte, A. Sutton, D. J. Langhout, and M. W. A. Verstegen. 1998a. Dietary protein affects nitrogen excretion and ammonia emission from slurry of growing–finishing pigs. *Livestock Production Science* 56: 181–191.

Canh, T. T., A. J. A. Aarnink, M. W. A. Verstegen, and J. W. Schrama. 1998b. Influence of dietary factors on the pH and ammonia emission of slurry from growing–finishing pigs. *Journal of Animal Sciences* 76: 1123–1130.

Clementson C. L., and K. E. Ileleji. 2009. Understanding bulk density variability of DDGS. In *Proceedings of the ASABE Annual International Meeting.* Reno, Nevada. June 2009.

Combs, G. F., and E. H. Bossard. 1969. Further studies on available amino acid content of corn distillers dried grains with soluble. *Proceedings of the 24th Distillers Feed Research Council,* pp. 53–58.

Corzo, A., E. T. Moran, Jr., and D. Hoehler. 2003. Arginine need of heavy broiler males: Applying the ideal protein concept. *Poultry Science* 82: 402–407.

Corzo, A., M. W. Schilling, R. E. Loar, V. Jackson, S. Kin, and V. Radharkrishnan. 2009. The effects of feeding distillers dried grains with soluble on broiler meat quality. *Poultry Science* 88: 432–439.

Couch, J. R., J. H. Trammell, A. Tolan, and W. W. Abbott. 1970. Corn distillers dried grains with solubles in low lysine diets for rearing broiler breeder replacement pullets. *Proceedings of the Distillers Feed Research Council* 25: 25–33. Cincinnati, OH.

Cromwell, G. L., K. L. Herkelman, and T. S. Stahly. 1993. Physical, chemical, and nutritional characteristics of distillers dried grains with solubles for chicks and pigs. *Journal of Animal Science* 71: 679–686.

Dale, N. M., and A. B. Batal. 2005. Distillers grains (DDGS): Focusing on quality control. *Poultry USA,* pp. 36–39. August 2005.

Day, E. J., B. C. Dilworth, and J. McNaughton. 1972. Unidentified growth factor source in poultry diets. *Proceedings of Distillers Feed Research Council Conference,* pp. 40–45. Cincinnati, OH.

Ergul, T., C. Martinez-Amezcua, C. M. Parson, B. Walters, J. Brannon, and S. L. Noll. 2003. Amino acid digestibility in corn distillers dried grains with solubles. *Poultry Science* 82(Suppl.1): 70.

Evonik Degussa. 2005. AminoDat 3.0., platinum version. Evonik Degussa, Hanau-Wolf-gang, Germany.

Fastinger, N. D., J. D. Latshaw, and D. C. Mahan. 2006. Amino acid availability and true metabolizable energy content of corn distillers dried grains with solubles in adult cecectomized roosters. *Poultry Science* 85: 1212–1216.

Fontaine, J., U. Zimmer, P. J. Moughan, and S. M. Rutherfurd. 2007. Effect of heat damage in an autoclave on the reactive lysine contents of soy products and corn distillers dried grains with solubles: Use of the results to check on lysine damage in common qualities of these ingredients. *Journal of Agricultural and Food Chemistry* 55: 10737–10743.

Ganesan, V., K. A. Rosentrater, and K. Muthukumarappan. 2007a. Dynamic water adsorption characteristics of distillers dried grains with solubles (DDGS). *Cereal Chemistry* 84: 548–555.

Ganesan, V., K. A. Rosentrater, and K. Muthukumarappan. 2007b. Modeling the flow properties of DDGS. *Cereal Chemistry* 84(6): 556–562.

Harms, R. H., R. S. Moreno, and B. L. Damron. 1969. Evaluation of distiller's dried grain with solubles in diets of laying hens. *Poultry Science* 48: 1652–1654.

Hughes, C. W., and S. M. Hauge. 1945. Nutritive value of distillers' dried solubles as a source of protein. *Journal of Nutrition* 30: 245–258.

Jensen, L. S., C. H. Chang, and S. P. Wilson. 1978. Interior egg quality: Improvement by distillers feeds and trace elements. *Poultry Science* 57: 648–654.

Jensen, L. S. 1978. Distillers feeds as sources of unidentified factors for laying hens. *Proceedings of the Distillers Feed Research Council* 33: 17–22. Louisville, KY.

Jensen, L. S. 1981. Value of distillers dried grains with solubles in poultry feeds. *Proceedings of the Distillers Feed Research Council* 36: 87–93. Cincinnati, OH.

Jensen, L. S., L. Falen, and C. H. Chang. 1974. Effect of distillers grains with solubles on reproduction and liver fat accumulation in laying hens. *Poultry Science* 53: 586–592.

Jung, B. Y., and A. B. Batal. 2009. The nutrient digestibility of high protein distillers dried grains and the effect of feeding various levels on the performance of laying hens. *Journal of Applied Poultry Research* 18: 741–751.

Jurgens, M. H., and K. Bregendahl. 2007. *Animal Feeding and Nutrition,* 10th ed. Dubuque, IA: Kendall/Hunt Publishing Company.

Kerr, B. J., and R. A. Easter. 1995. Effect of feeding reduced protein, amino acid supplemented diets on nitrogen and energy balance in grower pigs. *Journal of Animal Science* 73: 3000–3008.

Keshavarz, K., and R. Austic. 2004. The use of low-protein, low-phosphorus, amino acid- and phytase- supplemented diets on laying hen performance and nitrogen and phosphorus excretion. *Poultry Science* 83: 75–83.

Kim, E. J., C. Martinez Amezcua, P. L. Utterback, and C. M. Parsons. 2008. Phosphorus bioavailability, true metabolizable energy, and amino acid digestibilities of high protein corn distillers dried grains and dehydrated corn germ. *Poultry Science* 87: 700–705.

Klasing, K. C., and R. E. Austic. 2003. Nutritional diseases. In *Diseases of Poultry,* 11th ed., ed. Y. M. Saif, pp. 1027–1053. Ames, IA: Iowa State Press.

Leeson, S., and J. D. Summers. 2005. *Commercial Poultry Production,* 3rd ed. Guelph, ON: University Books.

Leeson, S., and J. D. Summers. 2001. *Nutrition of the Chicken,* 4th ed. Guelph, ON: University Books.

Leeson, S., and L. Caston. 2004. Enrichment of eggs with lutein. *Poultry Science* 83: 1709–1712.

Leeson, S., G. Diaz, and J. D. Summers. 1995. *Poultry Metabolic Disorders and Mycotoxins.* Guelph, ON: University Books.

Leske, K. L., O. Akavanichan, T. K. Cheng, and C. N. Coon. 1991. Effect of ethanol extract on nitrogen-corrected true metabolizable energy for soybean meal with broilers and roosters. *Poultry Science* 70: 892–895.

Leytem, A. B., P. Kwanyuen, and P. Thacker. 2008. Nutrient excretion, phosphorus characterization, and phosphorus solubility in excreta from broiler chicks fed diets containing graded levels of wheat distillers grains with solubles. *Poultry Science* 87: 2505–2511.

Lilburn, M. S., and L. S. Jensen. 1984. Evaluation of corn fermentation solubles as a feed ingredient for laying hens. *Poultry Science* 63: 542–547.

Loar II, R. E., J. S. Mortiz, J. R. Donaldson, and A. Corzo. 2010. Effects of feeding distillers dried grains with soluble to broilers from 0 to 28 days posthatch on broiler performance, feed manufacturing efficiency and selected intestinal characteristics. *Poultry Science* 89: 2242–2250.

Loar II, R. E., M. W. Schilling, C. D. Mc Daniel, C. D. Coufal, S. F. Rogers, K. Karges, and A. Corzo. 2010. Effect of dietary inclusion level of distillers dried grains with soluble on layer performance, egg characteristics, and consumer acceptability. *Journal of Applied Poultry Research* 19: 30–37.

Lumpkins, B., A. Batal, and N. Dale. 2005. Use of distillers dried grains plus solubles in laying hen diets. *Journal of Applied Poultry Science* 14: 25–31.

Lumpkins, B. S., A. B. Batal, and N. M. Dale. 2004. Evaluation of distillers dried grains with solubles as a feed ingredient for broilers. *Poultry Science* 83: 1891–1896.

Lumpkins, B. S., and A. B. Batal. 2005. The bioavailability of lysine and phosphorus in distillers dried grains with solubles. *Poultry Science* 84: 581–586.

Mack, S., D. Bercovici, G. De Groote, B. Leclercq, M. Lippens, M. Pack, J. B. Schutte, and S. van Cauwenberghe. 1999. Ideal amino acid profile and dietary lysine specification for broiler chickens of 20 to 40 days of age. *British Poultry Science* 40: 257–265.

Manley, J. M., R. A. Voitle, and R. H. Harms. 1978. The influence of distillers dried grains with solubles (DDGS) in the diet of turkey breeder hens. *Poultry Science* 57: 726–728.

Martinez Amezcua, C., and C. M. Parsons. 2007. Effect of increased heat processing and particle size on phosphorus bioavailability on corn distillers dried grains with solubles. *Poultry Science* 86: 331–337.

Martinez-Amezcua, C., C. M. Parsons, and D. H. Baker. 2006. Effect of microbial phytase and citric acid on phosphorus bioavailability, apparent metabolizable energy, and amino acid digestibility in distillers dried grains with solubles in chicks. *Poultry Science* 85: 470–475.

Martinez-Amezcua, C., C. M. Parsons, and S. L. Noll. 2004. Content and relative bioavailability of phosphorus in distillers dried grains with solubles in chicks. *Poultry Science* 83: 971–976.

Martinez-Amezcua, C., C. M. Parsons, V. Singh, R. Srinivasan, and G. S. Murthy. 2007. Nutritional characteristics of corn distillers dried grains with solubles as affects by the amount of grains versus solubles and different processing techniques. *Poultry Science* 86: 2624–2630.

Matterson, L. D., J. Tlustohowicz, and E. P. Singsen. 1966. Corn distillers dried grains with solubles in rations for high-producing hens. *Poultry Science* 45: 147–151.

Morrison, F. B. 1954. *Feeds and Feeding: A Handbook for the Student and Stockman,* 21st ed. Ithaca, NY: The Morrison Publishing Company.

National Research Council. 1994. *Nutrient Requirements of Poultry,* 9th rev. ed. Washington, DC: National Academy Press.

Noll, S. L., and J. Brannon, 2006. Inclusion levels of corn distillers grains with solubles and poultry byproduct meal in market turkey diets. *Poultry Science* 85 (Suppl. 1): 106–07.

Noll, S., and J. Brannon. 2009. Market turkey response to dietary protein as an energy source is protein the new "fat"? *Gobbles* 66: 4 13–14.

Noll, S. L., C. Parsons, and W. Dozier III. 2007a. Formulating poultry diets with DDGS, how far can we go? In *Proceedings of the 5th Mid-Atlantic Nutrition Conference,* pp. 91–99.

Noll, S. L., J. Brannon, and C. M. Parsons. 2006. Nutritional value of corn distiller dried grains with solubles: Influence of solubles addition. *Poultry Science* 85 (Suppl. 1): 204.

Noll, S. L., J. Brannon, and C. M. Parsons. 2007b. Nutritional value of corn distiller dried grains with solubles: Influence of solubles addition. *Poultry Science* 86(Suppl. 1): 204.

Noll, S. L., V. Stangeland, G. Speers, C. Parsons, and J. Brannon. 2002. Utilization of canola meal and distillers grains with solubles in market turkey diets. *Poultry Science* 81(Suppl. 1): 92.

Ouart, M. D., D. E. Bell, D. M. Janky, M. G. Dukes, and J. E. Marion. 1988. Influence of source and physical form of Xanthophyll pigment on broiler pigmentation and performance. *Poultry Science* 67: 544–548.

Pahm, A. A., C. S. Schere, J. E. Pettigrew, D. H. Baker, C. M. Parsons, and H. H. Stein. 2009. Standardized amino acid digestibility in cecectomized roosters and lysine bioavailability in chicks fed distillers dried grains with soluble. *Poultry Science* 88: 571–578.

Parsons, C. M., D. H. Baker, and J. M. Harter. 1983. Distillers dried grains with solubles as a protein source for the chick. *Poultry Science* 62: 2445–2451.

Parsons, C. M., C. Martinez, V. Singh, S. Radhadkrishnan, and S. Noll. 2006. Nutritional value of conventional and modified DDGS for poultry. *Proceeding of the Multistate Poultry Feeding and Nutrition,* Indianapolis, IN.

Pineda, L., S. Roberts, B. Kerr, R. Kwakkel, M. Verstegen, and K. Bregendahl. 2008. Maximum dietary content of corn dried distiller's grains with solubles in diets for laying hens: Effects on nitrogen balance, manure excretion, egg production, and egg quality. Iowa State University Animal Industry Report 2008, Iowa State University. Available online: www.ans.iastate.edu/report/air/. Accessed September 2008.

Potter, L. M. 1966. Studies with distillers feeds in turkey rations. *Proceedings of the Distillers Feed Research Council* 21: 47–51. Cincinnati, OH.

Roberson, K. D., J. L. Kalbfleisch, W. Pan, and R. A. Charbeneau. 2005. Effect of corn distiller's dried grains with solubles at various levels on performance of laying hens and yolk color. *International Journal of Poultry Science* 4(2): 44–51.

Roberson, K. D. 2003. Use of dried distillers grains with solubles in growing–finishing diets of turkey hens. *International Journal of Poultry Science* 2: 389–393.

Roberson, K. D., J. L. Kalbfleisch, W. Pan, and R. A. Charbeneau. 2005. Effect of corn distiller's dried grains with solubles at various levels on performance of laying hens and egg yolk color. *International Journal of Poultry Science* 4(2): 44–51. 2005.

Roberts, S. A., H. Xin, B. J. Kerr, J. R. Russell, and K. Bregendahl. 2007a. Effects of dietary fiber and reduced crude protein on ammonia emission from laying-hen manure. *Poultry Science* 86: 1625–1632.

Roberts, S. A., H. Xin, B. J. Kerr, J. R. Russell, and K. Bregendahl. 2007b. Effects of dietary fiber and reduced crude protein on nitrogen balance and egg production in laying hens. *Poultry Science* 86: 1716–1725.

Runnels, T. D. 1966. The biological nutrient availability of corn distillers dried grains with soluble in broiler feeds. *Proceeding of the 21st Distillers Feed Research Council,* pp. 11–15.

Runnels, T. D. 1968. Effective levels of distillers feeds in poultry rations. *Proceedings of the Distillers Feed Research Council* 23: 15–22. Cincinnati, OH.

Scheideler, S. E., M. Masa'dah, and K. Roberson. 2008. Dried distillers grains with solubles in laying hens ration and notes about mycotoxins in DDGS. In *Pre-Show Nutrition Symposium, Midwest Poultry Federation Convention.* St. Paul, MN. March 18–20, 2008.

Schilling, M. W., V. Battula, R. E. Loar II, V. Jackson, S. Kin, and A. Corzo. 2010. Dietary inclusion level effects of distillers dried grains with soluble on broiler meat quality. *Poultry Science* 89: 752–760.

Scott, M. L. 1970. Twenty-five years of research on distillers feeds for broilers. *Proceeding of the 25th Distillers Feed Research Council,* pp. 19–24.

Shurson, G., L. Johnston, and J. Goihl. 2007. Do selected flow agents improve DDGS flowability in commercial systems? *Distillers Grains Quarterly* 22–25.

Sibbald, I. R. 1976. A bioassay for true metabolizable energy of feed stuffs. *Poultry Science* 55: 303–308.

Sibbald, I. R. 1986. The T.M.E. system of feed evaluation: Methodology, feed composition data and bibliography. *Technical Bulletin 1986-4E.* Ottawa, ON: Animal Research Centre, Agriculture Canada.

Spiehs, M. J., M. H. Whitney, and G. C. Shurson. 2002. Nutrient database for distiller's dried grains with soluble produced from new ethanol plants in Minnesota and South Dakota. *Journal of Animal Sciences* 80: 2639.

Stein, H. H., M. L. Gibson, C. Pedersen, and M. G. Boersma. 2006. Amino acid and energy digestibility in ten samples of distillers dried grain with solubles fed to growing pigs. *Journal of Animal Sciences* 84: 853–860.

Summers, J. D. 1993. Reducing nitrogen excretion of the laying hen by feeding lower crude protein diets. *Poultry Science* 72: 1473–1478.

Swiatkiewicz, S., and J. Korleski. 2006. Effect of maize distillers dried grains with solubles and dietary enzyme supplementation on the performance of laying hens. *Journal of Animal and Feed Sciences.* 15: 253–260.

Thacker, P. A., and G. P. Widyaratne. 2007. Nutritional value of diets containing graded levels of wheat distillers grains with solubles fed to broiler chicks. *Journal of the Science of Food and Agriculture* 87: 1386–1390.

University of Minnesota, Department of Animal Science. 2008a. Overview. Distillers grains by-products in livestock and poultry feeds. Available online: www.ddgs.umn.edu/overview.htm. Accessed May 13, 2008.

University of Minnesota, Department of Animal Science. 2008b. Proximate analysis of distiller's dried grains with solubles (DDGS). Nutrient profiles—Comparison tables. Distillers grains by-products in livestock and poultry feeds. Available online: http://www.ddgs.umn.edu/profiles-current.htm. Accessed September 2, 2008.

U.S. Grains Council. 2008. DDGS user handbook. Available online: http://www.grains.org/ddgs-information/217-ddgs-user-handbook. Accessed June 17, 2008.

Waldroup, P. W., and K. R. Hazen. 1979. Examination of corn dried steep liquor concentrate and various feed additives as potential sources of a Haugh unit improvement factor for laying hens. *Poultry Science* 58: 580–586.

Waldroup, P. W., J. A. Owen, B. E. Ramsey, and D. L. Whelchel. 1981. The use of high levels of distillers dried grains with solubles in broiler diets. *Poultry Science* 60: 1479–1484.

Waldroup, P. W., Z. Wang, C. Coto, S. Cerrate, and F. Yan. 2007. Development of a standardized nutrient matrix for corn distillers dried grains with solubles. *International Journal of Poultry Science* 6: 478–783.

Wang, Z., S. Cerrate, C. Coto, F. Yan, and P. W. Waldroup. 2007a. Effect of rapid and multiple changes in level of distillers dried grain with solubles (DDGS) in broiler diets on performance and carcass characteristics. *International Journal of Poultry Science* 6: 725–731.

Wang, Z., S. Cerrate, C. Coto, F. Yan, and P. W. Waldroup. 2007b. Use of constant or increasing levels of distillers dried grains with solubles (DDGS) in broiler diets. *International Journal of Poultry Science* 6: 501–507.

Wang, Z., S. Cerrate, C. Coto, F. Yan, and P. W. Waldroup. 2007c. Utilization of distillers dried grains with solubles (DDGS) in broiler diets using a standardized nutrient matrix. *International Journal of Poultry Science* 6: 470–477.

Wang, Z., S. Cerrate, C. Coto, F. Yan, F. P. Costa, A. Abdel-Maksoud, and P. W. Waldroup. 2008. Evaluation of corn distillers dried grains with solubles in broiler diets formulated to be isocaloric at industry energy levels or formulated to optimum density with constant 1% fat. *International Journal of Poultry Science* 7(7): 630-637.

Zhang, Y., J. Caupert, P. M. Imerman, J. L. Richard, and G. C. Shurson. 2009. The occurrence and concentration of mycotoxins in U.S. distillers' dried grains with solubles. *Journal of Agricultural and Food Chemistry* 57(20): 9828–9837.

Part IV

Further Uses

16 Feeding DDGS to Finfish

Michael L. Brown, Travis W. Schaeffer, Kurt A. Rosentrater,
Michael E. Barnes, and K. Muthukumarappan

CONTENTS

16.1 INTRODUCTION

As discussed in the previous chapter, the use of distillers dried grains with solubles (DDGS) as an ingredient for poultry diets is increasing. There is also growing interest in using DDGS in aquafeeds. The availability of DDGS has spurred research to determine its use in other monogastric animal feeds, such as aquafeeds for commercial finfish production. Globally, aquaculture has been growing at a fairly rapid pace (currently 8.5% per year) over the past two decades (Delgado et al., 2003; Tacon and Metian, 2008) and is recognized as the fastest growing food production sector of agriculture in the United States (FAO, 2009). The dramatic growth in this industry has primarily been driven by declines in marine capture fisheries, increases in the human population, and a per capita increase in the consumption fish and shellfish products. Landing data and population assessments for wild-caught fish and shellfish over the last five decades reveal that most capture fisheries throughout the world have reached capacity, likely peaking in the mid-1990s (FAO, 2009; NMFS, 2009). In general, about two-thirds of the world's commercial fish stocks are currently fished at or above their capacity, with another 10% overharvested to the point where populations will require years to recover (FAO, 2009). Further, the global demand for fish and shellfish products has not diminished, but has steadily increased to about 21 kg/y/person (FAO, 2009). Similarly, the U.S. per capita seafood consumption has shown increases during the last two decades, reaching 4.8 billion pounds in 2004. Thus, with increasing population size and rising consumption of fish and shellfish products worldwide, wild-caught fisheries will be unable to accommodate rising demand, supporting an even greater need for aquaculture production.

Growth of aquaculture and other industries (e.g., other monogastric and ruminant livestock feed applications) has also placed an increased demand on desirable marine fish meal resources (Tacon and Metian, 2008), particularly for carnivorous finfish and marine shrimp aquafeeds. Fish meal (not fishery waste meal) is a high-protein feed ingredient found in most aquaculture diets and is processed by drying or cooking, and grinding fish or shellfish (Hertrampf and Piedad-Pascual, 2000). Fish meal is the traditional protein source in finfish feeds. This diet constituent provides an important nutritional base in aquaculture diets because of its well-balanced profile of amino acids, essential fatty acids, digestible energy (DE), vitamins, and minerals (Abdelghany, 2003). However, the reliability of fish meal supplies and increasing costs of meal have encouraged researchers and feed manufacturers to seek lower cost, nutritive fish meal replacements.

Reducing feed costs and increasing sustainability of fish meal stocks require protein replacement with more cost-efficient protein sources. Formulation of cost-effective diets requires accurate information on the amino acid composition and digestibility of ingredients. For example, diets containing corn gluten meal may be deficient in EAA arginine, lysine, and tryptophan. Similarly, diets containing animal protein byproducts, such as feather meal, poultry (byproduct) meal, and meat and bone meal may be deficient in several amino acids or have variable digestibility. Some of the plant sources investigated for fish meal protein replacement include soybean meal (Shiau et al., 1990; Webster et al., 1992), corn gluten meal (Wu et al., 1996, 1997), lupins (Fontaínhas-Fernandes et al., 1999), rapeseed (Jauncey and Ross, 1982; Davies et al., 1990), and cottonseed meal (Jauncey and Ross, 1982; Rinchard et al., 2002). However, replacement of fish meal with plant-based meals frequently results in decreased performance (e.g., Mbahinzirek et al., 2001; Sklan et al., 2004; Gatlin et al., 2007), requiring alternative ingredients or various supplements to satisfy dietary requirements.

Due to recent increases in biofuel production, fuel-based distillers coproducts derived from corn (maize) have become broadly available. Distillers dried grains with solubles (DDGS) is a coproduct of predominantly dry grind fuel ethanol processing, and to a smaller extent the manufacture of beverage alcohols (i.e., distilleries) (Rosentrater and Muthukumarappan, 2006). In early 2010, 201 operating ethanol plants produced a total of 51 billion L of ethanol, with an estimated industry expansion of nearly 4.9 billion L (RFA, 2010). In 2005, however, the industry had a capacity of 97 manufacturing plants, which produced only 15.8 billion L of ethanol (RFA, 2010). With exponential growth in fuel ethanol production in recent years, significant quantities of distillers grains are now produced. Compared to raw corn, corn meal, and other corn flour products, DDGS often contains up to three times the normal concentration of nutrients, due to starch consumption by yeast metabolism during fermentation (Chevanan et al., 2005, 2008). DDGS typically contains 26.8% to 33.7% protein (Rosentrater and Muthukumarappan, 2006; Wu et al., 1997), which closely matches crude protein requirements of commercially produced fishes such as tilapia *Oreochromis* spp. (Lim, 1989) and channel catfish *Ictalurus punctatus* (Lovell, 1989b; Chevanan et al., 2005). Also, DDGS does not contain certain antinutritional factors (e.g., trypsin inhibitors and gossypol) that are present in soybean meal (Wilson and Poe, 1985; Shiau et al., 1987) and cottonseed meal (Jauncey and Ross, 1982; Robinson, 1991). However, DDGS usually contains lower amounts of the essential amino acids lysine and methionine compared to fish meal (Cheng and Hardy, 2004). Thus, given the current and projected availability of DDGS there may be opportunities to replace fish meal (partially or completely, with supplements) with DDGS to provide least-cost, nutritionally complete diets.

The commercially cultured finfish species for which most DDGS feeding trials have been conducted include Nile tilapia, channel catfish, and rainbow trout *Oncorhynchus mykiss*. Similarly, these species compose the bulk of food finfish production in the United States (USDA-NASS, 2004). This chapter discusses relevant species characteristics and provides an overview of nutrient requirements and deficiency diseases as they relate to these important commercial aquaculture species. We summarize the results of feeding trials conducted with aquafeeds formulated with fuel-based DDGS and discuss feed processing aspects associated with DDGS.

16.2   OVERVIEW OF SPECIES CHARACTERISTICS

16.2.1   TILAPIA

Tilapia are considered tropical species endemic to freshwaters in Africa, Jordan, and Israel; however, they can be cultured in almost any type of production system including both fresh and saltwater in tropical, subtropical, and temperate climates (Lim and Webster, 2006). Due to their fast growth, efficient use of aquatic foods, omnivorous nature, resistance to handling and diseases, ease of reproduction in captivity, and tolerance of a wide range of environmental conditions, tilapia have become one of the most successful aquaculture species worldwide (Lim, 1989; Lim and Webster, 2006). Currently, tilapia production ranks eighth in the world, increasing from 800,000 metric tons in 1996 to 1.82 million metric tons in 2006 (FAO, 2002, 2006).

Generally, "tilapia" refers to two genera containing the most cultured species (Lim, 1989). *Tilapia* are macrophagous, substrate spawners with redbreast tilapia *T. rendalli* and redbelly tilapia *T. zilli* species being used in aquaculture (Lim, 1989). *Oreochromis* are microphagous, mouth brooders with Mozambique tilapia *O. mossambicus*, Nile tilapia *O. niloticus*, and blue tilapia *O. aureus* species used in practical aquaculture (Lim, 1989). Currently, Nile tilapia and various hybrids are most commonly used in aquaculture (Green, 2006). In general, tilapia can tolerate a pH range of 5 to 11, dissolved oxygen concentrations down to 0.1 mg/L, and unionized ammonia concentration of 2.4 mg/L; however, most tilapia are unable to survive water temperatures below 8°C to 12°C and stop feeding around 16°C (Lim, 1989).

16.2.2 CHANNEL CATFISH

Channel catfish (*Ictalurus punctatus*) are a member of the Ictaluridae family and are native to North America (Hargreaves and Tucker, 2004). Commercial scale production of channel catfish began in the 1950s, largely due to research conducted at Auburn University (Stickney, 1996; Kelly, 2005). Due to many desirable culture traits (i.e., captive spawning, rapid growth, acceptance of supplemental feeds, resistance to diseases, and general ease of extensive and intensive culture) channel catfish have become the top aquaculture fish produced in the United States (Lovell, 1989a; Hargreaves and Tucker, 2004; Kelly, 2005; FAO, 2006). Most channel catfish production occurs in earthen levee ponds in Mississippi, Alabama, Arkansas, and Louisiana (Kelly, 2005). Channel catfish grow rapidly, from 10 to 500 g in 6 months if water temperatures remain above 23°C, and can tolerate daily and seasonal variations in water quality and temperature (Lovell, 1989a). Since 1986, production of channel catfish has doubled from 150,000 tons (Lovell, 1989a) to just over 300,000 tons in 2003, which represents about half of the total finfish aquaculture production in the United States (Hargreaves and Tucker, 2004).

16.2.3 RAINBOW TROUT

Originally native to the Pacific coast of North America from Northern Mexico to Alaska, and also the Kamchatka peninsula of Russia (Quinn, 2005), rainbow trout (*Oncorhynchus mykiss*) are now widely distributed throughout the United States and in every other continent except Antarctica (AGMRC, 2009; FAO, 2010). Rainbow trout are also known as steelhead in their anadromous form, and steelhead was probably the original source for today's hatchery stocks of rainbow trout (Behnke, 1992). Rainbow trout belong to the family Salmonidae, one of the most important global commercial aquaculture groups of fish (Knapp et al., 2007).

Compared to tilapia or channel catfish, rainbow trout require colder water, with optimum water temperatures reported by Dwyer et al. (1981, 1983) from 13°C to 16°C, and Jobling (1994) from 16.5°C to 17.2°C. Rainbow trout are easily cultured, grow rapidly, and are highly tolerant of crowding (AGMRC, 2009). However, they require high levels (>5 mg/L) of dissolved oxygen (Westers, 2001). They are also relatively intolerant of decreased water quality, and require unionized ammonia levels below 0.60 mg/L (Thurston and Russo, 1983).

Over 6 million tons of rainbow trout were produced worldwide in 2007, an increase of over 1 million tons from 2005 (FAO, 2010). In 2008, nearly 25 million kg of rainbow trout were produced at private fish farms in the United States, with an additional 167 million rainbow trout grown by governmental agencies for release into public recreational fisheries (USDA-NASS, 2007).

16.3 OVERVIEW OF NUTRITION REQUIREMENTS

16.3.1 DIETARY ENERGY

Defining nutrient requirements for a species is complex and tedious because dietary needs can vary with size and age, temperature, and growth rate; however, minimum values for diet constituents have been described for important aquaculture fishes. Energy is an important property of nutrients that is released during metabolic oxidation of proteins, carbohydrates, and lipids (NRC, 1993; Li et al., 2004; Lim and Webster 2006). Generally, proteins and lipids provide the primary sources of energy and are given first priority in diet formulation due to their higher cost in aquaculture feeds, followed by major minerals and vitamins (Lovell, 1989a; Lim and Webster, 2006). Energy excesses, deficiencies, or a suboptimal nutrient:energy ratio can result in a reduction in growth (e.g., Lovell, 1989a). If insufficient amounts of energy exist, proteins will be used for energy and result in reduced growth, while an excess of energy can limit feed consumption and reduce use of protein and other nutrients (NRC, 1993; Lim and Webster, 2006). Highly unbalanced ratios can lead to undesirable deposits of body fat and off flavors (Lovell, 1989a; Li et al., 2004). However, sufficient energy is needed to allow somatic (i.e., muscle tissue) growth.

Dietary energy (DE) in fishes is described as gross energy (GE) or its components, digestible energy (DE), and metabolizable energy (ME) (Lovell, 1989a). GE is the amount of heat released by completely oxidizing a compound into carbon dioxide, water, and other gases (Lovell, 1989a; Li et al., 2004). DE is the difference between GE of food consumed and energy lost in feces and can be determined through direct or indirect means (Lovell, 1989a; NRC, 1993). ME is DE less the energy lost from excretion through gills and urinary wastes, but is more difficult to determine than DE (Lovell, 1989a). For initial feed formulation purposes, GE can be indirectly estimated using composition values of each ingredient multiplied by physiological fuel values of 5.64, 9.44, and 4.11 kcal/g for proteins, lipids, and carbohydrates, respectively (NRC, 1993). These gross estimates can be further refined to estimate DE by applying general nutrient digestibilities (%) for the species. For example, trout nutrient digestibility values for protein, fat and carbohydrate are about 75%, 85%, and 40%, respectively. Thus, available energy (kcal/g, approximate DE) values for a trout feed would be about 3.9 (5.64 kcal/g × 0.7), 8.5 (9.44 kcal/g × 0.9), and 1.6 (4.11 kcal/g × 0.4) kcal/g for proteins, lipids, and carbohydrates, respectively (Piper et al., 1982). These approximate DE values can then be used to gauge the total DE and protein-to-energy ratios of a particular feed composition.

16.3.1.1 Tilapia

Cultured tilapia digest GE from most commercial feed ingredients and are similar to channel catfish in assimilating starch carbohydrate in cereal grains such as corn and whole wheat, but are more efficient than channel catfish and rainbow trout in digesting complex carbohydrates found in highly fibrous feedstuffs. Tilapia, like most fishes, digest proteins and fats more readily than carbohydrates (Lim, 1989). All three types of energy (GE, DE, and ME) in relation to the dietary level of protein (protein/energy ratio) have been reported for Nile tilapia (Lim and Webster, 2006). Nile tilapia fry (mean initial weight = 12 mg) achieved the best growth with a protein/GE ratio of 110 mg/kcal (Table 16.1) (El-Sayed and Teshima, 1992). Kubaryk (1980) indicated optimum protein/DE ratios of 108 to 120 mg/kcal resulted in the best growth performance of Nile tilapia fry

TABLE 16.1
Optimum Dietary Protein-to-Energy Ratios for Various Tilapia Species

Species	Fish size (g)	Protein/Energy Ratio (mg/kcal)	Energy Type	Reference
O. niloticus	0.012	110	Gross	El-Sayed and Teshima (1992)
	1.7	120	Digestible	Kubaryk (1980)
O. niloticus[a]	2.9	75	Metabolizable	Fineman-Kalio and Camacho (1987)
O. aureus	2.5	123	Digestible	Winfree and Stickney (1981)
	7.5	106	Digestible	Winfree and Stickney (1981)
T. zilli	50	103	Digestible	El-Sayed (1987)
O. niloticus × *O. aureus*	0.6	111	Digestible	Santigo and Laron (1991)
O. niloticus × *O. aureus*[b]	1.6	68–104	Metabolizable	Shiau and Huang (1990)

Source: Lim, C. E., and C. D. Webster. *Tilapia: Biology, Culture, and Nutrition,* eds. C. E. Lim, and C. D. Webster, pp. 469–501. The Haworth Press, Inc., 2006.

[a] Cultured in seawater at 18 to 50 ppt salinity.
[b] Cultured in seawater at 32 to 34 ppt salinity.

(mean initial weight = 1.7 g) over a range of dietary protein levels. Fineman-Kalio and Camacho (1987) reported Nile tilapia (mean initial weight = 2.9 g) cultured in seawater (18 to 50 ppt salinity) showed the best performance when fed a diet with a protein/ME ratio of 75 mg/kcal; however, based on economic analysis they stated a diet using a protein/ME ratio of 66.7 mg/kcal would be more profitable. Other tilapia species and hybrids have shown similar protein/energy ratios (Table 16.1).

16.3.1.2 Channel Catfish

Absolute energy requirements are not well established for channel catfish diets, but estimates have been made by measuring weight gain (WG) or protein gain of catfish fed diets containing known amounts of energy (Li et al., 2004). Because DE is easier to calculate, no ME energy values have been determined (Li et al., 2004). Generally, energy requirements for catfish are expressed as a ratio of DE/crude protein or digestable protein/digestible energy (DP/DE) (Li et al., 2004). Table 16.2 lists several studies that have indicated optimum DP/DE ratios for channel catfish, while Table 16.3 lists protein/DE ratios for channel catfish of various sizes. Protein:DE ratios range from 7.4 to 12 kcal/g (Li et al., 2004; Robinson and Li 2007). A study by Robinson and Li (2007), supported this range by indicating a DE/protein ratio of 8.5 to 9.5 kcal/g for juvenile channel catfish (mean initial weight = 5.2 g). Gatlin et al., (1986) reported that the maintenance requirement for fingerling

TABLE 16.2
Digestible Protein (DP) to Digestible Energy (DE) Ratios for Various Sizes of Channel Catfish

Weight (g)	DP (%)	DE (kcal/g)	Final DP/DE (mg/kcal)	Reference
10	27.0	2.78	97	Mangalik (1986)
34	28.8[a]	3.07[a]	94	Garling and Wilson (1976)
266	27.0	3.14	86	Mangalik (1986)
526	22.2	2.33	95	Page and Andrews (1973)
600	24.4[a]	3.05[a]	81	Li and Lovell (1992)

Source: National Research Council. *Nutrient Requirements of Fish.* National Academy Press, 1993.
[a] Digestible protein and energy were estimated from ingredient composition of the diet.

TABLE 16.3
Protein to Digestible Energy (DE) Ratios for Various Sizes of Channel Catfish

Weight (g)	Protein (g/100 g fish/d)	DE (kcal/100 g fish d)	Protein/DE Ratio (mg/kcal)
3	1.64	16.8	97
10	1.11	11.4	97
56	0.79	9.0	88
198	0.52	6.1	85
266	0.43	5.0	86

Source: Mangalik, A. Dietary energy requirements of channel catfish. Doctoral Dissertation. Auburn, Alabama: Auburn University, 1986; Lovell, T. *Nutrition and Feeding of Fish,* ed. T. Lovell, pp. 11–71. Van Nostrand Reinhold, 1989a.

channel catfish (initial weight = 8.3–10 g) was 15 to 17.3 kcal energy/kg body weight (BW) per day and that 99.75 kcal/kg BW per day resulted in the highest growth rates. An adequate DE level in commercial channel catfish feeds is considered to be 9 to 10 kcal/g crude protein (Li et al., 2004). Increasing above this range may increase fat deposition, while low energy values may lead to slow growth (Li et al., 2004).

16.3.1.3   Rainbow Trout

The absolute energy requirements of rainbow trout have also been investigated (Cho and Kaushik, 1990; NRC, 1993; Rodehutscord and Pfeffer, 1999). Between 3700 and 4200 kcal of DE is required to produce one kilogram of rainbow trout (Piper et al., 1982; Kim and Kaushik, 1992). Daily maintenance energy expenditures for 150 g rainbow trout at 18°C are approximately 22 kcal/kg/d, with larger fish (300 g) at lower temperatures (15°C) requiring only 14 kcal/kg/d (Guillaume et al. 2001). Energy requirements in relation to digestible protein (DP/DE) range from 76 to 105 mg/kcal (Cho and Kaushik, 1985; Cho and Woodward, 1989; Cho, 1990, 1992; Higgs et al., 1995; Iwama, 1996; Azevedo et al., 2004b), with a value of 95 recommended by NRC (1993). At DP:DE ratios of 84 or less, protein retention can be increased and nitrogenous excretion decreased if dietary energy from nonprotein sources is also increased (Médale et al., 1995; Steffens et al., 1999; Encarnação et al., 2004; Eliason et al., 2007). At higher DP:DE ratios, such as 125, protein utilization in rainbow trout is adversely impacted (Eliason et al. 2007). DE and DP in rainbow trout have been shown to be related to rearing temperatures, increasing with temperature increases from 5°C to 15°C (Watanabe et al., 1996a, 1996b; Azevedo et al., 1998). This relationship was not observed by Cho and Kaushik (1990) however. Fish size also influences energy requirements in salmonids such as rainbow trout, as the DP/DE ratio was observed to decrease as fish size increased by Cho and Kaushik (1990) and Einen and Roem (1997).

16.3.2   Protein and Amino Acids

Proteins consist of a large group of complex, organic compounds that contain carbon, hydrogen, oxygen, nitrogen, and small amounts of iron; most proteins also contain sulfur (Lovell, 1989a; Li et al., 2004; Lim and Webster, 2006). Proteins exist and function in many different ways, including: formation and maintenance of tissues, regulatory substances (e.g., enzymes, hormones, and plasma globulins), and energy sources (Li et al., 2004). Optimum levels of protein for finfish diets range from approximately 25% to 50%, depending upon the species and life stage. However, an excess or deficiency in protein can have negative effects on fish and feeding costs. Inadequate amounts of protein cause slowing or cessation of growth and/or weight loss from protein transition to vital tissues (Lim and Webster, 2006). Only a partial amount of protein will be converted to new tissue when excess protein exists, while the remainder is converted to energy or excreted (Lim and Webster 2006). Also, excessive protein can increase feed costs because protein is one of the most expensive feed ingredients (Coyle et al., 1994). Several studies have indicated that a number of factors, including: size and age of fish, protein quality, nonprotein energy levels, water temperature, salinity, feed allowance, management practices, and amount of natural food in a culture system affect the dietary protein requirements (NRC, 1993; Lim and Webster, 2006; Robinson and Li, 2007).

Proteins consist of amino acids linked together by peptide bonds; most protein sources contain 18 amino acids, but some may have 22 to 26 (Lim and Webster, 2006). These amino acids can be separated into two nutritional groups: *nonessential*, those that an animal can synthesize at necessary levels to support maximum growth; and *essential*, those an animal cannot synthesize in sufficient quantities to support maximum growth (Lovell, 1989a). Bony (teleost) fishes require ten essential amino acids (EAA): arginine, histidine, isoleucine, leucine, lysine, methionine, threonine, tryptophan, phenylalanine, and valine (Lim, 1989; Lovell, 1989a; Li et al., 2004; Lim and Webster, 2006). Although fish can adequately synthesize most nonessential amino acids (NEAA), their presence in diets has a nutritional contribution (Lim and Webster, 2006). Fish require a source of nitrogen source for synthesis of NEAA. The EAA should compose 50% of the total amount of available amino acids (EAA:NEAA = 1:1). Tables 16.4, 16.5, and 16.6 provide known quantitative amino acid requirements for juvenile Nile tilapia, channel catfish, and rainbow trout, respectively.

16.3.2.1 Tilapia

Tilapia require a continuous supply of protein for maintenance, growth, and other physiological functions (Lim and Webster, 2006). According to El-Sayed and Teshima (1992), Nile tilapia fry

TABLE 16.4
Amino Acid Requirements for Juvenile Nile tilapia

Amino Acid	Protein in Diet (%)	Requirement as Percentage of Dietary Protein[a]	Requirement as Percentage of Dry Diet[a]
Arginine	28	4.20	1.18
Histidine	28	1.72	0.48
Isoleucine	28	3.11	0.87
Leucine	28	3.39	0.95
Lysine	28	5.12	1.43
Methionine[b]	28	2.68	0.75
Phenylalanine[c]	28	3.75	1.05
Threonine	28	3.75	1.05
Tryptophan	28	1.00	0.28
Valine	28	2.80	0.78

Source: National Research Council. *Nutrient Requirements of Fish.* National Academy Press, 1993.

[a] Santiago and Lovell (1988).
[b] Cystine 0.54% of dietary protein, 0.15% of dry diet.
[c] Tyrosine 1.79% of dietary protein, 0.5% of dry diet.

TABLE 16.5
Amino Acid Requirements for Juvenile Channel Catfish

Amino Acid	Protein in Diet (%)	Requirement as Percentage of Dietary Protein	Requirement as Percentage of Dry Diet	Reference
Arginine	24	4.3	1.0	Robinson et al. (1981)
Histidine	24	1.5	0.4	Wilson et al. (1980)
Isoleucine	24	2.6	0.6	Wilson et al. (1980)
Leucine	24	3.5	0.8	Wilson et al. (1980)
Lysine	24	5.1	1.2	Wilson et al. (1977)
	30	5.0	1.5	Robinson et al. (1980b)
Methionine[a]	24	2.3	0.6	Harding et al. (1977)
Phenylalanine[b]	24	5.0	1.2	Robinson et al. (1980a)
Threonine	24	2.0	0.5	Wilson et al. (1978)
Tryptophan	24	0.5	0.12	Wilson et al. (1978)
Valine	24	3.0	0.71	Wilson et al. (1980)

Source: National Research Council. *Nutrient Requirements of Fish.* National Academy Press, 1993; Li, M. H. et al. *Biology and Culture of Channel Catfish,* eds. C. S. Tucker, and J. A. Hargreaves, pp. 279–323. Elsevier B. V., 2004.

[a] In the absence of dietary cystine.
[b] Diet contained 0.3% tyrosine.

TABLE 16.6
Amino Acid Requirements for Juvenile Rainbow Trout

Amino Acid	Protein in Diet (%)	Requirement as Percentage of Dietary Protein	Requirement as Percentage of Dry Diet	Reference
Arginine	33	4.7	1.6	Cho et al. (1989)
	35	4.0	1.4	Kim et al. (1983), Kim et al. (1992b)
	36	3.3	1.2	Kaushik (1979)
	45	3.6	1.6	Walton et al. (1986)
	47	5.9	2.8	Ketola (1983)
Lysine	35	3.7	1.3	Kim and Kayes (1982), Kim et al. (1992b)
	45	4.2	1.9	Walton et al. (1984a)
	47	6.1	2.9	Ketola (1983)
Methionine	35	1.4	0.5	Kim et al. (1992a)
	35	2.9[a]	1.0[a]	Kim et al. (1984)
	35	3.0[b]	1.1[b]	Rumsey et al. (1983)
	41	1.5	0.6[d]	Cowey et al. (1992)
	46.4	2.2[c]	1.0[c]	Walton et al. (1982)
Methionine and Cystine (combined)	35	2.3	0.8	Kim et al. (1992a)
Phenylalanine[e]	35	2.0	0.7	Kim (1993)
Threonine	40	2.7	1.1	Bodin et al. (2008)
Tryptophan	35	0.6	0.2	Kim et al. (1987)
	42	1.4	0.6	Poston and Rumsey (1983)
	55	0.5	0.3	Walton et al. (1984b)

Sources: National Research Council. *Nutrient Requirements of Fish.* National Academy Press, 1993; Kim, K. I. *Aquaculture* 151, 3–7, 1997.

[a] Diet contained 0.5% cystine.
[b] Diet contained 0.3% cystine.
[c] Diet lacked cystine.
[d] Diet contained 0.16% cystine.
[e] Diet contained tyrosine.

(mean initial weight = 12 mg) fed diets containing 45% protein produced the best growth rates (Table 16.7). Similarly, Siddiqui et al. (1988) indicated protein levels of 40% and 30% resulted in the best growth responses for Nile tilapia fry (mean initial weight = 0.8 g) and advanced juveniles (mean initial weight = 40 g), respectively (Table 16.7). Other tilapia species have similar requirements to Nile tilapia (Table 16.7). However, De Silva et al. (1989) performed a least cost dietary protein study on Nile, Mozambique, blue, and redbelly tilapia and showed that dietary protein levels of 34% to 36% provided maximum growth of 1 to 5 g tilapia, while the most cost-effective protein levels occurred from 25% to 28%.

Tilapia have a requirement for sulfur-containing amino acids, which can be met by methionine alone or an appropriate mixture of methionine and cystine, a NEAA (Lim and Webster 2006). Likewise, tyrosine, a NEAA, will reduce the amount of phenylalanine required by tilapia (NRC, 1993).

16.3.2.2 Channel Catfish

Channel catfish require a certain amount of nonspecific nitrogen for normal growth and survival (Li et al., 2004; Robinson and Li, 2007). Crude protein levels in commercial feeds usually range from 45% to 50% for fry, 32% to 36% for fingerlings, and 26% to 32% for grow-out diets. Page and Andrews (1973) stated that juvenile channel catfish (mean initial weight = 14 g) showed increased

TABLE 16.7
Dietary Protein Requirements for Several Tilapia Species Cultured in Freshwater

Species	Protein Source	Size (g)	Protein Requirement (%)	Reference
O. niloticus	Fish meal	0.8	40	Siddiqui et al. (1988)
	Fish meal	40	30	Siddiqui et al. (1988)
	Casein	0.012	45	El-Sayed and Teshima (1992)
O. aureus	Casein/egg albumin	Fry–2.5	56	Winfree and Stickney (1981)
	Casein/egg albumin	2.5–7.5	34	Winfree and Stickney (1981)
	Soybean or fish meal	0.3–0.5	36	Davis and Stickney (1978)
O. mossambicus	Fish meal	Fry	50	Jauncey and Ross (1982)
	Fish meal	0.5–1.0	40	Jauncey and Ross (1982)
	Fish meal	6–30	30–35	Jauncey and Ross (1982)
T. zilli	Casein	1.3–3.5	35	Mazid et al. (1979)

Source: Lim, C. E., and C. D. Webster. *Tilapia: Biology, Culture, and Nutrition,* eds. C. E. Lim, and C. D. Webster, pp. 469–501. The Haworth Press, Inc., 2006.

WG as protein levels were increased from 25% to 35%, while the reverse occurred for adult channel catfish (mean initial weight = 114 g). Another study indicated maximum growth of swim-up stage channel catfish fry occurred with a protein level of 58%, while minimum protein requirements were 55% for 0.2 g channel catfish and 46% to 50% for 3 to 5 g channel catfish (Winfree and Stickney, 1984). Variables for such a range in protein requirements include: size of fish, daily feed allowance, amount of nonprotein energy in diet, protein quality, water temperature, and amount of natural food available (Lovell, 1989b). In general, a minimum dietary protein concentration of 24% is needed for maximum growth of advanced fingerling channel catfish to marketable size when fed to satiation (Li et al., 2004). Robinson and Li (2007) indicated that channel catfish fed diets with 26% dietary protein or less have a reduction in processed yield. Therefore, to obtain optimum growth, processing yield, and body composition, channel catfish grow-out diets likely require a minimum dietary protein concentration of 28% (Li et al., 2004; Robinson and Li, 2007).

Channel catfish can synthesize methionine and phenylalanine from cystine and tyrosine, respectively (NRC, 1993; Li et al., 2004). Cystine can replace or spare about 60% of the methionine requirement on a molar sulfur basis, while tyrosine can spare about 50% of the phenylalanine requirement (NRC, 1983; Lovell, 1989b; Li et al., 2004). Both of these NEAA can only be synthesized from essential amino acid precursors (NRC, 1983).

16.3.2.3 Rainbow Trout

In general, rainbow trout require from 38% to 45% dietary protein (Satia, 1974; Wilson, 1989; Kim et al., 1991; NRC, 1993). Piper et al. (1982) recommended a variation in protein levels, roughly corresponding to the three growth stanzas of rainbow trout (Dumas et al., 2007), with higher levels of protein in starter diets (45%–55%), compared to lesser amounts in feeds to juveniles (35%–50%), or larger fish (30%–40%). Cho and Cowey (1991) also suggested increased protein levels in feed for rainbow trout fry compared to grower diets. The 40% protein requirement can be roughly split into two functions: 24% provides the necessary amino acids, while 16% of the protein is available for energy needs (Kim, 1997). Several studies have indicated that the portion of the protein available for energy may be spared through the use of less-expensive lipid sources, thereby decreasing the total protein percentage of diet. Yigit et al. (2002), Chaiyapechara et al. (2003), and Morrow et al. (2004) all indicated increased rainbow trout growth rates with low protein/high lipid diets. However, other studies (Azevedo et al., 2004a, 2004b; Eliason et al., 2007), while observing no negative effects on growth and feed efficiency from decreased dietary protein levels, have observed changes in carcass composition. Kim (1997) stated that dietary protein levels of up to 25% are required if other energy sources with ME values similar to protein are included in the feed. The EAA for rainbow trout was suggested by NRC (1993) and tested by Green and Hardy (2002).

16.3.3 Lipids and Fatty Acids

Dietary lipids provide the only source of essential fatty acids needed by fish and are sources of highly digestible energy (Lim and Webster, 2006). Lipids also play several physiological roles by assisting in the absorption of fat-soluble vitamins and maintaining membrane flexibility and permeability of cellular structures in fishes (Lovell, 1989a; Lim and Webster, 2006). Common lipid sources in fish feeds are fish and seed oils and animal fats. Although essential fatty acid requirements are not well defined across species, most freshwater fishes require at least 0.5% linoleic (18:2 [ω-6]) and 0.5% linolenic (18:3 [ω-3]) (both common in seed oils) in their diets (NRC, 1993). Other requirements for eicosapentaenoic acid (EPA, 20:5 [ω-3]) and docosahexaenoic acid (DHA, 22:6 [ω-3]) have been identified for specific fishes (NRC, 1993). Both EPA and DHA are recognized to have positive effects on human health and certain disease therapies whereas long chain ω-6 fatty acids, such as arachidonic acid (ARA, 20:4 [ω-6]), are likely detrimental to human health. Fish do not naturally produce EPA or DHA, but rather concentrate these fatty acids from their food sources.

16.3.3.1 Tilapia

Tilapia do not require or tolerate high lipid levels as do salmonids (Lim, 1989); Jauncey and Ross (1982) indicated depressed growth of hybrid tilapia when dietary lipid levels exceeded 12%. Chou and Shiau (1996) demonstrated that 12% dietary lipid was optimal for growth, but 5% was a minimum lipid level for growth and survival.

Tilapia do have a requirement for linoleic fatty acid, which are highly variable but common in soybean and corn oils (Lim, 1989; Lim and Webster, 2006). Tilapia demonstrated better growth performance when fed diets containing corn and soybean oil (~59% and 50% linoleic fatty acids, respectively) as compared to beef tallow (~47% oleic acid, 18:1 [ω-9]) (Takeuchi et al., 1983a; Lim, 1989; Lim and Webster, 2006). Optimum dietary levels of essential fatty acids have been determined for two tilapia species (Lim, 1989; Lim and Webster, 2006). The dietary requirement of linoleic fatty acid for Nile tilapia was estimated to be 0.5% (Takeuchi et al., 1983b), while no firm estimate exists for a dietary requirement of linolenic fatty acid, such as found in flax oil (~58%) (Table 16.8) (Lim and Webster, 2006). Similarly, Kanazawa et al. (1980) indicated redbelly tilapia (mean initial weight = 0.5 g) require ω-6 fatty acids compared to ω-3 and estimated a 1% dietary requirement of 18:2 [ω-6] or ARA fatty acids (Table 16.8). Essential fatty acid requirements for blue tilapia have not been determined; however, studies have indicated a need for them (Lim and Webster, 2006). Similarly, essential

TABLE 16.8
Essential Fatty Acid Requirements for Young Tilapia

Species	Essential Fatty Acid	Requirement (% in Diet)	Reference
O. niloticus	18:2 [ω-6] or 20:4 [ω-6]	0.5	Takeuchi et al. (1983b)
	18:3 [ω-3]	Unknown	Takeuchi et al. (1983b)
O. aureus	18:2 [ω-6] and 18:3 [ω-3]	≥1.0	Stickney and McGeachin (1983)
T. zilli	18:2 [ω-6] or 20:4 [ω-6]	1.0	Kanazawa et al. (1980)
O. niloticus × *O. aureus*	18:2 [ω-6] and 18:3 [ω-3]	Unknown	Chou et al. (2001)

Source: Lim, C. E., and C. D. Webster. *Tilapia: Biology, Culture, and Nutrition,* eds. C. E. Lim, and C. D. Webster, pp. 469–501. The Haworth Press, Inc., 2006.

fatty acid requirements for various hybrids have not been determined, but Chou and Shiau (1999) and Chou et al. (2001) have suggested that ω-3 fatty acids along with ω-6 fatty acids are needed for maximum growth (Lim and Webster, 2006). Tilapia diets deficient in ω-6 and ω-3 fatty acids exhibit the following symptoms: poor appetite, retarded growth, and swollen, pale, and fatty livers (Lim, 1989).

16.3.3.2 Channel Catfish

Lipid requirements have been not precisely defined for channel catfish (Li et al., 2004). Channel catfish can tolerate moderate levels of dietary lipids; a dietary level up to 16% does not appear to negatively affect growth or feed efficiency (Li et al., 2004). However, excessive lipid levels in diets contribute to off flavor and increased fat deposition in body tissues and also cause difficulty in processing pellets during extrusion (Li et al., 2004). Typical lipid levels in commercial catfish feeds used for growing advanced fingerlings to market size do not exceed 6% (Li et al., 2004). Up to 75% of the lipid contained in commercial diets is usually contributed by ingredients, with the remainder applied as a top dressing to the feed (Kelly, 2005).

Similar to other fishes, channel catfish require certain essential fatty acids to achieve optimum growth and health (Li et al., 2004). In contrast to tilapia, channel catfish require ω-3 essential fatty acids including: linolenic acid, EPA, and/or DHA (NRC, 1993). Satoh et al. (1989) reported the essential fatty acid requirements of channel catfish (mean initial weight = 4.7 g) to be 1% to 2% linolenic acid or 0.5% to 0.75% EPA and/or DHA.

16.3.3.3 Rainbow Trout

Diets with a lipid content, as a percent of dietary dry weight, from 13% to 15% have been used successfully for rainbow trout culture (Takeuchi et al., 1978; Chaiyapechara, 2003; Geurden et al., 2006). The optimal lipid level was reported as 18% by Guillaume et al. (2001). However, because of the protein-sparing effect of additional lipids (Lee and Putnam, 1973; Ruohonen et al., 1998; Steffens et al., 1999; Azevedo et al., 2004a; Guerden et al., 2006), more recent commercial salmonid feeds have increased dietary lipid levels to 40% (Caballero et al., 2002). For example, Forsman and Ruohonen (2009) showed a positive protein-sparing effect with dietary lipid levels up to 40%. Thus, the lipid component of rainbow trout diets must be considered in conjunction with the DP/DE ratio (Guillaume et al., 2001). Lipid percentages of 50% are likely inappropriate in rainbow trout diets (Forsman and Ruohonen, 2009).

Rainbow trout have an essential fatty acid requirement (NRC 1993) and linolenic acid at 1.0% was recommended by Castell et al. (1972). Watanabe et al. (1984) showed the importance of essential fatty acids for successful rainbow trout reproduction and recommended dietary linolenic concentrations of 0.8%. Takeuchi and Watanabe (1977) determined the essential fatty acid requirement of rainbow trout to be 20% of their dietary lipid as linolenic acid, or 10% as EPA and DHA.

The source of lipids in diets affects the fatty acid profile present in rainbow trout tissues (Chaiyapechara et al., 2003; Kiron et al., 2004; Rinchard et al., 2007; Trenzado et al., 2008), which is important given the benefits to human health due to the consumption of ω-3 highly unsaturated fatty acids (Moreno and Mitjavila, 2003). However, essential fatty acids can be overfed. Rainbow trout fed greater than the required amount of ω-3 highly unsaturated fatty acids experienced decreased growth (Takeuchi and Watanabe, 1979). In addition, Trenzado et al. (2007) also observed decreased resistance to the negative effects of stress in rainbow trout fed elevated dietary levels (30.5 g/kg) of highly unsaturated fatty acids. Some of the negative effects of feeding diets high in ω-3 highly unsaturated fatty acids to rainbow trout may be ameliorated by the inclusion of vitamin E (Puangkaew et al., 2005). The ratio of polyunsaturated to saturated fatty acids in rainbow trout diets is also important, with a decreased ratio associated with decreased fatty acid digestibility (Caballero et al., 2002).

16.3.4 CARBOHYDRATES

In general, due to the natural aquatic environment where carbohydrates are scarce or inconsistently available, digestive and metabolic systems of fish metabolize proteins and lipids for energy better than carbohydrates (Lovell, 1989a). Studies indicate fish can maintain satisfactory growth and not exhibit any symptoms of nutritional diseases or growth deficiencies when fed carbohydrate-free diets (Lim 1989) and, therefore, carbohydrates are considered to be nonessential (Lovell 1989a). However, carbohydrates can serve several practical feed processing and fish metabolic functions, including pellet binders, precursors to metabolic intermediates and nucleic acids needed for growth, and a sparing effect on dietary protein utilization as an inexpensive source of energy (NRC, 1993; Lim and Webster, 2006). Carbohydrates that influence these functions include starch, sugars, and fiber (crude, neutral detergent, soluble, and insoluble).

16.3.4.1 Tilapia

No specific requirements for carbohydrates have been recommended for tilapia (Lim and Webster, 2006), but tilapia are largely herbivorous and are expected to utilize carbohydrates more efficiently than cultured piscivorous fishes (NRC, 1993; Lim and Webster, 2006). Carbohydrate utilization is affected by source, other diet constituents, feeding frequency, and size (El-Sayed, 2006). Anderson et al. (1984) found that Nile tilapia (mean initial weight = 2 g) exhibited higher growth rates when fed diets containing glucose, sucrose, dextrin, and starch compared to carbohydrate-free diets; WG increased when dietary carbohydrates were raised from 0% to 40%. Similarly, Wee and Ng (1986) reported improved growth performance and feed conversion efficiency of Nile tilapia (mean initial weight = 4 g) with increasing levels of cassava (~35% carbohydrate) up to 60% of diet. Anderson et al. (1984) observed that juvenile tilapia growth, food conversion, and survival were better with feeds containing ≤5% fiber (α-cellulose) and protein utilization decreased at excessive (10%) fiber levels. El-Sayed and Garling (1988) found that redbelly tilapia (mean initial weight = 1.85 g) appeared to utilize dextrin as efficiently as Nile tilapia, but more efficiently than channel catfish.

16.3.4.2   Channel Catfish

Channel catfish are omnivorous and do not have a defined carbohydrate requirement, but utilize carbohydrates more efficiently (higher intestinal amylase activity) than most coldwater and marine fishes (Robinson and Li, 2002; Li et al., 2004). According to Page and Andrews (1973), up to 25% corn can be used as an effective energy source in channel catfish diets. Garling and Wilson (1977) indicated dextrin levels of 14% to 28% resulted in the best gains and feed conversions for channel catfish (initial length = 2.5 to 7.5 cm). Within certain lipid:carbohydrate ratios, channel catfish were able to utilize polysaccharides (dextrin and starch) as an energy alternative to lipids; however, monosaccharides and disaccharides were not as useful (Wilson and Poe, 1987). Typical commercial catfish feeds contain 25% or more soluble carbohydrates, while an additional 3% to 6% is present as crude fiber and unavailable to channel catfish (Li et al., 2004).

16.3.4.3   Rainbow Trout

Rainbow trout can tolerate up to 12% to 20% carbohydrates in their diet, although this value is greatly diminished at lower water temperatures (Barrows and Hardy, 2001). Carbohydrates are not considered an essential component of trout diets, but can have a protein-sparing effect similar to lipids if used judiciously (Kim and Kaushik, 1992). Glucose is highly digestible in rainbow trout, with apparent digestibility coefficients approaching 99% (Singh and Nose, 1967; Hilton et al., 1982). However, when glucose and sucrose were fed at concentrations greater than 20%, growth was negatively affected at lower dietary protein levels (Bergot, 1979a, 1979b; Luquet, 1971). Storebakken et al. (1998) reported that apparent digestibility coefficients of protein and lipids was greater in diets containing 100 g of carbohydrates (glucose or dextrin, partially hydrolyzed starch) per kg compared to diets with 200 g/kg. Hilton and Atkinson (1982) noted a negative relationship between glucose and WG when glucose was used to replace the lipids in rainbow trout diets.

Starch digestibility is inversely related to dietary inclusion (Kim and Kaushik, 1992; Brauge et al., 1994) and feed intake (Bergot and Breque, 1983) in rainbow trout. A similar pattern with decreased digestibility was observed when dextrin was increased in rainbow trout diets (Singh and

Nose, 1967). However, if starch is gelatinized, apparent digestibility coefficients increase to up to 90% (Bergot and Breque, 1983). In diets containing either raw starch, dextrin, or simple sugars (glucose or maltose), rainbow trout growth decreased as the carbohydrate source became more complex (Hung and Storebakken, 1995). Diets with 23% starch resulted in significantly decreased rainbow trout growth rates, compared to diets with only 7% starch (Krogdahl et al., 2004). However, Page et al. (1999) observed no negative effects on rainbow trout immune response with increasing levels of dietary starch. Dietary fiber, such as cellulose and nonstarch polysaccharides, in general has a negative impact on rainbow trout nutrition (Ovrum-Hansen and Storebakken, 2007; Glencross, 2009). Storebakken et al. (1998) observed reductions in fat digestibility with diets containing greater than 100g/kg of indigestible carbohydrates. Removal of fiber may improve digestibility in rainbow trout diets (Glencross et al., 2004).

16.3.5 VITAMINS

Vitamins are essential for metabolic processes in fish and are required in trace amounts from the diet or other external sources for normal growth, reproduction, and health (NRC, 1993; Lim and Webster, 2006). Vitamins are classified into two categories, water soluble and fat soluble (NRC, 1993). Eight of the water-soluble vitamins (B_1 [thiamin], B_2 [riboflavin], B_3 [niacin], B_5 [pantothenic acid], B_6 [pyroxidine], B_7 [biotin], B_9 [folic acid], and B_{12} [cyanocobalamin]) are only needed in trace amounts and are collectively called the water-soluble vitamin B complex; these vitamins primarily serve coenzyme function roles (Lovell, 1989a). Choline, inositol, and ascorbic acid (vitamin C) are also water-soluble vitamins but are needed in larger quantities; these have functions other than as coenzymes (Lovell, 1989a; NRC, 1993). Fat-soluble vitamins A, D, E, and K function independent of enzymes or may have coenzyme roles and are absorbed along with fats during digestion (NRC, 1993).

16.3.5.1 Tilapia

Most vitamin requirements for tilapia are met when culture occurs extensively in ponds where natural food items exist; however, diets must be supplemented to sustain normal growth and health in intensive systems (Lim, 1989; Lim and Webster, 2006; Shiau and Lin, 2006). Some vitamins can be synthesized by tilapia; thereby, reducing the dietary need (Lim, 1989; Lim and Webster, 2006). Lovell and Limsuwan (1982) reported that Nile tilapia (mean initial weight = 7.1 g) produce enough vitamin B_{12} by intestinal synthesis to support normal growth and erythropoiesis such that dietary supplementation is probably unnecessary. Similarly, Peres et al. (2004) indicated Nile tilapia (mean initial weight = 5.8 g) probably synthesize enough inositol for normal growth, feed efficiency, immune function, and disease resistance, but not enough to prevent alteration of lipid metabolism. This fact, along with inositol found in many common feed ingredients, indicates that practical diets likely contain sufficient levels of this vitamin to meet the metabolic requirements of Nile tilapia (Peres et al., 2004; Lim and Webster, 2006).

While several vitamins can be internally produced, most must be met from the diet. Deficiencies of vitamins in diets can cause symptoms ranging from depressed appetite to severe tissue deformities (Lovell, 1989a). Table 16.9 lists the known vitamin requirements and deficiency symptoms for tilapia species.

16.3.5.2 Channel Catfish

Channel catfish usually meet vitamin requirements through natural food items found in extensive pond culture systems. However, vitamin inputs from feed ingredients and natural food items have not been thoroughly considered due to a lack of information on the bioavailability of these vitamins (Li et al., 2004). Therefore, commercial catfish feeds are generally supplemented with a vitamin premix containing all the essential vitamins in sufficient quantities to meet all requirements (Li et al., 2004).

Most of the quantitative vitamin requirements for channel catfish have been determined (Table 16.10) (NRC, 1993). Channel catfish demonstrate *de novo* synthesis of inositol within the liver

TABLE 16.9
Vitamin Requirements and Deficiency Symptoms for Various Tilapia Species

Vitamin/Species	Requirement (mg/kg Diet)	Deficiency Symptoms[a]	Reference
Thiamin			
O. niloticus	4.0	Light coloration, nervous disorder, low hematocrit and red blood cell count, and increased serum pyruvate.	Lim et al. (2000)
O. mossambicus × O. niloticus	2.5		Lim et al. (1991)
Riboflavin			
O. aureus	6.0	Lethargy, high mortality, loss of color, fin erosion, short-body dwarfism, and lens cataracts.	Soliman and Wilson (1992b)
O. mossambicus × O. niloticus	5.0		Lim et al. (1993)
Pyridoxine			
O. mossambicus	5.0–11.7	Nervousness, convulsion, caudal fin erosion, mouth lesions, high mortality, and reduced hepatic alanine aminotransferase.	Oyetayo et al. (1985)
O. mossambicus × O. niloticus	3.0		Lim et al. 1995
O. niloticus × O. aureus	1.7–9.5[b]		Shiau and Hsieh (1997)
O. niloticus × O. aureus	15.0–16.5[c]		Shiau and Hsieh (1997)
Pantothenic acid			
O. aureus	6.0	Hemorrhage, sluggishness, high mortality, anemia, and hyperplasia of gill lamellae epithelial cells.	Roem et al. (1991)
O. aureus	10.0		Soliman and Wilson (1992a)
Niacin			
O. niloticus × O. aureus	26.0[d]	Hemorrhages, deformed snout, gill edema, and skin, fin, and mouth lesions.	Shiau and Suen (1992)
O. niloticus × O. aureus	121.0[e]		Shiau and Suen (1992)
Biotin			
O. niloticus × O. aureus	0.06	Low tissue biotin, and reduced hepatic pyruvate carboxylase and acetyl CoA caboxylase activities.	Shiau and Chin (1999)
Folic acid			
O. niloticus	0.5–1	Reduced feed intake and efficiency.	Lim and Klesius (2001)
O. niloticus × O. aureus	0.82		Shaiu and Huang (2001)
Vitamin B$_{12}$			
O. niloticus	Not required		Lovell and Limsuwan (1982)
O. niloticus × O. aureus	Not required		Shiau and Lung (1993)
Inositol			
O. niloticus	Not required	Anemia, fin erosion, dark skin coloration, slow gastric emptying, and decreased cholinesterase and certain aminotransferase activities.[f]	Perres et al. (2004)
O. niloticus × O. aureus	400		Shiau and Su (2005)
Choline			
O. aureus	Not required	Poor survival and reduced blood triglyceride, cholesterol, and phospholipids concentration.	Roem et al. (1990a) Roem et al. (1990c)
O. niloticus × O. aureus	1000		Shiau and Lo (2000)
Ascorbic acid			
O. niloticus	50	Lordosis, scoliosis, poor wound healing, hemorrhage, anemia, exophthalmia and gill and operculum deformity.	Abdelghany (1996)
O. niloticus	420		Soliman et al. (1994)
O. aureus	50		Stickney et al. (1984)
O. spilurus	100–200[g]		Al-Amoudi et al. (1992)
O. niloticus × O. aureus	79		Shiau and Jan (1992)

continued

TABLE 16.9 (continued)
Vitamin Requirements and Deficiency Symptoms for Various Tilapia Species

Vitamin/Species	Requirement (mg/kg Diet)	Deficiency Symptoms[a]	Reference
Vitamin A			
O. niloticus	5000 (IU/kg)	High mortality, restlessness, abnormal swimming,	Saleh et al. (1995)
O. niloticus ×	5850–6970	blindness, skin, fin, and eye hemorrhages, and mucus	Hu et al. (2006)
O. aureus	(IU/kg)	secretion.	
Vitamin D			
O. aureus	Not required	Reduced hemoglobin, hepatosomatic index, and	O'Connell and Gatlin (1994)
O. niloticus ×	374.8 (IU/kg)	alkaline phosphatase activity.	Shiau and Hwang (1993)
O. aureus			
Vitamin E[h]			
O. niloticus	50–100	Skin hemorrhage, impaired erythropoiesis, muscle	Satoh et al. (1987)
O. aureus	10–25	degeneration, seroid in liver and spleen, and	Roem et al. (1990b)
O. niloticus ×	42–44	abnormal skin coloration.	Shiau and Shiau (2001)
O. aureus			
Vitamin K			
O. niloticus ×	5.2	Low plasma prothrombin concentration.	Lee (2003)
O. aureus			

Sources: Lim, C. E., and C. D. Webster. *Tilapia: Biology, Culture, and Nutrition,* eds. C. E. Lim, and C. D. Webster, pp. 469–501. The Haworth Press, Inc., 2006; Shiau, S. Y., and Y. H. Lin. 2006. Vitamin requirements of tilapia—A review. In *Advances in Aquaculture Nutrition VIII,* eds. L. E. C. Suárez, D. R. Marie, M. T. Salazar, M. G. N. López, D. A. V. Cavazos, A. C. P. Cruz, and A. G. Ortega, pp. 129–138. Monterrey, Mexico: Aquaculture Nutrition International, Symposium 8.

[a] Anorexia, poor growth, and poor feed efficiency are common deficiency signs; therefore, they are not listed.
[b] 28% protein diet.
[c] 36% protein diet.
[d] For glucose diet.
[e] For dextrin diet.
[f] Symptoms seen in other fish species (NRC, 1993).
[g] Requirement given in mg of L-ascorbyl-2-sulfate/kg in diet.
[h] 5%–6% dietary lipid levels.

and intestinal tract in sufficient quantities to meet their requirement (NRC, 1993; Li et al., 2004). In the presence of cobalt, synthesis of vitamin B_{12} by intestinal microorganisms has been demonstrated in sufficient quantities to meet normal growth and erythropoiesis (Li et al., 2004), but dietary supplementation is generally needed to prevent anemia (NRC, 1993). Lovell and Buston (1984) indicated that channel catfish (mean initial weight = 2 g) produce biotin through intestinal synthesis, but not in sufficient amounts. However, high concentrations of biotin are found in corn and soybean feed ingredients; therefore, diets containing these ingredients do not require supplemental biotin (Lovell and Buston, 1984; NRC, 1993; Li et al., 2004). Conflicting studies have occurred over the requirement of vitamin K in channel catfish diets. Dupree (1966) reported that hemorrhaging occurred when channel catfish (initial weight = 7 to 12 g) were fed diets containing 4.0 mg menadione (synthetic vitamin K)/100 kg of diet. However, Murai and Andrews (1977) could not demonstrate a need for vitamin K based on channel catfish fingerlings (mean initial weight = 20 g) fed diets deficient in vitamin K for 30 weeks. Therefore, vitamin K is generally supplemented to ensure adequacy (Li et al., 2004).

16.3.5.3 Rainbow Trout

Rainbow trout relying on artificial diets under intensive rearing conditions have more rigid vitamin requirements than carnivorous species reared extensively with access to supplemental prey items (De Silva and Anderson, 1995). Additionally, vitamin synthesis by gastrointestinal microorganisms is

TABLE 16.10
Vitamin Requirements and Deficiency Symptoms for Channel Catfish

Vitamin	Requirement (mg/kg Diet)	Deficiency Symptoms[a]	Reference
Thiamin	1	Loss of equilibrium, nervousness, and dark skin color.	Murai and Andrews (1978b)
Riboflavin	6	Short-body dwarfism.	Serrinni et al. (1996)
	9		Murai and Andrews (1978a)
Pyridoxine	3	Greenish-blue coloration, tetany, nervous disorders, and erratic swimming.	Andrews and Murai (1979)
Pantothenic acid	10	Clubbed gills, emaciation, anemia, and eroded epidermis.	Murai and Andrews (1979)
	15		Wilson et al. (1983)
Niacin	7.4	Skin and fin lesions, exophthalmia, deformed jaws, and anemia.	Ng et al. (1997)
	14		Andrews and Murai (1978)
Biotin	Required	Hypersensitivity, skin depigmentation, and reduced liver pyruvate.	Robinson and Lovell (1978)
Folic acid	1.5	Anemia.	Duncan and Lovell (1991)
Vitamin B$_{12}$	Required	Anemia.	Limsuwan and Lovell (1981)
Inositol	Not required		Burtle and Lovell (1989)
Choline	400	Fatty liver, and hemorrhagic kidney and intestine.	Wilson and Poe (1988)
Ascorbic acid	60	Scoliosis, lordosis, internal and external hemorrhage, fin erosion, and reduced bone collagen formation.	Lim and Lovell (1989)
	45		Robinson (1990)
	11		El Naggar and Lovell (1991)
Vitamin A	1000–2000 (IU/kg)	Exophthalmia, edema, hemorrhagic kidney, and skin depigmentation.	Dupree (1970)
Vitamin D	500 (IU/kg)	Low body ash, calcium, and phosphorus.	Lovell and Li (1978)
	1000 (IU/kg)		Andrews et al. (1980)
	250 (IU/kg)		Brown (1988)
Vitamin E	25 (IU/kg)	Muscular dystrophy, exudative diathesis, skin depigmentation, erythrocyte hemolysis, splenic and pancreatic hemosiderosis, fatty liver, and ceroid deposition.	Murai and Andrews (1974)
	50 (IU/kg)		Wilson et al. (1984)
Vitamin K	Required	Hemorrhagic skin.	Dupree (1966)
	Not Required		Murai and Andrews (1977)

Source: National Research Council. *Nutrient Requirements of Fish.* National Academy Press, 1993; Li, M. H. et al. *Biology and Culture of Channel Catfish,* eds. C. S. Tucker, and J. A. Hargreaves, pp. 279–323. Elsevier B. V., 2004.

[a] Anorexia, reduced weight gain, and mortality are common deficiency signs; therefore, they are not listed.

minimal in rainbow trout (Hepher, 1988), unlike channel catfish (Limsuwan and Lovell, 1981; Burtle and Lovell, 1989; Li et al., 2004) and tilapia (Lovell and Limsuwan, 1982). Although some vitamins are present in feed ingredients, vitamin premixes containing the total vitamin needs of rainbow trout are used during feed formulation and manufacturing (Barrows and Hardy 2001). The vitamin requirements and symptoms of vitamin deficiencies for rainbow trout are listed in Table 16.11.

Barrows et al. (2008) recommended increasing levels of selected vitamins (A, B$_1$, B$_2$, B$_5$, B$_6$, B$_9$, D, and E) compared to NRC (1993) recommendations, prior to extrusion of feed for young, fast-growing rainbow trout. Elevated vitamin levels are also suggested for rainbow trout undergoing rearing stress. For example, the vitamin C dietary value of 50 mg/kg was recommended by NRC (1993) for optimal growth of rainbow trout. However, much higher values (up to 1000 mg/kg) have to been shown to improve rainbow trout disease resistance (Navarre and Halver, 1989; Anggawati-Satyabudhy et al., 1989; Wahli et al., 2006) and immune function (Hardie et al., 1993; Verlhac and

TABLE 16.11
Vitamin Requirements and Deficiency Symptoms for Rainbow Trout

Vitamin	Requirement (mg/kg Diet)	Deficiency Symptoms[a]	Reference
Biotin	0.05–0.25	Degenerative gill lamellae, skin lesions, muscle atrophy, spastic convulsion, reduced liver enzymes, liver lipid infiltration, pancreatic acinar cell degeneration.	McLaren et al. (1947)
	0.08		Woodward and Frigg (1989)
	0.14		Woodward and Frigg (1989)
Choline	50–100	Fatty liver, exopthalmia, extended abdomen, hemorrhagic kidney and intestine.	McLaren et al. (1947)
	714–813		Rumsey (1991)
Folate	1.0	Lethargy, anemia, dark skin coloration.	Cowey and Woodward (1993)
Myoinositol	250–500		McLaren et al. (1947)
Niacin	1–5	Lesions and hemorrhages in skin and fins, anemia, photosensitivity.	McLaren et al. (1947)
	10		Poston and Wolfe (1985)
Pantothenic acid	10–20	Clubbed gills, distended operculum, atrophied pancreatic acinar cells, mortality.	McLaren et al. (1947)
	20		Cho and Woodward (1990)
Riboflavin	5–15	Lethargy, dark pigmentation, spinal deformities, fin hemorrhage and erosion, photophobia, corneal vascularization, eye hemorrhage, reduced activity of erythrocyte glutathione reductase.	McLaren et al. (1947)
	6		Takeuchi et al. (1980)
	3		Hughes et al. (1981)
	2.7		Amezaga and Knox (1990)
Thiamin	1–10	Nervous disorders, loss of equilibrium, convulsions, hyperirritability, low kidney and erythrocyte transketolase activity.	McLaren et al. (1947)
	1		Morito et al. (1986)
Vitamin A	2500 (IU/kg)	Skin pigmentation, exophthalmia, eye and lens displacement, corneal thinning, retinal degeneration, edema, ascites.	Kitamura et al. (1967)
Vitamin B_6	1–10	Nervous problems, erratic swimming, convulsions, hyperirritability.	McLaren et al. (1947)
	2		Woodward (1990)
	3–6		Woodward (1990)
Vitamin B_{12}	0.21	Microcytic hypochromic anemia, fragmented erythrocytes.	Phillips et al. (1964b)
Vitamin C	250–500	Lethargy, hemorrhagic exophthalmia, intramuscular hemorrhage, distorted gill filaments, lordosis, scoliosis, ascites, anemia.	McLaren et al. (1947)
	100		Halver et al. (1969)
	40		Hilton et al. (1978)
Vitamin D	1600–2400 (IU/kg)	Impaired calcium homeostasis, tetany of skeletal muscle, increased liver lipid.	Barnett et al. (1982)
Vitamin E	30 (IU/kg)	Skin pigmentation, anemia, ascites, abnormal erythrocytes, muscular dystrophy, edema.	Woodall et al. (1964)
	25		Hung et al. (1980)
	100		Watanabe et al. (1981)
	50		Cowey et al. (1983)
Vitamin K	50		Halver (1972)

Sources: National Research Council. *Nutrient Requirements of Fish.* National Academy Press, 1993; De Silva, S. S., and T. A. Anderson. 1995. *Finfish Nutrition in Aquaculture.* Chapman and Hall. 162, 1995; Guillaume, J. et al. *Nutrition and Feeding of Fish and Crustaceans,* pp. 145–165. Praxis, 2001.

[a] Anorexia, reduced weight gain, and mortality are common deficiency signs; therefore, they are not listed.

Gabaudan, 1994; Verlhac et al., 1996). At relatively high concentrations, both vitamins C and E may modulate the stress response in rainbow trout reared in crowded conditions (Trenzado et al. 2008). Vitamin E requirements increase with increasing amounts of dietary polyunsaturated fatty acids in trout diets (Watanabe et al., 1981; Cowey et al., 1983). Trenzado et al. (2007) observed the best rearing performance when rainbow trout received 275.6 mg/kg vitamin E in diets containing 12.5 g/kg highly unsaturated fatty acids. Puangkaew et al. (2005) recommended only moderate amounts of vitamin E (100 mg/kg) in response to high dietary lipid levels, and Kiron et al. (2004) also observed minimal benefit when rainbow trout received more than 100 mg/kg vitamin E.

16.3.6 Minerals

Fish require minerals to maintain normal life processes (NRC, 1993); however, not all minerals found within a fish are essential in its diet (Lovell, 1989a). Minerals are important components of hormones and enzymes and have numerous other functions, including formation of skeletal structures, electron transfer, regulation of acid–base equilibrium, and osmoregulation (NRC, 1993). Minerals can be placed into two categories, major and trace (Lovell, 1989a). Major minerals are calcium, phosphorus, magnesium, sodium, potassium, chlorine, and sulfur, while trace minerals include iron, iodine, manganese, copper, cobalt, zinc, selenium, fluorine, and chromium (Lovell, 1989a; NRC, 1993).

Only limited attention has been given to studying mineral requirements of fish. Formulating mineral-free diets, overcoming tissue stores of minerals, and the exchange of ions and dissolved minerals from the water create problems in mineral nutrition research (Lovell, 1989a; NRC, 1993). Nine quantitative mineral requirements have been determined for several fish species (NRC, 1993), while 22 minerals have been determined to be necessary for one or more animal species (Lovell, 1989a). Mineral deficiencies can cause numerous symptoms, including anorexia, poor growth, loss of equilibrium, high mortality, hypochromic microcytic anemia, low hepatic iron content, low hematocrit and hemoglobin content, and lethargy (Lim and Webster, 2006).

16.3.6.1 Tilapia

Little information exists regarding mineral requirements for tilapia (Lim and Webster, 2006); however, several studies have estimated some mineral requirements for Nile tilapia (Table 16.12). Watanabe et al. (1988) listed Nile tilapia mineral requirements of 0.8% to 1.0%, 10 mg/kg, 12 mg/kg, and 3 to 4 mg/kg for phosphorus, zinc, manganese, and copper, respectively (Table 16.12). Dabrowska et al. (1989) reported dietary magnesium levels of 0.59 to 0.77 g/kg diet for optimum performance of Nile tilapia (mean initial weight = 10.4 g) but suggested that the magnesium requirement will increase with high-protein diets. Elhamid and Ghomin (1994) found that the minimum dietary zinc requirement of Nile tilapia (mean initial weight = 8.4 g) was 30 mg/kg dry diet, as determined by WG and feed efficiencies. To maintain normal erythropoiesis, Nile tilapia require a minimum of 60 mg/kg diet of available iron (Lim and Webster, 2006). Mineral requirements for other tilapia species have also been determined (Table 16.12).

16.3.6.2 Channel Catfish

Due to the previously mentioned difficulties in mineral nutrition research of fish, not all mineral requirements for channel catfish have been determined either (Table 16.13). One of the most important minerals needed in channel catfish diets is phosphorus due to the species high physiological requirement, low levels of dissolved phosphorus in natural waters, and low amounts of phosphorus in plant-based feed ingredients (Lovell, 1989b). The physiological requirement for calcium is also high, but channel catfish, along with other fish species, typically obtain sufficient calcium from their aqueous environment to meet this need (Lovell, 1989b; NRC, 1993; Li et al., 2004). Feeds containing low levels of animal byproducts may be deficient in trace minerals; therefore, mineral premixes are generally added to most commercial catfish feeds (Lovell, 1989b). Table 16.13 lists mineral requirements and deficiency signs common in channel catfish (Li et al., 1989b).

TABLE 16.12
Mineral Requirements for Juvenile Tilapia

Mineral	Species	Requirement (unit/kg Diet)	Reference
Major			
Calcium	*O. aureus*	7.0 g	Robinson et al. (1987)
	O. aureus	7.5 g	O'Connell and Gatlin (1994)
Phosphorus	*Tilapia*[a]	0.5%	NRC (1993)
	O. niloticus	0.8–1.0 g	Watanabe et al. (1988)
	O. aureus	5.0 g	Robinson et al. (1987)
Potassium	*O. niloticus* × *O. aureus*	2.0–3.0 g	Shiau and Hsieh (2001)
Magnesium	*Tilapia*[a]	0.06%	NRC (1993)
	O. niloticus	0.59–0.77 g	Dabrowska et al. (1989)
	O. aureus	0.50–0.65 g	Reigh et al. (1991)
Chlorine		Not tested	NRC (1993)
Sodium		Not tested	NRC (1993)
Trace			
Manganese	*O. aureus*	12.0 mg	Watanabe et al. (1988)
	O. mossambicus	1.7 mg	Ishac and Dollar (1967)
Iron	*O. niloticus*	60 mg	Kleemann et al. (2003)
	O. niloticus × *O. aureus*	150–160 mg	Shiau and Su (2003)
Zinc	*Tilapia*[a]	20 mg	NRC (1993)
	O. niloticus	10 mg	Watanabe et al. (1988)
	O. niloticus	30.0 mg	Elhamid Eid and Ghomin (1994)
Copper	*O. niloticus*	3–4 mg	Watanabe et al. (1988)
Iodine	*Tilapia*[a]	0.001 g	Jauncey and Ross (1982)
Selenium	*Tilapia*[a]	0.004 g	Jauncey and Ross (1982)

Source: Jauncey, K., and B. Ross. *A Guide to Tilapia Feeds and Feeding.* Institute of Aquaculture, University of Stirling, 1982; National Research Council. *Nutrient Requirements of Fish.* National Academy Press, 1993; Lim, C. E., and C. D. Webster. *Tilapia: Biology, Culture, and Nutrition,* eds. C. E. Lim, and C. D. Webster, pp. 469–501. The Haworth Press, Inc., 2006.)

[a] Indicates suggested requirement from various studies.

16.3.6.2 Rainbow Trout

Many of the mineral requirements of rainbow trout have been identified, and are listed in Table 16.14. However, the amounts of some minerals, such as calcium, are still undetermined (Fontagné et al., 2009), and other mineral requirements may vary. For example, the dietary phosphorous requirement of rainbow trout determined by Ketola and Richmond (1994) was between 0.34% and 0.51%. However, McDaniel et al. (2005) observed that phosphorous concentrations of 1.0% increased rainbow trout growth at high dissolved oxygen levels, but Fontagné et al. (2009) noted substantial mortality in rainbow trout at dietary phosphorous levels of 2.2%. The size of the rainbow trout also affects the dietary phosphorous requirement (Lellis et al., 2004). Several plant feedstuffs contain antinutritional factors, such as phytates that are strong chelators, which reduce bioavailability of proteins and minerals (Hardy and Shearer, 1985; Satoh et al., 1987a, 1987b, 1987c). As these factors increase, trout mineral requirements increase as well (Apines et al., 2003). The effects of these antinutritional factors may be mitigated by using amino acid chelated elements to provide dietary

TABLE 16.13
Mineral Requirements and Deficiency Symptoms for Channel Catfish

Mineral	Requirement (unit/kg Diet)	Deficiency Signs[a]	Reference
Major			
Calcium[b]	Not required	Reduced bone ash.	Li et al. (2004)
Phosphorus	0.45%	Reduced bone ash, calcium, and phosphorus.	Lovell (1978)
	0.3–0.4g[c]		Li et al. (2004)
Potassium	0.26 g	None.	Wilson and El Naggar (1992)
Magnesium	0.04%	Sluggishness, muscle flaccidy, and reduced	Gatlin et al. (1982)
	0.02%	body magnesium.	Lim and Kleisus (2003)
Chlorine	Required	Not determined.	NRC (1993)
Sodium	Required	Not determined.	NRC (1993)
Trace			
Manganese	2.4 mg	None.	Gatlin and Wilson (1984a)
Iron	20–30 mg	Reduced hemoglobin, hematocrit, erythrocyte count, reduced serum iron, and reduced transferring saturation levels.	Gatlin and Wilson (1986a)
Zinc	20 mg	Reduced serum zinc, serum alkaline phosphatase activity, bone zinc, and calcium concentrations.	Gatlin and Wilson (1983)
Copper	5 mg	Reduced heart cytochrome c oxidase, and reduced hepatic copper-zinc superoxide dimutase activities.	Gatlin and Wilson (1986b)
Iodine	1.1[d]	Not determined	NRC (1993)
Selenium	0.25	Reduced liver and plasma selenium-dependent glutathione peroxidase activity.	Gatlin and Wilson (1984b)

Source: Lovell, T. *Nutrition and Feeding of Fish,* ed. T. Lovell, pp. 11–71. Van Nostrand Reinhold, 1989a; National Research Council. *Nutrient Requirements of Fish.* National Academy Press, 1993; Li, M. H. et al. *Biology and Culture of Channel Catfish,* eds. C. S. Tucker, and J. A. Hargreaves, pp. 279–323. Elsevier B. V., 2004.

[a] Anorexia, reduced weight gain, and mortality are common deficiency signs; therefore, they are not listed.
[b] Deficiency cannot be demonstrated in water containing sufficient ambient calcium.
[c] Requirement expressed on available basis.
[d] Estimated value.

minerals. Compared to inorganic (sulfate) sources, amino acid chelated elements produced greater rainbow trout growth and whole body contents of zinc (Hardy et al., 1987; Apines et al., 2003), manganese (Satoh et al., 2001; Apines et al., 2003), and copper (Apines et al., 2003), suggesting amino acid chelation could decrease the feed-inclusion level of these minerals. Selenium requirements may vary depending on the effects of various husbandry stressors on rainbow trout, with optimal dietary selenium supplementation as high as 8 mg/kg (Rider et al., 2009).

16.4 DDGS COMPOSITION RELEVANT TO FINFISH AND FEEDING VALUE

16.4.1 COMPOSITION AND VARIABILITY OF DDGS

Many fish feeding studies have been conducted in the past with DDGS, but most of these have used DDGS from beverage-alcohol production. This type of DDGS (which is often produced with multiple feedstocks such as wheat, corn, rye, etc.) differs from DDGS produced at modern, corn-based

TABLE 16.14
Mineral Requirements and Deficiency Symptoms for Rainbow Trout

Mineral	Requirement	Deficiency Signs[a]	Reference
Major			
Calcium[b]	0.03%–0.24%	Kyphosis, reduced vertebral size, skeletal	Guillaume et al. (2001)
	1.0%	abnormalities.	Fontagné et al. (2009)
Phosphorus	0.5%–0.8%	Reduced food conversion, poor bone	Ogino and Takeda (1978)
	0.34%–0.54%	mineralization.	Ketola and Richmond (1994)
Potassium	Required	Convulsion, tetany.	NRC (1993)
Magnesium	0.04%–0.06%	Sluggishness, nephrocalcinosis, cataracts, degeneration of muscle fibers, pyloric caeca epithelium, and gill filaments, skeletal deformity, reduced bone mineralization.	Ogino et al. (1978) Knox et al. (1981) Shearer (1989)
Chlorine	Required	Not determined.	NRC (1993)
Sodium	Required	Not determined.	NRC (1993)
Trace			
Manganese	13 m/kg	Skeletal deformities, poor egg hatchability, decreased Cu–Zn superoxide dismutase activity in cardiac muscle and liver, suppressed Ca and Mn vertebrae concentrations.	Ogino and Yang (1980)
Iron	60	Hypochromic microcytic anemia.	Desjardins (1985)
Zinc	15–30 mg/kg	Reduced growth, short-body dwarfism, cataracts, fin erosion.	Ogino and Yang (1978)
Copper	3 mg/kg	Reduced heart cytochrome c oxidase and liver Cu–Zn-superoxide dismutase activity.	Ogino and Yang (1980)
Iodine	Required	Thyroid hyperplasia.	NRC (1993)
Cobalt	Required	Not determined.	Watanabe et al. (1997)
Selenium	0.15–0.38	Muscular dystrophy, exudative diathesis, reduced glutathione peroxidase activity.	Hilton et al. (1980)

Source: Desjardins, L. M. The effect of iron supplementation on diet rancidity and the growth and physiological response of rainbow trout. M.S. Thesis. Ontario, Canada: University of Guelph, 1985; National Research Council. *Nutrient Requirements of Fish.* National Academy Press, 1993; Ketola, H. G., and M. E. Richmond. *Transactions of the American Fisheries Society* 123, 587–594, 1994; Watanabe, T. et al. *Aquaculture* 151, 185–207, 1997; Guillaume, J. et al. *Nutrition and Feeding of Fish and Crustaceans,* pp. 145–165. Praxis, 2001; Fontagné, S. et al. *Aquaculture* 297, 141–150, 2009.

[a]   Anorexia, reduced weight gain, and mortality are common deficiency signs; therefore, they are not listed.

[b]   Deficiency cannot be demonstrated in water containing sufficient calcium.

fuel ethanol plants. Proximate composition of DDGS differs among corn varieties and, more importantly, has varied with changes in ethanol production technology (Table 16.15). DDGS has a protein content of up to ~34%, which is about 50% of the crude protein content found in menhaden fish meal (~65% protein content), a common fish meal source used in fish feeds (NRC, 1993). Crude lipid content in menhaden fish meal is similar to DDGS lipid content (~11%), although fatty acid profiles differ. The quality of the raw corn grain and processing conditions contribute to differences in nutrient densification and quality of DDGS (Chevanan, 2005).

Several studies have analyzed the chemical composition of DDGS (Table 16.15) to estimate the potential nutritional value as a feedstock (Chevanan, 2005; Rosentrater and Muthukumarappan, 2006). Cromwell et al. (1992) reported a protein level of 26.9% in DDGS, while Lodge et al. (1997) found protein and lipid levels for DDGS at 29.2% and 11.4%, respectively. Belyea et al. (2004) analyzed DDGS samples from 1997 to 2001 and indicated an increase in protein and lipid levels from 28.3% and 10.9% to 33.3% and 12.6%, respectively in 2001, with a mean protein and lipid level of 31.3% and 11.9%,

TABLE 16.15
Typical Ranges of Chemical Composition of Corn-based DDGS

Property	Reported Values (%)[a]	Reported Values (%)[b]	Menhaden Fish meal (%)[c]
Dry matter	86.2–93.0	91	90
Protein	26.8–33.7	27.0	65
Fat	3.5–12.8	9.3	10
Ash	2.0–9.8	6.4	19
Carbohydrate	39.2–61.9	9.1	
Crude fiber	5.4–10.6		
Total dietary fiber	24.2–39.8		
Starch	4.7–5.9		
Nitrogen free extract	33.8–54.0		
Neutral detergent fiber	25.0–51.3		
Acid detergent fiber	8.0–21.0		

Source: Rosentrater, K. A., and K. Muthukumarappan. *International Sugar Journal,* 108, 648–657, 2006; National Research Council [NRC]. *Nutrient Requirements of Fish.* Washington DC: National Academy Press, 1993.

[a] Rosentrater and Muthukumarappan (2006).

[b] NRC (1993).

[c] Various commercial products.

respectively. Similarly, Spiehs et al. (2002) analyzed DDGS samples from 1997 to 1999 and found protein and lipid concentrations of 30.2% and 10.9%, which were higher than levels reported (28.1% protein and 8.2% lipid) for DDGS derived from older processes. These studies indicate an increasing trend in DDGS protein and lipid levels as ethanol production technologies evolve, but composition variability should be anticipated as processes, along with the different technologies used among ethanol production facilities, advance. Please refer to Chapter 8 for more information on the composition of DDGS.

Compared to corn, nutrient concentrations in DDGS are increased about threefold as a result of the fermentation process; however, the amino acid profile of DDGS (Table 16.16) remains similar to whole corn (Chevanan, 2005). Comparing the EAA requirements of tilapia, channel catfish, and rainbow trout (Tables 16.4 through 16.6) with the average DDGS profile (Table 16.16) it appears that lysine and methionine are the most deficient. Further, overheating in the feed extrusion process (depending on temperatures and processing times used) may lead to an additional reductions in the lysine concentration and digestiblity and partial destruction of cystine; therefore, accounting for deficiencies and processing losses in these and other EAA may require that supplements be used to meet the amino acid requirements of fishes (Chevanan, 2005).

One of the more promising attributes of DDGS for application in fish feeds is that it does not contain antinutritional factors often found in other plant protein sources, such as protease (trypsin) inhibitors in soybean meal (Wilson and Poe, 1985; Shiau et al., 1987) or gossypol found in cottonseed meal (Jauncey and Ross, 1982; Robinson, 1991). However, DDGS is highly fibrous, about 5% to 10% crude fiber, which is indigestible non-nutritive bulk in comparison to menhaden meal fiber, which is only about 0.7%. DDGS can also bind with water-soluble nutrients, thereby reducing nutrient digestibility. Enzyme technologies may improve these nutritional issues.

Compared to the raw grain, DDGS provides many vitamins (Table 16.17) and minerals (Table 16.18). As such, DDGS has a threefold increase in riboflavin, niacin, pantothenic acid, folic acid, and choline compared to corn (Hertrampf and Piedad-Pascual, 2000).

Similarly, DDGS can provide minerals within a diet; however, there is often wide variation in the mineral content of DDGS (Hertrampf and Piedad-Pascual, 2000). DDGS provides a potential

TABLE 16.16
Typical Ranges of Amino of Acids of corn-based DDGS

Amino Acid	Reported Values (%)[a]	Reported Values (%)[b]	Menhaden Fish meal (%)[c]
Arginine	0.9–2.2	1.12	3.7
Cystine	0.4–0.8	0.46	0.6
Histidine	0.6–1.0	0.64	1.7
Isoleucine	0.9–1.5	1.09	2.6
Leucine	2.4–4.0	2.89	4.5
Lysine	0.5–1.1	0.65	5.0
Methionine	0.5–0.8	0.50	1.8
Phenylalanine	1.3–1.7	1.39	2.5
Threonine	0.8–1.3	0.98	2.4
Tryptophan	0.2–0.3	0.10	5.2
Tyrosine	0.8–1.0	0.99	1.9
Valine	1.3–1.8	1.50	3.1
Alanine	1.8		
Aspartic acid	1.8		
Glutamic acid	4.6		
Glycine	1.0		
Hydroxyproline	0.2		
Proline	2.6		
Serine	1.4		

[a] Rosentrater, K. A., and K. Muthukumarappan. *International Sugar Journal* 108, 648–657, 2006.
[b] National Research Council. *Nutrient Requirements of Fish*. National Academy Press, 1993.
[c] Various commercial products.

TABLE 16.17
Typical Vitamin Composition of DDGS

Vitamin	Reported Value (mg/kg)[a]	Reported Value (mg/kg)[b]	Menhaden Fish meal (mg/kg)[c]
Biotin	1.04	0.77	0.1
Choline	2548	2551	3112
Folacin	0.73	0.90	0.1
Niacin	88.5	72	55
Pantothenic acid	13.83	13.9	8.6
Pyridoxine	4.60	5.00	4.7
Riboflavin	8.39	8.3	4.8
Thiamin	5.9	2.8	0.6
Vitamin E	40.7	39.1	
Vitamin A	1363 (IU/kg)		
Vitamin D	600 (IU/kg)		

Source: Hertrampf, J. W. and F. Piedad-Pascual. Distillery by-products. In *Handbook on Ingredients for Aquaculture Feeds*, pp. 115–124. Boston, Massachusetts: Kluwer Academic Publishers 2000; National Research Council [NRC]. *Nutrient Requirements of Fish*. Washington DC: National Academy Press, 1993.

[a] Hertrampf and Piedad-Pascual (2000).
[b] National Research Council. *Nutrient Requirements of Fish*. National Academy Press, 1993.
[c] Various commercial products.

TABLE 16.18
Typical Mineral Composition of Corn-based DDGS

Mineral	Reported Values (%)[a]	Reported Values (%)[b]	Menhaden Fish meal (%)[c]
Calcium	0–0.5	0.14	3.8
Phosphorus	0.4–1.0	0.66	2.6
Potassium	0.5–1.3	0.40	1.1
Magnesium	0.1–0.4	0.16	1.9
Sulfur	0.3–1.1	0.35	
Sodium	0–0.5	0.52	0.7
Chlorine	0.1–0.4	0.16	
Zinc	38.0–312.1 mg/kg	80 mg/kg	98.9 mg/kg
Manganese	9.0–49.5 mg/kg	22.8 mg/kg	48.6 mg/kg
Copper	3.0–13.5 mg/kg	52.8 mg/kg	3.3 mg/kg
Iron	68.0–295.0 mg/kg	236 mg/kg	904 mg/kg
Selenium		0.35 mg/kg	1.9 mg/kg

[a] Rosentrater, K. A., and K. Muthukumarappan. *International Sugar Journal* 108, 648–657, 2006.
[b] National Research Council. *Nutrient Requirements of Fish*. National Academy Press, 1993.
[c] Various commercial products.

source of phosphorus, principally in the phytate-phosphorous form at about 0.4% to 1.5% phytate. Phytate is the principal form of phosphorus found in most plant-based feed ingredients; however, much of the phosphorus contained within phytate is in a form not be readily digested by monogastrics. Furthermore, phytate is a strong chelator that binds with protein and mineral nutrients reducing their availability. Recent research indicates that microbial phytase (e.g., *Aspergillus*) may be added to the stillage during ethanol production to catalyze phytate hydrolysis, thereby improving DDGS-based phosphorous availability in feeds (Noureddini, 2009). Other primary minerals, such as calcium, chlorine, and potassium, are present in low amounts and require supplementation (NRC, 1993; Hertrampf and Piedad-Pascual, 2000). Trace mineral, including zinc, iron, manganese, and copper, concentrations are considerably lower in DDGS than in fish meal, but can be easily met with minor element supplementation.

16.4.2 FINFISH FEEDING TRIALS WITH DDGS-BASED FEEDS

Several feeding studies have investigated the inclusion of DDGS as a feed ingredient in tilapia, channel catfish, and rainbow trout diets. As mentioned previously, several studies have used beverage-based DDGS, which differs from fuel-based DDGS. An overview of fuel-based, corn-derived DDGS studies is presented below, by species, in chronological order.

16.4.2.1 Tilapia

A series of tilapia fry studies using fuel-based DDGS were investigated by Wu et al. (1994, 1996, 1997). Wu et al. (1994) observed that tilapia fry (mean initial weight = 30 g) fed a commercial feed (36% crude protein, fish meal base) had lower WG than fish fed diets containing either corn gluten meal (18%) or DDGS (29%) formulated at 32% or 36% crude protein. In a similar feeding trial, Wu et al. (1996) evaluated DDGS levels at 35% and 49% of the total diet fed to Nile tilapia fry (mean initial weight = 0.4 g) in an 8-week feeding trial, and found that both diets had good WG, feed conversion ratios (FCR), and protein efficiency ratios (PER). Because amino acid deficiencies may result from lower protein diets containing coproducts, supplements may be needed to provide a sufficient EAA profile. Thus, Wu et al. (1997) used feeds containing composite coproducts (corn gluten meal, corn gluten feed, and corn DDGS) that were supplemented with soy flour and/or synthetic lysine and tryptophan (formulated to contain 28% or 32% protein). The DDGS were included at 63%, 77.75%, and 82.23% of the total diet. Diets were fed to Nile tilapia fry (mean initial weight = 0.5 g) and compared to results obtained from a commercial diet. In general, WG was correlated with protein content and the lysine:protein ratio. Within given protein levels, FCR and PER were comparable, indicating there was a potential for using all-plant diets for tilapia fry.

Several studies have been conducted to assess utilization of DDGS in fingerling and larger sized tilapia. Coyle et al. (2004) included DDGS (source unknown) at 30% of total diet fed to hybrid tilapia *O. niloticus* × *O. aureus* (mean initial weight = 2.7 g) and concluded that diets containing DDGS, meat and bone meal, and soybean meal appeared to provide an effective, economical diet. Shelby et al. (2008) assessed growth and immune function of tilapia (mean initial weight = 6.7 g) fed formulated diets composed of 0%, 30%, or 60% DDGS (fuel-based, 32.5% crude protein) in combination with graded amounts of soy and corn meals and with or without a lysine supplement. That study showed that WG and FCR were negatively affected in the 60% DDGS diet without the lysine supplement. In a postfeeding trial challenge using *Streptococcus iniae* (1×10^5 CFU, intraperitoneal injection), serological measurements and survival did not differ among treatments.

Abo-State et al. (2009) conducted a Nile tilapia (mean initial weight = 2 g) feeding trial using DDGS (unknown source) to replace soybean meal (0%, 25%, 50%, 75%, and 100% replacement) and with (150 mg/kg) or without phytase. They found highest growth performance and feed utilization in 0%, 25%, and 50% DDGS diets containing phytase.

Schaeffer (2009) conducted feeding trials using fuel-based DDGS as a fish meal replacement for larger fish. Six diets were formulated to contain 0%, 17.5%, 20%, 22.5%, 25%, and 27.5% DDGS (29% crude protein, with no supplements) and fed to Nile tilapia (initial weight = 34.9 g). The reference diet (0% DDGS) resulted in higher WG, FCR, and PER than experimental diets, with the exception of better FCR and PER for the 17.5% DDGS diet. Apparent digestibility did not significantly differ among diets. In a subsequent feeding trial, six 32% protein diets were formulated to contain 20%, 25%, and 30% DDGS with and without 0.125% probiotic, and fed to Nile tilapia (initial weight = 43.6 g). No differences were detected among treatments for WG, FCR, or PER. Following a stress challenge, there was no apparent increase in survival or decrease in plasma cortisol levels of fish fed probiotic feeds. In total, results of the above feeding trials indicate good potential for replacing fish meal with DDGS, particularly if appropriate supplements are used to offset deficiencies in tilapia feeds.

16.4.2.2 Channel Catfish

Several studies have investigated the use of DGS in channel catfish diets. Tidwell et al. (1990) incorporated DGS (replacing corn and soybean meals) at 10%, 20%, and 40% of the total diet in fingerling (initial weight = 1.5 g) diets. They observed similar WG, survival, FCR, and PER for fish fed a diet containing 40% DGS compared to the control diet (containing 0% DDGS). Webster et al. (1991) used DGS levels of 35% and 70%, and determined that DGS could be added up to 35% in channel catfish diets without affecting growth (initial weight = 10 g); however, addition of 0.4% L-lysine allowed for 70% DGS to be used without adversely affecting growth. Another study indicated that diets using a combination of 35% DGS and varying concentrations of soybean meal could totally replace fish meal in channel catfish diets without adversely affecting growth or survival (mean initial weight = 11 g) (Webster et al., 1992a). A study on the winter feeding and growth of channel catfish (mean initial weight = 515 g) using DGS inclusion levels of 35%, 54.75%, and 90% found that 90% DGS was sufficient for growth without the addition of supplemental amino acids (Webster et al., 1992b). In a cage rearing study, Webster et al. (1993) found that WG and FCR were not different for catfish (initial weight = 33 g) fed diets containing 10%, 20%, or 30% DGS compared to a commercial catfish diet. They also observed that carcass composition and organoleptic properties did not differ among treatments.

In two pond feeding trials, Robinson and Li (2008) evaluated combinations of fuel-based DDGS, soybean, and cottonseed meals to assess soybean meal replacement in fingerling catfish diets. Those studies showed that 30% to 40% DDGS (29% protein), with a lysine supplement, could be used to replace soybean meal without affecting performance (growth, feed conversion). Those findings concur with previous cage and tank studies that replaced soybean meal with DGS in catfish diets (e.g., Webster et al., 1992a). Similarly, Zhou et al. (2010) evaluated fuel-based DDGS as a replacement for soybean and corn meals in juvenile hybrid catfish (channel catfish × blue catfish *I. furcatus*). They observed that diets containing 30% fuel-based DDGS supported good growth, protein retention, and feed conversion. These studies suggest that relatively high levels of DDGS can be used in channel catfish diets without adversely affecting survival, growth, or feed utilization.

16.4.2.3   Rainbow Trout

Phillips (1949) may have been the first to incorporate DDGS in rainbow trout diets. Subsequently, Sinnhuber (1964) successfully used 3% DDGS, and Phillips et al. (1964a) incorporated 21% dried distillers solubles into rainbow trout feeds. Further experimentation focused on other salmonids (Fowler and Banks, 1976; Hughes 1987), with no negative repercussions found when distillers grain products composed up to 8% of the salmonid diet. The apparent digestion coefficients for protein in distillers dried solubles in rainbow trout were reported as 71.9% by Smith et al. (1980) and for DDGS as 90.4% by Cheng and Hardy (2004b).

Cheng and Hardy (2004b) conducted a 42-day feeding trial, with 50 g rainbow trout receiving DDGS at 7.5%, 15%, and 22.5% (equivalent to replacing 25%, 50%, and 75% of the fishmeal component), with or without additional lysine and methionine supplementation. FCR results were not significantly different among the trout receiving the fish meal control diet and those receiving either 15% DDGS without supplemental amino acids or any of the DDGS diets that had additional lysine and methionine; only the 7.5% and 22.5% DDGS diets led to significantly higher FCR than that observed with the fish meal control. There was no difference in rainbow trout survival among any of the treatments. In general, trout body fat increased and moisture decreased with increasing dietary DDGS inclusion levels. Cheng and Hardy (2004b) stated that DDGS could replace up to 50% of the dietary fish meal (15% inclusion), and, if lysine and methionine are supplemented, DDGS could replace 75% of the fish meal (22.5% inclusion).

Cheng and Hardy (2004a) recommended the supplementation of phytase in rainbow trout diets containing DDGS. Apparent digestibility coefficients in diets containing 30% DDGS were significantly improved for calcium, magnesium, total phosphorus, and a number of essential and nonessential amino acids. Cheng and Hardy (2004a) also noted no negative effects on rainbow trout WG, FCR, survival, body composition, or apparent nutrient retention when the amount of trace mineral premix was reduced in diets containing 15% DDGS and 500 FTU/kg phytase supplementation.

Stone et al. (2005) compared diets containing both DDGS and corn gluten meal that were made either by cold pelleting or extrusion and found that the extrusion did not improve rainbow trout growth, feed conversion, or nutrient retention in the plant-based diets. Fish receiving the fish meal control diets performed significantly better than those receiving any of the DDGS and corn gluten meal diets.

One other experiment indirectly examined the use of DDGS in trout diets. Rainbow trout receiving diets containing 18.5% DDGS, 17.5% soybean meal, and 17.5% fish meal, in conjunction with the use of a methionine hydroxyl analogue, performed similar to trout fed 18.5% DDGS and 35% fish meal (Cheng et al. 2003).

Determining the optimum level of DDGS to use in a diet targeted for a specific species/age is one challenge; the other is determining appropriate processing conditions with which to produce water stable, durable, floating feed pellets. Extrusion processing is the most common way to achieve these goals.

16.5 EXTRUSION PROCESSING AND PROPERTIES OF DDGS-BASED FEEDS

Extrusion is a common processing method for the food and feed industries. It combines several unit operations into one machine, including pumping, mixing, cooking, shaping, and forming. Extruders are high temperature, short time bioreactors that can transform a variety of raw materials into modified intermediate and finished products. Advantages of the extrusion process include (1) continuous high throughput processing; (2) processing of materials with a range of moisture contents, from relatively dry to very wet materials; (3) improved textural and flavor characteristics; (4) control of the thermal changes that occur in the material; and (5) expanded products that are stable in water (i.e., will not disintegrate) and can float on the surface of water. All of these are important for the production of fish feeds.

Although various extruders exist, there are essentially two basic types: single screw (Figure 16.1) and twin screw (Figure 16.2). The goals of the following discussion are to introduce the reader to the generalities of extrusion processing, and to review prior work with extrusion of DDGS fish feeds. The discussion will encompass both types of extruders.

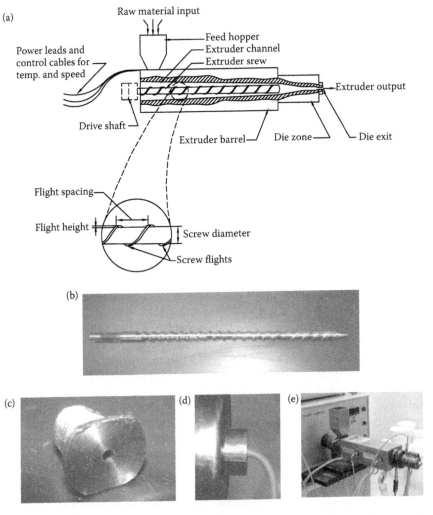

FIGURE 16.1 An example of a single screw laboratory-scale extruder: (a) a schematic depicting key aspects of the extruder; (b) extruder screw (single element); (c) extruder die insert; (d) extrudate exiting the extruder die during processing; and (e) view of a fully assembled laboratory-scale extruder.

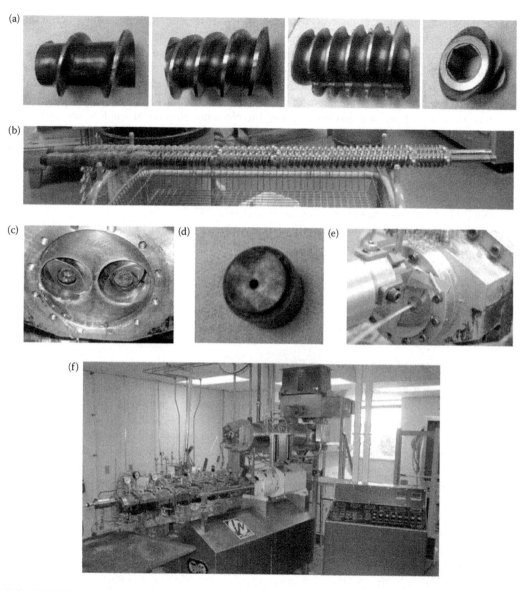

FIGURE 16.2 An example of a twin-screw pilot-scale extruder: (a) various screw elements; (b) the screw is formed by attaching several elements together in various configurations; (c) screws installed within the extruder barrel; (d) extruder die insert; (e) extrudate exiting the die during processing (rotary cutter also shown); (f) the entire extruder during operation, note the feed hopper and preconditioner are also shown.

Essentially, extruders are machines that accept incoming raw feed material, convey this material down a chamber using a flighted screw (or screws), apply heat to this material as it is conveyed (through frictional viscous dissipation, by heated jackets surrounding the extruder barrel, and by direct steam injection), and then force this dough (which usually becomes a plasticized mass due to the heating and mixing) out of a small opening (or openings), known as a die. Product shaping and forming occur at the die exit, due to expansion because of pressure release and moisture evaporation effects, and due to the type of die exit and shearing knives used to cut the outgoing product (Fellows, 1996; Gould, 1996).

The operation of extruders depend upon the pressure build up prior to exiting the die, slip at the barrel wall, and the degree to which the screw is filled. Based on the direction of screw rotation, twin screw extruders are divided into counter rotating and corotating. Considering screw configuration and degree of intermeshing, twin-screw extruders can be further divided into fully intermeshing, partially intermeshing, and nonintermeshing.

Extruders are generally divided into different zones, depending on the temperature and screw elements present. Usually single screw extruders are divided into a feeding zone, transition zone, metering zone, and die zone. In twin-screw extruders, more options are available, because the entire screw section can consist of combinations of conveying elements, kneading elements, reverse screw elements (to restrict flow), and additional conveying elements. Similar to single screw extruders, twin-screw extruders also have pressure and die zones. For both types of extruders, heater coils and/ or jackted heaters and thermocouples are typically provided to accurately control the temperatures in these various zones. Temperatures in feed zones are usually low (sometimes at ambient levels), in order to prevent plugging and back flow, and the barrel temperature usually increases as the material travels down the screw. The temperature at the die is often maintained at levels up to 130°C, although for heat sensitive ingredients this can be as low as 80°C.

Twin screw extruders offer more flexibility and provide better mixing. Single screw extruders are widely used to manufacture fish feeds due to lower capital investment and operating costs. In fact, the cost of installing a twin-screw extruder can be up to twice as much as for a single screw extruder. A thorough comparison between single and twin-screw extruders can be found in Harper (1981).

The extrusion process and the final extrudate products are affected by many processing parameters, including extrusion time, screw rotational speed, extrusion temperature, die opening size and shape, and raw ingredient moisture content, particle size, and composition.

16.5.1 Processing Conditions and Effects

Extrusion cooking is a complex process and is widely regarded as an art more than a science. There are a number of variables that contribute to final product quality. A very brief discussion regarding the most important factors and their effects on the extrusion process and on the resulting extrudate products will follow. The processing parameters that have the greatest effect on extrusion are screw speed, barrel temperature, and die diameter. Specifically, the effects of these parameters on apparent viscosity of the dough melt, the energy input required, and product expansion will be discussed.

Feed rate, also known as mass flow rate, or throughput, is a measure of the production capacity during processing. The through rate of an extruder depends on the type of the screw elements, operating screw speed, type of feeding elements, and feed moisture contents used. As screw speed increases, the apparent viscosity of the dough melt decreases. This occurs because food and feed doughs behave pseudoplastically, which means their non-Newtonian nature exhibits a decrease in viscosity as shear strain rate (rate of material deformation) increases (e.g., they are called shear thinning fluids due to this behavior). Typically, the specific mechanical energy input (SME) is an indication of the viscous dissipation of mechanical energy, which is provided by the screw drive shaft, into the dough due to frictional resistance (Marsman et al., 1995). The SME, in fact, quantifies the competing effects of viscosity changes due to changes in screw speed (i.e., the pseudoplastic behavior), and the resulting change in the torque that is required to convey the dough through the extruder at an increased shaft speed. Typically, it has been noted that as screw speed increases, SME also increases, due to the changes in energy input to the screw being of a greater order of magnitude than the decrease in torque associated with the decrease in apparent viscosity due to the shear thinning behavior of the non-Newtonian material (Mercier et al., 1989). A decrease in apparent viscosity also leads to a more expanded product upon exiting the die (Chen et al., 1979; Kokini et al., 1992; Mercier et al., 1989). Product expansion can be quantified by measuring extrudate diameter vis-à-vis the extruder die exit diameter. Conversely, expansion can also be quantified using unit density, which is defined as the mass of an extrudate for a given volume. The greater the expansion, the lower the unit density.

An increase in extruder barrel temperature generally leads to a decrease in dough apparent viscosity. This decrease in viscosity requires less energy input (SME) into the drive shaft to turn the screw at a given speed. As mentioned previously, a decrease in viscosity produces a more expanded product upon die exit. Furthermore, a higher temperature produces more water evaporation upon die exit, also leading to a more expanded product (Chen et al., 1979; Mercier et al., 1989; Kokini et al., 1992).

It appears that die size has little effect on apparent viscosity in the screw channel. But, as die diameter increases, SME required decreases, due to the lower pressure buildup behind the die prior to exit (i.e., less force is required to push the dough through the die, and thus less force is required to turn the screw). As the pressure drop across the die decreases, the resulting extrudates expand less, due to less water evaporation upon die exit (Mercier et al., 1989).

16.5.2 Feed Ingredients and Effects

The ingredient properties that have the greatest effects on extrusion are moisture content, particle size, and chemical composition. In particular, the amount of water will play an important role in both processing behavior and final product quality. An increased moisture content of the feed/dough will produce a lower apparent viscosity in the dough, and thus, it also produces a lower required SME. As moisture content increases, however, product and die temperatures and pressures decrease, thus producing less expansion upon die exit (Aguilera and Kosikowski, 1976; Chen et al., 1979; Mercier et al., 1989). However, it has been noted that an increase in moisture will often lead to an improvement in product durability (the ability of a feed pellet to resist mechanical breakage and attrition during storage and transport) and water stability (which is the ability of a pellet to remain intact in water for an extended period of time) (Rolfe et al., 2001).

Particle size distribution also plays an important role in final product quality. Usually a finely ground feed will result in a product with good water stability, water absorption index, expansion properties, and floatability. But, as ingredient particle size increases, apparent viscosity decreases, which leads to a decrease in SME. Also, increased particle size can result in increased product expansion, and thus a decrease in product bulk density (Garber et al., 1997). Greater particle size may also lead to increased fracturing of the extrudates.

Chemical composition also impacts the processing conditions and resulting product quality. The amount of protein, lipid, fiber, and starch in the ingredient mix all contribute. Fiber is generally a non-interacting component; thus the higher the fiber content of an ingredient mix, the lower the expansion, cohesiveness, durability, and water stability. Starch, however, is well known for its ability to produce expansion and cohesiveness. The extent of starch gelatinization and protein denaturation that occur during processing will impact the resulting durability and water stability, as well as other physical properties (Case et al., 1992; Ibanoglu et al., 1996; Thomas et al., 1999). During extrusion processing of high starch ingredients, an elastic melt is formed inside the barrel, which results in a more expanded final product (Ilo et al., 1996; Alves et al., 1999; Lin et al., 2000). But during extrusion processing of high-protein ingredients, a plastic melt is formed, which ultimately results in a more porous and textured final extruded product, but which has limited expansion (Cumming et al., 1973; Gwiazda et al., 1987; Singh et al., 1991; Sandra and Jose, 1993). It has also been noted that increasing the lipid content can reduce the degree of starch gelatinization, and thus product expansion upon exiting the die (Lin et al. 1997). Lipids also act as a lubricant, and decrease processability and expansion.

16.5.3 Extrusion Studies with DDGS

As discussed previously, there have been a number of studies on the use of DDGS in various fish diets over the years, but until recently there was no information on the actual processing of DDGS into fish feeds. After 2007, several studies have appeared in the literature that examined the effects of both processing conditions as well as feed ingredient compositions upon resulting extrudate (i.e., pellet) properties, and which aimed to maximize the quantity of DDGS in various fish feed blends while maintaining product quality. These studies are summarized in Table 16.19. Figure 16.3 illustrates extrudates that were produced in a few of the studies. Most of this work examined processing using single screw extruders; one study was conducted with a commercial-scale twin-screw extruder.

TABLE 16.19
Key Properties in Work Published To Date on Processing of Corn-based DDGS into Aquafeeds

DDGS (%)	Binder	Unit Density (g/cm³)	Durability (%)	Citation
Single screw extrusion				
20	None	0.96	89.00	Chevanan et al. (2008a)
30		0.93	65.00	
40		0.93	56.00	
20	Whey	1.05	94.04	Chevanan et al. (2009)
30		1.07	94.02	
40		1.06	93.52	
40	Whey	0.88–1.03	85.06–97.99	Chevanan et al. (2007a)
20	Cassava starch	0.78	81.6	Kannadhason et al. (2009a)
30		0.88	84.2	
40		0.86	86.1	
20	Corn	0.90	85.3	
30		0.94	75.7	
40		0.91	62.7	
20	Potato starch	0.79	81.8	
30		0.88	85.3	
40		0.90	87.4	
20	Corn	1.03	70.9	Rosentrater et al. (2009a)
25		1.01	90.8	
30		1.02	69.7	
20	Tapioca starch	0.94	90.2	Kannadhason et al. (2009b)
25		0.93	95.8	
30		0.99	84.1	
20	Potato starch	0.85	89.1	Rosentrater et al. (2009b)
25		0.97	96.4	
30		0.93	82.4	
Twin-screw extrusion				
20	Whey	0.24	97.76	Chevanan et al. (2007c)
40		0.34	97.60	
60		0.61	96.85	

Note: Unit density and durability values reported in this table are mean values from each of the studies listed.

Overall, the studies found that the high levels of fiber in the DDGS were problematic, especially as higher concentrations of DDGS were used. Various binding materials appeared to help, however. And processing conditions, including moisture, also impacted pellet quality. Each study will be briefly discussed below, with primary emphases placed on pellet durability and unit density (as these are key to viable feeds); the reader is referred to the original articles for more detailed information.

An initial study was conducted by Chevanan et al. (2008a). Three isocaloric (3.5 kcal/g) blends, with net protein adjusted to 28% (wb), were formulated with 20%, 30%, and 40% (wb) DDGS; the balance of each blend consisted of soy flour, corn flour, fish meal, mineral, and vitamin mix. Blends were conditioned to 15%, 20%, and 25% (wb) moisture content prior to processing, and no binders were used. Processing was conducted using a single screw (compression ratio of 3:1) extruder (barrel length of 317.5 mm; length-to-diameter ratio of 20:1), with a die length/diameter of 13 mm/2.7 mm (thus L/D = 4.81), using screw speeds of 100 rpm, 130 rpm, and 160 rpm, and

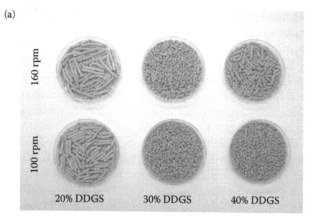

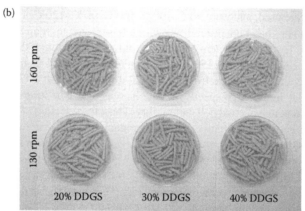

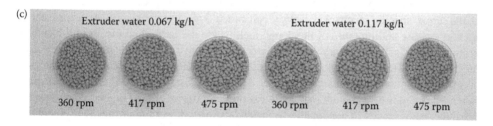

FIGURE 16.3 Examples of extrudates from some of the DDGS processing studies: (a) single screw processing of blends without binder resulted in many noncohesive extrudates; (b) single screw processing of blends using whey as a binder resulted in better structures; (c) twin-screw processing of 20% DDGS blends (data unpublished) resulted in optimal extrudates. (Part (a) from Chevanan, N. et al. *Cereal Chemistry* 84(4), 389–398, 2007a; Part (b) from Chevanan, N. et al. *Food and Bioprocess Technology* 2, 177–185, 2009.)

processing temperatures of 90°C, 90°C, and 120°C (corresponding to the feed, transition, and die sections, respectively). The extruder used for this study is shown in Figure 16.1. All extrudates had unit densities less than 1.0 g/cm^3, which indicates that they all floated. Unfortunately, the durability values were low, and ranged from 56% to 89% (which indicates that they fell apart relatively easily). The authors found that increasing the DDGS content from 20% to 40% resulted in a 37.1% and 3.1% decrease in extrudate durability and unit density, respectively. Increasing the screw speed from 100 rpm to 160 rpm resulted in a 20.3% increase in durability. And increasing the moisture content from 15% to 25% (wb) resulted in a 28.2% increase in durability, but an 8.3% decrease in unit density.

A follow-up study was then conducted by Chevanan et al. (2009), using the same blend formulations and extruder setup. In this study, however, 5% whey was added to the ingredient blends to improve binding. Compared to the previous research, the durability and unit density of the resulting extrudates were found to increase substantially, which was specifically due to the addition of the whey. All durability values were greater than 92% (indicating cohesive products which held together well), but unit densities were all greater than 1.05 g/cm^3 (which indicates that the extrudates did not float). Overall, increasing the moisture content of the blends from 15% to 25% resulted in an increase of 8.7% in durability. This study demonstrated that ingredient moisture content and screw speed are critical considerations when producing extrudates with feed blends containing DDGS; but further work was necessary to produce floating feeds.

To further understand the influence of processing conditions on extrudate quality, Chevanan et al. (2007a) then investigated the effects of die dimensions, barrel temperature profile, and blend moisture content, using the same extruder, but with a single feed blend of 40% (wb) DDGS (using the same balance of ingredients as previously; 5% whey was added for binding as well). Seven unique circular dies were used (ranging in diameter from 2.0 mm to 6.0 mm, with L/D varying from 3.33 to 10.0); three blend moisture contents were used, and consisted of 15%, 20%, and 25% (wb); three extruder temperature profiles were used, and were 90°C–100°C–100°C, 90°C–120°C–120°C, and 90°C–140°C–140°C. Resulting extrudates had high durability values (all were greater than 85%). Some floated (i.e., had unit density values less than 1.0 g/cm^3), but some did not. They found that increasing the moisture content of the ingredient mix from 15% to 25% resulted in an 11.6% increase in pellet durability, whereas increasing the die temperature from 100°C to 140°C resulted in a 17.0% and 5.9% decrease in unit density and pellet durability, respectively. And increasing the L/D ratio of the die resulted in a slight increase in both unit density and pellet durability. Thus, the selection of die geometry, temperature, and moisture content levels were all critical for producing DDGS-based fish feeds.

Chevanan et al. (2008b) extended this work and investigated the effects of DDGS level (20%, 30%, and 40%, wb), blend moisture content (15%, 20%, and 25%, wb), barrel temperature profile (90°C–100°C–100°C, 90°C–130°C–130°C, and 90°C–160°C–160°C), and screw speed (80, 100, 120, 140, and 160 rpm) on various extrusion processing parameters. They used the same balance of ingredients (with 5% whey as a binder), and the same single screw extruder, with a 2.7-mm-diameter circular die (L/D of 4.81). Processing parameters that were measured included mass flow rate, net torque required to turn the screw, specific mechanical energy consumed during processing (SME), apparent viscosity, and temperature and pressure of the dough melt inside the barrel and die. For all blends, as the temperature profile increased, mass flow rate exhibited a slight decrease, die pressure decreased, and apparent viscosity exhibited a slight decrease as well. The net torque requirement, specific mechanical energy consumption, and apparent viscosity decreased as screw speed increased, but mass flow rate increased. As moisture content increased, die pressure decreased. At higher temperatures in the barrel and die, the viscosity of the dough was lower, leading to lower torque and SME. Increasing the DDGS content, on the other hand, resulted in a higher mass flow rate and decreased pressure inside the die. Again, they found that processing temperature and moisture content levels are critical for processing of DDGS-based ingredient blends.

Experimental results from these previous studies were then combined by Chevanan et al. (2007b). Various multiple linear regression and neural network models were developed to explain the effects of DDGS inclusion, blend moisture content, extruder temperature profile, screw speed, and die dimension on resulting extrusion processing parameters and extrudate properties. In general, the regression and neural network models predicted the extrusion processing parameters with better accuracy than they did the extrudate properties. Regression models for extrudate properties, using three predictor variables (moisture, temperature, and die L/D), resulted in R^2 values ranging between 0.29 and 0.84. Regression models using three (moisture, temperature, and die L/D), and six (moisture, temperature, die length, die diameter, die L/D, and screw speed) input variables predicted extrusion processing parameters with R^2 values of 0.56 to 0.97 and 0.75 to 0.97, respectively. Neural network models (using three, five, and six input variables), on the other hand, had better performance, and predicted extrusion processing parameters with R^2 values of 0.82 to 0.98, 0.86 to 0.99, and 0.90 to 0.99, respectively. With the regression modeling, even though increasing the number of input variables from three to six resulted in better R^2 values, there was no decrease in the coefficient of variation between the measured and predicted variables. On the other hand, the neural network models developed with six input variables resulted in more accurate predictions, with reduced coefficients of variation and standard errors. Because of the ability to produce accurate results with reduced variation and standard error, neural network modeling has greater potential for developing robust models for extrusion processing than does regression modeling. Thus more work needs to be done to develop this tool.

To build upon the previous work, Kannadhason et al. (2009a) examined multiple protein levels and multiple starch sources, in addition to multiple DDGS levels. Feed blends were formulated with DDGS (20%, 30%, and 40% wb), the balance of which consisted of ingredients similar to those used in the prior studies, except cassava, corn, and potato starches were used as binders (instead of whey), and three protein levels were used (28%, 30%, and 32% wb). All blends had a moisture content of 20% wb. The same single screw extruder was used, with a 3.0-mm diameter circular die, using a temperature profile of 90°C–120°C–120°C, and screw speed of 130 rpm. All extrudates had durability values greater than 62%, and most were higher than 81%. Unit density values ranged from 0.78 g/cm³ to 0.94 g/cm³. For all three starch sources, increasing the DDGS level resulted in a significant increase in sinking velocity. Moreover, as DDGS and protein levels increased, unit density and pellet durability also increased for cassava and potato starch extrudates. For the cassava starch blends, a 20% level of DDGS and 28% level of protein exhibited better expansion and floatability. But the corn starch blends at 40% DDGS and 32% protein exhibited better durability.

Rosentrater et al. (2009a) pursued this in more depth. Ingredient blends were formulated using three levels each of DDGS (20%, 25%, and 30% db), protein (30%, 32.5%, and 35% db), and feed moisture content (25%, 35%, and 45% db), along with appropriate quantities of corn starch, soybean meal, fish meal, whey, vitamin, and mineral mix. The blends were extruded with the same single screw extruder, using three screw speeds (100, 150, and 200 rpm) and three extruder barrel temperature profiles (80°C–90°C–100°C, 80°C–100°C–125°C, and 80°C–125°C–150°C), and a 2.9-mm-diameter circular die (L/D of 3.19). Extrudate unit density values ranged from 1.01 g/cm³ to 1.03 g/cm³ (thus some extrudates floated while others sank), whereas durability ranged from 69.7% to 90.8%. As moisture increased, unit density increased by 16.7%, but as protein increased, it decreased by 1.4%. Increasing the DDGS levels from 20% to 25% db, protein content from 30% to 32.5% db, feed moisture content from 25% to 35% db, processing temperature from 100°C to 125°C, and screw speed from 100 rpm to 150 rpm increased the durability values by 28.1%, 18.1%, 31.8%, 6.6%, and 32.2%, respectively; all of these decreased as these independent variables increased to their highest levels, however.

Furthermore, Kannadhason et al. (2009b), examined similar blends (except tapioca starch was used as a binder instead of corn starch), using the same extruder and the same processing conditions as Rosentrater et al. (2009a). Extrudate unit densities ranged from 0.93 g/cm³ to 0.99 g/cm³, while durability values ranged from 84.1% to 95.8%. Thus, compared to the corn starch experiments, the extrudates from this study had somewhat better durability. Increasing the DDGS levels from 20% to 30% db, protein content from 30% to 35% db, feed moisture content from 25% to 45% db, and processing temperature from 100°C to 150°C decreased the resulting durability values by 7.5%, 16.2%, 17.2%, and 16.6%, respectively.

Rosentrater et al. (2009b) used the same extruder, same processing conditions, and similar feed blends (except potato starch was used as a binder instead of corn starch) as Rosentrater et al. (2009a). In this study, unit densities ranged from 0.85 g/cm³ to 0.97 g/cm³, while durability values ranged from 82.4% to 96.4%. Thus, these pellets were somewhat better in terms of structural cohesion compared to the corn starch/DDGS extrudates of Rosentrater et al. (2009a). Again, some extrudates floated while others did not. Increasing the DDGS levels from 20% to 30% (db), protein content from 30% to 35% (db), feed moisture content from 25% to 45% (db), and processing temperature from 100°C to 150°C decreased the durability values by 7.5%, 10.7%, 4.0%, and 16.8%, respectively, but in a curvilinear fashion.

Small scale single screw extrusion processing is a very effective tool for proof of concept and quantifying general behaviors, but extrusion at pilot and commercial scales must also be done in order to truly replicate the production of DDGS-based feeds that can be used in commercial aquaculture settings. Chevanan et al. (2007c) pursued this. Feed blends containing DDGS at three levels (20%, 40%, and 60%, wb) were formulated with soy flour, corn flour, fish meal, vitamin mix, and mineral mix to achieve a net protein content of 28% (similar to the blends used by Chevanan et al., 2009). These blends were extruded in a Wenger TX-52 twin-screw extruder (see Figure 16.2), which had two 52-mm diameter screws (each of which consisted of 25 individual screw segments), a barrel length-to-diameter ratio of 25.5:1, and dual 3.175-mm-diameter circular dies. During processing, two levels of blend moisture content were attained (15% and 19%, wb), and two screw speeds were used (350 rpm and 420 rpm). Unit density values for the resulting extrudates ranged from 0.24 g/cm³ to 0.61 g/cm³ (the low values were a consequence of the high expansion, which ranged from 5% to 66%), and thus all pellets floated in water—which was one of the goals of the study. Durability values ranged from 96% to 98%; hence these pellets were highly resistant to breakage during storage and transport—which was another goal of the study. Increasing the DDGS content from 20% to 60% resulted in a 36.7% decrease in the radial expansion, leading to a 159% increase in the unit density; but even the 60% DDGS pellets were better than those produced by small scale extrusion. Thus, all pellets were of very high quality, for all DDGS levels and for all processing conditions used. Consequently, it was determined that DDGS could be included at a rate of up to 60% (using these formulations) and viable extruded floating feeds could be produced.

16.6 CONCLUSIONS

Several factors (e.g., human population growth, food production needs, declining marine stocks) are driving the need to incorporate sustainable materials in finfish feeds. Increased biofuel production, improving chemical composition of DDGS, and growing availability and converge to make this coproduct a logical plant-based candidate for fish meal replacement in aquaculture feeds. The use of DDGS in aquaculture feeds will continue to expand as research continues to explore approaches to satisfy fish nutrition needs with plant-based alternatives. Important considerations include not only proximate composition but also essential amino and fatty acids, DE, vitamins, and minerals. Also, changes in biofuel production practices will alter DDGS composition and will need to be considered in formulations, requiring that various supplements be used to satisfy all nutritional requirements.

The body of literature on extrusion processing of aquaculture feed blends containing DDGS is growing. The studies done to date emphasize the importance of the balance between ingredient composition (which is a function of the target fish species), the physical properties of the raw feed blends, the type of extruder, and processing conditions that are used. It is also important to understand how all of these can affect the growth response and performance in actual fish—which is an area that still needs more attention to most effectively use DDGS as a commercial fish feed ingredient. Overall, continuing research investigations to determine optimal uses of DDGS in feed production systems should ultimately aid in sustaining livestock and companion animal markets. It thus appears that the use of DDGS as an aquafeed ingredient is promising. Using DDGS as ingredients for other nonlivestock animals is a topic explored in the following chapter.

REFERENCES

Abdelghany, A. E. 1996. Growth response of Nile tilapia, *Oreochromis niloticus* to dietary L-ascorbic acid, L-ascorbyl-2-sulfate, and L-ascorbyl-2-polyphophate. *Journal of the World Aquaculture Society* 27: 449–455.

Abdelghany, A. E. 2003. Partial and complete replacement of fish meal with gambusia meal in diets for red tilapia *Oreochromis niloticus* × *O. mossambicus*. *Aquaculture Nutrition* 9: 145–154.

Abo-State, H. A., A. M. Tahoun, and Y. A. Hammouda. 2009. Effect of replacement of soybean by DDGS combined with commercial phytase on Nile tilapia (*Oreochromis niloticus*) fingerlings growth performance and feed utilization. *American–Eurasian Journal of Agricultural and Environmental Sciences* 5: 473–479.

Al-Amoudi, M. M., A. M. N. El-Nagar, and B. M. El-Nouman. 1992. Evaluation of optimum dietary requirement of vitamin C for the growth of *Oreochromis spifurees* fingerlings in water from the Red Sea. *Aquaculture* 105: 165–173.

Alves, R. M. L., M. V. E. Grossmann, and R. S. S. Silva. 1999. Gelling properties of extruded yam (*Dioscorea alota*) starch. *Food Chemistry* 67: 123–127.

Agricultural Market Resource Center (AGMRC). 2009. *Trout Profile*. Washington, DC: U.S. Department of Agriculture. Available online: www.agmrc.org/commodities__products/aquaculture/trout_profile.cfm. Accessed July 10, 2010.

Aguilera, J. M., and F. V. Kosikowski. 1976. Soybean extruded product: A response surface analysis. *Journal of Food Science* 41(3): 647–651.

Amezaga, M. R., and D. Knox. 1990. Riboflavin requirements in on-growing rainbow trout, *Oncorhynchus mykiss*. *Aquaculture* 88: 87–98.

Anderson, J., A. J. Jackson, A. J. Matty, and B. S. Capper. 1984. Effects of dietary carbohydrate and fiber on the tilapia, *Oreochromis niloticus* (Linn). *Aquaculture* 37: 303–314.

Andrews, J. W., and T. Murai. 1978. Dietary niacin requirements of channel catfish. *Journal of Nutrition* 108: 1508–1511.

Andrews, J. W., and T. Murai. 1979. Pyridoxine requirements of channel catfish. *Journal of Nutrition* 109: 533–537.

Andrews, J. W., T. Murai, and J. W. Page. 1980. Effects of dietary cholecalciferol and ergocalciferol on catfish. *Aquaculture* 19: 49–54.

Anggawati-Satyabudhy, A. M., B. G. Grant, and J. E. Halver. 1989. Effect of L-ascorbyl phosphate (AsPP) on growth and immunoresistance of rainbow trout (*Oncorhynchus mykiss*) to infectious hematopoietic necrosis (IHN) virus. In *Third International Symposium on Feeding and Nutrition in Fish, Toba, Japan*, eds. M. Takeda, and T. Watanabe, pp. 411–426. Tokyo, Japan: Tokyo University of Fisheries. August 28–September 1.

Apines, M. J. S., S. Satoh, V. Kiron, T. Watanabe, and T. Akoi. 2003. Availability of supplemental amino acid-chelated trace elements in diets containing tricalcium phosphate and phytate to rainbow trout *Oncorhynchus mykiss*. *Aquaculture* 225: 431–444.

Azevedo, P. A., C. Y. Cho, S. Leeson, and D. P. Bureau. 1998. Effects of feeding level and water temperature on growth, nutrient and energy utilization and waste outputs of rainbow trout (*Oncorhynchus mykiss*). *Aquatic Living Resources* 11: 227–238.

Azevedo, P. A., S. Leeson, C. Y. Cho, and D. P. Bureau. 2004a. Growth and feed utilization of large size rainbow trout (*Oncorhynchus mykiss*) and Atlantic salmon (*Salmo salar*) reared in freshwater: Diet and species effects, and responses over time. *Aquaculture Nutrition* 10: 401–411.

Azevedo, P. A., S. Leeson, C. Y. Cho, and D. P. Bureau. 2004b. Growth, nitrogen and energy utilization of juveniles from four salmonid species: Diet, species, and size effects. *Aquaculture* 234: 393–414.

Barnett, B. J., C. Y. Cho, and S. J. Slinger. 1982. Relative biopotency of ergocalciferol and cholecalciferol and the role of and requirement for vitamin D in rainbow trout (*Salmo gairdneri*). *Journal of Nutrition* 112: 2011–2019.

Barrows, F. T., and R. W. Hardy. 2001. Nutrition and feeding. In *Fish Hatchery Management*, 2nd ed., ed. G. A. Wedemeyer, pp. 483–559. Bethesda, Maryland: American Fisheries Society.

Barrows, F. T., G. Gaylord, W. M. Sealey, L. Porter, and C. E. Smith. 2008. The effect of vitamin pre-mix in extruded plant-based and fish meal based diets on growth efficiency and health of rainbow trout, *Oncorhynchus mykiss*. *Aquaculture* 283: 148–155.

Behnke, R. J. 1992. Native Trout of Western North America. Monograph 6, pp. 172–174. Bethesda, Maryland: American Fisheries Society.

Belyea, R. L., K. D. Rausch, and M. E. Tumbleson. 2004. Composition of corn and distillers dried grains with solubles from dry grind ethanol processing. *Bioresource Technology* 94: 293–298.

Bergot, F. 1979a. Carbohydrate in rainbow trout diets: Effects of the level and source of carbohydrate and the number of meals on growth and body composition. *Aquaculture* 18: 157–167.

Bergot, F. 1979b. Problemes particuliers poses par l'utilisation des glucides chez la truite are-en-ciel. *Annales de la nutrition et de l'alimentation* 33: 247–257.

Bergot, F., and J. Breque. 1983. Digestibility of starch by rainbow trout: Effects of physical state of starch and the intake level. *Aquaculture* 34: 203–212.

Bodin, N., M. Mambrini, J. -B Wauters, T. Abboudi, W. Ooghe, E. Le Boulenge, Y. Larondelle, and X. Rollin. (2008) Threonine requirements for rainbow trout (*Oncorhynchus mykiss*) and Atlantic salmon (*Salmo salar*) at the fry stage are similar. *Aquaculture* 274: 353–365.

Bothast, R. J., and M. A. Schlicher. 2005. Biotechnological processes for conversion of corn into ethanol. *Applied Microbiology and Biotechnology* 67: 19–25.

Brauge, C., F. Medale, and G. Corraze. 1994. Effects of dietary carbohydrate levels on growth, body composition and glycemia in rainbow trout, *Oncorhynchus mykiss*, reared in seawater. *Aquaculture* 123: 109–120.

Brown, B. P. 1988. Vitamin D requirement of juvenile channel catfish reared in calcium-free water. Doctoral Dissertation. College Station, Texas: Texas A&M University.

Burtle, G. J., and R. T. Lovell. 1989. Lack of response of channel catfish (*Ictalurus punctatus*) to dietary myo-inositol. *Canadian Journal of Fisheries and Aquatic Sciences* 46: 218–222.

Caballero, M. J., A. Obach, G. Rosenlund, D. Montero, M. Gisvold, and M. S. Izquierdo. 2002. Impact of different dietary lipid sources on growth, lipid digestibility, tissue fatty acid composition and histology of rainbow trout, *Oncorhynchus mykiss*. *Aquaculture* 214: 253–271.

Case, S. E., D. D. Hamann, and S. J. Schwartz. 1992. Effect of starch gelatinization on physical properties of extruded wheat and corn based products. *Cereal Chemistry* 69(4): 401–404.

Castell, J. D., R. O. Sinnhuber, J. H. Wales, and D. J. Lee. 1972. Essential fatty acids in the diet of rainbow trout (*Salmo gairdneri*). Growth, feed conversion and some gross deficiency symptoms. *Journal of Nutrition* 102: 77–86.

Chaiyapechara, S., M. T. Casten, R. W. Hardy, and F. M. Dong. 2003. Fish performance, fillet characteristics, and health assessment index of rainbow trout (*Oncorhynchus mykiss*) fed diets containing adequate and high concentrations of lipid and vitamin E. *Aquaculture* 219: 715–738.

Chen, A. H., Y. C. Jao, J. W. Larkin, and W. E. Goldstein. 1979. Rheological model of soy dough in extrusion. *Journal of Food Processing Engineering* 2(4): 337–342.

Cheng, Z. J., R. W. Hardy, and M. Blair. 2003. Effects of supplementing methionine hydroxyl analogue in soybean meal and distillers dried grain-based diets on the performance and nutrient retention of rainbow trout [*Oncorhynchus mykiss* (Walbaum)]. *Aquaculture Research* 34: 1303–1310.

Cheng, Z. J., and R. W. Hardy. 2004a. Effect of microbial phytase supplementation in corn distiller's dried grain with solubles on nutrient digestibility and growth performance of rainbow trout, *Oncorhynchus mykiss*. *Journal of Applied Aquaculture* 15: 83–100.

Cheng, Z. J., and R. W. Hardy. 2004b. Nutritional value of diets containing distiller's dried grain with solubles for rainbow trout, *Oncorhynchus mykiss*. *Journal of Applied Aquaculture* 15: 101–113.

Chevanan, N., K. A. Rosentrater, and K. Muthukumarappan. 2005. Utilization of distillers dried grains for fish feed by extrusion technology—A review. *ASAE Annual International Meeting/Paper Number 056025*. Tampa, Florida.

Chevanan, N., K. A. Rosentrater, and K. Muthukumarappan. 2007c. Twin screw extrusion processing of feed blends containing distillers dried grains with solubles. *Cereal Chemistry* 84(5): 428–436.

Chevanan, N., K. A. Rosentrater, and K. Muthukumarappan. 2008a. Effect of DDGS, moisture content, and screw speed on the physical properties of extrudates in single screw extrusion. *Cereal Chemistry* 85(2): 132–139.

Chevanan, N., K. A. Rosentrater, and K. Muthukumarappan. 2008b. Effects of processing conditions on single screw extrusion of feed ingredients containing DDGS. *Food and Bioprocess Technology* 3: 111–120.

Chevanan, N., K. Muthukumarappan, K. A. Rosentrater, and J. Julson. 2007a. Effect of die dimensions on extrusion processing parameters and properties of DDGS based extrudates. *Cereal Chemistry* 84(4): 389–398.

Chevanan, N., K. Muthukumarappan, and K. A. Rosentrater. 2007b. Neural network and regression modeling of extrusion processing parameters and properties of extrudates containing DDGS. *Transactions of the ASABE* 50(5): 1765–1778.

Chevanan, N., K. Muthukumarappan, and K. A. Rosentrater. 2009. Extrusion studies of aquaculture feed using distillers dried grains with solubles and whey. *Food and Bioprocess Technology* 2: 177–185.

Cho, C. Y., and S. J. Kaushik. 1985. Effects of protein intake on metabolizable and net energy values of fish diets. In *Nutrition and Feeding of Fish,* eds. C. B. Cowey, A. M. Mackie, and J. G. Bell, pp. 95–117. London: Academic Press.

Cho, C. Y., S. J. Kaushik, and B. Woodward. 1989. Dietary arginine requirement for rainbow trout. *Federation of American Societies for Experimental Biology Journal* 3: 2459. (Abstr.)

Cho, C. Y., and B. Woodward. 1989. Studies on the protein-to-energy ratio in diets for rainbow trout (*Salmo gairdneri*). In *Proceedings of the Eleventh Symposium on Energy Metabolism, Lunteren 1988*, pp. 37–40. Wageningen, the Netherlands: European Association for Animal Production Publication 43.

Cho, C. Y. 1990. Fish nutrition, feeds and feeding: With special emphasis on salmonid aquaculture. *Food Reviews International* 6: 333–357.

Cho, C. Y., and B. Woodward. 1990. Dietary pantothenic acid requirements of young rainbow trout (*Oncorhynchus mykiss*). *Federation of American Societies for Experimental Biology Journal* 4: 3747. (Abstr.)

Cho, C. Y., and C. Cowey. 1991. Rainbow trout, *Oncorhynchus mykiss*. In *Handbook of Nutrient Requirements of Finfish*, ed. R. P. Wilson, pp. 131–144. Boca Raton, Florida: CRC Press,

Cho, C. Y. 1992. Feeding systems for rainbow trout and other salmonids with reference to current estimates of energy and protein requirements. *Aquaculture* 100: 107–123.

Cho, C. Y., and S. J. Kaushik. 1990. Nutritional energetics in fish: energy and protein utilization in rainbow trout (*Salmo gairdneri*). *World Review of Nutrition and Dietetics* 61: 132–172.

Chou, B. S., and S. Y. Shiau. 1999. Both n-6 and n-3 fatty acids are required for maximal growth of juvenile hybrid tilapia. *North American Journal of Aquaculture* 61: 13–20.

Chou, B. S., S. Y. Shiau, and S. S. O. Hung. 2001. Effect of dietary cod liver oil on growth and fatty acids of juvenile hybrid tilapia. *North American Journal of Aquaculture* 63: 277–284.

Cowey, C. B., J. W. Adron, and A. Youngson. 1983. The vitamin E requirement of rainbow trout (*Salmo gairdneri*) given diets containing polyunsaturated fatty acids derived from fish oil. *Aquaculture* 30: 85–93.

Cowey, C. B., C. Y. Cho, J. G. Sivak, J. A. Weerheim, and D. D. Stuart. 1992. Methionine intake in rainbow trout (*Oncorhynchus mykiss*), relationship to cataract formation and the metabolism of methionine. *Journal of Nutrition* 122: 1154–1163.

Cowey, C. Y., and B. Woodward. 1993. The dietary requirement of young rainbow trout (*Oncorhynchus mykiss*) for folic acid. *Journal of Nutrition* 123: 1594–1600.

Coyle, S. D., G. J. Mengel, J. H. Tidwell, and C. D. Webster. 2004. Evaluation of growth, feed utilization, and economics of hybrid tilapia, *Oreochromis niloticus* × *Oreochromis aureus*, fed diets containing different protein sources in combination with distillers dried grains with solubles. *Aquaculture Research* 35: 365–370.

Cromwell, G. L., K. L. Herkelman, and T. S. Stahly. 1993. Physical, chemical, and nutritional characteristics of distillers dried grain with solubles for chicks and pigs. *Journal of Animal Science* 71: 679–686.

Cumming, D. B., D. W. Stanley, and J. M. Deman. 1973. Fate of water soluble soy protein during thermoplastic extrusion. *Journal of Food Science* 38(2): 320.

Dabrowska, H., K. Meyer-Burgdorff, and K. D. Gunther. 1989. Interaction between dietary protein and magnesium level in tilapia (*Oreochromis niloticus*). *Aquaculture* 76: 277–291.

Davis, A. T., and R. R. Stickney. 1978. Growth response of *Tilapia aurea* to dietary protein quality and quantity. *Transactions of the American Fisheries Society* 107: 479–483.

Desjardins, L. M. 1985. The effect of iron supplementation on diet rancidity and the growth and physiological response of rainbow trout. M.S. Thesis. Ontario, Canada: University of Guelph.

De Silva, S. S., and T. A. Anderson. 1995. *Finfish Nutrition in Aquaculture*. London, United Kingdom: Chapman and Hall. p. 162.

Dumas, A., J. France, and D. P. Bureau. 2007. Evidence of three growth stanzas in rainbow trout (*Oncorhynchus mykiss*) across life stages and adaptation of the thermal-unit growth coefficient. *Aquaculture* 267: 139–146.

Duncan, P. L., and R. T. Lovell. 1991. Effect of folic acid on growth, survival and hematology in channel catfish (*Ictalurus punctatus*). 22nd Annual Conference of the World Aquaculture Society. San Juan, Puerto Rico.

Dupree, H. K. 1966. Vitamins essential for growth of channel catfish, *Ictalurus punctatus*. *Bureau of Sport Fisheries and Wildlife*, Technical Paper No. 7: 2–12. Washington DC.

Dupree, H. K. 1970. Dietary requirement of vitamin A acetate and beta carotene. In *Progress in Sport Fishery Research,1969*, Resource Publication No. 88. pp. 148–150. Washington DC: Bureau of Sport Fisheries and Wildlife.

Dwyer, W. P., C. E. Smith, and R. G. Piper. 1981. Rainbow trout growth efficiency as affected by temperature. *Information Leaflet 18*. Bozeman, Montana: U.S. Fish and Wildlife Service Fish Cultural Development Center.

Dwyer, W. P., C. E. Smith, and R. G. Piper. 1983. Steelhead trout growth efficiency as affected by temperature. *Information Leaflet 27*. Bozeman, Montana: U.S. Fish and Wildlife Service Fish Cultural Development Center.

Einen, O., and A. J. Roem. 1997. Dietary protein/energy ratios for Atlantic salmon in relation to fish size: Growth, feed utilization and slaughter quality. *Aquaculture Nutrition* 3: 115–126.

Elhmind, E. A., and S. I. Ghonim. 1994. Dietary zinc requirement of fingerling *Oreochromis niloticus*. *Aquaculture* 119: 259–264.

Eliason, E. J., D. A. Higgs, and A. P. Farrell. 2007. Effect of isoenergetic diets with different protein and lipid content on the growth performance and heat increment of rainbow trout. *Aquaculture* 272: 723–736.

El Naggar, G. O., and R. T. Lovell. 1991. L-Ascorbyl-2-monophosphate has equal antiscorbutic activity as L-ascorbic acid but L-acorbyl-2-sulfate is inferior to L-ascorbic acid for channel catfish. *Journal of Nutrition* 121: 1622–1626.

El-Sayed, A. -F. M. 1987. Protein and energy requirement of *Tilapia zilli*. Doctoral Dissertation. East Lansing, Michigan: Michigan State University.

El-Sayed, A. -F. M. 2006. *Tilapia Culture*. Wallingford, United Kingdom: CABI Publishing.

El-Sayed, A. -F. M., and D. L. Garling, Jr. 1988. Carbohydrate-to-lipid ratios in diets for *Tilapia* fingerlings. *Aquaculture* 73: 157–163.

El-Sayed, A. -F. M., and S. Teshima. 1992. Protein and energy requirements of Nile tilapia, *Oreochromis niloticus*, fry. *Aquaculture* 103: 55–63.

Encarnação, P., D. de Lange, Z. M. Rodehutscord, D. Hoehler, W. Bureau, and D. P. Bureau. 2004. Diet digestible energy content affects lysine utilization, but not dietary lysine requirements of rainbow trout (*Oncorhynchus mykiss*) for maximum growth. *Aquaculture* 235: 569–586.

Fellows, P. J. 1996. *Food Processing Technology Principles and Practice*. Abingdon, England: Woodhead Publishing Limited.

Fineman-Kalio, A. S., and A. S. Camacho. 1987. The effects of supplemental feeds containing different protein: energy ratios on the growth and survival of *Oreochromis niloticus* (L.) in brackish water ponds. *Aquaculture and Fisheries Management* 18: 139–149.

Fontagné, S., N. Silva, D. Bazin, A. Ramos, P. Aguirre, A. Surget, A. Abrantes, S. J. Kaushik, and D. M. Power. 2009. Effects of dietary phosphorous and calcium level on growth and skeletal development in rainbow trout (*Oncorhynchus mykiss*) fry. *Aquaculture* 297: 141–150.

Fontaínhas-Fernandes, A., E. Gomes, M. A. Reis-Henriques, and J. Coimbra. 1999. Replacement of fish meal by plant proteins in the diet of Nile tilapia: Digestibility and growth performance. *Aquaculture International* 7: 57–67.

Food and Agricultural Organization of the United Nations [FAO]. 2002. The state of the world fisheries and aquaculture, part 1. Rome, Italy: Food and Agriculture Organization, Available online: www.fao.org/docrep/005/y7300e/y7300e04.htm. Accessed July 10, 2010.

Food and Agricultural Organization of the United Nations [FAO]. 2006. The state of the world fisheries and aquaculture. Rome, Italy: Food and Agriculture Organization, Available online: www.fao.org/docrep/009/a0699e/a0699e00.HTM. Accessed July 10, 2010.

Food and Agricultural Organization of the United Nations [FAO]. 2009. The state of the world fisheries and aquaculture. Rome, Italy: Food and Agriculture Organization, Available online: www.fao.org/docrep/011/i0250e/i0250e00.HTM. Accessed July 10, 2010.

Food and Agricultural Organization of the United Nations [FAO]. 2010. Cultured aquatic species information programme: Rainbow trout fact sheet. Rome, Italy: FAO Fisheries and Aquaculture Department, Food and Agriculture Organization, Available online: www.fao.org/fishery/culturedspecies/Oncorhynchus_mykiss/en. Accessed July 10, 2010.

Forsman, A., and K. Ruohonen. 2009. Dynamics of protein and lipid intake regulation of rainbow trout studied with a wide range of encapsulated diets and self-feeders. *Physiology and Behavior* 96: 85–90.

Fowler, L. G., and J. L. Banks. 1976. Fish meal and wheat germ substitutes in the Abernathy diet. *Progressive Fish Culturist* 38: 127–130.

Garber, B. W., H. E. Huff, and F. Hsieh. 1997. Influence of particle size on the twin-screw extrusion of corn meal. *Cereal Chemistry* 74: 656–661.

Garling, Jr.,D. L., and R. P. Wilson. 1976. Optimum dietary protein-to-energy ratios for channel catfish fingerlings, *Ictalurus punctatus*. *Journal of Nutrition* 106: 1368–1375.

Garling, Jr.,D. L., and R. P. Wilson. 1977. Effects of dietary carbohydrate-to-lipid ratios on growth and body composition of fingerling channel catfish. *The Progressive Fish-Culturist* 39: 43–47.

Gatlin, III, D. M., W. E. Poe, and R. P. Wilson. 1986. Protein and energy requirements of fingerling channel catfish for maintenance and growth. *Journal of Nutrition* 116: 2121–2131.

Gatlin, III, D. M., E. H. Robinson, and W. E. Poe. 1982. Magnesium requirement of fingerling channel catfish. *Journal of Nutrition* 112: 1181–1187.

Gatlin, D. M., III, and R. P. Wilson. 1983. Dietary zinc requirement of fingerling channel catfish. *Journal of Nutrition* 114: 630–635.

Gatlin, III, D. M., and R. P. Wilson. 1984a. Studies on manganese requirement of fingerling channel catfish. *Aquaculture* 41: 85–92.

Gatlin, III, D. M., and R. P. Wilson. 1984b. Dietary selenium requirement of fingerling channel catfish. *Journal of Nutrition* 114: 627–633.

Gatlin, III, D. M., and R. P. Wilson. 1986a. Characterization of iron deficiency and the dietary iron requirement for fingerling channel catfish. *Aquaculture* 52: 191–198.

Gatlin, III, D. M., and R. P. Wilson. 1986b. Dietary copper requirement of fingerling channel catfish. *Aquaculture* 54: 277–285.

Gatlin, III, D. M., F. T. Barrows, P. Brown, K. Dabrowski, T. G. Gaylord, R. W. Hardy, et al. 2007. Expanding the utilization of sustainable plant products in aquafeeds: a review. *Aquaculture Research* 38: 551–579.

Glencross, B. D., C. G. Carter, N. Duijster, D. E. Evans, K. Dods, P. McCafferty, W. E. Hawkins, R. Maas, and S. Sispas. 2004. A comparison on the digestive capacity of Atlantic salmon (Salmo salar) and rainbow trout (*Oncorhynchus mykiss*) when fed a range of plant protein products. *Aquaculture* 273: 333–346.

Glencross, B. 2009. The influence of soluble and insoluble lupin non-starch polysaccharides on the digestibility of diets fed to rainbow trout (*Oncorhynchus mykiss*). *Aquaculture* 294: 256–261.

Gould, W. A. 1996. *Unit Operations for the Food Industries*. Baltimore, MD: CTI Publications, Inc.

Green, B. W. 2006. Fingerling production systems. In *Tilapia: Biology, Culture, and Nutrition*, eds. C. Lim, and C. D. Webster, pp. 181–210. Binghamton, New York: The Haworth Press, Inc.

Green, J. A., and R. W. Hardy. 2002. The optimum dietary essential amino acid pattern for rainbow trout (*Oncorhynchus mykiss*), to maximize nitrogen retention and minimize nitrogen excretion. *Fish Physiology and Biochemistry* 27: 97–108.

Guerden, I., E. Gondouin, M. Rimbach, W. Koppe, S. Kaushik, and T. Boujard. 2006. The evaluation of energy intake adjustments and preferences in juvenile rainbow trout fed increasing amounts of lipid. *Physiology and Behavior* 88: 325–332.

Guillaume, J., S. Kaushik, P. Bergot, and R. Métailler. 2001. *Nutrition and Feeding of Fish and Crustaceans*, pp. 145–165. Chichester, United Kingdom: Praxis.

Gwiazda, S., A. Noguchi, and K. Saio. 1987. Microstructural studies of texturized vegetable protein products: Effects of oil addition and transformation of raw materials in various sections of a twin screw extruder. *Food Microstructure* 6: 57–61.

Halver, J. E. 1972. The vitamins. In *Fish Nutrition*, ed. J. E. Halver, pp. 29–103. New York: Academic Press.

Halver, J. E., L. M. Ashley, and R. R. Smith.1969. Ascorbic acid requirements of coho salmon and rainbow trout. *Transactions of the American Fisheries Society* 90: 762–771.

Harding, D. E., O. W. Allen, Jr., and R. P. Wilson. 1977. Sulfur amino acid requirement of channel catfish: L-Methionine and L-cystine. *Journal of Nutrition* 107: 2031–2035.

Hardie, L. J., M. J. Madsen, T. C. Fletcher, and J. C. Secombes. 1993. In vitro addition of vitamin C affects rainbow trout lymphocyte responses. *Fish and Shellfish Immunology* 3: 207–219.

Hardy, R. W., and K. D. Shearer. 1985. Effects of dietary calcium phosphate and zinc supplementation on whole body zinc concentration of rainbow trout (*Salmo gairdneri*). *Canadian Journal of Fisheries and Aquatic Sciences* 42: 181–184.

Hardy, R. W., C. V. Sullivan, and A. M. Koziol. 1987. Absorption, body distribution, and excretion of dietary zinc by rainbow trout (*Salmo gairdneri*). *Fish Physiology and Biochemistry* 3: 133–143.

Hargreaves, J. A., and C. S. Tucker. 2004. Industry development. In *Biology and Culture of Channel Catfish*, eds. C. S. Tucker, and J. A. Hargreaves, pp. 1–13. Amsterdam, the Netherlands: Elsevier B. V.

Harper, J. M. 1981. *Extrusion of Foods*, Vol. I. Boca Raton, FL: CRC Press.

Hepher, B. 1988. *Nutrition of Pond Fishes*. New York: Cambridge University Press.

Hertrampf, J. W. and F. Piedad-Pascual. 2000. Distillery by-products. In *Handbook on Ingredients for Aquaculture Feeds*, pp. 115–124. Boston, Massachusetts: Kluwer Academic Publishers.

Higgs, D. A., J. S. McDonald, C. D. Levnings, and B. S. Dosanjh. 1995. Nutrition and feeding habits in relation to life history stage. In *Physiological Ecology of Pacific Salmon*, eds. C. Groot, L. Margolis, and W. C. Clark, pp. 161–135. Vancouver, Canada: UBC Press.

Hilton, J. W., C. Y. Cho, and S. J. Slinger. 1978. Effect of graded levels of supplemental ascorbic acid in practical diets fed to rainbow trout (*Salmo gairdneri*). *Journal of the Fisheries Research Board of Canada* 35: 431–436.

Hilton, J. W., P. V. Hodson, and S. J. Slinger. 1980. The requirement and toxicity of selenium in rainbow trout (*Salmo gairdneri*). *Journal of Nutrition* 110: 2527–2535.

Hilton, J. W., J. L. Atkinson, and S. J. Slinger. 1982. Maximum tolerable level, digestion, and metabolism of D-glucose (cerelose) in rainbow trout (*Salmo gairdneri*) reared on a practical trout diet. *Canadian Journal of Fisheries and Aquatic Sciences* 39: 1229–1234.

Hu, C. J., S. M. Chen, C. H. Pan, and C. H. Huang. 2006. Effects of dietary vitamin A or β-carotene concentrations on growth of juvenile tilapia, *Oreochromis niloticus* × *Oreochromis aureus*. *Aquaculture* 253: 602–607.

Hughes, S. G., G. L. Ramsey, and J. G. Nichum. 1981. Riboflavin requirements of fingerling rainbow trout. *Progressive Fish Culturist* 43: 167–172.

Hughes, S. G. 1987. Distillers products in salmonid diets. *Proceedings of the Distillers Feed Conference* 42: 27–31.

Hung, S. S. O., C. Y. Cho, and S. J. Slinger. 1980. Measurement of oxidation in fish oil and its effect on vitamin E nutrition of rainbow trout, *Salmo gairdneri*. *Canadian Journal of Fisheries and Aquatic Sciences* 37: 1248–1253.

Ibanoglu, S., A. Paul, and D. H. George. 1996. Extrusion of tarhana: Effect of operating variables on starch gelatinization. *Food Chemistry* 57(4): 541–544.

Ilo, S., E. U. Tomschik, U. Berghofer, and N. Mundigler. 1996. The effects of extrusion operating conditions on the apparent viscosity and properties of extrudates in twin screw extrusion cooking of maize grits. *Lebensmittel-Wissenschaft und-Technologie* 29: 593–598.

Ishac, M. M., and A. M. Dollar. 1967. Studies on manganese uptake in *Tilapia mossambica* and *Salmo gairdnerii*. *Hydrobiologia* 31: 572–584.

Iwama, G. K. 1996. Growth of salmonids. In *Principles of Salmonid Aquaculture,* Vol. 29, eds. W. Pennell, and B. A. Barton, pp. 467–515. Developments in Aquaculture and Fisheries Science. Amsterdam, The Netherlands: Elsevier.

Jauncey, K., and B. Ross. 1982. *A Guide to Tilapia Feeds and Feeding.* Stirling, Scotland: Institute of Aquaculture, University of Stirling.

Jobling, M. 1994. *Fish Bioenergetics.* London, United Kingdom: Chapman and Hall.

Kanazawa, A., S. Teshima, M. Sakamoto, and M. A. Awal. 1980. Requirements of *Tilapia zilli* for essential fatty acids. *Bulletin of the Japanese Society of Scientific Fisheries* 46: 1353–1356.

Kannadhason, S., K. Muthukumarappan, and K. A. Rosentrater. 2009a. Effect of starch sources and protein content on extruded aquaculture feed containing DDGS. *Food and Bioprocess Technology* DOI.10.1007/s 11947-008-0177-4.

Kannadhason, S., K. Muthukumarappan, and K. A. Rosentrater. 2009b. Effects of ingredients and extrusion parameters on aquafeeds containing DDGS and tapioca starch. *Journal of Aquaculture Feed Science and Nutrition* 1(1): 6–21.

Kaushik, S. 1979. Application of a biochemical method for the estimation of amino acid needs in fish. Quantitative arginine requirements of rainbow trout in different salinities. In *Finfish Nutrition and Fishfeed Technology,* eds. K. Tiews and J. E. Halve, pp. 197–201. Berlin, Germany: Heenemann GmbH.

Kelly, A. M. 2005. Channel catfish culture. In *American Fisheries Society, Symposium 46,* eds. A. M. Kelly, and J. Silverstein, pp. 251–272. Bethesda, Maryland.

Ketola, H. G., 1983. Requirement for dietary lysine and arginine by fry of rainbow trout. *Journal of Animal Science* 56: 101–107.

Ketola, H. G., and M. E. Richmond. 1994. Requirement of rainbow trout for dietary phosphorous and its relationship to the amount discharged in hatchery effluents. *Transactions of the American Fisheries Society* 123: 587–594.

Kim, K. I. 1993. Requirement for phenylalanine and replacement value of tyrosine for phenylalanine in rainbow trout (*Oncorhynchus mykiss*). *Aquaculture* 113: 243–250.

Kim, K. I. 1997. Re-evaluation of protein and amino acid requirements of rainbow trout. (*Oncorhynchus mykiss*). *Aquaculture* 151: 3–7.

Kim, K. I., T. B. Kayes, and C. H. Amundson. 1992a. Requirements for sulfur amino acids and utilization of D-methionine by rainbow trout (*Oncorhynchus mykiss*). *Aquaculture* 101: 95–103.

Kim, K. I., T. B. Kayes, and C. H. Amundson. 1992b. Requirements for lysine and arginine by rainbow trout (*Oncorhynchus mykiss*). *Aquaculture* 106: 333–334.

Kim, K. I., T. B. Kayes, and C. H. Amundson. 1991. Purified diet development and reevaluation of the dietary protein requirement of fingerling rainbow trout. (*Oncorhynchus mykiss*). *Aquaculture* 96: 57–67.

Kim, K. I., and T. B. Kayes. 1982. Test diet development and lysine requirement of rainbow trout. *Federation of American Societies for Experimental Biology* 41: 716. (Abstr.)

Kim, K. I., T. B. Kayes, and C. H. Amundson. 1983. Protein and arginine requirements of rainbow trout. *Federation of American Societies for Experimental Biology* 42: 2198. (Abstr.)

Kim, K. I., T. B. Kayes, and C. H. Amundson. 1984. Requirements for sulfur containing amino acids and utilization of D-methionine by rainbow trout. *Federation of American Societies for Experimental Biology* 43: 3338. (Abstr.)

Kim, K. I., T. B. Kayes, and C. H. Amundson. 1987. Effects of dietary tryptophan levels on growth, feed/gain carcass composition and liver glutamate dehydrogenase activity in rainbow trout (Salmo gairdneri). *Comparative Biochemistry and Physiology* 88B: 737–741.

Kim, K. I., T. B. Kayes, and C. H. Amundson. 1991. Purified diet development and reevaluation of the dietary protein requirement of fingerling rainbow trout. (*Oncorhynchus mykiss*). *Aquaculture* 96: 57–67.

Kiron, V., J. Puangkaew, K. Ishizaka, S. Satoh, and T. Watanabe. 2004. Antioxidant status and nonspecific immune responses in rainbow trout (*Oncorhynchus mykiss*) fed two levels of vitamin E along with three lipid sources. *Aquaculture* 234: 361–379.

Kitamura, S., T. Suwa, S. Ohara, and K. Nakagawa. 1967. Studies on vitamin requirements of rainbow trout. 3. Requirements for vitamin A and deficiency symptoms. *Bulletin of the Japanese Society of Scientific Fisheries* 33: 1126–1131.

Knapp, G., C. A. Roheim, and J. L. Anderson. 2007. *The Great Salmon Run: Competition between Wild and Farmed Salmon.* Washington, DC: Traffic North America, World Wildlife Fund.

Knox, D., C. B. Cowey, and J. W. Adron. 1981. Studies on the nutrition of salmonids fish. The magnesium requirement of rainbow trout (*Salmo gairdneri*). *British Journal of Nutrition* 45: 137–148.

Kokini, J. L., C. N. Chang, and L. S. Lai. 1992. The role of rheological properties on extrudate expansion. In *Food Extrusion Science and Technology*, ed. J. L. Kokini, C. T. Ho, and M. V. Karewe, pp. 631–652. New York, New York: Marcel Dekker, Inc.

Krogdahl, Å., A. Sundby, and J. J. Olli. 2004. Atlantic salmon (*Salmo salar*) and rainbow trout (*Oncorhynchus mykiss*) digest and metabolize nutrients differently. Effects of water salinity and dietary starch level. *Aquaculture* 229: 335–360.

Kubaryk, J. M. 1980. Effect of diet, feeding schedule and sex on food consumption, growth and retention of protein and energy by tilapia. Doctoral Dissertation. Auburn, Alabama: Auburn University.

Lee, D. J., and G. B. Putnam. 1973. The response of rainbow trout to varying protein/energy ratios in a test diet. *Journal of Nutrition* 103: 916–922.

Lee, J. Y. 2003. Vitamin K requirements of juvenile hybrid tilapia (*Oreochromis niloticus* × *O. aureus*) and grouper (*Epinephelus malabaricus*). M.S. Thesis. Keelung, Taiwan: National Taiwan Ocean University.

Lellis, W. A., F. T. Barrows, and R. W. Hardy. 2004. Effects of phase feeding dietary phosphorous on survival, growth, and processing characteristics of rainbow trout *Oncorhynchus mykiss*. *Aquaculture* 242: 607–616.

Li, M., and R. T. Lovell. 1992. Comparison of satiate feeding and restricted feeding of channel catfish with various concentrations of dietary protein in production ponds. *Aquaculture* 103: 165–175.

Li, M. H., E. H. Robinson, and B. B. Manning. 2004. Nutrition. In *Biology and Culture of Channel Catfish*, eds. C. S. Tucker, and J. A. Hargreaves, pp. 279–323. Amsterdam, the Netherlands: Elsevier B. V.

Lim, C. 1989. Practical feeding—Tilapias. In *Nutrition and Feeding of Fish*, ed. T. Lovell, pp. 163–183. New York, New York: Van Nostrand Reinhold.

Lim, C., and R. T. Lovell. 1978. Pathology of the vitamin C deficiency syndrome in channel catfish (*Ictalurus punctatus*). *Journal of Nutrition* 108: 1137–1146.

Lim, C., and P. H. Klesius. 2003. Influence of dietary levels of magnesium on growth, tissue mineral content and resistance of channel catfish challenged with *Edwardsiella ictaluri*. *Journal of the World Aquaculture Society* 34(1): 18–28.

Lim, C., and P. H. Klesius. 2001. Influence of dietary levels of folic acid on growth response and resistance of Nile tilapia (*Oreochromis niloticus*) to *Streptococcus iniae*. In *Book of Abstracts*, p. 150. Kaohsiung, Taiwan: 6th Asian Fisheries Forum.

Lim, C. E., and C. D. Webster. 2006. Nutrient requirements. In *Tilapia: Biology, Culture, and Nutrition*, eds. C. E. Lim, and C. D. Webster, pp. 469–501. Binghamton, New York: The Haworth Press, Inc.

Lim, C., B. LeaMaster, and J. A. Brock. 1991. Thiamin requirement of red hybrid tilapia grown in seawater. In *Abstracts*, p. 39. World Aquaculture Society 22nd Annual Conference and Exposition, San Juan, Puerto Rico. Baton Rouge, Louisiana: World Aquaculture Society.

Lim, C., B. LeaMaster, and J. A. Brock. 1993. Riboflavin requirement of fingerling red hybrid tilapia grown in seawater. *Journal of the World Aquaculture Society* 24: 451–458.

Lim, C., B. LeaMaster, and J. A. Brock. 1995. Pyridoxine requirement of fingerling red hybrid tilapia grown in seawater. *Journal of Applied Aquaculture* 5: 49–60.

Lim, C., M. M. Barros, P. H. Klesius, and C. A. Shoemaker. 2000. Thiamin requirement of Nile tilapia, *Oreochromis niloticus*. In *Book of Abstracts*, p. 201. Aquaculture America 2000, New Orleans, Louisiana. Baton Rouge, Louisiana: World Aquaculture Society.

Limsuwan, T., and R. T. Lovell. 1981. Intestinal synthesis and absorption of vitamin B12 in channel catfish. *Journal of Nutrition* 111: 2125–2132.

Lin, S, F. Hsieh, and H. E. Huff. 1997. Effects of lipids and processing conditions on degree of starch gelatinization of extruded dry pet food. *Lebensmittel-Wissenschaft und-Technologie* 30(1): 754–761.

Lin, S., H. E. Huff, and F. Hsieh. 2000. Texture and chemical characteristics of soy protein meat analog extruded at high moisture. *Journal of Food Science* 65: 264–269.

Lodge, S. L., R. A. Stock, T. J. Klopfenstein, D. H. Shain, and D. W. Herold. 1997. Evaluation of corn and sorghum distillers byproducts. *Journal of Animal Science* 75: 37–43.

Lovell, R. T. 1978. Dietary phosphorus requirement of channel catfish. *Transactions of the American Fisheries Society* 107: 617–621.

Lovell, T. 1989a. The nutrients. In *Nutrition and Feeding of Fish*, ed. T. Lovell, pp. 11–71. New York: Van Nostrand Reinhold.

Lovell, T. 1989b. Practical feeding—Channel catfish. In *Nutrition and Feeding of Fish*, ed. T. Lovell, pp. 145–162. New York: Van Nostrand Reinhold.

Lovell, R. T., and J. C. Buston. 1984. Biotin supplementation of practical diets for channel catfish. *Journal of Nutrition* 114: 1092–1096.

Lovell, R. T., and Y. P. Li. 1978. Essentiality of vitamin D in diets of channel catfish (*Ictalurus punctatus*). *Transactions of the American Fisheries Society* 107: 809–811.

Lovell, R. T., and T. Limsuwan. 1982. Intestinal synthesis and dietary nonessentiality of vitamin B_{12} for *Tilapia nilotica*. *Transactions of the American Fisheries Society* 11: 485–490.

Luquet, P. 1971. Efficatite des proteins en relation evac leur taux d'incorporationdans l'alimentation de la truite arc-en-ciel. *Annales d'hydrobiologie* 2: 175–186.

Mangalik, A. 1986. Dietary energy requirements of channel catfish. Doctoral Dissertation. Auburn, Alabama: Auburn University.

Marsman, G. J. P., H. Gruppen, D. J. van Zuilichem, J. W. Resink and A. G. J. Voragen. 1995. The influence of screw configuration on the in vitro digestibility and protein solubility of soybean and rapeseed meals. *Journal of Food Engineering* 26(1): 13–28.

Mazid, A. M., Y. Tanaka, T. Katayama, A. M. Rahman, K. L. Simpson, and C. O. Chichester. 1979. Growth response of *Tilapia zilli* fingerlings fed isocaloric diets with variable protein levels. *Aquaculture* 18: 115–122.

Mbahinzirek, G. B., K. Dabrowski, K. J. Lee, D. El-Saidy, and E. R. Wisner. 2001. Growth, feed utilization and body composition of tilapia fed with cottonseed meal-based diets in a recirculating system. *Aquaculture Nutrition* 7: 189–200.

McDaniel, N. K., S. H. Sugiura, T. Kehler, J. W. Fletcher, R. M. Coloso, P. Weis, and R. P. Ferraris. 2005. Dissolved oxygen and dietary phosphorous modulate utilization and effluent partitioning of phosphorous in rainbow trout (*Oncorhynchus mykiss*) aquaculture. *Environmental Pollution* 138: 350–357.

McLaren, B. A., E. Keller, D. J. O'Donnell, and C. A. Elvehjem.1947. The nutrition of rainbow trout.1. Studies of vitamin requirements. *Archives of Biochemistry and Biophysics*. 15: 169–178.

Médale, F., C. Brauge, F. Vallée, and S. J. Kaushik. 1995. Effects of dietary protein/energy ratio, ration size, dietary energy source and water temperature on nitrogen excretion in rainbow trout. *Water Science Technology* 31: 185–194.

Mercier, C., P. Linko, and J. M. Harper. 1989. *Extrusion Cooking*. St. Paul, Minnesota: American Association of Cereal Chemists, Inc.

Moreno, J. J., and M. T. Mitjavila. 2003. The degree of unsaturation of dietary fatty acids and the development of atherosclerosis. *Journal of Nutritional Biochemistry* 14: 182–195.

Morito, C. L. H., D. H. Conrad, and J. W. Hilton.1986.The thiamin deficiency signs and requirements of rainbow trout (*Salmo gairdneri*, Richardson). *Fish Physiology and Biochemistry* 1: 93–104.

Morrow, M. D., D. Higgs, and C. J. Kennedy. 2004. The effects of diet composition and ration on biotransformation enzymes and stress parameters in rainbow trout, *Oncorhynchus mykiss*. *Comparative Biochemistry and Physiology, Part C* 137: 143–154.

Murai, T., and J. W. Andrews. 1974. Interactions of dietary α-tocopherol, oxidized menhaden oil and ethoxyquin on channel catfish (*Ictalurus punctatus*). *Journal of Nutrition* 104: 1416–1431.

Murai, T., and J. W. Andrews. 1977. Vitamin K and anticoagulant relationships in catfish diets. *Bulletin of the Japanese Society of Scientific Fisheries* 43: 785–794.

Murai, T., and J. W. Andrews. 1978a. Thiamin requirement of channel catfish fingerlings. *Journal of Nutrition* 108: 176–180.

Murai, T., and J. W. Andrews. 1978b. Riboflavin requirement of channel catfish fingerlings. *Journal of Nutrition* 108: 1512–1517.

Murai, T., and J. W. Andrews. 1979. Pantothenic acid requirement of channel catfish fingerlings. *Journal of Nutrition* 109: 1140–1142.

National Research Council [NRC]. 1983. *Nutrient Requirements of Warmwater Fishes and Shellfishes*. Washington DC: National Academy Press.

National Research Council [NRC]. 1993. *Nutrient Requirements of Fish*. Washington DC: National Academy Press.

Navarre, O., and J. E. Halver. 1989. Disease resistance and humoral antibody production in rainbow trout fed high levels of vitamin C. *Aquaculture* 79: 207–221.

Ng, W. K., G. Serrini, Z. Zhang, and R. P. Wilson. 1997. Niacin requirement and inability of tryptophan to act as a precursor of NAD^+ in channel catfish *Ictalurus punctatus*. *Aquaculture* 152: 273–285.

Noureddini, D. J. 2009. Degradation of phytates in distiller's grains and corn gluten feed by *Aspergillus niger* phytase. *Applied Biochemistry and Biotechnology* 159: 11–23.

O'Connell, J. P., and D. M. Gatlin. 1994. Effects of dietary calcium and vitamin D_3 on weight gain and mineral composition of the blue tilapia (*Oreochromis aureus*) in low-calcium water. *Aquaculture* 125: 107–117.

Ogino, C., and H. Takeda. 1978. Requirements of rainbow trout for dietary calcium and phosphorus. *Bulletin of the Japanese Society of Scientific Fisheries* 44: 1019–1022.

Ogino, C., and G. Y. Yang. 1979. Requirement of rainbow trout for dietary zinc. *Bulletin of the Japanese Society of Scientific Fisheries* 44: 1015–1018.

Ogino, C., and G. Y. Yang. 1980. Requirements of carp and rainbow trout for dietary manganese and copper. *Bulletin of the Japanese Society of Scientific Fisheries* 46: 455–458.

Ogino, C., F. Takashima, and J. Y. Chiou. 1978. Requirements of rainbow trout for dietary magnesium. *Bulletin of the Japanese Society of Scientific Fisheries* 44: 1105–1108.

Ovrum-Hansen, J., and T. Storebakken. 2007. Effects of dietary cellulose level on pellet quality and nutrient digestibilities in rainbow trout (*Oncorhynchus mykiss*). *Aquaculture* 272: 458–465.

Oyetayo, A. S., C. C. Thornburn, A. J. Matty, and A. Jackson. 1985. Pyridoxine and survival of tilapia (*Sarotherodon mossambicus Peters*). In *Proceedings of the 4th Annual Conference of the Fisheries Society of Nigeria (Fison)*, pp. 223–230. Nigeria: Port Harcourt.

Page, G. I., K. M. Hayworth, R. R. Wade, A. M. Harris, and D. P. Bureau. 1999. Non-specific immunity parameters and the formation of advanced glycosylation end-products (AGE) in rainbow trout, *Oncorhynchus mykiss* (Walbaum), fed high levels of dietary carbohydrates. *Aquaculture Research* 30: 287–297.

Page, J. W., and J. W. Andrews. 1973. Interactions of dietary levels of protein and energy on channel catfish (*Ictalurus punctatus*). *Journal of Nutrition* 103: 1339–1346.

Peres, H., C. Lim, and P. H. Klesius. 2004. Growth, chemical composition and resistance to Streptococcus iniae challenge of Nile tilapia (*Oreochromis niloticus*) fed graded levels of dietary inositol. *Aquaculture* 235: 423–432.

Piper, R. G., I. B. McElwain, L. E. Orme, J. P. McCraren, L. J. Fowler, and J. R. Leonard. 1982. *Fish Hatchery Management*. Washington, DC: U.S. Fish and Wildlife Service.

Phillips, A. M. 1949. Fisheries Research Bulletin No. 13, Cortland Hatchery Report No. 18, Cortland, NY.

Phillips, A. M., G. L. Hammer, J. P. Edwards, and H. F. Hosking. 1964a. Dry concentrates as complete trout foods for growth and egg production. *Progressive Fish-Culturist* 26: 155–159.

Phillips, A. M., Jr., H. A. Podoliak, H. A. Poston, D. L. Livingston, H. E. Booke, E. A. Pyle, and G. L. Hammer.1964b. The production of anemia in brown trout. *Fisheries Research Bulletin* 27: 66–70. Albany, New York: State of New York Conservation Department.

Poston, H. A., and G. L. Rumsey. 1983. Factors affecting dietary requirement and deficiency signs of L-trypotophan in rainbow trout. *Journal of Nutrition* 113: 2568–2577.

Poston, H. A., and M. J. Wolfe. 1985. Niacin requirement for optimum growth, feed conversion and protection for rainbow trout, *Salmo gairdneri* Richardson, from ultraviolet-B-irradiation. *Journal of Fish Diseases* 8: 451–460.

Puangkaew, J., V. Kiron, S. Satoh, and T. Watanabe. 2005. Antioxidant defense of rainbow trout (*Oncorhynchus mykiss*) in relation to dietary n-3 unsaturated fatty acids and vitamin E contents. *Comparative Biochemistry and Physiology, Part C.* 140: 187–196.

Quinn, T. P. 2005. *The Behavior and Ecology of Pacific Salmon and Trout*. Bethesda, Maryland: American Fisheries Society, pp. 19–20.

Rasco, B. A., A. E. Hashisaka, F. M. Dong, and M. A. Einstein. 1989. Sensory evaluation of baked foods incorporating different levels of distillers' dried grains with solubles from soft white winter wheat. *Journal of Food Science* 54: 337–342.

Reigh, R. C., E. H. Robinson, and P. B. Brown. 1991. Effects of dietary magnesium on growth and tissue magnesium content of blue tilapia, *Oreochromis aureus*. *Journal of the World Aquaculture Society* 22: 192–200.

Rider, S. A., S. J. Davies, A. N. Jha, A. A. Fisher, J. Knight, and J. W. Sweetman. 2009. Supra-nutritional dietary intake of selenite and selenium yeast in normal and stressed rainbow trout (*Oncorhynchus mykiss*): Implications on selenium status and health responses. *Aquaculture* 295: 282–291.

Rinchard, J., G. Mbahinzirek, K. Dabrowski, K. J. Lee, M. A. Garcia-Abiado, and J. Ottobre. 2002. Effects of dietary cottonseed meal protein level on growth, gonad development and plasma sex steroid hormones of tropical fish tilapia *Oreochromis* sp. *Aquaculture International* 20: 11–28.

Rinchard, J., S. Czesny, and K. Dabrowski. 2007. Influence of lipid class and fatty acid deficiency on survival, growth, and fatty acid composition in rainbow trout juveniles. *Aquaculture* 264: 363–371.

Robinson, E. 1990. Re-evaluation of the ascorbic acid (vitamin C) requirement of channel catfish (*Ictalurus punctatus*). *Federation of American Societies for Experimental Biology* 4: 3745 (Abstr.)

Robinson, E. H. 1991. Improvement of cottonseed meal protein with supplemental lysine in feeds for channel catfish. *Journal of Applied Aquaculture* 1: 1–14.

Robinson, E. H., and M. H. Li. 2002. Channel catfish, Ictalurus punctatus. In *Nutrient Requirements and Feeding of Finfish for Aquaculture*, eds. C. D. Webster, and C. E. Lim, pp. 293–318. Wallingford United Kigdom: CABI Publishing.

Robinson, E. H., and M. H. Li. 2007. *Catfish Protein Nutrition* (revised). Mississippi Agricultural and Forestry Experimental Station, Bulletin 1159. Stoneville, Mississippi.

Robinson, E. H., and M. H. Li. 2008. Replacement of soybean meal in channel catfish, *Ictalurus punctatus*, diets with cottonseed meal and distiller's dried grains with soluble. *Journal of the World Aquaculture Society* 39: 521–527.

Robinson, E. H., and R. T. Lovell. 1978. Essentiality of biotin for channel catfish *Ictalurus punctatus* fed lipid and lipid-free diets. *Journal of Nutrition* 108: 1600–1605.

Robinson, E. H., D. LaBomascus, P. B. Brown, and T. L. Linton. 1987. Dietary calcium and phosphorus requirements of *Oreochromis aureus* reared in calcium-free water. *Aquaculture* 64: 267–276.

Robinson, E. H., R. P. Wilson, and W. E. Poe. 1980a. Total aromatic amino acid requirement, phenylalanine requirement and tyrosine replacement value for fingerling channel catfish. *Journal of Nutrition* 110: 1805–1812.

Robinson, E. H., R. P. Wilson, and W. E. Poe. 1980b. Re-evaluation of the lysine requirement and lysine utilization by fingerling channel catfish. *Journal of Nutrition* 110: 2313–2316.

Robinson, E. H., R. P. Wilson, and W. E. Poe. 1981. Arginine requirement and apparent absence of a lysine-arginine antagonist in fingerling channel catfish. *Journal of Nutrition* 111: 46–52.

Roem, A. J., C. C. Kohler, and R. R. Stickney. 1990a. Inability to detect a choline requirement for the blue tilapia, *Oreochromis aureus*. *Journal of the World Aquaculture Society* 21: 238–240.

Roem, A. J., C. C. Kohler, and R. R. Stickney. 1990b. Vitamin E requirements of the blue tilapia, *Oreochromis aureus* (Steindachner), in relation to dietary lipid level. *Aquaculture* 87: 115–164.

Roem, A. J., C. C. Kohler, and R. R. Stickney. 1990c. Vitamin requirements of blue tilapias in a recirculating water system. *The Progressive Fish-Culturist* 52: 15–18.

Roem, A. J., R. R. Stickney, and C. C. Kohler. 1991. Dietary pantothenic acid requirement of blue tilapia. *The Progressive Fish-Culturist* 53: 216–219.

Rodehutscord, M., and E. Pfeffer. 1999. Maintenance requirement for digestible energy and efficiency of utilisation of digestible energy for retention in rainbow trout, *Oncorhynchus mykiss*. *Aquaculture* 179: 95–107.

Rolfe, L. A., H. E. Huff, and F. Hsieh. 2001. Effects of particle size and processing variables on the properties of an extruded catfish feed. *Journal of Aquatic Food Product Technology* 10(3): 21–33.

Rosentrater, K. A. 2006. Some physical properties of distillers dried grains with solubles (DDGS). *Applied Engineering in Agriculture* 22: 589–595.

Rosentrater, K. A., and K. Muthukumarappan. 2006. Corn ethanol coproducts: Generation properties, and future prospects. *International Sugar Journal* 108: 648–657.

Rosentrater, K. A., K. Muthukumarappan, and S. Kannadhason. 2009a. Effects of ingredients and extrusion parameters on aquafeeds containing DDGS and corn starch. *Journal of Aquaculture Feed Science and Nutrition* 1(2): 44–60.

Rosentrater, K. A., K. Muthukumarappan, and S. Kannadhason. 2009b. Effects of ingredients and extrusion parameters on aquafeeds containing DDGS and potato starch. *Journal of Aquaculture Feed Science and Nutrition* 1(1): 22–38.

Ruohonen, K., J. Vielma, and D. J. Grove. 1998. Growth and food utilisation of rainbow trout (*Oncorhynchus mykiss*) fed low-fat herring and dry diets enriched with fish oil. *Aquaculture* 163: 275–283.

Rumsey, G. L. 1991. Choline-betaine requirements of rainbow trout (*Oncorhynchus mykiss*). *Aquaculture* 95: 107–116.

Rumsey, G. L., J. W. Page, and M. L. Scott. 1983. Methionine and cystine requirements of rainbow trout. *Progressive Fish-Culturist* 45: 139–143.

Saleh, G., W. Eleraky, and J. M. Gropp. 1995. A short note on the effects of vitamin A hypervitaminosis and hypovitaminosis on health and growth of *Tilapia nilotica* (*Oreochormis niloticus*). *Journal of Applied Ichthyology* 11: 382–385.

Sandra, H. P. F., and A. G. A. Jose. 1993. Effect of phospholipid on protein structure and solubility in the extrusion of lung proteins. *Food Chemistry* 47: 111–119.

Santiago, C. B., and M. A. Laron. 1991. Growth response and carcass composition of redbelly tilapia fry fed diets with varying protein levels and protein to energy ratio. In *Fish Nutrition Research in Asia,* ed. S. S. De Silva, pp. 55–62. Manila, Philippines: Asian Special Publication No. 5.

Santiago, C. B., and R. T. Lovell. 1988. Amino acid requirements for growth of Nile tilapia. *Journal of Nutrition* 118: 1540–1546.

Satia, B. P. 1974. Quantitative protein requirements of rainbow trout. *Progressive Fish-Culturist* 36: 80–85.

Satoh, S., M. J. Apines, T. Tsukioka, V. Kiron, T. Watanabe, and S. Fujita. 2001. Bioavailability of amino acids chelated and glass embedded manganese to rainbow trout, *Oncorhynchus mykiss*, fingerlings. *Aquaculture Research* 32S: 18–25.

Satoh, S., W. E. Poe, and R. P. Wilson. 1989. Effect of dietary n-3 fatty acids on weight gain and liver polar lipid fatty acid composition of fingerling channel catfish. *Journal of Nutrition* 119: 23–28.

Satoh, S., T. Takeuchi, and T. Watanabe. 1987. Requirement of tilapia for alpha tocopherol. *Nippon Suisan Gakkaishi* 53: 119–124.

Satoh, S., K. Izume, K. Takeuchi, and T. Watanabe. 1987a. Availability to rainbow trout of zinc contained in various types of fish meals. *Nippon Suisan Gakkaishi* 53: 1861–1866.

Satoh, S., K. Tabata, K. Izume, K. Takeuchi, and T. Watanabe. 1987b. Effect of dietary tri-calcium phosphate on availability of zinc to rainbow trout. *Nippon Suisan Gakkaishi* 53: 1199–1205.

Satoh, S., K. Takeuchi, and T. Watanabe. 1987c. Availability to rainbow trout of zinc in white fish meal and of various zinc compounds. *Nippon Suisan Gakkaishi* 53: 595–599.

Schaeffer, T. W. 2009. Performance of Nile tilapia and yellow perch fed diets containing distillers dried grain with solubles and extruded diet characteristics. M.S. Thesis. Brookings, SD: South Dakota State University.

Serrini, G., Z. Zhang, and R. P. Wilson. 1996. Dietary riboflavin requirement of fingerling channel catfish (*Ictalurus punctatus*). *Aquaculture* 139: 285–290.

Shearer, K. D. 1989. Whole body magnesium concentration as an indicator of magnesium status in rainbow trout (*Salmo gairdneri*). *Aquaculture* 77: 201–210.

Shelby, R. A., C. Lim, M. Yildrem-Aksoy, and P. H. Klesius. 2008. Effect of distillers dried grains with soluble incorporated diets on growth, immune function and disease resistance in Nile tilapia (*Oreochromis niloticus* L.). *Aquaculture* 39: 1351–1353.

Shiau, S. Y., and Y. H. Chin. 1999. Estimation of the dietary biotin requirement of juvenile hybrid tilapia. *Oreochromis niloticus × O. aureus. Aquaculture* 170: 71–78.

Shiau, S. Y., and H. L. Hsieh. 1997. Vitamin B sub(6) requirements of tilapia *Oreochromis niloticus × O. aureus* fed two dietary protein concentrations. *Fisheries Sciences* 63: 1002–1007.

Shiau, S. Y., and J. F. Hsieh. 2001. Quantifying the dietary potassium requirement of juvenile hybrid tilapia (*Oreochromis niloticus × O. aureus*). *British Journal of Nutrition* 85: 213–218.

Shiau, S. Y., and S. L. Huang. 1990. Influence of varying levels with two protein concentrations in diets for hybrid tilapia (*Oreochromis niloticus × O. aureus*) reared in seawater. *Aquaculture* 91: 143–152.

Shiau, S. Y., and S. L. Huang. 2001. Dietary folic acid requirement for maximal growth of juvenile tilapia, *Oreochromis niloticus × O. aureus. Fisheries Science* 67: 655–659.

Shiau, S. Y., and J. Y. Hwang. 1993. Vitamin D requirements of juvenile hybrid tilapia, *Oreochromis niloticus × O. aureus. Nippon Suisan Gakkaishi* 59: 553–558.

Shiau, S. Y., and F. L. Jan. 1992. Dietary ascorbic acid requirement of juvenile tilapia *Oreochromis niloticus × O. aureus. Nippon Suisan Gakkaishi* 58: 671–675.

Shiau, S. Y., and Y. H. Lin. 2006. Vitamin requirements of tilapia—A review. In *Advances in Aquaculture Nutrition VIII*, eds. L. E. C. Suárez, D. R. Marie, M. T. Salazar, M. G. N. López, D. A. V. Cavazos, A. C. P. Cruz, and A. G. Ortega, pp. 129–138. Monterrey, Mexico: Aquaculture Nutrition International, Symposium 8.

Shiau, S. Y., and P. S. Lo. 2000. Dietary choline requirements of juvenile hybrid tilapia, *Oreochromis niloticus × O. aureus. Journal of Nutrition* 130: 100–103.

Shiau, S. Y., and C. Q. Lung. 1993. No dietary vitamin B_{12} required for juvenile tilapia *Oreochromis niloticus × O. aureus. Comparative Biochemistry and Physiology* 105: 147–150.

Shiau, S. Y., and L. F. Shiau. 2001. Re-evaluation of the vitamin E requirements of juvenile tilapia (*Oreochromis niloticus × O. aureus*). *Animal Science* 72: 529–534.

Shiau, S. Y., and L. W. Su. 2003. Ferric sulfate is half effective as ferrous sulfate in meeting iron requirement for juvenile tilapia, *Oreochromis niloticus + O. aureus*. In *Book of Abstracts*, p. 719. World Aquaculture, Salvador, Brazil. Baton Rouge, Louisiana: World Aquaculture Society.

Shiau, S. Y., and S. L. Su. 2005. Juvenile tilapia (*Oreochromis niloticus × O. aureus*) requires dietary *myo*-inositol for maximal growth. *Aquaculture* 243: 273–277.

Shiau, S. Y., and G. S. Suen. 1992. Estimation of the niacin requirement for tilapia fed diets containing glucose or dextrin. *Journal of Nutrition* 122: 2030–2036.

Shiau, S. Y., J. L. Chuang, and G. L. Sun. 1987. Inclusion of soybean meal in tilapia (*Oreochromis niloticus × O. aureus*) diets at two protein levels. *Aquaculture* 65: 251–261.

Shiau, S., S. Lin, S. Yu, A. Lin, and C. Kwok. 1990. Defatted and full-fat soybean meal as partial replacements for fish meal in tilapia (*Oreochromis niloticus × O. aureus*) diets at low protein level. *Aquaculture* 86: 401–407.

Siddiqui, A. Q., M. S. Howlander, and A. A. Adam. 1988. Effects of dietary protein levels on growth, feed conversion and protein utilization in fry and young Nile tilapia, *Oreochromis niloticus. Aquaculture* 70: 63–73.

Singh, R. P., and T. Nose. 1967. Digestibility of carbohydrates in young rainbow trout. *Bulletin of Freshwater Fisheries Research Laboratory* (Japan) 17: 21–25.

Singh, R. K., S. S. Nielsen, and J. V. Chambers. 1991. Selected characteristics of extruded blends of milk protein raffinate or nonfat dry milk with corn flour. *Journal of Food Processing and Preservation* 15: 285–302.

Sinnhuber, R. O. 1964. Pelleted fish food. *Feedstuffs* 36(28): 16.

Sklan, D., T. Prag, and I. Lupatsch. 2004. Apparent digestibility coefficients of feed ingredients and their prediction in diets for tilapia *Oreochromis niloticus* × *Oreochromis aureus* (Telepstie, Cichlidae). *Aquaculture Research* 35: 358–364.

Smith, R. R., M. C. Peterson, and C. Allred. 1980. Effect of leaching on apparent digestion coefficients of feedstuffs for salmonids. *Progressive Fish-Culturist* 42: 195–199.

Soliman, A. K., and R. P. Wilson. 1992a. Water-soluble vitamin requirements of tilapia. 1. Pantothenic acid requirement of blue tilapia, *Oreochromis aureus. Aquaculture* 104: 121–126.

Soliman, A. K., and R. P. Wilson. 1992b. Water-soluble vitamin requirements of tilapia. 2. Riboflavin requirement of blue tilapia, *Oreochromis aureus. Aquaculture* 104: 309–314.

Soliman, A. K., K. Jauncey, and R. J. Roberts. 1994. Water-soluble vitamin requirements of tilapia: ascorbic acid (vitamin C) requirement of Nile tilapia, *Oreochromis niloticus* (L.). *Aquaculture and Fisheries Management* 25: 269–278.

Spiehs, M. J., M. H. Whitney, and G. C. Shurson. 2002. Nutrient database for distiller's dried grains with solubles produced from new ethanol plants in Minnesota and South Dakota. *Journal of Animal Science* 80: 2639–2645.

Steffens, W., B. Rennert, M. Wirth, and R. Krüger. 1999. Effect of two lipid levels on growth, feed utilization, body composition an some biochemical parameters of rainbow trout, *Oncorhynchus mykiss* (Walbaum 1792). *Journal of Applied Ichthyology* 12: 159–164.

Stickney, R. R. 1996. *Aquaculture in the United States: A Historical Review*. Wiley, New York.

Stickney, R. R., and R. B. McGeachin. 1983. Responses of *Tilapia aurea* to semipurified diets of differing fatty acid composition. In *Proceedings of the International Symposium on Tilapia in Aquaculture*, eds. L. Fishelson, and Z. Yaron, pp. 346–355. Tel Aviv, Israel: Tel Aviv University.

Stickney, R. R., R. B. McGeachin, D. H. Lewis, J. Marks, R. S. Sis, E. H. Robin, and W. Wurts. 1984. Response of *Tilapia aurea* to dietary vitamin C. *Journal of the World Aquaculture Society* 15: 179–185.

Stone, D. A., R. W. Hardy, F. T. Barrows, and Z. J. Cheng. 2005. Effects of extrusion on nutritional value of diets containing corn gluten meal and corn distiller's dried grain for rainbow trout *Oncorhynchus mykiss. Journal of Applied Aquaculture* 17: 1–20.

Storebakken, T., K. D. Shearer, S. Refstie, S. Lagocki, and J. McCool. 1998. Interactions between salinity, dietary carbohydrate source and carbohydrate concentration on the digestibility of macronutrients and energy in rainbow trout (*Oncorhynchus mykiss*). *Aquaculture* 163: 347–359.

Tacon, A. G. J., and M. Metian. 2008. Global overview of the use of fish meal and fish oil in industrially compounded aquafeeds: Trends and future prospects. *Aquaculture* 285: 146–158.

Takeuchi, T., and T. Watanabe. 1977. Dietary levels of methyl laurate and essential fatty acid requirement of rainbow trout. *Bulletin of the Japanese Society of Scientific Fisheries* 43: 893–898.

Takeuchi, T., and T. Watanabe. 1979. Effect of excess amounts of essential fatty acids on growth of rainbow trout. *Bulletin of the Japanese Society of Scientific Fisheries* 45: 977–982.

Takeuchi, T., S. Satoh, and W. Watanabe. 1983a. Dietary lipid suitable for practical feeding of *Tilapia nilotica. Bulletin of the Japanese Society of Scientific Fisheries* 49: 1361–1365.

Takeuchi, T., S. Satoh, and W. Watanabe. 1983b. Requirement of *Tilapia nilotica* for essential fatty acids. *Bulletin of the Japanese Society of Scientific Fisheries* 49: 1127–1134.

Takeuchi, L., and T. Takeuchi, and C. Ogino. 1980. Riboflavin requirements in carp and rainbow trout. *Bulletin of the Japanese Society of Scientific Fisheries* 46: 733–737.

Takeuchi, T., T. Watanabe, and C. Ogino. 1978. Optimum ratio of protein to lipid in diets of rainbow trout. *Bulletin of the Japanese Society of Scientific Fisheries* 44: 683–685.

Thomas, M., T. H. J. H. Paul, V. V. Ton, J. V. Z. Dick, and F. B. V. P. Antonius. 1999. Effect of process conditions during expander processing and pelleting on starch modification and pellet quality of tapioca. *Journal of the Science of Food and Agriculture* 79: 1481–1494.

Thurston, R. V., and R. C. Russo. 1983. Acute toxicity or ammonia to rainbow trout. *Transactions of the American Fisheries Society* 112: 696–704.

Tidwell, J. H., C. D. Webster, and D. H. Yancey. 1990. Evaluation of distillers grains with solubles in prepared channel catfish diets. *Transactions of Kentucky Academy of Science* 51: 135–138.

Trenzado, C. E., M. de la Higuera, and A. E. Morales. 2007. Influence of dietary vitamins E and C and HUFA on rainbow trout (*Oncorhynchus mykiss*) performance under crowding conditions. *Aquaculture* 263: 249–258.

Trenzado, C. E., A. E. Morales, and M. de la Higuera. 2008. Physiological changes in rainbow trout held under crowded conditions and fed diets with different levels of vitamins E and C and highly unsaturated fatty acids (HUFA). *Aquaculture* 277: 293–302.

USDA-NASS (National Agriculture Statistics Service). 2004. *2002 Census of Aquaculture,* Vol. 3, Special Studies Part II. Washington, DC: U.S. Department of Agriculture. Available online: www.agcensus.usda. gov/Publications/2002/index.asp. Accessed July 10, 2010.

USDA-NASS (National Agricultural Statistics Service). 2007. *Trout Production.* Washington, DC: United States Department of Agriculture. Available online: www.agcensus.usda.gov/Publications/2007/index. asp. Accessed July 10, 2010.

Verlhac, V., and J. Gabaudan. 1994. Influence of vitamin C on the immune system of salmonids. *Aquaculture and Fisheries Management* 25: 21–36.

Verlhac, V., J. Gabadudan, A. Obach, W. Schuep, and R. Hole. 1996. Influence of dietary glucan and vitamin C on non-specific and specific immune responses of rainbow trout (*Oncorhynchus mykiss*). *Aquaculture* 143: 132–133.

Wahli, T., R. Frischknecht, M. Schmitt, J. Gabaudan, V. Verlhac, and W. Meier. 2006. A comparison of the effect of silicone coated ascorbic acid and ascorbyl phosphate on the course of ichthyophthirirosis in rainbow trout, *Oncorhynchus mykiss* (Walbaum). *Journal of Fish Diseases* 18: 347–355.

Walton, M. J., C. B. Cowey, and J. W. Adron. 1984a. The effect of dietary lysine levels on growth and metabolism of rainbow trout (*Salmo gairdneri*). *British Journal of Nutrition* 52: 115–122.

Walton, M. J., R. M. Coloso, C. B. Cowey, J. W. Adron, and D. Knox. 1984b. The effects of dietary tryptophan levels on growth and metabolism of rainbow trout (*Salmo gairdneri*). *British Journal of Nutrition* 51: 279–287.

Walton, M. J., C. B. Cowey, R. M. Coloso, and J. W. Adron. 1986. Dietary requirements of rainbow trout for tryptophan, lysine and arginine determined by growth and biochemical measurements. *Fish Physiology and Biochemistry* 2: 161–169.

Watanabe, T., V. Kiron, and S. Satoh. 1997. Trace minerals in fish nutrition. *Aquaculture* 151: 185–207.

Watanabe, T., S. Satoh, and T. Takeuchi. 1988. Availability of minerals in fish meal to fish. *Asian Fisheries Science* 1: 75–195.

Watanabe, T., F. Takeuchi, M. Wada, and R. Vehara. 1981. The relationship between dietary lipid levels and α-tocopherol requirement of rainbow trout. *Bulletin of the Japanese Society of Scientific Fisheries* 47: 1463–1471.

Watanabe, T., T. Takeuchi, M. Saito, and K. Nishimura. 1984. Effect of low protein-high calorie or essential fatty acid deficient diet on reproduction of rainbow trout. *Bulletin of the Japanese Society of Scientific Fisheries* 50: 1207–1215.

Watanabe, T., T. Takeuchi, S. Satoh, and V. Kiron. 1996a. Digestible crude protein contents of various feedstuffs determined with four freshwater species. *Fisheries Science* 62: 278–282.

Watanabe, T., T. Takeuchi, S. Satoh, and V. Kiron. 1996b. Methodological influences and mode of calculation. *Fisheries Science* 62: 288–292.

Webster, C. D., J. H. Tidwell, and D. H. Yancey. 1991. Evaluation of distiller's grains with solubles as a protein source in diets for channel catfish. *Aquaculture* 96: 179–190.

Webster, C. D., J. H. Tidwell, and L. S. Goodgame. 1993. Growth, body composition, and organoleptic evaluation of channel catfish fed diets containing different percentages of distillers' grains with solubles. *The Progressive Fish-Culturist* 55: 95–100.

Webster, C. D., J. H. Tidwell, L. S. Goodgame, D. H. Yancey, and L. Mackey. 1992a. Use of soybean meal and distillers grains with solubles as partial or total replacement of fish meal in diets for channel catfish, *Ictalurus punctatus*. *Aquaculture* 106: 301–309.

Webster, C. D., J. H. Tidwell, L. S. Goodgame, J. A. Clark, and D. H. Yancey. 1992b. Winter feeding and growth of channel catfish fed diets containing varying percentages of distillers grains with solubles as a total replacement of fish meal. *Journal of Applied Aquaculture* 1: 1–4.

Wee, K. L., and L. T. Ng. 1986. Use of cassava as an energy source in pelleted feed for the tilapia, *Oreochromis niloticus* L. *Aquaculture and Fisheries Management* 17: 129–138.

Westers, H. 2001. Production. In *Fish Hatchery Management,* 2nd ed., pp. 31–90. Bethesda, Maryland: American Fisheries Society.

Wilson, R. P. 1989. Amino acid and proteins. In *Fish Nutrition,* 2nd ed., ed. J. E. Halver, pp. 111–151. San Diego, California: Academic Press.

Wilson, R. P., and G. El Naggar. 1992. Potassium requirement of fingerling channel catfish, *Ictalurus puncta-tus*. *Aquaculture* 108: 169–175.

Wilson, R. P., and W. E. Poe. 1985. Effects of feeding soybean meal with varying trypsin inhibitor activities on growth of fingerling channel catfish. *Aquaculture* 46: 19–25.

Wilson, R. P., and W. E. Poe. 1988. Choline nutrition of fingerling channel catfish. *Aquaculture* 68: 65–71.

Wilson, R. P., P. R. Bowser, and W. E. Poe. 1983. Dietary pantothenic acid requirement of fingerling channel catfish. *Journal of Nutrition* 113: 2124–2134.

Wilson, R. P., P. R. Bowser, and W. E. Poe. 1984. Dietary vitamin E requirement of fingerling channel catfish. *Journal of Nutrition* 114: 2053–2058.

Wilson, R. P., D. E. Harding, and D. L. Garling, Jr. 1977. Effect of dietary pH on amino acid utilization and the lysine requirement of fingerling channel catfish. *Journal of Nutrition* 107: 166–177.

Wilson, R. P., O. W. Allen, Jr., E. H. Robinson, and W. E. Poe. 1978. Tryptophan and threonine requirements of fingerling channel catfish. *Journal of Nutrition* 108: 1595–1599.

Wilson, R. P., W. E. Poe, and E. H. Robinson. 1980. Leucine, isoleucine, valine and histidine requirements of fingerling channel catfish. *Journal of Nutrition* 110: 627–633.

Winfree, R. A., and R. R. Stickney. 1981. Effect of dietary protein and energy on growth, feed conversion efficiency and body composition of *Tilapia aurea*. *Journal of Nutrition* 111: 1001–1012.

Winfree, R. A., and R. R. Stickney. 1984. Starter diets for channel catfish: effect of dietary protein on growth and carcass composition. *The Progressive Fish-Culturist* 46: 79–86.

Woodall, A. N., L. M. Ashley, J. E. Halver, H. S. Olcott, and J. van der Veen. 1964. Nutrition of salmonid fishes. XIII. The α-tocopherol requirement of chinook salmon. *Journal of Nutrition* 84: 125–135.

Woodward, B. 1990. Dietary vitamin B6 requirements of young rainbow trout (*Oncorhynchus mykiss*). *Federation of American Societies for Experimental Biology Journal* 4: 3748 (Abstr.)

Woodward, B., and M. Frigg. 1989. Dietary biotin requirements of young trout (*Salmo gairdneri*) determined by weight gain hepatic biotin concentration and maximal biotin-dependent enzyme activities in liver and white muscle. *Journal of Nutrition* 110: 54–60.

Wu, Y. V., R. R. Rosati, D. J. Sessa, and P. B. Brown. 1994. Utilization of protein-rich ethanol co-products from corn in tilapia feed. *Journal of American Oil Chemists Society* 71: 1041–1043.

Wu, Y. V., R. R. Rosati, and P. B. Brown. 1996. Effect of diets containing various levels of protein and ethanol coproducts from corn on growth of tilapia fry. *Journal of Agricultural and Food Chemistry* 44: 1491–1493.

Wu, Y. V., R. R. Rosati, and P. B. Brown. 1997. Use of corn-derived ethanol products and synthetic lysine and tryptophan for growth of tilapia (*Oreochromis niloticus*) fry. *Journal of Agricultural and Food Chemistry* 45: 2174–2177.

Yigit, M., O. Yardim, and S. Koshlo. 2002. The protein sparing effects of high lipid levels in diets for rainbow trout (*Oncorhynchus mykiss*, w. 1792) with special reference to reduction of total nitrogen excretion. *Israeli Journal of Aquaculture Bamidgeh* 54: 79–88.

Zhou, P., W. Zhang, D. A. Davis, and C. Lim. 2010. Growth response and feed utilization of juvenile hybrid catfish fed diets containing distiller's grains with soluble to replace a combination of soybean meal and corn meal. *North American Journal of Aquaculture* 72: 298–303.

17 Feeding DDGS to Other Animals

Kurt A. Rosentrater

CONTENTS

17.1 INTRODUCTION

Historically distillers dried grains with solubles (DDGS) has primarily been fed to beef, dairy, swine, and poultry animals (Chapters 12 through 15). It also appears to be a viable ingredient for aquafeeds (Chapter 16). It should be appropriate as a feed ingredient for other animals as well. There have only been a few published studies to date, however, on the efficacy of multispecies feeding of DDGS (or other ethanol coproducts). These limited reports include sheep and companion animals (but only horses and dogs). So in terms of conclusive feeding recommendations for these animals, there are still many unknowns. No book on DDGS would be complete without at least a brief overview of this information.

17.2 SHEEP

There were approximately 5.6 million sheep in the United States at the beginning of 2010. Historically, they have been raised primarily for wool and meat, and in recent years their use for leather has increased as well (ERS, 2010). Due to declining wool revenues, though, the number of sheep in the United States has been steadily declining since the late 1800s (Figure 17.1), especially during the latter half of the twentieth century. Even so, they are still an important segment of our agricultural economy. As sheep are ruminants, they can readily digest feeds that are high in fiber, such as forages and DDGS.

To date, there have been only a few studies published on the use of DDGS (or any other ethanol coproducts) in sheep diets (Table 17.1 provides a brief synopsis of these). These previous studies have examined the use of coproducts up to 100% of the diet: some used ad libitum; others used specific feeding levels. Most of these studies, however, have examined the use of DDGS and DDG only as a minor dietary constituent, not a primary component. Furthermore, the majority of these studies have not quantified the efficacy of DDGS, but have focused on animal metabolism and nutrient flow through the digestive system. While some have examined protein, amino acids, and energy, as well as rumen function (including pH and volatile fatty acid levels), other have determined digestibility and degradability of various chemical components. Overall, these studies have agreed that DDGS does effectively serve as a protein supplement when used with diets that primarily rely on low- to

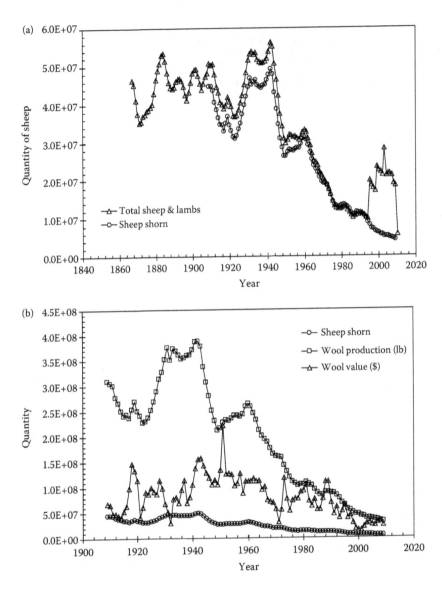

FIGURE 17.1 Historic U.S. sheep data: (a) sheep numbers; (b) wool production. (From NASS, *National Statistics—Sheep and Lambs*. National Agricultural Statistics Service, United States Department of Agriculture: Washington, DC, 2010.)

moderate-quality roughages (i.e., wheat straw, corn stover, bromegrass hay, switchgrass, silage) or cereal grains such as barley or corn.

Only five studies have specifically examined the efficacy of coproducts as protein sources in sheep diets (Vipond et al., 1995; Held, 2006; Huls et al., 2006; Flores et al., 2008; Gutierrez Zetina et al., 2009). Only a few of these though, have provided reliable feeding recommendations. For growing lambs, an inclusion level of corn-based DDGS of 15% has been recommended (Gutierrez Zetina et al., 2009); 1 kg of malt DDG/day/10 kg live weight has been recommended for lamb finishing diets, but this should be supplemented with 0.3 kg/day of cereal grain (Vipond et al., 1995). For pregnant ewes, recommendations are a maximum of 1 kg of malt DDG/day/25 kg live weight (Vipond et al., 1995). The malt DDG in these studies was a coproduct of a whiskey distillery. Held (2006), on the other hand, recommends for ewe rations corn-based DDGS inclusion of up to 20% in

TABLE 17.1

Overview of Work Published to Date on Use of Ethanol Coproducts in Sheep Diets

Animal Type	Coproduct Inclusion (%)	Crude Protein (%)	NDF (%)	ADF (%)	Energy (kcal/g)	Average Daily Gain (g/day)	Reference
Primary protein source							
Lambs, growing, mixed, 25 kg	0 15 30 (corn DDGS)					284 285 221	Gutierrez Zetina et al. (2009)
Lambs, growing, unspecified, 33 kg, 43 kg	7.5 15.0 (corn DDGS)						Flores et al. (2008)
Lambs, finishing, mixed, 35 kg	Up to 100 (malt DWG)	23.5	61.8	24.7	2.63	100–206	Vipond et al. (1995)
Lambs, finishing, wethers, 43 kg	22.9	14.6			3.4	290	Huls et al. (2006)
Ewes, pregnant, 73 kg	Up to 100 (malt DWG)	23.2	63.9	27.1	2.72		Vipond et al. (1995)
Ewes	Up to 20%						Held (2006)
Nutrient supplement only							
Lambs, growing rams, 25–28 kg	0 7.2 16.4 25.6 32.8 (malt DDG)	5.9 13.8 15.1 16.3 24.4	34.6 37.5 40.9 44.6 55.5			182–193	Richardson et al. (2003)
Lambs, finishing, unspecified, 31 kg	40 (corn DDGS)	17.3	26.0			310	Lodge et al. (1997)
Lambs, growing rams, 33 kg	0 9.7 11.3 19.4 (corn DDG)	10.9 12.9 12.8 12.0				15–26	Willms et al. (1991)
Lambs, growing, wethers, 35 kg	12.8 (corn DDG)	26.0– 30.1					Nakamura et al. (1994)

continued

TABLE 17.1 (continued)
Overview of Work Published to Date on Use of Ethanol Coproducts in Sheep Diets

		Diet Composition					
Animal Type	Coproduct Inclusion (%)	Crude Protein (%)	NDF (%)	ADF (%)	Energy (kcal/g)	Average Daily Gain (g/day)	Reference
Lambs, growing, wethers, 57 kg	100 g/day (corn DDGS)	29.5			3.11		Archibeque et al. (2008)
Lambs, growing, wethers, 61 kg	0		47.4	30.9			Willms et al. (1991)
	10.2		52.4	31.8			
	13.3		53.0	32.1			
	21.9		55.1	33.4			
	30.5 (corn DDG)		55.9	34.1			
Sheep, rams, 77 kg	0	26.8	46.4	11.8	4.66		Carvalho et al. (2005)
	15						
	30						
	45 (corn DDG)						
Sheep, wethers, 79 kg	16.4	14.6			2.48		Richardson et al. (2003)
	6.0 (malt DDG)	14.5			2.41		

Note: DDG is distillers dried grains; DDGS is distillers dried grains with solubles; DWG is distillers wet grains.

a forage-based diet (ad libitum or limit fed). Furthermore, Huls et al. (2006) determined that using up to 22.9% DDGS in a grain-based diet as partial replacement for corn and soybean meal had no negative consequences in terms of growth performance, muscle yield, and dressing, nor did it result in acidosis or bloating.

As with other livestock (and as discussed in previous chapters), when the level of DDGS in sheep diets increases past an optimum level, which has not yet been determined, the growth performance may decline (i.e., slower weight gain, lower dry matter intake, etc.) compared to control diets. Successful use of DDGS, of course, will depend upon the other dietary ingredients that are used in entire diets. Definitive research is yet to be done, and hopefully will be forthcoming in the near future. Additionally, sheep are particularly sensitive to urinary calculi, and with the high phosphorus content in DDGS, the risk for this development may be increased.

17.3 DOGS AND CATS

Many Americans have pets, especially dogs and cats. Petfood thus constitutes a potential market for many feed ingredients, including DDGS. In 2008, for example, U.S. petfood sales were over $17 billion (Pet Food Institute, 2010). According to the American Veterinary Medical Association (AVMA, 2007), there were approximately 72 million dogs, in more than 43 million households in the United States in 2007, and there were nearly 82 million cats in approximately 37.5 million homes.

No published studies have reported to date the use DDGS in cat foods. But there has been one study published on the use of DDGS in dog diets (Table 17.2). Allen et al. (1981) found that low

TABLE 17.2

Overview of Work Published to Date on Use of Ethanol Coproducts in Canine Diets

Animal Type	DDGS Inclusion (%)	Diet Composition			Digestibly of Dry Matter (%)	Reference
		Crude Protein (%)	ADF (%)	Energy (kcal/g)		
Adult,	0				83.6	Allen et al. (1981)
2 years	4				83.4	
12 female,	6				82.3	
19.4 kg	8				82.1	
	(DDGS)					
Adult,	0				84.8	Allen et al. (1981)
2 years	8.9				83.6	
12 female,	15.7				79.9	
19.4 kg	(DDGS)					
Adult,	0	24.4		4.9	74.9	Allen et al. (1981)
1.5 years	13.1	24.3		5.0	74.0	
12 female,	26.1	24.0		5.0	69.6	
17.8 kg	(DDGS)					
Puppies,	0	15.6	3.3	4.8	85.7	Allen et al. (1981)
5 months	14.1	16.3	8.6	4.9	81.2	
12 female,	(DDGS)					
7.1 kg						

Note: DDGS is distillers dried grains with solubles.

levels of distillery-based DDGS (i.e., up to 8%) resulted in no negative effects vis-à-vis dry matter and starch digestibility. At DDGS levels greater than 16.1%, however, dry matter digestibility and energy digestibility decreased, but protein digestibility did not appear to be impacted.

As discussed in de Godoy et al. (2009), challenges associated with using DDGS in petfoods include high fiber content, amino acid deficiencies (especially lysine, arginine, and other essential amino acids), and compositional variability (as discussed in other chapters). And, as with traditional livestock, potential mycotoxin contamination is also a concern. Until these issues are overcome (perhaps with some of the new production processes, such as fractionation), the use of DDGS in commercial petfoods will likely be relatively low.

17.4 HORSES

According to AVMA (2007), there were approximately 7.3 million horses in the United States in 2007 in more than 2 million households across the nation. The number of horses and ponies in the United States declined considerably since the advent of the automobile and farm tractor (Figure 17.2); with numbers rising again since the 1990s. This has been due, in large measure, to the increased use of horses for recreation.

Even though horses have the ability to digest high-fiber feed materials, there have been only a few studies published to date on the use of DDGS (Table 17.3). These studies have focused primarily on either bluegrass hay (Leonard et al., 1975) or alfalfa hay (Bonoma et al., 2008) based diets, and have examined the potential to partially replace corn and/or soybean meal feed supplements with DDGS (but not to use DDGS as the major feed constituent). Distillery-based DDGS appeared to increase feed palatability, at least up to a 20% inclusion level (Pagan, 1991), and distillery-based

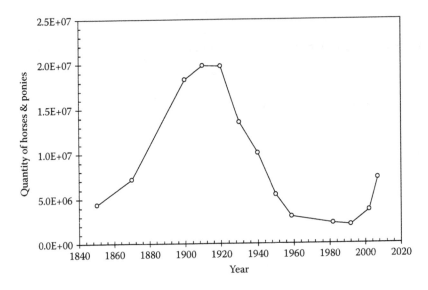

FIGURE 17.2 Historic U.S. horse and pony data. (From Kentucky Equine Research, *Equine Review, HW 37*, 2007.)

TABLE 17.3
Overview of Work Published to Date on Use of Ethanol Coproducts in Equine Diets

		Diet Composition						
Animal Type	**DDGS Inclusion (%)**	**Crude Protein (%)**	**NDF (%)**	**ADF (%)**	**Energy (kcal/g)**	**Digestibly of Dry Matter (%)**	**Average Daily Gain (g/day)**	**Reference**
Weanlings, 12 fillies, 4 colts, 276 kg	15 (corn DDGS)	16.3	32.9	22.2	2.60	51.09	610	Bonoma et al. (2008)
4 mature geldings, 400 kg	0 5 10 (DDGS)	7.5 7.5 7.4			1.84 1.85 1.75	44.05 40.05 41.39		Leonard et al. (1975)
12 mature horses, 460 kg	0 9.1 18.2 (DDGS)	9.2 9.7 11.5			1.91 1.93 1.59	43.01 42.14 41.54		Leonard et al. (1975)
Varying ages	0 5 10 20 (DDGS)							Pagan (1991)

Note: DDGS is distillers dried grains with solubles.

DDGS appeared to stimulate cellulose digestion (Leonard et al., 1975). Further, Hill (2002) found that feeding behavior depended on the physical form of the distillery-based DDGS itself (i.e., dry vs. wet). Inclusion levels of corn-based DDGS greater than 15% are not recommended (Bonoma et al., 2008) due to decreased digestibility of dry matter, protein, and fiber compared to control diets. Furthermore, the relatively fine particle size of DDGS may be a limitation for horses, as it is with other animals.

17.5   CONCLUSIONS

It appears that DDGS may be a suitable feed ingredient for animals other than traditional livestock (which have been the highest consumers of DDGS to date). More research is necessary in order to reliably formulate feeds for these types of animals, however. As the ethanol industry continues to grow, this should behoove animal researchers to pursue more extensive DDGS feeding trials, as there appears to be considerable room for growth in these feed sectors.

REFERENCES

Allen, S. E., G. C. Fahey, Jr., J. E. Corbin, J. L. Pugh, and R. A. Franklin. 1981. Evaluation of byproduct feed-stuffs as dietary ingredients for dogs. *Journal of Animal Science* 53(6): 1538–1544.

Archibeque, S. L., H. C. Freetly, and C. L. Ferrell. 2008. Feeding distillers grains supplements to improve amino acid nutriture of lambs consuming moderate-quality forage. *Journal of Animal Science* 86: 691–701.

AVMA. 2007. Market research statistics—U.S. pet ownership. American Veterinary Medical Association. Available online: www.avma.org/reference/marketstats/ownership.asp. Accessed February 18, 2011.

Bonoma, T. A., A. C. Brogren, K. H. Kline, and K. M. Doyle. 2008. Effects of feeding distiller's dried grains with solubles on growth and feed efficiency of weanling horses. *Journal of Equine Veterinary Science* 28(12): 725–727.

Carvalho, L. P. F., D. S. P. Melo, C. R. M. Pereira, M. A. M. Rodrigues, A. R. J. Cabrita, and A. J. M. Fonseca. 2005. Chemical composition, in vivo digestibility, N degradability and enzymatic intestinal digestibility of five protein supplements. *Animal Feed Science and Technology* 119: 171–178.

de Godoy, M. R. C., L. L. Bauer, C. M. Parsons, and G. C. Fahey, Jr. 2009. Select corn coproducts from the ethanol industry and their potential as ingredients in pet foods. *Journal of Animal Science* 87: 189–199.

ERS. 2010. *Sheep and Wool: Background.* Economic Research Service, United States Department of Agriculture: Washington, DC. Available online: www.ers.usda.gov/Briefing/Sheep/. Accessed February 18, 2011.

Flores, A. I., J. R. Sewell,, L. L. Berger, M. J. Cecava, P. H. Doane, J. L. Dunn, M. K. Dyer, A. H. Grusby, and N. A. Pyatt. 2008. Digestibility of corn replacement pellets in growing lamb diets. Poster No. 8. *Corn Utilization and Technology Conference,* June 2–4, 2008, Kansas City, MO.

Gutierrez Zetina, A., J. R. Orozco Hernandez, I. J. Ruiz Garci, and J. J. Olmos Colmenero. 2009. Effect of level of spent corn from the ethanol industry and lamb sex on performance. *Journal of Veterinary Advances* 8(3): 595–597.

Held, J. 2006. Feeding soy hulls and dried distillers grain with solubles to sheep. ExEx 2052. South Dakota State University Cooperative Extension Service. South Dakota State University: Brookings, SD. Available online: agbiopubs.sdstate.edu/articles/ExEx2052.pdf. Accessed February 18, 2011.

Hill, J. 2002. Effect of level of inclusion and method of presentation of a single distillery by-product on the processes of ingestion of concentrate feeds by horses. *Livestock Production Science* 75: 209–218.

Huls, T. J., A. J. Bartosh, J. A. Daniel, R. D. Zelinsky, J. Held, and A. E. Wertz-Lutz. 2006. Efficacy of dried distiller's grains with solubles as a replacement for soybean meal and a portion of the corn in a finishing lamb diet. *Sheep & Goat Research Journal* 21: 30–34.

Kentucky Equine Research. 2007. Changes in the horse industry. *Equine Review, HW 37.* Available online: www.ker.com/library/EquineReview/. Accessed February 18, 2011.

Leonard, T. M., J. P. Baker, and J. Willard. 1975. Influence of distillers feeds on digestion in the equine. *Journal of Animal Science* 40(6): 1086–1090.

Lodge, S. L., R. A. Stock, T. J. Klopfenstein, D. H. Shain, and D. W. Herold. 1997. Evaluation of wet distillers composite for finishing ruminants. *Journal of Animal Science* 75: 37–43.

Nakamura, T., T. J. Klopfenstein, and R. A. Britton. 1994. Evaluation of acid detergent insoluble nitrogen as an indicator of protein quality in nonforage proteins. *Journal of Animal Science* 72: 1043–1048.

NASS. 2010. *National Statistics—Sheep and Lambs.* National Agricultural Statistics Service, United States Department of Agriculture: Washington, DC. Available online: www.nass.usda.gov/. Accessed February 18, 2011.

Pagan, J. D. 1991. Distillers dried grains as an ingredient for horse feeds: palatability and digestibility study. *Proceedings of the Distillers Feed Research Council Conference* 1991: 83–86.

Pet Food Institute. 2010. Manufacturers. Available online: www.petfoodinstitute.org. Accessed February 18, 2011.

Richardson, J. M., R. G. Wilkinson, and L. A. Sinclair. 2003. Synchrony of nutrient supply to the rumen and dietary energy source and their effects on the growth and metabolism of lambs. *Journal of Animal Science* 81: 1332–1347.

Vipond, J. E., M. Lewis, G. Horgan, and R. C. Noble. 1995. Malt distillers grains as a component of diets for ewes and lambs and its effects on carcass tissue lipid composition. *Animal Feed Science and Technology* 54: 65–79.

Willms, C. L., L. L. Berger, N. R. Merchen, and G. C. Fahey. 1991. Effects of supplemental protein source and level of urea on intestinal amino acid supply and feedlot performance of lambs fed diets based on alkaline hydrogen peroxide-treated wheat straw. *Journal of Animal Science* 69: 4925–4938.

18 Using DDGS as a Food Ingredient

Kurt A. Rosentrater

CONTENTS

18.1 INTRODUCTION

Over the years, the benefits of diets containing high levels of dietary fiber have become well known, including lowering of serum cholesterol levels, blood pressure, risk of heart disease, risk of various cancers, as well as weight loss and control (Burkitt, 1977; Anderson et al., 1987; Anderson et al., 1994; Mehta, 2005; Anderson et al., 2009; Chawla and Patil, 2010). Recently, dietary regimes that also contain low levels of nonfiber carbohydrates (i.e., sugars and starches), such as the Adkins Diet (Atkins, 1992) and South Beach Diet (Agatston, 2003), have become popular (Angelich and Symanski, 2004; Sloan, 2004; Hursh and Martin, 2005). Not only do diets that contain high fiber and low starch/sugar promote weight loss/control, but research into glycemic response and resulting satiety indicates that these diets also have substantial health benefits for both diabetic patients and those suffering from obesity (Brand-Miller, 2004; Gross et al., 2004; Hofman et al., 2004; Layman and Baum, 2004; Li et al., 2003; Rendell et al., 2005). These benefits are true not only for blood sugar control in diagnosed patients, but they also help towards the prevention of obesity and diabetes onset. Because distillers grains is high in fiber and protein, and low in starch, it has potential for use in these types of dietary approaches. Because it is gluten free, it may also be appropriate for those who suffer from Celiac Disease as well. And, as discussed in previous chapters, distillers dried grains with solubles (DDGS) is a very low-cost ingredient.

18.2 POTENTIAL BENEFITS OF DDGS AS A FOOD INGREDIENT

Distillers grains is a low-starch, high-fiber material that can be used in various food products as a partial substitute for wheat-based (especially all-purpose) flour. For consumers in general, the American Dietetic Association has noted the importance of consuming foods from the carbohydrate group, especially those made with whole grains (vs. those that are highly refined and consequently have high starch contents) (American Dietetic Association, 2002). Individuals with diabetes (type 2), however, are especially encouraged to consume foods containing fiber, such as whole grains, fruits, and vegetables. High intakes of dietary fiber have been shown to result in metabolic benefits

399

vis-à-vis glycemic control, hyperinsulinemia, and plasma lipids (American Dietetic Association, 2002; Buttriss and Stokes, 2008). Dietary fiber intake of 50 g/day has been shown to slow digestion, gastric emptying, and the absorption of glucose, which can help regulate both immediate postprandial glucose metabolism as well as long-term glucose control—both of which are important for individuals with type 2 diabetes (Lafrance et al., 1998; American Dietetic Association, 2002). In addition to normalizing (and regulating) blood glucose levels within the body, which in turn can regulate insulin levels, dietary fiber has also been shown to lower cholesterol, which can help prevent heart disease as well as type 2 diabetes (American Dietetic Association, 2002). The American Dietetic Association (2002) has recommended a total dietary fiber intake of 20 g/day to 35 g/day for adults. Because of its composition, DDGS can be used to help achieve these goals. For example, Bechen (2008) has shown that DDGS can have dietary fiber contents of approximately 34 g/100 g and protein contents of nearly 31 g/100 g (dry basis).

Fiber has also been shown to help maintain colon health and decrease the incidence of colon cancer. Diets that are high in fiber are typically digested at a slower rate, and thus increase the physiological feeling of "fullness," which can lead to lower caloric intake, and ultimately a decreased propensity for obesity (American Dietetic Association, 2002). Stool weight increases as fiber intake increases, and the fiber tends to normalize defecation frequency, with a gastrointestinal transit time of 2 to 4 days (American Dietetic Association, 2002). The American Association of Cereal Chemists (2001) defines dietary fiber as "the edible parts of plants or analogous carbohydrates that are resistant to digestion and absorption in the human small intestine with complete or partial fermentation in the large intestine. Dietary fiber includes polysaccharides, oligosaccharides, lignin, and associated plant substances. Dietary fibers promote beneficial physiological effects including laxation, and/or cholesterol attenuation, and/or blood glucose attenuation" (American Association of Cereal Chemists, 2001).

If fiber is one side of the diabetic equation, then simple carbohydrates (such as sugar and starch) are the other. Monitoring the glycemic index (GI) of foods is a practice that is often used to assess the simple carbohydrate content in a food prior to consumption; this helps diabetic patients to control the food they ingest, and thus the resulting blood glucose levels they will experience (American Dietetic Association, 2006). The GI quantifies the blood glucose response of a food for a period of 2 h after consumption (i.e., postprandial), compared to the response due to a reference standard (which is most often 50 g of pure glucose). Figure 18.1 illustrates typical blood glucose curves over time. The GI is computed based on the ratio of the area under the test food's curve to that of the

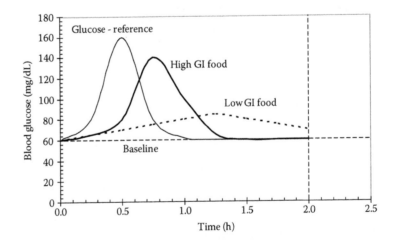

FIGURE 18.1 Illustration of blood glucose curves over time, comparing a reference (glucose) to a high glycemic index (GI) food and a low GI food. Note that 2 h is the cutoff for the GI calculation, and the baseline is the blood glucose level before consuming any food.

reference standard, and is expressed as a percentage. Foods with a low GI (i.e., high fiber, high protein materials) are considered better for diabetics than foods with a high GI (i.e., high sugar or starch), since the glucose is expelled at a slower rate over a longer period of time (Jones, 2002).

Intake of foods with high GI has been shown to be highly related to a greater risk of type 2 diabetes (Schulze et al., 2004). Miller et al. (2006) found that consumption of lower GI foods was associated with decreased incidence of diabetes, as well as better glucose control within diabetics, reduced serum lipid levels, improved insulin levels, and a lower risk of colon cancer. Short-term studies completed by Arora and McFarlane (2005) concluded that a low carbohydrate diet resulted in lower HbA1c levels, greater glycemic control, lower postprandial glucose levels, and improved insulin sensitivity. Foods higher in starch, however, increased postprandial glucose levels and insulin needs—an insulin dependent diabetic would need to use an increased quantity of injected insulin, while a non-insulin dependent diabetic would need to restrict the quantity of high starch foods consumed. Long-term consumption of low GI foods has also been associated with increased satiety and body weight control (Bloomgarden, 2004; Schultze et al. 2004; Ostman, 2006).

As DDGS is low in starch, but high in fiber, it may be well suited for this type of dietary application. But no studies have yet been conducted to examine this potential. Toward this end, Bechen (2008) conducted a study to assess the efficacy of using DDGS as a component of the diabetic diet. Eight participants, ages 18 to 25, were fed baked samples of all-purpose flour, whole wheat flour, and DDGS; blood samples were then drawn from each participant every 30 min for 2 h after consumption. After collection, blood glucose levels were measured, and glucose response over time was calculated. Results showed a clear difference in glucose levels among the three types of food samples (Figure 18.2). The DDGS resulted in the lowest glucose response, even better than whole wheat flour, and thus appeared to be a highly effective food ingredient for diabetic diets. Further testing is warranted, however, to more fully quantify DDGS's benefits for diabetics.

Celiac Disease (or gluten sensitive enteropathy, GSE), on the other hand, is categorized by sensitivity to gluten. Gluten is a protein composite that consists of prolamin and glutelin, and is primarily present in wheat, rye, triticale, and barley. Oats may also be problematic, due to possible contamination by wheat. Gluten is often intimately linked to the starch, and plays a key role in dough rheology and thus texture of resulting baked products. Celiac Disease is actually a food intolerance linked to an autoimmune disorder, not a food allergy. Approximately 1 in 133 people in the United States

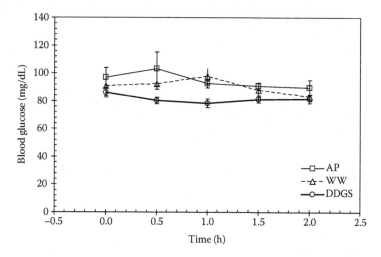

FIGURE 18.2 Clinical results of blood glucose levels over time (mean levels of eight participants) comparing all-purpose (AP) flour, whole wheat (WW) flour, and DDGS; error bars represent ±1 standard error of the mean. (Adapted from Bechen, A. The glycemic response of distillers dried grains as a human food ingredient. M.S. Thesis. Brookings, SD: South Dakota State University, 2008.)

have been diagnosed with some form of this condition (Mahan and Escott-Stump, 2004). At this point in time, the only known treatment is a gluten-free diet. The Mayo Clinic (2006) advises that the following flours and starches are safe for consumption by individuals with Celiac Disease: arrowroot, corn, potato, rice, soy, and tapioca. The restrictive nature of this disease can negatively impact the quality of life for many. A study by Lee and Newman (2003) concluded that 86% of survey respondents who had this diagnosis considered themselves negatively affected, especially in terms of their ability to dine out. It is true that corn products are referred to as containing "gluten," however corn proteins differ from wheat in that they lack gliadin, and thus do not pose a problem for gluten-intolerant people. No human trials have yet been performed using DDGS in diets for Celiac patients; but corn-based DDGS (i.e., from fuel ethanol production) is a gluten-free material, and would naturally fit the needs of this dietary approach.

Beyond the potential use of DDGS for diabetic and Celiac patients, some evidence has arisen in recent years that ethanol coproducts may also have beneficial effects in the prevention of heart disease (Sethi and Diaz, 2007). Sethi (2004, 2007) investigated the effects of wheat-based DDG on ischemic heart disease (known as myocardial ischemia, which is a result of reduced blood flow to the heart muscle, and is often caused by coronary artery disease) in rats. In this study, three rats were fed either DDG ($n = 3$) or a control diet ($n = 3$), the study ran for 21 days, and then heart cells from each rat were subjected to mineral oil layering to simulate the onset of ischemia. After 2 h, cells from the control rats had only about 3% survival, while the cells from the DDG-fed rats had nearly 13% survival ($p < 0.05$). Unfortunately, this study did not examine why this survival effect occurred, or what specifically about the DDGS was influential. And wheat-based DDGS will invariably behave differently compared to corn-based DDGS.

Perhaps one reason for this observed effect was the presence of phytosterols in the DDG. Phytosterols are chemicals that are naturally present in vegetable oils, grains, and cereals. They are important because they can reduce and even prevent the absorption of cholesterol in the intestinal tract (Gylling and Miettinen, 2005). Phytosterols include several distinct compounds, such as sitosterol, campesterol, stigmasterol, cycloartenol, sitostanol, and campestanol. Ostlund (2002) provides a thorough review of phytosterols in foods, their properties, and importance in human nutrition. According to Moreau et al. (1996) and Kochhar (1983), corn contains high concentration of three types of phytosterols; these include ferulate esters, long chain fatty acyl esters, and free phytosterols. Because of the fuel ethanol production process, DDGS should contain at least three times the levels found in unprocessed corn kernels (see Chapter 5 for details about the production process).

In fact, a few studies have been conducted in recent years to quantify phytosterol levels in DDGS. For example, Singh et al. (2002) measured phytosterol contents of DDGS samples from three fuel ethanol plants. They found that ferulate phytosterol esters were present in quantities between 43.16 mg/100 g (of dry DDGS) and 47.35 mg/100 g; free phytosterols ranged from 67.98 mg/100 g to 71.41 mg/100 g; phytosterol fatty acyl esters were between 178.22 mg/100 g and 230.27 mg/100 g; and total phytosterol contents ranged from 292.97 mg/100 g to 342.95 mg/100 g.

Along similar lines, Leguizamon et al. (2009) quantified total phytosterol (which consisted of campesterol, stigmasterol, and sitosterol) content in corn-based DDGS that was collected from one fuel ethanol plant, and found up to 134 mg/100 g (of dry DDGS), which was nearly 670% higher than that found in the raw corn itself (~20 mg/100 g of dry corn). In terms of the lipids that were present in the DDGS, campesterol was found at levels up to 245 mg/100 g (of lipids); stigmasterol was up to 88 mg/100 g; and sitosterol was observed up to 817 mg/100 g.

Correspondingly, Winkler et al. (2007) examined corn-based DDGS from one fuel ethanol plant. After using various extraction methods (i.e., Soxhlet extraction vs. accelerated solvent extraction, using hexane, ethanol, or supercritical CO_2), phytosterol and ferulate phytosterol esters were quantified. Total phytosterols (consisting of campesterol, campestanol, stigmasterol, sitosterol, sitostanol, $\Delta 5$-avenasterol, and $\Delta 7$-stigmastenol) in the DDGS ranged from 192 to 291 mg/100 g (of dry DDGS); whereas the ferulate phytosterol esters ranged from 35 to 53 mg/100 g.

Further, Winkler-Moser and Vaughn (2009) examined corn-based DDGS collected from one fuel ethanol plant. The oil was extracted from the DDGS and then concentrated using molecular distillation. Total phytosterols (consisting of steryl ferulate, steryl fatty acid esters, and free sterols) were observed at levels of 1620 mg/100 g (of lipids) in the DDGS oil, whereas the concentrated distillate had levels up to 9320 mg/100 g (a 575% increase). Steryl ferulates were measured at 400 mg/100 g in the oil, and 1040 mg/100 g in the distillate (which amounted to a 260% increase). Moreover, they also examined the performance of this distillate as an antioxidant in sunflower oil that had been stripped of its original tocopherols and phytosterols. They found that at levels between 0.1% and 1%, oxidative stability was improved, and tocopherol, tocotrienol, and steryl ferulates exhibited delayed degradation over time. For more information on lipid composition in DGGS, refer to Chapter 9.

Another potential benefit to using DDGS in human foods is the presence of residual nucleic acids from the yeast during ethanol production. Nucleic acids have been shown to have some health benefits in terms of boosting immune functions (Hong et al., 2004; Feng et al., 2007). A detailed discussion about residual yeast in DDGS is provided in Chapter 8.

Overall, it appears that DDGS may offer potential health benefits from several vantage points. It would be prudent, however, to conduct further research to validate these initial findings, especially in terms of clinical trials. Furthermore, a challenge to the proposition of using DDGS in human foods is the resulting functionality, performance, and ultimate consumer acceptance of food products that are made with DDGS, as will be discussed next.

18.3 USING DDGS IN FOODS

The author has investigated the use of corn-based DDGS from fuel ethanol production plants in a variety of food products, and has achieved varying levels of success. A few examples will be discussed below. Figure 18.3 illustrates some results from baking bread using DDGS as a partial substitute for flour. For these experiments, two types of flour were used (all-purpose flour and whole wheat flour), and two types of DDGS were used (traditional, unmodified DDGS (i.e., regular fat DDGS; fat content ~10% db) that was collected from a fuel ethanol plant, and low-fat DDGS, which had been

FIGURE 18.3 Examples of breads containing DDGS as a partial substitute for all-purpose flour (denoted here as wheat) or whole wheat flour (denoted here as whole wheat). RF denotes regular fat (i.e., unmodified) DDGS; LF denotes low-fat DDGS. Noted percentages indicate flour substitution level, not the actual quantity of the DDGS in each blend (author's unpublished data).

subjected to solvent extraction to remove lipids (final fat content ~2% db)). Each of these types of DDGS was used at five substitution levels: 0%, 5%, 10%, 15%, 20%, and 25%. The experiments were conducted using the same formulation (varied only according to type of flour and DDGS substitution rate), using the same bread machine in order to standardize the testing methodology. As shown in Figure 18.3, there were considerable changes in the resulting bread loaves. For example, as DDGS increased, the product volume decreased, the color appeared darker, the cell sizes decreased, and the internal structure became coarser/grainier. These visual observations were born out by the data (Figure 18.4). For both types of DDGS and both types of flour, as the DDGS level increased, Hunter L (brightness) decreased, whereas Hunter a and b values increased (i.e., the loaves became redder and yellower). Additionally, the expansion decreased (the resulting loaf height decreased), and the structural stiffness (compressive strength) increased. Figure 18.5 shows a magnified view of the internal cell structure. Overall, at least for bread, it appears that the use of DDGS as a flour additive/substitute can be viable up to about 15%; in some cases even 20% did not produce an objectionable appearance or texture. Clearly, however, substitution of DDGS over 20% appeared to be less desirable.

Gluten offers structure to baked products by trapping gases, which allows dough to rise during fermentation. Weak gluten structure can result in excessive expansion and uneven texture, whereas strong gluten structure can result in decreased expansion and small loaf volume (Schofield and Booth, 1983). Corn-based DDGS is high in fiber and protein, but is devoid of the gluten-forming glutenin and gliadin proteins (Lilliard, 2000; Wang et al., 2004). So, depending on the level of DDGS substitution

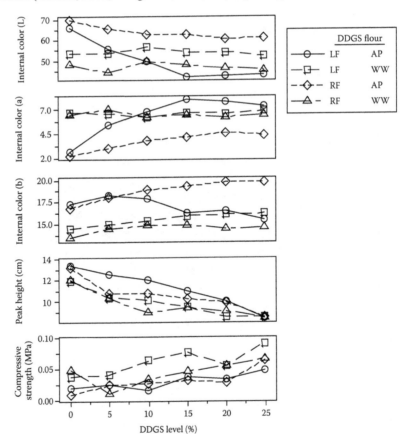

FIGURE 18.4 Effects of adding DDGS as a partial substitute for all-purpose flour (AP) or whole wheat flour (denoted here as WW) on the resulting physical properties of bread loaves. RF denotes regular fat (i.e., unmodified) DDGS; LF denotes low-fat DDGS. Noted percentages indicate flour substitution level, not the actual quantity of the DDGS in each formulation (author's unpublished data).

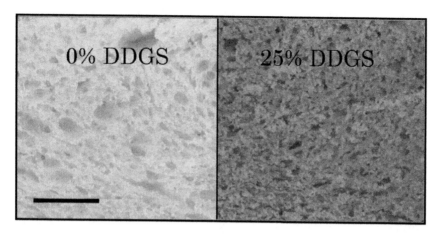

FIGURE 18.5 Comparison of bread cross sections illustrates the effect of DDGS substitution for all-purpose flour. Cell sizes decreased, the texture became grainier, and the color darkened as DDGS level increased. The scale bar indicates a length of approximately 1 cm (author's unpublished data).

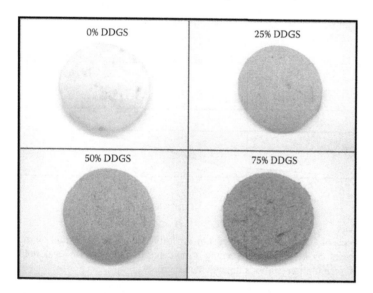

FIGURE 18.6 Examples of basic sugar cookies containing DDGS as a partial substitute for all-purpose flour. Noted percentages indicate flour substitution level, not the actual quantity of the DDGS in each blend (author's unpublished data).

for flour, bread fermentation may be hindered due to the dilution of gluten-forming proteins, thus altering the quality of the final product. The structure may also be compromised due to potential cutting of gluten strands by the fiber particles. Furthermore, increased concentrations of fiber will often absorb more water (Dreese and Hoseney, 1982); it is thus possible that increased water absorption can negatively affect gluten formation (Fennema, 1996), as additional water may disrupt molecular interactions and the formation of disulphide bonds that hold the gluten complex together.

Baked products that rely on chemical leavening (such as cookies, quick breads, and flat breads) instead of yeast (e.g., yeast breads), may be better suited for inclusion of DDGS, because leavening does not require carbohydrates (i.e., starch or sugars) to proceed; only a chemical reaction (generally an acid working with a base) is required to produce carbon dioxide bubbles, and thus expansion in the product. Figure 18.6 illustrates some results from experimental cookies using DDGS (unmodified DDGS from a commercial fuel plant). For these experiments, a simple cookie formulation was used (Method 10-50D;

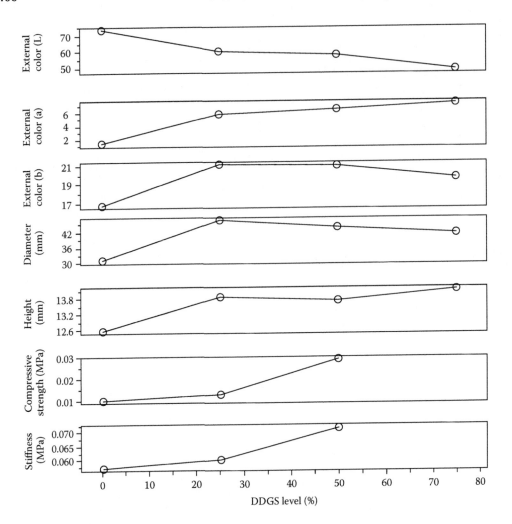

FIGURE 18.7 Effects of adding DDGS as a partial substitute for all-purpose flour on the resulting physical properties of basic sugar cookies. Noted percentages indicate flour substitution level, not the actual quantity of the DDGS in each formulation (author's unpublished data).

AACC, 2000), using corn-based DDGS at four flour substitution levels (0%, 25%, 50%, 75%). As with the bread baking trials, as the DDGS level increased, the resulting products became darker in appearance. Additionally, the cookies became redder and yellower (Figure 18.7); the diameter and height (and thus volume) remained favorable; but strength and stiffness (which result in texture) increased. Expansion and shape were mostly unaffected due to the use of chemical leavening instead of yeast. But, as DDGS increased, flavor declined (as measured by taste panel). Overall, it appeared that up to 50% DDGS was visually acceptable, although the flavor became intense over 25% DDGS.

Figure 18.8 illustrates some results from further baking trials (again, following AACC Method 10-50D) using two types of DDGS (commercial unmodified, and low-fat DDGS) in combination with various nonwheat flours (thus all resulting cookies were completely gluten free). As shown, some formulations behaved dramatically better than others. For example, the rice and bean flours were most promising, as they had good expansion and spread, good volume and shape, and good color, even up to 75% DDGS substitution. The tapioca and potato starch cookies were of moderate, yet acceptable structure. The corn meal and potato flour cookies, however, had very limited functionality, as evidenced by a lack of expansion, spread, or cohesion. In general, the low-fat DDGS seemed to produce cookies that were slightly darker than those that contained the unmodified DDGS.

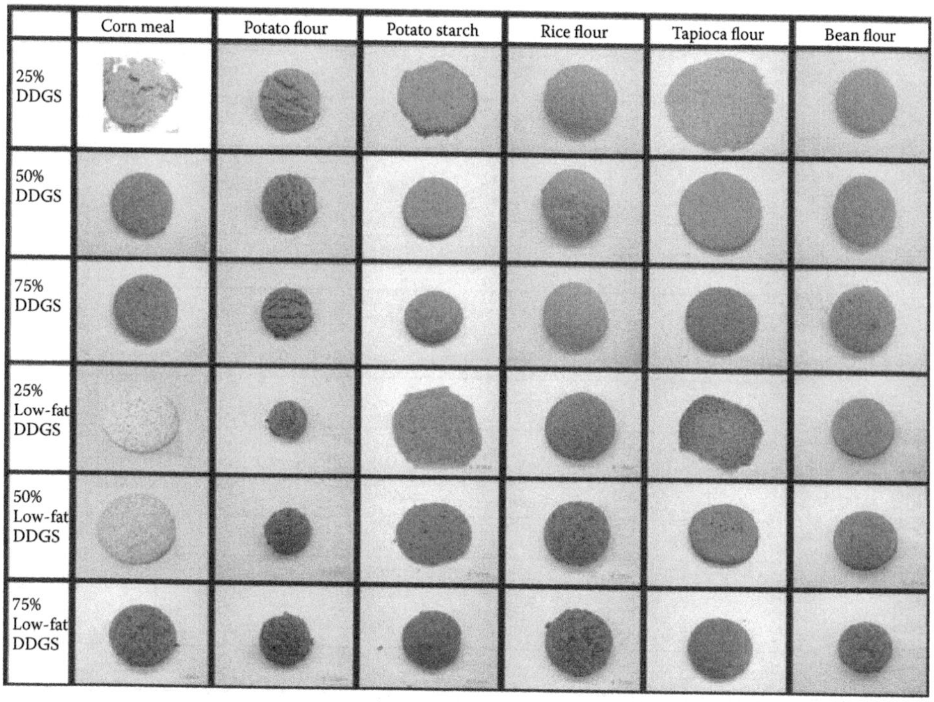

	Corn meal	Potato flour	Potato starch	Rice flour	Tapioca flour	Bean flour
25% DDGS						
50% DDGS						
75% DDGS						
25% Low-fat DDGS						
50% Low-fat DDGS						
75% Low-fat DDGS						

FIGURE 18.8 Examples of cookies containing DDGS as a partial substitute for various gluten-free flours. Noted percentages indicate flour substitution level, not the actual quantity of the DDGS in each blend. Control cookie shown in Figure 18.6 (author's unpublished data).

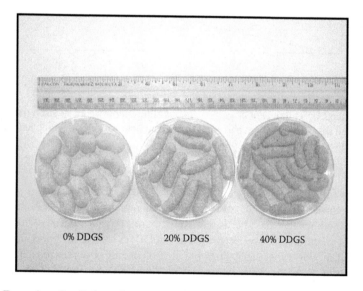

FIGURE 18.9 Examples of puffed snacks containing DDGS as a partial substitute for corn flour (author's unpublished data).

Extruded food products have also been shown to be promising. Figure 18.9 illustrates some extruded snacks that were produced using corn flour with DDGS substituted at three levels: 0%, 20%, and 40%. The 0% DDGS resulted in products that many consumers are familiar with—a traditional corn snack puff. As DDGS level increased, the product became darker, redder, bluer, and the puffing decreased, as evidenced by a decrease in product diameter (Figure 18.10). In fact,

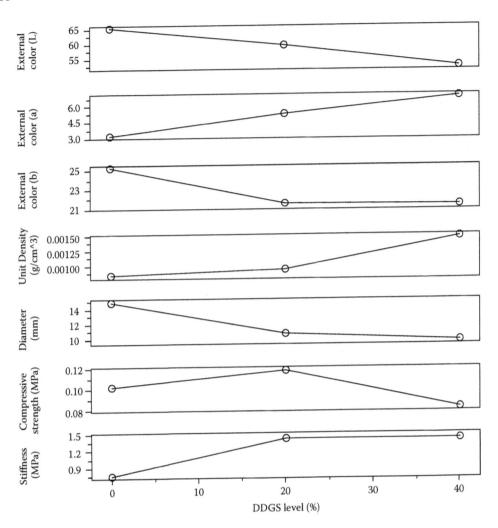

FIGURE 18.10 Effects of adding DDGS as a partial substitute for all-corn flour on the resulting physical properties of extruded snack puffs. Noted percentages indicate flour substitution level, not the actual quantity of the DDGS in each formulation (author's unpublished data).

at 40% DDGS, the expansion was only 65.5% that of the 0% DDGS extrudates, although the 20% blend exhibited expansion of 71.7% that of 0% DDGS. By increasing the DDGS, the resulting structural integrity declined because the fiber is relatively nonreactive, and does not bind well with other ingredients; moreover, the starch level decreased as the DDGS increased (starch is essentially a glue that binds extruded products together). Thus the size of the air pores decreased (Figure 18.11) as DDGS level increased. Concomitantly, as expansion decreased, the unit density of each extrudate increased, the overall strength appeared to decline, but the stiffness increased.

From the author's trials (considering both published and unpublished data), it appears that the optimal level of DDGS in human foods (at least the products that have been tested) lies between 15% and 25%, depending upon the composition of the DDGS itself, whether lipids have been extracted from the DDGS or not, and upon the specific food product. The ability to use DDGS also depends upon the other ingredients in the food matrix, all of which will interact with the DDGS particles and potentially increase or even counteract some of the negative functionality that DDGS possesses.

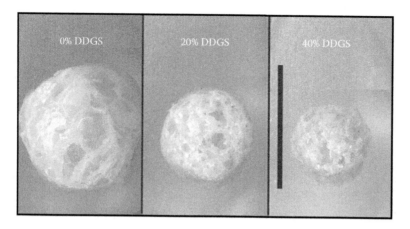

FIGURE 18.11 Extrudate cross sections illustrate the effect of DDGS addition. As DDGS level increased, overall expansion as well as cell sizes decreased, and color darkened. The scale bar indicates a length of approximately 15 cm (author's unpublished data).

18.4 PREVIOUS STUDIES

As with many food and organic processing residue and byproduct streams, feeding ethanol coproducts to animals is a logical means to attain full use (and the concomitant economic returns) of these materials, because they contain high levels of various nutrients (as discussed in previous chapters). Over the years, numerous research studies have been conducted in order to optimize their use in livestock feed rations; comprehensive reviews of this research can be found in other chapters in this book as well. Many previously published studies have, in fact, also examined coproduct use as high protein/high fiber/low starch functional ingredients in human foods. And several of these prior investigations have been compiled and reviewed by Rosentrater and Krishnan (2006).

Thus the concept of using DDGS as a food ingredient is not novel. For example, during the 1980s, 23 scientific studies were conducted and 47 food products were investigated; in the 1990s, only nine studies were published and 11 products were examined; in the 2000s, however, only three studies and six products were investigated.

Ironically, as the fuel ethanol industry grew, interest in using coproducts in human foods declined—due, in large measure, to the challenges that they pose (they are not readily used at high levels, for a variety of reasons, which will be discussed shortly). Most prior studies have almost exclusively focused on using dried coproducts (i.e., DDG and/or DDGS), most used grains other than corn as a substrate for fermentation, and most of these coproducts were from beverage distilleries (as opposed to corn-based fuel plants). Even so, these results are a good base from which to work vis-à-vis future product development research; much can be gleaned from the prior research. Previously published research has examined coproduct use in breads (Table 18.1), various baked products (Table 18.2), cookies (Table 18.3), pasta and extruded products (Table 18.4), and various other food products (Table 18.5). Additionally, there have been several patents (Table 18.6) that have been issued as well. Most of the older studies used either beverage-based DDGS or DDGS that was produced in the authors' laboratories; thus most prior studies were conducted on food-grade DDGS, not the one from a dry grind fuel ethanol plant. The food studies in recent years are most likely to have used DDGS from fuel ethanol plants, which are not food-grade quality, so before these results can be used commercially, they will likely require regulatory approval.

As illustrated in these tables, the use of distillers grains has generally negatively impacted the resulting organoleptic quality of food products as substitution (generally for wheat flour) rates increased, regardless of what food product was investigated. In terms of color, most food products became darker in appearance as the level of DDG or DDGS increased. Moreover, most of the

TABLE 18.1
Examples of Previous Studies that Have Examined Ethanol Coproducts in Breads

Original Flour	Coproduct Used	Ethanol Feedstock	Flour Substitution Level (%)	Functionality Results	Taste Panel Results	Citation
All-purpose flour	DDGS	Soft white winter wheat	0, 30	Darker in appearance	Acceptable to good	Rasco et al. (1987a)
Bread flour	DDGS	Soft white winter wheat	0, 20	High concentrations of dietary fiber, iron, calcium, zinc, especially soluble minerals	—	Rasco et al. (1990a)
Bread flour (hard red winter wheat)	DDG, DDGS	Barley, red wheat, soft white winter wheat	0, 4, 8	As DDG increased, slightly darker appearance; decreased loaf volume; increased water absorption	—	Rasco et al. (1990b)
Bread flour, whole wheat flour	DDG	Barley, corn, rye	0, 5, 10, 15, 20	As DDG increased, poor dough development, loaf volume decreased, darker in appearance	As DDG increased, bitter flavor, harsher mouthfeel; 10% DDG was still acceptable	Brochetti et al. (1991)
Wheat flour	DDG	Sorghum (brown, white, white waxy, yellow)	0, 5, 10, 15	As DDG increased, darker in appearance, loaf volume decreased	Acceptable up to 10% DDG	Morad et al. (1984)
Wheat flour	DDGS	Soft white winter wheat	0, 20	Darker in appearance, reduced loaf volume, increased loaf density, unacceptable crumb color	—	Rasco and Dong (1991)
Wheat flour	DDGS	Soft white winter wheat	0, 12.5, 25	Darker in appearance	Less acceptable as DDGS increased due to off-flavor	Rasco et al. (1989)
White flour, whole wheat flour	DDGS	Corn and other cereal grains	0, 10, 20	As DDGS increased, increased water absorption, decreased dough stability, reduced loaf volume and crumb quality, darker in appearance	—	Tsen et al. (1983)
Whole wheat flour	DDGS	Soft white winter wheat	0, 30	Darker in appearance	Acceptable to good	Rasco et al. (1987a)
All-purpose flour, bread flour	DDGS	Corn	0, 25, 50	As DDGS increased, color and texture decreased	As DDGS increased, acceptability decreased	Saunders (2008)

Note: DDG is distillers dried grains; DDGS is distillers dried grains with solubles.

TABLE 18.2
Examples of Previous Studies that Have Examined Ethanol Coproducts in Various Baked Products

Type of Product	Original Flour	Coproduct Used	Ethanol Feedstock	Flour Substitution Level (%)	Functionality Results	Taste Panel Results	Citation
Muffins— oatmeal	All-purpose flour	DDGS	Barley, corn, rye	0, 5, 15, 36	Slightly darker color	Acceptable	Abbott et al. (1991)
Yeast rolls	Whole wheat flour	DDGS	Barley, corn, rye	0, 33	Slightly darker color	Acceptable	Abbott et al. (1991)
Muffin—corn	All-purpose flour	DDG	Barley, corn, rye	20, 30	Greater water and oil absorption; darker in appearance	Acceptable up to 20%	Brochetti and Penfield (1989)
Donuts—spiced	All-purpose flour	DDG	Barley, corn, rye	20, 30	Greater water and oil absorption; darker in appearance	Acceptable up to 20%	Brochetti and Penfield (1989)
Hush puppies	All-purpose flour	DDG	Barley, corn, rye	10, 20	Greater water and oil absorption; darker in appearance	Acceptable up to 20%	Brochetti and Penfield (1989)
Muffins	All-purpose flour	DDG (bleached and defatted)	Barley	0, 15	Cooked volume increased, but spread decreased; water absorption increased	Poor flavor compared to control	Dawson et al. (1985)
Carrot coconut bread	Whole wheat flour	DDGS	Barley, corn, rye	0, 40	Darker in appearance; decreased volume	All acceptable	O'Palka et al. (1989)
Dinner rolls	Whole wheat flour	DDGS	Barley, corn, rye	0, 17, 33	Darker in appearance; decreased volume	All acceptable	O'Palka et al. (1989)
Nut rolls	Whole wheat flour	DDGS	Barley, corn, rye	0, 33	Darker in appearance; decreased volume	All acceptable	O'Palka et al. (1989)

continued

TABLE 18.2 (continued)
Examples of Previous Studies that Have Examined Ethanol Coproducts in Various Baked Products

Type of Product	Original Flour	Coproduct Used	Ethanol Feedstock	Flour Substitution Level (%)	Functionality Results	Taste Panel Results	Citation
Muffins—oatmeal	All-purpose flour and oats	DDGS	Barley, corn, rye	0, 33	Darker in appearance; increased volume	All acceptable	O'Palka et al. (1989)
Banana bread	All-purpose flour	DDGS	Soft white winter wheat	0, 30	As DDGS increased, darker in appearance	Good flavor	Rasco et al. (1987a)
Cinnamon rolls	All-purpose flour	DDGS	Soft white winter wheat	0, 12.5, 25	As DDGS increased, darker in appearance	Acceptable, but stronger flavor	Rasco et al. (1989)
Muffins—Wheat	Wheat flour	DDGS	Grain (undefined)	0, 10, 15, 20	Texture acceptable up to 10%	Appearance and flavor acceptable up to 10%	Reddy et al. (1986b)
Chipathi (flat bread)	Whole wheat flour	DDG	Corn	0, 5, 7, 10	As DDG increased, water requirements increased		Ahmed (1995)
Chipathi (flat bread)	Whole wheat; whole white wheat; bread flour	DDG	Corn	0, 10, 20	As DDG increased, color and texture decreased	As DDG increased, acceptability decreased	Arra et al. (2008)

Note: DDG is distillers dried grains; DDGS is distillers dried grains with solubles.

TABLE 18.3
Examples of Previous Studies That Have Examined Ethanol Coproducts in Cookies

Type of Cookie	Original Flour	Coproduct Used	Ethanol Feedstock	Flour Substitution Level (%)	Functionality Results	Taste Panel Results	Citation
Bar	Bleached soft white wheat	DDG	Grain (undefined)	0, 15, 25	Darker appearance; decreased width and thickness	Acceptable, but control still had better taste	Tsen et al. (1982)
Chocolate chip	Wheat flour	DDG	Sorghum (brown, white, white waxy, yellow)	0, 5, 10, 15	Darker in appearance; spread decreased at higher DDG	Acceptable	Morad et al. (1984)
Chocolate chip	All-purpose flour	DDGS	Soft white winter wheat	0, 15, 30	Darker in appearance	As good as control	Rasco et al. (1987a)
Chocolate chip	All-purpose flour	DDGS	Soft white winter wheat	0, 12.5, 25	Darker in appearance	All levels had acceptable flavor	Rasco et al. (1989)
Chocolate chip	Unbleached soft white wheat	DDG	Grain (undefined)	0, 15, 25	Darker appearance; decreased width and thickness	Acceptable, but control had better taste	Tsen et al. (1982)
Molasses	Wheat flour	DDG	Sorghum (brown, white, white waxy, yellow)	25, 50	Darker in appearance; spread decreased at higher substitution	Acceptable	Morad et al. (1984)
Molasses—raisin	All-purpose flour	DDG	Barley, corn, rye	0, 10, 20, 30	Greater water and oil absorption; darker in appearance	Acceptable up to 30%	Brochetti and Penfield (1989)
Oatmeal	All-purpose flour	DDG, Bleached DDG, Defatted DDG	Barley	15	Lipid composition degraded during ethanol processing; bleaching worsened lipid damage	Acceptable flavor; defatted DDG better than DDG and defatted bleached DDG	Dawson et al. (1984)
Spice	Unbleached soft white wheat	DDG	Grain (undefined)	0, 15, 25	Darker appearance; decreased width and thickness	Acceptable, but control had better taste	Tsen et al. (1982)
Sugar	Wheat flour	DDG	Sorghum (brown, white, white waxy, yellow)	0, 5, 10, 15	Darker in appearance; spread decreased at higher DDG	Acceptable	Morad et al. (1984)
Sugar	Wheat flour	DDG, DDGS	Barley, red wheat, soft white winter wheat	0, 2, 4, 8	Darker appearance; variable spread	—	Rasco et al. (1990b)
Sugar	Unbleached soft white wheat	DDG	Grain (undefined)	0, 15, 25	Darker appearance; decreased width and thickness	—	Tsen et al. (1982)

Note: DDG is distillers dried grains; DDGS is distillers dried grains with solubles.

TABLE 18.4
Examples of Previous Studies that Have Examined Ethanol Coproducts in Pasta and other Extruded Products

Type of Product	Original Flour	Coproduct Used	Ethanol Feedstock	Flour Substitution Level (%)	Functionality Results	Taste Panel Results	Citation
Extruded product	Corn, rice, potato, wheat	DDG	Hard red spring wheat	0, 10, 20, 40	As DDG increased, darker in appearance; unit density and longitudinal expansion increased; radial expansion decreased	—	Kim et al. (1989a)
Extruded product	Corn grits, degerminated corn, potato, rice, wheat	DDG	Barley, corn, oat, rye, sorghum, wheat	0, 20, 50, 100	As DDG increased, unit density and longitudinal expansion increased; radial expansion decreased	—	Kim et al. (1989b)
Extruded product	Corn meal	DDGS	Corn	0, 10, 20, 30	As DDGS increased, water absorption, solubility, expansion, and brightness decreased	—	Shukla et al. (2005)
Extruded product	Corn, potato, rice, wheat	DDG	Grain (undefined)	0, 10, 20, 30, 40	As DDG increased, expansion decreased	Acceptable up to 20%; poor quality greater than 20%	Wampler and Gould (1984)
Pasta	Whole wheat durum semolina	DDG	Hard red spring wheat	0, 25, 50	As DDG increased, darker appearance; cooked weight decreased; lower water absorption	Appearance, flavor, texture acceptable at 25%, but unacceptable at 50%	Maga and Van Everen (1989)
Spaghetti	Durum semolina	DDG	Hard red wheat, hard white wheat, rye, sorghum	0, 5, 10, 15, 20, 30	Cooking quality acceptable, but lower as DDG increased; darker appearance	Poor sensory qualities at levels greater than 15%	Van Everen et al. (1992)
Spaghetti	Durum semolina	DDG	Corn	0, 5, 10, 15	As DDG increased, cooking losses increased, firmness decreased	Flavor and texture decreased; over 10% was less acceptable	Wu et al. (1987)

Note: DDG is distillers dried grains; DDGS is distillers dried grains with solubles.

TABLE 18.5
Examples of Previous Studies that Have Examined Ethanol Coproducts in Various other Types of Foods

Type of Product	Original Flour	Coproduct Used	Ethanol Feedstock	Flour Substitution Level (%)	Functionality Results	Taste Panel Results	Citation
Beef stew	Modified starch	DDGS	Grain (undefined)	0, 1, 2, 3	—	Acceptable flavor, appearance, and mouthfeel	Reddy et al. (1986a)
Blended ingredients (food aid)	Corn, soy	DDG	Corn	0, 5, 10	As DDG increased, became darker in appearance	Flavor quality was poor; solvent extraction improved flavor	Bookwalter et al. (1984)
Blended ingredients (food aid)	Corn, soy	DDG	Corn	0, 5, 7.5	As DDG increased, became darker in appearance	Flavor quality was poor; solvent extraction improved flavor	Bookwalter et al. (1988)
Blended ingredients (food aid)	Corn, soy	DDG, DDGS	Corn	0, 5, 10	Acceptable digestibilities during rat feeding trials	—	Wall et al. (1984)
Chili	—	DDGS	Grain (undefined)	0, 1, 2, 3	—	Acceptable flavor, appearance, and mouthfeel	Reddy et al. (1986a)
Deep-fried fish batter	All-purpose flour	DDGS	Hard red wheat, soft white wheat, corn	0, 25	Coatings with DDGS were darker than control	Corn DDGS was more acceptable than control; wheat DDGS was worse than control	Rasco et al. (1987b)
Granola	—	DDG	Barley	7.5	—	Flavor not as good	Dawson et al. (1987)
Granola bar	—	DDG	Barley	2.4	—	Flavor not as good	Dawson et al. (1987)
Hot dog sauce	—	DDGS	Grain (undefined)	0, 1, 2, 3	—	Acceptable flavor, appearance, and mouthfeel	Reddy et al. (1986a)

Note: DDG is distillers dried grains; DDGS is distillers dried grains with solubles.

TABLE 18.6
Patents Related to the Use of Ethanol Coproducts in Food

Coproduct Used	Ethanol Feedstock	Summary	Recommended Use	Citation
DDGS DWG	Wheat	Fermentation residues (moisture content >50%) were combined with 0.05% (g/100 g of mixture) to 5.0% sodium bicarbonate, 0.05% to 5.0% amino acid (a combination of lysine and tyrosine), and 10% to 50% potato starch; The mixture was then dried to form a food ingredient similar to "bran"	Baked products • bread • noodles	Reddy and Stoker (1993)
DDG DDGS DWG	Cereal grains (such as barley, corn, rice, sorghum, wheat)	High-quality distillers grains (i.e., light in color) were used as a source of salt flavoring or as a sodium potentiating agent (i.e., a flavor enhancer that increases the perception of sodium) at inclusion levels less than 10%	Baked products bread • pizza crust • crackers • Cereal • Cheese sauce Meat analog Processed cheese Salad dressing Seasoning mix Stuffing mix	Silver et al. (2008)
WS	Cereal grains (such as barley, corn, milo, rice, rye, sweet sorghum, wheat)	Prior to fermentation, hydrolyzed slurry was pH adjusted to between 4.0 and 5.0 using citric, malic, acetic, lactic, tartaric, fumaric, hydrochloric, or succinic acid (all of which result in acceptable taste properties and are nontoxic) instead of sulfuric acid; After fermentation, WS is then neutralized to pH between 5.0 and 8.0, using NaO, NaOH, KO, KOH, CaO or CaOH; The WS was then dried at $T \leq 77°C$ until a final moisture between 5% and 10% was attained	Brownies (35%) Cookies (35%) Yeast breads (30%–50%) Quick breads (30%)	Rasco and McBurney (1989)
WS	Cereal grains (such as barley, corn, millet, oats, sorghum, rice, rye, wheat)	Whole stillage was removed prior to fermentation; solids were then removed from whole stillage until a solids content between 30% and 50% was achieved, with a residual solids between 10% and 30%; solids were dried at $T \leq 170°F$	Baked products Diet supplements Meat extenders RTE cereals Salad toppings Snack foods Yogurt	Thacker and Dodgin (1991)
WS	Cereal grains	WS was concentrated under pressure (100 mbar to 800 mbar) in a high-pressure homogenizer to between 15% and 19% solids content; Yogurt- or butter-innoculated milk was then added to the WS, and allowed to ferment at T of 38°C to 48°C for 10 to 20 h, or T of 18°C to 24°C for 20 to 36 h; Additional milk was then added to produce $T <15°C$ Resulting product could be used directly or spray dried	Ready to consume products such as milk, fruit, or vegetable preparations	Tolle (2004)

Note: DDG is distillers dried grains; DDGS is distillers dried grains with solubles; DWG is distillers wet grains; T is temperature; WS is whole stillage.

studies found a decreased functionality as the level of distillers grains increased; this often included decreased volume and expansion during baking, increased moisture absorption (due to higher fiber levels), as well as negative impacts on texture and mouth feel. Further, for almost all food products investigated, relatively high levels of distillers grains had a definite negative impact on resulting flavor: most products were typically rated as marginally acceptable to not acceptable due to off-flavors. Poor flavor has been shown to improve, though, by bleaching and/or solvent extracting lipids from the DDG/DDGS prior to use. These processes remove fatty acids (which can hold the fermentation odors and flavors) (Dawson et al., 1985); these steps can also neutralize flavors imparted by organic salts resulting from the ethanol production process. On the other hand, almost all food products had improved nutrient profiles when ethanol coproducts were used as a partial flour replacement: both protein and fiber levels increased substantially as coproduct level increased. Even so, with humans, odor, flavor, and appearance are essential to ultimate acceptance.

Unfortunately, the poor success of the food development studies over the last 30 years has discouraged significant new research efforts into food uses for distillers grains. And not surprisingly, there are very few commercial food products available that use ethanol processing coproducts as ingredients. In fact, there are only two commercial products currently available: sour mash pancake mix (www.atasteofkentucky.com/shop/sour-mash-bourbon-pancake-p-139.html) and sour mash bread mix (www.atasteofkentucky.com/shop/sour-mash-bourbon-bread-p-140.html); both of which are distillery-based, not fuel ethanol-based.

18.5 RESEARCH NEEDS

Based upon the discussion of previous research, what needs to be done to move this concept forward? Essentially, there are two major thrusts that should be pursued. First, new research needs to be conducted on corn-based fuel ethanol coproduct in terms of their use as food ingredients. Second, new research needs to examine potential options to help overcome the aforementioned challenges to effective use (i.e., poor flavor, odor, texture, functionality).

It is true that the few studies that have been conducted with ethanol-based coproducts have indicated that they may have challenges similar to the distillery-based coproducts (which have been by far the majority of materials studied to date). But, as discussed in previous chapters, the U.S. fuel ethanol industry has grown and matured over the years, and many improvements and process modifications have been developed and deployed. And as the dry grind manufacturing process has gained prevalence, even more technology enhancements have been developed. Modern dry grind plants produce distillers grains with different nutrient contents and physical properties vis-à-vis those produced by their predecessors—the corn wet mills of the 1970s and 1980s, and the beverage distilleries. As this industry continues to expand, "next generation" coproducts and fractionated coproducts (which are discussed in other chapters) are becoming more widely available as well. Ethanol plants have become increasingly interested in expanding the utilization of distillers grains beyond the traditional use as livestock feed. The caveat, of course, is that any alternative use of distillers grains must be profitable for the ethanol manufacturer. Part of this equation will consist of constructing and operating plants that will produce food-grade products (which is not inexpensive); the other side of the equation will be finding market outlets for the food-grade distillers grains. Thus, to convert these new types of coproducts into value-added food ingredients, additional research work needs to be pursued.

In order for viable ingredients (i.e., food-grade distillers grains, fractions thereof, or potentially other coproducts) to be successfully manufactured and marketed from fuel ethanol plants, considerable research efforts are needed. According to Rosentrater (2006), six key aspects are essential to achieving this objective: (1) analysis of current commercial fuel DDGS streams and/or specific fractions for food-grade applicability, especially nutritional contents and chemical levels, including major constituents, vitamins, minerals, nucleic acids, pigments, heavy metals, and toxic or other deleterious compounds that may be present; (2) developing methods for processing and upgrading

DDGS streams and/or specific fractions into high-quality ingredient streams, including, possibly, (a) various pretreatments, such as separation/fractionation/concentration of proteins, fibers, lipids, or other compounds, (b) washing, cleaning, and other quality upgrading steps, (c) bleaching, (d) deodorizing, (e) drying, (f) sterilizing, (g) milling (i.e., into a finer particle size, in order to more closely resemble flour), (h) assessing storage stability and preservation, and (i) analysis of any residues that may result from these upgrading steps; (3) determining enhancements that may be necessary to improve flavor, odor, texture, functionality, and ultimately the utilization potential in food products; (4) demonstrating the viability and acceptability of these enhanced food ingredients in specific food products, such as breads, bakery goods, noodles, pastas, or other low starch/high protein/high fiber foods; (5) quantifying storability, shelf life, and preservation behavior of these resulting food products (e.g., determining lipid oxidation and protein solubility over time); and, finally, (6) sensory analysis and acceptability testing of these resulting food products.

At this point in time, a few of these ideas have actually been pursued, at least cursorily. Saunders et al. (2008) reviewed several methods that may be appropriate for bleaching or removing color pigments (particularly carotenoids) from DDGS. They determined that the use of hydrogen peroxide (to bleach), solvent extraction (to remove fat-soluble pigments), and/or the addition of lipoxygenases (to enzymatically reduce pigments) may be the most promising avenues for reducing the color of DDGS. Saunders (2008) then examined ethanolic extraction of DDGS using two particle sizes (0.38 or 0.33 mm), up to three distinct extractions, three extraction times (30, 60, or 90 min), and three ethanol concentrations (5, 10, or 15 mL/g), and found that up to 48% of the total pigments could be removed from the DDGS. A few ethanol plants have also examined solvent extraction, but on a commercial basis, in order to remove corn oil for the production of biodiesel. This has successfully led to low-fat DDGS, which has improved flavor, odor, and color (Figure 18.12) because fatty acids and other compounds that have entrained the "fermentation flavor" have been removed.

Grinding can also improve the color of DDGS. This phenomenon occurs because the outer surfaces of the particles are darker than the interiors, due to the DDGS production process, specifically coating by CDS as well as browning (Maillard as well as denaturation) that results from drying. Grinding thus releases the lighter colored materials inside the particles and dilutes the intensity of the DDGS color. This improvement in brightness can be seen both before and after baking. For example, Figure 18.13 illustrates the color improvement due to grinding of DDGS (at three specific particle size ranges); this improvement is still visible even after baking for 10 min at 200°C (to

FIGURE 18.12 Commercial DDGS samples. The sample on the left is traditional DDGS (lipid content ~10%); that on the right has been subjected to solvent extraction (lipid content ~2%) (author's unpublished data).

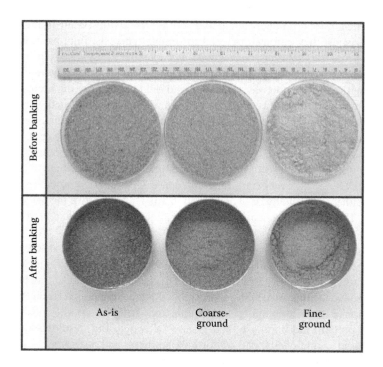

FIGURE 18.13 The finer the particle size, the lighter the color becomes. When heated (i.e., during baking—at 200°C for 10 min in this case), the DDGS darkens, but again, the finer the particle size the lighter the final color. Hunter L–a–b data as a function of particle size is provided in Figure 18.14 (author's unpublished data).

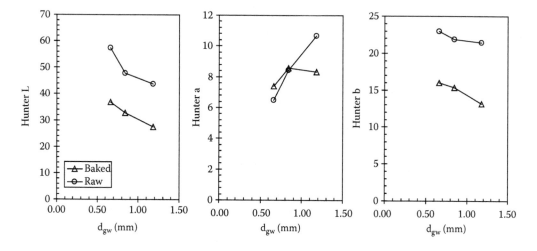

FIGURE 18.14 Grinding improves the color of DDGS, even after baking (at 200°C for 10 min in this example). Hunter L denotes lightness/darkness; a denotes red (+)/green (–); b denotes yellow (+)/blue (–) (author's unpublished data).

simulate a cookie baking test). Resultant Hunter color values are provided in Figure 18.14, which corroborates the visual observations.

Another promising avenue is the use of dough additives. A variety of conditioners, hydrocolloids, and enzymes can be added to ingredient blends to overcome shortcomings in flour quality and improve overall functionality (Azizi and Rao, 2004). They are often used to impart various functional properties to food products, including gelling, thickening, emulsifying, stabilizing, and foaming characteristics; they can also be used to inhibit syneresis, improve water retention and resulting texture properties (Rosell et al., 2007). For example, they can help increase the quality of baked products by increasing dough strength, increasing rates of hydration, mixing tolerance, crumb strength, loaf volume, shelf stability, and reducing shortening requirements. Specific examples of commonly used additives include guar gum, hydroxypropylmethylcellulose (HPMC), K-carrageenan, sodium alginate, sodium stearoyl lactylate (SSL), vital gluten, and xanthan gum (see, e.g., Collar et al., 1998; Rosell et al., 2001; Azizi and Rao, 2004; Guarda et al., 2004; Ribotta et al., 2004; Barcenas and Rosell, 2005; Caballero et al., 2006). Products high in fiber generally exhibit increased water absorption (Dreese and Hoseney, 1982), which will contribute to decreased baking functionality. As DDGS has a very high fiber content, it will readily absorb water. Thus various dough additives, either individually or in combination, may help improve the behavior of DDGS as a food ingredient. Unfortunately, not all additives are gluten free, and would not be appropriate for products targeted at Celiac diets.

Toward this end, Saunders (2008) examined the effect of adding SSL at levels of 0%, 0.15%, and 0.3% to ingredient mixes containing either all-purpose flour or bread flour, with three DDGS (fuel based) substitution levels (0%, 25%, and 50%). The qualities of the resulting bread loaves were enhanced for those blends that used SSL, and exhibited increasing dough strength, rates of hydration, mixing tolerance, crumb strength, loaf volume, and shelf life. As the substitution of DDGS increased, so did protein, moisture, ash, and Hunter "a" value. But, resulting loaf height, width, and length decreased. Additionally, cell uniformity, softness, and acceptability declined as well. Overall, 25% to 50% DDGS substitution appeared to have a negative effect on all physical features of the bread (even though the nutrient content was enhanced). Additional work, however, has found that DDGS substitution levels between 0% and 20% did not result in negative effects, either physically or organoleptically. And optimal DDGS substitution probably lies between 15% and 20%.

Even if milling, the use of additives, or pigment reduction can improve the functionality of the DDGS, it may still not work well in all baked products. In fact, it will probably have a higher probability of success if used in chemically leavened products compared to those that require yeast, as the available carbohydrate fractions have already been mostly consumed during the ethanol production process.

18.6 SAFETY AND REGULATION

A major barrier that impedes the potential use of distillers grains in human foods is the fact that currently fuel ethanol-based coproducts are not approved by the FDA for use as food ingredients, and thus they are currently not considered safe for consumption by humans. It is true that beverage alcohol production must strictly use only approved additives during processing, and must follow all food-grade requirements that are specified by the FDA and the Code of Federal Regulations; thus beverage-based distillers grains is already food-grade ethanol-based. But this is not the case for fuel coproducts; not all of the materials that are added during processing are approved for food use. Moreover, most fuel ethanol plants do not follow the specified Good Manufacturing Practices that are also required to produce food products (Title 21, Code of Federal Regulations, Part 110 (FDA, 2010a)). Even though distillers grains from fuel ethanol plants are required to be safe for livestock consumption, and thus to humans (who consume the livestock meat, milk, eggs, etc., as food), they are not automatically considered food-grade materials.

If a fuel ethanol manufacturer is interested in marketing distillers grains as a food ingredient, then that company will have to consult with the FDA (information about this process can be found in Title 21, Code of Federal Regulations, Parts 170–190 (FDA, 2010a)). During this consultative process, the manufacturer must provide extensive information about the safety of the proposed food ingredient (in this case DDGS), such as identity, proposed use, use level, intended technical effects in foods, and safety data, including chemical, microbiological, and toxicological information.

TABLE 18.7
Safety Issues and Regulatory Concerns Related to the Use of Ethanol Coproducts in Human Foods

Incoming grain
- Heavy metal content
- Pesticide content
- Mycotoxin content
- Dioxin content
- Other potential contaminants
- Genetically modified traits

Processing additives
- Enzymes (generally alpha-amylase and gluco-amylase)
- Yeast (generally *Sacchharomyces cerevisiae*)
- Urea (used as a protein source for fermentation)
- Ammonia or ammonium hydroxide (added during the slurry process prior to cooking)
- pH adjusters (such as sulfuric acid or caustic solutions [e.g., Clean-in-Place detergents and sanitizers such as sodium hydroxide])

Processing aids
- Antifoamers (used during fermentation)
- Antibiotics (currently virginiamycin is the only antibiotic in the United States that has been given "non-objection" status, although others have been detected in DDGS samples in recent years)
- Boiler chemicals (which prevent scaling and aid in steam production for the plant)

Other issues of concern
- Cleaning solutions (used in pipes, tanks, heat exchangers, etc.)
- Antifoulers (used in evaporators)
- Recycled process water (some plants only recycle backset, others also recycle water from anaerobic digesters, boilers, etc.)
- Salts
- The concentration of potential contaminants in the distillers grains

Note: Based upon discussions with the U.S. FDA (Food and Drug Administration).

Moreover, detailed information about the manufacturing facility, manufacturing methods, all ingredients used to manufacture that food ingredient, as well as the controls used in its production must also be provided. This means that for fuel ethanol plants, information must be provided on a host of issues, including (but not limited to) the items listed in Table 18.7. Some of the additives and chemicals that are used in a given plant may indeed be food grade, but many are not, which would thus preclude the use of the resulting distillers grains as a human food ingredient. Of course, the ability to produce a food-grade material will be specific to each plant, because each is unique vis-à-vis processing steps, additives, chemicals, and operational conditions used. More details about the information requirements can be found at the FDA's website (FDA, 2010b), especially the information found in the guidance for industry (FDA, 2010c), to which the reader is referred.

Distillers grains can only be approved as safe for human consumption if they meet all of the CFR's safety requirements, throughout the entire manufacturing process, beginning with the raw grain itself. In other words, any random lot of DDGS cannot be upgraded to food-grade status, regardless of the processes that may be applied to it.

18.7 CONCLUSIONS

Using ethanol coproducts in human foods is not a new concept. Considerable work over the years has examined the substitution of distillery-based DDG and/or DDGS for wheat flour in a multitude of baked products, although research on the use of fuel ethanol coproducts is quite limited. As the fuel ethanol industry evolves, and newer coproducts have become available, coupled with the shear volume of distillers grains that are now produced, the possibility of using these coproducts as high fiber/low starch/high protein human food ingredients should be considered. DDGS appears to offer multiple potential health benefits. But, there is still a substantial need to overcome the challenges associated with taste, smell, and functionality, in order to result in food products with qualities that are organoleptically and visually acceptable to the average consumer. Additionally, regulatory issues and food safety concerns need to be addressed.

REFERENCES

AACC. 2000. Method 10-50D—Baking quality of cookie flour. *Approved Methods of the American Association of Cereal Chemists,* 10th ed. St. Paul, MN: American Association of Cereal Chemists.

Abbott, J., J. O'Palka, and C. F. McGuire. 1991. Dried distillers' grains with solubles: Particle size effects on volume and acceptability of baked products. *Journal of Food Science* 56(5): 1323–1326.

Agatston, A. 2003. *The South Beach Diet.* Emmaus, PA: Rodale, Inc.

Ahmed, J. 1995. Utilization of corn distillers' grains in chipathi. M.S. Thesis. Brookings, SD: South Dakota State University.

American Association of Cereal Chemists. 2001. The definition of dietary fiber. *Cereal Foods World* 46: 112–129.

American Dietetic Association. 2002. Position of the American Dietetic Association: Health implications of dietary fiber. *Journal of the American Dietetic Association* 102(7): 993–1000.

American Dietetic Association. 2006. Nutrition care manual. Available online: www.nutritioncaremanual.org. Accessed December 21, 2009.

Anderson, J. W., B. M. Smith, and N. J. Gustafson. 1994. Health benefits and practical aspects of high-fiber diets. *American Journal of Clinical Nutrition* 59: 1242S–1247S.

Anderson, J. W., N. J. Gustafson, C. A. Bryant, and J. Tietyen-Clark. 1987. Dietary fiber and diabetes: a comprehensive review and practical application. *Journal of the American Dietetic Association* 87(9): 1189–1197.

Anderson, J. W., P. Baird, R. H. Davis Jr., S. Ferreri, M. Knudtson, A. Koraym, V. Waters, and C. L. Williams. 2009. Health benefits of dietary fiber. *Nutrition Reviews* 67(4): 188–205.

Angelich, A. P. R., and E. V. Symanski. 2004. Challenges in formulating low-carb bread products. *Cereal Foods World* 49(6): 326–330.

Arora, S. K., and S. I. McFarlane. 2005. The case for low carbohydrate diets in diabetes management. *Nutrition & Metabolism* 2(16): 1–9.

Arra, S., P. G. Krishnan, and K. A. Rosentrater. 2009. Fortifying chapathis, an Asian whole wheat unleavened bread, using corn distillers dried grains. *Poster No. 029-36, IFT Annual Meeting and Expo,* June 7, 2009. Anaheim, CA. Chicago, IL: Institute of Food Technologists.

Atkins, R. C. 1992. *Dr. Atkins' New Diet Revolution.* New York: Avon.

Azizi, M. H., and G. V. Rao. 2004. Effect of surfactant gel and gum combinations on dough rheological characteristics and quality of bread. *Journal of Food Quality* 27(5): 320–336.

Barcenas, M. E., and C. M. Rosell. 2005. Effect of HPMC addition on the microstructure, quality and aging of wheat bread. *Food Hydrocolloids* 19(6): 1037–1043.

Bechen, A. 2008. The glycemic response of distillers dried grains as a human food ingredient. M.S. Thesis. Brookings, SD: South Dakota State University.

Bloomgarden, Z. T. 2004. Type 2 diabetes in the young: The evolving epidemic. *Diabetes Care* 27(4): 998–1010.

Bookwalter, G. N., K. Warner, and Y. V. Wu. 1988. Processing corn distillers' grains to improve flavor: storage stability in corn–soy-milk blends. *Journal of Food Science* 53(2): 523–526.

Bookwalter, G. N., K. Warner, J. S. Wall, and Y. V. Wu. 1984. Corn distillers' grain and other by-products of alcohol production in blended foods. II. Sensory, stability, and processing studies. *Cereal Chemistry* 61(6): 509–513.

Brand-Miller, J. C. 2004. Postprandial glycemia, glycemic index, and the prevention of type 2 diabetes. *American Journal of Clinical Nutrition* 80: 243–244.

Brochetti, D., and M. P. Penfield. 1989. Sensory characteristics of bakery products containing distillers' dried grain from corn, barley, and rye. *Journal of Food Quality* 12: 413–426.

Brochetti, D., M. P. Penfield, and M. F. Heim-Edelman. 1991. Yeast bread containing distillers' dried grain: dough development and bread quality. *Journal of Food Quality* 14(4): 331–344.

Burkitt, D. 1977. Food fiber: benefits from a surgeon's perspective. *Cereal Foods World* 22(1): 6–9.

Buttriss, J. L., and C. S. Stokes. 2008. Dietary fibre and health: an overview *British Nutrition Foundation Nutrition Bulletin* 33: 186–200.

Caballero, P. A., M. Gomez, and C. M. Rosell. 2006. Bread quality and dough rheology of enzyme-supplemented wheat flour. *Zeitschrift fur Lebensmittel-Untersuchung und-Forschung A* 224(5): 525–534.

Chawla, R., and G. R. Patil. 2010. Soluble dietary fiber. *Comprehensive Reviews in Food Science and Food Safety* 9: 178–196.

Collar, C., E. Armero, and J. C. Martinez. 1998. Lipid binding of formula bread doughs. *Zeitschrift fur Lebensmittel-Untersuchung und Forschung A* 207(2): 110–121.

Dawley, L. J., and J. R. Dawley. 2003. *High Protein Corn Product production and Use. U.S. Patent No. 2003/0232109A1.* United States Patent and Trademark Office: Washington, DC. Available online: www.uspto.gov/. Accessed February 18, 2011.

Dawson, K. R., I. Eidet, J. O'Palka, and L. L. Jackson. 1987. Barley neutral lipid changes during the fuel ethanol production process and product acceptability from the dried distillers grains. *Journal of Food Science* 52(5): 1348–1352.

Dawson, K. R., J. O'Palka, N. W. Hether, L. Jackson, and P. W. Gras. 1984. Taste panel preference correlated with lipid composition of barley dried distillers' grains. *Journal of Food Science* 49: 787–790.

Dawson, K. R., R. K. Newman, and J. O'Palka. 1985. Effects of bleaching and defatting on barley distillers grain used in muffins. *Cereal Research Communications* 13(4): 387–391.

Dreese, V. G., and R. C. Hoseney. 1982. Baking properties of the bran fraction from brewer's spent grains. *Cereal Chemistry* 59(2): 89–91.

FDA. 2010a. *CFR—Code of Federal Regulations Title 21.* U.S. Food and Drug Administration: Washington, DC. Available online: www.accessdata.fda.gov/scripts/cdrh/cfdocs/cfcfr/cfresearch.cfm. Accessed February 18, 2011.

FDA. 2010b. *U.S. Food and Drug Administration Homepage.* U.S. Food and Drug Administration: Washington, DC. Available online: www.fda.gov. Accessed February 18, 2011.

FDA. 2010c. *Questions and Answers about the Petition Process.* U.S. Food and Drug Administration: Washington, DC. Available online: www.fda.gov/Food/GuidanceComplianceRegulatoryInformation/GuidanceDocuments/FoodIngredientsandPackaging/ucm078136.htm. Accessed February 18, 2011.

Feng, J., X. Liu, Z. R. Xu, Y. P. Lu, and Y. Y. Liu. 2007. Effects of fermented soybean meal on intestinal morphology and digestive enzyme activities in weaned piglets. *Digestive Diseases and Sciences* 52(8): 1845–1850.

Fennema, O. R. 1996. *Food Chemistry,* 3rd ed. New York: Marcel Dekker Inc.

Gross, L. S., L. Li, E. S. Ford, and S. Liu. 2004. Increased consumption of refined carbohydrates and the epidemic of type 2 diabetes in the United States: an ecologic assessment. *American Journal of Clinical Nutrition* 79: 774–779.

Guarda, A., C. M. Rosell, C. Benedito, and M. J. Galotto. 2004. Different hydrocolloids as bread improvers and antistaling agents. *Food Hydrocolloids* 18: 241–247.

Gylling, H., and T. A. Miettinen. 2005. The effect of plant stanol- and sterol-enriched foods on lipid metabolism, serum lipids and coronary heard disease. *Annals of Clinical Biochemistry* 42: 254–263.

Hofman, Z., J. D. E. van Drunen, C. de Later, and H. Kuipers. 2004. The effect of different nutritional feeds on the postprandial glucose response in healthy volunteers and patients with type II diabetes. *European Journal of Clinical Nutrition* 58: 1553–1556.

Hong, K. -J., C. -H. Lee, and S. W. Kim. Aspergillus oryzae GB-107 fermentation improves nutritional quality of food soybeans and feed soybean meals. *Journal of Medicinal Food* 7(4): 430–435.

Hursh, H., and J. Martin. 2005. Low-carb and beyond: the health benefits of inulin. *Cereal Foods World* 50(2): 57–60.

Jones, J. M. 2002. *Contradictions and Challenges: A look at the Glycemic Index.* Wheat Foods Council: Parker, CO. Available online: www.wheatfoods.org/. Accessed December 21, 2009.

Kim, C. H., J. A. Maga, and J. T. Martin. 1989a. Properties of extruded blends of wheat dried distiller grain flour with other flours. *International Journal of Food Science and Technology* 24(4): 373–384.

Kim, C. H., J. A. Maga, and J. T. Martin. 1989b. Properties of extruded dried distiller grains (DDG) and flour blends. *Journal of Food Processing and Preservation* 13(3): 219–231.

Kochhar, S. P. 1983. Influence of processing on sterols of edible vegetable oils. *Progress in Lipid Research* 22: 161–188.

Lafrance, L., R. Rabasa-Lhoret, D. Poisoon, F. Ducros, and J. L. Chiasson. 1998. The effects of different gly-caemic index foods and dietary fiber intakes on glycaemic control in type 1 patients with diabetes on intensive insulin therapy. *Diabetic Medicine* 15(11): 972–978.

Layman, D. K., and J. I. Baum. 2004. Dietary protein impact on glycemic control during weight loss. *The Journal of Nutrition* 134(4): 968S–973S.

Lee, A., and J. Newman. 2003. Celiac diet: Its impact on quality of life. *Journal of the American Dietetic Association* 103(11): 1533–1535.

Leguizamon, C., C. L. Weller, V. L. Schlegel, and T. P. Carr. 2009. Plant sterol and policosanol characterization of hexane extracts from grain sorghum, corn and their DDGS. *Journal of the American Oil Chemists Society* 86: 707–716.

Li, J., T. Kaneko, L.-Q. Qin, J. Wang, Y. Wang, and A. Sato. 2003. Long-term effects of high dietary fiber intake on glucose tolerance and lipid metabolism in GK rats: Comparison among barley, rice, and cornstarch. *Metabolism* 52(9): 1206–1210.

Lilliard, D. A., G. L. Christen, and J. S. Smith. 2000. *Food Chemistry: Principles and Applications,* pp. 131–148. West Sacramento, CA: Science Technology System.

Maga, J. A., and K. E. Van Everen. 1989. Chemical and sensory properties of whole wheat past products supple-mented with wheat-derived dried distillers grain (DDG). *Journal of Food Processing and Preservation* 13: 71–78.

Mahan, L. K., and S. Escott-Stump. 2004. *Krause's Food, Nutrition, & Diet Therapy,* 11th ed., ed. Y. Alexopoulos. Philadelphia, PA: Saunders Press.

Mayo Clinic. 2006. Celiac disease. Available online: http://www.mayoclinic.com/health/celiac-disease/DS00319/DSECTION=9.-. Accessed December 21, 2009.

Mehta, R. S. 2005. Dietary fiber benefits. *Cereal Foods World* 50(2): 66–71.

Miller, C., R. A. Gabbay, J. Dillon, J. Apgar, and D. Miller. 2006. The effect of three snack bars on glycemic response in healthy adults. *Journal of the American Dietetic Association* 106(5): 745–748.

Morad, M. M., C. A. Doherty, and L. W. Rooney. 1984. Utilization of dried distillers grain from sorghum in baked food systems. *Cereal Chemistry* 61(5): 409–414.

O'Palka, J., I. Eidet, and J. Abbott. 1989. Use of sodium bicarbonate and increased liquid levels in baked prod-ucts containing sour mash corn dried distillers' grains. *Journal of Food Science* 54(6): 1507–1514.

Ostlund, Jr., R. E. 2002. Phytosterols in human nutrition. *Annual Reviews of Nutrition* 22: 533–549.

Ostman, E. M., A. H. Frid, L. C. Groop, I. M. E. Bjorck. 2006. A dietary exchange of common bread for tai-lored bread of low glycemic index and rich in dietary fibre improves insulin economy in young women with impaired glucose tolerance. *European Journal of Clinical Nutrition* 60: 334–341.

Power, R. F. 2006. Methods for Improving the Nutritional Quality of Residues of the Fuel, Beverage Alcohol, Food and Feed Industries. *U.S. Patent No. 2006/0233864A1.* United States Patent and Trademark Office: Washington, DC. Available online: www.uspto.gov/. Accessed February 18, 2011.

Rasco, B. A., and F. M. Dong. 1991. Baking and storage stability properties of high fiber breads containing comparable levels of different fiber ingredients. *Journal of Food Processing and Preservation* 15(6): 433–442.

Rasco, B. A., A. E. Hashisaka, F. M. Dong, and M. A. Einstein. 1989. Sensory evaluation of baked foods incor-porating different levels of distillers' dried grains with solubles from soft white winter wheat. *Journal of Food Science* 54(2): 337–342.

Rasco, B. A., and W. J. McBurney. 1989. Human Food Product Produced from Dried Distillers' Spent Cereal Grains and Solubles. *U.S. Patent No. 4,828,846,* United States Patent and Trademark Office: Washington, DC. Available online: http://www.uspto.gov/. Accessed February 18, 2011.

Rasco, B. A., G. Rubenthaler, M. Borhan, and F. M. Dong. 1990b. Baking properties of bread and cookies incor-porating distillers' or brewer's grain from wheat or barley. *Journal of Food Science* 55(2): 424–429.

Rasco, B. A., S. E. Downey, and F. M. Dong. 1987a. Consumer acceptability of baked goods containing distill-ers' dried grains with solubles from soft white winter wheat. *Cereal Chemistry* 64(3): 139–143.

Rasco, B. A., S. E. Downey, F. M. Dong, and J. Ostrander. 1987b. Consumer acceptability and color of deep-fried fish coated with wheat or corn distillers' dried grains with solubles (DDGS). *Journal of Food Science* 52(6): 1506–1508.

Rasco, B. A., S. S. Gazzaz, and F. M. Dong. 1990a. Iron, calcium, zinc, and phytic acid content of yeast-raised breads containing distillers' grains and other fiber ingredients. *Journal of Food Composition and Analysis* 3(1): 88–95.

Reddy, J. A. and R. Stoker. 1991. Bakery Product from Distiller's Grain. *U.S. Patent No. 5,225,228,* United States Patent and Trademark Office: Washington, DC. Available online: http://www.uspto.gov/. Accessed February 18, 2011.

Reddy, N. R., F. W. Cooler, and M. D. Pierson. 1986a. Sensory evaluation of canned meat-based foods supplemented with dried distillers grain flour. *Journal of Food Quality* 9(4): 233–242.

Reddy, N. R., M. D. Pierson, and F. W. Cooler. 1986b. Supplementation of wheat muffins with dried distillers grain flour. *Journal of Food Quality* 9(4): 243–249.

Rendell, M., J. Vanderhoof, M. Venn, M. A. Shehan, E. Arndt, C. S. Rao, G. Gill, R. K. Newman, and C. W. Newman. 2005. Effect of a barley breakfast cereal on blood glucose and insulin response in normal and diabetic patients. *Plant Foods for Human Nutrition* 60: 63–67.

Ribotta, P. D., G. T. Perez, A. E. Leon, and M. C. Anon. 2004. Effect of emulsifier and guar gum on micro structural, rheological and baking performance of frozen bread dough. *Food Hydrocolloids* 19: 93–99.

Rosell, C. M., J. A. Rojas, and C. Benedito. 2001. Influence of hydrocolloids on dough rheology and bread quality. *Food Hydrocolloids* 15: 75–81.

Rosell, C. M., C. Collar, and M. Haros. 2007. Assessment of hydrocolloid effects on the thermo-mechanical properties of wheat using the Mixolab. *Food Hydrocolloids* 21: 452–462.

Rosentrater, K. A. 2006. Expanding the role of systems modeling: considering byproduct generation from biofuel production. *Ecology and Society* 11(1): 1–12.

Rosentrater, K. A., and P. G. Krishnan. 2006. Incorporating distillers grains in food products. *Cereal Foods World* 51(2): 52–60.

Saunders, J. A. 2008. Analysis of physical, chemical, and functionality properties of distillers dried grains with solubles (DDGS) for use in human foods. M.S. Thesis. Brookings, SD: South Dakota State University.

Saunders, J. A., K. A. Rosentrater, and P. Krishnan. 2008. Potential bleaching techniques for corn distillers grains. *Journal of Food Technology* 6(6): 242–252.

Schofield, J. D., and M. R. Booth. 1983. Wheat proteins and their technological significance. In *Developments in Food Proteins-2*, ed. B. J. F. Hudson. London: Applied Science Publishers.

Schulze, M. B., S. Liu, E. B. Rimm, J. E. Manson, W. C. Willett, and F. B. Hu. 2004. Glycemic index, glycemic load, and dietary fiber intake and incidence of type 2 diabetes in younger and middle-age women. *The American Journal of Clinical Nutrition* 80: 348–356.

Sethi, R. 2004. Beneficial effect of distiller's grain in cardiovascular disease. *U.S. Patent No. 2004/0234630 A1*. United States Patent and Trademark Office: Washington, DC. Available online: http://www.uspto. gov/. Accessed February 18, 2011.

Sethi, R. 2007. Beneficial effects of DDG in ischemic heart disease. *FASEB Journal* 21(6): 1b377.

Sethi, R., and D. Diaz. 2007. Distillers grains as a dietary additive for the prevention of cardiovascular disease. *Distillers Grains Quarterly* 4: 24–27.

Shukla, C. Y., K. Muthukumarappan, and J. L. Julson. 2005. Effect of single-screw extruder die temperature, amount of distillers' dried grains with solubles (DDGS), and initial moisture content on extrudates. *Cereal Chemistry* 82(1): 34–37.

Silver, R. S., K. E. Petrofsky, P. H. Brown, C. A. Leduc, and Y. -Y. Pai. 2008. DDGS as a Low-Cost Flavor Enhancer and Sodium Reduction Enabler. *U.S. Patent No. 2008/0160132A1*, United States Patent and Trademark Office: Washington, DC. Available online: http://www.uspto.gov/. Accessed February 18, 2011.

Singh, V., R. A. Moreau, K. B. Hicks, R. L. Belyea, and C. H. Staff. 2002. Removal of fiber from distillers dried grains with solubles (DDGS) to increase value. *Transactions of the ASAE* 45(2): 389–392.

Sloan, A. E. 2004. The low-carb diet craze. *Food Technology* 58(1): 16.

Thacker, R. S., and B. A. Dodgin. 1991. Process for the Co-production of Ethanol and an Improved Human Food Product from Cereal Grains. *U.S. Patent No. 5,061,497*, United States Patent and Trademark Office: Washington, DC. Available online: http://www.uspto.gov/. Accessed February 18, 2011.

Tolle, M. 2004. *U.S. Patent No. 6,706,294*, Method for Producing Foodstuffs, Dietetic Foodstuffs and Food Additives on the Basis of Grain Stillage. United States Patent and Trademark Office: Washington, DC. Available online: http://www.uspto.gov/. Accessed February 18, 2011.

Tsen, C. C., J. L. Weber, and W. Eyestone. 1983. Evaluation of distillers' dried grain flour as a bread ingredient. *Cereal Chemistry* 60(4): 295–297.

Tsen, C. C., W. Eyestone, and J. L. Weber. 1982. Evaluation of the quality of cookies supplemented with distillers' dried grain flours. *Journal of Food Science* 47: 684–685.

Van Everen, K., J. A. Maga, and K. Lorenz. 1992. Spaghetti products containing dried distillers grains. *Developments in Food Science* 29: 551–563.

Wall, J. S., Y. V. Wu, W. F. Kwolek, G. N. Bookwalter, and K. Warner. 1984. Corn distillers' grains and other by-products of alcohol production in blended foods. I. Compositional and nutritional studies. *Cereal Chemistry* 61(6): 504–509.

Wampler, D. J., and W. A. Gould. 1984. Utilization of distillers' spent grain in extrusion processed doughs. *Journal of Food Science* 49: 1321–1322.

Wang, X., S. Choi, and W. L. Kerr. 2004. Water dynamics in white bread and starch gels as affected by water and gluten content. *Lebensmittel-Wissenschaft und-Technologie* 37(3): 377–384.

Winkler, J. K., K. A. Rennick, F. J. Eller, and S. F. Vaughn. 2007. Phytosterol and tocopherol components in extracts of corn distiller's dried grain. *Journal of Agricultural and Food Chemistry* 55: 6482–6486.

Winkler-Moser, J. K., and S. F. Vaughn. 2009. Antioxidant activity of phytochemicals from distillers dried grain oil. *Journal of the American Oil Chemists Society* (in press): DOI 10.1007/s11746-009-1439-7.

Wu, Y. V., V. L. Youngs, K. Warner, and G. N. Bookwalter. 1987. Evaluation of spaghetti supplemented with corn distillers' dried grains. *Cereal Chemistry* 64(6): 434–436.

Part V

Emerging Uses

19 Using DDGS in Industrial Materials

Nicholas R. DiOrio, Robert A. Tatara, Kurt A.
Rosentrater, and Andrew W. Otieno

CONTENTS

19.1 INTRODUCTION

As discussed in the previous chapter, the use of distillers dried grains with solubles (DDGS) as an ingredient for human foods is one method of potentially adding value to distillers grains; use in industrial materials is another. The success of the plastics industry has significantly impacted the environment; the relative inertness of disposable (e.g., one-time use) products, along with products that have exceeded their usefulness and are discarded, can lead to pollution. Solid plastic waste must be dealt with on a large scale. Some plastics are recycled, although the effort requires collection, handling, sorting, cleaning, and remanufacture preparation. There is substantial cost associated with recycling and not all plastics are recyclable. Another option is incineration, a process in which the plastic is burned as fuel to generate electricity. However, issues such as hazardous gas emissions, global warming, greenhouse gases, ash disposal, and heavy metals make incineration a difficult and costly option. A third option is landfill disposal. This also has limitations as valuable area is taken and improper landfill operation releases contaminants that may enter local groundwater and pollute the water supply. Use of biobased or "green" plastics as well as other materials that are biodegradable can minimize environmental effects.

In the effort to manufacture biobased plastics and polymers, two options are available: process biomaterials directly into 100% biobased plastics or use a biomaterial in combination with a synthetic material. Examples of the former option include films made from biomaterials such as alginic acid, cellulose, chitin, curdlan, lignin, soy protein, starch, xanthan, xylan, whey, and zein, and foams derived from starch, corn stover, and soybean oil (Rosentrater and Otieno, 2006). For the latter option, flax, palm, sisal, jute, and DDGS fibers have reinforcing potential in natural and synthetic polymers. It should be noted that approximately one-third of DDGS is fiber; the presence of fiber in the dried solid content in combination with a plastic resin may improve some mechanical properties by creating a biocomposite similar to wood–plastic composites. Generally, reinforcing agents consist of relatively long fibers combined with a resin matrix during composite processing to enhance mechanical strength. Primarily tensile, compressive, impact, and bending strengths are all greatly increased. Reinforcing materials range from common and inexpensive glass to exotic polymer or carbon fibers.

Biomaterials can also be blended with synthetic (or natural) polymers as simple filler instead of reinforcement. Currently, many plastic products utilize low-cost materials as fillers. Ideally filler is added in a concentration that allows the product to retain sufficient mechanical strength, physical properties, and final quality. Certain fillers may improve a plastic's other properties, such as thermal stability, color, and opacity but the primary benefit is in replacing the higher-cost polymer resin. Common fillers include clay, talc, ground limestone, carbon black, marble dust, glass, paper, wood flour, and metals, which are often added in concentrations ranging from 10% to 50% (by weight). Recently, to conserve petroleum resources and enhance biodegradability, biobased fillers are receiving increased attention. Examples include sugar cane, palm, jute, sisal, lignin, flax, grasses, bamboo, starch, chicken feathers, soy protein, and cellulose. This trend is consistent with recent U.S. governmental policies that prioritize the procurement of materials having significant biobased content: this program is targeted to increase to a 50% biobased level over the next several decades.

Adding certain biological materials as fillers to plastics can enhance any existing biodegradability or provide biodegradability where none had previously existed. One potential biofiller is DDGS. In fact, several studies have been conducted recently that have investigated the use of DDGS in various plastics applications. Table 19.1 provides an overview of most of these studies. Some have investigated production-oriented operations, such as compression and injection molding; others have pursued small-scale hand or laboratory forming methods. They will be discussed in more depth through the course of the chapter.

Overall, DDGS as raw industrial material provides an additional market for the agriculture industry and is a higher-value utilization opportunity for these processing coproducts.

19.2   COMBINING DDGS WITH BIOPLASTICS

Most advantageously would be to combine DDGS with a biobased resin. This maximizes the "green" content and biodegradation potential. Accordingly, one potential resin for DDGS is polylactic acid (PLA), a biocompatible and biodegradable polymer. To investigate this, test specimens were prepared with PLA and 0%, 10%, 20%, 30%, 40%, by weight, corn-derived DDGS from an ethanol plant in the Midwest United States. The specimens were injection molded, and subsequent mechanical testing included tensile strength, hardness, stiffness, and ductility. To prepare PLA resin for molding required drying. Per the resin manufacturer's recommendation, the PLA was spread in a tray to a depth of 1/2 in (13 mm) and dried between 140°F to 170°F (60°C to 77°C) for 7 to 10 h in a 6142 Btu/h (1800 W) tray dryer. DDGS was similarly dried. The PLA and DDGS were dried separately and blends were mixed immediately. The coarse powder grains were mixed by weight, with PLA pellets and placed in the hopper of the injection molding machine to be molded. This process was adequate for pure PLA as well as the 30% and 40% DDGS blends. But the dried mixtures with 10% and 20% DDGS content did not injection mold well; instead the DDGS appeared to form clumps in the molder and could not be injected into the closed mold. Thus for 10% and 20% blends, both the DDGS and PLA were not dried but directly molded without any difficulty.

The hydraulic injection molder used for this study was a three-platen, 27.5-ton (245-kN), with a screw diameter of 1 in (25 mm), and a maximum injection pressure of 22,300 psi (154 MPa). The mold for producing the test specimens had one tensile bar cavity and one bending bar. The tensile bars were manufactured according to ASTM D638-03, while the bending bars were consistent with ASTM D747-02. To maintain a constant mold body temperature, a small water bath circulated a water and ethylene glycol solution in a U-shaped pattern inside the mold body. The bath's capacity was 1.5 gal (5.7 L). Table 19.2 presents the complete injection molding conditions. The blends were molded at nearly the same conditions.

Once molded, the tensile and bending bars (Figure 19.1) were conditioned at 73.4°F ± 36°F (23°C ± 2.0°C) and 50% ± 5% humidity for a minimum of 40 h. Tensile pulls were performed with a

TABLE 19.1
Overview of Work Published to Date on Use of DDGS in Biocomposites

Process	Polymer Matrix	DDGS Inclusion Level (%)	Tensile Strength (MPa)	Tensile Modulus at Yield (MPa)	Composite Properties Flexural Strength at Yield (MPa)	Flexural Modulus (MPa)	Elongation at Yield (%)	Citation
Compression molding	Phenolic resin	0–75	4.3–28.5	841–2632	—	—	0.54–1.13	Tatara et al. (2009)
	Phenolic resin	0–90	7.7–32.1	1076–2878	—	—	0.73–1.25	Tatara et al. (2007)
	Castor oil-based polyurethane	50–75	11.6–16.0	249–312	—	—	5.2–13.8	Wu and Mohanty (2007)
	Phenolic resin glue	0–50	—	—	0.15–0.38	0.006–0.035	—	Cheesbrough et al. (2008)
Injection molding	Polyethylene	20–30	14.25–16.95	392–425	20.2–20.6	416–452	—	Julson et al. (2004)
	Polyethylene, zein	25–50	12.6–19.0	—	—	—	—	Vogel and Grewell (2008)
	Polypropylene	20–30	24.90–27.80	697–736	41.7–42.6	1020–1110	—	Julson et al. (2004)
Film formation	Lipid and zein extraction	100	7–42	—	—	—	0.4–3.3	Xu et al. (2009)
Foam	MDI	33–50	—	—	—	—	—	Yu et al. (2008)

TABLE 19.2
Typical Injection Molding Conditions for PLA Composites

Zone 1 (hopper)	377 (192)	°F (°C)
Zone 2 (middle of screw)	392 (200)	°F (°C)
Zone 3 (injection end)	386 (197)	°F (°C)
Zone 4 (nozzle)	386 (197)	°F (°C)
Shot size	0.505 (14.3)	oz (g)
Cooling time	50	s
Cycle time	60	s
Clamping pressure	2611 (18.0)	psi (MPa)
Injection pressure	22,300 (154)	psi (MPa)
Injection time	4	s
Injection speed	1.0 (25)	in/s (mm/s)
Inlet mold circulating water temperature	82.5 (28.1)	°F (°C)
Outlet mold circulating water temperature	82.4 (28.0)	°F (°C)

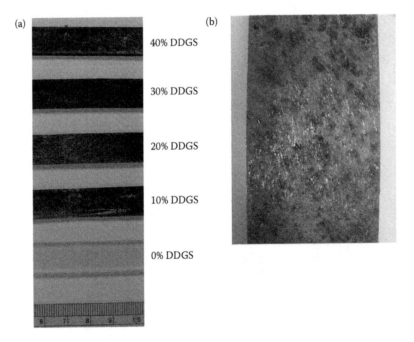

FIGURE 19.1 DDGS/PLA composites manufactured by injection molding: (a) comparison of each of the blends shows that as the DDGS level increases, the resulting product becomes darker, due to the color of the DDGS scale bar indicates cm; (b) 10× magnification illustrates how the DDGS particles are fairly uniformly dispersed within the composite matrix.

5500-lb$_f$ (24.5-kN) tester, while a 6-in-lb$_f$ (0.68-J) stiffness tester was used to determine flexural modulus. Three tensile bar specimens were tested and values averaged for each blend that was produced. Flexural modulus was also obtained by averaging three tests using the bending bars for each blend. Shore D surface hardness measurements were taken at five different locations on both sides of a bending bar and a tensile bar; this was replicated three times.

Generally the engineering stress–strain curves exhibited a linear region at low loading, from which Young's modulus was computed. The data displayed a traditional yield point with breakage occurring shortly afterwards with little extra extension. The yield also corresponded to the

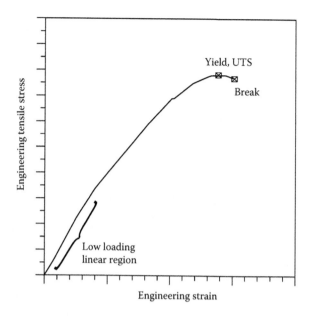

FIGURE 19.2 Typical stress–strain relationship for DDGS/PLA composites.

TABLE 19.3
Effect of DDGS Concentration on Mechanical Properties of PLA Composites

Property	DDGS Level (%)				
	0	10	20	30	40
Ultimate tensile stress (UTS), psi	8894	6481	4097	3385	3290
(MPa)	(61.3)	(44.7)	(28.2)	(23.3)	(22.7)
Break stress, psi	8754	6436	3448	3051	2887
(MPa)	(60.4)	(44.4)	(23.8)	(21.0)	(19.9)
Young's modulus, psi	270,690	304,130	278,160	251,830	305,140
(MPa)	(1866)	(2097)	(1918)	(1736)	(2104)
Flexural modulus, psi	342,730	326,890	338,140	338,070	300,870
(MPa)	(2363)	(2254)	(2331)	(2331)	(2074)
Elongation to UTS, %	5.1	3.3	2.3	2.0	2.0
Elongation to break, %	5.4	3.4	2.8	2.3	2.2
Hardness–Shore D	79	77	77	77	78

maximum loading (ultimate tensile stress) of the material. Qualitatively, Figure 19.2 presents a typical stress–strain relationship for the blends.

For each blend, average values for the ultimate tensile strength (UTS), break stress, Young's modulus, flexural modulus, percent elongation to UTS, percent elongation to break, and hardness are provided in Table 19.3. As the DDGS content increased, the blend's tensile strength decreased. UTS ranged from 8894 to 3290 psi (61.3 to 22.7 MPa) as DDGS content increased from 0% to 40%. The presence of DDGS had a slightly greater effect on the break stress, which decreased to 2887 psi (19.9 MPa) at the maximum DDGS level. Generally, the break stress was 90% of the yield for the blends. The stiffness of the specimens, as measured by Young's modulus, was slightly greater when DDGS was present, although a definite trend was unclear. The flexural modulus (relative stiffness), on the other hand, was mostly unchanged until 40% DDGS, when it was 12% less. The elongation at the maximum stress behaved similar to the tensile strength. It ranged from 5.1% to 2.0% as DDGS content increased from 0% to 40%. Shore D surface hardness was slightly less when DDGS was introduced into the PLA.

It appears that mixing DDGS with a bioplastic such as PLA may be promising. Up to 40% DDGS still provided adequate mechanical strength for many applications, although, it was substantially lower than pure PLA. Certainly the low-cost feature of DDGS and its potential to accelerate biodegradability makes it attractive, as most bioplastics tend to be expensive compared to their petroleum-based counterparts.

Along this line, another set of experiments investigated use of a bioplastic by blending corn-derived DDGS with a commercially available biodegradable corn-starch plastic resin. The DDGS was a tan color, coarse powder, and obtained from an Illinois commercial dry grind ethanol plant. Using standard AOAC methods, the DDGS had a protein content of 28.3%, a fiber content of 7.9%, a fat content of 11.0%, an ash content of 5.0%, and a nitrogen-free extract of 48.0% (on a dry basis). Moisture content was 10% to 12% (on a dry basis). Two treatments of DDGS were tested: flaked and ground powder. The flaked DDGS was coarser with larger particle size, and was actually the natural, raw state when produced in an ethanol plant, while the powdered DDGS was obtained after grinding the flakes into smaller particles. As a control, samples with no DDGS (100% corn-starch resin) were also tested. The blends were mechanically tested for tensile strength, material stiffness, ductility, and hardness. In addition to the mechanical properties, other properties such as biodegradability, surface finish, color, and general appearance were evaluated, as these qualitative aspects are also important. The results were extrapolated to optimize the blending and molding conditions needed to produce commercially viable products.

To process tensile test specimens, injection molding was conducted with a cycle time of approximately 0.5 min. Other injection molding parameters varied from blend to blend, and an optimal set of molding conditions was chosen for each 10% or 25%, flaked or ground blend. Some of the runs required water cooling of the mold. After processing, the specimens were placed in a constant temperature/humidity chamber for conditioning at 73.4°F ± 3.6°F (23°C ± 2°C) with 50% ± 5% relative humidity, for at least 40 h prior to tensile testing, following the prescribed ASTM procedure.

All specimens were pulled with a 5500-lb$_f$ (24.5-kN) tensile tester until breakage. For all the data points, ultimate engineering tensile strength, Young's modulus, and percent elongation to break were based on an arithmetic mean of three samples. Surface hardness was measured with a Durometer-type Shore D scale indentor; each test specimen was sampled by averaging eight readings at various locations, then averaging over three samples. Biodegradability tests were 20 weeks in duration and performed per ASTM D5988-03 and ASTM D6400-04.

Typically all specimens displayed moderate strength until yielding, where the material elongated without an increase in tensile loading, at or very near the UTS. After the yield point, the material continued to elongate at diminishing loading until break. Hardness measurements were consistent throughout each specimen and the blends readily biodegraded. Figure 19.3a presents the UTS for the DDGS/corn-starch blends, including the 100% starch control case. It can be seen that, as expected with any filler, the presence of DDGS lowered the material's strength. Using 25% and 10% DDGS reduced the corn-starch's strength by 45% and 25%, respectively. There was not significant difference between blending flaked (raw) or ground DDGS. The stiffness, as measured through Young's modulus and shown in Figure 19.3b, decreased 15% for 25% raw or ground DDGS compared to the control. But 10% DDGS increased the stiffness approximately 10%. There was no significant difference between the raw and ground filler. The flexibility of the biomaterial is given in terms of its elongation to break. Figure 19.3c indicates that all corn-starch formulations were rather brittle, similar to the PLA and DDGS blends, with less than 5% elongation. No consistent trends were found, although 25% raw or ground DDGS showed less flexibility than the control. However, the raw 10% DDGS blend exhibited better strain than the 0% DDGS case; it is possible that the larger DDGS particle size held the corn-starch matrix together longer. The 10% ground blend was

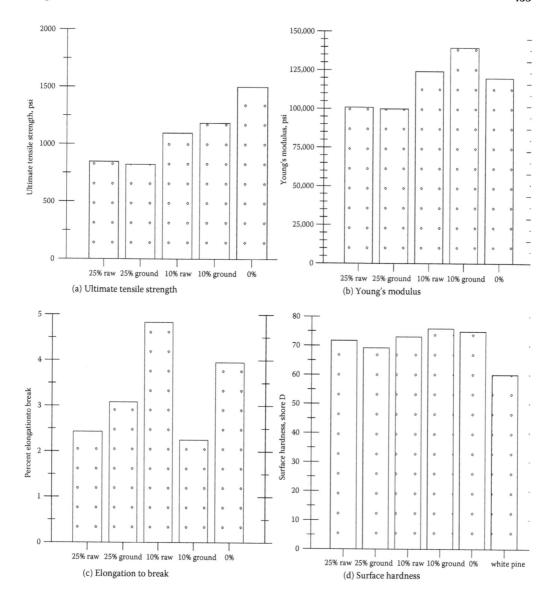

FIGURE 19.3   Mechanical properties of DDGS/corn-starch blends.

less ductile than pure corn-starch. Figure 19.3d presents the hardness, and indicates that the DDGS blends were only slightly softer than pure corn-starch; no appreciable difference existed between raw and ground DDGS. A value for white pine is also shown in the figure; all corn-starch blends were somewhat harder than the wood, but slightly softer than PLA blends. Figure 19.4 demonstrates the accelerated biodegradability as the DDGS content increased. For these tests, a 50% ground and a 35% raw blend were also available. Here the relative biodegradability is directly correlated with the rate of mass loss. Therefore a doubling in the y-axis value indicates that a specimen will decompose in soil in half the time. Clearly, 25% or more DDGS greatly increased biodegradability compared to the pure corn-starch material and white pine wood. Raw and ground additive conditions exhibited similar performance.

As DDGS contains protein and cellulosic fiber, it seems amendable to direct processing into a biobased plastic. But due to low protein content, any resulting DDGS bioplastic would probably be relatively weak. Thus one proposed method is to combine raw or ground DDGS with zein protein

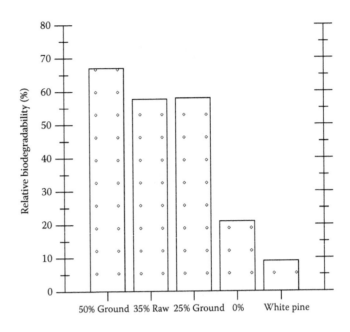

FIGURE 19.4 Biodegradability (%) of DDGS/corn-starch blends.

to form a better composite material (Vogel et al., 2007). In that study, DDGS/zein mixtures (by weight) of 0/100, 25/75, 50/50, 75/25, and 100/0 were formulated. However, fabricating compounds with high concentration of DDGS through traditional methods such as injection molding and extrusion is impractical. Also, such methods may damage the fiber component of the grains. Thus the researchers used ultrasonic energy to compact and thermally fuse the blends into tensile bars that were then tested for ultimate tensile stress. It was found that 3 to 4 s was a reasonable compaction time; a shorter time led to incomplete fusion while longer times thermally degraded the blends. Due to the low bulk density of the materials, the compaction took place layer by layer with two or three needed to fully fill the mold cavity. Using optimal compaction time and amplitude, 25% ground DDGS generally exhibited the greatest UTS, just over 1450 psi (10 MPa). Figure 19.5 shows the various effects of DDGS pretreatment and concentration, with a compaction time of 5 s and ultrasonic amplitude of 0.0011 in (28 μm), with the ground DDGS blends somewhat outperforming the raw DDGS. No difference existed between 50% and 75% DDGS, which had 50% to 60% the strength of the 25% blends.

The efforts with PLA and corn-starch, coupled with the growing ethanol industry and the resultant increase in supply of distillers grains as coproducts, suggest that there may be commercial opportunity. There are many current applications of plastics and composites that utilize fillers and extenders such as wood flour and other cellulosic matter. Therefore, the shift to use of distillers grains in plastics seems viable and reasonable. It is also plausible that some of the core components within the distillers grains such as the oil, fiber, and lignin may play a role beyond that of merely filler. It may be that these ingredients positively impact the ability to mold or extrude plastic products as well as provide certain contributions to material properties.

Any proposed products using "green" plastics with biofillers may also offer marketing benefits to manufacturers. The biomass presence may add more natural characteristics to plastics and plastic-based products. Many of these already use fillers such as wood flour (refined sawdust). However, wood flour is increasing in price as wood becomes more of an endangered resource. DDGS, from easily grown corn, is a potential economical substitute. Thus an immediate application of DDGS is as a wood flour substitute or alternative. One example is a biodegradable golf tee (Figure 19.6a) that was manufactured from a DDGS corn-starch polymer blend. Field trials and qualitative inspection

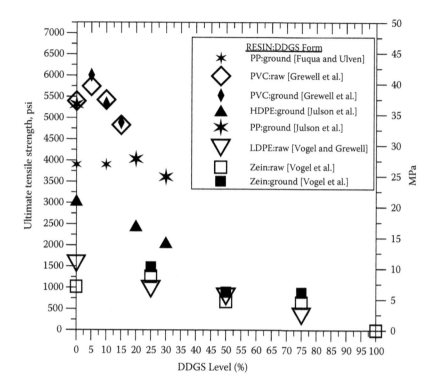

FIGURE 19.5 Ultimate tensile strength for various plastic matrices with DDGS.

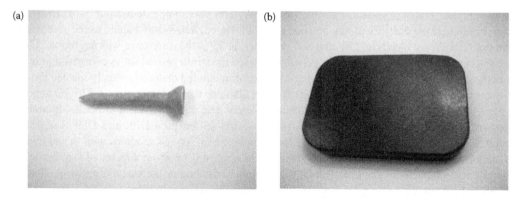

FIGURE 19.6 Potential commercial applications of DDGS-based composites: (a) golf tee manufactured by injection molding a blend of DDGS and corn-starch; (b) compression molding a 50% DDGS/50% phenolic resin blend has resulted in commercially viable composite for tile applications.

of the golf tees showed a smooth surface, typical of injection-molded parts, along with performance properties comparable to existing products in the marketplace. Figure 19.6b shows an example of a 50% DDGS/50% phenolic resin blend that may be suitable for tile applications.

19.3 COMBINING DDGS WITH THERMOPLASTICS

Since DDGS may only provide marginal value as reinforcement in a 100% bioplastic, there has been considerable effort expended in investigating the combination of DDGS with conventional, petroleum-based resins. Biobased filler materials or composites can replace a portion of a petrochemical plastic resin, which is usually derived from natural gas or crude oil. This will thus conserve nonrenewable petroleum resources while simultaneously adding a "green" component to a plastic material or product. Several studies have provided engineering data for these types of applications.

Vogel and Grewell (2008) injection molded test specimens blended from combinations of DDGS, zein, and either high density polyethylene (HDPE) or low density polyethylene (LDPE). Before injection molding, pure DDGS, or DDGS/zein in a 50/50 (by weight) mixture, pellets were extruded at 176°F to 194°F (80°C to 90°C). These were then simply mixed with pellets of HDPE or LDPE and injection molded at 266°F and 248°F (130°C and 120°C), respectively. Additionally, to better combine the grains with the resin, some of the DDGS or DDGS/zein pellets were precompounded with LDPE pellets in an extruder to formulate pellets consisting of the resin and DDGS that were then injection molded. UTS, Young's modulus, and elongation at maximum stress data were obtained for 50% HDPE and 50% DDGS/zein; since the ratio of DDGS to zein was 1:1, this amounts to a 25/25/50 DDGS/zein/HDPE blend. The introduction of DDGS significantly decreased the tensile strength of the specimens from 2760 to 1830 psi (19.0 to 12.6 MPa) and the DDGS particles acted as filler rather than reinforcement. Similarly, Young's modulus decreased 40% and elongation 15% in the presence of the biofiller. For LDPE, using the precompounded pellets that combined resin and DDGS did not greatly affect the maximum tensile strength compared to just mixing DDGS and polyethylene pellets at the injection molding machine and, as expected, the strength was reduced as filler content increased. Compared with only DDGS, the effect of blending DDGS with zein into the LDPE was not consistent; results indicated 25% decreased strength for the 50/50 DDGS-zein/LDPE blend, while the 25/75 and 75/25 blends exhibited 10% and 50% increases, respectively. Figure 19.5 presents UTS versus DDGS-only data from injection molding of the mixed LDPE and DDGS pellets. On the other hand, stiffness was significantly improved with this biofiller, especially at 25% loading, where Young's modulus doubled from the pure LDPE value. This is reasonable based on the fact that the fiber component of the DDGS is stiffer than the LDPE molecular chain structure. As before, it is not clear if DDGS alone or DDGS with zein yielded higher modulus. But precompounded pellets led to much higher Young's modulus than the mixed-only condition. For extension at the UTS, DDGS severely reduced the ductility of any blended material. Here, DDGS-only blends were more flexible than the DDGS/zein blends at 25% and 50% biofiller; at 75%, the difference was negligible. Use of precompounded DDGS or DDGS/zein pellets increased the strain several times compared to results from the mixed-pellet process, but even using the precompounded material severely limited the ductility compared to 100% LDPE. This research effort also saw the introduction of 1% to 2% nanoclay to the LDPE/DDGS-only and LDPE/DDGS/zein blends. The nanoclay was a mineral, aluminosilicate, and it is postulated that its layered structure could stiffen the LDPE and DDGS molecular structure. Testing did confirm a large (2-1/2 times) increase in Young's modulus with the addition of the nanoclay. Interestingly, the elongation drastically improved by about a factor of ten, but tensile strength was relatively unchanged. The DDGS/zein/nanoclay composite was more flexible than the DDGS-only blend, while Young's modulus was independent of type of filler.

Grewell et al. (2008) created polyvinyl chloride (PVC) and corn-based DDGS composites for use in window frames. The DDGS added some strength, as long as its concentration remained under 10% (by weight); over this, the DDGS acted as filler only and mechanical properties declined. Treatments to the DDGS included grinding and caustic solution. DDGS was dried to a moisture level of 8% to 10%, mixed with PVC, extruded, and pelletized. DDGS was provided raw (unground), or ground to a particle size corresponding to ASTM Mesh No. 35 prior to compounding. To investigate the benefit of pretreating with NaOH, some 10% DDGS batches were immersed in the 5% caustic solution for 1 h prior to drying and pellet compounding. Also, to increase impact strength a 2% concentration of poly (VA-co-ethylene) was compounded into some blends. Optimal extrusion conditions included barrel zone temperatures ranging from 226°F to 338°F (130°C to 170°C), extruder die at 266°F (130°C), and a screw rotational speed of 11 rpm. The compounded pellets were fed to a 22-ton (200-kN) injection molding machine to form conventional ASTM tensile specimens. Injection molder barrel temperatures were held at 320°F to 338°F (160°C to 170°C). Ten test samples were prepared for each composite blend. Tensile tests indicated that DDGS provided a maximum reinforcing effect at 5%; a 7% and 11% enhancement of UTS was noted with raw and ground DDGS, respectively. At 10% DDGS content, UTS was unchanged, while at 15% DDGS tensile strength decreased 10% for both ground and raw DDGS. Figure 19.5 includes the effect of varying DDGS content with maximum tensile stress. The authors theorized that the presence of DDGS, especially at higher levels, disrupted the PVC phase. This is especially relevant to the ductility of the composites as measured by the elongation to break, which is extremely limited with any DDGS. For 5% raw and ground DDGS, the extensions were only 19% and 35%, respectively, while the control value was 123%. DDGS contents of 10% and 15% further reduced the tensile strain to 12% and 10% for raw filler and 26% and 16% for ground DDGS. Caustic pretreatment had a minor, negative effect on tensile and elongation properties, while the poly (VA-co-ethylene) additive had insignificant effects on the mechanical properties regardless of DDGS particle size. Overall, these results indicate that blending PVC with up to 15% DDGS is of some value as reinforcement or filler, depending on level used, if the DDGS is inexpensive compared to the petrochemical resin.

In another study, polypropylene (PP) and HDPE were blended with different natural fibrous materials: big blue stem grass, mixed pinewood, soybean hulls, and DDGS from ethanol processing (Julson et al., 2004); these biobased materials were added at 20% and 30% levels. The DDGS was ground to 40 mesh and dried at 140°F (60°C) for 16 h, bringing moisture down to 1.5%–1.8%. The screened DDGS was then mixed with resin and extruded into compounded pellets for further processing. Extruder zone temperatures ranged from 230°F to 329°F (110°C to 165°C) and 284°F to 401°F (140°C to 205°C) for HDPE and PP, respectively. For each treatment, five test specimens were injection molded at 302°F to 329°F (150°C to 165°C) for HDPE, and 365°F to 383°F (185°C to 195°C) for PP. Mechanical and melt flow index tests were conducted. The presence of 20% and 30% DDGS in PP lowered the UTS by about 25% and 30%, respectively, as seen in Figure 19.5. Also in Figure 19.5, for the HDPE-based blends, UTS decreased about 20% and 30% for 20% and 30% DDGS, respectively. At the same mesh size, the tensile strength of the DDGS/PP composite was similar to that of the ground soybean hulls, and approximately 20% less than the big blue stem grass or pinewood blends; this trend in strength reduction grew to 30% in the DDGS/HDPE blend. Clearly, DDGS acted as filler and did not offer increased strength as in a true composite. Consistent with typical fillers, there was a small elevation in stiffness. For HDPE, Young's modulus was 9% and 18% greater for 20% and 30% DDGS, respectively. The effect was 0% with PP and 20% DDGS, but increasing to 30% DDGS raised PP stiffness by just 6%. Furthermore, in this study flexural tests also demonstrated that soybean hulls and DDGS fillers were inferior to pinewood or big blue stem grass. Compared to pure resin, the flexural modulus decreased about 15% in the case of DDGS/PP; values for the polyethylene mixtures were unchanged. Similarly, flexural strength diminished 20% and 5% to 10% for DDGS/PP and DDGS/HDPE blends, respectively; it is postulated that the DDGS may have clumped and thus was not able to provide fibrous reinforcement. Melt flow index tests were also performed and shrinkage measurements showed only slightly less shrinkage in all the biocomposites compared to the pure resin control. Notched impact strength was reduced by nearly half when DDGS was introduced to PP, although this effect was comparable to the behavior of adding pinewood. On the other hand, the combination of HDPE and DDGS raised impact strength by 80%. But for unnotched impact, the addition of biofiller severely reduced the ability of the blends to resist sudden energy absorption; this effect was proportional to the amount of biofiller present. About 75% of toughness was lost when DDGS was blended with PP; this loss grew to about 90% for the DDGS with HDPE.

Fuqua and Ulven (2008) used a ball mill to fractionate DDGS into fibers with a maximum length of 0.0008 in (20 μm) although material clumping was observed. The fibrous grains, at 10% by weight, were then pelletized with PP in an extruder. The extruder speed was 350 rpm; the residence time was limited to 30 to 40 s to minimize thermal degradation; and zone temperatures ranged from 351°F (177°C) at the pellet hopper to 383°F (195°C) at the die. Some pellet batches were prepared with DDGS pretreated with three different chemical coupling agents to enhance molecular bonding;

the pretreatment additives were employed up to 5%, by weight. Depending on the specific coupling agent, the pretreatment technique differed and involved various levels of solvent dissolving, mixing, filtering, heating, and drying. Pellets were injection molded to produce conventional test specimens, five per each condition. The injection molding zone temperatures ranged from 379°C to 421°F (193°C to 216°C) with 399°F (204°C) at the injection nozzle tip. From tensile testing, strength and Young's modulus measurements were obtained and compared to the 100% PP baseline of 3900 psi (26.9 MPa) tensile strength and 206,000 psi (1420 MPa) modulus. Results demonstrated that at 10% concentration DDGS fibers had no measurable effect on tensile stress (Figure 19.5) while the modulus increased by 7%. On average, using DDGS with the various coupling additives indicated a slight (6%) enhancement in tensile strength and 14% greater Young's modulus, but this required significant effort in the pretreatments. The best tensile strength improvement was 9% with a maleic anhydride grafted PP additive.

19.4   COMBINING DDGS WITH THERMOSETTING RESINS

Another promising approach is to use DDGS with thermosetting resins. To fully take advantage of low-cost DDGS, a simple process was devised in which DDGS was mixed with phenolic powder to produce a molding compound that was then cured through a compression molding process (Tatara et al., 2009). The phenol and formaldehyde resin was a commercial-grade powder consisting of 91.5% phenolic with 8.5% hexa curing agent without any other additives or fillers that are typically found in conventional molding compounds. The DDGS was obtained from a dry grind ethanol plant in eastern South Dakota, and had (on a dry basis) 12.3% moisture content, protein content of 27.6%, fiber content of 11.1%, lipid content of 9.3%, ash content of 4.2%, and other carbohydrates of 47.8%. The DDGS was utilized in its raw, untreated form. DDGS was blended with resin at four weight levels (0%, 25%, 50%, and 75%), and then compression molded at 1.0, 2.5, or 3.5 tons/in² (13.8, 34.5, or 48.3 MPa) and 315°F, 345°F, or 375°F (157°C, 174°C, or 191°C). The DDGS contained volatiles that release immediately upon heating. Traditional compression molding did not vent all these gases, so all specimens were produced with cold (i.e., room temperature) pressing up to molding pressure, then about 30 min of heating was needed to bring the mold from room temperature up to the curing condition. A design of experiments approach was used to systematically evaluate the effects of each parameter, and for every combination of DDGS concentration, pressure, and temperature, three test specimens were molded and tested to quantify mechanical properties including tensile yield strength, Young's modulus, percent elongation, and hardness, as well as the physical and performance properties of density, water absorption, and biodegradability. It was found that compression pressure and temperature each had little effect on the resulting properties. The presence of DDGS, on the other hand, greatly influenced all of the tested properties.

For the 0% and 25% DDGS cases, the resulting specimens were brittle, with yield and break occurring simultaneously; however, with 50% and 75% DDGS levels, there was some actual yielding and ductility observed (but not in every case). It is postulated that this added ductility resulted from the fibrous DDGS bearing proportionally more of the tensile loading. Tensile yield strengths ranged from 2102 psi (14.5 MPa) to 621 psi (4.3 MPa), while Young's modulus ranged from 333,000 psi (2296 MPa) to 122,000 psi (841 MPa) as the DDGS content increased from 25% to 75%. At 25% DDGS, the tensile yield strength was approximately half that of the pure resin case. Higher levels of DDGS biofiller further reduced the strength to nearly one-quarter (at 50% DDGS), and under one-sixth (at 75% DDGS) the baseline. Young's modulus was also reduced from that of 100% resin, but less dramatically. At 25% DDGS, a 10% to 15% stiffness reduction was noted, while 50% and 75% filler caused 50% and 70% decreases in Young's modulus, respectively. Percent elongation was restricted about 50% by inclusion of the DDGS, although the trend was not consistent. Surface hardness indicated a general softening as DDGS increased; the Shore D values were about 10% and 25% lower for 25% DDGS and 50% to 75% DDGS, respectively. Although this data analysis was

FIGURE 19.7   A 3 in (76 mm) diameter disk composed of 90% DDGS and 10% phenolic resin.

limited to a maximum of 75% DDGS, specimens having 90% DDGS content were also compression molded and exhibited reasonable strength and stiffness; Figure 19.7 shows a 3 in (76 mm) disk composed of 90% DDGS and 10% phenolic resin.

In terms of physical properties, for the 25% DDGS case water absorption ranged from 1.6% to 7.9% as the immersion time lengthened from 2 h to one week. The water uptake increased for the 50% blend up to 30% after 1 week, while 75% DDGS demonstrated only a slight increase in water absorption compared to that at 50%. Overall, the presence of DDGS created considerable porosity in the material, which also improved biodegradability, which was noted as mass loss. With the pure phenolic resin exhibiting no biodegradability, a 9% mass loss was observed with 25% DDGS, 25% with 50% DDGS, and 40% with 75% DDGS. Of course, the rather elevated water absorption of the blends contributed to the significant biodegradability of the thermoset-based material as well. Again, commercial applicability is essential to the long-term success of DDGS-based biocomposites. Figure 19.6b shows an example of a tile made from compression molding blends consisting of 50% DDGS and 50% phenolic resin. This tile had dimensions of 3 in (76 mm) length, 2 in (51 mm) width, and 0.4 in (10 mm) thickness, had a very smooth surface finish, and was very durable.

In another approach, distillers dried grains (DDG) was liquefied in an acidic reaction to form molecules rich with biopolyols; their functional hydroxyl groups were then cross-linked in the formation of polyurethane foams with the goal of creating a more biodegradable foamed material (Yu et al., 2008). The method required the separation and purification of the biopolyols from the reacted DDG. This study showed that rigid and flexible foams were possible. Foaming was achieved with a conventional isocyanate and water reaction, accompanied by the release of carbon dioxide. Foam test samples, 0.8 in (20 mm) cubes, were buried in soil and after 10 months experienced a 12.6% mean weight loss. The presence of natural proteins and fats, as well as some degree of incomplete cross-linking in the foams, is postulated to have enhanced biodegradability.

19.5   MACHINABILITY OF DDGS-FILLED PHENOLIC COMPOSITES

It is important to characterize the resulting physical and mechanical properties when using DDGS (or other biological fillers) in polymer matrices, as discussed in the previous sections of this chapter. It is also important to examine the manufacturability as well as the final quality of the biocomposites themselves vis-à-vis manufacturing processes. Many manufacturing operations, and combinations thereof, are necessary in order to process industrial materials into finished products, and each of these will impact the finished product (Creese, 1999; Kalpakjian and Schmid, 2001; Geng, 2004).

One of the most common manufacturing processes, especially for finishing operations, is machining. It is therefore important to establish the ease with which a material can be machined (which is also known as machinability). This is usually quantified by several factors such as surface finish, tool life, and power consumption during the machining process (Wyatt and Trmal, 2006). Machining can include operations such as cutting (e.g., turning, boring, drilling, milling, planing, shaping, sawing, grinding, etc.); it also encompasses finishing operations (e.g., honing, lapping, polishing, burnishing, deburring, surface treating, etc.). Machinability depends, in part, on the physical and mechanical properties of the materials being manufactured; it also depends on the types of tools used, tool geometry, tool wear, the nature of the material's matrix, and the product's propensity to chip or flake (Lou et al., 1999). Some of these variables can be controlled very well; others cannot.

In addition to machinability, several other parameters are typically assessed to quantify product quality. These generally include how closely a final product matches specifications, including length (i.e., required dimensional tolerances) for size, shape, and thickness (especially the minimum dimension), functionality requirements, physical and mechanical property requirements, surface finish requirements, and surface roughness, which quantifies the small-scale geometric deviations along a product's surface. Surface roughness is a good indicator of the machinability of a product, because it can be used to quantify the behavior of cutting processes and their effects on surface quality (Lou et al., 1999; Savage and Chen, 1999).

Products containing biomaterials (such as DDGS) may require more finishing operations to achieve specific dimensional tolerances and surface finishes (e.g., biomaterials may result in rougher surfaces during manufacture). Manufacturing biocomposites may also result in products that are more prone to chipping or flaking than traditional products (due to the heterogeneous nature of the shapes and sizes of biological particles). Further, biocomposites may be more softer and more brittle compared to traditional plastic products, and may not have sufficient mechanical strength. All of these potential issues may reduce the machinability of the biocomposite, which in turn may lead to poor final product quality. Thus it is important to test potential biocomposites not only for mechanical strength, but also machinability and surface quality. Even though this type of data is important, to date, however, almost nothing has been published regarding the machining of biocomposites, including those utilizing DDGS.

The final portion of this chapter discusses a series of machinability tests that were conducted on the compression molded DDGS/phenolic resin blends that were discussed in earlier parts of this chapter (Figure 19.8). Three types of specimens were compression molded: 100% (0% DDGS), 75% (25% DDGS), and 50% (50% DDGS) phenolic resin. After molding, the samples were machined on a computer numerically controlled (CNC) milling machine. The machining process involved cutting a rectangular slot, or channel, across the surface of each specimen (Figure 19.9) using a ½ in (12.7 mm) diameter carbide two-fluted end mill. Six slots were machined onto each sample (three on each side). A new end mill was used after every specimen to reduce the effects of potential tool wear. The depth of cut was kept constant at 0.08 in (2 mm).

Cutting speed and feed rate have long been known to be key parameters during milling operations (Alauddin, 1996). In this experiment, the cutting speed was varied between 394 and 525 ft/min (120 and 160 m/min) at feed rates between 8 and 12 in/min (203 and 305 mm/min). Table 19.4 lists the selected cutting conditions. Two machining factors were considered; the range of values for each factor was set at three different levels; and each treatment was replicated twice. Therefore the total number of tests conducted was $3^2 \times 2 = 18$ runs. Table 19.5 shows the full experimental design; all treatment combinations were tested using a full factorial approach.

The resulting surface roughness was measured at the bottom of each channel using a profilometer (Figure 19.10). The effects of the independent variables on the resulting surface roughness were then assessed with Analysis of Variance using a level of significance (α) of 5% (i.e., a level of confidence of 95%) to test for main and interaction effects.

FIGURE 19.8 Compression molded DDGS/phenolic resin specimen used for machinability experiments.

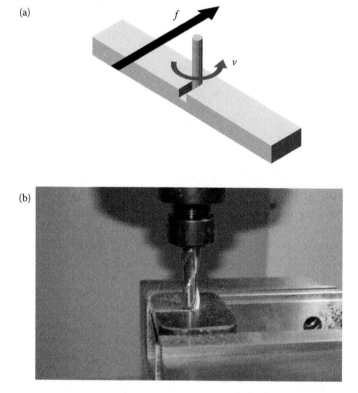

FIGURE 19.9 (a) Illustration of machining a workpiece with an end mill, where f denotes the feed rate, and v denotes the cutting speed and (b) actual machining test in operation.

Table 19.6 shows the main effects of each of the independent variables on the resulting surface roughness. As the level of DDGS in the composites increased, the surface roughness significantly increased. Additionally, the variability in the data increased as well, which indicates that the composites had an increased tendency to chip and flake as the end mill passed through the specimen. Cutting speed, however, did not have any significant effect on the resulting surface roughness. Feed rate, on the other hand, caused a small increase in roughness as the rate increased. Interactions between all

TABLE 19.4
Compression Molded DDGS/Phenolic Resin Blends were Examined for Machinability with Two Milling Variables, Each at Three Levels

Parameter	Low	Medium	High
Cutting speed, v, ft/min (m/min)	394 (120)	459 (140)	525 (160)
Feed rate, f, in/min (mm/min)	8 (203)	10 (254)	12 (305)

TABLE 19.5
Milling Experiments were Conducted on Compression Molded DDGS/Phenolic Resin Blends Following a 3^2 Factorial Design

Experimental Treatment	Cutting speed, v ft/min (m/min)	Feed rate, f in/min (mm/min)
1	394 (120)	8 (203)
2	394 (120)	10 (254)
3	394 (120)	12 (305)
4	459 (140)	8 (203)
5	459 (140)	10 (254)
6	459 (140)	12 (305)
7	525 (160)	8 (203)
8	525 (160)	10 (254)
9	525 (160)	12 (305)

Note: Each treatment was replicated (i.e., milled) twice on independent samples.

independent variables are given in Table 19.7. DDGS and feed rate exhibited significant effects individually, although no interactions appeared to be significant among any of the independent variable combinations. Examining the effects of DDGS and feed rate simultaneously (Figure 19.11) reflects these trends, and illustrates the somewhat curvilinear nature of these relationships.

Although an initial effort has been made to quantify the machinability of phenolic composites filled with DDGS, further research is warranted in order to refine the relationships between surface texture and cutting speed and feed rate, as well as other machinability parameters. Additional experiments should be carried out to examine other machining variables, including cutting force measurements (i.e., how much power is consumed during machining to achieve desired results), the performance of operations such as drilling, grinding, polishing, etc., and the determination of overall machinability indices and predictors.

19.6 CONCLUSIONS

Most data prove that DDGS is a suitable filler but does not offer composite reinforcing strength. From Figure 19.5, it is obvious that tensile stress diminishes as more DDGS is added, generally regardless of the resin matrix. But the varying strength of the base polymer makes comparisons difficult. To facilitate comparison, UTS values have been normalized based on the UTS of each pure resin, and these transformed data are displayed in Figure 19.12. This approach clearly illustrates the effect of DDGS content. When the ratio of UTS with DDGS to UTS at 0% DDGS is greater than 1.0, the presence of DDGS adds strength to the blend and is acting as reinforcement. Values below 1.0, on the other hand, indicate that tensile strength is actually degraded and DDGS

(a)

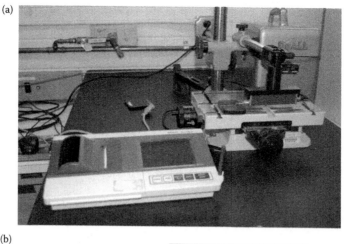

(b)

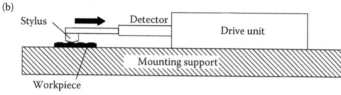

FIGURE 19.10 (a) Surface profilometer measuring roughness of DDGS/phenolic resin sample after machining and (b) schematic illustrating use of the profilometer.

TABLE 19.6
Main Effects of Each Independent Variable upon Surface Roughness of Compression Molded DDGS/Phenolic Resin Blends. Values Underneath and in Brackets Indicate ±1 Standard Deviation

DDGS (%)	0	25	50
Roughness, micro inches (μm)	76 (1.93)[a]	137 (3.49)[b]	193 (4.91)[c]
	[0.32]	[0.89]	[1.85]
Speed, ft/min (m/min)	**394 (120)**	**459 (140)**	**525 (160)**
Roughness, micro inches (μm)	141 (3.58)[a]	132 (3.35)[a]	134 (3.40)[a]
	[1.41]	[1.83]	[1.86]
Feed, in/min (mm/min)	**8 (203)**	**10 (254)**	**12 (305)**
Roughness, micro inches (μm)	124 (3.14)[a]	136 (3.45)[ab]	147 (3.74)[b]
	[1.78]	[1.64]	[1.66]

Note: Differing letters within a given row indicate significant differences ($\alpha = 0.05$, LSD) between surface roughness values for a given independent variable.

---

is thus considered filler. From Figure 19.12, 5%–10% DDGS with PVC exhibits only minor strength enhancement. Only the zein/25% DDGS combination has reinforcing benefit, but it should be noted that pure zein is a low-strength material. In all the other cases, DDGS does not promote composite strength. The natural polymers, PLA and corn-starch, show similar behavior and the DDGS reduces strength at a greater rate than with the synthetic resins PP and polyethylene. Interestingly, phenolic blends follow the general trend too. But over half the strength of the base resin still exists up to

TABLE 19.7
Interaction Effects between Independent Variables on Surface Roughness of Compression Molded DDGS/Phenolic Resin Blends

Source	Pr > F*
DDGS	0.0001
Speed	0.5807
DDGS × Speed	0.3676
Feed	0.0466
DDGS × Feed	0.9620
Speed × Feed	0.8976
DDGS × Speed–Feed	0.9717

* Pr denotes probability.

(a)

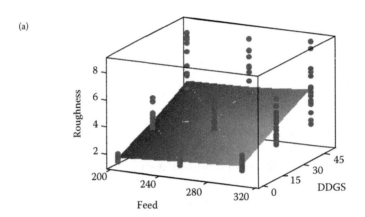

(b)

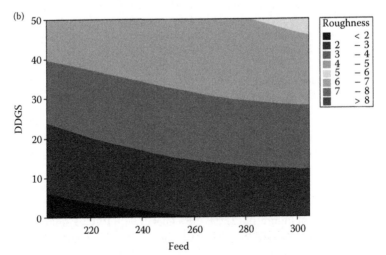

FIGURE 19.11 Surface roughness (μm) as a function of DDGS inclusion (%) and feed rate (mm/min): (a) surface plot; (b) contour plot.

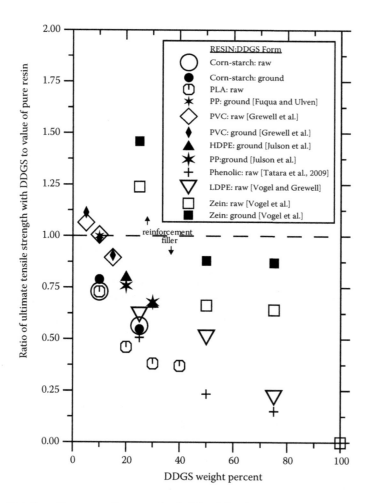

FIGURE 19.12 Relative effect on tensile strength of adding DDGS to a plastic matrix.

approximately one-third DDGS loading. This remaining mechanical strength may be adequate for a wide range of parts and products.

Except for the case of zein, no difference is shown between raw (natural) and ground DDGS. With zein, the ground form performs much better than the raw state. Generally, Figure 19.12 implies that the effect on tensile strength of adding DDGS is the same regardless of the composition of the base resin or DDGS pretreatment; this holds for natural, synthetic, thermoplastic, and thermosetting resins. Therefore, improvement needs to focus on better bonding at the molecular level between resin and DDGS. It is true that use of DDGS as a filler in industrial materials is a promising method of adding value to distillers grains. Using DDGS as a source of bioenergy is another; this topic will be explored in the following chapter.

REFERENCES

Alauddin, M., M. A. E. Baradie, and M. S. J. Hashmi. 1996. Optimization of surface finish in end milling Inconel 718. *Journal of Materials Processing Technology* 56(1–4): 54–65.

Cheesbrough, V., K. A. Rosentrater, and J. Visser. 2008. Properties of distillers grains composites: A preliminary investigation. *Journal of Polymers and the Environment* 16: 40–50.

Creese, R. C. 1999. *Introduction to Manufacturing Processes and Materials*. New York, NY: Marcel Dekker.

Fuqua, M. A., and C. A. Ulven. 2008. Preparation and characterization of polypropylene composites reinforced with modified lignocellulosic corn fiber. American Society of Agricultural and Biological Engineers, ASABE Paper 084364.

Geng, H. 2004. *Manufacturing Engineering Handbook*. New York: McGraw-Hill.

Grewell, D., G. Srinivasan, and M. Baboi. 2008. Feasibility study of the use of DDGS plastic composites. Society of Plastics Engineers ANTEC 2008: 1638–1641.

Julson, J. L., G. Subbarao, D. D. Stokke, H. H. Gieselman, and K. Muthukumarappan. 2004. Mechanical properties of biorenewable fiber/plastic composites. *Journal of Applied Polymer Science* 93: 2484–2493.

Kalpakjian, S., and S. R. Schmid. 2001. *Manufacturing Engineering and Technology*, 4th Ed. Upper Saddle River, NJ: Prentice Hall.

Lou, M. S., J. C. Chen, and C. M. Li. 1999. Surface roughness prediction technique for CNC end-milling. *Journal of Industrial Technology* 15(1): 1–6.

Rosentrater, K. A., and A. W. Otieno. 2006. Considerations for manufacturing bio-based plastic products. *Journal of Polymers and the Environment* 14: 335–346.

Savage, M. D., and J. C. Chen. 1999. Effects of tool diameter variations in on-line surface roughness recognition system. *Journal of Industrial Technology* 15(4): 1–7.

Tatara, R. A., K. A. Rosentrater, and S. Suraparaju. 2009. Design properties for molded, corn-based DDGS-filled phenolic resin. *Industrial Crops and Products* 29: 9–15.

Tatara, R. A., S. Suraparaju, and K. A. Rosentrater. 2007. Compression molding of phenolic resin and corn-based DDGS blends. *Journal of Polymers and the Environment* 15: 89–95.

Vogel, J., and D. Grewell. 2008. Impact study of substituting non degradable plastics by biodegradable composites (DDGS and zein) and the potential of nanoclays as reinforcement. Society of Plastics Engineers ANTEC 2008: 1633–1637.

Vogel, J., M. T. Montalbo, G. Srinivasan, and D. Grewell. 2007. Potential ultrasonic compaction of zein and distiller's dry grain (DDGS). Society of Plastics Engineers ANTEC 2007: 1057–1062.

Wyatt, D. E., and G. J. Trmal. 2006. Machinability: Employing a drilling experiment as a teaching tool. *Journal of Industrial Technology* 22(1): 1–12.

Wu, A., and A. K. Mohanty. 2007. Renewable resource based biocomposites from coproduct of dry milling corn ethanol industry and castor oil based biopolyurethanes. *Journal of Biobased Materials and Bioenergy* 1(2): 257–265.

Xu, W., N. Reddy, and Y. Yang. 2009. Extraction, characterization and potential applications of cellulose in corn kernels and distillers' dried grains with solubles (DDGS). *Carbohydrate Polymers* 76: 521–527.

Yu, F., Z. Le, P. Chen, Y. Liu, X. Lin, and R. Ruan. 2008. Atmospheric pressure liquefaction of dried distillers grains (DDG) and making polyurethane foams from liquefied DDG. *Applied Biochemistry and Biotechnology* 148: 235–243.

20 Using DDGS as a Feedstock for Bioenergy via Thermochemical Conversion

Kurt A. Rosentrater

CONTENTS

20.1 INTRODUCTION

Over the last several years there has been growing interest in producing bioenergy from many biomass feedstocks, including dry grind ethanol coproducts. In fact, many have asked about the possibility of "burning" distillers dried grains with solubles (DDGS). More specifically, Morey et al. (2006, 2009) proposed that ethanol plant efficiencies and energy balances could be improved if the coproducts (or at least a portion thereof) are used for heat and/or electricity generation at the plants themselves. This could be achieved with a variety of thermochemical conversion technologies. This chapter will briefly discuss these topics. Although not intended to be a comprehensive review of thermochemical processing, it is intended to provide only a very basic introduction to the major concepts.

Combustion, gasification, and pyrolysis are three techniques (Figure 20.1) that can be used to generate bioenergy from various biomass feedstocks. Biomass can be readily transformed into energy, primarily in the form of heat or electric power (e.g., stationary generation). Table 20.1 provides a few examples of research results for biomass conversion into bioenergy via the thermochemical route found in the literature. This is a very incomplete listing, but it should provide the reader a feel for the diversity of conversion processes and potential feedstocks that are being examined and developed.

Combustion, which essentially is the conversion of biomass into heat, occurs with a sufficient level of oxygen, and can result in flame temperatures up to, and even greater than, 2000°C (Brown, 2003). It can be accomplished in a variety of equipment: grate-fired, suspension, and fluidized bed combustors, furnaces, and boilers, for example. One of the prerequisites for combustion is an excess of air/oxygen. Figure 20.2 depicts, at least in a very generic form, several potential processes for thermochemical conversion of biomass by combustion.

Gasification, which is the conversion of biomass into flammable gas using an atmosphere that is deficient in oxygen (only about one-thirds of the oxygen required for complete combustion is supplied), generally at temperatures between 750°C and 850°C (Brown, 2003). It can be accomplished by a variety of gasifiers, which most commonly have updraft, downdraft, or fluidized bed

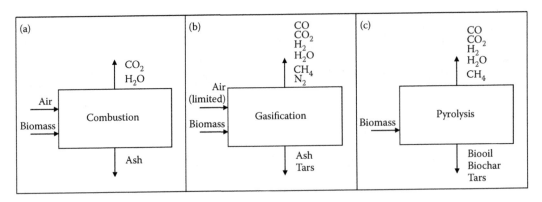

FIGURE 20.1 Simplified flow chart indicating inputs and outputs for various thermochemical conversion processes, including (a) combustion; (b) gasification; and (c) pyrolysis.

configurations (Figure 20.2). The synthesis gas (i.e., syngas) produced by gasifiers is often rich in hydrogen and carbon monoxide, and can be combusted and used to drive electric generation equipment, or to power boilers, which results in process steam (heat). Additionally, gasification results in an ash byproduct that must be disposed of—often as fertilizer or soil amendments.

Thermochemical conversion also includes pyrolysis, where the biomass is heated in the absence of oxygen, generally at temperatures between 200°C and 500°C (Brown, 2003). This results in a bio-oil, which can be converted into fuel, along with residual solids (known as char), and various gases (such as methane, carbon monoxide, and carbon dioxide).

Additionally, the products produced via thermochemical conversion contain high concentrations of organic compounds, and thus are useful as concentrated sources of substrates for further utilization (i.e., conversion into fuels and chemicals). The results from many studies are available in the literature (e.g., more information regarding the conversion of several types of biomass can be found in Table 20.1). The reader is referred to these sources for more details about thermochemical conversion in general.

20.2 PROPERTIES RELEVANT TO BIOENERGY PRODUCTION

The ability to utilize a material via thermochemical conversion is impacted by the composition of that material as well as the energy content. As discussed by Brown (2003) and Giuntoli et al. (2009), ash, especially the elements potassium, sodium, silicon, chlorine, sulfur, and proteins (due to nitrogen) can limit the use of many biomass feedstocks in industrial thermochemical conversion processes due to the ease with which these elements form complexes, which can subsequently lead to deleterious effects such as slagging, equipment fouling, reduced conversion efficiencies, and increased emissions such as NO_x. The energy content of a material is quantified via the heating value. Most often the higher heating value (HHV) is reported, as opposed to the lower heating value (LHV). HHV is defined as the full energy content of a material, and it accounts for the water formed and condensed during combustion, and thus the latent enthalpy of condensation. The LHV, on the other hand, is also a measure of energy content, but at the end of combustion the water is still in vapor form, and has not condensed (Brown, 2003; ASABE, 2006).

Although there is an emerging interest in converting ethanol coproducts via thermochemical processes into energy, there has been only a little research in the area to date (Table 20.2) regarding properties that are relevant to bioenergy production. This limited data indicates that, on a dry basis, the major ethanol coproducts (e.g., DDGS, DDG [distillers dried grains], DWG [distillers wet grains], and CDS [condensed distillers solubles]) have energy contents between approximately 20 and 22 MJ/kg. Additionally, all but CDS have a fairly low ash content.

An important question that must be asked is: how do ethanol coproducts compare to other materials in terms of potential for thermochemical conversion? Fossil-based fuels, such as charcoal, coal,

TABLE 20.1
A Few Examples of Thermochemical Conversion of Biomass

Biomass	Reactor Type	Operating Conditions Time	Temperature (°C)	Yields (%) Misc.	O₂	CO	CO₂	CH₄	H₂	Reference
Combustion										
Rapeseed meal and softwood bark pelletized (10% and 30%)	Fluidized bubbling bed	8 h	800	5 kg fuel						Eriksson et al. (2009)
Rapeseed meal and softwood bark pelletized (10% and 30%)	Grate combustor	24 h	1200–1250							
Pure rapeseed meal	Powder burner	6 h	900–950		$32.2\text{–}33.51$					Lawrence et al. (2009)
Coal and dairy biomass	Boiler burner rated at 29 kW		1092			0–12,000 ppm	17–21			Sathitruangsak et al. (2009)
Coal and rice husk	Short-combustion-chamber fluidized bed combustor	8 h								
Switchgrass	Electric-resistance heated; Bunte-Baum softening temperature testing instrument		500–700							Szemmelveisz et al. (2009)
Sunflower seed shell			420–750							
Hemp			400–725							
Gasification										
Almond shells	Fluidized bed reactor — Sand		800			33.2	11.7	11.5	43.6	Rapagna and Latif (1997)
	Dolomite					24	14.1	6.4	55.5	
	Olivine					23	16.9	7.9	52.2	
Pine sawdust	Air-steam gasification in fluidized bed at atmospheric pressure and indirect heat		700		0.71	42.89	20.51	9.12	21.48	Lv et al. (2003)
			800		0.25	37.73	18.55	7.46	32.10	
			900		0.49	33.42	19.36	6.10	39.40	
Pine sawdust	Air-steam gasification in fluidized bed		700	Fuel yield 1.43–2.57 Nm³/kg biomass	0.17–0.34	37.25–40.59	16.75–21.18	6.63–7.79	30.47–31.62	Lv et al. (2004)
			800							
			900							

continued

TABLE 20.1 (continued)
A Few Examples of Thermochemical Conversion of Biomass

Biomass	Operating Conditions				Yields (%)					Reference
	Reactor Type	Time	Temperature (°C)	Misc	O_2	CO	CO_2	CH_4	H_2	
Food waste	Carbonized in stainless steel autoclave and then subjected to a fluidized bed reactor		300	Solid 30.1g; Oil 1.1 g; Gas 3700 mL						Ko et al. (2001)
Radish				Solid 9.0 g; Oil 0.55 g; Gas 925 mL						
Cellulose Xylan, and lignin	Downdraft fixed-bed gasifier		900			35.5 mol%	27 mol%		28.7 mol%	Hanaoka et al. (2005)
Rice hull	Fluidized-bed gasifier		700–800	Heating range 11.1–12.1 MJ/m^3		25 mol%	36 mol%		32 mol%	Boateng et al. (1992)
Sugar cane residue	Cyclone gasifier; feeding rate of 39 and 46 kg/h		820–850	4.7 MJ/Nm^3 dry gas						Gabra et al. (2001)
Texas lignite, Wyoming coal, dairy biomass, feedlot biomass	10 kW batch type, counter current, fixed-bed reactor			Syngas		0.119–0.23 mol%	0.161–0.155 mol%	0.005–0.020 mol%	0.072–0.186 mol%	Gordillo et al. (2009)
Wet pine sawdust	Quartz tube reactor placed in electric furnace; gas collected after passing through a serpentine cooler and cotton fiber									Guoxin et al. (2008)
Switchgrass with an MC of 20%	Pressurized, oxygen blown, fluidized bed gasifier			Fischer–Tropsch (F–T) fuels (diesel and gasoline blendstocks), dimethyl ether (DME), hydrogen, electricity (coproduct)						Larson et al. (2009)

Feedstock	Process	Time	Temp (°C)	Products					Reference
Corn DDGS	Gasification with oxygen in a continuous down flow fixed-bed microreactor; oxygen/nitrogen ratio 0.08–0.2 (optimum was 0.08)	15–45 min; optimum 30 min	700–900; optimum 900			57	29	13	Tavasoli et al. (2009)
Wheat DDGS						50	30	11	
Rice husk	Cyclone air gasification					18.60–22.00	15.81–18.01	5.03–7.10	Sun et al. (2009)
Pyrolysis									
Dry sugarcane bagasse	Batch pyrolysis reactor	90 min				422.460–954.06 cm³	687–1672.51 cm³	219.19–1551.48 cm³	Arni et al. (2010)
Sugarcane bagasse pretreated with water									
Waste fish oil	Continuous pyrolysis	17 s	525	Biogas 15.85%, bio-oil 72.83%, coke solid 11.32%	1.56–2.61	12.28–13.43	9.67–11.16	7.61–4.81	Wiggers et al. (2009)
Corncob	Feed of 45–50 kg/h, the bio-syngas was then put through additional treatments		400	Gas yield of 40–45 Nm³/h, dimethyl ether yield of 124.3–203.8 kg/m³/h		9.04–11.6	26.3–27.8	23.7–25.6	Li et al. (2009)
Hazelnut shell	Non-isothermal pyrolysis experiments were performed under nitrogen flow of 40 mL/min; temp was increased from ambient to 900°C at a rate of 40°C/min.; trial was terminated once 900°C was reached		<900						Haykiri-Acma and Yaman (2010)
Peat									
A. Lignite									
S. Lignite									
Bituminous coal									
Anthracite									

TABLE 20.2
Overview of Work Published to Date on Ethanol Coproduct Properties Relevant to Thermochemical Conversion

Coproduct	n	Higher Heating Value (HHV) (BTU/lb)	(MJ/kg)	Ash (%, db)	Reference
DDGS	—	9422	21.87	4.13	AURI (2009)
	30	9133.51	21.20	—	Bhadra et al. (2010)
	4	9349	21.75	3.89	Morey et al. (2009)
DDG	—	9848	22.86	2.24	Auri (2009)
	5	9288.61	21.56	—	Bhadra et al. (2010)
DWG	5	9438	21.95	2.58	Morey et al. (2009)
CDS	5	8482	19.73	7.02	Morey et al. (2009)

Note: n is the number of samples analyzed for that particular property; all properties reported on a dry basis (db); DDGS is distillers dried grains with solubles; DDG is distillers dried grains; DWG is distillers wet grains; CDS is condensed distillers solubles.

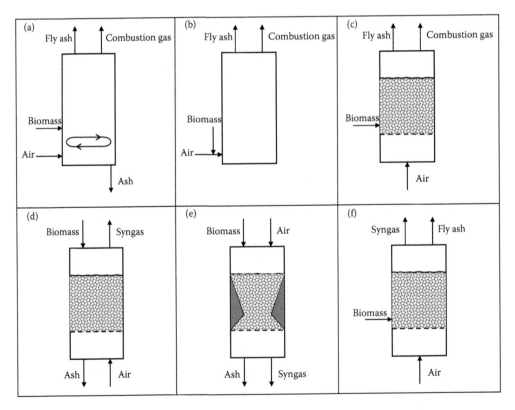

FIGURE 20.2   Thermochemical conversion of biomass can be accomplished by several types of processes. Three conventional systems for combustion include (a) grate; (b) suspension; and (c) fluidized bed; three typical systems for gasification include (d) countercurrent; (e) concurrent; and (f) fluidized bed. (Based, in part on, McKendry, P., *Bioresource Technology* 83, 47–54, 2002, and Brown, R. C., *Biorenewable Resources: Engineering New Products from Agriculture.* Iowa State Press, 2003 [in part].)

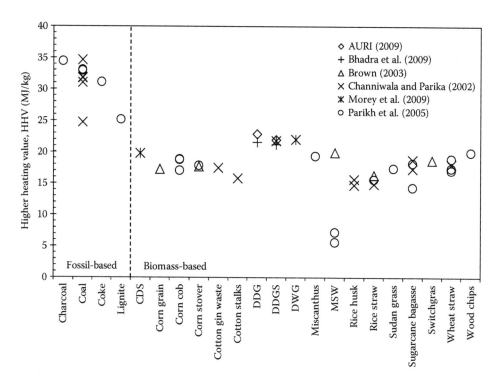

FIGURE 20.3 Previous research has found that ethanol coproducts are at the higher end of biomass in terms of energy content, but they are still lower than fossil-based fuels. (All products reported on a dry basis.) CDS is condensed distillers solubles; MSW is municipal solid waste; DDG is distillers dried grains; DDGS is distillers dried grains with solubles; DWG is distillers wet grains.

coke, and lignite are the benchmarks, and generally have HHV values greater than 25 MJ/kg, and often approach 30 to 35 MJ/kg (Figure 20.3). These traditional fuels are on the high end of the spectrum in terms of energy content. Most biomass feedstocks are generally substantially lower (i.e., less than 25 MJ/kg) in energy. Even so, there is growing interest in using them as sources of energy, especially at biorefineries, in general, and at ethanol plants in particular, as they are readily available sources of fuel. A few examples are depicted in Figure 20.3 (although it is not comprehensive, it does provide a reference base for the discussion). Compared to other biomass fuels, DDGS, DDG, and DWG are all on the higher end of energy contents vis-à-vis other biological materials. This indicates that ethanol coproducts could indeed prove to be viable sources of energy, especially for use in ethanol plants for process heat and/or electricity. This could be, at least in part, due to the relatively high lipid and protein contents of the various ethanol coproducts compared to other biomass materials (Table 20.3).

In terms of bioenergy production, one of the key issues regarding biomass is that it can exhibit a range of potential energy contents; this is reflected in some of the values shown in Figure 20.3. Biomass materials are, by nature, heterogeneous mixtures of organic (i.e., protein, lipid, carbohydrate) as well as inorganic (i.e., mineral) constituents. And, a given material will have variability in both physical and chemical properties due to a number of factors, including the type of biomass, the hybrid grown, the growing location, the quality of the growing season (for instance droughts, floods, pest infestations [i.e., insects, microorganisms, or diseases]), which will ultimately affect the resulting quality (e.g., composition) and quantity (e.g., yield) of the biomaterials available for use. For the ethanol industry, these factors all impact the corn grain that is delivered to each plant, and thus the coproducts that result from processing.

To account for some of this variability, several previous studies have developed potential empirical correlations to predict heating values of various fuel sources' based on specific chemical constituents,

TABLE 20.3
Compositional Data for Ethanol Coproducts and Other Biomass Materials

Material	Moisture (%, db)	Protein (%, db)	Fat (%, db)	NDF (%, db)	Ash (%, db)	C (%, db)	N (%, db)	Reference
DDGS	5.1	29.9	10.5	36.7	12.8			Bhadra et al. (2010)
						46.0	4.4	Unpublished (author's data)
DDG	2.6	30.9	8.9	31.4	10.9			Bhadra et al. (2010)
DWG	34.4	28.6	11.1	33.8	13.3			Bhadra et al. (2010)
Corn grain					1.27	44.0	1.24	Brown (2003)
		7.82	1.65	38.33	4.04			McCann et al. (2007)
Corn cob					1.36	46.58	0.47	Brown (2003)
Corn stover					5.58	43.65	0.61	Brown (2003)
Cotton gin waste					1.61	42.66	0.18	Parikh et al. (2005)
Cotton stalks					17.3	39.47	1.2	Parikh et al. (2005)
MSW					12.00	47.60	1.20	Brown (2003)
Rice husk					21.24	38.5	0.45	Channiwala and Parikh (2002)
Rice straw					20.38	35.68	0.28	Channiwala and Parikh (2002)
Sudan grass					8.65	44.6	1.21	Parikh et al. (2005)
Sugarcane bagasse					3.2	45.48	0.15	Channiwala and Parikh (2002)
Switchgrass					3.61	47.45	0.74	Brown (2003)
Wheat straw					13.5	45.5	1.8	Parikh et al. (2005)
Wood chips					0.10	48.1	0.08	Parikh et al. (2005)

such as carbon, hydrogen, sulfur, oxygen, nitrogen, chlorine, phosphorus, total ash, organic matter, volatile matter, and/or combinations thereof. Some of these studies include Raveendran and Ganesh (1996), Demirbas (1997), Channiwala and Parikh (2002), Parikh et al. (2005), Sheng and Azevedo (2005), and Thipkhunthod et al. (2005). The reader is referred to these for more in-depth information. Of these correlations, some are specific to certain types of biomass; others are general in nature.

For example, according to Brown (2003), a useful generic relationship is:

$$HHV = 0.4571 \times C - 2.70 \tag{20.1}$$

where HHV is Higher Heating Value (MJ/kg) and C is carbon content (%, dry basis).

Or, according to Parikh et al. (2005):

$$HHV = 0.196 \times C + 14.119 \tag{20.2}$$

In the author's laboratory (data unpublished), the carbon content of 80 DDGS samples collected from ethanol plants in South Dakota were measured in duplicate using a C/N analyzer (LECO Corporation, St. Joseph, MI). From this analysis, a minimum carbon content of 45.37% (dry basis), and a maximum value of 46.99% were found. According to Equation 20.1, the predicted HHV for these samples should range from 18.18 to 18.78 MJ/kg; this was somewhat low compared to experimental findings, however (as shown in Table 20.2). On the other hand, according to Equation 20.2, the HHV should range from 18.90 to 23.33 MJ/kg; this was actually a better estimate for the author's DDGS samples. No one has yet attempted to develop equations for predicting the HHV of DDGS as a function of chemical constituents, although this would be valuable, as it could be used to help develop and deploy effective thermochemical conversion systems for ethanol coproducts.

20.3 THERMOCHEMICAL CONVERSION RESEARCH

Even though ethanol coproducts are relatively unexplored in terms of fuel properties, a few conversion investigations have already been conducted and have examined the gasification behavior of DDGS. For example, Hu and Yuan (2008) examined a laboratory-scale downdraft fixed-bed gasifier system using DDGS as a feedstock, with syngas production ranges between 2.8 ft³/min and 5.6 ft³/min (0.08 m³/min and 0.16 m³/min), and found that the resulting syngas was 19% H_2, 20% CO, and 51% N, but only 1.3% CH_4 (methane). Additionally, their system resulted in a tar concentration of 4.2 g/Nm³. In a similar study, Tavasoli et al. (2009) also examined the conversion of DDGS using a small-scale continuous down-flow fixed-bed gasifier. Reaction time was varied between 15 and 45 min, temperature was set between 700°C and 900°C, and O_2/N_2 ratio was varied between 0.08 and 0.2. They determined that optimal operating conditions for this system occurred near 900°C, with an O_2/N_2 ratio of 0.08, using a reaction time of 30 min. The resulting syngas was 10% H_2, 55% CO, and 29% CO_2, but only 4.5% CH_4. And carbon conversion efficiencies between 45% and 50% could be achieved. Kumar et al. (2009) examined the conversion of DDGS using a laboratory-scale fluidized bed gasifier that used both steam and air. This system was investigated using temperatures between 650°C and 850°C, steam-to-biomass ratios between 0 and 14.3, and equivalence ratios (i.e., the ratio of air that was supplied to the air required for total theoretical combustion of the DDGS) between 0.07 and 0.29. The resulting syngas had a CO content less than 7% (for all processing conditions), CH_4 content varied from approximately 13% up to nearly 24%, while H_2 varied between approximately 3% to 16%. Carbon conversion efficiencies up to 90% could be achieved; higher temperatures and higher equivalence ratios resulted in better efficiencies, due to greater molecular conversions. Data from this system was used to develop a computer model, which was then used to simulate the gasification of DDGS based upon elemental composition changes (Kumar et al., 2008). Unfortunately, the model did not predict final gas content well, nor did it adequately predict gas composition, and thus required further refinement.

Along another vein, Wang et al. (2007) subjected DDGS to sequential fluid extraction, using supercritical CO_2, subcritical water, and then supercritical ethanol as solvents, in order to remove lipids, ash, and other soluble components prior to thermochemical conversion. The remaining residues were then subjected to thermogravimetric analysis in order to quantify pyrolysis (up to 650°C) and combustion (up to 850°C) behavior. At the end of pyrolysis, approximately 27% of the extracted residue remained, while at the end of combustion, only 5.5% of the residue was not decomposed. However, this analysis did not result in the production of bio-oil or biochar, both of which are key to the overall pyrolysis process.

Only one study (Mullen et al., 2010) has investigated the use of DDGS as a feedstock for actual pyrolysis. This study used a bubbling fluidized bed system to convert DDGS from the barley-to-ethanol process (i.e., not corn-based DDGS). The authors found that the resulting bio-oil had an energy content greater than 30 MJ/kg, and a yield of nearly 70% could be achieved. In terms of conversion efficiencies, 76.6% of the C, 59.7% of the H, 10.4% of the O, and 79.4% of the N, respectively, was converted into bio-oil. The balance of each constituent was converted into biochar, noncondensable gases, and water. The biochar had a resulting energy content of 21.5 MJ/kg.

Overall, it appears that ethanol coproducts have relatively good properties vis-à-vis biomass fuels, and they have been proven to be amenable to conversion into bioenergy, at least on small scales. The potential costs and economic returns, if implemented on an industrial scale, however, must be examined in order to determine the value of these technologies to the industry.

20.4   ECONOMICS OF THERMOCHEMICAL CONVERSION

A few studies have actually tried to answer these questions in recent years. For example, Morey et al. (2006) examined the economics of four options for converting DDGS into process heat and electricity at ethanol plants: (1) no conversion, (2) process heat only, (3) combined heat and power—using combustion/gasification—to meet the plant's needs, and (4) combined heat and power for the plant as well as sale of excess electricity to the electric grid. They examined potential energy cost savings, and the net energy utilized from the corn. They found that current technology (i.e., the purchase of electricity and natural gas) only produces a net value of 9.25 MJ/L of ethanol, whereas implementing combined heat and power technology for the plant's use as well as sale to the grid resulted in a net value of 23.9 MJ/L of ethanol produced—a considerable net energy increase of 14.65 MJ/L. They also determined that, for an ethanol plant that produced 150 million L/y, at a DDGS sales price of $80/ton, a cost savings of $8.6 million/y could be achieved if the natural gas price was $6/GJ, whereas at $10/GJ, then a cost savings of nearly $15.1 million/y could be achieved. For the same natural gas price ranges, at a DDGS sales price of $100/ton, however, the cost savings to the plant ranged from $6.3 to $12.8 million/y.

Wang et al. (2009) furthered this research and developed a computer model to simulate the implementation of these combined heat and power (i.e., fluidized bed combustion and gasification) systems for a 190 million L/y corn dry grind ethanol plant. They modeled all of the unit operations, process flows, energy consumption and generation, chemical reactions, and emissions. They utilized two ethanol coproducts in their study: CDS was combusted with corn stover, while DDGS was gasified. For each of these scenarios, they found that the electricity that could be sent to the electrical grid resulted in 0.4 kWh/L of ethanol produced. They concluded that systems that only produced steam for the production process (i.e., process heat) were probably the best choice for ethanol plants, because they had somewhat better thermal efficiencies compared to systems that produced heat only or systems that produced heat and electricity, and then sent excess electricity to the electrical grid. They also point out that when combusting the syngas that is produced, pollution control equipment will be necessary to reduce emissions, which could be elevated due to high nitrogen levels in the coproducts.

Tiffany et al. (2009) examined the capital and operational costs of implementing DDGS gasification (i.e., fluidized bed gasification) or combustion of CDS mixed with corn stover (i.e., fluidized bed combustion), at ethanol plants at scales of 190 million L/y and 380 million L/y. Again, they examined scenarios where the biomass conversion was used for (1) process heat, (2) process heat and electricity, and (3) process heat and electricity, with the sale of excess electricity to the electrical grid. They also accounted for equipment and technologies that would be necessary in order to control NO_x, SO_x, and HCl emissions. For combusting the CDS at a plant that produces 190 million L/y, they determined that, to produce process heat only, installation costs would be $0.74/L of ethanol, but for combined heat and power with the sale of excess electricity it would be $0.92/L. At an ethanol plant that produces 380 million L/y, however, the economies of scale would reduce the installation costs to $0.60/L and $0.75/L, respectively. The option of gasifying the DDGS had installation costs that were very similar to those of the CDS combustion. For a 190 million L/y ethanol plant, both options resulted in rates of return that ranged from 5.8% to 8.0% (compared to a return of 3.7% for a conventional ethanol plant using natural gas and electricity), but for a 380 million L/y plant, the rates of return ranged from 9.9% to 12.6% (compared to a return of 7.1% for a conventional plant of the same size).

Wang et al. (2009) also conducted an analysis of the capital and operational costs of implementing thermochemical conversion in a 189 million L/y ethanol plant, but examined the use of DWG instead. This coproduct was at a moisture content 64.7%. In their system, the DWG was dried at a rate of 41.1 t/h, and then gasified in a fluidized bed gasifier; the resulting syngas was then used to produce process heat (steam was generated in a boiler) and electricity (using a combined heat and power fuel cell generator). They found that using DWG resulted in an energy cost of $0.101/L of ethanol, which was 107% the cost of current technology (i.e., purchasing natural gas and electricity).

Tiffany et al. (2009) indicate that, at least at the time of publication, so far only one commercial fuel ethanol plant (185 million L/y) in Minnesota has implemented a fluidized bed reactor and utilizes CDS as a feedstock. This plant produces process heat (i.e., steam) at a rate of approximately 45,400 kg/h; this rate of generation meets all of the plant's needs for steam, it still produces DDG for sale as animal feed, and it also sells the ash byproduct from the fluidized bed for fertilizer. Further, Lemke (2008) indicates that two additional plants in Minnesota have installed gasifiers for the conversion of DDGS as well as other biomass materials.

It is worthwhile to consider price information for both natural gas and DDGS. Figure 20.4 shows the price history for DDGS and natural gas from January 2001 through July 2009 (103 months of data). On an absolute basis the price of natural gas has been relatively stable over this time frame, and ranged from $3.42/ft³ to $12.18/ft³. DDGS, on the other hand, has experienced some fairly substantial price fluctuations, and ranged from $61/ton to $165/ton. On a relative scale (maximum/minimum), natural gas has experienced a ratio of ~2.7 compared to that of ~3.6 for DDGS over time.

But, this type of comparison does not provide an accurate picture, because each material contains a different amount of energy. Transforming this data and plotting it on a per unit of energy basis (Figure 20.5), the price comparison becomes much more complex. Because DDGS can exhibit variability, three energy contents are provided to encompass the data presented earlier (Table 20.2): 16 MJ/kg, 20 MJ/kg, and 24 MJ/kg. Note that this price comparison is based only upon energy that is available in the fuel itself; it does not consider any capital or operational expenditures that may be necessary to convert the DDGS into heat or power at the ethanol plant. As shown in this figure, if DDGS contained 24 MJ/kg, a considerable price advantage compared to natural gas emerges for most of the last 103 months. At even at the lowest energy content, the price for DDGS-based energy was fairly competitive with the price of natural gas. Furthermore, by examining the absolute difference in price between natural gas and DDGS (on a per unit of energy basis, $/MJ) (Figure 20.6), it appears that, depending on the DDGS energy content, for most of the last several years there would

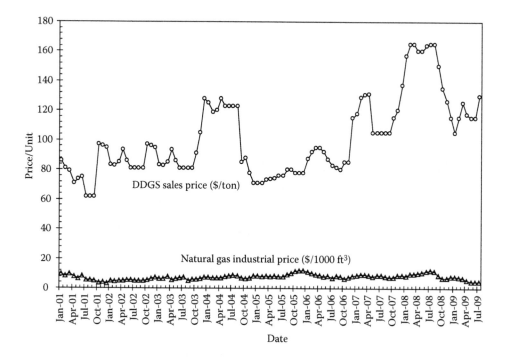

FIGURE 20.4 Monthly sales prices between January 2001 and July 2009 for DDGS and natural gas. (Based on EIA, *Natural Gas Summary,* Energy Information Administration, United States Department of Energy: Washington, DC, 2009.)

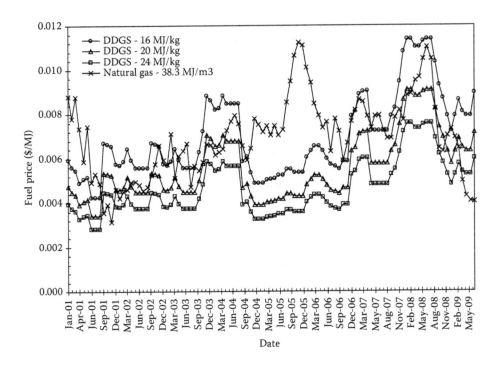

FIGURE 20.5 Comparison between natural gas and DDGS for use as fuel, on a monthly basis between January 2001 and July 2009, on a per unit of energy basis. All comparisons made in terms of $/MJ. Three levels of energy content were assumed for DDGS (16, 20, and 24 MJ/kg) in order to provide a range of fuel properties.

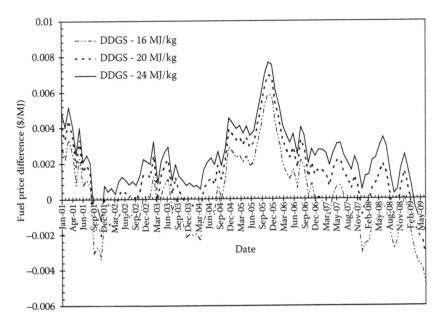

FIGURE 20.6 On a monthly basis between January 2001 and July 2009, it appears that the use of DDGS as a fuel source would have been cost effective compared to using natural gas for the majority of the time, especially if the DDGS energy content was 24 MJ/kg. The higher the energy content of the DDGS, the better the price advantage. Negative values of price difference indicate that natural gas was less expensive than DDGS as a source of energy. Fuel price difference = price of natural gas − price of DDGS ($/MJ).

have been a cost advantage to using DDGS as a source of fuel vis-à-vis natural gas. If the DDGS only contained 16 MJ/kg, 48% of the time DDGS had a cost advantage; at 20 MJ/kg DDGS was less expensive than natural gas 76% of the time; and at 24 MJ/kg DDGS was cheaper 93% of the time.

20.5   BYPRODUCTS OF THERMOCHEMICAL CONVERSION

Investigating conversion processes and quantifying the associated economics if implemented in various industrial settings are necessary considerations. Additionally, it is important to understand what will be left when the conversion processes are complete—the ash byproducts, or process residues, in other words. Unfortunately, almost nothing has been investigated in this regard to date in terms of ethanol coproducts.

In fact, no investigations have examined ash byproducts from the thermochemical conversion of any types of ethanol coproducts. Giuntoli et al. (2009) did characterize barley-based DDGS as a thermochemical fuel, and examined its conversion behavior. They conducted a TGA (thermogravimetric analysis) to quantify pyrolysis kinetics (up to 900°C), as well as FTIR (Fourier transform infrared spectroscopy) analysis, in order to characterize the evolution of nitrogen-based compounds, which can be NO_x precursors. The DDGS that they examined had a HHV of 19.8 MJ/kg (which was similar to corn-based DDGS values, Table 20.2), and an ash content of 7.1% (which was almost twice as much as corn-based DDGS, Table 20.2). They determined that nearly 75% of that ash was composed of SiO_2, K_2O, and P_2O_5. During pyrolysis, barley-based DDGS exhibited behavior similar to other lignocellulosic materials: at 280°C the hemicellulose showed peak decomposition, whereas the cellulose exhibited peak decomposition at 330°C. When the process reached 900°C, nearly 80% of the DDGS had decomposed, and only about 20% of the original DDGS remained as residue. In terms of nitrogen-based compounds, they found three primary components: NH_3 exhibited a peak release at nearly 328°C; HCN was released at 664°C; and HNCO was released at 407°C. These occurred due to the decomposition of proteins—amino acids specifically. These findings indicated that NO_x remediation would be necessary if thermochemical conversion processes for DDGS were used commercially.

Moreover, to date no work has yet examined the utilization of the residues resulting from DDGS thermochemical conversion. Although Tiffany et al. (2009) indicated that the ethanol plant that has installed a fluidized bed does sell the ash residues as fertilizer. Also, Edwards and Anex (2007) did examine use of DDGS as a binder (at levels between 0% and 45% inclusion) for pelleting of gasifier ash (which was produced using corn stover as a feedstock for gasification). The goal was to use this ash as a fertilizer and to improve its pelleting performance using some type of binder. They found that using 30% DDGS as a binder produced the strongest pellets, with a crushing strength of 3829 g; this compared to only 1906 g for the 0% DDGS pellets.

20.6   CONCLUSIONS

Considering the use of ethanol coproducts as a source of fuel for the generation of process heat and/or electricity appears to be a viable alternative for these materials. Previous research has shown that ethanol coproducts have relatively high energy content (compared to other biomass materials), conversion research has proven their effectiveness, and economic analyses indicate positive energy balances. So, all things considered, it appears that ethanol coproducts may be a viable source of fuel for the production of heat and power at ethanol plants. It will indeed be interesting to see if this avenue for coproduct utilization will be ubiquitously adopted by the industry, or if it will only be implemented at a few plants.

REFERENCES

Arni, S. A., B. Bosio, and E. Arato. 2010. Syngas from sugarcane pyrolysis: an experimental study for fuel cell applications. *Renewable Energy* 35: 29–35.

ASABE. 2006. ANSI/ASABE S593—Terminology and definitions for biomass production, harvesting and collection, storage, processing, conversion, and utilization. *ASABE Standards 2006—Standards, Engineering Properties, Data.* St. Joseph, MI: American Society of Agricultural and Biological Engineers.

AURI. 2009. *AURI Fuels Initiative: Agricultural Renewable Solid Fuels Data.* Waseca, MN: Agricultural Utilization Research Institute. Available online: http://www.auri.org/research/fuels.pdf.

Bhadra, R., K. Muthukumarappan, and K. A. Rosentrater. 2010. Chemical and physical properties of distillers dried grains with solubles (DDGS) relevant to value added uses. *Cereal Chemistry* 87(5): 439–447.

Boateng, A. A., W. P. Walawender, L. T. Fan, and C. S. Chee. 1992. Fluidized-bed steam gasification of rice hull. *Bioresource Technology* 40: 235–239.

Brown, R. C. 2003. *Biorenewable Resources: Engineering New Products from Agriculture.* Ames, IA: Iowa State Press.

Channiwala, S. A., and P. P. Parikh. 2002. A unified correlation for estimating HHV of solid, liquid, and gaseous fuels. *Fuel* 81: 1051–1063.

De Kam, M. J., R. V. Morey, and D. G. Tiffany. 2009. Integrating biomass to produce heat and power at ethanol plants. *Applied Engineering in Agriculture* 25(2): 227–244.

Demirbas, A. 1997. Calculation of higher heating values of biomass fuels. *Fuel* (76)5: 431–434.

Edwards, K. A., and R. P. Anex. 2007. *Recycling Nutrients from Biofuel Production: Pelletizing Gasifier Ash for Fertilizer.* ASABE Paper No. 076078. St. Joseph, MI: American Society of Agricultural and Biological Engineers.

EIA. 2009. *Natural Gas Summary.* Washington, DC: Energy Information Administration, United States Department of Energy: tonto.eia.doe.gov/dnav/ng/ng_sum_top.asp. Accessed January 10, 2009.

Eriksson, G., H. Hedman, D. Bostrom, E. Pettersson, R. Backman, and M. Ohman. 2009. Combustion characterization of rapeseed meal and possible combustion applications. *Energy and Fuels* 23: 3930–3939.

Gabra, M., E. Pettersson, R. Backman, and B. Kjellstrom. 2001. Evaluation of cyclone gasifier performance for gasification of sugar cane residue—part 2: Gasification of cane trash. *Biomass and Bioenergy* 21: 371–380.

Giuntoli, J., W. de Jong, S. Arvelakis, H. Spliethoff, and A. H. M. Verkooijen. 2009. Quantitative and kinetic TG-FTIR study of biomass residue pyrolysis: dry distiller's grains with solubles (DDGS) and chicken manure. *Journal of Analytical and Applied Pyrolysis* 85: 301–312.

Gordillo, G., K. Annamalai, and N. Carlin. 2009. Adiabatic fixed-bed gasification of coal, dairy biomass, and feedlot biomass using an air-stream mixture as an oxidizing agent. *Renewable Energy* 34: 2789–2797.

Guoxin, H., H. Hao, and L. Yanhong. 2008. The gasification of wet biomass using $Ca(OH)_2$ as CO_2 absorbent: The microstructure of char and absorbent. *International Journal of Hydrogen Energy* 33: 5422–5429.

Hanaoka, T., S. Inoue, S. Uno, T. Ogi, and T. Minowa. 2005. Effect of woody biomass components on air-steam gasification. *Biomass and Bioenergy* 28: 69–76.

Haykiri-Acma, H. and S. Yaman. 2010. Interaction between biomass and different rank coals during co-pyrolysis. *Renewable Energy* 35: 288–292.

Hu, M., and W. Yuan. 2008. *Development of a Unique Downdraft Gasifier System for Low Bulk Density Biomass Materials.* ASABE Paper No. 083774. St. Joseph, MI: American Society of Agricultural and Biological Engineers.

Ko, M. K., W. Y. Lee, S. B. Kim, K. W. Lee, and H. S. Chun. 2001. Gasification of food waste with steam in fluidized bed. *Korean Journal of Chemical Engineering* 18(6): 961–964.

Kumar, A., H. Noureddini, D. D. Jones, and M. A. Hanna. 2008. *Modeling of Fluidized Bed Steam-Air Gasification of Biomass.* ASABE Paper No. 084523. St. Joseph, MI: American Society of Agricultural and Biological Engineers.

Kumar, A., K. Eskridge, D. D. Jones, and M. A. Hanna. 2009. Steam-air fluidized bed gasification of distillers grains: Effects of steam to biomass ratio, equivalence ratio and gasification temperature. *Bioresource Technology* 100: 2062–2068.

Larson, E. D., H. Jin, and F. E. Celik. 2009. Large-scale gasification-based coproduction of fuels and electricity from switchgrass. *Biofuels, Bioproducts, and Biorefining* 3: 174–194.

Lawrence, B., K. Annamalai, J. M. Sweeten, and K. Heflin. 2009. Cofiring coal and dairy biomass in a 29 kW furnace. *Applied Energy* 86: 2359–2372.

Lemke, D. 2008. Redefining Ag wastes as coproducts. *Biocycle* 49(4): 42.

Li, Y., T. Wang, X. Yin, C. Wu, L. Ma, H. Li, and L. Sun. 2009. Design and operation of integrated pilot-scale dimethyl ether synthesis system via pyrolysis/gasification of corncob. *Fuel* 88: 2181–2187.

Lv, P., J. Chang, Z. Xiong, H. Huang, C. Wu, and Y. C. Zhu. 2003. Biomass air-steam gasification in a fluidized bed to produce hydrogen-rich gas. *Energy and Fuels* 17(3): 677–682.

Lv. P. M., Z. H. Xiong, J. Chang, X. Z. Wu, Y. Chen, and J. X. Zhu. 2004. An experimental study on biomass air-steam gasification in a fluidized bed. *Bioresource Technology* 95: 95–101.

McCann, M. C., W. A. Trujillo, S. G. Riordan, R. Sorbet, N. N. Bogdanova, and R. S. Sidhu. 2007. Comparison of the forage and grain composition from insect-protected and glyphosate-tolerant MON 88017 corn to conventional corn (*Zea mays* L.). *Journal of Agricultural and Food Chemistry* 55: 4034–4042.

McKendry, P. 2002. Energy production from biomass (part 2): Conversion technologies. *Bioresource Technology* 83: 47–54.

Morey, R. V., D. G. Tiffany, and D. L. Hatfield. 2006. Biomass for electricity and process heat at ethanol plants. *Applied Engineering in Agriculture* 22(5): 723–728.

Morey, R. V., D. L. Hatfield, R. Sears, D. Haak, D. G. Tiffany, and N. Kaliyan. 2009. Fuel properties of biomass feed streams at ethanol plants. *Applied Engineering in Agriculture* 25(1): 57–64.

Mullen, C. A., A. A. Boateng, K. B. Hicks, N. M. Goldberg, and R. A. Moreau. 2010. Analysis and comparison of bio-oil produced by fast pyrolysis from three barley biomass/byproduct streams. *Energy Fuels* 24: 699–706.

Parikh, J., S. A. Channiwala, and G. K. Ghosal. 2005. A correlation for calculating HHV from proximate analysis of solid fuels. *Fuel* 84: 487–494.

Rapagna, S., and A. Latif. 1997. Steam gasification of almond shells in a fluidized bed reactor: The influence of temperature and particle size on product yield and distribution. *Biomass and Bioenergy* 12(4): 281–288.

Rapagna, S., N. Jand, A. Kiennemann, and P. U. Foscolo. 2000. Steam-gasification of biomass in a fluidized-bed of olivine particles. *Biomass and Bioenergy* 19(3): 187–197.

Raveendran, K., and A. Ganesh. 1996. Heating value of biomass and biomass pyrolysis products. *Fuel* 75(15): 1715–1720.

Sathitruangsak, P., T. Madhiyanon, and S. Soponronnarit. 2009. Rice husk co-firing with cola in a short-combustion-chamber fluidized-bed combuster (SFBC). *Fuel* 88: 1394–1402.

Sheng, C., and J. L. T. Azevedo. 2005. Estimating the higher heating value of biomass fuels from basic analysis data. *Biomass and Bioenergy* 28: 499–507.

Sun, S., Y. Zhao, F. Ling, and F. Su. 2009. Experimental research on air staged cyclone gasification of rice husk. *Fuel Processing Technology* 90: 465–471.

Szemmelveisz, K., I. Szucs, A. B. Palotas, L. Winkler, and E. G. Eddings. 2009. Examination of the combustion conditions of herbaceous biomass. *Fuel Processing Technology* 90: 839–847.

Tavasoli, A., M. G. Ahangari, C. Soni, and A. K. Dalai. 2009. Production of hydrogen and syngas via gasification of the corn and wheat dry distiller grains (DDGS) in a fixed-bed micro reactor. *Fuel Processing Technology* 90: 472–482.

Tavasoli, A., M. G. Ahangari, C. Soni, and A. K. Dalai. 2009. Production of hydrogen and syngas via gasification of the corn and wheat dry distillers grains (DDGS) in a fixed-bed micro reactor. *Fuel Processing Technology* 90: 472–482.

Thipkhunthod, P., V. Meeyoo, P. Rangsunvigit, B. Kitiyanan, K. Siemanond, and T. Rirksomboon. 2005. Predicting the heating value of sewage sludges in Thailand from proximate and ultimate analyses. *Fuel* 84: 849–857.

Tiffany, D. G., R. V. Morey, and M. J. De Kam. 2009. Economics of biomass gasification/combustion at fuel ethanol plants. *Applied Engineering in Agriculture* 25(3): 391–400.

USDA ERS. 2009. *Feed Grains Database: Yearbook Tables.* Washington, DC: United States Department of Agriculture, Economic Research Service. www.ers.usda.gov/data/feedgrains. Accessed January 10, 2009.

Wang, L., A. Kumar, C. L. Weller, D. D. Jones, and M. A. Hanna. 2007. *Co-production of Chemical and Energy Products from Distillers Grains Using Supercritical Fluid Extraction and Thermochemical Conversion Technologies.* ASABE Paper No. 076064. St. Joseph, MI: American Society of Agricultural and Biological Engineers.

Wang, L., M. A. Hanna, C. L. Weller, and D. D. Jones. 2009. Technical and economical analyses of combined heat and power generation from distillers grains and corn stover in ethanol plants. *Energy Conversion and Management* 50: 1704–1713.

Wiggers, V. R., A. Wisniewski, Jr., L. A. S. Madureira, A. A. Chivanga Barros, and H. F. Meier. 2009. Biofuels from waste fish oil pyrolysis: Continuous production in a pilot plant. *Fuel* 88: 2135–2141.

21 Using DDGS as a Feedstock for Bioenergy via Anaerobic Digestion

Conly L. Hansen

CONTENTS

21.1 INTRODUCTION

The production of corn-based ethanol in the United States is dramatically increasing, and consequently, so is the quantity of byproduct materials generated from this processing sector. These coproduct streams are currently often utilized as livestock feed, which is a route that provides ethanol processors with a revenue source and increases the profitability of the production process. With the construction and operation of many new plants in recent years, the residue has much potential for making energy. This option holds the promise of economic benefit for corn processors, especially if the livestock feed market in some areas of the county becomes saturated with byproduct feeds. Anaerobic digestion has been successfully utilized to produce methane from a variety of food and organic processing residues, but has not yet been used much in the ethanol industry. This chapter will outline the origin of major ethanol coproducts, explain the anaerobic digestion process, and talk about the potential for utilization in the ethanol industry.

In the production of bioethanol from cereals, the dry matter is divided into three almost equivalent fractions: one-third is converted into ethanol, one-third into carbon dioxide, and one-third into residues. After fermentation and removal of the ethanol with fractional distillation, the remaining slurry, called whole stillage, is centrifuged to remove solids from the water. The centrate, called thin stillage, is concentrated in evaporators to make condensed distillers' solubles (CDS), commonly known as syrup, which contains relatively high concentrations of Na, K, and P. The centrifuged solids can be dried alone, often in rotary drums, to produce distillers dried grains (DDG), but are typically added back to the CDS and this mixture is then dried to make distillers' dried grains with solubles (DDGS), which is traditionally utilized by direct feeding to livestock containing a mixture of crude fat, protein, and fiber (Bonnardeaux, 2007). Refer to Chapter 5 for more information on processing of corn to ethanol and coproducts.

With increasing energy costs, combustion of DDGS and other ethanol coproducts has been considered for energy production (Morey, et al., 2006). The energy thus produced would reduce the cost of corn-based ethanol production relative to using natural gas, and would also increase the net energy value of the ethanol produced (Hirl, 2006). Refer to Chapter 20 for information on thermochemical conversion of ethanol coproducts into bioenergy.

The method of energy recovery from ethanol coproducts to be emphasized in this chapter is anaerobic digestion to produce biogas that is mostly methane. This would help to reduce energy usage from fossil fuels to make ethanol. Anaerobic digestion has more potential for energy recovery than combustion because whole stillage can be treated directly, which would eliminate the energy required for centrifuges, evaporators, and driers to convert whole stillage to DDGS (Cassidy, 2008). The sheer quantity of DDGS (for every liter [gallon] of alcohol distilled off, there are 13 liters [gallons] of stillage to be discharged), indicates the potential for use of coproducts for energy via anaerobic digestion and relaxes pressure on the livestock feed market.

21.2 OVERVIEW OF ANAEROBIC DIGESTION

Anaerobic digesters convert solid and liquid biomass wastes using microorganisms operating in an environment without the presence of oxygen to produce combustible gas (biogas) that is similar to natural gas and a liquid effluent stream containing solids that retain the nutrients originally in the influent. The effluent can thus be used as a fertilizer. The anaerobic digestion process can be divided into hydrolysis, acidogenesis, acetogenesis, and methanogenesis (Figure 21.1). Each of the steps has its own unique type of microorganisms (Kalia, 2007).

1. **Hydrolysis:** During hydrolysis, complex organic polymer chains such as carbohydrates, fats, and proteins are broken down into lower molecular weight compounds—simple sugars, fatty acids, and amino acids. These reactions tend to depend on extracellular enzymes, for example cellulase, amylase, protease, and lipase. This step is normally a rate-limiting step for the anaerobic treatment of insoluble wastes such as biomass.

2. **Acidogenesis:** In the second step of anaerobic digestion, the fermentation/acidification of soluble substrate into more oxidized intermediates occurs by acidogens. They reduce the low molecular weight compounds into volatile fatty acids (VFAs) (or so called carboxylic acids), alcohols, propionate, butyrate, carbon dioxide, ammonia, and hydrogen. The metabolism of acidogenic bacteria demands a low partial pressure of hydrogen

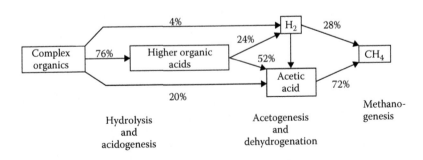

FIGURE 21.1 Outline of the steps of anaerobic digestion. (Adapted from McCarty, P. L. *Anaerobic Digestion 1981*, ed. D. E. Hughes, pp. 3–22. Elsevier Biomedical Press, 1982. With permission.).

($\leq 10^{-6}$M) so hydrogen must be utilized by other groups of microorganisms or otherwise removed from solution. Some of the acidogenic reactions are thermodynamically endergonic under standard conditions so once again the optimal metabolism of acidogenic bacteria requires other microorganisms with which the acidogens can share energy. This step is the fastest step in the anaerobic conversion of complex material in liquid phase digestion. Acidogens are fast-growing bacteria, with minimum doubling time of around 40 min.

3. **Acetogenesis:** The third major step in the anaerobic digestion process is acetogenesis, which is basically a hydrogen-producing, acetate-forming step. Although some acetate (20%) and H_2 (4%) are directly produced by acidogenic fermentation, both products are primarily derived from the acetogenesis and dehydrogenation of higher VFAs with acetogenic bacteria (FAO, 1997). About 76% of organic matter is degraded this way (Figure 21.1) and can be oxidized further to methanogenic substrates by homoacetogenic and hydrogenogenic bacteria. In the acetogenesis process, bacteria further digest the remnants of the previous process into hydrogen (H_2), acetic acid (CH_3COOH), and carbon dioxide(CO_2). Hence, acetogenesis is a key process in the mineralization of organic matter in the methanogenic environment. The acetogens are slow growing and depend on low partial pressure of hydrogen conditions for making their conversion energy-yielding. These reactions occur in syntrophic association, where several anaerobic microorganisms can share the energy available in the bioconversion of a molecule of organic matter to methane (CH_4) and carbon dioxide and thus can achieve intermediate reactions that are endergonic under standard conditions.

4. **Methanogenesis:** Methanogens catabolize acetate and a carbon compound to biogas, which consists of methane, carbon dioxide, water vapor, ammonia, and hydrogen sulfide. Methanogens are unique among prokaryotes and are classified as members of archaea, a group of phylogenetically different microorganisms. These obligate anaerobes are very sensitive. Most methanogens can utilize hydrogen and carbon dioxide as growth substrates and some of them can also catabolize acetate and one carbon compound to methane. Other possible methanogenic substrates like formate ($CHOO^-$), methanol (CH_3OH), carbon monoxide (CO), and methylamines are of minor importance in most anaerobic digestion processes.

About 28% of total methane is produced via the hydrogenotrophic pathway where hydrogen from fermentation and acetogenic reactions is converted to methane by hydrogen-consuming methanogens. Doubling time for hydrogenotrophic methanogens is 6 h. Although a significant amount of hydrogen is produced in acidogenesis/acetogenesis, hydrogen is rarely detectable in the final biogas product because it is utilized by methane-forming and other microorganisms.

Hydrogen-consuming acetogenic bacteria are present in small numbers in an anaerobic digester. They produce acetate from carbon dioxide and hydrogen and, therefore, compete for hydrogen with the methanogenic bacteria. Also, the synthesis of propionate from acetate, as well as production of longer chain VFA, occurs to a limited extent in anaerobic digestion.

21.3 ADVANTAGES OF ANAEROBIC TREATMENT

Lettinga and Hulshoff Pol (1991) list anaerobic digestion as the heart of a very promising environmental protection and resource recovery concept for the following reasons:

1. Anaerobic treatment is significantly cheaper than aerobic treatment with respect to the treatment of medium and high strength wastewater ([COD] chemical oxygen demand > 1500 mg/L).

2. Energy is produced instead of used. This is especially of interest when treating highly concentrated waste.

3. Less space is required for an anaerobic treatment plant compared to an aerobic treatment plant.
4. Anaerobic treatment technology has relatively low equipment cost.
5. Anaerobic treatment is more suitable for the treatment of wastewaters of seasonal industries, because anaerobic sludge can be preserved without serious deterioration in activity or settleability, provided the temperature is maintained below 15°C.
6. Available anaerobic treatment systems can be applied at small as well as large scale.
7. The anaerobic treatment technology does not require the import of expensive equipment.
8. Anaerobic process stability can be easily achieved.
9. Waste biomass disposal costs are low.
10. Nitrogen and phosphorus supplementation costs are low.
11. Energy can be recovered, thus providing ecological and economical benefits.
12. Management requirement is low.
13. Off-gas air pollution can be eliminated.
14. Foaming of surfactant containing wastewaters can be avoided.
15. Aerobic nonbiodegradable organics can be degraded anaerobically.
16. Chlorinated organic toxicity levels can be reduced.
17. Seasonal treatment can be provided.

21.4 TYPES OF ANAEROBIC REACTORS

There are many types of anaerobic digesters. Some feature short hydraulic (liquid) retention times or HRTs. In order to do this, solids (living microorganisms) must either be retained in the digester or collected from the effluent and recycled back to the digester. This is because anaerobic microorganisms are slow growing unlike most aerobic microorganisms and would otherwise "wash out" of the digester meaning they cannot grow fast enough to replace themselves when the solid retention time (SRT) is short. By decoupling the SRT and the HRT, the liquid wastewater can be processed faster. HRT is the time water is retained within the digester and is equal to reactor volume divided by the average volumetric flowrate. In many instances a short HRT will reduce capital and operation costs. There may be some advantages for a simple design like a complete mix digester where SRT is equal to HRT. The simple designs are generally reliable and easily managed; although modern controls permit hands off management of more complex designs that decouple HRT and SRT. Some of the more common digester types are given below (Ma, 2002). The following anaerobic digesters will be discussed in this section: continuous stirred tank reactor (CSTR), anaerobic contact reactor, upflow anaerobic sludge blanket reactor (UASB), anaerobic filter, anaerobic sequencing batch reactor (ASBR), plug flow reactor, and the induced bed reactor (IBR).

21.4.1 CONTINUOUS STIRRED TANK REACTOR

In a CSTR, wastewater is fed into a closed continuously mixed vessel operated in either continuous or batch feed mode. Figure 21.2 shows a diagram for the CSTR. The biogas produced is collected from the top of the reactor and the sludge is discharged through the bottom of the digester. In the case of a continuous system, the untreated wastewater (influent) is introduced from the bottom of the reactor and the treated wastewater (effluent) is discharged from the top of the reactor. Three stirring methods can be used in this system: mechanical stirring, biogas recycling, and treated wastewater recycling. The organic loading rate (OLR) is 2–3 kg COD/m³-day (0.12–0.19 lb_m/ft³-day) for mesophilic treatment and 5–6 kg COD/m³-day (0.31–0.37 lb_m/ft³-day) for thermophilic treatment. The advantage of this type of reactor is that it can be used to directly treat wastewater with high content of floating solid and large particles. Its disadvantage, however, is that the activated sludge (living bacteria) is difficult to keep in the digester, thus resulting in relatively low treatment efficiency.

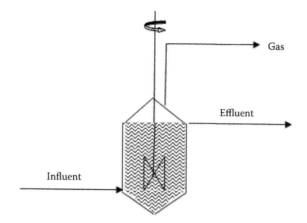

FIGURE 21.2   Continuously fed stirred tank reactor (CSTR).

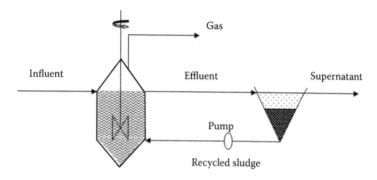

FIGURE 21.3   Anaerobic contact reactor.

21.4.2   ANAEROBIC CONTACT REACTOR

This system consists of a digester and a settling tank (Figure 21.3). The activated settled sludge is pumped back to the digester. The advantages of this system are that (1) the concentration of the sludge in the digester can be kept high (normally 10–15 g/l) by sludge recycling; (2) the OLR is high (normally 2–10 kg/m³-day (0.12–0.6 lb$_m$/ft³-day); (3) the HRT is short; (4) it can directly treat waste-water with higher content of floating solids and larger particles; (5) there are relatively low plugging problems; and (6) the quality of the effluent is often good. Its disadvantages, however, are that (1) it requires some equipment (capital expense) for settling, recycling, and degassing; (2) solid and liquid are difficult to separate in the settling tank; and (3) significant energy is required for mixing.

21.4.3   UPFLOW ANAEROBIC SLUDGE BLANKET REACTOR

Figure 21.4 shows a diagram for the UASB. No external power is used for mixing and the wastewa-ter is introduced from the bottom of the reactor whereas the gas and effluent are collected from the top of the reactor. Retention of sludge is greatly aided by formation of granules that are basically clumps of living microorganisms. Once granules are formed influent COD can be very high; some have operated at over 50 g/L. The OLR can also be very high or at least 30 kg/m³-day (1.9 lb$_m$/ft³-day). Upflow liquid velocity is 0.72 to 0.96 m/day (2.4–3.14 ft/day) and HRTs are often less than 1 day. A concentration of NH_4–N more than 1000 mg/L will inhibit granulation.

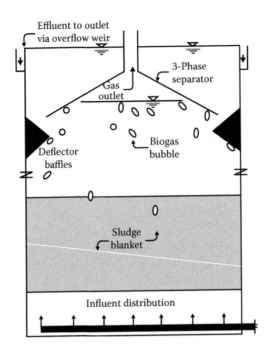

FIGURE 21.4   Upflow anaerobic sludge blanket reactor (UASB).

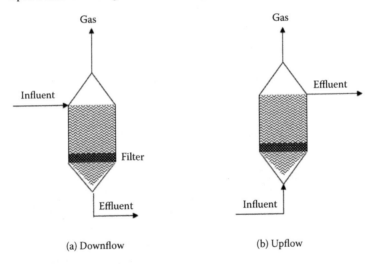

(a) Downflow   (b) Upflow

FIGURE 21.5   Anaerobic filter.

21.4.4 ANAEROBIC FILTER

The anaerobic filter (also called anaerobic fixed film reactor) is illustrated in Figure 21.5. In this reactor, anaerobic bacteria adhere to the surface of a supporting media to form a biofilm. When wastewater passes through the supporting medium, the bacteria in the biofilm degrade the organics in the wastewater. Biogas is produced and collected from the top of the filter. Anaerobic filters can be divided into two types: (1) upflow anaerobic filter in which wastewater is fed from the bottom of the filter and treated wastewater is discharged from the top of the filter and (2) downflow filter, in which wastewater is fed from the top of the filter and the treated water is discharged from the bottom of the filter.

The advantages of anaerobic filters include the following: (1) biofilm retention time can be up to 100 days; (2) the OLR can be up to 16 kg COD/m^3-day; (3) it can absorb peak loading rates;

(4) the organic removal rate is high; (5) sludge recycling and stirring are not required; and (6) the period of start-up or restart-up is short. The disadvantages include the following: (1) The filter can become plugged when the wastewater contains floating matter or significant concentration (>12%) of suspended solids; and (2) the filter is difficult to clean.

21.4.5 ANAEROBIC SEQUENCING BATCH REACTOR

The ASBR shown in Figure 21.6 is a suspended growth reactor. Solids separation occurs within the reactor rather than in an external clarifier. It is used for separating the SRT from the HRT. It also has the ability of processing large volumes of wastewater with a short HRT while maintaining a long SRT. The ASBR includes four processes: filling, reacting, settling, and decanting. Settling time typically ranges from 10 to 30 min. The reactor can achieve efficient treatment of relatively dilute wastewater and can be used at low temperature. The OLR is relatively high compared to other reactors.

21.4.6 PLUG FLOW REACTOR

The simplest explanation of a plug flow digester is the velocity profile of a fluid flowing in a conduit (Figure 21.7). In plug flow, the velocity of the fluid is assumed to be constant across any cross section of the pipe perpendicular to the axis of the pipe. The plug flow model assumes there is no boundary layer adjacent to the inner wall of the pipe. A plug flow digester requires minimal maintenance and is suitable for substrate with about 10% solids. The HRT will commonly be in the 20–25 day range and OLR will be moderate compared to other types of digesters.

21.4.7 INDUCED BED REACTOR

The IBR relies on biomass retention in the digester to provide an accumulation of active solids, similar to the operational principles first proposed by Dr. Gatze Lettinga (Lettinga et al., 1980) for his UASB reactor design (Figure 21.8). UASBs operate on the principle of separation of SRT and HRT via granulated biomass retention. The active biomass acts as an effective filter, metabolizing available carbon to CO_2 and CH_4. In the UASB, influent enters the reactor from the bottom via a diffuser to provide equal distribution through the sludge blanket and minimize short circuiting.

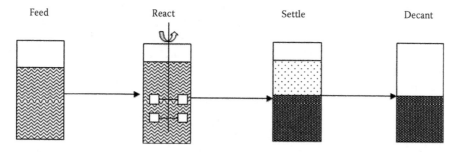

FIGURE 21.6 Anaerobic sequencing batch reactor (ASBR).

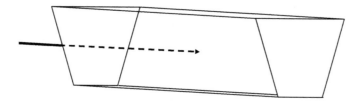

FIGURE 21.7 Schematic of plug flow digester.

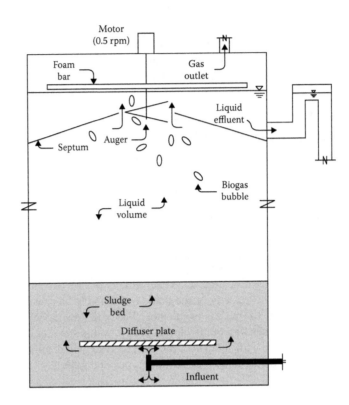

FIGURE 21.8   Induced bed reactor (IBR).

The differences between IBRs and UASBs are derived from the need to address the high solids content of the influent. The influent of the IBR is typically a single line double jet directed at a diffuser plate and bottom of the tank to provide some diffusion of the influent into the sludge bed. This configuration is dictated by the high solids content of the influent and the need to prevent the solids from settling in and plugging the influent line. The thickened sludge area in the IBR is near the bottom of the vessel and thus is called a bed rather than a blanket. The upflow velocity in the IBR is much less than in the UASB and the device for separating solids from gas at the top of the vessel is very robust with an auger filling the hole in the septum. The purpose of the auger is to help prevent bacteria from readily escaping while at the same time pulling large solid particles, that is, greater than 30–40 mm, to the outlet. The IBR is capable of handling solids concentration up to the point it would become a plug flow digester. OLR rate is similar to the UASB and equal to at least 30 kg/ m^3-day. The HRT is most often in the range of 3–8 days.

21.5   ANAEROBIC DIGESTION OF ETHANOL COPRODUCTS

Experiments were conducted at Utah State University to assess the potential for using ethanol-processing residue streams as feedstocks for anaerobic digestion (Rosentrater et al., 2006). Laboratory testing of anaerobic digestibility was conducted on DDG, DDGS, whole stillage, thin stillage, and CDS. Testing of anaerobic digestibility was conducted and biochemical methane yields were measured for each. Some of the results are shown in Figure 21.9. They indicate that whole stillage and thin stillage produced acceptable levels of methane, but CDS was the most promising coproduct. It yielded a biomethane potential (BMP) of 253 mL gas/g sample (4 ft^3/lb_m), which is high, and thus should be examined further in follow-up studies.

Ultrasonic pretreatment of ethanol coproducts appears to make a difference. Hearn (2009) found that treating DDGS, solids, and thin stillage, under various ultrasonic conditions improved anaerobic

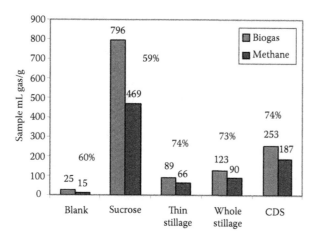

FIGURE 21.9 Cumulative biogas and methane, with average methane %, produced in the BMP assay using various ethanol coproduct substrates.

digestibility compared to untreated samples. The amplitude of ultrasonic treatment ranged from 52.8 μm_{pp} to 160 μm_{pp} and the time was varied from 10 to 50 s. For samples consisting of solid/liquid suspensions of DDGS, solids showed a significant decrease in particle size and an associated increase in the surface-area-to-volume ratio. Treatment of DDGS resulted in a 23% average increase in methane biogas production.

Cassidy et al. (2008) found that approximately 10 million BTUs (British thermal units) of energy could be produced from the biological conversion of one ton (907 kg) of distillers' syrup (CDs) to methane. Shin et al. (1991) reported that anaerobic digestion of distillery (barley and sweet potato) wastewater in a two-phase upflow anaerobic sludge blanket (UASB) system resulted in a daily methane production of 0.28 L/g COD added under mesophilic condition. The UASB reactor was also used for mesophilic anaerobic digestion of stillage from distilleries using beet sugar, cane molasses, or corn.

Thin stillage contains relatively high fats, oils, and grease (FOG). FOG, which is common at high concentrations in thin stillage from dry mills without fractionation of the corn grain, can accumulate in mesophilic digesters by forming a foam layer, and cause operating problems by washing out active biomass (Jeganathan and Bassi, 2006). This is not a problem in thermophilic digesters due to sufficient solubilization and degradation of FOG at higher temperatures (Reimann et. al., 2002). Application of thermophilic digestion would only require cooling the stillage to less than 70°C (158°F), which occurs naturally during temporary stillage storage. Agler et al. (2008) studied the applicability of an integrated method of thermophilic anaerobic digestion of thin stillage from dry mill corn grain-to-ethanol plants by utilizing ASBRs. They estimated the methane yield by total COD loading rates and removal rates. The estimated methane yield was 0.245 L/g (3.9 ft³/lb$_m$) of COD added (approximately equals to 0.35 L/g (5.6 ft³/lb$_m$) of volatile solids (VS) added) after reaching sustainable operating performance. They also suggested that methane generated from thermophilic anaerobic digestion of corn thin stillage could replace 51% of natural gas consumed at a conventional dry mill and improve the net energy balance ratio from 1.26 to 1.70. Schaefer and Sung (2008) tested anaerobic digestion of corn ethanol thin stillage at thermophilic temperature (55°C (131°F)) using two completely stirred tank reactors. A significant reduction of VS (89.8%) was observed at the 20-day HRTs. Methane yield ranged from 0.6 to 0.7 L/g (9.6 to 11.2 ft³/lb$_m$ VS added during steady-state operation.

21.6 CONCLUSIONS

Anaerobic digestion of corn-based ethanol coproduct streams holds the promise of economic benefit for corn processors, especially if the livestock feed market in some areas of the county becomes saturated with these byproducts. Advantages of anaerobic digestion of ethanol coproduct streams include: renewable energy is produced, relatively low equipment cost, anaerobic process stability can be easily achieved, waste biomass disposal costs are low, and off-gas air pollution can be eliminated. Popular types of anaerobic digesters include completely stirred, plug flow, sequencing batch and various types of sludge retention or recycle reactors. High rate anaerobic digesters must either retain the slow growing anaerobic microorganisms (UASB, IBR, SBR) or recycle them back into the main reaction chamber (contact process). Care should be taken when choosing an anaerobic digester for ethanol coproducts to select one that is not easily plugged if significant amounts of solids are included in the influent stream. Experiments were conducted wherein whole stillage, thin stillage and CDS were anaerobically digested. All produced acceptable amounts of methane. CDS was the most promising coproduct. It produced 253 mL of gas containing 70% methane per gram of sample loaded, which is relatively high for anaerobic substrates.

REFERENCES

Agler, M. W., M. Garcia, E. Lee, M. Schlicher, and L. T. Angenent. 2008. Thermophilic anaerobic digestion to increase the net energy balance of corn grain ethanol. *Environmental Science & Technology* 42: 6723–6729.

Bonnardeaux, J. 2007. Potential uses for distillers grains. Department of Agriculture and Food Western Australia 3 Baron-Hay Court South Perth WA, 6151.

Cassidy, D. P., P. J. Hirl, and E. Belia 2008. Methane production from ethanol co-products in anaerobic SBRs. *Water Science & Technology* 58(4): 789–793.

FAO. 1997. *Renewable Biological Systems for Alternative Sustainable Energy Production* (FAO Agricultural Services Bulletin—128), ed. K. Miyamoto. Rome, Italy: Food and Agriculture Organization of the United Nations Corporate Document Repository.

Hearn, C. J. 2009. Particle size characterization of ultrasonic treatment of dry milling coproducts for enhanced biofuel production. Graduate thesis. Iowa State University, 2009.

Hirl, P. J. 2006. Process for ethanol production and for energy recovery. U.S. Patent Application No. 20070141691, filed November 21, 2006.

Jeganathan, J. N., and G. Bassi. 2006. A. long-term performance of high-rate anaerobic reactors for the treatment of oily wastewater. *Environmental Science & Technology* 40: 6466–6472.

Kalia, V. C. 2007. Microbial Treatment of Domestic and Industrial Wastes for Bioenergy Production. Applied Microbiology (e-Book): http://nsdl.niscair.res.in/bitstream/123456789/650/1/DomesticWaste.pdf Published by NISCAIR, CSIR, New Delhi.

Lettinga, G., A. F. M. van Velsen, S. W. Hobma, W. de Zeeuw, and A. Klapwijk. 1980. Use of the upflow sludge blanket (USB) reactor concept for biological wastewater treatment, especially for anaerobic treatment. *Biotechnology and Bioengineering* 22(4): 699–734.

Lettinga, G., and L. W. Hulshoff Pol. 1991. Application of modern high anaerobic treatment processes for wastewater treatment. In *New Developments in Industrial Wastewater Treatment*, ed. A. Turkman, and O. Uslu, pp. 33–64. Netherlands: Kluwer Academic Publishers.

Ma, Y. 2002. Fundamental and applied studies on anaerobic biotechnology for the treatment of high strength cheese processing waste. Ph.D. Dissertation, Utah State University, Logan, UT.

McCarty, P. L. 1982. *Anaerobic Digestion 1981*, ed. D. E. Hughes, pp. 3–22. Amsterdam, New York, Oxford: Elsevier Biomedical Press.

Morey, R. V., D. L. Hatfield, R. Sears, and D. G. Tiffany. 2006. Characterization of feed streams and emissions from biomass gasification/combustion at fuel ethanol plants. American Society of Agricultural and Biological Engineers. Paper Number: 064180.

Reimann, I., A. Klatt, and H. Markl. 2002. Treatment of fat-containing wastewater from the food processing industry with a thermophilic microorganism. *Chemie Ingenieur Technik* 74: 508–512.

Rosentrater, K. A., H. R. Hall, and C. L. Hansen. 2006. Anaerobic digestion potential for ethanol processing residues. American Society of Agricultural and Biological Engineers. Paper Number: 066167.

Schaefer, S. H., and S. Sung. 2008. Retooling the ethanol industry: Thermophilic anaerobic digestion of thin stillage for methane production and pollution prevention. *Water Environment Research* 80(2): 101–108.

Shin, H. -S., B. U. Bae, J. J. Lee, and B. C. Paik. 1991. Anaerobic digestion of distillery wastewater in a two-phase UASB system. *Water Science & Technology* 25(7): 361–371.

22 Dry Grind Coproducts as Cellulosic Ethanol Feedstock

Nathan S. Mosier

CONTENTS

22.1 INTRODUCTION

Corn grain is the staple feedstock for fuel ethanol production in the United States, accounting for more than 95% of fuel ethanol production (Saunders and Rosentrater, 2009). First-generation ethanol biofuel production from corn breaks down the starch portion of the grain into glucose, which is then fermented to ethanol. While improved efficiencies in the U.S. fuel ethanol industry have increased yields of ethanol near the theoretical maximum for corn starch, converting residual biomass possesses the opportunity for further increasing ethanol yields from a bushel of corn by as much as an additional 10%–14% (Kim et al., 2008a).

From a dry mill corn ethanol plant, the unfermented solids recovered by centrifugation from the aqueous slurry following ethanol distillation (commonly termed as distillers wet grains [DWG] [also known as wet cake]) are composed of three major components: (1) protein; (2) oil; and (3) carbohydrates (Table 22.1). When combined with the dissolved and suspended solids remaining in the aqueous fraction from the centrifuge (thin stillage), the resulting coproduct (distillers dried grains with solubles (DDGS), has considerable value as animal feed. However, the carbohydrate portion of this coproduct may have higher value as a cellulosic feedstock for additional fuel production. Most of the carbohydrates are represented by cellulose, residual unconverted starch, and hemicellulose. In total, these carbohydrates make up >40% wt. of the total dry matter content of these solids, on par with the protein content.

Carbohydrates represent a significant portion of the caloric value of distillers grains for animal feed in ruminant animals. However, neither cellulose nor hemicellulose is digestible by monogastrics such as swine and poultry, thus limiting the use of DDGS in monogastric feed rations. In methods for assessing animal forages, crude fiber more-or-less is equivalent to the cellulose fraction of wet cake (Belyea et al., 2004). The high amount of crude fiber present in DDGS and DWG is

TABLE 22.1
Composition of Wet Cake and Selected Other Cellulosic Feedstocks by Cellulosic Biofuels Analysis Methods

Dry Matter	Wet Cake[a] (%)	Corn Stover[b] (%)	Switchgrass[c] (%)
Ether extractives (Oil)	10.5	8.5	8.1
Crude protein	40.1	2.3	1.4
Cellulose	13.8	34.4	35.0
Starch	6.5	0.0	0.0
Hemicellulose (pentosans)	22.9	27.0	25.3
Lignin	N/A	17.0	21.4
Ash	2.2	6.1	3.3
Total dry matter (mass closure)	96.0	95.3	94.5

Source: Kim et al. *Bioresource Technology*, 99(12), 5165–5176, 2008a; Mosier et al. *Bioresource Technology*, 96, 1986–1993, 2005c; Kim et al. *Bioresource Technology*, 99, 5206–5215, 2008c.

[a] Kim et al., 2008a
[b] Mosier et al., 2005c
[c] Kim et al., 2008c

responsible, in part, for restricting its recommended inclusion in feed rations for swine and poultry. However, using technology under development for commercialization of cellulosic ethanol production, these carbohydrates could be utilized for ethanol production, thereby, lowering the fiber content in the residual feed product and increasing ethanol yields per bushel of corn.

22.2 CONVERTING CELLULOSE AND HEMICELLULOSE TO FUEL ETHANOL

Cellulose and hemicellulose are polymers of simple sugars that, when liberated by hydrolysis, can be fermented by microorganisms to produce fuel ethanol. Technologies in research, development, and at various stages on the path to commercialization exist for converting cellulose to fuel ethanol. Below is an outline of these technologies as they may be applied to converting the crude fiber in dry grind ethanol coproducts into fuel ethanol.

Cellulosic feedstocks are primarily comprised of three components, cellulose, hemicellulose, and lignin, all of which are found in the cell walls of plants (Wright, 1988). Cellulose and hemicellulose are carbohydrate polymers that act as structural components in the plant cell wall with lignin, a complex heterogeneous polymer, acting as the glue that binds them together and a water-resistant sealant (Wiselogel et al., 1996). Cellulose is a linear polysaccharide with 4000 to 8000 glucose units linked together through β-1,4 glycosidic bonds (Wright, 1988).

On the other hand, hemicellulose is a highly substituted heteropolysaccharide. Unlike cellulose, hemicellulose composition varies from species to species with similarities within families of plants. The predominant hemicellulose in grasses and cereals, such as corn, is (D-Glucurono)-L-arabino-D-xylan (GAX). GAX has a β-1,4-linked xylan backbone with primarily glucuronosyl and arabinosyl side chains (Dhugga, 2007). When hydrolyzed, the five-carbon sugars (pentoses) released (xylose and arabinose) are approximately 40% of the total carbohydrates available in lignocellulosic biomass.

The sugars released from both cellulose and hemicellulose are suitable for fermentation into ethanol. The remaining one-fifth of the biomass composition is predominantly a lignin-like material consisting of aromatics. The fibrous product from grain-to-ethanol processing (distillers grains) lacks true lignin, which is derived from the dehydration of three aromatic alcohols, and instead is elevated in protein content. This is contrasted to other cellulose-rich feedstocks such as corn stover and switchgrass, which have significant amount of lignification and relatively low crude protein contents (Table 22.1). Lignin both in nature and in a cellulosic ethanol production process significantly hinders

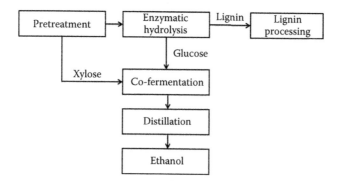

FIGURE 22.1   Biomass to ethanol production process. (Adapted from Wright, J. D., *Chemical Engineering Progress* 84(8), 62–74, 1988. With permission.)

the action of enzymes to hydrolyze the cell wall polysaccharides into fermentable sugars (Zhu et al., 2008). This requires pretreatment of the cellulosic material, often using harsh temperatures and/or chemicals, to improve the rates and yields of sugars from the feedstock (Mosier et al., 2005b).

An overview of the steps required for concerting cellulosic biomass to ethanol is shown below in Figure 22.1. Similar to starch-based grain-to-ethanol production technology, cellulose conversion requires a pretreatment step, analogous to starch liquefaction, to increase the reactivity of the polysaccharides to the subsequent enzymatic hydrolysis step (saccharification). The downstream fermentation and distillation steps are also very similar to the technology used to produce grain ethanol, allowing for the potential to integrate the processing cellulosic feedstocks, such as the fiber component of DDGS, into ethanol existing corn-to-ethanol facilities.

22.2.1   PRETREATMENT

One of the major hurdles associated with converting cellulosic biomass to ethanol is the extraction of fermentable (e.g., hydrolyzed) sugars. To facilitate their liberation, the cellulosic biomass is first pretreated. Pretreatment is necessary to disrupt the lignin/hemicellulose/cellulose matrix and to modify the structure of the biomass to ease access of the enzymes that catalyze the breakdown of carbohydrate polymers into fermentable sugars (Mosier et al., 2005b). Pretreatment can also result in extraction of hemicellulose fraction into the liquid stream (Wright, 1988). Numerous pretreatment techniques have been developed, and some are being commercialized (Mosier et al., 2005b). In all pretreatment technologies under commercial development, a combination of heat and chemicals (sometimes just water) is employed to disrupt the cellulose–hemicellulose–lignin complex within the plant cell walls to improve the accessibility of hydrolytic enzymes to the polysaccharides. This is accomplished by increasing the porosity of the material, partial or complete fractionation of the cell wall polymers (lignin, hemicelluloses, and cellulose), and altering the chemical and physical inter and intra actions among the major components of the cell wall. For a review of several pretreatment technologies see Mosier et al. (2005b).

22.2.2   ENZYMATIC HYDROLYSIS

The second step is required to release fermentable sugars from cellulosic biomass by hydrolysis. Following pretreatment, blends of enzymes are added to hydrolyze the cellulose and hemicellulose. Enzymatic hydrolysis is considered one of the most economically challenging steps, along with pretreatment, in producing ethanol (Lynd et al., 2002). To be cost competitive with conventional fuels and corn-based ethanol, enzyme costs still need to be further reduced per gallon of ethanol production.

The complexity of bonds and structures associated with plant cell walls necessitates a carefully balanced mixture of enzymes to achieve optimal sugar yields. This is only part of the reason that the enzyme loading, and thus cost, for releasing sugars from lignocellulose is higher than required for starch. Another reason is that the specific activity (amount of enzyme required to release sugars per unit mass) is lower for cellulases and hemicellulases compared to amylases. For a thorough review of enzymatic hydrolysis of lignocellulose for biofuel production see Lynd et al. (2002).

22.2.3 FERMENTATION

The processing of cellulosic biomass feedstocks through pretreatment and enzymatic hydrolysis results in the release of monomeric sugars, primarily glucose and xylose. The ultimate goal is to use microorganisms to ferment all of the sugars into ethanol at high yields, at fast rates, and at high final concentrations of ethanol (Aristidou, 2007). For example, *S. cerevisiae*, the yeast commonly used for industrial grain ethanol fermentations meets most of these criteria, but it lacks the ability to ferment xylose to ethanol. For herbaceous feedstocks, especially from grasses and crop residues (maize, rice, wheat, etc.) or potential bioenergy crops such as switchgrass, xylose comprises nearly 40% of the available fermentable sugars (Table 22.1). For economical conversion of these feedstocks to biofuel, xylose must also be fermented to ethanol.

To improve microorganisms for the fermentation of cellulosic biomass hydrolysates, genetic engineering techniques have been applied to either increase the number and type of sugars that can be fermented to ethanol (e.g., *S. cerevisiae*) or improve the yield and titer of ethanol produced by a microorganism that naturally ferments a wide range of sugars (e.g., *E. coli*) (Ho et al., 1998; Dien et al., 2003, and Chu and Lee, 2007). Many of the engineered strains are now being further developed for commercial use by cellulose-ethanol companies by improving the robustness of these organisms to inhibitors present in or produced by the processing of cellulosic feedstocks.

22.2.4 DISTILLATION AND ETHANOL RECOVERY

The recovery of the ethanol resulting from the fermentation of both pentose and hexose sugars uses the same unit operations and processing technology currently practiced in the corn-to-ethanol industry. However, distillation of dilute ethanol–water mixtures continues to be unattractive from both economic and energy consumption perspectives. One of the advantages of utilizing the fiber fraction of corn grain is the ability to utilize some of the same or similar product recovery equipment for both the starch and the cellulose conversion processing chains. Additionally, there is the potential to increase the final ethanol titer by either blending more dilute fiber sugar or ethanol streams with more concentrated parallel corn starch-derived streams. Both approaches leverage processing equipment, utilities, and other infrastructure to produce cellulosic ethanol collocated with grain ethanol production (Perkis et al., 2008).

22.3 COPRODUCTS OF DRY GRIND AS A CELLULOSIC ETHANOL FEEDSTOCK

Kim et al. (2008b) described possible modifications to dry grind corn-to-ethanol processes where the wet cake is further processed and the resulting sugar streams are recycled to increase ethanol yield per bushel of corn. Poet, ADM (Abbas et al., 2009), and Aventine Renewable Energy (Mosier et al., 2005a) have examined the commercialization of similar process modifications to enhance grain ethanol production, including modified dry grind technologies that fractionate the fiber from the other grain components at the beginning of the process (Singh and Eckhoff, 2001). It is reported that prefractionation results in a high fiber fraction similar to corn fiber, a coproduct of corn wet milling (Dien et al., 1997; Weil et al., 1998; Saha and Bothast, 1999; Mosier et al., 2005a). Back-end fractionation of fiber from ethanol coproducts has also been attempted (for more information, refer to Chapter 25). Figure 22.2 illustrates one example of how cellulose and starch processing could be integrated into a single plant.

In these modified dry grind processes, the wet cake from the centrifugation of heavy stillage undergoes pretreatment and saccharification separate from the starch processing of the grain (Ladisch et al., 2008). For example, the Kim et al. study (2008b) applies a hot-water pretreatment

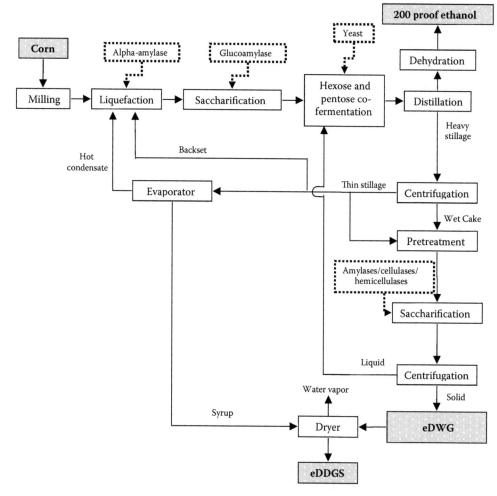

FIGURE 22.2 Example of integrating cellulose conversion from wet cake into existing dry grind corn ethanol plant. eDDGs = enhanced distillers dried grains with solubles; eDWG = enhanced distillers wet grains. (From Kim, Y. et al. *Bioresource Technology*, 99(12), 5177–5192, 2008b. With permission.)

method utilizing the thin stillage from the corn ethanol process as the pretreatment medium in order to reduce the overall plant water usage and to control pH during the reaction step.

After pretreatment, the solids are then hydrolyzed by a mixture of amylases, cellulases, and hemicellulases to release both glucose and xylose from the solids. The saccharified stream is then centrifuged and the liquid stream is then recycled directly to the corn fermentation (Figure 22.2). As discussed by Kim et al., other possible integration scenarios are possible. The report also discusses numerous constraints imposed by conjoining cellulosic and corn ethanol processes, such as the impact of recycling process streams on final ethanol titers and throughput of downstream processing equipment (distillation, etc.). Addressing these issues requires a detailed engineering analysis of energy and mass flows. In addition, special consideration of water use and recycling, bioreactor(s) capacity, utility and distillation capacity is needed to successfully integrate processing cellulose from wet cake in a dry grind ethanol. In addition, if the pentoses (xyloses) in the wet cake are to be fermented into ethanol, genetically modified (GM) yeasts are required to convert pentoses into ethanol. Assuring separation of the GM yeasts from DDGS is not likely to be feasible. This may require a strict segregation of xylose-fermenting GM yeasts from non-GM yeasts, which may necessitate separate fermentation, distillation, and other unit operations.

In the Kim et al. study (2008b), a computer process model was developed that included material balances formulated for each unit operation throughout the entire process. Using processing assumptions and data derived from laboratory-scale experiments, the model was used to determine overall ethanol yield, compositions of the new ethanol coproduct, enhanced distillers dried grains with solubles (eDDGS) mass of eDDGS per gal of ethanol produced, sugar, final ethanol concentrations in the fermentation tank, and accumulation of potential fermentation inhibitors in the process for several cases. In addition to addressing the technical feasibility of the project, model results were also used to inform an economic analysis of these cellulose–starch hybrid processing scenarios (Perkis et al., 2008).

The model analysis concluded that hybrid cellulose–starch grain process could yield 14% more ethanol per bushel of corn than the conventional process (assuming laboratory yields of 90% at each conversion step are maintained). The final coproduct (eDDGS) was predicted to have higher protein content than DDGS (40%–43% compared to 28%–32%), however the total solids sent to the drier was less since a significant fraction of the fiber was converted to additional ethanol. Perkis et al. (2008) modified a previously developed financial model for dry grind ethanol facilities to include the integration of extracting the sugars from the cellulose in the DDGS and examined the financial impact of these process changes. A 32% increase in net present value (NPV) for the overall operation was calculated from adding the cellulosic ethanol process to a 100 million gallon ethanol plant, assuming an enzyme cost of $0.20 for each additional gallon of ethanol produced.

Even though eDDGS had higher protein levels, current pricing of DDGS appears to be more sensitive to the amino acid profile than the total protein level, and the eDDGS had lower lysine (a key amino acid) levels. To determine the possible value, a swine feed ration with 15% inclusion of DDGS was used to model the potential value of the eDDGS (Thaler, 2002). Due to the lower lysine content, additional supplementation was required compared to the DDGS case. In balance, the model predicted no increase in the value of the eDDGS. In the context of the whole process economics, this resulted in decreased revenue from eDDGS due to less coproduct being formed due to its conversion to ethanol. However, the increased revenue from higher ethanol yields was greater than this loss of coproduct revenue plus the sum of all added costs, which include higher capital costs, larger loan payments, and increased operating costs. From the assumptions used for this analysis (Perkins et al., 2008), the breakeven ethanol price for the pretreatment process was roughly $1.95 per gallon of ethanol compared to $1.99 per gallon for the conventional dry grind process. They concluded that any improvements that added value to the eDDGS or lowered the enzyme cost had the most significant impact on improving the economics of integrating converting cellulose from DDGS to ethanol in a dry grind ethanol plant.

22.3.1 Pretreatment of Wet Grains or DDGS

As described above, the fiber component of ethanol coproducts lack true lignin that is present in most other cellulosic feedstocks (e.g., corn stover). Lignin is one of the most significant barriers to rapid rates and high yields from saccharification of cellulosic feedstocks (Yang and Wyman, 2004). Kim et al. (2008c) showed that the native DDGS digests rapidly compared to other biomass feedstocks, such as corn stover, in the absence of pretreatment prior to the addition of enzymes. Digestion of untreated DDGS leveled off after 72 h with a final glucose yield of 76% of theoretical (measured at 168 h) (Figure 22.3). By contrast, enzymatic hydrolysis of unpretreated corn stover particles (53–75 μm) at the same enzyme loading (15 FPU/g glucan) resulted in a 3× less (only 25%) glucan conversion after 168 h of hydrolysis (Zeng et al., 2006).

Pretreatment improves the rates and yields of glucose from DDGS, as it does for other cellulosic feedstocks. Kim et al. (2008c) compared the effects of liquid hot water (LHW) and AFEX pretreatment (Alizadeh et al., 2005; Teymouri et al., 2005; Bals, 2006). Maximum yield as well as rate of hydrolysis was significantly improved by pretreatment (Figure 22.3). LHW pretreatment at 160 °C for 20 min increased the ultimate yield of glucose to 98% at 72 h of hydrolysis and increased the initial hydrolysis rate by 10×, with hydrolysis leveling off at 5 h. AFEX pretreatment also resulted

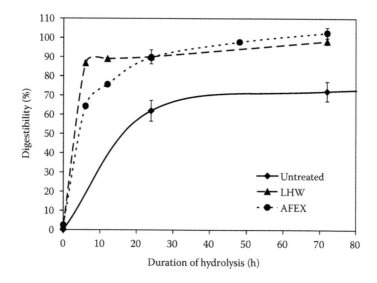

FIGURE 22.3 Glucan digestibility of untreated DDGS versus DDGS pretreated by liquid hot water pretreatment (LHW). (From Kim, Y. et al. *Bioresource Technology*, 99, 5206–5215, 2008c. With permission.)

in a significantly increased hydrolysis rate and an enhanced glucose yield near stoichiometric yield within 72 h. In a separate study of acid-catalyzed pretreatment, the optimal conditions (140°C for 20 min with 3.27 wt.% H_2SO_4) dissolved 77% of the carbohydrates (65% of the glucan and 93% of the xylan) in distillers grains as a mixture of monomeric and oligomeric sugars with only negligible degradation (Tucker et al., 2004). With the addition of cellulase and amylase, the glucose yield increased to >92% when pretreated with dilute acid.

In contrast, LHW pretreatment released only 2.9% of the total glucan and avoided any detectable degradation of glucose. During the enzymatic hydrolysis, yields of xylose and arabinose from the hemicellulose fraction were much lower than glucose yields (44% compared to >95% for glucose). Xylose yields for AFEX-treated DDGS were also low, obtaining less than 20% yield after 72 h of hydrolysis. This is because neither LHW nor AFEX pretreatments end-hydrolyse hemicellulose to monomeric sugars, which is advantageous in avoiding their degradation during pretreatment. Therefore, the hemicellulose must be enzymatically hydrolyzed as well as the cellulose in order to release the fermentable sugars.

Dien et al. (2008) tested several commercial enzyme preparations in combination with commercial cellulase to improve saccharifaction yields for LHW- and AFEX-pretreated DDGS. They found that adding commercial grade pectinase and feruloyl esterase preparations caused arabinose and xylose yields to increase. While DDGS contains very little pectin, commercial pectinase has rather high hemicellulose activity. Feruloyl esterase hydrolyzes a key side chain of GAX hemicellulose, ferulic acid, which cross-links separate hemicellulose chains in the plant cell wall (Saulnier et al., 2001; Bunzel et al., 2005). For LHW-pretreated wet cake, the xylose and arabinose yields were increased to a maximum of 82% and 70% respectively with the addition of these hemicellulytic enzymes. For AFEX-pretreated wet cake, the xylose and arabinose yields were improved to 81% and 99% (Dien et al., 2008). This illustrates the importance of matching the correct enzyme activities with each pretreatment method for achieving high sugar yields.

The ratio of solids to liquid has a strong effect on the yields that can be achieved by enzymatic saccharifaction. Higher solids-to-liquid ratios tend to lower the achievable yields due to a combination of factors not yet completely understood, but which are known to include product feedback inhibition of the enzymes, concentration of enzyme inhibitors and deactivators released by pretreatment and saccharifaction, and poor mass transfer of enzyme and sugars to/from the substrates (Jørgensen et al., 2006).

22.3.2 FERMENTATION OF PRETREATED WET GRAINS

Kim et al. (2008c) fermented the resulting sugars generated from the enzymatic hydrolysis of concentrated pretreated DWG (20% solids) using a nonrecombinant *Saccharomyces* yeast strain ATCC 4124. Fermentation byproducts such as glycerol, lactic acid, and acetic acid were elevated compared to standard dry grind ethanol fermentations because the thin stillage (backset) was substituted for much of the water during pretreatment.

The prepared hydrolysate of the LHW treated contained 120 g/L glucose, 21 g/L xylose, 47 g/L glycerol, 4.0 g/L lactic acid, and 2.5 g/L acetic acid. The hydrolysate of AFEX-treated DWG contained 75 g/L glucose, 9 g/L xylose, 12 g/L glycerol, 1.1 lactic acid, and 1.4 g/L acetic acid. Over 95% of the glucose was fermented within 24 h for the LHW-treated WDG hydrolysate, resulting in 60 g/L ethanol. Glucose was completely consumed within 6 h for the hydrolysate of AFEX-treated DWG, resulting in 45 g/L ethanol. For these fermentations, the yield of ethanol was approximately 75% of the theoretical based on the glucan in the starting material for the LHW-treated DWG, and approximately 80% of theoretical based upon the glucan in the starting material for the AFEX-treated DWG. For comparison, Tucker et al. (2004) reported that sulfuric acid pretreated DWG resulted in complete fermentation of glucose (15 g/L initial concentration) by *S. cerevisiae* D5A within 12 h to ethanol at approximately 80% of theoretical yield. These results indicate that the hexose sugars from pretreated DWG are readily fermented to ethanol by conventional distillery yeast.

22.3.3 FEED ANALYSIS OF ENHANCED DDGS

Pretreatment and fermentation of the wet cake can be expected to alter the composition of the DDGS. This is expected to enhance the resulting feed value because removal of the fiber fraction will enrich the modified feed for protein content. Both Kim et al. (2008c) and Tucher et al. (2004) collected the unfermented solids remaining after pretreatment, saccharifaction, and fermentation for testing as potential animal feed. In the Kim et al. (2008c) paper, the fermented slurry was dried to produce a potential animal feed product similar to DDGS. Tucher et al. (2004) filtered and centrifuged the undissolved solids present in the fermentation slurry and dried them to produce a potential animal feed product. In this case, the condensed distillers thin soluble solids from evaporating the thin stillage were not included in the feed product. In both studies, the feed products were subjected to proximate analyses consistent with DDGS (Table 22.2).

TABLE 22.2
Feed Analysis Results of DDGS and Enhanced DDGS

Compositions (%)	DDGS[a]	DDGS[b] (from LHW)	DDGS[b] (from AFEX)	DDGS[c] (from H$_2$SO$_4$)
Moisture	10.4	6.6	11.5	4.2
Crude protein	28.3	41.2	50.8	57.8
Crude fat	14.5	14.7	7.2	14.6
Crude fiber[d]	7.6	2.9	0.5	3.9
Ash	4.8	5.3	6.0	1.4

Source: Thiex, N. *Journal of AOAC International,* 92(1), 61–73, 2009; Kim et al. *Bioresource Technology,* 99, 5206–5215, 2008c; Tucker et al. *Applied Biochemistry and Biotechnology,* 113–116, 1139–1159. 2004; AOAC Official Method 978.10, 2006.

Note: Results are expressed on a dry matter basis (wt./wt. %).

[a] Thiex, 2009
[b] Kim et al., 2008c
[c] Tucker et al., 2004
[d] AOAC Official Method 978.10, 2006

The protein level in the new DDGS, after all three cellulosic processes, increased by at least 50%. The crude protein from the sulfuric acid pretreated DWG was the highest at 57.8% dry matter. However, the mass balance performed by Tucker et al. (2004) concluded that this was only 56% of the original protein mass of the feedstock (DWG). The remainder of the protein was either suspended or dissolved in the thin stillage and not recovered by centrifugation or was degraded by the pretreatment. Furthermore, the state of the remaining protein was not determined. No mass balance on protein was performed by Kim et al. (2008c), so recovery numbers are not available for comparison.

Both Kim et al. (2008c) and Tucker et al. (2004) performed analyses of the amino acids in the protein of the new DDGS. The overall amino acid profiles of the DDGS from all three pretreated DWG closely matched DDGS from the traditional dry grind process with a few exceptions. Lysine content was found to be lower in the DDGS from LHW- and sulfuric acid-treated DWG. Aspartic and glutamic acids were higher than traditional DDGS in DDGS from both LHW- and AFEX-treated DWG (no comparison for the acid pretreated DWG was possible). This indicated that the pretreatment process may significantly damage the protein and may affect the protein quality. More telling, the summation of the amino acids as measured in the traditional DDGS closely matched the estimated protein by the Kjeldahl method (25.6% vs. 28.3%) but the summations of amino acids from the DDGS after pretreatment were significantly lower than the Kjeldahl measurements of total nitrogen (31.6% for LHW and 37.6% for AFEX compared to 41.2% and 50.8%, respectively). A comparison between total nitrogen and amino acid profile was not possible from the data published by Tucker et al. (2004).

Tucker et al. (2004) also reported results from a limited feeding trial for the DDGS from the sulfuric acid processed DWG. No such feeding trial was reported by Kim et al. (2008c). The Tucker et al. (2004) feeding trial was of 350 recently hatched male commercial turkey poults. After a starting diet of the control ration for 3 days, 350 from 420 initial poults were selected (lightest and heaviest 70 poults were removed from the trial). The 350 test poults were randomized between the control diet (corn, soybean meal, and meat bone meal) and four levels of inclusion of the new DDG (5%–20%) into the control feed as a replacement for corn and soybean meal. The poults were fed for an additional 16 days and weighed periodically throughout the trial and at the end. The preliminary results were promising. The poults did not separate out the acid-treated DDG from the base feed ration during feeding. No mortality was reported for any of the groups. A small but significant (−6%) difference in total weight gain was observed, but only at the highest two levels of inclusion (15% and 20%). These data all indicated that DDG or DDGS from pretreated DWG has potential value as an animal feed coproduct. However, the quality of the protein must be assessed in these new coproducts. Traditional methods for determining protein by proximate analysis (Kjeldahl nitrogen) may overestimate the available amino acids that can be degraded to other non-nutritive nitrogen containing compounds during pretreatment. Any proposed process to integrate further processing of the DDGS to extract cellulosic sugars must carefully examine the potential impact on the quality of this valuable coproduct.

22.4 SUMMARY

The fibrous coproduct of corn dry grind ethanol production (distillers grain or corn fiber) has potential as a cellulosic feedstock that is readily available, easily harvested, stored, and transported, and could be processed in existing corn milling facilities. Distillers grain has the potential of increasing ethanol yields per bushel of corn by up to 14% while still yielding a high-protein byproduct, which appears to be an animal feed of similar performance to DDGS, but with the benefit of much lower fiber content. This opens greater possibilities for inclusion of low-fiber, high-protein DDGS into monogastric diets (poultry and swine) at higher levels than practical for traditional DDGS. Use of the fiber is one method for producing additional biofuel. Extraction and use of the lipids is another. This will be discussed in the following chapter.

REFERENCES

Abbas, C., T. P. Binder, K. E. Beery, M. J. Cecava, P. H. Doane, D. P. Holzgraefe, and L. P. Solheim. 2009. Process for the production of animal feed and ethanol and novel animal feed. *U.S. Patent 7,494,675.*

Alizadeh, H., F. Teymouri, T. I. Gilbert, and B. E. Dale. 2005. Pretreatment of switchgrass by ammonia fiber explosion (AFEX). *Applied Biochemistry and Biotechnology* 121–124: 1133–1141.

Aristidou, A. A. 2007. Application of metabolic engineering to the conversion of renewable resources to fuels and fine chemicals: Current advance and future prospects. *Fermentation Microbiology and Biotechnology.* E. M. T. El-Mansi. Boca Raton, FL: CRC Press.

Bals, B., B. E. Dale, and V. Balan, V. 2006. Enzymatic hydrolysis of distiller's grain and solubles (DDGS) using ammonia fiber expansion pretreatment. *Energy Fuels* 20: 2732–2736.

Beguin, P., and J. P. Aubert. 1994. The biological degradation of cellulose. *Fems Microbiology Reviews* 13(1): 25–58.

Belyea, R. L., K. D. Rausch, and M. E. Tumbleson. 2004. Composition of corn and distillers dried grains with solubles from dry grind ethanol processing. *Bioresource Technology* 94: 293–298.

Bunzel, M., J. Ralph, C. Funk, and H. Steinhart. 2005. Structural elucidation of new ferulic acid-containing phenolic dimers and trimers isolated from maize bran. *Tetrahedron Letters* 46: 5845–5850.

Collins, T., C. Gerday, and G. Feller. 2005. Xylanases, xylanase families and extremophilic xylanases. *Fems Microbiology Reviews* 29(1): 3–23.

Chu, B. C. H., and H. Lee. 2007. Genetic improvement of Saccharomyces cerevisiae for xylose fermentation. *Biotechnology Advances* 25(5): 425–441.

Dien, B. S., M. A. Cotta, and T. W. Jeffries. 2003. Bacteria engineered for fuel ethanol production: Current status. *Applied Microbiology and Biotechnology* 63(3): 258–266.

Dien, B. S., R. B. Hespell, L. O. Ingram, and R. J. Bothast. 1997. Conversion of corn milling fibrous co-products into ethanol by recombinant Escherichia coli strains KO11 and SL40. *World Journal of Microbiology & Biotechnology* 13: 619–625.

Dien, B. S., E. A. Ximenes, P. J. O'Bryan, M. Moniruzzaman, X. Li, V. Balan, B. E. Dale, and M. A. Cotta. 2008. Enzymatic extraction of sugars from AFEX and liquid hot-water pretreated distiller's grains and their conversion to ethanol. *Bioresource Technology* 99(12): 5216–5225.

Din, N., H. G. Damude, N. R. Gilkes, R. C. Miller, R. A. J. Warren, and D. G. Kilburn. 1994. C-1-C-X revisited—Intramolecular synergism in a cellulase. *Proceedings of the National Academy of Sciences of the United States of America* 91(24): 11383–11387.

Dhugga, K. S. 2007. Maize biomass yield and composition for biofuels. *Crop Science* 47(6): 2211.

Henrissat, B., H. Driguez, C. Viet, and M. Schulein. 1985. Synergism of cellulases from Trichoderma-reesei in the degradation of cellulose. *Bio-Technology* 3(8): 722–726.

Ho, N. W. Y., Z. Chen, and A. P. Brainard. 1998. Genetically engineered Saccharomyces yeast capable of effective cofermentation of glucose and xylose. *Applied and Environmental Microbiology* 64(5): 1852–1859.

Jørgensen, H., J. Vibe-Pedersen, J. Larsen, and C. Felby. 2006. Liquefaction of lignocellulose at high-solids concentrations. *Biotechnology and Bioengineering* 96(5): 862–870.

Kim, Y., N. S. Mosier, R. Hendrickson, T. Ezeji, H. Blaschek, B. S. Dien, M. A. Cotta, B. E. Dale, and M. R. Ladisch. 2008a. Composition of corn dry-grind ethanol byproducts: DDGS, wet cake, and thin stillage. *Bioresource Technology* 99(12): 5165–5176.

Kim, Y., N. S. Mosier, and M. R. Ladisch 2008b. Process simulation of modified dry grind ethanol plant with recycle of pretreated and enzymatically hydrolyzed distillers' grains. *Bioresource Technology* 99(12): 5177–5192.

Kim, Y., R. Hendrickson, N. S. Mosier, M. R. Ladisch, B. Bals, V. Balan, and B. E. Dale. 2008c. Enzyme hydrolysis and ethanol fermentation of liquid hot water and AFEX pretreated distillers' grains at high-solids loadings. *Bioresource Technology* 99: 5206–5215.

Ladisch, M., M. Dale, W. Tyner, N. S. Mosier, Y. Kim, M. A. Cotta, B. S. Dien, H. Blaschek, E. Laurenas, B. Shanks, J. Verkad, C. Schell, and G. Petersen. 2008. Cellulose conversion in dry grind ethanol plants. *Bioresource Technology* 99(12): 5157–5159.

Lynd, L. R., P. J. Weimer, W. H. van Zyl, and I. S. Pretorius. Microbial cellulose utilization: Fundamentals and biotechnology. *Microbiology and Molecular Biology Reviews* 66(3): 506–577.

Mosier, N. S., R. Hendrickson, M. Brewer, N. Ho, M. Sedlak, R. Dreshel, G. Welch, B. S. Dien, A. Aden, and M. R. Ladisch. 2005a. Industrial scale-up of pH-controlled liquid hot water pretreatment of corn fiber for fuel ethanol production. *Applied Biochemistry and Biotechnology* 125: 77–97.

Mosier, N. S., C. Wyman, B. E. Dale, R. Elander, Y. Lee, Holtzapple, and M. R. Ladisch. 2005b. Features of promising technologies for pretreatment of lignocellulosic biomass. *Bioresource Technology* 96: 673–686.

Mosier, N. S., R. Hendrickson, N. Ho, M. Sedlak, and M. R. Ladisch. 2005c. Optimization of pH controlled liquid hot water pretreatment of corn stover. *Bioresource Technology* 96: 1986–1993.

Perkis, D., W. E. Tyner, and R. Dale. 2008. Economic analysis of a modified dry grind ethanol process with pretreatment and recycle. *Bioresource Technology* 99(12): 5243–5249.

Polizeli, M. L. T. M., A. C. S. Rizzatti, R. Monti, H. F. Terenzi, J. A. Jorge, and D. S. Amorim. 2005. Xylanases from fungi: Properties and industrial applications. *Applied Microbiology and Biotechnology* 67(5): 577–591.

Reese, E. T., R. G. H. Siu, and H. S. Levinson. 1950. The biological degradation of soluble cellulose derivatives and its relationship to the mechanism of cellulose hydrolysis. *Journal of Bacteriology* 59(4): 485–497.

Saha, B. C., and R. J. Bothast. 1999. Pretreatment and enzymatic saccharification of corn fiber. *Applied Microbiology and Biotechnology* 76: 67–77.

Saulnier, L., and J. F. Thibault. 1999. Ferulic acid and diferulic acids as components of sugar-beet pectins and maize bran heteroxylans. *Journal of the Science of Food and Agriculture* 79: 396–402.

Saunders, J. A., and K. A. Rosentrater. 2009. Survey of US fuel ethanol plants. *Bioresource Technology* 100: 3277–3284.

Shallom, D., and Y. Shoham. 2003. Microbial hemicellulases. *Current Opinion in Microbiology* 6(3): 219–228.

Shimada, M., and M. Takahashi. 1991. Wood and cellulosic chemistry, Ch. 13. In *Biodegradation of Cellulosic Materials*, vol. iii, p. 1020. New York: Marcel Dekker.

Singh, V., and S. R. Eckhoff. 2001. Process for recovery of corn coarse fiber (pericarp). *U.S. Patent 6,254,914*.

Teeri, T. T. 1997. Crystalline cellulose degradation: New insight into the function of cellobiohydrolases. *Trends in Biotechnology* 15(5): 160–167.

Teeri, T. T., A. Koivula, M. Linder, G. Wohlfahrt, C. Divne, and T. A. Jones. 1998. Trichoderma reesei cellobiohydrolases: Why so efficient on crystalline cellulose? *Biochemical Society Transactions* 26(2): 173–178.

Teymouri, F., L. Laureano-Perez, H. Alizadeh, and B. E. Dale. 2005. Optimization of the ammonia fiber explosion (AFEX) treatment parameters for enzymatic hydrolysis of corn stover. *Bioresource Technology* 96: 2014–2018.

Thaler, B. 2002. Use of distillers dried grains with solubles (DDGS) in swine diets. *Extension Extra* 2035, Cooperative Extension Service, South Dakota State University.

Thiex, N. 2009. Evaluation of analytical methods for the determination of moisture, crude protein, crude fat, and crude fiber in distillers dried grains with solubles. *Journal of AOAC International* 92(1): 61–73.

Tucker, M. P., N. J. Nagle, E. W. Jennings, K. N. Ibsen, A. Aden, Q. A. Nguyen, K. H. Kim, and S. Noll. 2004. Conversion of distiller's grain into fuel alcohol and a higher-value animal feed by dilute-acid pretreatment. *Applied Biochemistry and Biotechnology* 113–116: 1139–1159.

Wiselogel, A., S. Tyson, and D. Johnson. 1996. Biomass feedstock resources and composition. In *Handbook on Bioethanol: Production and Utilization*, ed. C. E. Wyman, pp. 105–118. Washington, DC: Taylor & Francis.

Weil, J. R., A. Sarikaya, S. L. Rau, J. Goetz, C. M. Ladisch, M. Brewer, R. Hendrickson, and M. R. Ladisch. 1998. Pretreatment of corn fiber by pressure cooking in water. *Applied Biochemistry and Biotechnology* 73: 1–17.

Wright, J. D. 1988. Ethanol from biomass by enzymatic hydrolysis. *Chemical Engineering Progress* 84(8): 62–74.

Yang, B., and C. E. Wyman. 2004. Effect of xylan and lignin removal by batch and flowthrough pretreatment on the enzymatic digestibility of corn stover cellulose. *Biotechnology and Bioengineering* 86(1): 88–95.

Zhu, L., J. P. O'Dwyer, V. S. Chang, C. B. Granda, and M. T. Holtzapple. 2008. Structural features affecting biomass enzymatic digestibility. *Bioprocess Technology* 99(9): 3817–3828.

23 Extraction and Use of DDGS Lipids for Biodiesel Production

Michael J. Haas

CONTENTS

23.1 INTRODUCTION

23.1.1 DRIVERS FOR THE USE OF RENEWABLE FUELS, AND A CONDENSED HISTORY OF BIODIESEL

In addition to the application of pretreatment technologies to convert fiber components of DDGS into substrates amendable to additional fermentation into ethanol, as described in the preceding chapter, uses are being developed for the oil components of DDGS as well. Chief among these is as a feedstock for the production of liquid transportation fuel, particularly "biodiesel."

Contemporary society is inescapably dependent on petroleum products. For over a century our technologies have been incentivized, designed, and optimized to operate on petroleum-based fossil fuels and chemical feedstocks. Within the past decade the secondary consequences of this societal choice have become increasingly clear, manifested in various forms of environmental upset, political tension, human suffering and death, and the diversion of human, financial, and material resources to military expenditures. This has increased the desire in many regions of the world to develop alternate sources for the energy and chemical feedstocks presently obtained from petroleum products. The facts that the world has passed the peak of new oil discovery, which will lead to reduced supplies and increasing prices (Government Accountability Office, 2007), and that petroleum use is growing in large developing countries further indicates the necessity of moving beyond fossil fuels and feedstocks.

The United States excels at agricultural productivity. Considering its strengths of soil and climate, including a well-developed agricultural production system, a strong agriculture-related private sector, and an extensive agricultural education and research infrastructure, it is unequalled in the world. As the need to develop new sources of energy and chemical feedstocks becomes clear and urgent, this system is being mobilized in the effort to replace petroleum. The use of domestically produced agricultural materials eases the necessary transition away from crude oil as a feedstock, reduces the emission of nonrecycled carbon as greenhouse gases, and has the potential to greatly reduce global political tensions and conflicts.

The use of triacylglycerol-based fuels as petroleum replacements in compression–ignition (diesel) engines has long been known, with early versions of this engine being fueled by peanut oil at the turn of the 20th century (Diesel, 1913). At that time, however, global production of fats and oils was insufficient to support the growth of a lipid-based liquid transportation fuel industry. The emerging petroleum industry occupied that niche. The recent growth of interest in the use of biobased fuels follows a century of engine optimization around petroleum. Thus, although the most economical option would be to fuel diesel engines with unmodified fats and oils, as when these engines were first invented, the engines have evolved and become specialized to the extent that at present they are generally unable to operate long term on these materials. The relatively high viscosities of triacyglycerols leads to poor in-cylinder atomization, inefficient combustion, fuel deposition, coking on fuel injectors and cylinder walls, and engine wear and failure (Ziejewski et al., 1986a, 1986b; Goering et al., 1987).

The transesterification of triacylglycerols converts their fatty acids from glycerol esters to simpler (e.g., methyl, ethyl) acyl esters. Viscosity is reduced approximately tenfold (Knothe et al., 2005, appendix A) and the resulting fuel, termed "biodiesel," performs well in unmodified diesel engines (Graboski and McCormick, 1998; Chase et al., 2000; Biobus Committee, 2003). This substantial compatibility of biodiesel with contemporary engines and fuel handling systems is a significant feature and has stimulated the growth of the biodiesel industry. Other biobased liquid diesel fuels are under development (Dry, 2002; Spath and Dayton, 2003; Kalnes et al., 2007) but none approach biodiesel in terms of demonstrated performance, functionality and compatibility with the existing fuel storage and handling infrastructure, regulatory approval, number and distribution of production sites, or output. Annual U.S. production of biodiesel rose from 2 to 650 million gal (7.6 million to 2.5 billion L) between 2000 and 2008 (National Biodiesel Board, 2010), taking biodiesel from a research topic to a component of the national fuel infrastructure. Production volumes are expected to continue to increase. The Renewable Fuels Standard of the 2007 United States Energy Policy Act (United States Government Printing Office, 2007) pledges further growth in the production of renewable fuels, to reach 36 billion gal (136 billion L) of biofuels annually by 2022. Biodiesel production is anticipated to exceed 1.5 billion gal (5.7 billion L) annually under this scenario.

23.1.2 CURRENT AND FUTURE LANDSCAPES REGARDING BIODIESEL FEEDSTOCKS

Annual U.S. production of fats and oils is approximately 36 billion lb (16 million metric tons), with 72% of this being vegetable oils and the balance animal fats and rendered greases (Schnepf, 2007). At approximately 20 billion lb (2.5 billion gal) (9.1 million metric tons, 9.5 billion L) annually, soybean oil production constitutes roughly 79% of the vegetable oil output, or 55% of total lipid production (Schnepf, 2007; United States Department of Agriculture, 2008a; United States Census Bureau, 2009). It is the nation's most abundant and least expensive refined edible oil, and thus also historically and presently the predominant material for biodiesel production. Soybean oil currently accounts for just less than 50% of annual biodiesel production. Edible animal fats and high-quality rendered greases make up the bulk of the remainder of the feedstock mix. To date the use of these lipids has not substantially impacted food prices (Europa Bio, 2007; National Biodiesel Board, 2008). However, prohibitive increases in the costs of oil for human consumption could result from further substantial increases in the use of these oils for biodiesel production. Even dedication of the entire soybean crop to biodiesel production would only displace about 6% of the nation's annual diesel fuel consumption (Hill et al., 2006). Consideration of existing biodiesel production targets and their comparison with current feedstock availability and with overall diesel fuel consumption has raised concerns about feedstock availability, fuel cost, food cost, and the environmental impact of fuel production agriculture.

Fuel affordability considerations also suggest the need to develop lower cost feedstocks for biodiesel production. Even by using soybean oil, the least expensive vegetable oil, feedstock expense renders the final fuel unable to compete economically with petroleum diesel. The current soy oil price of nearly $0.60/lb ($1.32/kg) amounts to a feedstock cost in the range of $4.25/gal of biodiesel. Nonfeedstock fuel production costs are approximately $0.40/gal (0.11/L of $4.25/gal) (Haas et al., 2006), resulting in a product that is unable to compete with U.S. petroleum diesel fuel on a price basis in the absence of government support. Higher value oils such as canola and edible corn oil give proportionately more expensive biodiesel costs.

Increased demand for the lowest cost refined oil, that is soybean oil, to meet increased fuel production goals may not cause substantial increases in oil production since more of the value of a soybean lies in its meal component than in its oil content. The use of the high-protein meal in animal nutrition is a greater stimulus to soybean production. U.S. meat production is not rising substantially (United States Department of Agriculture, 2008b). It is therefore difficult, at best, to determine the degree to which an increase in oil demand would increase soybean production. By the same token, animal fats and waste greases are byproducts of the animal and food preparation industries, respectively, and therefore the production of these lower cost feedstocks is not amenable to efforts to increase output for the sake of biofuel use.

Thus, to reduce feedstock and biofuel prices, to increase biodiesel supplies, and to minimize the impact of biodiesel production on the availability of triacyglycerols for human consumption, there is interest in the use of inedible fats and oils as feedstocks for biodiesel production. Corn oil present in the mixtures generated during the fermentation of corn to ethanol is one such material.

The United States used approximately 3.9 billion bu. of corn for ethanol production in 2009 (Economic Research Service, 2010). This generated approximately 30 million tons (27 million metric tons) of DDGS as a byproduct (Rosentrater, 2009). The typical market value of this material, and of the corn oil that it contains, is $0.15/lb ($0.33/kg), approximately half the value of refined soybean oil. This postfermentation corn oil could be an economically attractive biodiesel feedstock given affordable technology for its conversion to high-quality biodiesel. With a typical oil content of 10 wt.%, the current output of DDGS has the potential to provide up to 6 billion lb (2.7 million metric tons) of oil, or a comparable amount of biodiesel (810 million gal; 3.1 billion L). This is comparable to or exceeds annual U.S. biodiesel output in recent years. Since corn oil recovered to date from postfermentation streams has not been of sufficient quality for human consumption, its use as a fuel would not impact the cost or availability of food.

23.2 BIODIESEL: CHEMISTRY, PRODUCTION, QUALITY

Following the demonstration in 1900 of the use of vegetable oils as engine fuels, there was little attention paid to the topic until the oil export embargos of the 1970s stimulated interest in the development of domestic energy sources. The resulting research initiatives began producing notable results in the early 1980s. Biodiesel is synthesized from acylglycerols via a relatively simple transesterification process involving reaction with a short chain alcohol, typically catalyzed by simple inorganic alkaline catalysts such as sodium or potassium hydroxide. The reaction (Figure 23.1) involves the sequential transfer of each of the three fatty acids of the acylglycerol from one of the hydroxyl groups of the trihydroxy compound glycerol to a monohydroxy alcohol, usually methanol. Di- and monoacyl glycerol esters are intermediates in the reaction series, which produces fatty acid methyl esters (FAME) and glycerol as final products. Freedman et al. (1984) published a seminal study of the chemistry of the reaction for fuel production. Working at temperatures below the boiling point of the reaction (i.e., 60 °C for methanol reactions), these authors identified effective catalyst concentrations for inorganic acid and base catalysts; showed alkaline catalysis to be superior to acid for vegetable oil feedstocks; identified effective concentrations for sodium hydroxide and sodium methylate and other catalysts; demonstrated effective transesterification by methanol, ethanol, and butanol; and showed that the reaction proceeded rapidly to high degrees of completion at a 2:1 initial molar ratio of alcohol:fatty acid, that is, a 6:1 ratio of alcohol:triglyceride. Mittelbach and Trathnigg (1990) and Komers et al. (2002) characterized the reaction kinetics, which are complicated by the facts that the oil and alcohol reactants have limited cosolubility, as do the glycerol and fatty acid

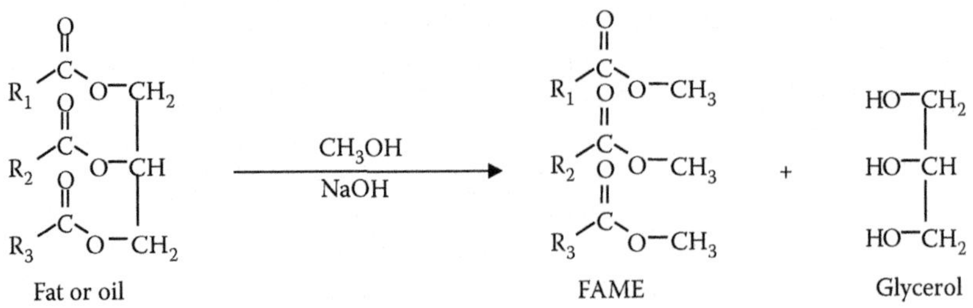

FIGURE 23.1 The alkali-catalyzed transesterification of a triacylglycerol to produce fatty acid methyl ester (FAME) and glycerol. R_x = fatty acid alkyl chain.

ester products. Initial reaction rates are limited by the mutual insolubilities of the reactants, and mixing plays a key role in initiating effective reaction until the population of intermediate di- and monacylglycerols reaches levels sufficient to emulsify the reaction effectively (Ma et al., 1999). Other investigators subsequently described alternate liquid- and solid-phase catalysts, supercritical reaction conditions, cosolvent addition to convert the reaction to a single phase system, alternate alcohols, various batch and continuous reactor geometries, and multiple methods to mix the reaction (Mittelbach and Remschmidt, 2004; Knothe et al., 2010). Investigations such as these could result in substantial changes to the methods for biodiesel production in the future. Presently however, the optimal conditions defined by Freedman et al. (1984) largely form the basis of contemporary industrial biodiesel production technology. A process diagram for a typical production facility is shown in Figure 23.2, although individual production facilities often implement alternative and/or proprietary technologies. For further information on the process shown in Figure 23.2, and a model for estimation of the capital and operational costs of this technology, see Haas et al. (2006).

Following the synthesis of FAME, and irrespective of the feedstock or chemical conversion process employed, unreacted alcohol is removed by heating and the glycerol byproduct is removed by gravity settling or centrifugation. Despite a proliferation of solid-phase adsorbants for "waterless biodiesel cleanup," the bulk of existing production facilities employ water washing to remove traces of glycerol, methanol, and catalyst from the biodiesel product (T. Kemper, DeSmet Ballestra North America; and M. Danzer, Renewable Energy Group, pers. comm.).

Alkali is an effective catalyst for the transesterification of acylglycerols, and is employed as described above when the feedstock is primarily composed of glycerides. However, alkali is not an acceptable catalyst for FAME production from feedstocks containing significant amounts of free fatty acids. These are converted to "soaps" under alkaline conditions, the term given to the cation salts of free fatty acids. Soaps cannot serve as an engine fuel, and also are insoluble in biodiesel. When a feedstock contains less than about 4–5 wt.% FFA, it can be practical to employ alkaline catalysis for FAME production, with the resulting soaps being removed and sold or disposed of. At higher FFA levels the elevated consumption of alkali to replace that lost to the soap fraction, and the attendant reduction in FAME yield compared to conversion of the FFA to FAME, are unacceptable. In such cases use can be made of the fact that inorganic acid readily catalyzes the esterification of FFA. However, it is a poor catalyst for the transesterification of acylglycerols. Thus, for high FFA feedstocks dual sequential reaction protocols have been developed wherein acid-catalyzed esterification of the FFA fraction is followed by alkali-catalyzed transesterification of the acylglycerol component to achieve high FAME yields and low residual FFA levels (Canacki and Van Gerpen, 2003).

It is inappropriate to discuss, contemplate, or conduct the production of fatty acid ester preparations for use as biodiesel without considering the aspect of product quality. Whereas fuel ethanol is a single chemical species necessarily produced by technologies that involve distillation, thereby generating a pure product, biodiesel is at best a mixture of multiple fatty acid esters. At worst, and much more typically, it is an undistilled product generated from feedstocks that may contain a myriad of oil-soluble materials found in vegetable oils, animal fats, or waste greases. Today's diesel

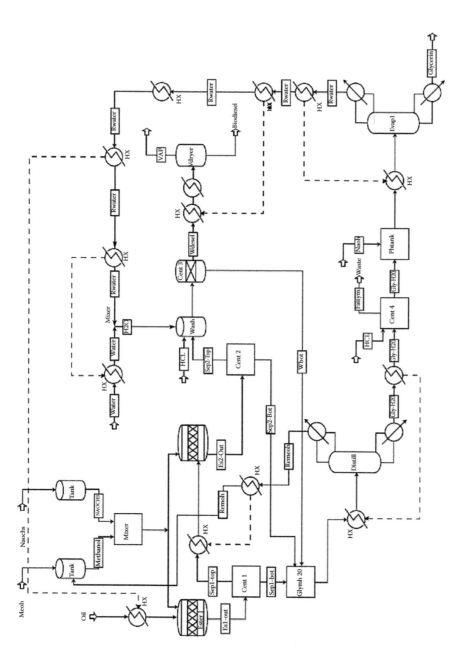

FIGURE 23.2 Process flowsheet for a typical process for the production of biodiesel from soybean oil. CENT: centrifuge; ES1, 2 OUT: posttransesterification ester mixture exiting reactors 1 and 2, respectively; FATTYM: free fatty acid waste stream; GLYH2O: glycerol–water stream, GLYMH2O: crude glycerol accumulation tank; HX: heat exchanger; dashed lines indicate heat transfer; MEOH: methanol; NAOCH3: sodium methoxide; PHTANK: pH adjustment tank, REMEOH: recovered methanol; RWATER: recovered water; SEP 1, 2 BOT: heavy (glycerol-rich) layer exiting centrifugal separators 1 and 2, respectively; SEP 1, 2 TOP: light layer (containing ester product) exiting centrifugal separators 1 and 2, respectively; VAP: water vapor; VDRYER: vacuum dryer; WBOT: aqueous stream recovered after a water wash of crude biodiesel, WDESEL: biodiesel–water mixture. Solid lines indicate the flow of liquid streams, dashed lines the flow of heat.

TABLE 23.1
Biodiesel Standard ASTM D6751-09

Property	Assay Method (ASTM)[a]	Maximum Allowed (Unless Stated)
1. Flash point (°C)	D 93	93 (minimum)
2. Alcohol control		
Meet one of:		
A. Methanol content (mass %)	EN 14110	0.2
B. Flash point (°C)	D 93	130 (min)
3. Water and sediment (vol %)	D 2709	0.05
4. Carbon residue (wt%)	D 4530	0.05
5. Sulfated ash (mass %)	D 874	0.020
6. Kinematic viscosity (mm²/s @ 40°C)	D 445	1.9–6.0
7. Sulfur (wt%)	D 5453	0.0015, 0.05[b]
8. Cloud point (°C)	D 2500	Report
9. Copper corrosion	D 130	3
10. Cetane no.	D613	47 (min)
11. Cold soak filterability (seconds)—for use	Annex A1	360
below 12°C	Annex A1	200
12. Acid number (mg KOH/g)	D 664	0.50
13. Free glycerin (wt%)	D 6584	0.020
14. Total glycerin (wt%)	D 6584	0.240
15. Oxidation stability (hours)	EN14112	3 (min)
16. Calcium plus magnesium (ppm)	EN14538	5
17. Sodium plus potassium (ppm)	EN 14538	5
18. Phosphorus (wt%)	D 4951	0.001
19. Reduced pressure distillation (temperature at 90% recovery, °C)	D 1160	360

Source: American Society for Testing and Materials. 2009. D6751-09 Standard specification for
biodiesel fuel (B100) blend stock for distillate fuels. West Conshohocken: ASTM.

[a] "D" prefixes indicate methods authorized by ASTM. "EN" prefixes designate methods released by
the European Committee for Standardization, Brussels, Belgium.

[b] The limits are for S15 (0.0015 wt.% = 15 ppm maximum sulfur) and S500 (500 ppm) Grade fuels,
respectively. The former is most often used for "on-road" transportation applications while the latter
is for "off-road" applications such as construction equipment.

engines, fuel systems, and emissions control technologies are sensitive to damage by any of a number of contaminants that can originate either in the feedstock or from the fuel production technology. To protect these systems, official standards have been developed and adopted in a number of countries. In the United States the accepted Standard is ASTM D6751, which was developed and released by the American Society for Testing and Materials (2009). An ever-evolving document, it presently specifies the evaluation of 19 fuel properties and establishes allowable ranges for the resulting values (Table 23.1). Not until these have been passed can a FAME preparation correctly be referred to as "biodiesel." The parameters that are tested and the results specified as constituting "acceptable" in the Official Standard were chosen either by extension from similar requirements that are in place for petroleum-based diesel fuel or after analysis of the causes of in-field failures of biodiesel. These standards have been tightened and expanded as the industry has become aware of necessary augmentations to improve the utility, reliability and performance of biodiesel. Continued refinement will occur over time as new needs are identified.

Historically there have been some instances of the marketing of FAME preparations that met some, but not all, of the parameters of D6751. The undesirability of such actions can hardly be overstated since, as with any new industry, performance problems caused by an off-specification fuel could brand all biodiesel as unreliable, thus threatening the survival and growth of the entire industry. In recognition of this fact, the National Biodiesel Board (NBB, see www.biodiesel.org), the representative umbrella organization for the U.S. biodiesel industry, has instituted a quality accreditation program. Known as BQ 9000®, this program certifies biodiesel producers, marketers, distributors, and analytical laboratories in the United States and Canada. The program is a unique combination of ASTM D6751 and a quality systems program that includes storage, sampling, testing, blending, shipping, distribution, and fuel management practices. Participation in the program has grown steadily. As of mid-2010, NBB reported that 77% of U.S. biodiesel production for the fiscal year beginning October 2009 had occurred in BQ9000 facilities. An ongoing commitment to quality is essential to the reliability of biodiesel, and to the survival of the biodiesel industry.

23.3   CONTEMPORARY TECHNOLOGIES FOR BIODIESEL PRODUCTION FROM POSTFERMENTATION CORN OIL

A comprehensive discussion of the lipid content of DDGS is presented in Chapters 8 and 9 of this text. The fatty acid compositions of corn oil (Gunstone, 2005) and corn oil recovered postfermentation (Winkler, 2007; Moreau, 2010) are quite similar to that of soy oil. They differ from soy oil mainly in lacking the approximately 7% linolenic acid (18 carbons long, three double bonds), and display a corresponding increase in oleic acid (18:1). A low linolenic acid content is desirable in a biodiesel since it imparts oxidative stability to the fuel. Thus, on the basis of its chemical composition and probable suitability as a fuel, there is interest in the production of biodiesel from corn oil. There has been very little reported research on the use of refined corn oil in biodiesel production to date, probably because the cost of this feedstock historically exceeds that of soy oil by approximately 20%. However, inedible corn oils, such as are present in postfermentation ethanol coproduct streams, are viable low-cost potential biodiesel feedstocks.

The dry grind process is the predominant technology in the United States for fuel ethanol production from corn, accounting for more than 80% of the output. In this method, described elsewhere in this text, dry corn is ground, slurried in water, and subjected to a saccharification/fermentation process that produces ethanol from the starch of the corn kernel. Refer to Chapter 5 for more details about ethanol production processes. The corn germ, which contains the bulk of the corn triacylglycerols, is thereby pulverized, along with the rest of the kernel, and subjected to the yeast fermentation process. The residual material after fermentation and removal of the ethanol product is termed "whole stillage," which is typically processed into products known as "distillers wet grains," with approximately 35% solids, and "thin stillage," which is about 8% solids. The latter is often evaporated into a syrup that is added to the distillers wet grains (also known as "wet cake") and dried, producing distillers dried grains with solubles, or DDGS. The majority of the corn oil triacylglycerols are not substantially degraded, hydrolyzed, or metabolized during fermentation, and exit with these streams. Approximately one-third of the input corn oil is located in the thin stillage, with the remaining two-thirds in the distillers wet grains. Recovery and sale of the oil would augment the income of an ethanol production plant, and could increase biodiesel supplies.

In addition to this potential value of the recovered corn oil, removal of the oil from DDGS can increase the utility of the resulting oil-free coproduct. Oil removal may facilitate the industrial scale handling and transport of DDGS by increasing flowability (Agricultural Utilization Research Institute, 2005). It may also reduce the very pronounced tendency of DDGS to compact tightly during, for example, rail transport. Such lodging necessitates additional labor to remove the material from railcars, a process sometimes so vigorous that the cars themselves are damaged. A reduction of the oil content of DDGS could also expand its use in swine, poultry, and ruminant nutrition (McElroy, 2007, Chpaters 12–15 of this book).

23.3.1 OIL RECOVERY VIA SOLVENT EXTRACTION

Solvent extraction, typically using hexane, is routinely applied to corn germ for the production of high-quality corn oil for human consumption (Moreau, 2005). At large production levels this is the method of choice for oil recovery (Kemper, 2005; Williams, 2005). Corn germ has an oil content of 45–50 wt.%, while that of DDGS is typically 10–13 wt.%. The economics of solvent extraction are less attractive when applied to materials with such a relatively low lipid content. Nonetheless, the use of solvent extraction for oil recovery from distillers grains has been described.

Because of its ready availability at an ethanol production facility, Singh and Cheryan (1998) explored the use of anhydrous ethanol at 50°C as a solvent for corn oil extraction. In single batchwise extractions, 50% of the oil in DDGS could be recovered. Multiple extractions or countercurrent extraction technology might improve oil recovery. The use of the recovered oil as a biodiesel feedstock was not investigated and has not been subsequently described, however.

The use of alkyl acetates, for example, ethyl, butyl, and isopropyl acetate, in the recovery of oil from DDGS has also been described (Randhava et al., 2008). In this work, multiple batch extractions were conducted, achieving recovery of greater than 90% of the oil. Again, neither oil quality nor use of the oil in biodiesel production was described.

Hexane extraction can recover greater than 90% of the resident oil from DDGS. This oil contains 5%–15% free fatty acids (Janes et al., 2008; Winkler-Moser and Vaughn, 2009). The possibility of producing biodiesel from the oil, including the use of degumming, alkaline refining, and bleaching to remove contaminants, has been discussed (Janes et al., 2008), although actual analytical data for FAME so produced was not presented.

The capital costs for installation of a hexane extraction facility are substantial, approximately U.S. $30 million for a facility able to handle the DDGS generated by a 100 million gal (379 million L)/year ethanol plant. In addition, the use of hexane, a federally regulated neurotoxin and air pollutant (Galvin, 1997), is not allowed in all regions of the nation. This limits the location of extraction facilities and can necessitate long distance transport of feedstock for extraction. Even in locations where allowed, obtaining regulatory approval can be an involved process. Hexane also presents special challenges in the working environment, leading to efforts to eliminate it from oilseed processing technologies (Johnson, 1997). Solvent extraction technology requires an essentially dry feedstock whereas a substantial number of contemporary ethanol facilities are configured to produce a DWG. Perhaps for these reasons the recovery of corn oil from distillers grains by solvent extraction has not been widely adopted by industry. This approach was recently implemented at the industrial scale (Watkins, 2007), although details concerning the facility, which is not presently functioning, are lacking. At this time only one commercial facility, in the People's Republic of China, is operational. The recent description (Kemper and Subieta, 2009) of the pelletizing of DDGS and the use of this material in a continuous hexane extraction format, with virtually complete oil recovery, suggests that there is still interest in this approach and that further advances may be forthcoming.

23.3.2 OIL RECOVERY VIA PHYSICAL METHODS

Approximately one-third of the oil in corn kernels entering fermentation partitions into the liquid phase and exits the operation in the thin stillage stream. The thin stillage is evaporated and concentrated into condensed distillers solubles (CDS), which is known as "syrup." Technologies are being developed and implemented to recover this oil prior to addition of the syrup to the DWG to produce DDGS. (For a further examination on new technologies being developed for the fractionation of DDGS, refer to Chapter 5). Corn oil recovered at this step of the ethanol process is considered inedible and does not compete with food oils, but is an excellent raw material feedstock for the production of biofuels. Presently the leading technology developer in this sector is GS CleanTech Corporation (Alpharetta, GA; formerly Veridium), a branch of the GreenShift Corporation (www.greenshift.com). In the initial process developed (Cantrell and Winsness, 2009) the thin stillage is heated and concentrated to a water content of between 60% and 85% by evaporation. The oil is then recovered by heating and centrifugation. Reported oil recovery rates are between 35% and 80% of that in the thin stillage, for an overall recovery of approximately 10%–26% of the oil that entered fermentation. Variations in recovery are associated with such facility-dependent parameters as fermentation conditions and duration and types of prefermentation enzymes employed. Following removal of DDGS fines by filtration or centrifugation, and additional cleanup (described below) this oil can be successfully converted into FAME that meet the ASTM specifications for biodiesel (P. Bush, pers. comm.). This proprietary process is currently being implemented at four dry grind ethanol plants (www.greenshift.com). Some other technologies for the fractionation of DDGS are discussed in Chapters 5 and 25.

In another study of FAME preparations synthesized from corn oil recovered from DDGS, analytical results indicated that the preparation met all ASTM specifications except for oxidative stability (Sanford et al., 2009). The induction time of the sample, as determined by the accepted biodiesel oxidation stability test, was 2.2 h, whereas a minimum 3-h induction time is specified (American Society for Testing and Materials, 2009). The authors of the study noted that since the work was done on a small scale there was a greater opportunity for excess air to be present in the system, leading to a greater depletion of endogenous antioxidants than would typically be observed in a production-scale reaction. For this reason they felt that the observed stability was probably lower than would be routinely observed for a FAME sample produced at industrial scale. This interpretation is reasonable. At any rate, it is probable that this defect would not prevent use of the product as biodiesel since induction time is readily increased by the accepted industry practice of adding an antioxidant. Antioxidants for this purpose are affordable and readily available in bulk quantities.

A further refinement of the GreenShift process described above involves the additional step of heating to between 110°C and 121°C under sufficient pressure to prevent boiling (Winsness and Cantrell, 2009). In various manifestations this treatment may be applied to whole, thin, or concentrated thin stillage. The additional step is described as a "pressure cooking", that causes hydrolysis, releasing oil bound to solids in the feedstock. Following cooking, the mixture is cooled, the pressure returned to ambient, and the oil is recovered by centrifugation. This modified process is reported to give recovery of as much as two-thirds of the oil in the thin stillage.

In more recent work a similar approach has been applied to wet cake or solids, which can be separated from thin stillage by centrifugation; thin stillage is added to the wet cake stream, which can then be subjected to a pressure and heat step. This releases a substantial portion of the oil bound in the wet cake. Following depressurization, cooling, and centrifugation, recoveries of as high as 80% of the total corn oil that had entered fermentation are reported (G. Barlage, P. Bush, personal communication).

The corn oil recovered by these postfermentation methods is not suitable for food use, due largely to the fact that it has a free fatty acid content between 10% and 20%, typically 13%–15%. These and other contaminants also complicate use of this oil for biodiesel production. The free fatty acid content is sufficiently high that FFA conversion to soap by the addition of alkali, and subsequent removal and disposal, is not economically attractive. A more desirable approach (Canacki and Van Gerpen, 2003) would be to apply the dual reaction technology mentioned above: (1) two or more consecutive acid-catalyzed esterification reactions, with intervening removal of the water coproduct of fatty acid esterification, to convert the free fatty acids to the FAME that comprise biodiesel; and (2) alkali-catalyzed transesterification to produce FAME from the acylglycerol component. This approach increases the overall FAME yield. Ethanol could be employed in place of methanol in this and all biodiesel production. To date, however the higher cost of ethanol has limited its use.

High degrees of conversion of free and glyceride-linked fatty acids into FAME can be achieved by these methods, resulting in preparations with ester contents in excess of 97 wt.%. However, a high ester content alone does not ensure that a sample will function satisfactorily as a biodiesel. Minor components can have a profound impact on the handling and performance of a biodiesel. This is pertinent in the case of corn oil recovered from postfermentation streams because they can contain species other than acylglycerols and fatty acids. Of importance to fuel-use considerations is the presence of substantial amounts of "wax." In strictest chemical terms, a wax is an ester formed between a fatty acid and a long-chain (C > 20) alcohol. The levels of these in corn DDGS are very low (Leguizamon et al., 2009). However in the context of oil recovered from ethanol fermentations the term wax refers to the presence of a variety of long-chain alkanes, esters and polyesters, as well as the esters of long-chain primary alcohols and fatty acids. There has been little analytical characterization of these in corn oils isolated from postfermentation product streams. In examining three corn oil samples recovered by centrifugation of thin stillage, Moreau et al. (2010) found free sterol and hyrdoxycinnamate steryl ester levels higher than those in oil recovered by the hexane extraction of corn germ. Also, the steryl- and wax ester levels were more than twice those in conventional refined, bleached corn oil obtained by hexane extraction of corn germ.

There has recently been concern regarding the impact on biodiesel performance of another class of molecules that may be found in the "wax" fraction in corn oil recovered from postfermentation streams: steryl glucosides (Pfalzgraf et al., 2007; Lee et al. 2007; Van Hoed et al., 2008; Moreau et al., 2008). These are sterol species to which a saccharide is appended via a glycosidic linkage. Typically present as trace components of vegetable oils, steryl glucosides have a lower solubility in fatty acid ester preparations than in vegetable oils, which causes them to precipitate upon cooling downstream of biodiesel production. The resulting solids can block filters in, for example, storage tank exit lines and engines, compromising performance or stopping engines outright due to fuel starvation. The potential for this problem was not realized until the biodiesel industry grew to a sufficient size that large volumes of the fuel were being stored and transported. Subsequent to identification of the problem and its causes, methods both to reduce its occurrence as well as to test a fuel to determine its acceptability in this regard were developed. Recently, ASTM D6751, the standard specification for biodiesel (American Society for Testing and Materials, 2009) was modified to include determination of the filter plugging tendency of a biodiesel as a result of species such as these, and their aggregates with other biodiesel components. The problem can usually be eliminated by chilling and filtration of the fuel (Lee et al., 2006; Danzer et al., 2007), a procedure that may overcome the majority of wax-related issues associated with recovered postfermentation corn oil. Moreau et al. (2003) reported sterol glucoside (SG) levels of 20 mg/kg in whole corn and thus there is reason to be concerned that these species may also be present in DDGS and in FAME preparations made from DDGS or postfermentation corn oil. However, to date there has been no reported investigation into this topic.

Methods have been developed for the removal of such contaminants from corn oil recovered from thin stillage (Hagberg and Smith, 2010). The described purification protocol was patterned after technologies used in the edible oil industry. It includes (1) degumming, wherein the product is neutralized and washed with phosphoric acid and water, with subsequent removal of phospholipids and other impurities as a "gum" fraction; (2) a series of acid-catalyzed esterifications in the presence of methanol to produce FAME from the free fatty acid fraction; (3) alkali addition to remove residual FFA as soaps; (4) treatment with silica, diatomaceous earth and bleaching clay, which reduce color content and remove contaminants; and (5) chilling and holding at 4°C, followed by filtration to remove precipitated waxes. The acylglycerols in the resulting corn oil can be converted to FAME by conventional alkaline transesterification, yielding an ester product that meets ASTM biodiesel specifications. Similarly, other processes employ a cold super-degumming operation, involving the addition of water and phosphoric or citric acids followed by chilling and centrifugation, to remove sterols and waxes prior to conversion of the oil into FAME (P. Bush, pers. comm.).

In the United States, a red dye is added to diesel fuel intended for use in off-road vehicles such as agricultural and construction equipment. This fuel is also allowed to have a sulfur content of 500 ppm, which is 33 times greater than that for "on-road" (e.g., transportation) fuel. Off-road fuel is taxed at a lower rate than its undyed counterpart, and there is a penalty for the use of the former in on-road applications. Recovered postfermentation corn oil has been reported to have low sulfur levels (4.6 ppm, Sanford et al., 2009), but has a substantially darker, redder color than conventional corn oil and gives a correspondingly darker colored biodiesel. This is consistent with the detection in corn oil recovered from thin stillage of carotenoid levels four times greater than seen in oil recovered by hexane extraction of corn kernels (Moreau et al., 2010). There has been some concern that this color would cause biodiesel produced from recovered corn oil to be mistaken for off-road fuel by a regulatory authority if detected in storage tanks, fuel handling systems, or vehicles intended for on-road fuel. However, the absorption spectra of the colored compounds in recovered corn oil and of the dyes used to color off-road diesel fuel are sufficiently different to allow for discrimination between the two. Therefore this color issue is not considered to be a significant barrier to the use of biodiesel produced from recovered corn oil as an on-road fuel (S. Howell, pers. comm.). If found undesirable, distillation could be employed to produce a lighter or colorless biodiesel.

23.3.3 Direct Production of Biodiesel from Lipid-Bearing Materials

Rather than extracting the glycerides from a biological material and converting the fatty acids in the resulting oil to alkyl esters in a separate transesterification, research over the past 25 years has established that transesterification can also be directly conducted in the parent lipid-bearing material. In this process, termed "in situ transesterification," the lipid is not initially extracted from the feedstock. Rather, the feedstock is incubated directly with an alcohol and an acidic or basic catalyst. Transesterification occurs in the feedstock and the resulting fatty acid esters diffuse into the surrounding liquid reaction medium, from which they can be recovered. Both the in situ transesterification approach and the conventional technology of first isolating a lipid feedstock and then transesterifying it to produce FAME begin with the same feedstock—for example soybeans or DDGS—and end with the same products: biodiesel, glycerin, and a low-lipid meal (Figure 23.3). In an in situ transesterification approach, DDGS rather than recovered corn oil from DDGS would be incubated with a mixture of sodium methylate in methanol (other small alcohols are also acceptable substrates) to produce fatty acid methyl (or other alkyl) esters. Relative to present biodiesel production technologies, this method offers the advantage of completely eliminating the technology and

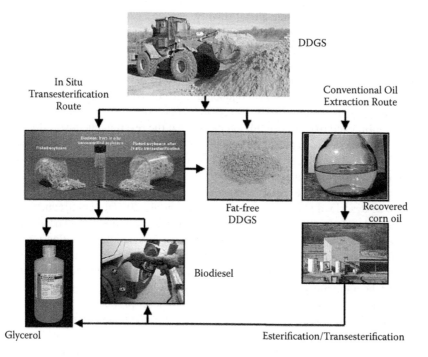

FIGURE 23.3 Simplified schematic representation of the routes of biodiesel production from DDGS by either conventional oil extraction/transesterification or by in situ transesterification.

cost of recovering lipid from its agricultural commodity or other source. Thus the regulatory and handling challenges associated with the use of hexane for oil extraction are avoided. To the degree that the protocol is applicable to feedstocks whose lipid contents are too low for economical oil recovery by solvent extraction, in situ transesterification may offer an economical and effective route to new supplies of biodiesel.

The first descriptions of this approach were by Harrington and d'Arcy-Evans (1985a, 1985b) who obtained high ester yields in acid-catalyzed transesterification of homogenized whole sunflower seeds in methanol. Some time later the approach was extended to the study of a range of acid and methanol levels in the direct transesterification of sunflower seeds (Siler-Marinkovic and Tomasevic, 1998), and other workers reported high yields in acid-catalyzed direct transesterification of rice bran (Ozgul-Yucel and Turkay, 1993, 2002, 2003). Substantial amounts of free fatty acids contaminated the product. This approach was also applied to ground soybeans, though only 20%–40% transesterification of the lipid pool was obtained (Kildiran et al., 1996), and to palm pulp (Obibuzor et al., 2003).

When applied to flaked soybeans, conditions for alkali-catalyzed transesterification were identified that gave greater than 90% of maximum theoretical FAME yield following 8-h reactions at room temperature and ambient pressure (Haas et al., 2004). Whereas in conventional transesterification of triacylglycerols a twofold molar excess of methanol over fatty acids is typical, optimal high-yield in situ transesterification of soybean flakes required a 181-fold molar excess when first described (Haas et al., 2004). It was subsequently determined that drying the substrate to less than 3 wt% moisture content reduced the molar methanol requirement to 76/1 while maintaining complete transesterification (Haas and Scott, 2007). Following a wash with aqueous alkali to reduce a slightly excessive free fatty acid content, this ester product met the ASTM biodiesel specifications (Haas and Scott, 2007). Subsequent work has shown that a simple physical pretreatment of the soybean flakes, such as extrusion, can reduce the molar requirement for methanol to ninefold over the fatty acid content (Haas et al., 2007a). Such reductions in solvent use are desirable because they lower the energy cost for the removal and recovery of unreacted alcohol.

It has also been shown that, at least in the case of soybeans, the lipid-free meal exiting the transesterification reaction retains its nutritional quality, and is an acceptable feed for trout (Barrows et al., 2008) and for chickens (Haas et al., 2009). Such uses for the coproduct meal are vital to the economics of any scheme for the recovery and use of the oil in DDGS.

The in situ transesterification technology for biodiesel production has also been successfully applied to DDGS (Haas et al., 2007b). Greater than 90% of maximum theoretical FAME production was achieved in 2–3 h in reactions conducted at 35°C. Higher reaction temperatures neither increased product yield nor substantially decreased reaction time. The fatty acid composition of the FAME product mimicked that of corn oil, indicating that the reaction chemistry does not discriminate against any substrate fatty acids. The majority of the protein remained in the DDGS following reaction. Free fatty acid and glyceride contents of the FAME product were low, only 0.32–0.40 wt.% of the preparation. The minimum methanol/fatty acid ratio for high level FAME yields in 35°C reactions using DDGS dried to a moisture content of 2.6% was 230. It is possible that pretreatment of DDGS such as described above for soybeans (Haas et al., 2007a) could reduce this value. The product was not subjected to the full panel of ASTM biodiesel quality tests. The virtually complete deoiling of the DDGS that was achieved in this manner eliminated the self-adhesion and poor flow properties of the parent material (Agricultural Utilization Research Institute, 2005).

An advantage of the in situ transesterification approach over that of recovering and separately transesterifying the oil in the thin stillage is that the former method gives higher FAME yields. This equates to a higher gross economic return from the biodiesel production process, since biodiesel sells for substantially more than does the meal coproduct of the reaction and such residual oil as it contains. Because it contains virtually no lipid, it is also possible that the meal generated by in situ transesterification could be included at higher blend rates in animal feeds. This could be especially useful in the case of dairy cattle and swine, which can accept only limited amounts of fat in their diets. In current applications within the area of animal nutrition in the oil in the DDGS provides an energy source to the animal. The impact of the removal of this oil by in situ transesterification remains to be determined.

To date, however, this approach has not been applied at the industrial scale for any feedstock. Further research toward a reduction in the methanol requirement will probably be necessary to achieve the economic competitiveness necessary to stimulate implementation of in situ transesterification technology at the industrial scale.

23.4 CONCLUSIONS

The production of renewable fuels from agricultural materials offers opportunities to reduce global warming, strengthen farm and rural economies, and reduce both the dependence on imported petroleum and the attendant political tensions and potential military conflicts. The oil resident in distillers grains produced by the fermentation of corn and other feedstocks to produce ethanol represents a potential feedstock for the synthesis of biodiesel. Given forecast growth in the production of ethanol from starch, the already substantial amount of oil residing in distillers grains could exceed amounts sufficient to produce over 300 million gal of biodiesel annually. This would represent a very significant amount of new biodiesel, significantly increasing production beyond the 500 to 700 million annual gal of recent years. To realize this goal, new technologies for biodiesel production from the oil in distillers grains have been developed and refined, and are now beginning to be implemented by the ethanol industry. These approaches include the transesterification of corn oil recovered from the thin stillage by centrifugation or from the postfermentation solids by solvent extraction, and the direct (in situ) transesterification of the oil in the DDGS. The deoiled DDGS that is a coproduct of these technologies could see expanded use as a component of animal feeds. Just as with the development of improved enzymes for biofuel production (Chapter 24), these efforts will expand supplies of renewable, domestically produced, liquid transportation fuels while reducing net fuel production costs.

ACKNOWLEDGMENTS

I appreciate the cooperation of M. Danzer (Renewable Energy Group, Ames IA, USA), P. Bollheimer (Bollheimer & Associates Inc, Memphis, TN, USA), G. Barlage and P. Bush (Greenshift Corp., Alpharetta, GA, USA), S. Howell (Marc-IV Consulting, Kearney, MO, USA), T. Kemper (DeSmet Ballestra North America, Marietta, GA, USA), and R. Moreau (ARS, USDA, Philadelphia, PA, USA) in providing background information for this work.

REFERENCES

Agricultural Utilization Research Institute. 2008. Distiller's dried grains flowability report. Waseca, Minnesota: Agricultural Utilization Research Institute. Available online: www.auri.org/research/Flowability_summary_10_17_05.pdf. Accessed February 22, 2011.

American Society for Testing and Materials. 2009. *D6751-09 Standard Specification for Biodiesel Fuel (B100) Blend Stock for Distillate Fuels.* West Conshohocken, Pennsylvania: ASTM.

Barrows, F. T., T. G. Gaylord, W. M. Sealey, M. J. Haas, and R. L. Stroup. 2008. Processing soybean meal for biodiesel production: effect of a new processing method on growth performance of rainbow trout, *Oncorhynchus mykiss. Aquaculture* 283: 141–147.

Biobus Committee. 2003. Biobus project final report: Biodiesel demonstration and assessment with the Societe de Transport de Montreal (STM). http://www.stm.info/English/info/a-biobus-final.pdf.

Canacki, M., and J. Van Gerpen. 2003. A pilot plant to produce biodiesel from high free fatty acid feedstocks. *Transactions of the ASAE* 46: 945–954.

Cantrell, D. F., and D. Winsness. 2009. Method of processing ethanol byproducts and related subsystems. *US Patent 7,601,858.*

Chase, C. L., C. L. Peterson, G. A. Lowe, P. Mann, J. A. Smith, and N. Y. Kado. 2000. *A 322,000 Kilometer (200,000 Mile) over the Road Test with HySEE Biodiesel in a Heavy Duty Truck.* SAE Technical Paper 2000-01-2647, Warrendale, Pennsylvania: Soc. Automotive Engineers.

Danzer, M. F., T. L. Ely, S. A. Kingery, W. W. McCalley, W. M. McDonald, J. Mostek, and M. L. Schultes. 2007. Biodiesel cold filtration process, *US Patent Application 20070175091.*

Diesel, E., 1937. Diesel- Der Mensch - Das Werk - Das Schicksal, Hanseatische Verlagsgesellschaft, Hamburg, as cited in Knothe, G. 2010. History of vegetable oil-based diesel fuels, in *The Biodiesel Handbook*, 2nd ed., eds. G. Knothe, J. Krahl, and J. Van Gerpen, pp. 5–19, Champaign, Illinois: AOCS Press.

Dry, M. E. 2002. High quality diesel via the Fischer-Tropsch process-a review. *Journal of Chemical Technology and Biotechnology* 77: 43–50.

Economic Research Service. 2010. *Agricultural Baseline Projections: U.S. Crops, 2009–2018.* Washington, DC: United States Department of Agriculture. http://www.ers.usda.gov/briefing/baseline/present2009.htm

Europa Bio. 2007. *Biofuels & Food.* Brussels: The European Association for Bioindustries. http://www.google.com/search?q=http%3A%2F%2Fwww.europabio.org%2FBiofuels%2FFood_Biofuels%2520factsheet.pdf&sourceid=ie7&rls=com.microsoft:en-US&ie=utf8&oe=utf8&rlz=1I7GGLD_en

Freedman, B., E. H. Pryde, and T. L. Mounts. 1984. Variables affecting the yields of fatty esters from transesterified vegetable oils. *Journal of the American Oil Chemists Society* 10: 1638–1643.

Galvin, J. B. 1997. Toxicity data for commercial hexane and hexane isomers. In *Technology and Solvents for Extracting Oilseeds and Nonpetroleum Oils,* eds. P. J. Wan, and P. J. Wakelyn, pp. 75–86. Champaign, Illinois: AOCS Press.

Government Accountability Office. 2007. *Crude Oil: Uncertainty about Future Oil Supply Makes It Important to Develop a Strategy for Addressing a Peak and Decline in Oil Production.* GAO-07-283, Washington, DC. Available online: www.gao.gov/new.items/d07283.pdf. Accessed February 22, 2011.

Goering, C. E., M. D. Schrock, K. R. Kaufman, M. A. Hanna, F. D. Harris, and S. J. Marley. 1987. *Evaluation of Vegetable Oil Fuels in Engines.* Proc. of the International Winter Meeting of the ASAE, Paper No. 87-1586. St. Joseph, Michigan: ASAE.

Graboski, M. S., and R. L. McCormick. 1998. Combustion of fat and vegetable oil derived fuels in diesel engines. *Progress in Energy and Combustion Science* 24: 125–164.

Gunstone, F. D. 2005. Vegetable oils. In *Bailey's Industrial Oil and Fat Products,* 6th Ed., Vol. 1: Edible Oil & Fats, ed. F. Shahidi, pp. 213–267. Hoboken, New Jersey: John Wiley & Sons, Inc.

Haas, M. J., F. F. Barrows, A. F. McAloon, W. C. Yee, and K. M. Scott, 2007a. *In Situ Transesterification: Process Modifications to Improve Efficiency, and Use of Spent Meal in an Aquaculture Diet.* Abstracts, 98th AOCS Annual Meeting & Expo, 24. Champaign, Illinois: American Oil Chemists Society.

Haas, M. J., A. J. McAloon,.W. C. Yee, and T. A. Foglia. 2006. A process model to estimate biodiesel production costs. *Bioresource Technology* 97: 671–678.

Haas, M. J., and K. M. Scott. 2007. Moisture removal substantially improves the efficiency of in situ biodiesel production from soybeans. *Journal of the American Oil Chemists Society* 84: 197–204.

Haas, M. J., K. M. Scott, W. N. Marmer, and T. A. Foglia. 2004. In situ alkaline transesterification: An effective method for the production of fatty acid esters from vegetable oils. *Journal of the American Oil Chemists Society* 81: 83–89.

Haas, M. J., K. M. Scott, T. A. Foglia, and W. N. Marmer. 2007b. The general applicability of in situ transesterification for the production of fatty acid esters from a variety of feedstocks. *Journal of the American Oil Chemists Society* 84: 963–970.

Haas, M. J., R. L. Stroup, D. Latshaw, and K. M. Scott. 2009. *In Situ Transesterification for Biodiesel Production: Investigating the Suitability of the Spent Meal as a Poultry Feed Component.* Abstracts, 100th AOCS Annual Meeting & Expo, 73. Urbana, Illinois: American Oil Chemists Society.

Hagberg, J., and N. Smith. 2010. *Corn Oil Extraction: Processes, Product Quality, Treatment & Economics.* Proc. 2nd Int. Congress on Biodiesel: The science and the technologies. Urbana, Illinois: AOCS Press.

Harrington, K. J., and C. D'Arcy-Evans. 1985a. Transesterification in situ of sunflower seed oil. *Industrial & Engineering Chemistry Product Research and Development* 24: 314–318.

Harrington, K. J., and C. D'Arcy-Evans. 1985b. A comparison of conventional and in situ methods of transesterification of seed oil from a series of sunflower cultivars. *Journal of the American Oil Chemists Society* 62: 1009–1013.

Janes, M., K. Bruinsma, T. Cooper, and D. Endres. 2008. Solvent extraction of oil from distillers dried grains and methods of using extraction products. *International Patent WO2008039859.*

Johnson, L. A. 1997. Theoretical, comparative, and historical analyses of alternative technologies for oilseeds extraction. In *Technology and Solvents for Extracting Oilseeds and Nonpetroleum Oils,* eds. P. J. Wan, and P. J. Wakelyn, pp. 4–47. Champaign, Illinois: AOCS Press.

Kalnes, T., T. Marker, and D. R. Shonnard. 2007. Green diesel: A second generation biofuels. *International Journal of Chemical Reactor Engineering* 5:Article A48. Available online: www.bepress.com/ijcre/vol5/A48. Accessed February 22, 2011.

Kemper, T. G. 2005. Oil extraction. In *Bailey's Industrial Oil & Fat Products,* Vol. 5, Edible Oil & Fat Products, Processing Technologies, ed. F. Shahidi, pp. 57–98. Hoboken, New Jersey: John Wiley & Sons, Inc.

Kemper, T., and A. T. Subieta. 2009. Method of reducing oil content in dry distillers grain with solubles (DDGS). *U.S. Patent Application 2009/0104325 A1.*

Kildiran, G., S. Ozgul-Yucel, and S. Turkay. 1996. In-situ alcoholysis of soybean oil. *Journal of the American Oil Chemists Society* 73: 225–228.

Knothe, G., J. Van Gerpen, and J. Krahl. 2010. *The Biodiesel Handbook,* 2nd ed., Urbana, Illinois: AOCS Press.

Komers, K., F. Skopal, R. Stloukal, and J. Machek. 2002. Kinetics and mechanism of the KOH-catalyzed methanolysis of rapeseed oil for biodiesel production. *European Journal of Lipid Science and Technology* 104: 728–737.

Lee, I., J. L. Mayfield, L. M. Pfalzgraf, L. Solheim, and S. Bloomer. 2006. Processing and producing biodiesel and biodiesel produced there from. *U.S. Patent Application 20070151146.*

Lee, I., L. M. Pfalzgraf, G. B. Poppe, E. Powers, and T. Haines. 2007. The role of sterol glucosides on filter plugging. *Biodiesel Magazine* 4: 105–112.

Leguizamon, C., C. L. Weller, V. L. Schlegel, and T. P. Carr. 2009. Plant sterol and policosanol characterization of hexane extracts from grain sorghum, corn and their DDGS. *Journal of the American Oil Chemists Society* 86: 707–716.

Ma, F., D. Clements, and M. A. Hanna. 1999. The effect of mixing on transesterification of beef tallow. *Bioresource Technology* 69: 289–293.

McElroy, A. K. 2007. Corn oil extraction opens new markets. Distillers Grains Quarterly. http://www.greenshift.com/pdf/DGQ_Q1_2007.pdf.

Mittelbach, M., and C. Remschmidt. 2004. *Biodiesel, The Comprehensive Handbook.* Graz, Austria: Martin Mittelbach, Publisher.

Mittelbach, M., and B. Trathnigg. 1990. Kinetics of alkaline catalysed methanolysis of sunflower oil. *Fat Science and Technology* 92: 145–148.

Moreau, R. A., M. J. Powell, and V. Singh. 2003. Pressurized liquid extraction of polar and nonpolar lipids in corn and oats with hexane, methylene chloride, isopropanol, and ethanol. *Journal of the American Oil Chemists Society* 80: 1063–1067.

Moreau, R. A. 2005. Corn oil in edible oil and fat products. In *Baileys Industrial Oil and Fat Products,* Vol. 2: Edible Oil and Fat Products: Edible Oils, ed. F. Shahidi, pp. 149–172. Hoboken, New Jersey: John Wiley & Sons, Inc.

Moreau, R. A., K. Hicks, D. Johnston, and N. Laun. 2010. The composition of crude corn oil recovered after fermentation via centrifugation from a commercial dry grind ethanol process. *Journal of the American Oil Chemists Society* 87: 895–902.

National Biodiesel Board. 2008. Biodiesel brings a lot to the table. Available online: www.biodiesel.org/resources/sustainability/pdfs/Food%20and%20FuelApril162008.pdf. Accessed February 22, 2011.

National Biodiesel Board. 2010. Estimate of U.S. Annual Biodiesel Production by Fiscal Year (Oct. 1 to Aug. 30). Available online: www.biodiesel.org/pdf_files/fuelfactsheets/Production_Graph_Slide.pdf. Accessed February 22, 2011. Jefferson City, Missouri: National Biodiesel Board.

Obibuzor, J. U., R. D. Abigor, and D. A. Okiy. 2003. Recovery of oil via acid-catalyzed transesterification. *Journal of the American Oil Chemists Society* 80: 77–80.

Ozgul, S., and S. Turkay. 1993. In situ esterification of rice bran oil with methanol and ethanol. *Journal of the American Oil Chemists Society* 70: 145–147.

Ozgul-Yucel, S., and S. Turkay. 2002. Variables affecting the yields of methyl esters derived from in situ esterification of rice bran oil. *Journal of the American Oil Chemists Society* 79: 611–613.

Ozgul-Yucel, S., and S. Turkay. 2003. FA monoalkylesters from rice bran oil by in situ esterification. *Journal of the American Oil Chemists Society* 80: 81–84.

Pfalzgraf, L., I. Lee, J. Foster, and G. Poppe. 2007. Effect of minor components in soy biodiesel on cloud point and filterability. *Inform* 18 (Suppl. 4): 17–21.

Randhava, S. S., R. L. Kao, S. G. Calderone, and A. S. Randhava. 2008. Oil recovery from dry corn milling ethanol production processes. *U.S. Patent Application 2008/0176298 A1.*

Rosentrater, K. A. 2009. Distiller's dried grains with solubles (DDGS)—A key to the fuel ethanol industry. *Inform* 20: 789–800.

Sanford, S. D., J. M. White, P. S. Shah, C. Wee, M. A. Valverde, and G. R. Meier. 2009. Feedstock and biodiesel characteristics report. Renewable Energy Group, Inc. Available online: www.regfuel.com. Accessed February 22, 2011.

Schnepf, R. 2007. Agriculture-based renewable energy production. CRS Report for Congress, Washington: Congressional Research Service, Library of Congress. Order Code RL32712. Available online: collinpeterson.house.gov/PDF/agenergy.pdf. Accessed February 22, 2011.

Siler-Marinkovic, S., and A. Tomasevic. 1998. Transesterification of sunflower oil in situ. *Fuel* 77: 1389–1391.

Singh, N., and M. Cheryan. 1998. Extraction of oil from corn distillers dried grains with solubles. *Transactions of the ASAE* 41: 1775–1777.

Spath, P. L., and D. C. Dayton. 2003. *Preliminary Screening—Technical and Economic Assessment of Synthesis Gas to Fuels and Chemicals with Emphasis on the Potential for Biomass-Derived Syngas,* NRELITP-5 10-34929. Golden, Colorado: National Renewable Energy Laboratory.

United States Census Bureau. 2009. Current Industrial Report: Fats and Oils: Production, consumption and stocks, 2008. Available online: http://www.census.gov/cir/www/311/m311k/m311k0813.xls. Accessed February 22, 2011.

United States Department of Agriculture. 2008a. Agricultural Statistics, Table 3-38.—Soybeans: Crushing, and oil and meal stocks, production, and foreign trade, United States, 1997–2006. Agricultural Statistics, pp. III–16. Available online: www.nass.usda.gov/Publications/Ag_Statistics/2008/Chap03.pdf. Accessed February 22, 2011.

United States Department of Agriculture. 2008b. Agricultural Statistics, Table 7-69.—Red Meat: Production by class of slaughter, United States, 1998–2007, Agricultural Statistics, pp. VII–44. Available online: www.nass.usda.gov/Publications/Ag_Statistics/2008/Chap07.pdf. Accessed February 22, 2011.

United States Government Printing Office. 2007. Energy Independence and Security Act of 2007. http://frwebgate.access.gpo.gov/cgi-bin/getdoc.cgi?dbname=110_cong_bills&docid=f:h6enr.txt.pdf.

Watkins, C. 2007. Two fuels from one kernel. *Inform* 18: 714–717.

Williams, M. A. 2005. Recovery of oils and fats from oilseeds and fatty materials. In *Bailey's Industrial Oil & Fat Products,* Vol. 5, Edible Oil & Fat Products, Processing Technologies, ed. F. Shahidi, pp. 99–189. Hoboken, New Jersey: John Wiley & Sons, Inc,.

Winkler, J. K., K. A. Rennick, F. J. Eller, and S. F. Vaughn. 2007. Phytosterol and tocopherol components in extracts of corn distiller's dried grain. *Journal of Agricultural and Food Chemistry* 55: 6482–6486.

Winkler-Moser, J. K., and S. F. Vaughn. 2009. Antioxidant activity of phytochemicals from distillers dried grain oil. *Journal of Agricultural and Food Chemistry* 86: 1073–1082.

Winsness, D., and D. F. Cantrell. 2009. Method of freeing the bound oil present in whole stillage and thin stillage. *U.S. Patent 7,608,729.*

Ziejewski, M., H. Goettler, and G. L. Pratt. 1986a. *Comparative Analysis of the Long-Term Performance of a Diesel Engine on Vegetable Oil Based Alternate Fuels.* SAE Technical Paper Series, No. 860301. Warrendale, Pennsylvania: Society of Automotive Engineers.

Ziejewski, M., H. Goettler, and G. L. Pratt. 1986b. *Influence of Vegetable Oil Based Alternate Fuels on Residue Deposits and Components Wear in a Diesel Engine.* SAE Technical Paper Series, No. 860302. Warrendale, Pennsylvania: Society of Automotive Engineers.

Part VI

Process Improvements

24 Improved and New Enzymes for Fuel Ethanol Production and Their Effect on DDGS

Milan Hruby

CONTENTS

24.1 INTRODUCTION

Enzymes are crucial in ethanol production from corn (*Zea mays*) and other starch sources. Enzymes in ethanol plants facilitate production of sugars used by yeast to yield ethanol through a fermentation process. They can also function as processing aids to reduce slurry viscosity, improve throughput, reduce deposits of organometallic salts or scale in processing equipment, reduce water usage and the overuse of other compounds such as acids and bases in ethanol production. Today, the ethanol produced in the United States comes mainly from plants employing dry milling technology (as discussed in Chapter 5). Besides carbon dioxide (CO_2) and ethanol, dry-grind plants also produce distillers dried grains with solubles (DDGS). DDGS is increasingly counted as an economically important product of the ethanol production process rather than a coproduct or even less desirable coproducts. Deliberate focus on high nutrient digestibility, absence or low levels of antinutrients, and reduction in variability of DDGS can result in a significant increase of ethanol plant's revenues. Importantly, this can be even more crucial at times when ethanol production gives lower return on investment due to low fuel prices or high input cost such as high corn and energy (e.g., natural gas) prices.

Research studies focusing on a trend of feeding higher levels of DDGS in animal feeds (e.g., Noll et al., 2007; Xu et al., 2010) provide confirmation that increased levels of DDGS can be used in diets of farm animals without a negative impact on performance. For information on uses of DDGS in various animals, refer to Chapters 12–15. Besides their domestic use, DDGS produced in the United States are also exported to other countries. Understanding what effect various enzymes in the ethanol production process can have on DDGS dietary value and uniformity can provide final users, whether domestic or in export markets, with a better focused DDGS application for a specific use. This, in the end, could result in a positive impact on economics of animal protein production more about feeding DDGS to animals is covered in Chapters 12 through 17.

Chapter 5 of this book covers subject of the dry-grind ethanol production in detail. This chapter will provide a brief review of enzymes employed by corn starch ethanol plants and their potential impact on DDGS value when used in animal feeds. Issues such as improved nutrient digestibility but also economic benefits of various DDGS products when different enzymes are used during production of ethanol will also be presented.

24.2   DDGS AND THEIR USE IN ANIMAL FEEDS: BENEFITS AND LIMITATIONS

Corn DDGS (~10% moisture) are a direct result of an ethanol production process. About 18 lb (8.16 kg) of DDGS are produced for every bushel (56 lb or 25.4 kg) of corn used in dry-grind ethanol plants. Besides a primary focus on ethanol production achieving about 2.7 gal (10.2 l) from one bushel of corn (Davis, 2001), CO_2 and distillers grains (wet or dry) are other important streams of most ethanol plants' revenues. It is estimated that around 31.5 million short tons of distillers grains were produced in the United States during 2009 (Deutscher, 2009). This is about a sixfold increase compared to only 10 years ago. Thus, it is not surprising that DDGS usage and export opportunities are growing proportionally.

The increase in DDGS production is obviously coupled with an enormous increase in ethanol production over the past 10 years. Expectedly, ethanol production is the main driver of ethanol plant profitability. Recently, however, there has been a stronger focus on quality of other products from the ethanol production process, namely DDGS. Improvement in DDGS nutritive value and uniformity can prove beneficial for ethanol plants, especially during periods of low ethanol prices and/or high costs of inputs including corn or other sources of fermentable substrate and energy such as natural gas or heating oil. It is apparent that continued focus on producing more uniform and higher quality DDGS can result in a better marketing opportunity for ethanol plants and consequently in an improvement of ethanol plant's revenues. Over the past 4 to 5 years, the use of DDGS in feeds of monogastric animals (namely pigs and poultry) increased by about 2.5 times. Over the same period, the feed production in the United States has been relatively stagnant. This suggests that increasingly higher levels of DDGS are used in poultry and pig feeds today than only few years ago. This is largely due to a greater body of available research focusing on the use of higher levels of DDGS in animal feeds (Noll et al., 2007; Xu et al., 2010). Additionally, it is also possible that animal nutritionists have frequently greater access to DDGS with higher and more predictable nutrient specification (i.e., higher nutrient digestibility and greater uniformity). This contributes to better and more uniform animal performance even when relatively high levels of DDGS are included in animal feeds. It also improves economics of animal production because nutrient safety margins, when formulating animal feeds with more uniform DDGS, can be relaxed. In the Unites States, it is not surprising to find today pig grower and finisher feeds containing over 30% DDGS. Many poultry feeds can achieve DDGS levels at 15% or more of total feed content. Obviously, in ruminants DDGS have been used at much greater inclusion for a long time since ruminants can deal better with a high fiber fraction present in distillers grains.

New-generation DDGS—from dry-grind ethanol plants, which have been constructed during the past 10 to 15 years—are highly desirable feed ingredient with a high level of protein, a good source of energy, phosphorus, and amino acids (Table 24.1). However, sometimes DDGS even from these new dry-grind plants can have a low and variable nutritive value. This can be caused by many reasons. Types and varieties of grains can greatly influence the value of DDGS. Furthermore, starting grain quality can be affected by factors such as soil conditions, fertilizer usage, weather, production, and harvesting methods (Miller, 2006) and storage. Moreover, a large portion of DDGS variability is caused by variation in the ethanol plant technologies and processes; temperatures used during liquefaction, types and levels of enzymes, throughput time, pH conditions, addition of solubles into distillers grains, and conditions (temperatures and length) during drying of DDGS. In a study conducted by Fastinger et al. (2006), five samples of DDGS were collected from different Midwestern USA ethanol plants and

TABLE 24.1

Nutrient Composition (As-Fed Basis) of New-Generation DDGS

Dry matter (%)	89	Lysine (%)	0.74
Crude protein (%)	27.2	Methionine (%)	0.49
Crude fat (%)	9.5	Cystine (%)	0.52
ADF (%)	14	Phenylalanine (%)	1.32
NDF (%)	38.8	Threonine (%)	1.01
DE (kcal/kg pig)	3529	Tryptophan (%)	0.21
ME (kcal/kg pig)	3197	Valine (%)	1.34
Arginine (%)	1.06	Calcium (%)	0.05
Histidine (%)	0.68	Phosphorus (%)	0.79
Isoleucine (%)	1.01	Available phosphorus (%)	0.71
Leucine (%)	3.18		

Source: Shurson, G. C. et al. Value and use of "new-generation" distiller's dried grains with solubles in swine diets. Presented at the 19th International Alltech Conf., Lexington, KY. May 13, 2003.

subjected to various analyses. The nutritive values for DDGS samples from this study are presented in Table 24.2. The variability among the samples highlights how difficult it may be for a nutritionist to predict accurately nutritive value of DDGS especially if sourcing DDGS from number of ethanol plants and a need to apply wider nutrient safety margins, thus increasing cost of feed production, when formulating feeds. While care is taken by many ethanol plants to analyze DDGS for various nutrients, it is extremely difficult and might be costly to evaluate digestibility of various nutrients for every batch of DDGS supplied to the feed producers. Table 24.2 also shows how a color-related parameter, lightness of DDGS, influenced by drying or the amount of added solubles, is not the most reliable indicator of DDGS dietary value (R^2 column in Table 24.2). A similar observation was supported by other researchers (e.g., Liu et al., 2008), underlying potential difficulty to predict quickly and reliably DDGS quality. More about variability in the production process is discussed in Chapter 5.

During the ethanol production process, starch is removed from corn or other starchy grains (sorghum, wheat, barley) by converting it into sugars. In ethanol plants, this is in general achieved with a varying degree of success. While the level of starch can be as low as 3% to 4% in some samples of DDGS, samples with 10% or higher starch levels are also supplied to the animal feed industry (Liu, 2008). This typically means that energy value of DDGS can be higher and the presence of less desirable compounds such as fiber (a potentially problematic antinutrient namely in diets of pigs and poultry) tends to be lower. On the other hand, it could be also concluded that low sugar levels in DDGS can have a positive correlation with, for example, higher amino acid digestibility since there is less opportunity for a development of Maillard reaction. Maillard reaction causes amino acids to form a strong matrix with reducing sugars after heat treatment resulting in impaired amino acid digestibility (Pahm et al., 2009). Naturally, ethanol plants try to extract as much starch as possible to increase ethanol yield from sugar fermentation, which could have an impact on DDGS, including amino acid digestibility. Furthermore, some of the starch present in corn will be resistant to enzymes and has characteristics similar to crude fiber (Xie et al., 2006), thereby reducing ethanol yield and potentially negatively affecting DDGS quality. This starch is known as resistant starch and various types of ethanol process can have a different impact on this portion of starch (Sharma et al., 2010) and DDGS quality.

The amount of condensed distillers solubles (CDS) blended with distillers wet grain to produce DDGS can vary significantly between and within ethanol plants and it can contribute further to

TABLE 24.2
Amino Acid, Crude Protein, Gross Energy, Neutral Detergent Fiber, and Acid Detergent Fiber Contents of the Five Sources of Dried Grains with Solubles (DDGS; As-Fed Basis)[a,b]

Item	1	2	3	4	5	R^2
			DDGS Source			
CP[c] (%)	28.2	28.3	29.8	27.3	27	0.01
GE[d] (kcal/kg)	4969	4895	4898	4888	4848	0
NDF[e] (%)	34.2	29.7	32.8	32.9	31.5	0.17
ADF[f] (%)	12	10.3	13	11	13.2	0.79
Essential amino acids (%)						
Arginine	0.96	1.06	1.08	1.06	0.86	0.65
Histidine	0.61	0.66	0.7	0.65	0.63	0.15
Isoleycine	0.84	1.03	1.09	0.99	0.96	0.19
Leucine	2.86	3.05	3.26	3.05	3.13	0
Lysine	0.51	0.75	0.7	0.76	0.48	0.86
Methionine	0.48	0.48	0.5	0.51	0.45	0.41
Phenylalanine	1.19	1.36	1.44	1.34	1.36	0.05
Threonine	0.89	0.98	1.03	1.01	0.84	0.56
Tryptophan	0.25	0.26	0.28	0.25	0.2	0.31
Valine	1.21	1.32	1.42	1.29	1.26	0.1
Nonessential amino acid (%)						
Alanine	1.8	1.85	1.99	1.85	1.84	0.01
Asparagine	1.66	1.72	1.77	1.71	1.46	0.47
Cysteine	0.46	0.48	0.5	0.5	0.45	0.48
Glutamine	4.39	3.89	3.94	3.68	4.01	0.42
Glycine	0.96	1.01	1.06	0.98	0.91	0.27
Proline	1.94	2.09	2.17	1.95	1.84	0.24
Serine	1.13	1.01	1.08	1.06	0.91	0.05
Tyrosine	0.86	0.98	1.01	1.03	0.96	0.29

Source: Fastinger, N. D., J. D. Latshaw, and D. C. Mahan. *Poultry Science,* 85, 1212–1216, 2006.

[a] Values represent a single analysis of each DDGS sample.

[b] R^2 indicates correlation between lightness value and nutrient composition.

[c] CP = crude protein.

[d] GE = gross energy.

[e] NDF = neutral detergent fiber.

[f] ADF = acid detergent fiber.

differences in levels of oil, protein, phosphorus (P), DDGS color, but also lysine digestibility as reported by Pahm et al. (2009). Finally, drying of distillers grains—time and temperature—will also introduce variability in DDGS nutritive value. It has been shown that amino acids and P availability can be affected to a great degree by varying temperatures used during DDGS drying. Martinez Amezcua et al. (2004) indicated that a substantial variability in phosphorus (P) bioavailability among different DDGS samples can be associated with increased heat processing (e.g., drying in the ethanol plant). The authors saw an increase in bioavailability of P in DDGS dried at higher temperatures. Interestingly, the authors commented that a DDGS sample containing the highest P bioavailability

TABLE 24.3
Amino Acid Concentrations (100% Dry Matter Basis as) of
DDGS Samples from Ten Midwest Ethanol Plants

Amino Acid	Mean (Range)(%)	Standard Deviation (%)
Lysine	0.88 (0.61–1.06)	0.13
Methionine	0.63 (0.54–0.73)	0.06
Threonine	1.14 (1.02–1.28)	0.09
Tryptophan	0.24 (0.18–0.34)	0.05

Source: Miller, P. S. et al. Corn distillers dried grains with solubles for swine, corn processing coproducts manual a review of current research on distillers grains and corn gluten. *University of Minnesota DDGS web site*, 2005.

value showed also a very low lysine digestibility potentially suggesting an increase in Maillard reaction—browning. This specific sample was dark brown in color, indicative of increased heat processing, which contributed to a significant destruction of lysine. It is not surprising that amino acids levels in DDGS can thus be highly variable, as presented by Miller et al. (2005) (Table 24.3).

The ethanol production process from corn and other starch sources varies from plant to plant. This affects nutritive value and uniformity of DDGS. To deal better with changes in input costs and ethanol market prices, some ethanol producers are trying to find a product niche or a new technology that gives their refinery a competitive advantage to deal better with possible market price fluctuations. For any ethanol plant, however, DDGS with uniform and high nutritive value may offer a greater economic value to the final users (e.g., nutritionists, animal feed producers) compared to DDGS produced without such a strong focus. Understanding that a specific process within an ethanol plant can be one of the largest influencing factors affecting DDGS quality and uniformity could help in critical evaluations of ethanol production systems and possibly lead to economically beneficial changes. Besides drying time and temperature, which can vary greatly, with temperatures typically over 127°C (260°F), other changes such as corn grind particle size, cooking, and fermentation time, can also have an impact on the final DDGS product.

Other quality parameters, such as screening for the presence of mycotoxin in corn could have a value since the ethanol refining does not destroy mycotoxins. The ethanol production actually increases mycotoxin concentration by about threefold in the final DDGS product (Wu and Munkvold, 2008). Thus, harvesting high moisture corn, which is exposed to delayed drying, can contribute to increase in a wide range of mycotoxins in DDGS. Many ethanol plants screen grains prior to accepting them for ethanol production and this will be happening more likely in the future with improvements in cost and speed of mycotoxin analyses. However, it is still up to the final user of DDGS to request assurance from DDGS producers that they have in place a reliable mycotoxin quality-control plan. Additional information on mycotoxin in DDGS is covered in Chapter 11, while information on chemical composition of DDGS, changes during dry grind processing, and various factors causing compositional variations is discussed in Chapters 3, 5, 7, and 8.

One specific factor in ethanol production that has been gaining more focus recently is the type of enzymes and their optimal use in the ethanol production process. This focus is not only for improving ethanol yield, but also for improving other areas of the ethanol plant, such as DDGS quality.

24.3 USE OF ENZYMES IN ETHANOL PLANTS AND EFFECTS ON DDGS QUALITY

Exogenous enzymes play a crucial role in the production of ethanol from corn (*Zea mays*) and other starch-containing ingredients. In the United States, corn is used as the main source of sugars for fermentation in the ethanol production process. In dry-grind plants, corn can be simply milled, mixed with water, cooked, and then fermented as a whole mash. Enzymes are added at various stages of this process. Additionally, some plants are starting to implement dry milling fractionation technology to produce more product streams at lower cost than a wet milling facility (RFA, 2009). This can involve partial degermination of corn before fermentation or the use of decortication with sorghum. Pretreatment of grains results in changes of the substrate available for fermentation and can also contribute to changes in DDGS nutritive value. In degermed corn, fat, P, and energy levels are typically reduced in the final DDGS product.

Hydrolysis of starch by enzymes is crucial in order to improve sugar availability for yeast in the fermenters. There are two types of starches found in corn—amylose and amylopectin. According to Takeda et al. (1988) amylose starches are small molecules (930 to 990) with an average degree of polymerization (2270–2550). They are composed of nearly equal numbers of branched (5.3 chains on average) and unbranched molecules. The amylopectins, on the other hand, are shorter with average chain lengths of 21–22. Amylopectin starch consists of branched glucose subunits. Corn varieties used today have been specifically designed with resulting effects on starch (and other compounds such as protein/amino acids and oil) to better meet needs of various industries widely using corn. For example, dent or normal corn starch contains about 27% amylose, with the remainder being amylopectin (Bothast and Schlicher, 2005). This type of corn is likely utilized to the greatest degree by the U.S. ethanol plants today. The entire focus of the ethanol production process is targeted towards achieving the most efficient starch hydrolysis, resulting in the highest possible alcohol yield. In the process of utilizing corn starch to provide substrate for yeast, to facilitate the actual fermentation process, several enzyme activities may be used at different points in the ethanol production process. From an animal nutritionist's point of view, types of enzymes used and conditions employed to achieve optimum performance for these enzymes within an ethanol plant can have a significant impact on the nutritive value and the uniformity of DDGS.

24.3.1 AMYLASES

Alpha-amylases (α-amylases) are enzymes which facilitate hydrolysis or breakdown of starch. Calcium (Ca) is required for activation and stability of alpha-amylase (Bush et al., 1989). By acting at random locations along the starch chain, α-amylase breaks down long-chain carbohydrates, ultimately yielding products such as maltotriose and maltose from amylose, or maltose, glucose, and dextrin from amylopectin. Because it can act anywhere on the starch substrate, α-amylase tends to be faster acting than other types of amylases (e.g., β-amylase) and it is used as a workhorse enzyme by ethanol production systems today. Alpha-amylases are typically included during the starch liquefaction stage of ethanol production. Corn can be simply ground, enzymatically hydrolyzed, and then fermented as a whole mash. This is probably the least capital intensive process to produce ethanol for industrial and fuel markets and it is employed by many dry-grind plants today. However, ethanol production can be also achieved from cornstarch that has been removed from most of the other components present in corn such as gluten, germ, and fiber.

More recently, there has been increased interest in lines of granular starch hydrolyzing enzymes, which can convert granular or uncooked starch to fermentable sugars on a continuous basis in a simultaneous saccharification and fermentation process, or SSF. Stargen™ of Genencor International, Inc. (Palo Alto, CA) is an example of such a product—a combination of amylase and another enzyme called glucoamylase. There are many potential advantages of this technology including energy savings due to the elimination of jet cooking, lower water and energy use, greater ethanol yield, and fewer chemicals used (Williams, 2006). For DDGS from such a process, it could mean less opportunity for protein and amino acid damage due to the lower processing temperatures.

For ethanol production from other starch sources, such as sorghum, a significant increase in ethanol yields can be achieved by subjecting the grain to decortication. Decortication results in the removal of hull and outer-layer pericarp prior to the fermentation. This optimizes starch hydrolysis, or digestion, and increases ethanol production. The decortication gives samples with higher starch levels and increases the amount of fermentable substrate, in turn resulting in higher fermentation yields. Research has indicated that the amount of unfermented starch in DDGS decreases as the degree of decortication increases (Corredor et al., 2006). Finally, new amylase-based enzyme products have been introduced recently, with a direct impact on DDGS quality. These products are able to hydrolyze starch across a wide range of pH levels allowing for a substantial reduction in sulfuric acid used during mainly the liquefaction process (e.g., Spezyme RSL, Genencor International, Palo Alto, CA). This new technology opportunity could help ethanol producers reduce input costs because of 25% to 50% reduction in sulfuric acid usage. DDGS from ethanol plants with this new enzyme could contain lower sulfur levels. Detailed review of sulfur toxicity in various animal species is available (Mineral tolerance of domestic animals, 1980) suggesting potential benefits in reducing sulfur acid level in ingredients used in animal feeds.

24.3.2 AMYLOGLUCOSIDASE

Amyloglucosidase, or glucoamylase, or γ-amylase, is responsible for hydrolysis of terminal glucose residue in the chain of glucose units. The amyloglucosidase enzymes are typically used during the starch saccharification phase. The main role of glucoamylase is to further hydrolyze substrates resulting from amylase action to provide a more suitable compound (i.e., glucose) for fermentation by yeast. As mentioned in the amylase section, the starch from corn is usually pressure cooked around 105°C (221°F) in the presence of alpha-amylase and then liquefied to dextrins at temperatures exceeding 90°C (194°F) using liquefactions enzymes (Kelsall and Lyons, 2003). The dextrinized mash is then cooled and saccharified at either 60°C (140°F) with glucoamylase or simply cooled to 32°C (89.6°F) and simultaneously saccharified with glucoamylase and fermented with yeast (SSF process) The fermentation usually takes up to 3 days and may involve the use of additional enzymes.

24.3.3 PROTEASES

Protease enzymes—either from bacterial or fungal sources—can be used at various stages of ethanol production. Typically, proteases, which hydrolyze proteins to yield peptides and free amino acids, can further improve fermentation of sugars. In addition, they can also facilitate the process of dewatering solids that remain after the ethanol has been extracted, as reported by McGinnis (2007). The article has also suggested that soaking corn in water (during wet milling process) and using protease increased starch recovery, making it available for use in various industries including corn starch ethanol production. For DDGS quality, this can mean samples with lower starch but also lower protein content and/or higher availability of amino acids. On the other hand, fiber levels in these samples can be particularly high, making DDGS from such a process less desirable for monogastric animals, but potentially a good source for ruminants. Vidal et al. (2009) has found that when using protease the residual starch in endosperm fiber after fermentation was reduced by 8% and 22% in dry fractionation with raw starch fermentation and dry fractionation with conventional fermentation process, respectively. The authors concluded that protease treatment can be used effectively to enhance grind processes. A dry-grind corn process modification called enzymatic milling (E-mill) employs protease together with starch-degrading enzymes during an incubation phase after soaking and grinding (Singh et al., 2005). The E-mill process helps to recover germ, pericarp fiber, and endosperm fiber, thus increases fermentation capacity but also improves the quality of DDGS. The authors found that when compared to a conventional process, the E-mill reduced DDGS fiber (reported as ADF) from 10.8% to 2.03% and increased protein from 28.5% to 58.5%. On the other hand, crude fat levels were reduced from 12.7% to 4.53%, mainly due to the removal of germ prior to the fermentation.

24.3.4 PHYTASES

Phytases, also known as myo-inositol hexakisphosphate phosphohydrolases or phytate phosphatases, break down phytate (phytic acid, myo-inositol hexa-phosphate, IP6) to release free phosphate molecules. A very high proportion of P in cereal grains is bound within phytate. The polyanionic phytate molecule has a tremendous capacity to chelate cations and form various nutrient–phytate complexes. In the ethanol production process, the presence of phytate can significantly reduce efficiency of ethanol production. Isaksen (2006) reported that incubating wheat or corn starch with alpha-amylase together with increased levels of phytic acid resulted in reduced activity of amylase, and the degradation of starch was almost halved at the highest tested concentration of phytic acid. The author also found that the negative effect of phytic acid on starch degradation was counteracted by increasing levels of Ca^{++} or by adding phytase. Ca is required for activation of amylase (Bush et al., 1989) and a potential chelation of Ca ion by phytate can reduce activity of alpha-amylase. Additionally, there is some evidence that phytate in the ethanol production process contributes to increased slurry viscosity, reduced throughput, increased deposits of scale in the processing equipment, and increased water usage (Shetty et al., 2008). By removing phytate from the ethanol production process, digestibility of P from DDGS by pigs and poultry could be further increased. The phosphorus concentration in DDGS is considerably greater than corn (0.89% vs. 0.28%) and it can also be highly variable (Figure 24.1). Shurson et al. (2004) suggested that pig available P in DDGS can be calculated as high as 80%. Pedersen et al. (2007) estimated that total tract digestibility of P from DDGS was only 59% implying considerable variability among various samples of DDGS. Since pigs and poultry do not produce significant amounts of endogenous phytase, using this enzyme in ethanol production could further improve P digestibility and more importantly make it uniformly present and predictable.

Besides phytate's effect on P availability, there is evidence that phytate contributes to an increase in maintenance requirement of animals (through endogenous losses of nutrients). This means that reducing phytate levels in DDGS with phytase use in ethanol plants could result in a reduction of protein and energy used by animals for maintenance, making these nutrients available instead for production of meat and eggs (Selle and Ravindran, 2007; Pirgozliev et al., 2007; Liu et al., 2008). In a study conducted by Hruby et al. (2007) the effect of phytase on different ethanol production

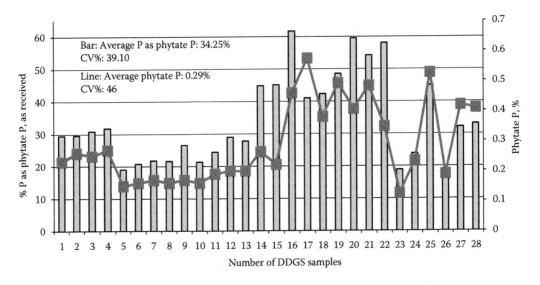

FIGURE 24.1 Phytate phosphorus levels expressed as percent of total P in DDGS from various U.S. ethanol plants. (Adapted from Danisco Animal Nutrition database.)

TABLE 24.4
The Effect of Phytase Addition in Ethanol Production Process on DDGS Nutritive Value

Parameter	DDGS w/o Phytase (control)	DDGS with Phytase	% Improvement	P Value
Dry matter digestibility, %	61.98	66.61	7.5	0.11
TME, kcal/kg DDGS (dmb)	3425	3641	6.3	0.04
Total amino acid digestibility, %	80.79	84.57	4.7	0.15
Essential AA digestibility, %	79.46	82.59	3.9	0.24
Nonessential AA digestibility, %	82.99	87.85	5.9	0.06
Lysine digestibility, %	71.87	75.94	5.7	0.24
Methionine digestibility, %	88.10	90.7	3.0	0.34
Threonine digestibility, %	73.24	79.68	8.8	0.05
Isoleucine digestibility, %	78.4	83.08	6.0	0.09
Leucine digestibility, %	84.69	88.76	4.8	0.04
Valine digestibility, %	78.24	83.12	6.2	0.10

Source: Hruby, M., J. K. Shetty, G. Chotani, T. Dodge, and C. N. Coon, *Journal of Animal Sciences.* 85, Suppl. 1/*Journal of Dairy Science,* 90, Suppl.1/*Poultry Science* 86, Suppl. 1: 397, 2007.
Essential amino acids: Arg, His, Iso, Leu, Lys, Met, Phe, Thr, Try, Val.
Nonessential amino acids: Ala, Asp, Cys, Glu, Ser, Tyr.
Data presented as a main effect of phytase treatment.
Impr. = improvement.

processes, that is, conventional hot cook and granular starch hydrolyzing enzyme processes, on the nutritive value of DDGS was evaluated. The researchers reported that almost no phytate P was detected in DDGS samples from the phytase-containing ethanol processes. DDGS derived from the phytase treatment had significantly higher true metabolizable energy (TME) and also digestibility of some amino acids (Table 24.4).

Economically, using the DDGS with improved nutritive values presented in Table 24.4 resulted in about U.S. $20 per metric tons (MT) greater value compared to DDGS from ethanol plants without phytase used in the process. Considering that during 2009, DDGS prices in the upper Midwest varied between U.S. $77/MT and 156/MT, the impact on ethanol plant's profitability could be significant if the benefit, due to improved nutrient digestibility, could be fully or partially realized by the ethanol producers.

24.3.5 NONSTARCH POLYSACCHARIDASES: CELLULASES, XYLANASES, AND BETA-GLUCANASES

DDGS is composed of unfermented polymeric sugars or fiber, which could potentially become a source of additional fermentable substrate to further improve ethanol yield. Fiber is considered an antinutritional compound, especially in diets of monogastric and aquatic species. It has been reported that fermentation of corn kernel fiber fraction can increase ethanol yield from a bushel of corn by about 10% and subsequently yield a higher-value and a higher-protein coproduct than is typically recovered in DDGS (Gulati et al., 1996). The opportunity to utilize DDGS as an additional source of fermentable sugars by the ethanol plant has the potential to increase overall ethanol yield in the current dry-grind ethanol production process. This could also have a positive impact on DDGS quality (i.e., less fiber) making it more beneficial, mainly for monogastric animals. For example, in a study conducted by Kim et al. (2007), controlled pH liquid hot water pretreatment and ammonia fiber expansion pretreatment were applied to enhance enzymatic digestibility of distillers grains.

Both pretreatment methods significantly increased the hydrolysis rate of DDGS versus unpretreated material. Fiber hydrolyzing enzymes, for example, cellulase and β-glucosidase, were also used as part of this process. Enhanced DDGS produced this process had higher crude protein levels and low crude fiber content compared to typical values found in commercial DDGS. Chapter 22 deals with this subject in detail.

24.3.6 COMBINATION OF ENZYMES

There is limited information regarding the use of various types of enzymes in combination at the same time—carbohydrases (amylase, glucoamylase, fiber-hydrolyzing enzymes), proteases, and phytases—in ethanol plants and their effects on DDGS nutritive values for animals. However, experience by the feed enzyme industry (enzymes added into animal feeds) suggests that possible benefits by the ethanol and DDGS industry could be realized if the use of various enzyme combinations is considered. For example, Rosen (2004), in his research review, found that average broiler feed conversion ratio (FCR or amount of feed per unit of gain) improvement in 17 published studies was 5.09% in diets supplemented with phytase and xylanase when in combination compared to 2.66% and 2.28% when supplementing xylanase or phytase alone, respectively. This suggests a somewhat synergistic effect for these two enzyme activities when supplemented in animal feeds. Similarly, Cowieson and Adeola (2005) saw a 10.4% improvement in FCR of broilers fed a corn-based diet with the addition of phytase, amylase, protease, and xylanase. However, they also saw "only" a 7.3% and 8.4% FCR improvement when using phytase alone or a mixture of amylase, protease, and xylanase on its own, respectively, resulting in a sub-additive effect of combining all the enzyme activities. Clearly, the substrate complexity both ethanol plants and gastrointestinal tract of animals must deal with calls for a multienzyme approach with possibly additive, synergistic, or sub-additive benefits depending on the type of substrate and the enzymes used.

24.4 SUMMARY

In February 2010, it was reported (Feedinfo News Service, 2010) that the U.S. ethanol industry produced approximately 30.5 million metric tons of distillers grains. Exports amounted to 18% of that total use. This is five times more than five years ago and strongly underlines the level of demand but also popularity this ingredient commands as a valuable source of nutrients in animal feeds. With both new and current ethanol plants looking for ways to maximize profitability, a consistent focus on DDGS quality, while improving efficiency of ethanol production, can result in a beneficial outcome for ethanol plants, DDGS marketers, and the final users of DDGS. Today's enzyme companies are employing research and development resources on a large scale to test and develop enzyme product candidates better equipped to deal with challenges of various substrates available to us in the process of ethanol production from corn to improve ethanol yield, reduce energy and other inputs, and increase production of various high quality coproducts. In many cases, these new enzyme products can be included price competitively offering additional benefits to ethanol plants, including higher DDGS quality and uniformity.

REFERENCES

Bothast R. J., and M. A. Schlicher. 2005. Biotechnological processes for conversion of corn into ethanol. *Applied Microbiology and Biotechnology* 67: 19–25.

Bush, D. S., L. Sticher, R. van Huystee, D. Wagner, and R. L. Jones. 1989. The calcium requirement for stability and enzymatic activity of two isoforms of barley aleurone alpha-amylase. *The Journal of Biological Chemistry* 264: 19392–19398.

Corredor, D. Y., S. R. Bean, T. Schober, and D. Wang. 2006. Effect of decorticating sorghum on ethanol production and composition of DDGS. *Cereal Chemistry* 83(1): 17–21.

Cowieson, A. J., and O. Adeola. 2005. Carbohydrases, protease, and phytase have an additive beneficial effect in nutritionally marginal diets for broiler chicks. *Poultry Science* 84: 1860–1867.

Davis, K. S. 2001. Corn milling, processing and generation of co-products. In proceedings: Minnesota Nutrition Conference and Minnesota Corn Growers Association Technical Symposium, Minnesota, September 11.

Deutscher, H. 2009. Economics of distillers grains. Ethanol Producer Magazine. June 2009. Available online: www.ethanolproducer.com/article.jsp?article_id=5653. Accessed May 16, 2010.

Fastinger, N. D., J. D. Latshaw, and D. C. Mahan. 2006. Amino acid availability and true metabolizable energy content of corn distillers dried grains with solubles in adult cecectomized roosters. *Poultry Science* 85: 1212–1216.

Gulati, M, K. Kohlmann, M. R. Ladisch, R. B. Hespell, and R. J. Bothast. 1996. Assessment of ethanol production options for corn products. *Bioresource Technology* 58: 253–264.

Hruby, M., J. K. Shetty, G. Chotani, T. Dodge, and C. N. Coon. 2007. Phytase in ethanol production process improves nutritive value of DDGS. *Journal of Animal Science* 85, Suppl.1/*Journal of Dairy Science* 90, Suppl.1/*Poultry Science* 86, Suppl.1: 397.

Isaksen, M. F. 2006. Phytic acid effect on endogenous alpha-amylase activity and starch degradation. *International Poultry Scientific Forum Abstracts* T111: 35–36.

Kelsall, D. R., and T. P. Lyons. 2003. Grain dry milling and cooking procedures: Extracting sugars in preparation for fermentation. In *The Alcohol Textbook: A Reference for the Beverage, Fuel and Industrial Alcohol Industries,* 4th ed., eds. K. A. Jacques, T. P. Lyons, and D. R. Kelsall, pp. 10–21. Nottingham, UK: Nottingham University Press.

Kim, Y., R. Hendrickson, N. S. Mosier, M. R. Ladisch, B. Bals, V. Balan, and B. E. Dale. 2007. Enzyme hydrolysis and ethanol fermentation of liquid hot water and AFEX pretreated distillers' grains at high-solids loadings. *Bioresource Technology* 99(12): 5206–5215.

Liu, K. 2008. Particle size distribution of distillers dried grains with solubles (DDGS) and relationships to compositional and color properties. *Bioresource Technology* 99(17): 8421–8428.

Liu, N., Y. J. Ru, F. D. Li, and A. J. Cowieson. 2008. Effect of diet containing phytate and phytase on the activity and messenger ribonucleic acid expression of carbohydrase and transporter in chickens. *Journal of Animal Science* 86: 3432–3439.

Martinez Amezcua, C., C. M. Parsons, and S. L. Noll. 2004. Content and relative bioavailability of phosphorus in distillers dried grains with solubles in chicks. *Poultry Science* 83: 971–976.

McGinnis, L. 2007. Enzymes boost ethanol production. *Agricultural Research Magazine.* April 2007. Available online: www.ars.usda.gov/is/pr/2007/070409.htm. Accessed February 20, 2010.

Miller, M. 2006. Scaling the golden DDGS mountain. *Pork Magazine.* December 2006. Available online: www.porkmag.com/directories.asp?pgID=728&ed_id=4578. Accessed January 5, 2010.

Miller, P. S., J. Uden, and D. E. Reese. 2005. Corn distillers dried grains with solubles for swine, corn processing co-products manual a review of current research on distillers grains and corn gluten. *University of Minnesota DDGS web site.* Available online: www.ddgs.umn.edu. Accessed September 10, 2009.

National Research Council Commitee on Animal Nutrition, 1980.

Noll, S. E., C. M. Parsons, and W. A. Dozier III. 2007. Formulating poultry diets with DDGS—How far can we go? Proceedings of the 5th Mid Atlantic Nutrition Conference, ed. N. G. Zimmerman. University of Maryland, College Park, March 28–29, 2007.

Pahm, A. A., C. Pedersen, and H. H. Stein. 2009. Standardized ileal digestibility of reactive lysine in distillers dried grains with solubles fed to growing pigs. *Journal of Agricultural and Food Chemistry* 57(2): 535–539.

Pedersen, C., M. G. Boersma, and H. H. Stein. 2007. Digestibility of energy and phosphorus in ten samples of distillers dried grains with solubles fed to growing pigs. *Journal of Animal Science* 85: 1168–1176.

Pirgozliev, V., O. Oduguwa, T. Acamovic, and M. R. Bedford. 2007. Diets containing *Escherichia coli*-derived phytase on young chickens and turkeys: Effects on performance, metabolizable energy, endogenous secretions, and intestinal morphology. *Poultry Science* 86: 705–713.

RFA. 2009. Growing Innovation. Ethanol Industry Outlook. Renewable Fuels Association. Washington, DC. Available online: http://www.ethanolrfa.org/page/-/objects/pdf/outlook/RFA_Outlook_2009.pdf.

Rosen, G. 2004. Admixture of exogenous phytases and xylanases in broiler nutrition. 6 pages on CD Proc. XXII World's Poultry Congress, Istanbul, Turkey: June 2004.

Selle, P. H., and V. Ravindran. 2007. Microbial phytase in poultry nutrition. *Animal Feed Science and Technology* 135: 1–41.

Sharma, V., K. D. Rausch, J. V. Graeber, S. J. Schmidt, P. Buriak, M. E. Tumbleson, and V. Singh. 2010. Effect of resistant starch on hydrolysis and fermentation of corn starch for ethanol. *Applied Biochemistry and Biotechnology* 160: 800–811.

Shetty, J. K., B. Paulson, M. Pepsin, G. Chotani, B. Dean, and M. Hruby. 2008. Phytase in fuel ethanol production offers economical and environmental benefits. *International Sugar Journal* 110 (1311): 2–12.

Shurson, G. C., M. J. Spiehs, J. A. Wilson, and M. H. Whitney. 2003. Value and use of "new generation" distiller's dried grains with solubles in swine diets. Presented at the 19th International Alltech Conf., Lexington, KY. May 13, 2003.

Shurson, G. C., M. J. Spiehs, and M. H. Whitney. 2004. The use of maize distiller's dried grains with solubles in pig diets. *Pig News and Information* 25(2): 75–83.

Singh, V., D. B. Johnston, K. Naidu, K. D. Rausch, R. L. Belyea, and M. E. Tumbleson. 2005. Comparison of modified dry-grind corn processes for fermentation characteristics and DDGS composition. *Cereal Chemistry* 82(2): 187–190.

Takeda, Y., T. Shitaozono, and S. Hizukuri. 1988. Molecular structure of corn starch. *Starch-Stärke* 40(2): 51–54.

US. 2009 Distillers Grains Exports Up 24 Per Cent. Feedinfo News Service. February 18th, 2010. (Accessed February 18, 2010). http://www.feedinfo.com/console/PageViewer.aspx?page=1704799&str=ETHANOL and PRODUCTION

Vidal, Jr., B. C., K. D. Rausch, M. E. Tumbleson, and V. Singh. 2009. Protease treatment to improve ethanol fermentation in modified dry grind corn processes. *Cereal Chemistry* 86(3): 323–328.

Williams, J. 2006. Break it down now. *Ethanol Producer Magazine* January 2006 issue.

Wu, F., and G. P. Munkvold. 2008. Mycotoxins in ethanol co-products: modeling economic impacts on the livestock industry and management strategies. *Journal of Agricultural and Food Chemistry* 56(11): 3900–3911.

Xie, X., Q. Liu, and S. W. Cui. 2006. Studies on the granular structure of resistant starches (type 4) from normal, high amylose and waxy corn starch citrates. *Food Research International* 39: 332–341.

Xu, G., S. K. Baidoo, L. J. Johnston, D. Bibus, J. E. Cannon, and G. C. Shurson. 2010. Effects of feeding diets containing increasing content of corn distillers dried grains with solubles to grower-finisher pigs on growth performance, carcass composition, and pork fat quality. *Journal of Animal Science* 88: 1398–1410.

25 Fractionation of DDGS Using Sieving and Air Classification

Radhakrishnan Srinivasan

CONTENTS

25.1 INTRODUCTION

25.1.1 MOTIVATION FOR FIBER SEPARATION FROM DDGS

"Dry grind processing is preferred over wet milling for fuel ethanol production from corn and other cereal grains because it requires less equipment and capital investment when establishing processing plants (Belyea et al., 2004). Currently, dry grind plants contribute to most of the U.S. ethanol production and it is expected that future corn-based ethanol production will arise mainly from construction of dry grind plants. Nearly 1 kg of distillers dried grains with solubles (DDGS) is produced per kg ethanol in a dry grind plant (Schilling et al., 2004). DDGS supply increases in proportion to the increase in ethanol production. There is a need to find new uses and make innovative products from DDGS" (Srinivasan et al., 2006).

Currently, DDGS is mainly used as feed for ruminants especially, dairy and beef cattle (Chapters 12 and 13). Swine and poultry (Chapters 14 and 15) are nonruminants and do not digest DDGS fiber very well. The inclusion levels of DDGS in swine and poultry diets are low due to the high fiber content in DDGS (Noll et al., 2001; Shurson, 2002). Therefore, separation of fiber would help to increase utilization of DDGS in poultry and swine diets. Fiber separation would result in products with higher protein and fat contents and lower fiber levels than regular DDGS and could increase the market value of DDGS due to the improved nutritional value, especially for nonruminants.

"Wu and Stringfellow (1986) reported that sieving produced DDGS fractions that were different in fiber and protein contents. Singh et al. (2002) observed that air classification could be a potential

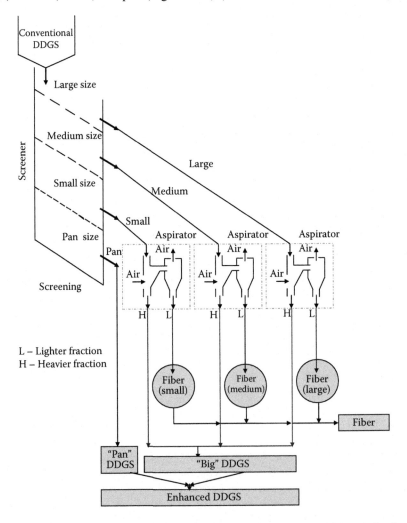

FIGURE 25.1 Schematic of Elusieve process for fiber separation from DDGS. In lab scale implementation of the Elusieve process, an elutriation column was used for air classification instead of an aspirator. (Reproduced from Srinivasan, R. et al. *Cereal Chemistry* 86, 393–397, 2009a.)

which comprises nearly 35 wt.% of original DDGS, has lower fiber (neutral detergent fiber; NDF) content and is a product by itself, called "Pan DDGS." Pan DDGS has 5% higher protein level than the conventional DDGS. Fiber is removed from the three biggest sizes by air classification, and the product obtained by mixing the remaining material is called "big DDGS," which comprises nearly 55 wt.% of original DDGS. Big DDGS has nearly the same protein content as conventional DDGS. The lighter fractions (fiber fractions) from air classification of the three size fractions are combined together as Elusieve fiber product, which comprises nearly 10 wt.% of original DDGS. In a variation of the Elusieve processing scheme, big DDGS and pan DDGS could be mixed together into a single product, enhanced DDGS. Enhanced DDGS has 1% to 3% higher protein content than conventional DDGS. "Currently, it is not clear whether industrial implementation of the Elusieve process would incorporate the scheme for production of three products (big DDGS, pan DDGS, and fiber) or the scheme for production of two products (enhanced DDGS and fiber)" (Srinivasan et al., 2009a).

25.2 ELUSIEVE PROCESSING, COMPOSITIONS OF FRACTIONS AND ECONOMICS

25.2.1 ELUSIEVE PROCESSING OF DDGS IN LAB SCALE

25.2.1.1 Procedure

In the first study on Elusieve processing by Srinivasan et al. (2005), air was blown into a 63 mm internal diameter (ID) column constructed of transparent perspex material to separate fiber from DDGS size fractions. In the next study by Srinivasan et al. (2008b), a 155 mm ID column constructed from commercial metal air ducts was used for elutriation. Material elutriated by the air was called "lighter fraction." Material that settled to the bottom of the elutriation column was called "heavier fraction." The elutriation apparatus in Srinivasan et al. (2008b) consisted of an elutriation column, an air blower for supplying air to the elutriation column, a surge box mechanism for controlling airflow, a vibratory feeder for feeding material into the column, and collection vessels for receiving the lighter and heavier fractions.

Commercial DDGS material (75 kg) was divided into three batches of 25 kg each. Each batch of DDGS was sieved using a vibratory screen (model ZS30S6666, SWECO Vibro-Energy Separator, Florence, KY). Screens were: 24T (869 µm), 34T (582 µm), 48T (389 µm), 62T (295 µm), and pan. "T" refers to tensil bolt cloth. The material retained on 24T was sieved by hand using a U.S. 7 mesh screen (2800 µm) to ensure that large lumps did not obstruct the flow of material when fed into the elutriation column. The six size fractions of material retained on these screens were referred as US7, 24T, 34T, 48T, 62T, and pan, based on their respective screen labels. Material (2 kg) was sieved on each screen for 1 h. Material passing through the sieve with a larger opening was collected and placed on the next smaller sieve size. Only one sieve was used at a time. The four intermediate size fractions, 24T, 34T, 48T, and 62T, were subjected to elutriation at three velocities. Due to low fiber content, elutriation was not carried out for US7 (>2800 µm) and pan (<295 µm) size fractions. Material was fed into the elutriation column at a rate of 5 g/min. Elutriation was carried out at three air velocities to obtain lighter fraction yields of 15, 20, and 25%.

Analogous to solvent selectivity in liquid extraction and relative volatility in distillation, NDF separation factor was defined to quantify fiber separation (Treybal, 1980; Srinivasan et al., 2005). The NDF separation factor (α_{NDF}) for elutriation was defined as the ratio of the NDF/nonNDF of the lighter fraction to the NDF/nonNDF of the heavier fraction.

$$\alpha_{NDF} = \frac{[NDF/(100-NDF)]_{Lighter\,fraction}}{[NDF/(100-NDF)]_{Heavier\,fraction}}$$

The NDF separation factor was indicative of the selectivity of air in carrying fiber rather than non-fiber at each air velocity. A large NDF separation factor indicates high selectivity.

Samples were analyzed for total nitrogen (AOAC, 2003, Method 990.03). Crude protein content was calculated as total N × 6.25. Samples were analyzed for crude fat using a hexane solvent extraction method described in Moreau et al. (1999). NDF content was determined using the procedure of Van Soest et al. (1991). Sample moisture contents were determined using a two-stage convection oven method (AACC International, 2000, Method 44-18).

25.2.1.2 Results

25.2.1.2.1 Sieving

The smallest size fraction, pan, contained 25.8% of the original DDGS (Table 25.1). The pan size fraction contained lower fiber, higher protein and fat contents than the original DDGS (Table 25.1).

25.2.1.2.2 Elutriation

Elutriation was effective in separating fiber from size fractions as indicated by higher NDF in the lighter fractions in conjunction with higher protein content in the heavier fractions (Table 25.2). For the 24T size fraction, elutriation at 1.88 m/s increased NDF from 33.2% to 61.9% in the lighter fraction, increased protein from 29.4% to 31.4% in the heavier fraction, and increased fat from 10.0% to 10.7% in the heavier fraction (Table 25.2; Srinivasan et al., 2008b). NDF separation factors were more than 1.0, which signified the separation of fiber (Table 25.2).

The coefficient of variability (COV) of lighter fraction yield was less than ±10% except at two velocities for 62T size fraction, ±21.2% and 13.9% at 1.06 and 1.20 m/s, respectively. COVs of NDF and fat content were less than ±10%.

25.2.1.2.3 Effect of Air Velocity on Fiber Separation

Increasing air velocity within a size fraction increased lighter fraction yield with a reduction in heavier fraction yield (Table 25.2). An increase in air velocity decreased NDF separation factors, because of the ability of air to carry nonfiber components at higher velocities (Table 25.2). For 24T size fraction, as air velocity increased from 1.88 to 2.37 m/s, lighter fraction yield increased from 15.3% to 28.2% with a decrease in NDF of lighter fraction from 61.9% to 52.2% and the protein

---

TABLE 25.1
Compositions of Material Retained on Each Screen After Sieving (Lab Scale)

Screen Label	Openings (μm)	wt.% Retained on Screen	Crude Protein* (% db)	Crude Fat (% db)	NDF (% db)
Original DDGS	NA	NA	34.7	10.1[a]	28.0[c,b]
US7	2800	1.7	34.4	11.6[a]	26.8[c]
24T	869	26.6	29.4	10.0[b]	33.2[a]
34T	582	18.3	27.7	9.0[c]	35.6[a]
48T	389	14.0	31.0	9.3[c]	32.2[b,a]
62T	295	13.6	36.0	10.4[b]	26.6[d,c]
Pan	0	25.8	41.4	11.6[a]	21.9[d]

Source: Reproduced from Srinivasan, R., and V. Singh. *Separation and Purification Technology* 61, 461–468, 2008b.

* Values for crude protein are means of two batches.

NA—Not applicable, NDF—Neutral detergent fiber. Values reported are means of three batches.

Values in the same column followed by same letter are not different (p < 0.05).

Coefficients of variability (COVs) of wt.% retained on screen were less than ±11% except for US7 screen (20.4%). COVs of NDF and crude fat were less than ±12%.

TABLE 25.2
Yield and NDF (% db) of Fractions from Elutriation of Size Fractions (Lab Scale)

Size Fraction	Velocity (m/s)	Yield (%) (L)	Yield (%) (H)	NDF L	NDF H	Separation Factor
24T (869 to 2800 μm)	0*	NA	NA	33.2[d]	33.2[a]	NA
	1.88	15.3[c]	84.7	61.9[a]	28.8[b]	4.0
	2.08	19.4[b]	80.6	56.9[b]	27.8[c,b]	3.4
	2.37	28.2[a]	71.8	52.2[c]	25.4[c]	3.2
34T (582 to 869 μm)	0*	NA	NA	35.6[c]	35.6[a]	NA
	1.60	16.8[c]	83.2	62.2[a]	30.1[b]	3.8
	1.75	23.7[b]	76.3	57.6[b,a]	27.5[b]	3.6
	1.92	31.8[a]	68.2	54.9[b]	27.3[b]	3.2
48T (389 to 582 μm)	0*	NA	NA	32.2[c]	32.2[a]	NA
	1.45	18.3[c]	81.7	54.3[a]	26.5[b]	3.3
	1.55	22.8[b]	77.2	51.8[b,a]	25.7[b]	3.1
	1.70	29.0[a]	71.0	47.0[b]	23.9[b]	2.8
62T (295 to 389 μm)	0*	NA	NA	26.6[c]	26.6[a]	NA
	1.06	8.2[c]	91.8	42.9[a]	25.5[a]	2.2
	1.20	17.3[b]	82.7	39.4[b,a]	24.5[a]	2.0
	1.33	31.7[a]	68.3	36.7[b]	24.8[a]	1.8

Source: Reproduced from Srinivasan, R., and V. Singh. *Separation and Purification Technology* 61, 461–468, 2008b.
* Values at 0 m/s denote initial material.
NDF—Neutral detergent fiber, H—Heavier fraction, L—Lighter fraction, NA—Not applicable.
Values are means of three batches. Values in the same column of a size fraction followed by the same letter are not significantly different (p < 0.05).

and fat content of heavier fraction increased from 31.4% to 32.9% and 10.7% to 11.5%, respectively (Table 25.2; Srinivasan et al., 2008b). The NDF separation factor decreased from 4.0 to 3.2 during this air velocity increase (Table 25.2).

As particle size of material fed to the elutriation column decreased, air velocity required to produce the same lighter fraction yield decreased (Table 25.2). For 24T size fraction, the lighter fraction yield at the air velocity of 1.88 m/s was 15.3% (Table 25.2). For 34T size fraction, air velocity needed to produce a similar yield (16.8%) was 1.60 m/s.

25.2.1.2.4 Compositions of Elusieve Products

The compositions (% db) of products calculated from Elusieve processing of two different commercial DDGS materials (DDGS-1 and DDGS-2) using the 63 mm ID column, reported in Srinivasan et al. (2005), shows that the protein content can be increased by 2.0% to 2.6% from original DDGS to enhanced DDGS (Table 25.3). The NDF of fiber product was substantially higher than original DDGS, by 33% to 37%.

25.2.2 ELUSIEVE PROCESSING OF DDGS IN PILOT SCALE

The Elusieve process was effective in separating fiber from commercial DDGS samples on a laboratory scale. In the laboratory-scale apparatus, processing was carried out in batch operation and air classification was carried out in an elutriation column that was custom built. In industrial-scale implementation of the Elusieve process, commercial sifters and aspirators would be used. The pilot

TABLE 25.3
Compositions (% db) and wt.% of Products Obtained from Elusieve Lab Scale Processing of Two Different Commercial DDGS Materials (DDGS-1 and DDGS-2)

Original DDGS Material	DDGS Product	wt.% of Original DDGS	Crude Protein	NDF
DDGS-1	Original DDGS	100	33.6	32.5
	Enhanced DDGS	88.3	36.2	28.1
	Fiber	11.7	14.0	65.7
DDGS-2	Original DDGS	100	32.9	33.6
	Enhanced DDGS	91.3	33.9	30.1
	Fiber	8.7	12.4	70.3

Source: Based on Results from Srinivasan, R. et al. *Cereal Chemistry* 82, 528–533, 2005 for elutriation at low air velocities that resulted in nearly 20% lighter fraction yield.
NDF—neutral detergent fiber.

plant work was done in order to determine its effectiveness using commercial equipment and to verify its operability in continuous mode (Srinivasan et al., 2009b). The pilot plant was assembled to evaluate fiber separation from DDGS on a continuous basis using a commercial sifter and aspirators. The sifter and aspirators were not custom made but were procured off-the-shelf from equipment manufacturers.

25.2.2.1 Procedure

A rectangular rotary sifter (Model 484, Gump, Savannah, GA) with a sieving area of 1.8 m² (19 ft²) per deck, consisting of three decks for stack sieving, was used to produce the four size fractions, which were denoted as "large," "medium," "small," and "pan." The opening size for screens was chosen such that each of the large, medium, and small size fractions would be nearly 20% by weight of the original DDGS. The fraction on the pan (i.e., the fraction of smallest particle size) is also denoted as "Pan DDGS."

The large, medium, and small size fractions were air classified using three multi-aspirators (Model VJ8X6, Kice, Wichita, KS). The multiaspirator comprises a material feeding section, through which the DDGS size fraction is fed, and an air-inlet section through which air is sucked into the aspirator by a fan. The fan for the multiaspirator that was used to aspirate the large size fraction was operated by a 1.1 kW (1.5 hp) motor and the fans in the other two aspirators were operated by 0.6 kW (0.75 hp) motors. A higher rating fan was used for the large size fraction because of the higher air velocities needed to separate fiber from large size fraction compared to the other fractions.

Commercial DDGS material was obtained from a local feed mill (Prairie Mills, Prairie, MS). The pilot plant was tested on three different DDGS materials in five different cases of processing batches. DDGS was gravity fed from a hopper to the sifter, through a manual gate valve, at a rate of 0.25 kg/s (1 ton/h).

25.2.2.2 Results

Lighter fractions from air classification of size fractions had higher fiber (NDF) content than corresponding heavier fractions. Trends were similar to those observed in laboratory scale studies (Srinivasan et al., 2005; Srinivasan et al., 2008b).

Out of the five processing cases carried out in the pilot scale work, the best implementation scenario was identified based on the highest revenue potential. The implementation scenario with highest revenue potential produced pan DDGS that had 4.8% higher protein content than original DDGS and big DDGS that had 0.7% lower protein content, but had higher fat and lower fiber contents, than original DDGS (Table 25.4) (Srinivasan et al., 2009b). The best implementation scenario for the Elusieve process would be to produce three products: (1) a product (pan DDGS) that would have 5% higher protein content than conventional DDGS (on a wet basis); (2) a product (big DDGS) that would have nearly the same protein content as conventional DDGS; and (3) a fiber product.

25.2.3 DDG (WITHOUT SOLUBLES) FIBER SEPARATION

"In the dry grind corn-to-ethanol process, distillers wet grains (DWG) and syrup (commonly known as condensed distillers solubles [CDS]) are blended and dried to produce DDGS. Some dry grind plants produce distillers dried grains (DDG) as a coproduct instead of DDGS. In these plants, syrup is not mixed with DWG and is sold as a liquid feed ingredient or used for energy generation by combustion. DDG is defined as the product obtained after removal of ethyl alcohol by distillation from yeast fermentation of a grain or a grain mixture by separating the resultant coarse grain fraction of the whole stillage and drying (AAFCO, 2002). The Elusieve process could be beneficial in separating fiber from DDG also. Fiber separation from DDG using the Elusieve process was evaluated. Elutriation of size fractions increased NDF in the lighter fractions, and NDF separation factors were more than 1.0. When DDG is separated via the Elusieve process, 11.9% would be obtained as Elusieve fiber and 88.1% would be obtained as enhanced DDG. Original DDG had NDF of 36.7% (db), while enhanced DDG would have NDF of 35.3% (db) and Elusieve fiber would have NDF of 57.3% (db). Thus, the Elusieve process can produce value-added products from both DDG and DDGS" (Srinivasan et al., 2008a).

25.2.4 COST FOR FRACTIONATION OF DDGS USING THE ELUSIEVE PROCESS

A processing rate of 2030 t/d (80,000 bu/d) of corn was taken as the base case; this plant would produce ethanol at a rate of 848 m³/d (224,000 gal/d or 76 million gal/y) and DDGS at a rate of 617 t/d (680 tons/d, 25.7 t/h or 56,670 lb/h). Three short sifters with a capacity of 10 t/h (22,000 lb/hr) each would be needed. Purchased equipment cost for sifters was obtained from an industrial manufacturer (Sweco, Florence, KY). Each sifter would cost 10 t/h; therefore total purchase cost for sifters would be $0.3 million (Table 25.5) (Srinivasan et al., 2006).

In order to ensure conservativeness of cost estimation, it was assumed that four size fractions would be air classified instead of just three. Four aspirators would be needed to separate fiber from the four largest size fractions and one aspirator would be needed for standby purposes. For the base case, with a corn processing rate of 2030 t/d (80,000 bu/d), each aspirator would have a capacity

TABLE 25.4
Composition (% wb), wt.% and Price of Products that Represent the Potential Implementation Scenario for the Elusieve Process; Obtained by Pilot Scale Processing

Product	wt.%	Protein	Fat	NDF	Ash	Moisture	Price Estimate ($/t)
Original DDGS	100.0	30.4	7.3	25.6	4.0	12.3	160
Enhanced DDGS	87.6	31.8	7.7	23.7	4.1	12.2	170
Big DDGS	54.9	29.7	8.5	23.8	4.0	12.0	160
Pan DDGS	32.8	35.2	6.2	23.5	4.3	12.6	195
Fiber	12.4	20.4	5.0	38.8	3.5	12.5	114

Source: Reproduced from Srinivasan, R. et al. *Cereal Chemistry* 86, 393–397, 2009a.

TABLE 25.5

Equipment and Associated Costs for Implementing the Elusieve Process in a Dry Grind Plant Processing Corn at a Rate of 2030 t/d (80,000 bu/d)

Type of Equipment	Sifter	Aspirator	Total
Capacity	10 t/h (22,000 lb/h)	12 t/h (25,000 lb/h)	NA
Units required for normal operation	3	4	NA
Units required for standby purpose	0	1	NA
Total units required	3	5	NA
Purchase cost per unit	$100,000	$26,000	NA
Purchase cost for plant	$0.3 million	$0.13 million	$0.43 million
Total capital investment	NA	NA	$1.4 million
Motor rating per unit	3.7 kW (5 hp)	11.2 kW (15 hp)	NA
Energy consumption	11.2 kW (15 hp)	44.7 kW (60 hp)	55.9 kW (75 hp)
Energy cost @ $0.05/kWh	$4,600/year	$18,300/year	$22,900/year ($0.02 million/year)
Total labor cost (2 man h/d @ $30/man-h)	NA	NA	$0.02 million/year
Total maintenance cost	NA	NA	$0.06 million/year
Total operating cost	NA	NA	$0.1 million year

Source: Reproduced from Srinivasan, R. et al. *Cereal Chemistry* 83, 324–330, 2006.
NA—Not applicable.

of 12,000 kg/h (12 t/h). Purchased equipment cost ($26,000 each) for aspirators was obtained from an industrial manufacturer (Kice Industries, Wichita, KS) (Srinivasan et al., 2006). Total purchase cost for aspirators would be $0.13 million (Table 25.5). Total capital investment was estimated as 3.25 times of purchased equipment cost (Peters and Timmerhaus, 1980). Working capital, building expenses, and service facilities were considered to be negligible because these are already available in dry grind plants. The Elusieve process has a small footprint and could be implemented in the DDGS warehouse. Total capital investment for the Elusieve process would be $1.4 million (Table 25.5).

For the base case, with a corn processing rate of 2030 t/d (80,000 bu/d), each sifter would be operated by a 3.7 kW (5 HP) motor and each aspirator would be operated by an 11.2 kW (15 HP) motor (Srinivasan et al., 2006). Energy costs @ $0.05/kWh would be $0.02 million/year (Table 25.5). Labor needed would be 2 man h/d costing $30/man-h, amounting to $0.02 million/y. Maintenance costs were estimated at $0.06 million/y. Total operating costs would be $0.1 million/y (Table 25.5). Using straight line depreciation over 15 years, processing cost would be 84 ¢/t of DDGS.

25.3 WORKING PRINCIPLES OF THE ELUSIEVE PROCESS

"Elutriation (or aspiration) separates particles based on the combined effect of density, size, and shape (Srinivasan, 2006; Srinivasan and Singh, 2008b). When air velocity is more than the terminal velocity of a particle, the particle is carried by air. Terminal velocity is the constant velocity attained by a falling particle when the upward drag on the particle and the buoyancy force balance the downward force of gravity (DRI, 2005). For spherical particles in turbulent flow conditions, the upward drag on the particle is proportional to the square of the particle velocity. When a particle falls from its position of rest, gravitational force on the particle is more than the buoyancy force; hence, the particle accelerates. As the particle accelerates, the upward drag on the particle increases; when the particle reaches its terminal velocity, the forces are balanced and there is no further acceleration.

Spherical particles with the same diameter but larger particle density have higher terminal velocity, as the drag force needed to balance the heavier mass is higher. Similarly, spherical particles with the same particle density but larger diameter have larger terminal velocity, as they have heavier mass.

Shape also affects particle terminal velocity. For particles with equivalent spherical diameter and the same particle density, flat-shaped particles have lower terminal velocity than spherical particles, as flat-shaped particles experience higher drag force. Particle shape is characterized by sphericity, which is the ratio of surface area of a sphere having the same volume as the particle to the actual surface area of the particle. Sphericity of a sphere-shaped particle is thus defined as 1.0. For non-spherical particles, particle size is characterized by equivalent spherical diameter (Srinivasan and Singh, 2008b).

To separate fiber from DDGS, air velocity must be greater than the fiber's terminal velocity, but less than the nonfiber terminal velocity. Terminal velocities of particles are governed by the density, size, and shape of the particles. The particle densities, equivalent spherical diameters, sphericities, and terminal velocities of fiber and nonfiber particles were determined to understand fiber separation from DDGS" (Srinivasan and Singh, 2008b).

Particle density of fiber was 1403 to 1470 kg/m^3, which was higher than particle density of non-fiber (1301 to 1337 kg/m^3). Sphericity of fiber (0.19 to 0.47) was lower than sphericity of nonfiber (0.72 to 0.92) indicating the flatness of fiber compared to nonfiber. Sphericity of fiber was higher for smaller size fractions compared to larger size fractions, which indicated fiber in smaller size fractions was not as flat as fiber in larger size fractions.

For larger size fractions, mass of fiber particles was lower than mass of smallest nonfiber particles, despite higher particle density of fiber, because of lower equivalent spherical diameter of fiber (Srinivasan and Singh, 2008b). For smaller size fractions, mass of fiber particles was comparable to mass of smallest nonfiber particles and the equivalent spherical diameter of fiber was comparable to equivalent spherical diameter of smallest nonfiber. The flatness of fiber particles (sphericity of 0.19 to 0.21) caused fiber particles to have a lower or comparable equivalent spherical diameter relative to nonfiber particles (sphericity of 0.72 to 0.92) because a flat particle passing through a screen has lower volume than a spherical particle passing through the same screen.

For larger size fractions, lower mass of fiber than smallest nonfiber particles explained fiber being carried selectively at lowest air velocity (Srinivasan and Singh, 2008b). For smaller size fractions, fiber was selectively carried at low air velocities, despite mass of fiber being marginally higher (by 7% to 15%) than mass of smallest nonfiber. For smaller size fractions, flatness of fiber (sphericity of 0.44 to 0.47) compared to nonfiber (sphericity of 0.72 to 0.92) explained fiber being carried selectively at low air velocity; flat particles experience higher drag force than spherical particles.

Thus, fiber particles were carried selectively in each size fraction, at low air velocities, as they had low terminal velocities due to their flat shape and low mass. NDF of lighter fractions was higher at lower air velocities for all three size fractions. Within each size fraction, nonfiber particles also are carried along with fiber at high air velocities because nonfiber particles with lower equivalent spherical diameter have the same terminal velocity as fiber particles with larger equivalent spherical diameter (Srinivasan and Singh, 2008b).

25.4   DDGS PRODUCTS FROM ELUSIEVE PROCESSING IN POULTRY DIETS

"Poultry nutrition studies have shown that Elusieve processing of DDGS enhances nutritional characteristics. Feeding experiments with precision-fed rooster assay using cecectomized roosters showed that pan DDGS had 7% higher true metabolizable energy (TMEn) than conventional DDGS (Kim et al., 2007). Feeding studies on broilers at 8% DDGS inclusion levels showed that birds fed with pan DDGS had significant differences in body weight (BW) ($p = 0.08$) compared to birds fed with conventional DDGS, and there was no significant difference in BW for birds fed with conventional of big DDGS (Loar et al., 2008). Further studies for determination of allowable inclusion levels of big, pan, and enhanced DDGS in broiler diets, from a nutritional perspective, are underway.

Despite the potential to increase inclusion levels of DDGS in poultry diets by fiber removal, there have been concerns about feed manufacturing characteristics at higher inclusion levels. Pellet-mill throughput, power consumption, and pellet quality were evaluated for broiler diets incorporating different levels (0%, 10%, and 20%) of conventional DDGS and DDGS products from the Elusieve process. Poultry oil contents were lower (1.5% to 1.6%) in diets comprising pan DDGS and the diet without DDGS than in other diets (2.2% to 3.1%). The feed throughput was not affected by inclusion levels or type of DDGS. Pellet quality (pellet durability index; PDI) for diets comprising pan DDGS (both 10% and 20% inclusion levels) was significantly better than PDI for diets comprising conventional DDGS, big DDGS, and the diet without DDGS. Better pellet quality of diets comprising pan DDGS could be due to lower quantity of poultry oil used, as well as compositional characteristics such as low fiber and high protein. Diets with big DDGS had similar pelleting characteristics as those with conventional DDGS. Pellet quality deteriorated at higher inclusion levels of conventional DDGS, big DDGS, and enhanced DDGS. Considering that pan DDGS would be included at higher inclusion levels in broiler diets, superior pellet quality of diets comprising pan DDGS is beneficial (Srinivasan et al. 2009a).

25.5 VARIATION OF THE ELUSIEVE PROCESS

Liu (2009) compared the procedure of air classification followed by sieving with the procedure of sieving followed by air classification for four different commercial DDGS materials. The air classification method used by Liu (2009) was winnowing, which utilizes a horizontal draft of air for separation. The study confirmed the effectiveness of the Elusieve process, sieving followed by air classification, for fractionation of DDGS (reported by Srinivasan et al., 2005). The study further found that the overall effect of air classification followed by sieving was the same as sieving followed by air classification. Liu (2009) recommended the procedure of the Elusieve process, sieving followed by air classification, as the choice of DDGS fractionation method, because it would need less time compared to air classification followed by sieving.

REFERENCES

AACC International. 2000. *Approved Methods of the AACC,* 10th ed. St Paul, MN: The American Association of Cereal Chemists.

AAFCO. 2002. *Official Publication of AAFCO.* Oxford, IN: The Association of American Feed Control Officials Incorporated.

AOAC. 2003. *Official Methods of the AOAC,* 17th ed. Gaithersburg, MD: The Association of Official Analytical Chemists.

Belyea, R. L., K. D. Rausch, and M. E. Tumbleson. 2004. Composition of corn and distillers dried grains with solubles from dry grind ethanol processing. *Bioprocess Technology* 94: 293–298.

Doner, L. W., H. K. Chau, M. L. Fishman, and K. B. Hicks. 1998. An improved process for isolation of corn fiber gum. *Cereal Chemistry* 75(4): 408–411.

DRI. 2005. Climate and weather terms glossary. Desert Research Institute. Available online: www.wrcc.dri.edu/ams/glossary.html. Accessed February 22, 2011. Reno, NV. October, 2005.

Kim, E., C. Parsons, V. Singh, and R. Srinivasan. 2007. Nutritional evaluation of new corn distillers dried grains with solubles (DDGS) produced by the enzymatic milling (E-Mill) and elusieve processes. *Poultry Science* 86(Suppl. 1): 397.

Liu, K. 2009. Fractionation of distillers dried grains with solubles (DDGS) by sieving and winnowing. *Bioresource Technology* 100: 6559–6569.

Loar, R., R. Srinivasan, M. Kidd, W. Dozier, and A. Corzo. 2009. Effects of fiber removal from distillers dried grains with solubles on the performance and carcass characteristics of male broilers. *Journal of Applied Poultry Research* 18(3): 494–500.

Moreau, R. A., V. Singh, S. R. Eckhoff, M. J. Powell, K. B. Hicks, and R. A. Norton. 1999. Comparison of yield and composition of oil extracted from corn fiber and corn bran. *Cereal Chemistry* 76(3): 449–451.

Noll, S., V. Stangeland, G. Speers, and J. Brannon. 2001. *Distillers Grains in Poultry Diets.* Bloomington, MN: 62nd Minnesota Nutrition Conference and Minnesota Corn Growers Association Technical Symposium.

Peters, M. S., and K. D. Timmerhaus. 1980. *Plant Design and Economics for Chemical Engineers,* 3rd ed. New York: McGraw-Hill.

Schilling, C. H., P. Tomasik, D. S. Karpovich, B. Hart, S. Shepardson, J. Garcha, and P. T. Boettcher. 2004. Preliminary studies on converting agricultural waste into biodegradable plastics. I. Corn distillers' dry grain. *Journal of Polymers and the Environment* 12: 257–264.

Shurson, G. C. 2002. *The Value and Use of Distiller's Dried Grains with Solubles (DDGS) in Swine Diets.* Raleigh, NC: Carolina Nutrition Conference.

Singh, V., R. A. Moreau, L. W. Doner, S. R. Eckhoff, and K. B. Hicks. 1999. Recovery of fiber in the corn dry grind ethanol process: a feedstock for valuable coproducts. *Cereal Chemistry* 76(6): 868–872.

Singh, V., R. A. Moreau, K. B. Hicks, R. L. Belyea, and C. H. Staff. 2002. Removal of fiber from distillers dried grains with solubles (DDGS) to increase value. *Transactions of the ASAE* 45: 389–392.

Srinivasan, R. 2006. *Separation of Fiber from Distillers Dried Grains with Solubles Using Sieving and Elutriation.* PhD dissertation. Urbana, IL: Department of Agricultural and Biological Engineering, University of Illinois.

Srinivasan, R., and V. Singh, 2006. Removal of fiber from grain products including distillers dried grains with solubles. US Patent 7,670,633 B2.

Srinivasan, R., and V. Singh, 2008a. Pericarp fiber separation from corn flour using sieving and air classification. *Cereal Chemistry* 85: 27–30.

Srinivasan, R., and V. Singh. 2008b. Physical properties that govern fiber separation from distillers dried grains with solubles (DDGS) using sieving and air classification. *Separation and Purification Technology* 61: 461–468.

Srinivasan, R., A. Corzo, K. Koch, and M. Kidd. 2009a. Effect of fiber separation from distillers dried grains with solubles (DDGS) on pelleting characteristics of broiler diets. *Cereal Chemistry* 86: 393–397.

Srinivasan, R., F. To, and E. Columbus. 2009b. Pilot scale fiber separation from distillers dried grains with solubles (DDGS) using sieving and air classification. *Bioprocess Technology* 100: 3548–3555.

Srinivasan, R., M. P. Yadav, R. L. Belyea, K. D. Rausch, L. E. Pruiett, D. B. Johnston, M. E. Tumbleson, and V. Singh. 2008b. Fiber separation from distillers dried grains with solubles (DDGS) using a larger elutriation apparatus and use of fiber as a feedstock for corn fiber gum. *Biological Engineering* 1: 39–49.

Srinivasan, R., R. A. Moreau, C. Parsons, J. D. Lane, and V. Singh. 2008a. Separation of fiber from distillers dried grains (DDG) using sieving and elutriation. *Biomass and Bioenergy* 32: 468–472.

Srinivasan, R., R. A. Moreau, K. D. Rausch, R. L. Belyea, M. E. Tumbleson, and V. Singh. 2005. Separation of fiber from distillers dried grains with solubles (DDGS) using sieving and elutriation. *Cereal Chemistry* 82: 528–533.

Srinivasan, R., V. Singh, R. L. Belyea, K. D. Rausch, R. A. Moreau, and M. E. Tumbleson. 2006. Economics of fiber separation from distillers dried grains with solubles (DDGS) using sieving and elutriation. *Cereal Chemistry* 83: 324–330.

Treybal, R. E. 1980. *Mass Transfer Operations,* 3rd ed. New York: McGraw-Hill.

Van Soest, P. J., J. B. Robertson, and B. A. Lewis. 1991. Methods for dietary fiber, neutral detergent fiber and non-starch polysaccharides in relation to animal nutrition. *Journal of Dairy Science* 74: 3583–3597.

Wu, Y. V., and A. C. Stringfellow. 1986. Simple dry fractionation of corn distillers dried grains and corn distillers dried grains with solubles. *Cereal Chemistry* 63: 60–61.

26 Concluding Thoughts— Toward Increasing the Value and Utility of DDGS

Kurt A. Rosentrater and KeShun Liu

Coproducts are one key to the economic viability of fuel ethanol production. As the industry has grown, the importance of distillers grains has also increased. Through the course of our discussions, we hope that it has become apparent that distillers grains, particularly distillers dried grains with solubles (DDGS), have become very important animal feed ingredients in the United States. The production level is expected to rise to nearly 40 million metric tons in the next few years, and as evidenced by increasing export levels (which are anticipated to consume almost one-fourth of all U.S. DDGS during 2010), DDGS is becoming a principal feed ingredient globally as well.

In this book, we have discussed various topics that are crucial to the use of DDGS, ranging from properties of raw grains themselves to their conversion and processing into ethanol and coproducts, from physical properties and chemical composition of the coproducts to mycotoxin levels, analytical methods, and various end-use options. We have also discussed several new and emerging opportunities for DDGS, including use as feeds for companion and aquatic animals, front-end and back-end fractionation processes, and conversion into bioenergy, plastic composites, and other value-added uses.

We hope that the information provided in this book has been informative, and that it can be used by the scientific and technical communities to move both the fuel ethanol and the livestock industries forward. We have tried to provide the most current information about this growing industry and answer key questions regarding the production, properties, and end uses of DDGS.

No doubt, there are still lingering controversies regarding using grains for fuel ethanol production. Some blame this industry for causing high commodity prices and food shortages; others argue that the energy balance is questionable between use of ethanol as a motor fuel and the resource inputs needed for processing. Still others contend that the ethanol industry is now mature, and that technological improvements will only be incremental from this point onward. There are still many questions that need to be answered and many innovations yet to be discovered and implemented. Some important questions to consider include:

1. What is the future of the corn-based ethanol industry?
2. What is the final level of industry equilibrium that will be achieved, and thus the long term production level of ethanol, DDGS, and other coproducts?
3. In terms of fractionation techniques that are being developed, what types of DDGS will be commercialized, by how many companies, and how soon will they be available?
4. What will be the resulting nutrient profiles of these products?
5. Will all ethanol plants eventually implement some type of fractionation, or will only a portion of them?
6. What is the best way to improve the quality control at each plant and reduce variability in the DDGS?
7. Can food-grade DDGS be manufactured and commercialized?

8. Should a valuable animal feed ingredient, such as DDGS, be used for industrial (nonfeed and nonfood) uses (e.g., fillers for bioplastics, substrates for bioenergy, etc.)?
9. How soon will cellulosic ethanol be economically viable and widely commercialized?
10. What will these new types of byproducts be composed of, and what will their physical properties be?
11. What are the potential uses for these byproducts?
12. What is the future of transportation fuels in general?
13. How does corn-based ethanol fit into this bigger picture?
14. Is corn-based ethanol only a transitional fuel, or will it be used for the long term?
15. How does it fit with cellulosic ethanol and other biofuels?
16. How will biofuels develop sustainably?
17. What are the effects upon the environment and upon ecosystems in the United States and abroad (especially land use change, fertilizer use, water use, biodiversity, invasive pests, etc.)?
18. What are the best means to mitigate negative environmental effects?

As the renewable fuels industry continues to grow, these questions will eventually be answered. Conversion of starch from corn and other grains into biofuels is one step on the path to sustainable energy independence, but it is not the only step on this journey. Corn-based ethanol plants will continue to be a cornerstone in the growing biorefining and biofuels industries. It is clear that DDGS and other coproducts will continue to play increasing roles in the feed, food, and industrial sectors for years to come, both domestically and internationally. Therefore, it is our hope that this book will not only provide timely information about distillers grains, but will also serve as a benchmark for coproducts coming from other renewable fuel sources.

Index

A

Acetogenesis, in anaerobic digestion, 467
Acid detergent fiber (ADF), 19–20, 67. *See also*
Detergent fiber
content of DDGS, 507
in livestock feeds, inclusion of, 199
Acid detergent insoluble crude protein (ADICP), 268
Acid detergent lignin, in livestock feeds, inclusion of, 199
Acidogenesis, in anaerobic digestion, 466–467
Acylated steryl glycosides (ASG), 183
Additives, 419–420
Aflatoxins, in feed ingredients, 208–209, 225, 227,
229–232. *See also* Mycotoxins
FDA action levels for, 208, 221
Agrinine, 21
Albumins, 64
Aleurone, 48
Alkali, in FAME production, 490
American Association of Cereal Chemists, 400
American Association of Feed Control Officials
(AAFCO), 99
American Dietetic Association, 399–400
American Feed Industry Association (AFIA), 26, 196
American Society for Testing and Materials (ASTM), 495
American Society of Animal Science Committee, 36
Amino acids, 302–303, 321–324, 348–349, 364
composition of cereal grains, 108
composition of distillers dried grains with solubles,
147–148, 507
during dry grind ethanol processing, 156–158
in livestock feeds, inclusion of, 203–204
standard, 203–204
tryptophan, 204
Ammonia fiber expansion (AFEX), 173
α-Amylase, 63, 64
β-Amylase, 63–64
γ-Amylase. *See* Amyloglucosidase
Amylases, in ethanol production, 510–511
Amylase-treated neutral detergent fiber (aNDF). *See also*
Detergent fiber
in livestock feeds, inclusion of, 199–200
Amyloglucosidase, in ethanol production, 511
Anaerobic contact reactor, 469
Anaerobic digestion
acetogenesis, 467
acidogenesis, 466–467
of ethanol coproducts, 472–473
hydrolysis, 466
methanogenesis, 467
steps of, 466
Anaerobic filter, 470–471
Anaerobic fixed film reactor, 470
Anaerobic reactors, types of, 468
anaerobic contact reactor, 469
anaerobic filter, 470–471
anaerobic sequencing batch reactor, 471
continuous stirred tank reactor, 468–469

induced bed reactor, 471–472
plug flow reactor, 471
upflow anaerobic sludge blanket reactor, 469
Anaerobic sequencing batch reactor (ASBR), 471, 473
Anaerobic treatment, advantages of, 467–468
Analytical methodology (quality standards), of distillers
dried grains with solubles, 193–213
for contaminants
antibiotic residues, 212–213
mycotoxins, 205–212
nutrients for inclusion in livestock feeds,
determination of
amino acids, 203–204
ash, 200
databases, 205
detergent fiber, 199–200
laboratory proficiency programs, for testing
distillers grains, 204–205
starch, 204
trace elements, 200–203
for trading purposes, 194–199
crude fat (oil) analysis, 195–196
crude fiber analysis, 196
crude protein analysis, 195
industry recommended methods, 196–199
moisture analysis, 195
Angle of difference, 134
Angle of fall, 134–135
Angle of repose, 124–126, 134–135
Angle of spatula, 134–136
Antibiotic residues, 212–213
AOAC International Research Institute, 194
Apparent total tract digestibility (ATTD), 301, 304
L-Arabinoes, 66
Ash, 19–20
in livestock feeds, inclusion of, 200
Aspergillus sp., 212, 220
Aspergillus kawachi, 63
Aspergillus niger, 63
Association of American Feed Control Officials, Inc.
(AAFCO), 36
Feed Check Sample Program, 204–205
Ingredient Definition Committee, 42
Laboratory Method and Services Committee, 213
Operations Manual, 42
Avenin, 64
Average daily feed intake (ADFI), 307
Average daily gain (ADG), 240, 241, 242, 244, 251, 257, 307

B

Baked products, containing ethanol coproducts, 401,
404–405, 411–412
Barley (*Hordeum vulgare*)
average proximate composition of, 57
caryopsis, 51
characteristics of, 55
crude protein and amino acid composition of, 108

Rice bran oil, 65
Rumen degradable protein (RDP), 266–267
Rumen-undegraded protein (RUP), 266, 269, 270, 277,
    288
Rye (*Secale cereale*)
    average proximate composition of, 57
    characteristics of, 56
    chemical composition of, 38
    dimension of, 51
    embryo and endosperm, chemical composition of, 58
    ethanol production from, 112–113
    sizes and shapes of, 51
    starch granules of, 60
    stillage after fermentation of, 106
    structure of, 50
    vitamin contents in, 39
    weight of, 51
    world production of, 107

S

Saccharification, 62
Saccharomyces cerevisiae, 81, 110, 238, 478
Seagrams Distillers Corporation, 40
Selenium, in livestock feeds, inclusion of, 202
sheep, feeding DDGS to, 391–394
S-induced polioencephalomalacia (sPEM), 250, 251
Single screw extrusion, 368, 370, 375–376
Sitostanyl ferulate, structure of, 183
β-Sitosterol, structure of, 183
Solid plastic waste, 429
Solid retention time (SRT), 468
Soluble dietary fiber (SDF), 298. *See also* Dietary fiber
Solvent extraction, of distillers dried grains with solubles,
    186–187
Sorghum (*Sorghum bicolor*)
    average proximate composition of, 57
    characteristics of, 56
    crude protein and amino acid composition of, 108
    dimension of, 51
    embryo and endosperm, chemical composition of, 58
    ethanol production from, 107
    fatty acids concentration in, 109
    sizes and shapes of, 51
    starch granules of, 60
    stillage after fermentation of, 106
    structure of, 50
    weight of, 51
    world production of, 107
Sorghum kernel oil, free fatty acid composition of,
    181–182
Sows, ethanol products inclusion in diets to, 310
Soybeans, in biodiesel production, 491
Specific mechanical energy input (SME), 370, 374
Standardized digestible (SDD) lysine, 322
Starch, 19
    in cereal grains
        chemistry, 59
        gelatinization, 61–63
        gel formation and retrogradation, 63
        starch-degrading enzymes, 63–64
    -degrading enzymes, 63–64
    granules, of cereal grains, 60–61
    in livestock feeds, inclusion of, 204

Starch-rich feedstocks
    composition of DDGS and stillage after fermentation
        of, 106
    for fuel ethanol, 103–107
        composition of, 105
    world production of, 107
Steryl ester, structure of, 183
Steryl ferulates, 402
Steryl glycosides (SG), 183
Streptococcus iniae, 366
6-in-lb$_f$ (0.68-J) stiffness tester, 432
Sulfur, in livestock feeds, inclusion of, 200–201
Surface roughness, 442, 434
Sweet potato (*Ipomoea batatas*)
    ethanol production from, 114
    stillage after fermentation of, 106
    world production of, 107
Swine, ethanol coproducts feeding, 297
    and diets, 298
        and carcass composition and quality, 309–310
        and live pig performance, 307–309
        to sows, 310
        to weanling pigs, 306–307
    energy concentration and digestibility, 305–306
    nutrient concentration and digestibility
        amino acids, 302–303
        carbohydrates, 298–301
        ether extract, 305
        phosphorus, 303–305
Switchgrass, composition of, 476
Syneresis, 63
Syrup. *See* Condensed distillers solubles (CDS)

T

T-2 toxins, 211, 220, 225–233
Thermal properties, of distillers dried grains with
        solubles, 127
Thermochemical conversion
    of biomass, 451–454
    byproducts of, 461
    economics of, 458–461
Thermochemical conversion research, 457
Thermogravimetric (TGA) analysis, 461
Thermoplastics, combining DDGS with, 437–440
Thermosetting resins, combining DDGS with, 440–441
Thin-layer chromatography (TLC), 222, 225
Threonine, 21
Tilapia, 343
    carbohydrates, 353
    dietary energy, 345–346
    feeding trials with DDGS-based feeds, 365–366
    lipids and fatty acids, 351–352
    minerals, 359–360
    proteins and amino acids, 348–349
    species characteristics, 343
    vitamins, 354–356
Tilapia rendalli, 343
Tilapia zilli, 343
Tip cap, 247
Tocopherols, 160, 184–185
    structure of, 180
Tocotrienols, 160, 184–185
    structure of, 180